CORE DYNAMICS

Treatise on Geophysics

CORE DYNAMICS

Editor-in-Chief

Professor Gerald Schubert
Department of Earth and Space Sciences and Institute of Geophysics and Planetary Physics, University of California Los Angeles, Los Angeles, CA, USA

Volume Editor

Dr. Peter Olson
Johns Hopkins University, Baltimore, MD, USA

ELSEVIER

AMSTERDAM • BOSTON • HEIDELBERG • LONDON • NEW YORK • OXFORD
PARIS • SAN DIEGO • SAN FRANCISCO • SINGAPORE • SYDNEY • TOKYO

Elsevier B.V.
Radarweg 29, 1043 NX Amsterdam, the Netherlands

First edition 2009

British Library Cataloguing in Publication Data
A catalogue record for this book is available from the British Library

Library of Congress Control Number: 2009929986

ISBN: 978-0-444-53457-6

Printed and bound in Spain

09 10 11 12 10 9 8 7 6 5 4 3 2 1

Contents

Preface vii

Contributors xi

Editorial Advisory Board xiii

1 Overview 1
P. Olson, *Johns Hopkins University, Baltimore, MD, USA*

2 Energetics of the Core 31
F. Nimmo, *University of California Santa Cruz, Santa Cruz, CA, USA*

3 Theory of the Geodynamo 67
P. H. Roberts, *University of California, Los Angeles, CA, USA*

4 Large-Scale Flow in the Core 107
R. Holme, *University of Liverpool, Liverpool, UK*

5 Thermal and Compositional Convection in the Outer Core 131
C. A. Jones, *University of Leeds, Leeds, UK*

6 Turbulence and Small-Scale Dynamics in the Core 187
D. E. Loper, *Florida State University, Tallahassee, FL, USA*

7 Rotational Dynamics of the Core 207
A. Tilgner, *University of Göttingen, Göttingen, Germany*

8 Numerical Dynamo Simulations 245
U. R. Christensen and J. Wicht, *Max-Planck-Institut für Sonnensystemforschung, Katlenburg-Lindau, Germany*

9 Magnetic Polarity Reversals in the Core 283
G. A. Glatzmaier and R. S. Coe, *University of California, Santa Cruz, CA, USA*

10 Inner-Core Dynamics 299
I. Sumita, *Kanazawa University, Kanazawa, Japan*
M. I. Bergman, *Simon's Rock College, Great Barrington, MA, USA*

11 Experiments on Core Dynamics 319
P. Cardin, *Université Joseph-Fourier, Grenoble, France*
P. Olson, *Johns Hopkins University, Baltimore, MD, USA*

12 Core–Mantle Interactions 345
B. A. Buffett, *The University of Chicago, Chicago, IL, USA*

Preface

Geophysics is the physics of the Earth, the science that studies the Earth by measuring the physical consequences of its presence and activity. It is a science of extraordinary breadth, requiring 10 volumes of this treatise for its description. Only a treatise can present a science with the breadth of geophysics if, in addition to completeness of the subject matter, it is intended to discuss the material in great depth. Thus, while there are many books on geophysics dealing with its many subdivisions, a single book cannot give more than an introductory flavor of each topic. At the other extreme, a single book can cover one aspect of geophysics in great detail, as is done in each of the volumes of this treatise, but the treatise has the unique advantage of having been designed as an integrated series, an important feature of an interdisciplinary science such as geophysics. From the outset, the treatise was planned to cover each area of geophysics from the basics to the cutting edge so that the beginning student could learn the subject and the advanced researcher could have an up-to-date and thorough exposition of the state of the field. The planning of the contents of each volume was carried out with the active participation of the editors of all the volumes to insure that each subject area of the treatise benefited from the multitude of connections to other areas.

Geophysics includes the study of the Earth's fluid envelope and its near-space environment. However, in this treatise, the subject has been narrowed to the solid Earth. The *Treatise on Geophysics* discusses the atmosphere, ocean, and plasmasphere of the Earth only in connection with how these parts of the Earth affect the solid planet. While the realm of geophysics has here been narrowed to the solid Earth, it is broadened to include other planets of our solar system and the planets of other stars. Accordingly, the treatise includes a volume on the planets, although that volume deals mostly with the terrestrial planets of our own solar system. The gas and ice giant planets of the outer solar system and similar extra-solar planets are discussed in only one chapter of the treatise. Even the *Treatise on Geophysics* must be circumscribed to some extent. One could envision a future treatise on Planetary and Space Physics or a treatise on Atmospheric and Oceanic Physics.

Geophysics is fundamentally an interdisciplinary endeavor, built on the foundations of physics, mathematics, geology, astronomy, and other disciplines. Its roots therefore go far back in history, but the science has blossomed only in the last century with the explosive increase in our ability to measure the properties of the Earth and the processes going on inside the Earth and on and above its surface. The technological advances of the last century in laboratory and field instrumentation, computing, and satellite-based remote sensing are largely responsible for the explosive growth of geophysics. In addition to the enhanced ability to make crucial measurements and collect and analyze enormous amounts of data, progress in geophysics was facilitated by the acceptance of the paradigm of plate tectonics and mantle convection in the 1960s. This new view of how the Earth works enabled an understanding of earthquakes, volcanoes, mountain building, indeed all of geology, at a fundamental level. The exploration of the planets and moons of our solar system, beginning with the Apollo missions to the Moon, has invigorated geophysics and further extended its purview beyond the Earth. Today geophysics is a vital and thriving enterprise involving many thousands of scientists throughout the world. The interdisciplinarity and global nature of geophysics identifies it as one of the great unifying endeavors of humanity.

The keys to the success of an enterprise such as the *Treatise on Geophysics* are the editors of the individual volumes and the authors who have contributed chapters. The editors are leaders in their fields of expertise, as distinguished a group of geophysicists as could be assembled on the planet. They know well the topics that had to be covered to achieve the breadth and depth required by the treatise, and they know who were the best of their colleagues to write on each subject. The list of chapter authors is an impressive one, consisting of geophysicists who have made major contributions to their fields of study. The quality and coverage achieved by this group of editors and authors has insured that the treatise will be the definitive major reference work and textbook in geophysics.

Each volume of the treatise begins with an 'Overview' chapter by the volume editor. The Overviews provide the editors' perspectives of their fields, views of the past, present, and future. They also summarize the contents of their volumes and discuss important topics not addressed elsewhere in the chapters. The Overview chapters are excellent introductions to their volumes and should not be missed in the rush to read a particular chapter. The title and editors of the 10 volumes of the treatise are:

Volume 1: Seismology and Structure of the Earth
Barbara Romanowicz
University of California, Berkeley, CA, USA
Adam Dziewonski
Harvard University, Cambridge, MA, USA

Volume 2: Mineral Physics
G. David Price
University College London, UK

Volume 3: Geodesy
Thomas Herring
Massachusetts Institute of Technology, Cambridge, MA, USA

Volume 4: Earthquake Seismology
Hiroo Kanamori
California Institute of Technology, Pasadena, CA, USA

Volume 5: Geomagnetism
Masaru Kono
Okayama University, Misasa, Japan

Volume 6: Crust and Lithosphere Dynamics
Anthony B. Watts
University of Oxford, Oxford, UK

Volume 7: Mantle Dynamics
David Bercovici
Yale University, New Haven, CT, USA

Volume 8: Core Dynamics
Peter Olson
Johns Hopkins University, Baltimore, MD, USA

Volume 9: Evolution of the Earth
David Stevenson
California Institute of Technology, Pasadena, CA, USA

Volume 10: Planets and Moons
Tilman Spohn
Deutsches Zentrum für Luft-und Raumfahrt, GER

In addition, an eleventh volume of the treatise provides a comprehensive index.

The *Treatise on Geophysics* has the advantage of a role model to emulate, the highly successful *Treatise on Geochemistry*. Indeed, the name *Treatise on Geophysics* was decided on by the editors in analogy with the geochemistry compendium. The *Concise Oxford English Dictionary* defines treatise as "a written work dealing formally and systematically with a subject." Treatise aptly describes both the geochemistry and geophysics collections.

The *Treatise on Geophysics* was initially promoted by Casper van Dijk (Publisher at Elsevier) who persuaded the Editor-in-Chief to take on the project. Initial meetings between the two defined the scope of the treatise and led to invitations to the editors of the individual volumes to participate. Once the editors were on board, the details of the volume contents were decided and the invitations to individual chapter authors were issued. There followed a period of hard work by the editors and authors to bring the treatise to completion. Thanks are due to a number of members of the Elsevier team, Brian Ronan (Developmental Editor), Tirza Van Daalen (Books Publisher), Zoe Kruze (Senior Development Editor), Gareth Steed (Production Project Manager), and Kate Newell (Editorial Assistant).

G. Schubert
Editor-in-Chief

Contributors

M. I. Bergman
Simon's Rock College, Great Barrington, MA, USA

B. A. Buffett
The University of Chicago, Chicago, IL, USA

P. Cardin
Université Joseph-Fourier, Grenoble, France

U. R. Christensen
Max-Planck-Institut für Sonnensystemforschung, Katlenburg-Lindau, Germany

R. S. Coe
University of California, Santa Cruz, CA, USA

G. A. Glatzmaier
University of California, Santa Cruz, CA, USA

R. Holme
University of Liverpool, Liverpool, UK

C. A. Jones
University of Leeds, Leeds, UK

D. E. Loper
Florida State University, Tallahassee, FL, USA

F. Nimmo
University of California Santa Cruz, Santa Cruz, CA, USA

P. Olson
Johns Hopkins University, Baltimore, MD, USA

P. H. Roberts
University of California, Los Angeles, CA, USA

I. Sumita
Kanazawa University, Kanazawa, Japan

A. Tilgner
University of Göttingen, Göttingen, Germany

J. Wicht
Max-Planck-Institut für Sonnensystemforschung, Katlenburg-Lindau, Germany

EDITORIAL ADVISORY BOARD

1 Overview

P. Olson, Johns Hopkins University, Baltimore, MD, USA

1.1	**A Scientific Journey to the Center of the Earth**	1
1.2	**State of the Core**	3
1.3	**The Search for a Dynamo Theory**	5
1.4	**Core Dynamics and the Geomagnetic Field**	11
1.5	**Core Energetics**	12
1.6	**Core Dynamics as a Heat Engine**	14
1.7	**Convection and Dynamo Action**	14
1.8	**Simulating the Geodynamo**	16
1.9	**Mantle Effects within the Core**	18
1.10	**The Dynamical Inner Core**	20
1.11	**Future Prospects and Problems**	20
1.12	**Additional References**	21
1.13	**Summary of the Chapters in This Volume**	21
1.13.1	Energetics of the Core	21
1.13.2	Theory of the Geodynamo	22
1.13.3	The Large Scale Flow in the Core	22
1.13.4	Thermal and Compositional Convection in the Core	23
1.13.5	Turbulence and Small-Scale Dynamics in the Core	23
1.13.6	Rotational Dynamics	24
1.13.7	Numerical Dynamo Simulations	24
1.13.8	Magnetic Polarity Reversals in the core	25
1.13.9	Inner Core Dynamics	26
1.13.10	Laboratory Experiments on Core Dynamics	26
1.13.11	Core–Mantle Interactions	27
References		28

1.1 A Scientific Journey to the Center of the Earth

For as long as man has speculated about the interior of the Earth, it has been presumed that there exists a central core. Centuries before the rise of modern science, philosophers, and theologians had concluded that the Earth has a hot region at its center, with properties distinct from all other parts of the planet. For nearly as long a time it has been known that the Earth is also magnetic, but the cause of the Earth's magnetism remained just as mysterious as the nature of the deep interior.

Scientific inquiry about the core grew from early investigations of the properties of the geomagnetic field, which began during the era of global exploration. Although the ancient Chinese deserve the credit for discovering Earth's magnetism, Gilbert (1600) was the first to demonstrate that the compass needle is controlled by a force originating within the Earth (**Figure 1**). He showed that the pattern of magnetic field lines on a uniformly magnetized sphere approximate the known directions of the compass needle over the Earth's surface. Three hundred and fifty years later, Sidney Chapman characterized Gilbert's demonstration as "the only successful experiment in the history of geomagnetism!" Later it was observed that Earth's magnetic field changes slowly with time. In his famous explanation for this secular variation, Halley (1683, 1692) proposed that the geomagnetic field has its origin near the Earth's center, in a region separated from the solid crust by a cavernous, fluid-filled shell. Halley (**Figure 2**) envisioned that both the crust and the central region or core rotate in the prograde sense, but the core spins slightly slower, causing the magnetic field to drift systematically westward as seen at the surface. Thus, two important and long-lasting concepts were born: the basic three-layer

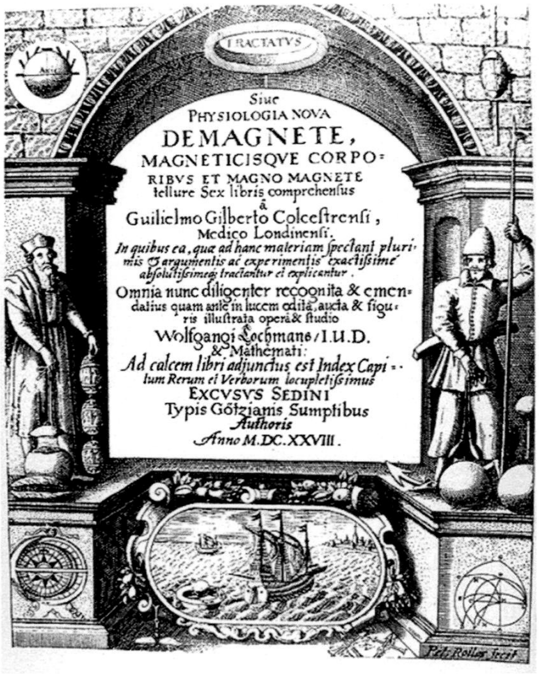

Figure 1 Title page from William Gilbert's 1600 treatise *De Magnete*.

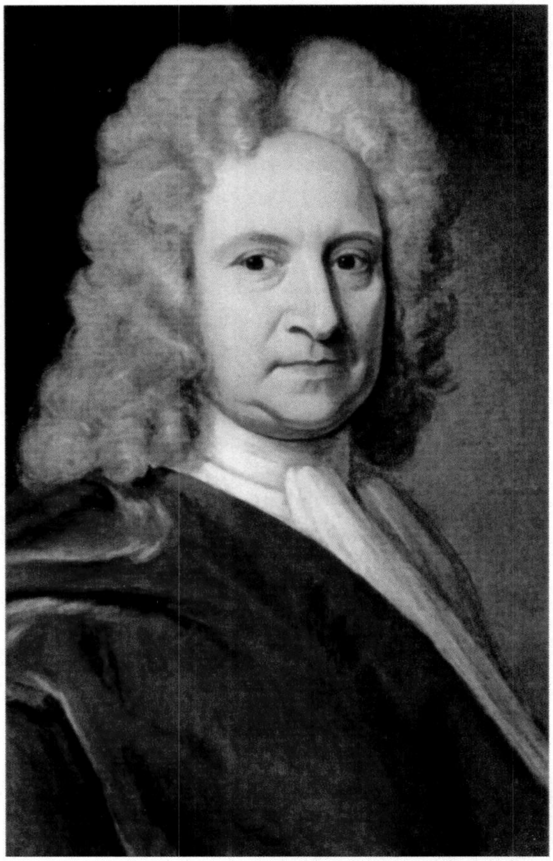

Figure 2 Edmond Halley (1656–1742) interpreted the geomagnetic secular variation.

model of Earth's interior (solid crust and mantle, liquid outer and solid inner core), and the association between the westward geomagnetic drift and westward motion of the fluid outer core with respect to other parts of the Earth system.

Halley's model implicitly assumed that the magnetic field originated in a solid inner core (Evans, 1988), akin to Gilbert's uniformly magnetized sphere. Subsequently, it was shown that Halley's model is at variance with the ferromagnetic properties of Earth materials, which lose their permanent magnetization at the Curie temperature at depths of a few tens of kilometers beneath the surface. However, by then the physical connection between magnetic fields and electric currents had been established, providing an alternative explanation for the geomagnetic field that relied on free electric currents rather than permanent magnetization.

The liquid (i.e., molten) state of the outer core was established during the early part of the twentieth century, but the roots of the idea can be traced back into antiquity. Several independent lines of scientific evidence appeared in the middle of the nineteenth century in favor of high temperatures in the Earth's deep interior, including the steep geothermal gradient measured in deep mines and petrologic discoveries that indicated that very high temperatures are needed to form most igneous rocks. However, the early estimates of the actual temperature variation through the deep Earth varied wildly, preventing any firm conclusion about the state of matter in the core.

Toward the end of the twentieth century, two competing models of the state of Earth's deep interior became prominent. One model assumed that the interior was solid (except for small melt regions below volcanoes) and also elastic, with a very high shear modulus, 'as rigid as steel', according to the Kelvin (1862) famous prescription. This model was supported by observations of the amplitudes of the tides (Darwin, 1879) and the period of Earth's free nutation, the Chandler wobble (Newcomb, 1892). The competing model held that the interior was largely fluid, an idea that was popular with geologists at that time, although it had prominent adherents within the physics community as well, for example, Ritter (1878), Poincaré

(1885, 1994), Arrhenius (1900), and even earlier, Franklin (1793).

The terms of this debate underwent a permanent shift with the publication by Wiechert (1897) of the first quantitative model of Earth structure. Wiechert's Earth model was based on all available astronomical and geodetic data, and featured a central metallic core surrounded by a rocky mantle. Wiechert (**Figure 3**) is often given credit for being the first to attribute both chemical and physical differences to the core and mantle, and to infer that the core–mantle boundary, the most significant discontinuity in the planet's interior, represents a change from silicates to iron, as well as a density jump. In any case, there is little doubt that his work launched the era of seismic exploration of the core. Within a decade, Oldham identified seismic P- and S-waves (1899) and interpreted the P-wave shadow as low velocity in a central core (Oldham, 1906). Shortly thereafter, Gutenberg (1912) determined the location of the core–mantle boundary, at depth of 2900 ± 20 km, consistent within his calculated uncertainty with the present-day value.

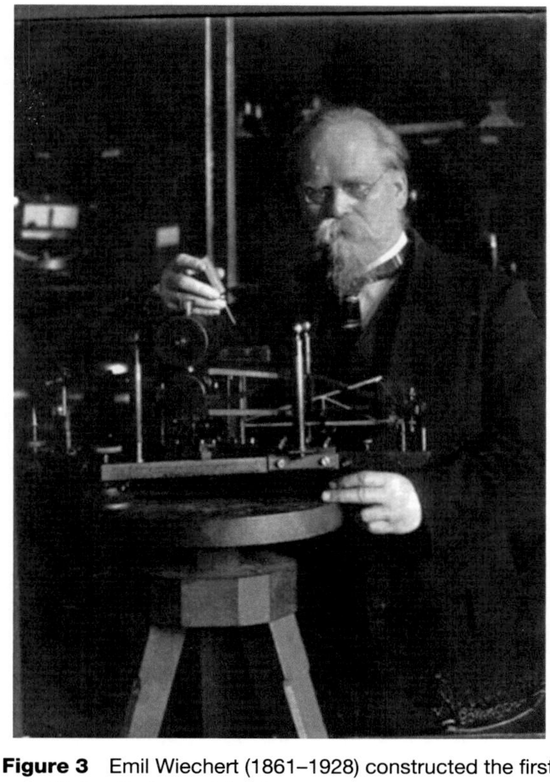

Figure 3 Emil Wiechert (1861–1928) constructed the first quantitative Earth model with a core.

Gutenberg's original Earth model included rigidity throughout the core, in spite of the fact that seismic shear wave transmission through the core had never been confirmed, a testimony to the lasting influence of Kelvin's ideas. Indeed, the fluidity of the core remained controversial for more than another decade. The issue was settled when Jeffreys (1926) showed that it was possible to reconcile seismic wave speeds with tidal and Chandler wobble observations using an Earth model with a liquid core of radius 3471 km, essentially the same as in the Gutenberg model. As Brush (1996) points out in his history of the exploration of the Earth's interior, Jeffreys' reputation became somewhat tarnished by his refusal to accept continental drift and the concept of mantle convection (Jeffreys, 1929). Ironically, Jeffreys made several fundamental contributions to the theory of convection in viscous fluids – the basic model for mantle convection – and he also contributed importantly to the acceptance of the geodynamo theory by demonstrating the outer core is liquid.

The final piece of the main radial structure of the core was provided by Lehmann (1936; **Figure 4**), who discovered the inner-core boundary, which she placed at 4970 km depth, or 1400 km radius (the currently preferred radius is about 1220 km). Following Lehmann's discovery, seismologists have succeeded in demonstrating the crystalline nature of the inner-core material. The study by Dziewonski and Gilbert (1971) of normal mode overtones excited by the 1964 Alaska earthquake provided an estimate of its average rigidity, and subsequent investigations have determined that the inner core is also anisotropic. **Table 1** gives the chronology of important milestones in this scientific journey.

1.2 State of the Core

A full review of the composition of the core is found of this treatise. Here we provide a brief summary of the major constituents of the core, for purposes of this chapter. **Table 2** lists some of the well-known physical properties of the core, **Table 3** lists some important thermodynamic and transport properties (which are generally less well-known), and **Figure 5** shows its basic radial structure.

The model of a predominantly iron core was firmly in place by the middle of the twentieth century, and was well-supported by evidence from seismology (Bullen, 1954) and mineral physics (Birch, 1952). One early argument that was often

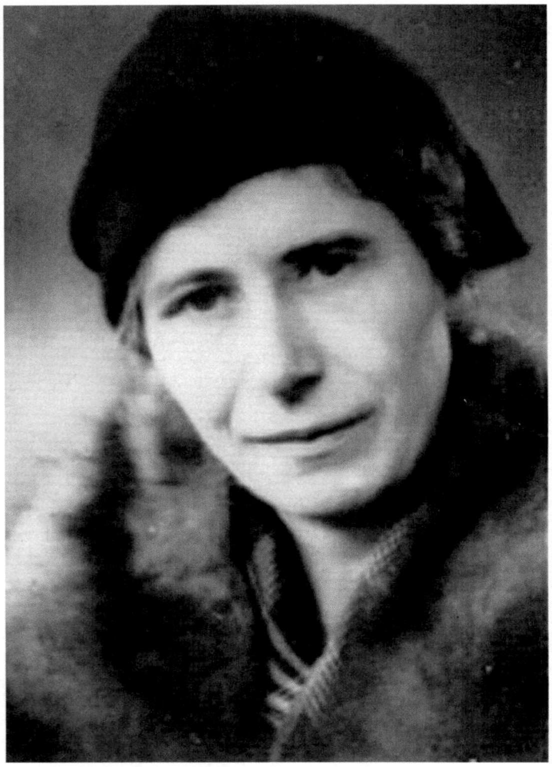

Figure 4 Inge Lehmann (1888–1993) discovered the inner core.

cited in support of an iron core was Birch's law (Birch, 1964), a linear relationship between density and bulk sound velocity with a coefficient proportional to the mean atomic weight of the material. Applications of Birch's law revealed that the mean atomic weight of the outer core and inner core is slightly less than iron, respectively, but are far too large to be explained by a phase transformation of lower mantle material. Although it is now recognized that the theoretical basis for Birch's law is weak, it was historically important because it seemed to demand a metallic core, rather than an oxide-rich or silica-rich one. Other metals might also be present in the core. The abundance of nickel in iron meteorites was historically used to argue that the core contains substantial Fe–Ni alloy. However, many of the important physical properties of nickel are indistinguishable from iron at core conditions, so the detection of nickel in the core is difficult.

The presence of lighter elements is quite important for the dynamics of the core. As described of this treatise, the abundance of the various candidate light elements has been a matter of intense controversy. Four decades of high-pressure measurements and theory have convincingly demonstrated Birch's interpretation that the outer core is less dense than pure iron at *in situ* conditions, even allowing for the expected density decrease due to melting, and therefore requires the addition of light elements. Candidate light elements frequently proposed to reconcile this density deficit include oxygen and sulfur (Poirier, 1994), and less frequently, silicon, hydrogen, and even helium. Sulfur has a known affinity for iron at low pressures, and high-pressure studies indicate that the oxygen is soluble in iron at core conditions (Alfe *et al.*, 2003). Both sulfur and oxygen are known to affect the phase diagram of iron substantially, including the melting point. Less is known about the effects of silicon in the core, and still less about hydrogen or helium.

Most detailed information on the present-day structure of the core comes from seismology. The variation of seismic wave velocity and density with depth through the fluid outer core shown in **Figure 5(b)** are basically consistent with a homogeneous mixture of iron and the lighter alloying elements listed above (Dziewonski and Anderson, 1981). Unfortunately, it has proved to be difficult to choose between the candidate light elements on the basis of their seismic properties alone. Some dilution of the solid inner core by lighter elements is also indicated, as the predicted density of solid iron is a few percent greater than the seismic properties indicate (Shearer and Masters, 1990; Masters and Gubbins, 2003). The actual density change at the inner-core boundary is 500–900 $\mathrm{kg\,m^{-3}}$, of which about one-third is due to solidification. The rest is likely due to differences in light element concentration between the inner and outer core. This small density difference has profound implications for the driving mechanism for flow in the core, and also for the power source of the geodynamo (Buffett, 2003). A related issue is the abundance of radioactive heat sources in the core. Among the possible radio isotopes, ^{40}K is the most likely to be energetically significant, because of its low-pressure affinity for iron sulfides. However, the amount of potassium in the core is still debated (Gessmann and Woods, 2002; Rama Murthy *et al.*, 2003). In addition to potassium, some uranium content has been proposed for the core, which could slightly augment the amount of internal heat production. As mentioned above, the

Table 1 Chronology

Year	Person(s)	Event
1000	Chinese	Discover lodestone south–north orientation
1600	W. Gilbert	Publication of *De Magnete*
1634	H. Gellibrand	Discovers geomagnetic secular variation
1683	E. Halley	Interprets secular variation
1820	H. Oersted, A. Ampere	Relate magnetism to electric currents
1831	M. Faraday	Introduces disk dynamo
1834	C. F. Gauss	Measures magnetic intensity
1836	C. F. Gauss	Spherical harmonic analysis
1897	E. Weichert	Iron-core Earth model
1906	R. Oldham	Observes core seismic waves
1906	B. Brunhes	Reversely magnetized rocks
1912	B. Gutenberg	Determines core radius
1919	J. Larmor	Proposes self-sustaining fluid dynamos
1933	T. Cowling	Antidynamo theorems
1936	I. Lehmann	Discovers inner core
1942	H. Alfven	MHD waves
1946	W. Elsasser	First MHD dynamo theory
1952	F. Birch	Composition of the core
1953	J. Jacobs	Present-day inner-core growth
1955	E. Parker	$\alpha\omega$-Dynamo effect
1958	A. Hertzenberg, G. Backus	Laminar kinematic dynamos
1961	J. Verhoogen	Inner-core freezing as core energy source
1963	F. Lowes & I. Wilkinson	Laboratory dynamo
1963	J. B. Taylor	Rotational dynamo constraint
1964	S. Braginsky	Asymptotic dynamo model Z
1966	M. Steenbeck *et al.*	α^2-Dynamo effect
1971	A. Dziewonski & F. Gilbert	Inner-core rigidity
1995	G. Glatzmaier & P. Roberts	Self-consistent reversing dynamo model
1995	A. Kageyama & T. Sato	Self-consistent compressible dynamo model
1996	Z. Song & P. Richards	Inner-core super-rotation
2000	A. Gailitis, U. Muller and others	First laboratory fluid dynamos

temperature distribution in the core appears to follow an adiabat. Based on melting point temperature measurements (Boehler, 1996), it is estimated that the temperatures at the core–mantle boundary and inner-core boundary are approximately 4000 and 5500 K, respectively, as shown in **Table 3**.

1.3 The Search for a Dynamo Theory

Until rather late in the twentieth century, 'core dynamics' was a rather narrowly defined subject. It was limited to topics drawn from classical mechanics, such as rotational and tidal deformation, and was pursued by a small number of theoretical physicists and applied mathematicians, G. Darwin, H. Poincaré (**Figure 6**), and H. Jeffreys among them. The full range of core dynamics, the subject of this volume, has received serious study only since the advent of the modern dynamo theory for the origin of the geomagnetic field.

After the time of Halley, relatively little work was done on the causes of the geomagnetic field until well into the twentieth century. Gauss (1832, 1839; **Figure 7**) conjectured that geomagnetic secular variation was a consequence of solidification of the crust, but evidently did not speculate much on the origin of the main field. Zollner (1871) constructed a model for geomagnetic secular variation that included several aspects of present-day dynamo theory, including electric currents induced in a liquid core, and interaction with the solid mantle. However, his ideas evidently were not taken up by others, and the subject languished in obscurity. Even in the middle decades of the twentieth century, leading theorists showed only passing interest in the physical origins of the geomagnetic field. For example, Chapman and Bartels (1940) treatise *Geomagnetism* devotes portions of just 7 pages (out of 1050 total) to this topic. They

Table 2 Known physical properties

Property	Notation	Units	Value
Core–mantle radius (mean)	r_o	m	3.480×10^6
Inner-core radius (mean)	r_i	m	1.22×10^6
Outer-core thickness	d	m	2.26×10^6
Core–mantle boundary area	A_o	m^2	1.52×10^{14}
Inner-core boundary area	A_i	m^2	1.87×10^{13}
Core–mantle boundary ellipticity	ϵ_o	nd	2.5×10^{-3}
Core volume	V	m^3	1.77×10^{20}
Inner-core volume	V_i	m^3	7.6×10^{18}
Outer-core volume	V_o	m^3	1.70×10^{20}
Core moment of inertia	I	kg m^2	9.2×10^{36}
Outer-core mass	M_o	kg	1.835×10^{24}
Inner-core mass	M_i	kg	9.68×10^{22}
Core density (mean)	ρ	kg m^{-3}	1.09×10^4
Inner-core density (mean)	ρ_i	kg m^{-3}	1.29×10^4
Core–mantle boundary gravity	g_o	m s^{-2}	10.68
Inner-core boundary gravity	g_i	m s^{-2}	4.40
Core–mantle boundary pressure	P_o	GPa	136
Inner-core boundary pressure	P_i	GPa	328
P-wave velocity below CMB	v_p	km s^{-1}	8.07
P-wave velocity above ICB	v_p	km s^{-1}	10.36
Poisson ratio, Inner core	ν_P	nd	0.44
Angular velocity of rotation	Ω	rad s^{-1}	7.292×10^{-5}
Free core nutation period	T_c	s	3.71×10^7
Magnetic dipole moment	m_d	A m^2	7.8×10^{22}
Magnetic dipole tilt	θ_d	deg	10.8
Magnetic intensity, CMB (rms)	B_{CMB}	mT	0.42
Magnetic dipole intensity, CMB (rms)	B_{CMB}^{dip}	mT	0.263

nd, not determined.

Table 3 Thermodynamic and transport properties

Property	Notation	Units	Range
Core–mantle boundary temperature	T_o	K	4000 ± 400
Inner-core boundary temperature	T_i	K	5500 ± 600
Outer-core temperature (mean)	T	K	4500 ± 500
Adiabatic temperature gradient, CMB	$-dT/dr_{ad}$	K km^{-1}	0.3 ± 0.1
Adiabatic temperature gradient, ICB	$-dT/dr_{ad}$	K km^{-1}	0.8 ± 0.2
Density jump, ICB	$\Delta\rho_i$	kg m^{-3}	700 ± 200
Light element density reduction, outer core	C_o	%	8 ± 2
Light element density reduction, inner core	C_i	%	4 ± 2
Thermal expansivity	α	K^{-1}	$1.4 \pm 0.5 \times 10^{-5}$
Light element expansivity	α_C	nd	0.7 ± 0.3
Specific heat	C_p	J kg^{-1} K^{-1}	850 ± 20
Latent heat, crystallization	L	J kg^{-1}	$1 \pm 0.5 \times 10^6$
Thermal conductivity	k	W m^{-1} K^{-1}	45 ± 10
Thermal diffusivity	κ	m^2 s^{-1}	$5 \pm 3 \times 10^{-6}$
Compositional diffusivity	D	m^2 s^{-1}	$1 \times 10^{-9 \pm 2}$
Kinematic viscosity, outer core	ν	m^2 s^{-1}	$1 \times 10^{-5 \pm 2}$
Kinematic viscosity, inner core	ν_i	m^2 s^{-1}	$1 \times 10^{10 \pm 3}$
Electrical conductivity	σ	A^2 kg^{-1} m^{-3} s^3	$5 \pm 2 \times 10^5$
Magnetic diffusivity	λ	m^2 s^{-1}	1.5 ± 0.5

nd, not determined.

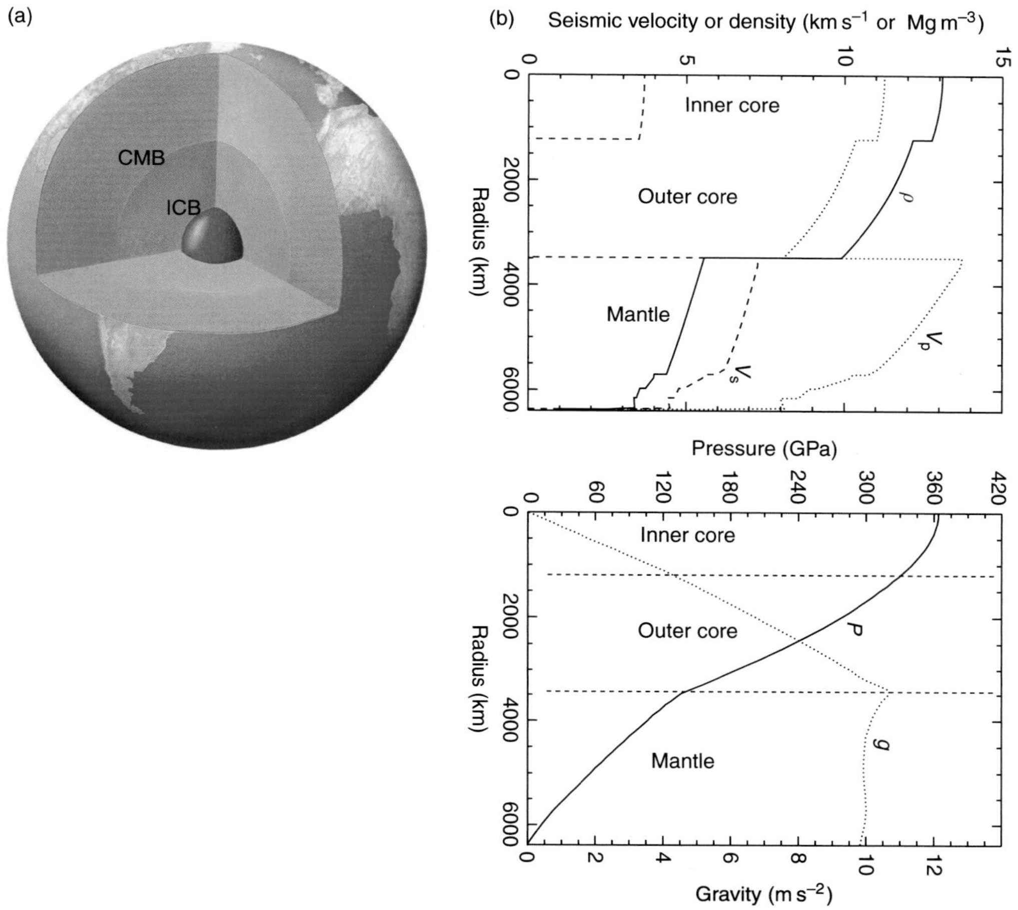

Figure 5 (a) Cutaway view showing the main layers of the Earth's interior, including the solid mantle (yellow), liquid outer core (orange), the solid inner core (red), and the core–mantle boundary (CMB) and the inner–outer core boundary (ICB). (b) Seismic Earth model PREM (Dziewonski and Anderson, 1981).

rejected permanent magnetization, ohmic decay of free electric currents, and gyromagnetism as primary causes. They also dismiss the self-sustaining dynamo mechanism, concluding, as did Schuster (1911) that the "difficulties which stand in the way of basing terrestrial magnetism on electric currents inside the Earth are insurmountable."

Evidently, Chapman and Bartels were responding to the suggestion in a short paper by Larmor (1919; **Figure 8**) that an internal circulation of a conducting fluid in the presence of a small magnetic field would induce an electric field, and if a suitable path for electric currents was created in the fluid, a magnetic field might be sustained indefinitely. In short, Larmor had proposed a self-sustaining fluid dynamo. Although Larmor's suggestion was primarily intended for application to the Sun (his paper refers to Hale's discovery of magnetic fields in sunspots), the idea seemed equally applicable to the Earth. However, it did not lead to much immediate progress on the geodynamo. Subsequently, Cowling (1933) showed that electromagnetic induction by a conducting fluid cannot maintain a steady axisymmetric field. This was the first of several antidynamo theorems that cast some doubt on the validity of the self-sustaing dynamo concept. In retrospect, it seems that Cowling's theorem was perhaps overinterpreted. A steady axisymmetric dynamo is probably an unphysical situation, and is certainly not applicable to the Earth. Nevertheless, many theorists of that era believed that Cowling's results implied that a general nonexistence proof of fluid dynamos would eventually be found.

The logjam started to break in the 1940s, beginning with a series of papers by Elsasser (1946, 1950; **Figure 9**), and E. C. Bullard (**Figure 10**) and colleagues (Bullard *et al.*, 1950; Bullard and Gellman, 1954),

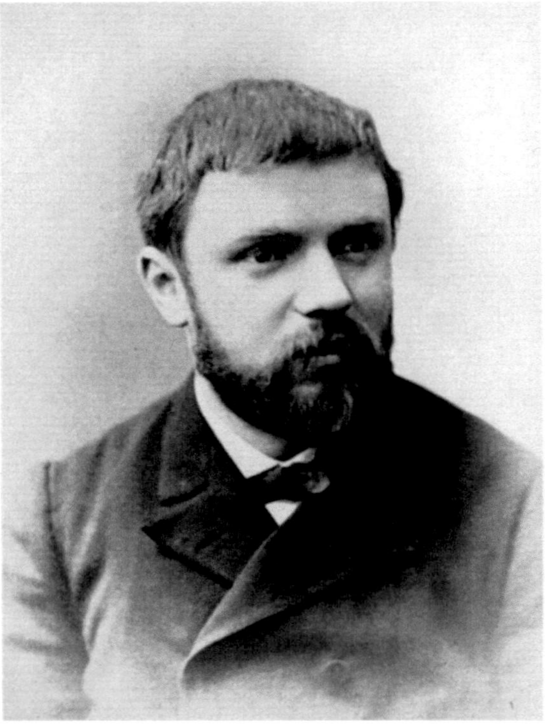

Figure 6 Henri Poincaré (1854–1912) developed the theory of rotating, precessing fluids.

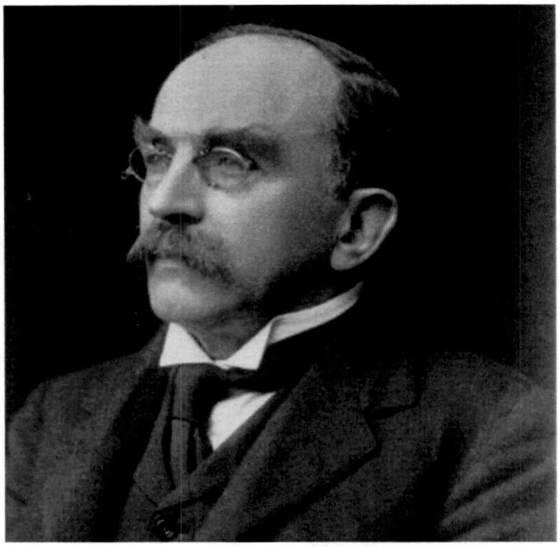

Figure 8 Joseph Larmor (1857–1942) proposed a self-sustaining fluid dynamo for the core.

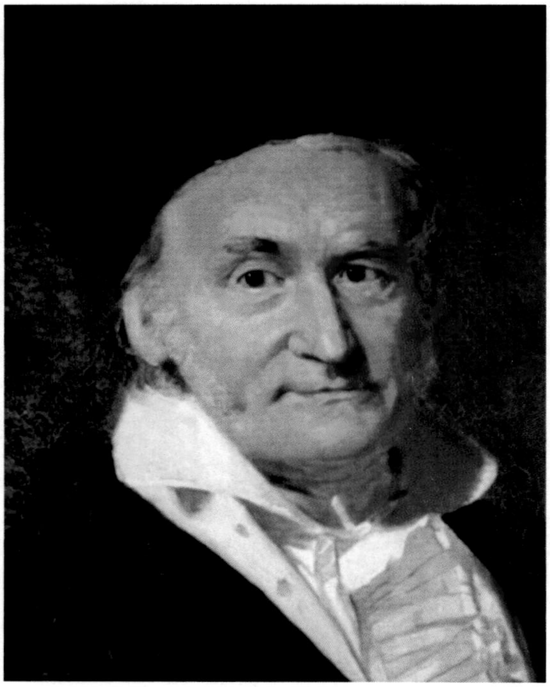

Figure 7 Carl Friedrich Gauss (1777–1855) father of geomagnetism.

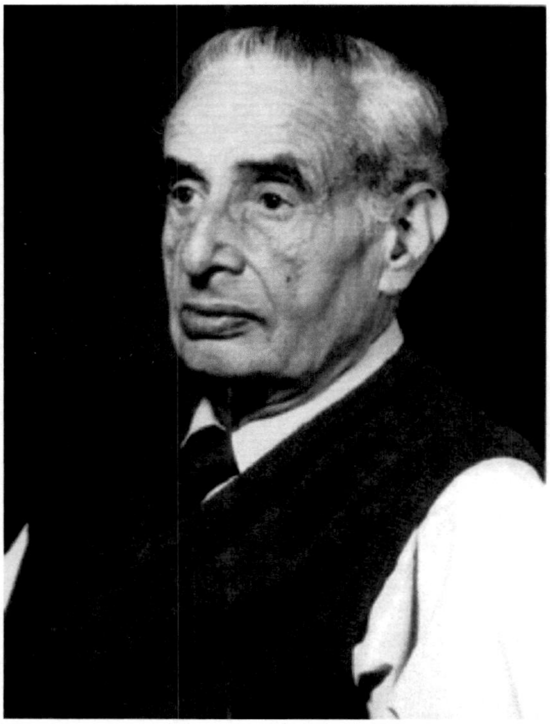

Figure 9 Walter M. Elsasser (1904–1991) pioneer in the dynamo theory of geomagnetism.

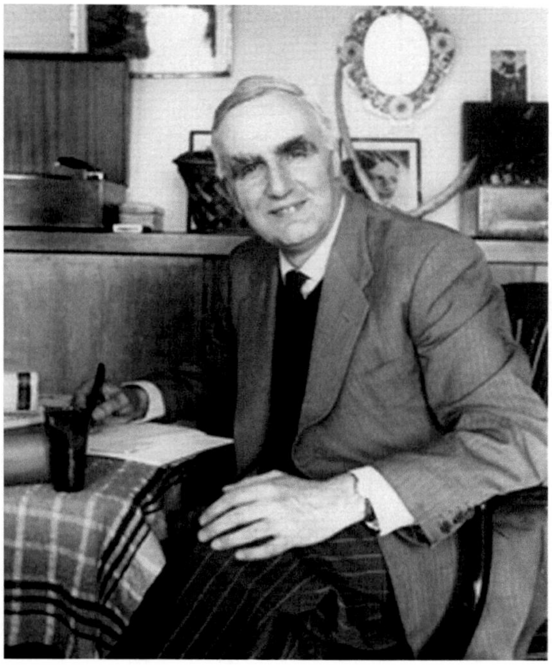

Figure 10 Edward C. Bullard (1907–1980), leading twentieth century geophysicist, the first to attempt numerical solution of the dynamo problem.

the first quantitative efforts to build a full magnetohydrodynamic (MHD) theory for the main geomagnetic field. Accomplishments in these early papers that have stood the test of time include the derivation of the magnetic induction equation for incompressible flow, representation of the magnetic field in terms of its poloidal and toroidal parts, amplification of toroidal field through interaction of toroidal shear flow with a poloidal (i.e., dipolar-type) field, selection rules for interaction between toroidal magnetic fields and poloidal flows, identification of outer-core convection as the main source of kinetic energy for the flow, and the definitions of several of the key dimensionless parameters that govern self-sustaining fluid dynamos, including the ratio of Lorentz to Coriolis forces, now called the Elsasser number. In addition, these early papers offered an explanation for the observed geomagnetic westward drift, based on conservation of angular momentum and core–mantle interaction.

The main shortcoming of these pioneering papers was that they failed to adequately prove the existence of self-sustaining dynamos, even kinematic ones. Elsasser's (1946) attempt at a theoretical demonstration fell short of completeness, and the early numerical efforts by Bullard and Gellman (1954) to construct kinematic dynamos in terms of a series of spherical harmonics have subsequently been shown to diverge when the series is extended (Liley, 1970; Gubbins, 1973). The first working theoretical examples of kinematic dynamos were by Backus (1958) and Herzenberg (1958). Although these were based on quite unrealistic flows (e.g., the Backus dynamo assumed periods of motion alternating with periods of stasis), they succeeded in demonstrating the existence of fluid dynamos, and with this, the limitations of Cowling's theorems. Several other examples of kinematic dynamos followed, in which the flow consists of large-scale, laminar-type motions amenable to analytical or simple numerical investigation (*see* Chapter 3), including an important experimental dynamo by Lowes and Wilkinson (1963) consisting of fluid and solid conductors. Although these idealized dynamos provided only limited insight into the dynamics of the geodynamo, they succeeded in defining some of the conditions necessary for magnetic field generation in the core.

Just as these successes were being registered, the direction of dynamo research began to shift away from large-scale kinematic models, following the introduction of mean-field dynamo concepts. The first mean-field theory for dynamo action was an early model by Parker (1955), who proposed that polodial magnetic field is induced in the core by the statistical action of smaller-scale convective vortices on the large-scale toroidal magnetic field. Parker used the term 'cyclonic' to describe the kinematics of convective flows in which the radial velocity correlates with the radial vorticity, a consequence of Earth's rotation. The term now used to describe such motions is 'helicity', defined as the inner product of vorticity and velocity vectors. Helical flows twist toroidal magnetic field lines into meridional planes, inducing a poloidal magnetic field from the original toroidal one.

This mechanism for inducing poloidal from toroidal magnetic field was recognized as a key ingredient missing from many of the failed efforts to produce kinematic dynamos. It became a particularly potent concept starting in the 1960s, with the introduction of 'mean field electrodynamics', a formal approach for calculating the large-scale induction properties of complex, small-scale flows (*see* Chapter 3). Mean field electrodynamics is implicit in the early work of Parker but the form in which it is used today came from astrophysics (Steenbeck *et al.*, 1967; Steenbeck and Krause, 1967). The basic idea is to represent the fluid velocity and the magnetic field as consisting of a mean and a fluctuating part, the latter part having zero average over some appropriately chosen length or time scales. Under appropriate conditions, the interaction between the small-scale velocity and magnetic fields generates a large-scale electric field that is parallel to the large-scale field. The proportionality between the induced electric field and the large-scale magnetic field is represented by the coefficient α, and the name given to this type of induction is the 'α-effect'. Dynamo action occurs if the α-coefficient is sufficiently large to overcome the inhibiting effects of magnetic diffusion. Roberts (1972) generates an extensive set of numerical calculations that showed how self-sustaining kinematic dynamos could be generated in conducting fluid spheres, using various distributions of the α-parameter, in combination with different types of large-scale flows. The nomenclature introduced by Roberts for the large-scale flows included ω for toroidal shear flows and m for meridional (poloidal) flows. Combinations of these parameters produced an assortment of dynamos. For example, α^2-dynamos, in which both the toroidal and poloidal magnetic fields are induced by the small-scale motion, $\alpha\omega$-dynamos, in which the poloidal fields is induced by the small-scale flow through the α-effect, but the toroidal field is induced by the ω-effect, and $\alpha\omega m$-dynamos, where all three effects are active. One advantage of this formalism is that kinematic dynamos exhibiting a wide variety of characteristics could easily be constructed, without explicitly specifying the details of the fluid motion, in essence by lumping all of the flow properties into a few parameters. It was subsequently shown that the α-effect is proportional in many rotating systems to the fluid helicity (Rädler, 1968; Moffatt, 1978), so that the kinematics of these dynamos could be related, at least indirectly, to real fluid motions.

With the successes of mean field electrodynamics, efforts to find dynamos driven entirely by large-scale flows waned in the 1970s. At about this same time, additional impetus for the study of mean field dynamos was provided by experimental and theoretical studies of thermal convection in rapidly rotating fluids. As discussed earlier, it is generally accepted that the most important fluid motions in the outer core for maintaining the geodynamo are convection – flows driven directly by thermal and compositional buoyancy forces. Convection in the core is surely affected by the Coriolis acceleration. Because of the small viscosity of liquid iron compounds, the Ekman number in the core is exceedingly small (the Ekman number is the ratio of viscous to Coriolis effects). In addition, the geomagnetic secular variation indicates that the fluid velocities in the core are of the order 10–20 km yr^{-1} (*see* Chapter 4). Assuming a length scale of a few hundred kilometers for these motions, the Rossby number implied for these motions is also rather small (the Rossby number is the ratio of vorticity in the fluid motion to planetary vorticity). Fritz Busse and colleagues (Busse, 1970; Busse and Carrigan, 1976) developed theoretical and experimental techniques to study the structure of convection in this, the so-called 'rapidly-rotating' limit. They demonstrated that rapidly rotating convection takes the form of small-scale columns or vortices that are nearly two dimensional (2D) and aligned parallel to the axis of rotation, as discussed in Chapter 11. Moreover, columnar convection is helical, with negative helicity in the Northern Hemisphere and positive helicity in the Southern Hemisphere (see Jones *et al.* (2000) and Chapter 5), and so is capable of dynamo action through the α^2-mechanism. Thus, the structure of rotating convection provides another important piece of the core dynamics puzzle.

While all of these advances in understanding the kinematics of dynamo action were taking place, only incremental progress was being made on the more difficult problem of dynamo equilibration. Dynamo equilibration involves the back-reaction of the magnetic field on the flow, through the Lorentz force. The Lorentz force is nonlinear in the magnetic field, and its action in the context of internally generated fields was a little-understood facet of MHDs. The main focus of efforts to understand dynamo equilibration mechanisms involved a theorem proved by Taylor (1963), showing that, for small Ekman and Rossby numbers, the torque exerted by the Lorentz force must vanish on cylinders coaxial with the rotation axis (*see* Chapter 3). This so-called Taylor constraint imposes some restrictions on the configuration and the strength of the internal magnetic field in such a dynamo. It was widely assumed that the Taylor constraint applies to the geodynamo, and therefore it determines the strength and the symmetry of the geomagnetic field inside the core. An alternative model of the field equilibration process was proposed by Braginsky (1964), the so-called 'model-Z', which assumed a nearly uniform magnetic field configuration in the core interior plus thin magnetic layers just below the core–mantle boundary.

The stumbling block for all magnetic field equilibration theories has been our lack of a full understanding of the Lorentz force in self-sustaining dynamos. The nonlinear character of the Lorentz force has precluded the development of full analytical solutions, except in a few, highly idealized theoretical cases, such as weakly nonlinear dynamo action in horizontally infinite plane layers. Convincing demonstrations of how dynamos in spherical geometry driven by fully developed convective flows subject to Lorentz forces equilibrate had to await the advent of 3D numerical models.

Full 3D, spherical MHD simulations were pioneered by astrophysicists (Gilman and Miller, 1981) for application to the Sun and magnetic stars. The first successful 3D MHD models applied to the geodynamo were made nearly simultaneously by Glatzmaier and Roberts (1995) and Kageyama and Sato (1995). Both models were driven by thermal convection, although former used the Boussinesq approximation whereas the later included fluid compressibility. Quickly thereafter, these models were extended to include the buoyancy derived from growth of the inner core, for example, the study by Glatzmaier *et al.* (1999) was based on the theory for thermo-chemical convection in a compressible fluid proposed by Braginsky and Roberts (1995). This modeling activity since has spread around the globe, with a wide variety of competing dynamo models in use (*see* Chapter 8). Further experimental confirmation of the dynamo theory came in the form of the first successful self-sustaining, purely fluid dynamos by Gailitis *et al.* (2003) and Muller *et al.* (2006) using liquid sodium (*see* Chapter 11).

1.4 Core Dynamics and the Geomagnetic Field

The study of the geomagnetic field is a rich topic in its own right, and is the subject in this treatise. An extensive description of the geomagnetic field and paleomagnetic field can also be found in Merrill *et al.* (1998) and the full theory of the geomagnetic field is given in Backus *et al.* (1996). Here it is appropriate to briefly list some of the main features of the geomagnetic field that we seek to explain in terms of the core's dynamics. First and foremost is the persistence of an internally generated geomagnetic field, not just the modern field, but a field sustained throughout most if not all of Earth history. The main geomagnetic field that originates in the core is at least 3.5 Ga, according to the rock record, and quite possibly is as old as the core itself. The present-day dipole moment, 7.8×10^{22} A m^2, is probably somewhat larger than the long-term average value, which may be around 6.5×10^{22} A m^2. Significantly, there is little evidence for a secular trend in the dipole moment over geologic time, although there is abundant evidence of fluctuations, as described below. For reference, the free decay time of the dipole field in the core is only about 20 ky. The existence of an ancient field, the lack of evidence of a decreasing dipole moment, and the inadequacy of permanent magnetization as a source for the field are three facts that demand there be a regeneration mechanism, that is, a dynamo theory.

The dominance of the dipole component is evident in the present-day field, where its energy exceeds that in any other spherical harmonic, even at the core–mantle boundary. This dominance becomes even stronger if time averages of the field are considered. The characteristic time constants of the nondipole terms in the spherical harmonic representation of the field are measured in centuries, so that when the field is averaged over a millenium or more it tends toward a dipolar state. Over somewhat longer time averages, of the order of 10 ky, the components of the dipole moment vector lying in the equatorial plane average out, and the field tends toward a geocentric axial dipole, an important reference state called GAD. The GAD representation is the basis of much of paleomagnetism and it appears to approximate the actual time-averaged field at about the 95% level. The main deviations from GAD in the time-averaged field in the last few million years are a small axial quadrupole term (with the same sign as the dipole) and a small 1–2% additional deviation that may be a nonaxial field, but is not fully resolved.

Despite its long-term persistence, the geomagnetic field exhibits fluctuations on a wide range of timescales, which are fully described of this treatise. Very briefly, these include the so-called geomagnetic jerks at the high-frequency end of the spectrum – probably the shortest time signal capable of traversing the weakly conducting mantle and still producing a measurable signal at the surface – and decade timescale fluctuations in tilt and other field parameters. The classic geomagnetic secular variation occurs on centuries timescales, including the present-day dipole moment decrease, which has been a feature of the geomagnetic field ever since Gauss' intensity measurements in the 1830s. Large amplitude, irregular dipole moment strength fluctuations occur on millenium timescales, and individual dipole polarity chrons have characteristic timescales of several hundred thousand years. The dipole polarity chrons are separated by polarity reversals, and are often subdivided into shorter subchrons, separated by brief dipole transitions called excursions. The duration of the polarity reversal process, typically 2–12 ky, defines another characteristic timescale of the dynamo. On the longest timescales, the frequency of polarity reversals changes with a 100–200 My time constant, the minimum in the cycle being marked by superchrons, lengthy intervals that are evidently devoid of polarity reversals. The relationship between these characteristics of polarity reversals and the dynamics of the core is the subject of Chapter 9.

1.5 Core Energetics

The sequence of events in the formation and evolution of the core is reasonably well established, even though the timing of events and important details of the processes remain uncertain. It is widely accepted that the core began to form during or shortly after the accretion from the solar nebula. Heat generated by large impacts during the accretion phase may have resulted in widespread melting in the near surface, allowing the metallic compounds to separate and sink to the center, as illustrated in **Figure 11**. The gravitational energy released in this process is enough to raise the core temperature by several thousand degrees, enough to ensure that the early core was entirely liquid. Along with Fe, the proto-core metal included Ni and some important light constituents, particularly S, O, and Si. Many of the abundant heat-producing elements such as Th and U were probably partitioned into what became the mantle and crust, but some heat producers, most notably ^{40}K, may have dissolved into the iron compounds and been swept into the core. The exact process by which the core segregated from the mantle remains uncertain, but regardless of its details, this was an extreme event capable of completely mixing the early core and setting the stage for its later evolution dominated by convection. Subsequently, as the cooling of the Earth proceeded, the temperature in the core became low enough for the solid inner core to begin to nucleate by crystallization at the center.

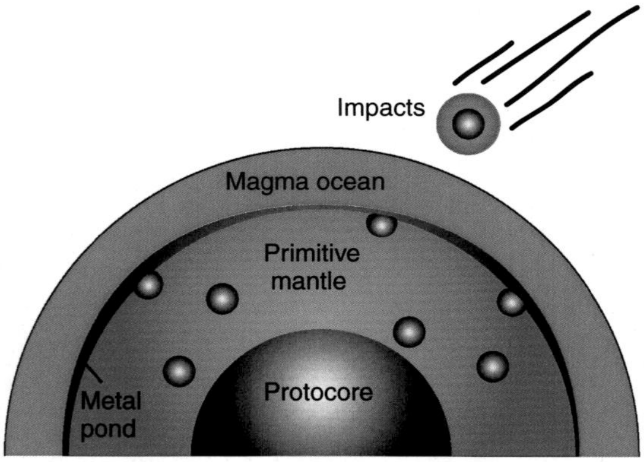

Figure 11 Schematic illustration of core formation. Iron droplets from impacting planetismals sink through a deep magma ocean and pond at the top of the high viscosity primitive mantle, forming large metal diapirs that descend into the protocore. From Wood BJ, Walter MJ, and Wade J (2006) Accretion of the Earth and segregation of its core. *Nature* 441: 825–833 (doi:10.1038/nature04763).

As described in Chapters 5, 6, and 7, the iron-rich outer-core liquid has a small enough viscosity to permit strong convective flows, waves, turbulence, and other types of fluid motion with typical velocities of 20 km y^{-1}. Together these fluid motions induce electric currents and magnetic fields, in the core which have persisted throughout Earth's history in spite of the electrical resistance of the iron and other core compounds. Collectively, the processes by which magnetic fields and electric currents are produced and maintained by core dynamics are called the 'geodynamo'. The geodynamo is fundamentally an MHD phenomenon, in which the kinetic energy of the outer-core fluid is converted into electromagnetic energy. This energy conversion is promoted by Earth's rotation, which structures the flow through the Coriolis acceleration, leading to sustained magnetic field production. The magnetic fields so produced exert forces back onto the core, altering the fluid motion and allowing the geodynamo to attain an overall statistical equilibrium in the face of a host of fluctuations and instabilities.

Since the energy for the geodynamo is ultimately derived from the overall cooling of the Earth (possibly agumented by radioactive heating in the core), the history of the geodynamo is linked to the thermal and chemical evolution of the core, and also to the thermal evolution of the mantle. The rejection of the light elements as the inner core crystallizes is one of the major sources of buoyancy for convection in the liquid outer core. Fluid enriched in light elements is expected to form ascending flows in the outer core, while the fluid that is relatively depleted in light elements is expected to form descending flows. Thermal buoyancy is another important source of outer-core convection. Thermal buoyancy originates near the core–mantle boundary, a consequence of heat extracted from the core by the cooler mantle. In addition, thermal buoyancy is also produced by inner-core growth, through release of latent heat of crystallization near the inner-core boundary. The relative contributions of thermal and compositional buoyancy to convection in the outer core are uncertain, but most estimates put the compositional effect larger than the thermal effect, by a factor of 2 or 3. However, this has not always been the situation. Earlier in Earth history the inner core was likely smaller and played a proportionally smaller role in driving outer-core convection, and prior to the nucleation of the inner core (which is variously estimated to have begun 1–3 Ga), the geodynamo would have had to rely on thermal convection alone.

An important factor in the energetics of the core is the control on the rate of core evolution imposed by the overlying mantle. The solid mantle has a larger thermal mass than the core, and its subsolidus convective overturn is much slower. Accordingly, heat transfer by the mantle is the rate-limiting process in the heat transfer from the core, and so is the rate limiter for convection in the core, inner-core growth, and the power available to the geodynamo. Recent estimates of the total heat flux from the core to the mantle are in the range of 6–14 TW, which correspond to 15–35% of the surface geothermal flux (see **Table 4**). This would seem more than adequate to power convection in the outer core, except that the adiabatic gradient in the outer core is quite large, nearly $0.8\,\mathrm{K\,km^{-1}}$, and owing to the large thermal conductivity of liquid iron, the heat conducted down the core adiabat is also large, about 4 TW. This conducted heat flow makes no direct contribution to outer-core convection, and represents a substantial tax on the geodynamo: the core must produce enough heat to maintain an adiabatic thermal profile just to ensure that the convection occurs throughout the outer core. In doing so, it uses up a relatively large fraction of its available energy.

Other complications in core–mantle thermal interactions arise because of the way the mantle draws heat from the core. The so-called D″ -layer, the 200 km thick layer at the base of the lower mantle above the core–mantle boundary is among the most heterogeneous regions in the Earth. As depicted in **Figure 12**, the lateral variations of seismic properties are larger in D″ than anywhere else in the lower mantle. In addition, the D″ layer is dynamically active. Its long wavelength structure is related to, and probably dictated by, the large-scale pattern of flow, density, and viscosity variations in the lower mantle (Schubert *et al.*, 2001). On shorter wavelengths, its own internal dynamics shape its structure. The D″ is generally considered to be a source of mantle plumes, hot upwelling structures that produce volcanic hot spots on the surface.

Table 4 Dynamical properties

Property	Notation	Units	Approximate range
Total heat flux, CMB	Q	TW	10 ± 4
Adiabatic heat flux, CMB	Q_{ad}	TW	4 ± 1
Buoyancy flux, outer core	F	$m^2\,s^{-3}$	$\sim 10^{-13}$
Core age	t_c	yr	4.4×10^9
Inner-core age	t_i	yr	$(0.5–3) \times 10^9$
Inner-core growth rate	dr_i/dt	$m\,yr^{-1}$	$(0.3–1) \times 10^{-3}$
Viscous diffusion time	t_ν	yr	$\sim 10^{10}$
Thermal diffusion time	t_κ	yr	$\sim 10^{11}$
Magnetic diffusion time	t_λ	yr	$\sim 10^5$
Dipole diffusion time	t_{dip}	yr	2×10^4
Circulation time	d/U	yr	100–300

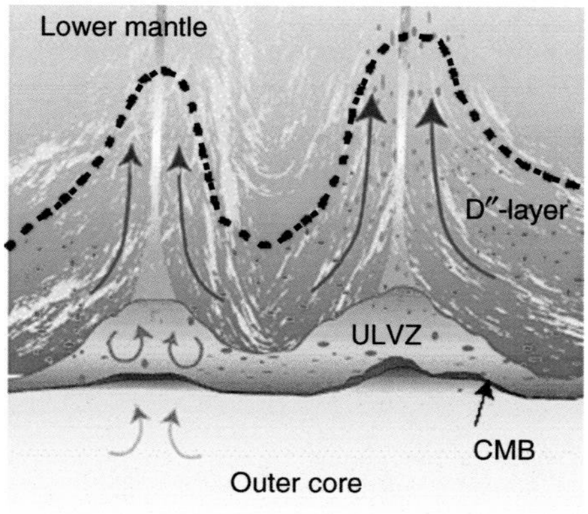

Figure 12 Schematic depiction of the D″ layer near the core–mantle boundary (CMB), illustrating various types of core–mantle interaction and small-scale dynamical processes that have been proposed for this region. The seismic ultralow velocity zone (ULV2) consists of thin lenses of material directly above the CMB that may be partially molten and may have a composition intermediate between the mantle and the core. Courtesy of E. Garnero.

Thermal plumes are expected to form in D″ by virtue of the large superadiabatic thermal gradient there, estimated to be in the neighborhood of $5\,\mathrm{K\,km^{-1}}$. The major volcanic hot spots – those with a large enough buoyancy flux to originate in D″ and traverse the whole mantle – are relatively small in number and are widely distributed over Earth's surface. This suggests that the spacing of plumes near the core–mantle boundary may be controlled by the large-scale pattern of mantle flow, although the heat flow from the core to the mantle in the neighborhood of the plumes may be controlled by the dynamics of the plumes themselves. In any event, a strong case can be made that the local heat flux from the core to the mantle is spatially variable over the core–mantle boundary, and this pattern of variability has changed with time in concert with changes in the pattern of convection in the lower mantle and changes in location and intensity of mantle plumes. Just how the core and the geodynamo respond to these mantle-induced thermal changes, and what effects this coupling has on the Earth system as a whole, are frontier research issues.

1.6 Core Dynamics as a Heat Engine

The overall operation of the dynamo can be thought of as a giant Carnot heat engine with several interacting parts. This analogy, which is fully described in Chapter 2, can also be found in several general treatments (Verhoogen, 1979; Roberts *et al.*, 2003; Gubbins *et al.*, 2003). Heat is extracted from the core by the mantle, providing a source of kinetic energy in the core that maintains the geomagnetic field. The sources of buoyancy for convection in the outer core have already been mentioned – thermal buoyancy due primarily to the secular cooling of the core, and compositional buoyancy derived from the growth of the solid inner core. Both of these sources are linked by the global energy flow in the core, whose path is the following: secular cooling of the core results in solidification at the inner-core boundary and growth of the inner core. Preferential fractionation of the light elements and compounds into the outer core reduces its mean density and lowers its gravitational potential energy with time (Loper, 1978). Since the adjustment toward lower potential energy requires convection to redistribute the light elements throughout the outer core, production of kinetic energy of fluid motion is necessary for this process to occur fast enough to keep up with the cooling. Part of this kinetic energy is then transformed into magnetic energy, the kinematic step in the dynamo process. This is called the gravitational dynamo mechanism. There are additional energy sources available to help drive the geodynamo, including latent heat released by inner-core crystallization (Verhoogen, 1961), and stirring of the core by tidal forces and by the precession of Earth's spin axis (Malkus, 1968; Tilgner, 2005; *see also* Chapter 7). The thermal–compositional effect is probably the most important energy source now, but the rotational effects may have played a proportionally larger role at times in the past.

Another important aspect of the Carnot heat engine analogy is the efficiency of the geodynamo. Fluid dynamos have intrinsically low thermodynamic efficiency. The primary energy dissipation process within the core is ohmic dissipation – ordinary Joule heating by electric currents flowing in the imperfectly conducting core material. The efficiency of the geodynamo can be expressed as the ratio of the total Joule heating in the core to the total convective heat transfer by the dynamo process. Theory indicates this ratio is about 0.1. Numerical dynamo models indicate the amount of Joule heating necessary to maintain the present-day geomagnetic field is between 0.1 and 1.0 TW. Accordingly, the convective heat flow in the core lies between 1 and 10 TW; to this we should add the heat flow down the core adiabat, which is about 4 TW, giving a total core heat loss of 6–14 TW. The lower end of this range does not present problems for the current energy budget of the Earth, but the high end does. In addition, the energy demands of the geodynamo become more acute further back in time. For these reasons, one of the primary objectives in core dynamics is to derive better constraints on the energy budget and energy flow in the geodynamo.

1.7 Convection and Dynamo Action

The finite electrical conductivity of the core means that the main dipole part of the geomagnetic has an e-fold decay time of about 20 000 years. The existence of the field over geological time requires continual regeneration on timescales shorter than this. The convection in the core is estimated to be sufficiently fast to induce large-scale magnetic fields on the required short timescales, and moreover, theoretical, experimental, and numerical studies of convection in rapidly rotating spherical shells give us valuable insight into how it works. Planetary rotation affects core convection through the so-called Taylor–Proudman constraint, where the local axis of fluid vorticity tends to align parallel (or anti-parallel) to Earth's spin axis (Taylor, 1923). The resulting convective motion, shown in **Figure 13**, is often called 'columnar', because the variation in flow structure parallel to the spin axis is smaller than the variations in perpendicular planes, that is, parallel to the equatorial plane. Theory indicates that the Lorentz force due to the geomagnetic field within the core distorts the columns in the equatorial plane, as does the compressibility of the core fluid and the curvature and roughness of the core–mantle and inner-core boundaries. Columnar convection includes secondary motions with components parallel to the spin axis. The combination of these secondary motions and the primary geostrophic motion makes convection helical, in which the vorticity and velocity vectors are correlated on average, being generally negative in the Northern Hemisphere of the outer core and generally positive in the Southern Hemisphere. This helical motion gives rise to the aforementioned α-effect through which poloidal magnetic field is induced from toroidal magnetic field, and vice versa. The forward and backward transformation of magnetic energy

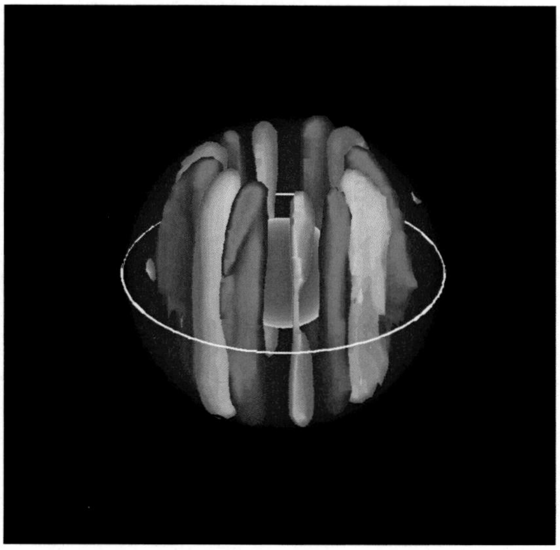

Figure 13 Columnar-style convection in a rotating spherical shell. Numerical model results show columnar concentrations of axial vorticity (red = positive, blue = negative vorticity).

between these two types of fields in the outer core is the key kinematic ingredient in the geodynamo. Helicity is not just a property of convection; other rotationally dominated flows, such as the flows induced by precession, are helical. Another induction mechanism is provided by large-scale toroidal flows, such as thermal winds and the flows associated with differential rotation of the inner core. The shear in these nearly horizontal flows converts poloidal magnetic field lines into toroidal magnetic field lines, much like windings on a ball of string, providing the so-called 'ω -effect'. The relative contributions of the α and ω effects in the geodynamo are uncertain, because the strength of the toroidal field in the core is unknown. If the toroidal field is only about as strong as the poloidal field in the core (1–10 mT), then the geodynamo is likely an α^2-type dynamo, and the ω effect due to large-scale shear flows does not control the field strength in the core. Much stronger field strengths in the core would imply a larger role for the ω effect, and the geodynamo would then be classified as $\alpha\omega$ or $\alpha^2\omega$. It is significant that both effects are functions of latitude, offering a fundamental explanation for why the geomagnetic field tends toward an axially symmetric dipole time-average configuration.

An important dynamical issue is the physics of magnetic field equilibration – what controls the geomagnetic field strength. In this context, theorists distinguish between two generic classes of dynamos, called 'weak' and 'strong', respectively. In strong field dynamos, the Lorentz force exerted by the magnetic field on the fluid is comparable to the Coriolis acceleration, whereas in weak-field dynamos the Lorentz force is secondary. The fluid motion in strong field dynamos is sometimes called 'magnetostrophic', indicating the tendency for the fluid to circulate around, rather than directly crossing, the magnetic field lines. In strong field dynamos, the Elsasser number is of order one. Assuming the geodynamo is a strong field dynamo, an Elsasser number of one implies an root mean square (rms) field of about 4 mT in the core. Strong-field dynamos tend to equilibrate by trading off magnetic and kinetic energy. Increases in magnetic energy increase the Lorentz force, reducing the kinetic energy of the fluid, which ultimately causes the field to decrease back toward its equilibrium strength. Conversely, decreases in magnetic energy decrease the Lorentz force, allowing the kinetic energy to increase, which induces a stronger magnetic field. This situation would seem to provide a stable equilibrium field strength (given a constant driving force for the convection), and may explain why the geomagnetic dipole moment has been stable for much of Earth's history.

However, the above simple argument fails to explain the most dramatic of geomagnetic field changes, the polarity reversals. Chapter 9 discusses what is known about the reversal process from theory and from numerical models. Mathematically, a simple change in the sign of the magnetic field (a full reversal) makes no difference in the governing MHD equations for the geodynamo, since the Lorentz force is quadratic and the induction equation is homogeneous in the magnetic field. In other words, there is no bias for polarity in an MHD dynamo; both polarities are equally likely. The facts that the paleomagnetic record shows no evidence for polarity bias, and also shows no evidence for differences in field structure with polarity, are strong points in favor of a MHD origin for the geodynamo.

The existence of occasional geomagnetic reversals, widely spaced in time and yet short in duration, suggested to some early investigators that they might possibly have an external trigger. However, theory and numerical experiments have convincingly shown that polarity reversals occur in dynamos through internal dynamics, without any external influence needed. Some dynamos reverse periodically, most notably the $\alpha\omega$-dynamos, which often show wave-like or oscillatory magnetic field behavior. Other dynamos reverse more irregularly, a consequence of chaotic fluctuations in fluid motion. It is likely that several different mechanisms are capable of producing polarity reversals in the geodynamo. It is also likely that there are many failed attempts at reversal, marked by events such as polarity excursions, for example. For these reasons, it is not clear that an individual reversal represents a very significant event in the evolution of the core.

Possibly more significant for core dynamics is the average frequency of reversals, which has changed appreciably through time. The average rate of reversals is about 4 per million years during the past few million years, but this rate has increased systematically since the 35 My long Cretaceous normal superchron, which ended about 83 Mo. Because dynamo theory indicates that the important dynamical timescales intrinsic to the geodynamo should be substantially less than hundreds of millions of years, many geodynamicists have looked to the mantle for the cause of long-term polarity reversal changes. As described in Chapter 8 of this volume, it is now well documented that the frequency of polarity reversals is sensitive to the overall vigor of convection in the core, and in addition, spatial variations in core convection, both of which the mantle controls. Additional evidence for a mantle control on the geodynamo comes from the tendency for some transition fields to follow statistically prevered paths during polarity change. A full understanding of the ways the core and mantle interact, and how these interactions affect the geodynamo and also the dynamics of the mantle, remain important goals for the future. Current ideas and possible new directions in this area are the topic of the Chapter 12 of this volume.

1.8 Simulating the Geodynamo

Along with the new discoveries of core structure coming from seismology, the delineation of the present-day and ancient geomagnetic field, and progress in fluid dynamo experiments, the most important advances in understanding core dynamics in this decade have come from numerical simulations of the dynamo process. Chapters 8 and 9 describe the rapid development, considerable successes, inherent limitations, and future prospects of this approach. Here it is appropriate just to describe the basic ideas behind these models and comment on their overall objectives.

Numerical dynamo models solve the conservation equations for fluid motion and magnetic field induction in rotating spherical shell geometries with homogeneous magnetic boundary conditions, to ensure that any sustained magnetic field is internally generated, without an external source. The field and the flow are free to evolve through mutual interaction. The goals are to obtain magnetic fields resembling the geomagnetic field, and to understand how the various interactions generate different aspects of geomagnetic field behavior.

As mentioned above, this approach has developed rapidly, starting with the breakthrough simulations of Glatzmaier and Roberts (1995) and Kageyama and Sato (1995), which quickly became a global activity. Numerical dynamos driven by Boussinesq convection in thick, rotating fluid shells commonly produce magnetic fields similar to the geomagnetic field, with a strong axial dipole component, secular variation, and occasional polarity reversals, as shown in **Figure 14**. Dynamo models with inner cores smaller than the present Earth predict different magnetic

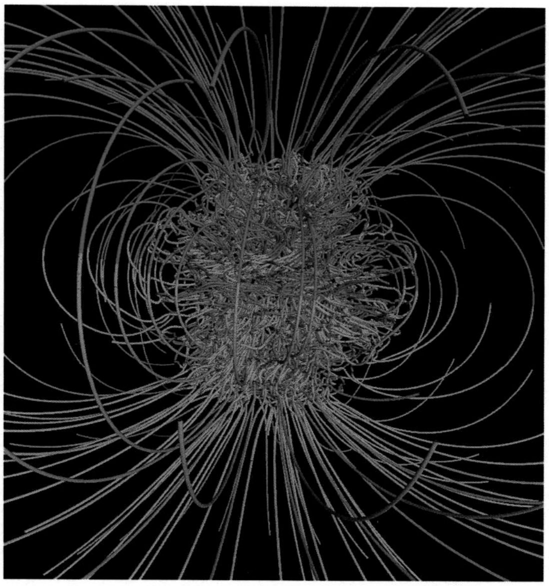

Figure 14 Magnetic field lines of force from a numerical dynamo model. Blue = inward directed, yellow = outward directed field. A dominantly dipolar magnetic field extends out from the core into the mantle. Complex magnetic field is induced inside the core by turbulent columnar convection. Courtesy of G. Glatzmaier.

field intensity at early stages of inner-core growth. These and other more recent high-profile modeling studies (e.g., Takahashi *et al.*, 2005) have led to the impression, in some quarters, that the dynamo problem has actually been solved. In truth, however, many fundamental problems remain in applying numerical dynamo model results directly to the Earth, not to mention the other magnetic planets, which have been swept under the rug, so to speak, in this initial burst of progress. Some of these problems stem from our limited knowledge of the core, but other arise because the dynamo models are far removed in parameter space from the Earth. Specifically, numerical dynamos rotate too slowly, are less turbulent, and have far too large viscosity (relative to electrical conductivity) compared to the core. In terms of the dimensionless parameters that control convective dynamos, this combination of factors means that the Rayleigh number is too small, and the Ekman and magnetic Prandtl numbers are too large in the dynamo models. Estimates of these and other dimensionless parameters important to core dynamics are listed in **Table 5**. As demonstrated in Chapter 8, prospect of direct numerical simulation with realistic values of these parameters is remote, because of the enormous temporal and spatial resolution such calculations would require.

As Chapter 8 describes, several different types of numerical models are now used for studying the geodynamo. Most dynamo models are driven by thermal convection, although some use a combination of thermal and compositional buoyancy, and there is substantial progress on precession-driven dynamos (*see* Chapter 7). In nearly all of these models, the poloidal magnetic field is generated by what is essentially a mean-field effect, through helicity produced by the convection. A part of the toroidal field is also generated by the convection, and some is generated by large-scale shear flows, particularly within the tangent cylinder of the inner

Table 5 Dimensionless parameters

Parameter	Notation	Definition	Approximate value
Radius ratio	r^*	r_i/r_o	0.35
Prandtl number	Pr	ν/κ	0.1–0.5
Magnetic Prandtl number	Pm	ν/λ	$\sim 10^{-5}$
Roberts number	q	κ/λ	$\sim 10^{-6}$
Ekman number	E	$\nu/\Omega d^2$	$\sim 3\times 10^{-14}$
Magnetic Ekman number	E_λ	$\lambda/\Omega d^2$	$\sim 4\times 10^{-9}$
Thermal Ekman number	E_κ	$\kappa/\Omega d^2$	$\sim 10^{-14}$
Rayleigh number, ΔT	Ra_T	$\alpha g \Delta T d^3/\kappa\nu$	$\sim 10^{20}$
Rayleigh number, q	Ra_q	$\alpha g q d^4/k\kappa\nu$	$\sim 10^{22}$
Dissipation number	Di	$\alpha g d/C_p$	0.3
Elsasser number	Λ	$\sigma B_{rms}^2/\rho\Omega$	~ 1
Elsasser number, CMB	Λ_{CMB}	$\sigma B_{CMB}^2/\rho\Omega$	~ 0.01
Reynolds number	Re	Ud/ν	$\sim 10^6$
Magnetic Reynolds number	Rm	Ud/λ	300–600
Peclet number	Pe	Ud/κ	$\sim 10^7$

Note: Ekman and Elsasser numbers are defined here without factors 2 in denominator.

core. Many numerical dynamos produce realistic large-scale magnetic fields: dominantly axial dipoles with Elsasser number $\Lambda \simeq 0.1$; nearly flat spectra at the core mantle boundary and realistic velocities (magnetic Reynolds number $Rm \simeq 100$–1000), even though the input Ekman number E, Rayleigh number Ra, Prandtl number Pr, and magnetic Prandtl number Pm are decidedly unrealistic. An example of an advanced dynamo model is shown in **Figure 14**. Some numerical dynamos have strong westward drift and large zonal velocities at low latitudes, although many do not have this feature. In many models, the most rapid westward drift is found at high latitudes, due to thermal wind-style flow within the inner-core tangent cylinder.

How do different boundary conditions affect numerical dynamos? The difference between no-slip and free-slip boundary conditions is relatively minor in dynamos with Newtonian (uniform) viscosity, although it is large in dynamos using hyperviscosity. In contrast, heat-flow boundary conditions have a strong effect, particularly nonuniform heat flow imposed at the core–mantle boundary. Large variations in core–mantle boundary heat flow tend to kill numerical dynamos. Moderate variations in boundary heat flow generate departures from axisymmetry in the time-average magnetic field, and also influence the frequency of polarity reversals. The so-called 'tomographic' heat-flow boundary condition assumes that core–mantle boundary heat flow variations are proportional to seismic velocity variations in the D″ layer. This boundary condition produces heterogeneity in numerical dynamos that can be compared with heterogeneity in the geomagnetic field, including concentrated flux spots at high latitudes, regions with westward drift, and regional differences in the time-average magnetic field and secular variation. In addition, calculations using this boundary condition suggest that thermal coupling between the mantle and the inner core is possible; the nonuniform heat flow at the core–mantle boundary is transmitted through the whole outer core, producing azimuthal variations in heat flow at the inner-core boundary. These characteristics are not found in numerical dynamos with uniform core–mantle boundary heat flow, which generate time-average magnetic fields with axial symmetry. Tomographic models do not explain all of the nondipole field however. In particular, they fail to explain the large quadrupole component in the paleomagnetic field.

Polarity reversals occur in numerical dynamos with large temporal fluctuations, particularly when the dipole field is weak and time variable. In some cases the polarity transition is short, comparable to the timescale of the transitions in the paleomagnetic record. However, some reversals in dynamo models are characterized by extremely long transition periods. Two broad classes of reversals have been found. In convection-dominated dynamo models, large temporal fluctuations are usually found at high Rayleigh number where the flow is strongly chaotic. In these dynamos the reversals are irregular (possibly random) in time and seem to develop from anomalously large reversed flux patches. It has been shown that the frequency of these irregular reversals is sensitive to thermal boundary conditions: boundary heat flow that is compatible with the heat flow pattern intrinsic to the dynamo stabilizes polarity, whereas incompatible boundary heat flow destablizes polarity (Glatzmaier *et al.*, 1999). Dynamo models with strong zonal flows also show polarity reversals, but at more regular intervals. This type of reversal may be produced by dynamo wave instabilities.

1.9 Mantle Effects within the Core

Many of the processes we associate with core dynamics and the geodynamo have timescales that are short compared to mantle dynamics timescales. **Figure 15**

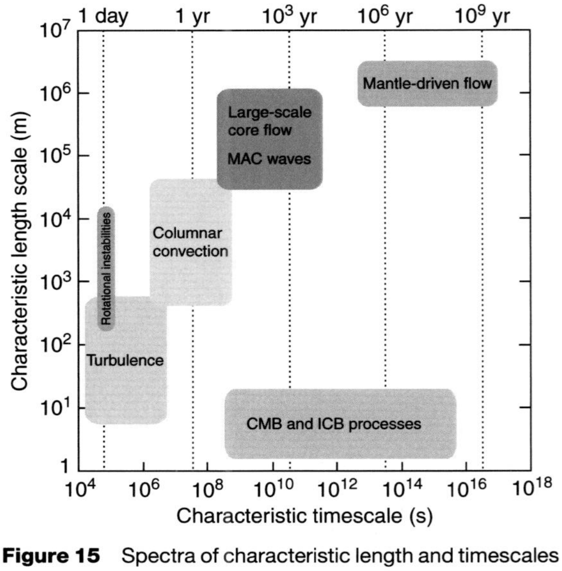

Figure 15 Spectra of characteristic length and timescales in core dynamics. MAC – Magnetic, Archimedean, Coriolis waves.

summarizes the space and timescales that characterize some of the important core dynamical processes. As discussed in Chapters 7, 4, and 12, inertia, Alfven, and buoyancy-driven waves have characteristic timescales ranging from 1 day to decades or centuries. The convective overturn time in the core is measured in centuries to millennia. The dipole free decay time is about 20 000 years, and the average duration of the last 10 dipole polarity chrons is about 250 000 years.

But in addition to these relatively short timescales connected with fluid dynamical processes in the core, the geodynamo shows evidence of much longer timescales. These more slowly evolving processes have often been attributed to the influence of the mantle on the core. The best-known example of this class of phenomena is the modulation in the frequency of magnetic polarity reversals seen in the record of sea-floor magnetization. A related phenomenon is the occasional magnetic superchron – lengthy intervals of time that appear to be devoid of reversal behavior. The reversal frequency modulation occurs on a 100–200 My timescale, which is very suggestive of the timescale for changes in mantle convection. Additional but less certain evidence comes from the persistence of nondipole field components and the tendency of the field to reverse polarity along paths that are confined in longitude. Since it seems unlikely on theoretical grounds that the core has such a long-term dynamic memory of its own, it is natural to look at the role of the mantle in the geodynamo process.

The lowermost region of the mantle is known to be among the most heterogeneous regions in the planet. Mantle heterogeneity that could directly affect the geodynamo includes lateral variations of temperature and heat flow near the core–mantle boundary, topography on the core–mantle boundary, and lateral variations in other physical and chemical properties, possibly including melts. As discussed in Chapter 12, this heterogeneity has the potential to influence the core in multiple ways, although a full exploration of the ramifications of each type of heterogeneity is yet to be made.

Thermal interaction with the mantle is expected to influence the core on very long timescales. Temperature differences of several hundred degrees are expected in the lower mantle, and these will cause substantial heat flow heterogeneity over the core–mantle boundary. It is possible that there are regions on the core–mantle boundary where the heat flow is less than the heat conducted down the core adiabat, and stable thermal stratification will prevail in these areas. In spite of the very small lateral temperature differences this boundary heterogeneity produces in the core, the circulation in the core will be strongly modified, through the addition of thermal wind-type flows and the suppression of convection in the stable regions.

The dynamical response of the core to mantle thermal heterogeneity is a rich subject, one that has received considerable attention in the past decades (e.g., Zhang and Gubbins, 1992) but the basic questions here are far from resolved. Numerical dynamo models indicate that mantle heterogeneity induces long-lasting, large-scale circulation in the outer core (*see* Chapter 12) and can influence the frequency and the character of polarity reversals (*see* Chapter 9). The fact that core–mantle boundary heterogeneity changes in both pattern and magnitude on 100–200 My timescales suggests a causal link between mantle convection and the long-term geodynamo variations, including reversal frequency modulation, superchrons, and low-frequency variations in the dipole field strength.

Chemical exchange is another type of long-term core–mantle interaction, that is even less well understood than thermal interactions. As discussed in Chapter 12, isotopic anomalies from volcanic hot spots have been interpreted by some geochemists as evidence for core contamination of mantle plumes originating near the core–mantle boundary. The seismic complexity of the core–mantle boundary region, which indicates heterogeneity over a wide range of spatial scales, is broadly consistent with the idea that chemical reactions, sedimentation, infiltration and compaction, and other processes active near the Earth's surface may have counterparts in the environment of the core–mantle boundary region.

Very different kinds of core–mantle interactions can occur on shorter timescales. Elevated electrical conductivity near the base of the mantle permits the geomagnetic secular variation to induce electric currents in the lower mantle that can couple the mantle and core electromagnetically (Rochester, 1962), and allow for exchange of angular momentum between the core and the mantle on decade and century timescales. Other short-term interactions include gravitational coupling between mantle density heterogeneity and density heterogeneity in the core, most notably the equatorial bulge of the inner core.

1.10 The Dynamical Inner Core

Perhaps the most unanticipated development in core dynamics in the past decade has come from realization of the importance of the solid inner core. Long thought to be homogeneous and dynamically passive, it is now seen to play a much more significant role in the evolution of the core and the geodynamo process. The core was recognized to be seismically anisotropic about 15 years ago, based on the travel times of seismic compressional waves, which travel a few percent faster along polar paths, compared with equatorial paths. Initially this anisotropy was seen as more or less uniform through the whole inner core, which led to an early suggestion that it could be a single crystal, possibly aligned by Earth's rotation. However, subsequently it has been shown that the anisotropy is not entirely rotationally aligned, and moreover, it is spatially heterogeneous through the inner core. In particular, anisotropy is weak or absent for a few hundred kilometers below the inner-core boundary, and even more surprising, it appears to vary in strength between the Eastern and Western Hemispheres of the inner core (Ishii and Dziewonski, 2003).

Several explanations have been offered to explain the inner-core anisotropy. The obvious one is textural anisotropy, the statistical alignment of crystals. Another possible explanation is partial melt fabric. Each of these explanations has its own merits, although both involve difficulties. There are also several competing theories for the origin of the anisotropy, which are discussed fully in Chapter 10 of this volume. Most of these explanations rely on the ability of the inner core to deform on timescales of tens of millions of years, which requires the solid material to have a small yield strength and a finite viscosity.

The most provocative interpretation of inner-core dynamics is its anomalous rotation. Numerous studies have used observations of small, secular changes in the travel times of seismic P-waves traversing the inner core to infer that the inner core is rotating relative to the crust and mantle. The initial reports suggested a relatively fast super-rotation, with the inner core rotating eastward relative to the crust by about one degree per year (Song and Richards, 1996). Subsequent investigations have demonstrated that this rate is too high. The observed splitting of certain normal modes that are sensitive to inner-core structure indicates little or no anomalous rotation, whereas follow-up studies based on the original P-wave technique still indicate a super-rotation, but only 0.2-0.3° yr^{-1}. Interestingly, the theoretical possibility of inner-core super-rotation had been known for more than a decade (*see* Chapter 10), and prograde anomalous inner-core rotation was observed in several of the earliest numerical dynamo models. Indeed, the pattern of core convection predicted by numerical dynamo models is fully consistent with inner-core super-rotation (*see* Chapter 8). However, there are some unresolved basic issues here. First, the uncertainty in the seismic observations does not entirely preclude an inner-core co-rotating with the rest of the Earth. In addition, it is theoretically possible that gravitational coupling locks the inner core to the mantle (*see* Chapter 12). This question will ultimately be settled, as ever better seismic data accumulate.

1.11 Future Prospects and Problems

It is impossible to know if the rapid progress we have witnessed in core dynamics over the past decade or two will continue at this same pace. Rather than attempt to forecast what lies ahead, for this field, it may be more useful to define some important goals for the future, in terms of key core dynamics questions. Most of the chapters in this volume have identified the key science questions particular to each area. Below is a partial list of broadly defined questions, by which future progress in core dynamics may be gauged.

- What is the present-day energy budget of the core?
- What is the path for energy flow in the present-day core?
- How old is the inner core?
- By what processes does the inner core formed, what accounts for its complex structure, and how has its formation affected the rest of the Earth?
- What is the strength of the geomagnetic field in the core, and what processes control its strength?
- How much energy is dissipated in maintaining the geodynamo?
- What is the relative importance of chemical versus thermal buoyancy in core convection?
- What are the characteristic length and time scales of core convection?

- What is the nature of turbulence in the outer core, and how does turbulence influence core dynamics at larger scales?
- How does dynamo action in numerical models differ from dynamo action in true liquid metals?
- Can numerical dynamos be scaled to Earth conditions?
- What is the relationship between the flow at the top of the core inferred from geomagnetic observations and the geodynamo process?
- What are the contributions of rotational instabilities to the geodynamo?
- Is the inner core rotationally locked to the mantle, or does it differentially rotate?
- What are the critical processes that lead to magnetic polarity changes and other forms of geomagnetic variation?
- What is the nature of long-term mantle control on core and geodynamo processes, such as the chemical and thermal evolution of the core and the frequency of magnetic polarity reversals?
- What are the characteristics of shorter-term core–mantle interactions?
- Which aspects of Earth's core dynamics are applicable to the cores in other planets?

1.12 Additional References

Although there is no published book like this volume, with comparable breadth and focus on the dynamics of the core, there are numerous works that treat the specific foundations of this subject. Several of these deserve special mention here. Monographs by Verhoogen (1979), Jacobs (1987), and the collected volumes edited by Dehant *et al.* (2003) and Jones *et al.* (2003) treat most aspects of core structure and energetics. The fundamentals of rotating fluid dynamics and geophysical turbulence can be found in Greenspan (1968) and Rhines (1979), respectively, and the fundamentals of dynamo theory in books by Moffatt (1978) and Davidson (2001). Stacey (1992) and Poirier (1988) are good sources of basic information on core properties, Schubert *et al.* (2001) describe the dynamics at the base of the mantle, and together Merrill *et al.* (1998) and Backus *et al.* (1996) give a comprehensive account of the geomagnetic and paleomagnetic fields as they relate to core dynamics. There have been numerous review papers on the recent advances in numerical dynamo models, including Dormy *et al.* (2000), Kono and Roberts (2002), Glatzmaier and Roberts (2002), and Glatzmaier (2002). Braginisky and Roberts (1995) have developed a complete set of equations for the core convection including thermal and compositional buoyancy.

1.13 Summary of the Chapters in This Volume

1.13.1 Energetics of the Core

Chapter 2 analyzes the gross energetics of the core, a subject of long-standing and on-going controversy. After reviewing the present state of the core, author Francis Nimmo derives the global balances of energy and entropy, which are critical to such basic issues as the growth rate of the solid inner core and the power necessary to maintain the geodynamo. Special attention is given in this chapter to the power requirements and the long-term consequences of inner-core growth. Growth of the inner core results in compositional convection and latent heat release, and the author concludes that the former is likely the most important energy source for the geodynamo. Geophysical considerations indicate that the present-day heat loss from the core is 6–14 TW. Nimmo shows how this heat flow translates into estimates of inner-core age, which vary from about 0.3 to around 2 Ga if the geodynamo is driven by compositional convection. Even when the uncertainties in core properties are taken into account, this range indicates that the inner core is far younger than the core itself. The lower end of this range puts the onset of inner-core nucleation well within the Paleozoic era, which would imply that the inner core has existed only for about two Wilson cycles of plate tectonics, and that it postdates the explosion of advanced life forms over the Earth's surface. The author concludes the chapter by considering the possible alternatives to the young inner-core scenario. The most important of these alternatives is an end-member model in which the core contains appreciable radioactive heat sources, most notably radioactive potassium ^{40}K, and the power requirements of the geodynamo remain low throughout inner-core growth. Such a model is marginally capable of extending the inner-core age to about 3 Ga, broadly consistent with some estimates obtained from rare-earth isotope measurements. A final resolution of this controversial yet central issue could come by detecting inner-core formation in the record of the paleomagnetic field, more precise modeling of core–mantle interaction, or some as yet unexploited geochemical evidence on the early core.

1.13.2 Theory of the Geodynamo

Long before there were 3D numerical dynamo simulations, and long before self-sustaining experimental sodium dynamos, the theoretical foundations for understanding the geodynamo were laid by applied mathematicians and physicists. The accomplishments of this dedicated group are all the more noteworthy in light of the fact that they were working without the benefits of realistic numerical or laboratory examples of fluid dynamos.

The important elements of dynamo theory are set out by Paul Roberts in Chapter 3. Roberts begins by defining what a fluid dynamo is, what is meant when a dynamo is 'self-sustaining', and the difference between kinematic dynamos, in which the fluid velocity is specified *a priori* and the magnetic field is then calculated, and MHD dynamos, in which the energy source is specified, and both the flow and the magnetic field are calculated simultaneously. Dynamo theory starts from the (pre) Maxwell equations for electromagnetic induction in a moving, conducting fluid. An important limiting case that is often used for intuition and practical applications is Alfven's frozen flux theorem, appropriate for a perfect electrical conductor.

Roberts then presents the basis for kinematic dynamos in both Cartesian and spherical geometries. Important results include a lower bound for the critical magnetic Reynolds number for dynamo action, plus a set of antidynamo theorems, including the famous theorems by Cowling, that reveal some of the necessary conditions for fluid dynamos to become self-sustaining. The chapter includes a full account of the very influential mean-field electrodynamics, the statistical treatment of the induction effects of turbulent fluid motion. This leads to several types of mean field dynamos, which derive their dynamo action from the α-effect, an induction property of convective turbulence, and the ω-effect, the induction property of large-scale shear flows, as discussed previously in this chapter.

Now that the study of the geodynamo has come to rely so heavily on large-scale numerical simulations, it is fair to ask what role classical dynamo theory plays in the present discussion. Roberts' summary provides a clear reply to this question. Because there are no true 'simple' fluid dynamos, analog models, be they numerical or experimental, offer only limited insight into the geodynamo on their own. By virtue of their inherent complexity, a substantial level of theoretical insight is needed to interpret the results of any numerical dynamo simulation, or any laboratory dynamo experiment, for that matter. Kinematic dynamo theory provides us with the fundamental concepts and a useful terminology, for analyzing and describing how complex fluid dynamos work.

1.13.3 The Large Scale Flow in the Core

The theory in other Chapters (Chapters 5, 7, and 6) and the results of numerical dynamo models presented in other Chapters (Chapters 8, 9) indicate that a broad spectrum of motions is likely in the fluid outer core, governed by a variety of possible force balances that include several possible driving mechanisms. In light of this inherent complexity in the geodynamo processes, there is little hope of deducing the nature of the flow in the core on the basis of theoretical considerations alone. For this reason, a substantial effort has been made over the past few decades to infer as many properties of the flow in the core as possible from other types of geophysical observations. By far the most success in this regard has come from analyzing the secular variation of the geomagnetic field on the core–mantle boundary. There is a full description of the properties of geomagnetic secular variation of the treatise.

In this volume we are primarily concerned with interpreting the images of flow near the top of the core, obtained from the geomagnetic secular variation, in terms of the types of motion that are important for core dynamics in general, and the geodynamo in particular. We are also interested in the evidence for interaction between the core and other parts of the Earth system as revealed by these motions. In Chapter 4, Richard Holme summarizes what we know about the actual fluid motions in the core, both in the present-day and in the past, as far back in time as the geomagnetic record allows us to infer them. Holme points out that geomagnetic secular variation has been used to 'trace' motions in the core for more than four centuries, starting with Halley's famous 1692 discovery of its westward drift, which Halley interpreted as the result of relative motion between magnetized regions in Earth's interior. Holme shows how the early idea of a uniformly westward drifting core, based on a preferred westward shift of anomalies in the core field with time, has been replaced by a far more complex picture of the core flow, which includes westward motion mostly confined to the Atlantic Hemisphere. However, the long-standing controversy over the geomagnetic westward drift and its interpretation provided impetus to develop the modern techniques for imaging core flow, which are based on the concept of frozen magnetic flux. Holme first reviews the frozen-flux hypotheses, including its assumptions and limitations, and then goes on to summarize the important aspects of the core flow that the applications of this technique have revealed. He shows how frozen flux reveals that an energetic, large-scale flow is present in the outer core, with characteristic velocities of 10–20 km yr^{-1}. This motion appears to be most vigorous beneath the Atlantic and Indian Oceans (since it is proportional to the geomagnetic secular variation, which is greatest there), and it includes several large-scale vortices and jets that are suggestive of the large-scale quasi-geostrophic vortices and jets seen in the general circulation of the atmosphere and ocean. Smaller-scale motions are doubtless also present (*see* Chapters 5 and 6), but are filtered from the geomagnetic secular variation by the electrical conductivity of the mantle and by permanent magnetization in the crust, and are therefore invisible at the surface. Wave motion in the core is another possible interpretation of the secular variation. Holme discusses several interesting interactions between the the core and the mantle related to core flow, including the decade scale variations in the rate of rotation. These variations involve angular momentum exchange between the core and mantle, and appear as nearly periodic variations in the zonally averaged part of the core flow. In addition, the core flow shows some response to the abrupt, transient changes in the core field, the class of phenomena called geomagnetic jerks.

1.13.4 Thermal and Compositional Convection in the Core

Convection in the liquid outer core is the primary energy source for the geodynamo, and also the primary mechanism for heat and mass transfer through the core. In Chapter 5, author Christopher Jones examines the complex fluid mechanics involved in convection in the core. Both thermal convection and compositional convection are important in the outer core, thermal convection produced by temperature variations arising at the inner-core boundary and especially at the core–mantle boundary, the compositional convection being a product of chemical fractionation of light elements such as sulfur and oxygen into the outer-core fluid accompanying crystallization of the inner core.

After introducing the basic physical principles involved in thermochemical convection in the core, Jones derives the basic equations governing the convection, with emphasis on the effects of compressibility, which are present in the outer core by virtue of the enormous (~200 GPa) increase in hydrostatic pressure between the core–mantle and inner–outer core boundaries. He then conducts a systematic tour of the physics of convection, starting with a model of high Rayleigh number convection in a nonrotating, electrically insulating fluid, then progressively adding in the effects of rotation and magnetic fields, ending up with a model of core convection in which all of the known physical ingredients are present and interacting with each other. The result is a model of core convection that is multiply anisotropic, with the flow jointly constrained by Earth's rotation, by the Lorentz force due to interaction between the induced magnetic field and associated electric currents, and other dynamical effects. Jones shows how the characteristic velocity, time, and length scales of this convection are all determined by internal dynamical balances involving subtle balances between Coriolis, Lorentz, buoyancy and inertial forces, and the spherical shell configuration of the outer core. He goes on to show how this rotating magnetoconvection model for the core dynamics fits into the numerical models of the geodynamo described in Chapter 8 on the large scale, and in the other extreme, how it fits with the concepts of smaller-scale core turbulence described in Chapter 6. Jones then considers how the particular deep Earth environment influences core convection, specifically, how the surrounding mantle with its substantial lateral heterogeneity can affect convection in the core. The control of core convection by the mantle is only partially understood, and is among the highest priorities for future expansion of knowledge in this whole subject.

1.13.5 Turbulence and Small-Scale Dynamics in the Core

Turbulence is ubiquitous in geophysical and planetary fluids, although it exists in a wide variety of forms. The causes and consequences of turbulence in the core is a relatively new subject in geodynamics, but with the growth of sophisticated numerical and laboratory models of the core dynamics, it is coming to play a more central role in our view of the geodynamo. In Chapter 6, David Loper gives a connected account of turbulence and the small-scale dynamics of the core. He starts by describing the special conditions for small-scale dynamics in the core, an environment dominated by the Coriolis and Lorentz forces due to planetary rotation and the core's magnetic field, respectively. This environment is affected by the great thickness of the outer-core fluid, where relatively weak buoyancy forces can generate turbulence even though the overall fluid velocity is small. The dominant form of energy dissipation in the core is ohmic (Joule heating) rather than viscous dissipation as in a normal fluid.

Interpreting the core as a turbulent MHD system, Loper finds that several dynamically distinct regimes are expected, including complex boundary layers near the inner and outer-core boundaries with overlaping magnetic, viscous, and thermal layers. Outside the boundary layers are regions dominated by buoyant plumes, especially beyond the inner-core boundary. He argues that the interior of the outer core should be well mixed, but there is the possibility of a stable layer near the top of the core, a consequence of incomplete mixing of lighter elements released by inner-core crystallization. He presents a model for each of these regions based on length- and time-scaling considerations. Lastly, he discusses some possible ways to parametrize core turbulence for use in global models of the core and the geodynamo.

1.13.6 Rotational Dynamics

Convection is not the only type of motion in the fluid outer core. There are a host of instabilities that arise because of irregularities in Earth's rotation. These are well documented in theory and experiments, for example, in the Chapter 11 of this volume. In Chapter 7, Andreas Tilgner summarizes the state of knowledge on rotation-induced core flows. He begins by deriving the equations of motion for a fluid rotating about a variable axis, in which the transverse or Poincaré acceleration appears in addition to the familiar Coriolis and centripetal accelerations. He summarizes the properties of inertial oscillations and the structure and transport properties of viscous Ekman layers.

The central problem in this subject is the flow induced by precession. Tilgner begins by presenting the classical solutions to the inviscid problem, including the Poincaré solution on which most subsequent studies were based. He then systematically introduces the complexities of a real fluid, first the effects of viscosity, then instabilities, tidal excitation and finally nonlinear interaction with convective flows. A long-standing question is whether precession can support a planetary dynamo by itself, without assistance from buoyancy-driven motions. Tilgner discusses the status of this debate, and points out the directions future work must take to supply a definitive answer.

1.13.7 Numerical Dynamo Simulations

The use of first-principles numerical models of self-sustaining fluid dynamos has been the most significant technical development in the study of the geodynamo in decades. The enormous progress in this arena, and the challenges for the future, are the subjects of Chapter 8. Ulrich Christensen and Johannes Wicht first review the governing equations for the geodynamo in their basic, Boussinesq form, including the fundamental dimensionless parameters that control the system. They also define the critical dimensionless output parameters that characterize each dynamo model solution and provide the basis for comparison with the geodynamo. Attention is paid to the choices of boundary conditions appropriate for the core–mantle boundary and the inner–outer-core boundary. The choice of boundary conditions is crucial for modeling specific aspects of the geodynamo behavior, particularly its interaction with the mantle, but also its interaction with the solid inner core.

In this chapter the challenges in dynamo modeling have already been discussed that stem from the fact that the physics of the geodynamo is characterized by extreme values of the governing dimensionless parameters, which are not accessible to direct numerical simulation. This situation has guided both the development of numerical dynamos and the techniques used to apply them to the Earth. Each of these aspects are discussed in detail in Chapter 8. Christensen and Wicht first summarize the simplifying approximations that are sometimes used in the models, such as representing the buoyancy force in convection in terms of just a single variable, imposed symmetries, and modifications to the inertial and diffusive terms in the equations. In terms of the numerical techniques, Christensen and Wicht focus on the spectral (or semispectral) approach, which has been the work-horse method to date, although they point out its inherent limitations and the need to continue development of alternative methods that will be more suitable for the massively parallel computer architectures in the future.

The authors then summarize the main classes of dynamo models found to date, describing their changes in structure as the main controlling parameters (Rayleigh and Ekman numbers) are systematically varied in the direction of Earth values. They point out that most convectively driven numerical dynamo models are basically of the α^2-type, although ω-effects from thermal wind flows are sometimes important contributors to the dynamo mechanics.

The rapid increase in the number of high-quality numerical dynamos in the literature has made it possible to derive scaling laws for the basic behavior of convective dynamos. This is an important development, since it provides a way to extrapolate numerical results with more modest input parameters to the extreme parameter regime of the geodynamo. Christensen and Wicht show a number of rather simple scaling laws for the main dynamo properties, that project reasonably well to the geodynamo, providing additional evidence that dynamo models are capturing important parts of the core's dynamics.

As for the prospects in dynamo modeling, it may be that the low-hanging fruit has, to a large extent, already been picked, and that future progress will hinge on solving a number of thorny physical problems that were conveniently swept under the rug in the rush of the past decade. Many of these problems involve the complicating but interesting effects discussed in other chapters of this volume. Christensen and Wicht end the chapter with a sober assessment of some of these problems, and where their solutions in the context of numerical dynamos might ultimately be found.

1.13.8 Magnetic Polarity Reversals in the core

The fact that the geomagnetic field has stably persisted for most of Earth's history, while exhibiting occasional polarity reversals, is among the strongest pieces of evidence in support of the dynamo theory. As discussed at length of this treatise, and as summarized in Chapter 9 of this volume, both the existence and the known characteristics of magnetic polarity reversals offer general confirmation of the dynamo interpretation, as well as severe challenges for any successful model of the geodynamo.

In Chapter 9, Gary Glatzmaier and Robert Coe examine the behavior of polarity reversals in numerical dynamo models in light of the evidence on the nature of reversals as seen in the paleomagnetic record. Key observations include the time required for polarity reversal (2 to about 12 ky), the characteristic duration of stable polarity chrons (100–1000 ky, but highly variable), the nature of the transition field, and the possibility of polarity bias, statistical differences between magnetic fields for normal (i.e., present-day-type) polarity and the reverse polarity configurations. They argue that nearly all records of reversals indicate reduced intensity for the transition field, but that the actual transition field structure is complex and may differ substantially between individual reversals, as evidenced by the variety of virtual geomagnetic pole (VGP) paths. They also point to the significance of large directional excursions, which appear to be more frequent than full polarity switches and may represent failed reversals in the core.

Dynamo models have shown convincingly that it is not necessary to have an external trigger for polarity reversals. There are several mechanisms that operate in numerical dynamos which can produce spontaneous polarity reversals, either in a semiregular way (i.e., nearly periodic reversals) or in a chaotic way (in which the reversal sequence is seemingly random in time). Glatzmaier and Coe analyze some specific numerical examples of each of these types of reversals. Their general conclusion is that chaotic, nearly random reversing dynamos better approximate the observed reversal behavior as measured by the main geomagnetic polarity timescale (i.e., the reversals that define the major polarity chrons). However, they also find that the presence of more frequent, short reversals implied by excursions may indicate that a more regular reversal mechanism also operates in the geodynamo.

The authors examine the role of reversals in illuminating the interaction between the core and the mantle, a recurring theme in the other chapters of this volume. The transition field structure in some (but not all) reversals seems to indicate some mantle control on the process, in that the VGP paths for these reversals tend to cluster within one of two longitude belts. As mentioned earlier in this chapter, one interpretation of this phenomenon is that it represents the tendency for the transition field to configure itself in accord with the large-scale heterogeneity of the lower mantle. As Glatzmaier and Coe show, the transition fields in numerical dynamo models are indeed sensitive to large-scale boundary heterogeneity. Preferred reversal paths do emerge in the statistical behavior of dynamo models with heterogeneous thermal boundary conditions, for example. However, like nearly all aspects of core–mantle interaction, the paleomagnetic data are still ambiguous and the numerical dynamo models are still too rudimentary to yield a firm conclusion on this issue.

The evidence for very long timescales in the polarity record, as typified by magnetic polarity superchrons and the 100–200 My modulation of the reversal frequency record, is of considerable theoretical interest for properly interpreting the longer-term thermal interaction of the core and the mantle. Here again, Glatzmaier and Coe point to several numerical studies that demonstrate dynamo model sensitivity to mantle heterogeneity, particularly thermal heterogeneity at the core–mantle boundary. Reversal frequency seems to be a function of the total heat loss at the core–mantle boundary, and also is sensitive to the pattern of that heat loss. The authors demonstrate that dynamo models generally show more frequent reversals with increased total heat flow, a consequence of increased flow complexity. They also present some preliminary evidence, drawn from the very few studies on this subject, that the pattern of core–mantle boundary heat flow can affect reversal frequency. Sensitivity of reversal frequency to the pattern of mantle dynamics offers a way to link the history of the paleomagnetic field to the history of mantle convection and plate tectonics – an exciting prospect indeed, but one that will require a large, sustained and focused effort before a full picture emerges.

1.13.9 Inner Core Dynamics

Chapter 10 summarizes the wealth of new information on the inner core, and discusses the various roles the inner core plays in core dynamics and the geodynamo. Ikuro Sumita and Michael Bergman first describe the current understanding of the phase diagram of the core, which is critical for properly interpreting the effects of inner-core growth, and for estimating the present-day thermal regime of the core. They then use the information from the phase diagram as the basis for a discussion of the small-scale processes near the inner–outer core boundary that affect and are affected by, the solidification. Solidification processes at the inner-core boundary naturally lead to their next topic, estimating the grain size and creep viscosity of the inner-core solid material. Viscosity and grain size are closely linked in the inner core (as they are in the crystalline mantle), and the value of inner-core viscosity controls the rate of creep deformation in dynamical processes such as the relaxation that accompanies heterogeneous inner-core growth and super-rotation. It also affects the ability of the inner core to convect on its own. Sumita and Bergman then describe the various models that have been advanced to explain the recent and most perplexing observations of the inner core, its heterogeneity, anisotropy, and (possible) anomalous rotation. These three are linked, since the lateral heterogeneity and the anisotropy of the inner core serve as the basis for detecting (and correctly interpreting) its possible anomalous rotation. They conclude that the evidence for anisotropy is unequivocal; the evidence for lateral heterogeneity is compelling (but likely subject to many future revisions), whereas the evidence for anomalous motion now indicates the super-rotation is quite small (a few tenths of degree per year at most) and there is still the possibility that it is zero.

1.13.10 Laboratory Experiments on Core Dynamics

Laboratory experiments always play a crucial role in fluid dynamics research, and core dynamics is no different in this respect. In Chapter 11, Philippe Cardin and Olson conduct a tour of the important experiments that have shaped our physical intuition about the flows in the core that produce the geodynamo and govern its evolution. Most, if not all, of the dynamical phenomena discussed elsewhere in this volume can be seen in their relatively pure forms in the framework of laboratory experiments. This chapter is possibly singular in all of geodynamics, in collecting photographic examples of all the types of flow thought to be significant in the core.

The chapter begins with a description of the fundamental flow structure in rotating fluids, the geostrophic column. Geostrophy is familiar in atmosphere and ocean dynamics, but this partiular class of structures, so-called Taylor columns, are especially important in the outer core because of its thick shell geometry and the dominance of the Coriolis acceleration over the effects of viscosity. Where geostrophic columns touch the core–mantle and inner–outer core boundaries, the no-slip condition leads to various shear boundary layers, including Ekman boundary layers. Similarly, where distinct geostrophic columns meet, free shear layers called Stewartson layers form. The most prominent Stewartson-type shear layer in the core is formed along the inner-core tangent cylinder – the imaginary cylinder parallel to the Earth's rotation axis and tangent to the inner-core equator. The strength of the inner-core tangent cylinder is an important element of the internal structure of the core, because the tangent cylinder wall tends to divide the outer core into dynamically separate fluid regions.

Complementing the theory of rotational instabilities in Chapter 7, this chapter discusses the experiments on precession-induced flows in spheres, spherical shells, and spheroids. These flows include the classic Poincaré-type flow (in which the outer core rotates about an axis that lags the mantle in precession), and the parametric instabilities that occur when precession effects reach finite amplitude. Experimental examples of the quasi-geostrophic turbulence discussed in Chapter 6 are shown in connection with these rotational instabilities.

Although rotational instabilities are likely in the core, the main source of energy for the geodynamo probably comes from thermo-chemical convection. The chapter includes a discussion of the clever experimental techniques that were developed to investigate thermal convection in rapidly rotating spherical shells, starting with the pioneering work by F. H. Busse and co-workers. Subjects include the nature of rotating convection at onset, and how the structures change as the convection becomes fully developed, rotating convection in liquid metals, and the differences between compositionally driven versus thermally driven flows.

Fluid mechanics in the core is strongly influenced by the geomagnetic field. As discussed in Chapter 5, the theoretical foundation here is classical MHDs of an electrically neutral but conducting fluid, subject to the combined effects of buoyancy, rotation, and its own internally generated magnetic field. An important simplification to this system uses an externally applied magnetic field in place of a self-sustained one; the resulting flow is called magnetoconvection. The fundamentals of rotating magnetoconvection as it occurs in laboratory fluid metals such as mercury, gallium, and molten sodium are described, with a focus on the differences between magnetic versus nonmagnetic convection in a rotating fluid.

Lastly, an account is given of the recent successes of, and the future prospects for, laboratory self-sustaining fluid dynamos. The design and the results from the Riga and Karlsruhe sodium dynamos are reviewed, both of which achieved self-sustaining behavior within the same year, 2000. These two dynamos used mechanical forcing to reach supercritical Reynolds numbers and helical-shaped ducts and channels to get the proper feedbacks between poloidal and toroidal field components. The next challenge is to produce a laboratory dynamo in an Earth-like geometry, with rotation included. The authors discuss the technical difficulties associated with scale and power requirements associated with this type of experiment, and close by summarizing the on-going efforts at labs around the world to make rotating spherical lab dynamos a reality.

1.13.11 Core–Mantle Interactions

In Chapter 12, Bruce Buffett examines a wide range of possible ways the mantle can affect core dynamics, and corresponding ways the core can affect the mantle. He separately considers the effects of thermal, mechanical, electromagnetic, and chemical interactions, and how each of these might be detected at the Earth's surface. As Buffett points out, the possible inventory of core–mantle interactions is quite large, comparable to the number and scope of interactions between the lithosphere and the hydrosphere. In terms of timescales, they range from about 1 day for rotational interactions to hundreds of millions of years for thermal and chemical interactions. The basic elements of core–mantle thermal interaction were referred to earlier in this chapter. Mantle dynamics has a controlling influence on the heat loss from the core, and as Buffett argues, both the total heat flow and the pattern of heat flow at the core–mantle boundary are expected to have a major influence on core dynamics. Furthermore, this influence is time dependent, because of the evolving thermal regime of the core–system, and also because the pattern and strength of mantle flow varies with time. During times when heat loss to the mantle is low, a thermally stably stratified layer might develop in the outer-core fluid below the core–mantle boundary. This layer could be destroyed by an episode of increased boundary heat flow, or through the accumulation of lower density, iron depleted fluid rising up from the inner-core boundary region. It is expected that high-frequency components of the geodynamo would be damped when the stable layer was present, resulting in a quieter geomagnetic field during those times. Conversely, at times when convection is vigorous below the core–mantle boundary, the geomagnetic field is expected to show more variability. The same reasoning applies to spatial variations of boundary heart flow, a consequence of the pattern of mantle convection. These variations produce lateral variations in the intensity of convection, and also larger-scale thermal wind flows, both of which tend to destroy the rotational symmetry of the geomagnetic field. Dynamo models indicate that the frequency and the character of magnetic polarity reversals are sensitive to boundary heat-flow heterogeneity, with some types of boundary heterogeneity acting to destabilize the dynamo (resulting in more frequent reversals) and other types of heterogeneity acting to stabilize it. The primary control on the character of reversals comes through longitudinal bias, that is, a tendency for the equatorial component of the dipole to amplify at longitudes where the heat flow is large, which in turn causes a statistical preference for the dipole to reverse along these longitudes. There is suggestive evidence for many of these thermal effects in the paleomagnetic and geomagnetic field behavior, but as Buffett explains, unambiguous identification of these effects requires better resolution of the paleomagnetic field and a more thorough comparison with dynamo models.

Electromagnetic core–mantle interactions described in Chapter 12 include the electromotive force exerted by the core field on the mantle through its interaction with electric currents in the mantle. The strength of this interaction depends on the electrical conductivity of the lower mantle, which is not very large unless it contains an appreciable amount of core material. The best-understood interaction of this type is the exchange of angular momentum between core and mantle which produces decade-scale length-of-day variations. Less well understood are electromagnetic interactions arising from lateral variations in mantle electrical conductivity. The effects of these interactions are expected to be most pronounced when the field is varying rapidly (as it is today) and particularly during polarity transitions. Mechanical interactions include the effect of core–mantle boundary topography and roughness on core dynamics, the large influence of seemingly small topographic variations being well known in ocean and atmosphere dynamics. Unfortunately, the core–mantle boundary topography has proved to be exceedingly difficult to image seismologically, so the importance of this interaction has yet to be quantified. Another class of mechanical interactions involve gravitational coupling between the solid inner core and the mantle, which limits the freedom of the inner core to rotate separately from the mantle.

Lastly, Buffett summarizes the evidence for and against active chemical interaction between the mantle and the core, including the mechanisms through which mass exchange across the core–mantle boundary might occur. This subject has recently been stimulated by the interpretation of certain trace element isotopes from mantle rocks as having a core signature. There are multiple ways that mass exchange might occur across the core–mantle boundary, so the real issue here is to determine which, if any, of them are large enough to produce an observable chemical tracer.

References

Alfe D, Gillan MJ, and Price GD (2003) Thermodynamics from first principles temperature and composition of the Earth's core. *Mineralogical Magazine* 67: 113–123.

Arrhenius S (1900) Zur Physik des Vulcanismus. *Geologiska Foreningens i Stockholm Forhandlingar* 22: 395–420.

Backus G, Parker R, and Constable C (1996) *Foundations of Geomagnetism*. Cambridge: Cambridge University Press.

Backus GE (1958) A class of self-sustaining spherical dynamos. *Annals of Physics* 4: 372–447.

Birch F (1952) Elasticity and constitution of the Earth's interior. *Journal of Geophysical Research* 57: 227–286.

Birch F (1964) Density and composition of mantle and core. *Journal of Geophysical Research* 69: 4377–4388.

Boehler R (1996) Melting temperature of the Earth's mantle and core: Earth's thermal structure. *Annual Review of Earth and Planetary Sciences* 24: 15–40.

Braginsky S and Roberts P (1995) Equations governing convection in Earth's core and the geodynamo. *Geophysical and Astrophysical Fluid Dynamics* 79: 1–97.

Braginsky SI (1964) Self-excitation of a magnetic field during the motion of a highly conducting fluid. *Soviet Physics JETP* 20: 726–735.

Brush SG (1996) *Nebulous Earth*. New York: Cambridge University Press.

Buffett BA (2003) The thermal state of Earth's core. *Science* 299: 1675–1677.

Bullard EC, Freedman C, Gellman H, and Nixon J (1950) The westward drift of the Earth's magnetic field. *Philosophical Transactions of the Royal Society of London Series A* 243: 67–92.

Bullard EC and Gellman H (1954) Homogeneous dynamos and terrestrial magnetism. *Proceedings of the Royal Society of London Series A* 247: 213–278.

Bullen KE (1954) Compositon of the Earth's outer core. *Nature* 174: 505.

Busse FH (1970) Thermal instabilities in rapidly rotating systems. *Journal of Fluid Mechanics* 44: 441–460.
Busse FH and Carrigan CR (1976) Laboratory simulation of thermal convection in rotating planets and stars. *Science* 191: 81–83.
Chapman S and Bartels J (1940) *Geomagnetism*. Oxford: Clarendon Press.
Cowling TG (1933) The magnetic field of sunspots. *Monthly Notices of the Royal Astronomical Society* 94: 39–48.
Darwin GH (1879) On the bodily tide of viscous and semi-elastic spheroids, and on the ocean tides upon a yielding nucleus. *Philosophical Transactions of the Royal Society of London* 170: 1–35.
Davidson PA (2001) *An Introduction to Magnetohydrodynamics*. New York: Cambridge University Press.
Dehant V, Creager K, Karato S, and Zatman S (2003) Earth's Core, Dynamics, Structure, Rotation. *American Geophysical Union Geodynamics Series 31*. Washington, DC: American Geophysical Union.
Dormy E, Valet JP, and Courtillot V (2000) Numerical models of the geodynamo and observational constraints. *Geochemistry Geophysics Geosystems* 1, doi:2000GC000062.
Dziewonski AM and Anderson DL (1981) Preliminary reference Earth model. *Physics of the Earth and Planetary Interiors* 25: 297–356.
Dziewonski AM and Gilbert F (1971) Solidity of the inner core of the Earth inferred from normal model observations. *Nature* 234: 465–466.
Elsasser WM (1946) Induction effects in terrestrial magnetism. Part II: The secular variation. *Physical Review* 70: 202–212.
Elsasser WM (1950) The Earth's interior and geomagnetism. *Reviews of Modern Physics* 22: 1–35.
Evans ME (1988) Edmond Halley, geophysicist. *Physics Today* 41: 41–45.
Franklin B (1793) Conjectures concerning the formation of the Earth. *Transactions of the American Philosophical Society* 3: 1–5.
Gailitis A, Lielausis O, Platacis E, Gerbeth G, and Stefani F (2003) The Riga Dynamo Experiment. *Surveys in Geophysics* 24: 247–267.
Gauss CF (1832) cited in 'Carl Friedrich Gauss und die Erforschung des Erdmanetismus', by Schering E, Abhand. Gesellschaft Wissenschaften Gottingen 34: 1–79.
Gauss CF (1839) Allgemeine Theorie des Erdmagnetismus. In: Gauss CF and Weber W (eds.) *Resultate aus den Beobachtungen des magnetischen Vereins*, pp. 1–57. Gottingen: Dieterichsche Buchhandlung.
Gessmann C and Wood B (2002) Potassium in the earth's core? *Earth and Planetary Science Letters* 200: 63–78.
Gilbert W (1600) *De Magnete*. London: P. Short.
Gilman PA and Miller J (1981) Dynamically consistent nonlinear dynamos driven by convection in a rotating spherical shell. *Astrophysical Journal Supplement Series* 46: 211–238.
Glatzmaier G (2002) Geodynamo simulations – How realistic are they? *Annual Review of Earth and Planetary Sciences* 30: 237–257.
Glatzmaier G, Coe R, Hongre L, and Roberts P (1999) The role of the Earth's mantle in controlling the frequency of geomagnetic reversals. *Nature* 401: 885–890.
Glatzmaier G and Roberts P (1995) A three-dimensional self-consistent computer simulation of a geomagnetic field reversal. *Nature* 337: 203–209.
Glatzmaier G and Roberts P (2002) Simulating the geodynamo. *Contemporary Physics* 38: 269–288.
Greenspan H (1968) *The Theory of Rotating Fluids*. Cambridge: Cambridge University Press.
Gubbins D (1973) Numerical solution of the kinematic dynamo problem. *Philosophical Transactions of the Royal Society of London Series A* 274: 493–521.
Gubbins D, Alfe D, Masters G, Price GD, and Gillan MJ (2003) Can the Earths dynamo run on heat alone? *Geophysical Journal International* 155: 609–622.
Gutenberg B (1912) Uber Erdbebenwellen, VIIA. *Gottingen Nachr* 125–176 (read 1912, published 1914).
Halley E (1683) A theory of the variation of the magnetical compass. *Philosophical Transactions of the Royal Society of London* 13: 208–221.
Halley E (1692) An account of the cause of the change of the variation of the magnetical needle, with an hypothesis of the structure of the internal parts of the Earth. *Philosophical Transactions of the Royal Society of London* 16: 563–578.
Herzenberg A (1958) Geomagnetic dynamos. *Philosophical Transactions of the Royal Society of London Series A* 250: 543–583.
Ishii M and Dziewonski AM (2003) Distinct seismic anisotropy at the center of the Earth. *Physics of the Earth and Planetary Interiors* 140: 203–217.
Jacobs JA (1987) *The Earth's Core,* 2nd edn. London: Academic Press.
Jeffreys H (1926) The rigidity of the Earth's central core. *Monthly Notices of the Royal Astronomical Society Geophysical Supplement* 1: 371–383.
Jeffreys H (1929) *The Earth: Its Origin, History and Physical Constitution,* 2nd edn. Cambridge: Cambridge University Press.
Jones C, Soward A, and Zhang K (eds.) (2003) *Earth's Core and Lower Mantle*, 218p. London and New York: Taylor and Francis.
Jones CA, Soward AM, and Mussa AI (2000) The onset of convection in a rapidly rotating sphere. *Journal of Fluid Mechanics* 405: 157–179.
Kageyama A and Sato T (1995) Computer simulation of a magnetohydrodynamic dynamo. II *Physics of Plasmas* 2: 1421–1431.
Kelvin L (1862) On the rigidity of the Earth. *Philosophical Transactions of the Royal Society of London* 153: 573–582.
Kono M and Roberts P (2002) Recent geodynamo simulations and observations of the geomagnetic field. *Reviews of Geophysics* 40: 1013.
Larmor J (1919) How could a rotating body such as the sun become a magnet? *British Association for the Advancement of Science* 87: 139–160.
Lehmann I (1936) P′, publications of the International geodetic & geophysical Union, Assosiation of seismology, Seria A, Travaux Scientifiques 14: 87–115.
Lilley FEM (1970) On kinematic dynamos. *Proceedings of the Royal Society of London Series A* 316: 153–167.
Loper D (1978) Some thermal consequences of the gravitationally powered dynamo. *Journal of Geophysical Research* 83: 5961–5970.
Lowes F and Wilkinson (1963) Geomagnetic dynamo: A laboratory model. *Nature* 198: 1158–1160.
Malkus VWR (1968) Precession of the earth as the cause of geomagnetism. *Science* 160: 259–264.
Masters G and Gubbins D (2003) On the resolution of density within the Earth. *Physics of the Earth and Planetary Interiors* 140: 159–167.
Merrill RT, McElhinny MW, and McFadden PL (1998) *The Magnetic Field of the Earth*. London: Academic Press.
Moffatt HK (1978) *Magnetic Field Generation in Electrically Conducting Fluids*. Cambridge: Cambridge University Press.
Müller U, Stieglitz R, and Horanyi S (2006) Experiments at a two-scale dynamo test facility. *Journal of Fluid Mechanics* 552: 419–440.
Newcomb S (1892) On the dynamics of the Earth's rotation, with respect to the periodic variation of latitude. *Monthly Notices of the Royal Astronomical Society* 52: 336–341.

Oldham RD (1906) The constitution of the Earth, as revealed by earthquakes. *Nature* 92: 684–685.
Parker EN (1955) Hydrodynamic dynamo models. *Astrophysical Journal* 122: 293–314.
Poincaré H (1885) Sur l'equilibre d'une masse fluide animee d'un mouvement de rotation. *Comptes Rendus de l' Academie des Sciences Paris* 100: 346–348.
Poincaré H (1910) Sur la precession des corps deormables. *Bulletin of the American Astronomical Society* 27: 321–356.
Poirier JP (1988) Transport properties of liquid metals and viscosity of the Earths core. *Geophysical Journal* 92: 99–105.
Poirier JP (1994) Light elements in Earth's outer core: A critical review. *Physics of the Earth and Planetary Interiors* 85: 319–337.
Rädler KH (1968) On the electrodynamics of conducting fluids in turbulent motion. II: Turbulent conductivity and turbulent permeability. *Zeitschrift für Naturforschung* 23a: 1851–1860.
Rama Murthy V, von Westrenen W, and Fei Y (2003) Experimental evidence that potassium is a substantial radioactive heat source in planetary cores. *Nature* 423: 163–165.
Rhines PB (1979) Geostrophic Turbulence. *Annual Review of Fluid Mechanics* 11: 401–441.
Ritter A (1878) Untersuchungen uber die Hole der Atmosphare und die Constitution gasforminger Weltkorper. *Annalen der Physik* 3,5: 405–445, 543–558.
Roberts PH (1972) Kinematic dynamo models. *Proceedings of the Royal Society of London Series A* 272: 663–698.
Roberts PH, Jones CA, and Calderwood AR (2003) Energy fluxes and ohmic dissipation in the earths core. In: Jones C, Soward A, and Zhang K (eds.) *Earths Core and Lower Mantle.*, pp. 100–129. London: Taylor and Francis.
Rochester MG (1962) Geomagnetic core-mantle coupling. *Journal of Geophysical Research* 67: 4833–4836.
Schubert G, Turcotte D, and Olson P (2001) *Mantle Convection in the Earth and Planets*. Cambridge: Cambridge University Press.
Schuster A (1911) *The Progress of Physics During 33 years (1875–1908)*. Cambridge: Cambridge University Press.
Shearer P and Masters G (1990) The density and shear velocity contrast at the inner core boundary. *Geophysical Journal International* 10: 491–498.
Song X and Richards P (1996) Seismological evidence for differential rotation of the earth's inner core. *Nature* 382: 221–224.
Stacey F (1992) *Physics of the Earth,* 3rd edn. Brisbane: Brookfield Press.
Steenbeck M, Krause F, and Radler K-H (1967) A calculation of the mean electromotive force in an electrically conducting fluid in turbulent motion, under the influence of Coriolis forces. *Zeitschrift für Naturforschung 21a*: 369–376.
Takahashi F, Matsushima M, and Honkura Y (2005) Simulations of a quasi-Taylor state geomagnetic field including polarity reversals on the Earth simulator. *Science* 309: 459–461.
Taylor GI (1923) The motion of a sphere in a rotating liquid. *Proceedings of the Royal Society of London Series A* 102: 180–189.
Taylor JB (1963) The magneto-hydrodynamics of a rotating fluid and the Earth's dynamo problem. *Proceedings of the Royal Society of London Series A* 274: 274–283.
Tilgner A (2005) Precession driven dynamos. *Physics of Fluids* 17: 034104.
Verhoogen J (1961) Heat balance of the Earth's core. *Geophysical Journal of the Royal Astronomical Society* 4: 276–281.
Verhoogen J (1979) *Energetics of the Earth*. Washington, DC: National Academy of Sciences.
Wiechert E (1897) Uber die Massenvertheilung im Innern der Erde, *Nachr. Ges ellschaft der Wissenschaften* Gottingen 221–243.
Wood BJ, Walter MJ, and Wade J (2006) Accretion of the Earth and segregation of its core. *Nature* 441: 825–833 (doi:10.1038/nature04763).
Zhang K and Busse F (1989) Convection driven magnetohydrodynamic dynamos in rotating spherical shells. *Geophysical Astrophysical Fluid Dynamics* 49: 97–116.
Zhang K and Gubbins D (1992) On convection in the Earth's core driven by lateral temperature variation in the lower mantle. *Geophysical Journal International* 108: 247–255.
Zollner F (1871) Uber den Ursprung des Erdmagnetismus und die magnetischen Bezienhungen der Weltkorper. *Gesellschaft der Wissenschaften, Mathematische-Physikalsche, Leipzig* 23: 479–575.

2 Energetics of the Core

F. Nimmo, University of California Santa Cruz, Santa Cruz, CA, USA

2.1	**Introduction**	33
2.2	**Core Structure and Magnetic Field Evolution**	34
2.2.1	Density	34
2.2.2	Composition	34
2.2.3	Temperature	35
2.2.4	Dynamo Behavior over Time	35
2.3	**Energy and Entropy Equations**	36
2.3.1	Preliminaries	37
2.3.2	Energy Balance	38
2.3.2.1	General expression	39
2.3.2.2	Core heat flow Q_{cmb}	40
2.3.2.3	Internal heating Q_R	40
2.3.2.4	Heat of reaction Q_H	40
2.3.2.5	Secular cooling and pressure heating Q_s, Q_P	41
2.3.2.6	Gravitational energy Q_g	41
2.3.2.7	Latent heat release Q_L	41
2.3.2.8	Pressure effect on freezing Q_{PL}	41
2.3.2.9	Summary	42
2.3.3	Entropy Terms	42
2.3.4	Example Core Structure	43
2.3.4.1	Specific heat (Q_s, E_s)	45
2.3.4.2	Radioactive heating (Q_R, E_R)	45
2.3.4.3	Latent heat (Q_L, E_L)	46
2.3.4.4	Gravitational contribution (Q_g, E_g)	46
2.3.4.5	Entropy of heat of solution (E_H)	46
2.3.4.6	Adiabatic contribution (Q_k, E_k)	46
2.3.4.7	Other contributions	46
2.3.5	Summary	47
2.4	**Present-Day Energy Budget**	47
2.4.1	Introduction	47
2.4.2	General Behavior	47
2.4.3	Core Properties	48
2.4.3.1	Density and gravity	49
2.4.3.2	Thermodynamic properties	49
2.4.3.3	Temperature and melting	50
2.4.3.4	Chemical properties	50
2.4.3.5	Ohmic dissipation	50
2.4.3.6	Internal heating	51
2.4.3.7	Present-day CMB heat flow	51
2.4.3.8	Geochemical constraints on IC age	52
2.4.3.9	Comparison with other models	52
2.4.4	Present-Day Behavior	53
2.5	**Evolution of Energy Budget through Time**	54
2.5.1	Introduction	54
2.5.2	Theoretical Models	55

2.5.3 Age of the IC 59
2.6 Summary and Conclusions 61
References 62

Nomenclature

c	mass fraction of light element
e	internal energy
f_{ad}	ratio of T_i to T_c
g, $\mathbf{g}$	acceleration due to gravity
h	internal heating rate
$\mathbf{i}$	flux vector of light element
k	thermal conductivity
$\mathbf{q}$	heat flux
r	radial distance
r_c	core radius
r_e	planetary radius
r_i	IC radius
s	entropy
t	time
$\mathbf{u}$	core contraction velocity
$\mathbf{v}$	local core fluid velocity
v_p, v_s	seismic P-wave and S-wave velocities
A	characteristic core lengthscale
B	characteristic core lengthscale
$\mathbf{B}$	magnetic field
C	characteristic core lengthscale
C_c	constant relating light element release to core growth
C_p	specific heat capacity
C_r	constant relating core growth to temperature change
D	lengthscale of adiabatic temperature variation
$\mathbf{E}$	electric field
E_g	entropy production rate due to potential energy
E_H	entropy production rate due to chemical heating
E_k	entropy production rate due to thermal conduction
E_L	entropy production rate due to latent heat
E_P	entropy production rate due to pressure heating
E_R	entropy production rate due to internal heating
E_s	entropy production rate due to secular cooling
E_α	entropy production rate due to molecular diffusion
E_Φ	entropy production rate due to Ohmic dissipation
$\tilde{E}_T$	total entropy production constant
G	gravitational constant
I_s	$\int \rho T \, dV$
I_T	$\int \rho/T \, dV$
K_0	compressibility at zero pressure
L	lengthscale of density variations
L_H	latent heats
L'_H	latent heat incorporating pressure effect
M_c	core mass
M_{oc}	outer core mass
P	pressure
P_c	pressure at the CMB
P_T	constant relating pressure change to temperature change
Q_{cmb}	CMB heat flow
Q_g	heat flow due to potential energy
Q_H	heat flow due to chemical reaction
Q_L	heat flow due to latent heat
Q_P	heat flow due to pressure heating
Q_R	heat flow due to radioactive heating
Q_s	heat flow due to secular cooling
$\bar{Q}_{cmb}$	mean core heat flow
$\tilde{Q}_T$	total heat flow constant
R_H	heat of reaction
$\mathbf{S}$	outward normal vector
T, T_a	temperature, adiabatic temperature
T_c	temperature at the CMB
T_i	temperature at the ICB
T_m	melting temperature
T_{m0}, T_{m1}, T_{m2}	coefficients describing melting curve
W_g	energy released since IC formation due to potential energy
W_L	energy released since IC formation due to latent heat
W_R	energy released since IC formation due to radioactive decay
W_s	energy released since IC formation due to secular cooling

α	thermal expansivity	τ_i	IC age
α_c	compositional thermal expansivity	τ'	deviatoric stress
γ	Gruneisen parameter	ψ	gravitational potential
λ	radioactive decay constant	Δc	compositional contrast across the ICB
μ	chemical potential		
μ_0	permeability of free space	ΔP	pressure difference from ICB to center of Earth
ρ	density		
ρ_{cen}	density at the center of the Earth	ΔT_c	change in T_c since IC formation
ρ_i	density at the ICB	$\Delta T'_c$	change in T_c prior to IC formation
ρ_m	mantle density	$\Delta\rho_c$	compositional density contrast
ρ_0	density at zero pressure	Φ	Ohmic dissipation
σ	electrical conductivity		

2.1 Introduction

Understanding the energy budget of the Earth's core is important for at least four reasons. First, it is this energy budget which determines the present-day viability of the Earth's dynamo and magnetic field. Second, the persistence of this field for at least 3.5 Gy places strong constraints on the thermal and energetic evolution of the core (and particularly, the growth of the inner core (IC)). Third, the core is a large reservoir of heat which is transferred to the mantle, at a rate controlled by the mantle; thus, the thermal evolution of the mantle both depends on and affects the energy budget of the core. Finally, an improved understanding of the way the Earth's dynamo has evolved will allow us to make better use of the fact that various other solar system bodies possess, or once possessed, apparently similar dynamos (e.g., Stevenson, 2003; Nimmo and Alfe, 2006).

The aim of this chapter is to summarize the contributing factors to the core's overall energy (and entropy) budgets; to quantify these factors, and their likely uncertainties; and to demonstrate how these results may be used to evaluate the behavior of the core, both now and in the past. The development followed below should allow a reader to start from general thermodynamic equations, and end up able to calculate the entropy and energy terms for an arbitrary core structure. Large parts of this chapter are based on previous works, particularly those of Braginsky and Roberts (1995), Buffett *et al.* (1996), Labrosse (2003), Roberts *et al.* (2003), Gubbins *et al.* (2003, 2004), and Nimmo *et al.* (2004). Other relevant chapters in this treatise include, Chapters 5, 6, 3, 8, 9, and 12. Such a treatment is inevitably lengthy; readers more interested in the application of the equations to the Earth than their derivation are advised to focus on Sections 2.4 and 2.5.

Perhaps the most robust conclusion is that the principal driving mechanism for the present-day dynamo is the release of one or more light elements at the inner core boundary (ICB) due to ongoing IC solidification. In the absence of such solidification, the core would have had to cool roughly 3 times as fast in order to maintain the same rate of dissipation in the dynamo (Section 2.4.4). The rate at which the core cools is controlled by the mantle, and the present-day heat flow at the core–mantle boundary (CMB) is estimated at 10 ± 4 TW, sufficient to maintain a present-day dynamo generating 1–5 TW of Ohmic dissipation. However, this range of core cooling rates implies that the IC is 0.37–1.9 Gy old (Section 2.5.3), much less than the age of the Earth. Prior to the formation of the IC, the rate of core cooling must have been faster, leading to early core temperatures likely implying widespread lower mantle melting. These high initial temperatures are reduced if the core contains radioactive potassium, which acts as an extra energy source for the dynamo and allows slower core cooling and less rapid inner-core growth. For the IC to be as old as 3.5 Gy requires a time-averaged core heat flux of 1.5–3.3 TW less than the present-day value; the presence of potassium in the core makes an ancient IC somewhat more plausible. Since geochemical arguments have been used to argue for such an ancient IC (e.g., Brandon *et al.*, 2003), there remains a currently unresolved disagreement between the geophysical and geochemical arguments, which will undoubtedly form the basis for future work.

The rest of the chapter is arranged as follows. Section 2.2 briefly summarizes core and dynamo parameters relevant to the later analysis. Section 2.3 gives the general energy and entropy equations describing core thermal evolution and dynamo generation. It also gives specific expressions for each of the mechanisms (e.g., latent heat release) contributing toward dynamo generation for a simple analytical model of core structure. Section 2.4 shows how these expressions may be used to establish general results for core and dynamo behavior. This section also derives specific results for the present-day energy and entropy budget of the core, based on recent models of core and mantle structure. Section 2.5 summarizes investigations of these budgets through time, focusing in particular on the age of the IC and the role of radioactive heating. Finally, Section 2.6 summarizes and concludes the chapter.

2.2 Core Structure and Magnetic Field Evolution

2.2.1 Density

The radially averaged density structure of the core may be derived directly from seismological observations. **Figure 1(a)** plots the inferred density structure of the core as a function of depth, and shows the increase with depth due to increasing pressure. The density discontinuity at the ICB arises because of two effects. First, solid core material is inherently denser than liquid core material at the same pressure and temperature (P, T) conditions. Second, the outer core contains more of one or more light elements than the IC (e.g., Poirier, 1994), and would therefore be less dense even if there were no phase change. This compositional density contrast has a dominant role in driving compositional convection in the core (see Section 2.3.3 below); unfortunately, its magnitude is uncertain by a factor of about 2.

Even the total density contrast across the ICB is somewhat uncertain. The two observations generally employed are normal mode data (e.g., Dziewonski and Anderson, 1981) and reflected body wave amplitude ratios (e.g., Shearer and Masters, 1990). Early application of these methods gave widely varying, and sometimes contradictory, results. However, a recent normal mode study (Masters and Gubbins, 2003) gives a total density contrast at the ICB of 640–1000 kg m^{-3}, which agrees rather well with the result of 600–900 kg m^{-3} obtained using body waves (Cao and Romanowicz, 2004), but is somewhat higher than the value obtained by Koper and Dombrovskaya (2005).

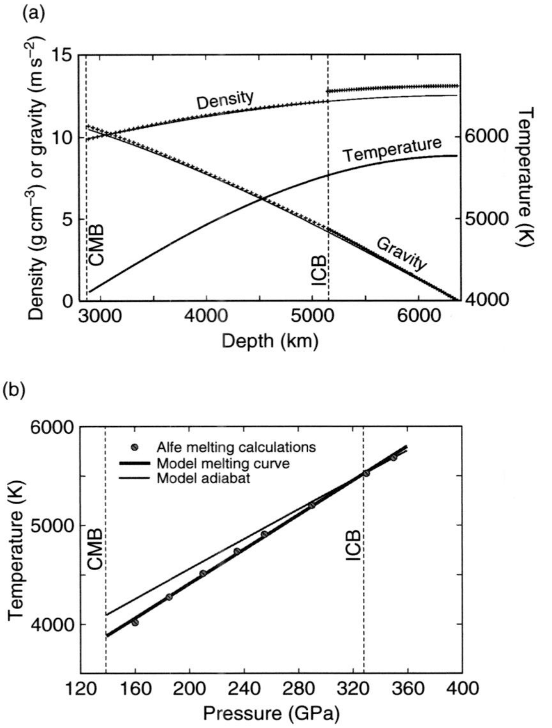

Figure 1 (a) Variation in density, gravity, and temperature with depth for the Earth's core. Crosses denote seismologically constrained values from PREM (Dziewonski and Anderson, 1981). Solid lines are analytical approximations to these observations, using the expressions given in Section 2.3.4 and the parameter values given in **Table 2**. CMB denotes core–mantle boundary and ICB inner core boundary. (b) Melting curve and adiabat as a function of pressure. Dots are computational melting results for pure iron from Alfe *et al.* (2004), with a reduction in temperature of 11% to account for the presence of the light element(s). Thick line is a least-squares fit to these data (eqn [51]). Thin line is the analytical adiabat (eqn [44]).

The density contrast, based on first-principles simulations, between pure solid and liquid Fe at the ICB is estimated at 1.6% (Laio *et al.*, 2000). This value contrasts with the 5–8% density contrast inferred from seismology to exist at the ICB. These results imply a compositional density contrast of 3.5–6.5%, or 400–800 kg m^{-3}, and may in turn be used to estimate the difference in light element(s) concentrations between inner and outer core.

2.2.2 Composition

Cosmochemical abundances leave no doubt that the bulk of the core is made of iron. However, it is also clear that the outer core is 6–10% less dense than pure liquid iron would be under the estimated P,T conditions (e.g. Poirier, 1994; Alfe *et al.*, 2002a). The IC also appears to be less dense than a pure iron composition would suggest (Jephcoat and Olson, 1987), though here the difference is smaller. Both the outer and inner core must therefore contain some fraction of light elements, of which the most common suspects are sulfur, silicon, and oxygen (Poirier, 1994). Furthermore, the outer core must contain a greater proportion of such elements than the IC. It is the expulsion of these light elements during IC crystallization that, to a large extent, drives the dynamo.

Using first-principles computations, Alfe *et al.* (2002a) find that oxygen does not tend to be retained within crystalline iron. It therefore partitions strongly into the outer core, and is most likely the element responsible for the compositional density contrast at the ICB. Conversely, S and Si have atomic radii similar to that of iron at core pressures, and thus substitute freely for iron in the solid IC. There is thus little difference between S/Si concentrations in the inner and outer core. Based on an assumed total ICB density contrast of 4.5% (575 kg m^{-3}), Alfe *et al.* (2002a) concluded that 2.8% (360 kg m^{-3}) of this density contrast arose from compositional variations. They suggested that the IC contains 8.5 ± 2.5% molar S/Si, while the outer core contains 10 ± 2.5% molar S/Si and 8 ± 2.5% molar O. A higher total density contrast would imply a higher molar fraction of O: for instance, a total density contrast of 7% (900 kg m^{-3}) would imply a compositional density contrast of 5.3% and 15% molar O in the outer core.

2.2.3 Temperature

As long as the core is vigorously convecting, its mean temperature profile will closely approximate that of an adiabat, except in very thin top and bottom boundary layers. Since the temperature at the ICB must equal the melting temperature of the core at that pressure, the temperature elsewhere in the core may be extrapolated from the ICB along an appropriate adiabat. Thus, determining the melting behavior of core material is crucial to establishing the temperature structure of the core.

The melting behavior of pure iron is difficult to establish: experiments at the P,T conditions required (e.g. Brown and McQueen, 1986; Yoo *et al.*, 1993; Boehler, 1993) are challenging, and computational (first-principles) methods (e.g. Laio *et al.*, 2000; Belonoshko *et al.*, 2000; Alfe *et al.*, 2002b) are time consuming and hard to verify. Furthermore, the presence of the light element(s) likely reduces the melting temperature from that of pure iron, but by an uncertain amount. These issues are discussed in detail elsewhere in this volume. Here, we will simply summarize what appears to be the most convincing set of computational results to date.

Based on first-principles calculations, Alfe *et al.* (2003) obtain a melting temperature of 6350 ± 500 K for pure iron at ICB pressures (330 GPa). They further use ideal solution theory to argue that the melting point depression due to the presence of oxygen is proportional to the oxygen concentration difference across the ICB, and obtain a temperature reduction of 700 ± 100 K. The predicted temperature at the ICB is therefore 5650 ± 600 K. Extrapolating from this point to other locations within the core depends on the adiabat, which involves further unknowns discussed below (Section 2.3.4).

The results of the calculations by Alfe *et al.* (2002a, 2002b, 2003) differ from previous calculations by Belonoshko *et al.* (2000) and Laio *et al.* (2000). However, this discrepancy arises mainly because of the different molecular dynamics techniques adopted; correcting for these differences, the results obtained are very similar (Alfe *et al.*, 2002a). The melting curve of Alfe *et al.* (2002a, 2002b) agrees well with the low-pressure diamond-anvil cell results of Shen *et al.* (1998) and Ma *et al.* (2004), though not those of Boehler (1993). Similarly, the curve agrees with the higher-pressure shock-wave results of Brown and McQueen (1986) and Nguyen and Holmes (2004), though not those of Yoo *et al.* (1993). Further discussion of the differing results may be found in Alfe *et al.* (2004).

Figure 1(b) shows the melting curve of Alfe *et al.* (2002a, 2002b), including a temperature reduction of 11% to account for the presence of the light element(s), and also shows a linear fit to the computational data. This fit gives a temperature at the ICB (P= 328 GPa) of 5520 K. **Figure 1(b)** also shows a hypothetical adiabat, and demonstrates that the temperature at the CMB, obtained by extrapolating down the adiabat, is 4100 K.

2.2.4 Dynamo Behavior over Time

The behavior of the Earth's magnetic field over time bears directly on core energetics (see reviews by

Valet (2003) and Jacobs (1998)). In particular, one might expect that the long-term behavior of the field would provide information on the evolution of the dynamo and core. In practice, however, as discussed below, the information is limited to the following: (1) a reversing, predominantly dipolar field has existed, at least intermittently, for at least the last 3.5 Gy; (2) the amplitude of the field does not appear to have changed in a systematic fashion over time.

One reason for this paucity of constraints is that the measurable magnetic field at the Earth's surface differs considerably from the field within the core. First, the short-wavelength surface field is dominated by crustal magnetic anomalies, which obscure the components of the dynamo field at wavelengths shorter than about 3000 km (e.g., Langel and Estes, 1982). Second, the toroidal component of the core's magnetic field has field lines which are parallel to the surface of the core. Thus, the toroidal component is not observable at the Earth's surface, though it is probably at least comparable in magnitude to the observable poloidal component (e.g., Jackson, 2003). As a result, the magnetic field that we can measure at the surface is different in both frequency content and amplitude from the field within the core. In particular, Ohmic heating is dominated by small-scale magnetic fields which are not observable at the surface.

Several other factors also make it difficult to relate the observed evolution of the geomagnetic field to the core's thermal evolution in any more detail. First, although current numerical simulations can now generate dynamos whose behavior resembles that of the Earth's magnetic field (see reviews by Busse (2000), Roberts and Glatzmaier (2000), Glatzmaier (2002), Kono and Roberts (2002), and Chapter 8), these models have not in general been used to explore how dynamo behavior (e.g., reversal frequency and field intensity) changes as a function of parameters such as core cooling rate or IC size. A notable exception is Roberts and Glatzmaier (2001), who found that increasing the IC size tended to result in a less axisymmetric field and (surprisingly) greater time variability. However, models with ICs 0.25 and 2 times the radii of the current IC both produced almost identical mean field amplitudes; a similar result was found by Bloxham (2000). Nor is it clear that changes in global variables, such as core cooling rate, will have a larger effect on the field behavior than local factors such as the heat flux boundary condition (e.g., Christensen and Olson, 2003).

Second, the present-day amount of dissipation actually generated by dynamo activity (Ohmic heating) is very uncertain, making a direct link between core thermal evolution and dynamo activity problematic (see Section 2.4.3.5). And finally, paleomagnetic observations are sparse at times prior to the oldest surviving oceanic floor (150 My BP), making identification of trends in field amplitude very difficult (see, e.g., Labrosse and Macouin, 2003).

Despite these difficulties, it is clear that, over timescales greater than a few thousand years, the mean position of the magnetic axis coincides with the rotation axis (Valet, 2003). Furthermore, the field appears to have remained predominantly dipolar over time (though, see Bloxham, 2000), and has apparently persisted for at least 3.5 Gy (McElhinny and Senanayake, 1980). The earliest documented apparent paleomagnetic reversal is at 3.2 Gy BP (Layer *et al.*, 1996). The magnetic field intensity has fluctuated over time, with the present-day magnetic field being anomalously strong (Selkin and Tauxe, 2000), and the field during the Mesozoic anomalously weak (Prevot *et al.*, 1990). The maximum field intensity appears never to have exceeded the present-day value by more than a factor of 5 (Valet, 2003; Dunlop and Yu, 2004). The pattern of magnetic reversals for the Proterozoic is well known, but not well understood. For instance, although reversals occur roughly every 0.25 My on average (Lowrie and Kent, 2004), there were no reversals at all in the period 125–85 Ma (e.g., Merrill *et al.*, 1996), for reasons which are obscure but may well have to do with the behavior of the mantle over that interval (e.g., Glatzmaier *et al.*, 1999).

In summary, the fact that a reversing dynamo has apparently persisted for >3.5 Gy can be used to constrain the energy budget of the core over time (see Section 2.5). Unfortunately, other observations which might potentially provide additional constraints, such as the evolution of the field intensity, are either poorly sampled or difficult to relate to the global energy budget, or both.

2.3 Energy and Entropy Equations

The Earth's dynamo is ultimately maintained by convection (either compositional or thermal), with the convective motions being modified by electromagnetic and rotational forces (see, e.g., Roberts and Glatzmaier, 2000). Since the dynamo has persisted over 3.5 Gy, core convection must have likewise persisted, which places constraints on the energy budget of the core. In this section, the basic equations which allow the different terms in the energy balance to be estimated are derived. It will be demonstrated that the Ohmic heating generated by dynamo activity does not appear in the energy equations. Thus, these equations are in general insufficient to determine whether or not a geodynamo will operate, though they do allow the evolution of parameters like the IC radius to be calculated. However, by deriving the equivalent entropy equations, in which the Ohmic heating term does appear, a criterion for geodynamo activity may be determined (see eqn [72]).

This section summarizes a large body of previous work on core thermodynamics. Pioneering works by Bullard (1950), Verhoogen (1961), and Braginsky (1963) were followed by studies focusing primarily on the entropy balance of the dynamo (Backus, 1975; Hewitt *et al.*, 1975; Gubbins, 1977; Loper, 1978; Gubbins *et al.*, 1979; Hage and Muller, 1979). More recent works include the monumental Braginsky and Roberts (1995) and contributions by Lister and Buffett (1995), Buffett *et al.* (1996), Buffett (2002), Lister (2003), Labrosse (2003), Roberts *et al.* (2003), Gubbins *et al.* (2003, 2004), and Nimmo *et al.* (2004).

The derivations given below, though not especially difficult, are somewhat lengthy. Individual terms contributing to the geodynamo are summarized in **Table 1**. The most important equations, those which summarize the entropy and energy balances, are given in eqns [32] and [39], respectively. These equations also make the underlying physics relatively easy to understand. Briefly, the principal sources of buoyancy capable of driving convection and a dynamo are either thermal (core cooling, latent heat release at the ICB, and radioactive decay) or compositional (light element release at the ICB). The thermal sources are less efficient at driving a dynamo (they generate less entropy) than the compositional sources. More rapid core cooling provides more entropy available to drive a dynamo; if the core cooling is too slow, convection ceases because the core is capable of losing its heat purely by conduction.

2.3.1 Preliminaries

The methods and notation adopted here are essentially reviews of Gubbins *et al.* (2003, 2004) and Nimmo *et al.* (2004). Alternative approaches which yield similar or identical results may be found in Buffett *et al.* (1996), Roberts *et al.* (2003), and Labrosse (2003). A useful set of simplified expressions is given by Lister (2003).

Because convection in the outer core is vigorous, the core fluid is assumed to be well mixed outside the thin boundary layers. This in turn implies that both the entropy and fraction of the light element are uniform in the fluid outer core. Over timescales longer than the convective transport timescale, the pressure field is assumed to average to hydrostatic:

$$\nabla P = \rho \nabla \psi \qquad [1]$$

where P is the pressure, ρ is the density, and ψ the gravitational potential. Although the hydrostatic balance is not precisely maintained (with important consequences for core convection), the difference from a hydrostatic reference state is negligible for globally averaged quantities. The nature of the reference state is discussed in more detail in Braginsky and Roberts (2002).

For an isentropic and isochemical outer core, the temperature gradient is adiabatic and obeys

$$\nabla T_a = \frac{\alpha \mathbf{g} T_a}{C_p} = \frac{\gamma \mathbf{g}}{\left(v_p^2 - \frac{4}{3} v_s^2\right)} \qquad [2]$$

where T_a is the temperature along an adiabat, α and C_p the thermal expansivity and specific heat capacity, respectively, $\mathbf{g}$ the acceleration due to gravity, γ is Gruneisen's parameter, and v_p and v_s the P- and S-wave seismic velocities. It should be noted that if the heat flux out of the top of the core is subadiabatic, a stable conductive layer may develop in the outer core (e.g., Loper, 1978; Labrosse *et al.*, 1997), while a similar effect may occur if the light element, rather than being well mixed, accumulates at the top of the core (e.g., Braginsky, 1999). Although in either of these cases the temperature gradient will not be adiabatic everywhere, in most of what follows it is assumed that the temperature is in fact adiabatic throughout.

For a two-component mixture, the thermodynamic relationship between the three state variables P, T (temperature), and the mass fraction of the light element, c, is given by

$$T\mathrm{d}s = \mathrm{d}e - \frac{P\mathrm{d}\rho}{\rho^2} - \mu \mathrm{d}c \qquad [3]$$

Here, $\mathrm{d}e$ and $\mathrm{d}s$ are the differentials of the internal energy and entropy, respectively, and μ is the chemical potential, where

$$\left(\frac{\partial \mu}{\partial T}\right)_{P,c} = -\left(\frac{\partial s}{\partial c}\right)_{P,T} \qquad [4]$$

Table 1 Summary of analytical expressions for entropy and energy terms

Term	Energy	Entropy
Secular cooling	$Q_s = -\int \rho C_p \frac{dT_c}{dt} dV$	$E_s = -\int \rho C_p \left(\frac{1}{T_c} - \frac{1}{T}\right) \frac{dT_c}{dt} dV$
Latent heat	$Q_L = -\frac{4\pi r_i^2 L_H T_i}{(dT_m/dP - dT/dP) g T_c} \frac{1}{} \frac{dT_c}{dt}$	$E_L = -\frac{4\pi r_i^2 L_H (T_i - T_c)}{(dT_m/dP - dT/dP) T_c g} \frac{1}{T_c} \frac{dT_c}{dt}$
Radioactive decay	$Q_R = \int \rho h \, dV$	$E_R = \int \rho h \left(\frac{1}{T_c} - \frac{1}{T}\right) dV$
Heat of reaction	$Q_H = \int \rho R_H \frac{Dc}{Dt} dV = 0$	$E_H = -\int \rho \frac{R_H}{T} \frac{Dc}{Dt} dV$
Compositional	$Q_g = \int \rho \psi \alpha_c \frac{Dc}{Dt} dV$	$E_g = \frac{Q_g}{T_c}$
Pressure heating	$Q_P = \int \alpha T P_T \frac{dT_c}{dt} dV$	$E_P = \frac{Q_P}{T_c} - \int \alpha P_T \frac{dT_c}{dt} dV$
Pressure freezing	$Q_{PL} = 4\pi r_i^2 L_H \frac{T'_m}{T'_m - T'} \frac{P_T}{g} \frac{dT_c}{dt}$	$E_{PL} = Q_{PL} \left(\frac{1}{T_c} - \frac{1}{T_i}\right)$

Note that the latent heat terms Q_L, E_L can incorporate the pressure freezing terms Q_{PL}, E_{PL} by substituting L'_H for L_H (eqn [30]). Also note that dT_c/dt has been substituted for DT/Dt (see text). Integrations are carried out over the entire core except for the compositional and heat of reaction terms.

and

$$\left(\frac{\partial \mu}{\partial P}\right)_{T,c} = -\frac{1}{\rho^2}\left(\frac{\partial \rho}{\partial c}\right)_{P,T} = \frac{\alpha_c}{\rho} \qquad [5]$$

and α_c is a dimensionless coefficient which indicates the sensitivity of the core density to the presence of the light element (Roberts *et al.*, 2003; Gubbins *et al.*, 2004):

$$\alpha_c = -\frac{1}{\rho}\left(\frac{\partial \rho}{\partial c}\right)_{P,T} \approx \frac{\Delta \rho_c}{\rho_i \Delta c} \qquad [6]$$

Here $\Delta\rho_c$ is the change in density across the ICB due to the change in light element concentration, Δc, across the same interface, and ρ_i is the density of the solid IC at the ICB. The parameter α_c depends on the properties of the light element and is assumed to be independent of the light element concentration (mass fraction), c.

The continuity equation for the core can be written

$$\frac{\partial \rho}{\partial t} + \nabla \cdot (\rho \mathbf{v}) = \frac{D\rho}{Dt} + \rho(\nabla \cdot \mathbf{v}) = 0 \qquad [7]$$

where $\mathbf{v}$ is the local fluid velocity. Similarly, conservation of mass of the light element can be written

$$\rho\frac{\partial c}{\partial t} + \rho \mathbf{v} \cdot \nabla c + \nabla \cdot \mathbf{i} = \rho\frac{Dc}{Dt} + \nabla \cdot \mathbf{i} = 0 \qquad [8]$$

where $\mathbf{i}$ is the flux vector of the light element. Both the compositional and heat flux vectors depend on the gradients of the three state variables, P, T, and s, according to the Onsager reciprocal relationships (Landau and Lifshitz, 1959):

$$\mathbf{q} = -k\nabla T + \mathbf{i}\left(\mu + \frac{\beta T}{\alpha_D}\right) \qquad [9]$$

$$\mathbf{i} = -\alpha_D \nabla \mu - \beta \nabla T \qquad [10]$$

where k is the thermal conductivity, α_D and β are material constants (defined in Gubbins *et al.*, 2004), and μ is related to P, T, and s via eqn [3].

Having established these preliminary expressions describing the reference state of the core, the individual terms in the core's energy and entropy budgets may now be derived.

2.3.2 Energy Balance

Although the individual terms in the core's energy budget can be quite complicated, the overall budget is actually quite simple to write down. The rest of this section demonstrates that the energy budget may be written as

$$Q_{cmb} = Q_s + Q_L + Q_g + Q_P + Q_H + Q_R \qquad [11]$$

Here, Q_{cmb} is the heat flow across the core–mantle boundary, the thermal energy extracted out of the core by the mantle. This energy arises from six main sources: Q_s, the secular cooling of the core; Q_L, the latent heat delivered as the IC solidifies; Q_g, the gravitational potential energy (or more properly compositional energy) associated with the release of the light element during IC solidification; Q_P, a small contribution due to the change in pressure during core cooling; Q_H, a contribution from chemical reactions which turns out to be negligible; and Q_R, a contribution from the decay of any radioactive elements within the core. The first four terms on the right-hand side (RHS) of the equation are all proportional to the rate of core cooling dT_c/dt, where T_c is the core temperature at the CMB (Gubbins *et al.*, 2003). Thus, in the absence of radioactive heating, the heat flow out of the core is directly proportional to the core cooling rate. Hence, a high CMB heat flow either requires a rapidly cooling core, or a large contribution from radioactive elements. In a situation lacking an IC, the only nonzero terms are Q_s, Q_P, and potentially Q_R. The same CMB heat flow would therefore require a more rapidly cooling core than a similar situation in which an IC existed.

An important aspect of eqn [11] is that heating due to dynamo activity (Joule heating) or viscous dissipation does not appear. This is because the dissipation only involves conversion of energy within the core (buoyancy forces generate kinetic and magnetic energy, which in turn are converted to heat via Ohmic and viscous dissipation). As a result, the global energy balance is not affected. In order to investigate the effect of dissipation, it is necessary to consider the entropy budget of the core. This approach allows investigation of the circumstances under which a dynamo will operate, and is the subject of Section 2.3.3.

2.3.2.1 General expression

A very general expression for energy conservation within the core may be written as follows (Gubbins *et al.*, 2003; see also Buffett *et al.*, 1996):

$$\begin{aligned}\frac{d}{dt}\int \rho e\, dV + \frac{d}{dt}\int \frac{1}{2}\rho \mathbf{v}^2 dV + \int \frac{\partial}{\partial t}\frac{\mathbf{B}^2}{2\mu_0} dV &= -\oint P\mathbf{v}\cdot d\mathbf{S} - \oint \frac{\mathbf{E}\times\mathbf{B}}{\mu_0}\cdot d\mathbf{S} \\ &\quad + \oint \mathbf{v}\cdot\tau'\cdot d\mathbf{S} - \oint \mathbf{q}\cdot d\mathbf{S} + \int \rho h\, dV \\ &\quad + \int \rho\mathbf{v}\cdot\nabla\psi\, dV \qquad [12]\end{aligned}$$

Here, ρ is density, e is internal energy, $\mathbf{v}$ is the core fluid velocity, $\mathbf{B}$ and $\mathbf{E}$ are magnetic and electric fields, μ_0 is the permeability of free space, P is the pressure, $\mathbf{S}$ is an outward normal surface vector, τ' is the deviatoric stress, $\mathbf{q}$ is the CMB heat flux, h the local volumetric heat generation, and ψ the gravitational potential. The left-hand side (LHS) of this equation gives the total rate of change of internal, kinetic, and magnetic energies, respectively. The surface integrals on the RHS give the work done on the surface by pressure and electromagnetic forces, surface tractions, and the heat flow across the boundary (Q_{cmb}). The final two volume integrals give the total heat generated and the work done against gravitational forces.

Several simplifications may be applied to this equation (Gubbins *et al.*, 2003, 2004). For the Earth's core there are two very different timescales implicit in eqn [12]: the short timescale (~1 year) associated with convective motions, and the much longer timescale (~100 My) associated with cooling and contraction of the core as a whole. Let the velocity associated with core contraction be denoted by $\mathbf{u}$. Then over timescales long compared to the convective timescale, but short compared to the core evolution timescale, the core will be well represented by the well-mixed, hydrostatic and isentropic basic state (see Section 2.3.1). Over this intermediate timescale, the total fluid velocity $\mathbf{v}$ may be assumed to average to the slow contractional velocity $\mathbf{u}$ (Gubbins *et al.*, 2003), while at the core boundary it is always true that $\mathbf{v}\cdot d\mathbf{S} = \mathbf{u}\cdot d\mathbf{S}$. Over this same timescale, it is assumed that the time fluctuations of kinetic and magnetic energy are negligible (see Braginsky and Roberts, 1995) and thus the second two terms on the LHS are zero.

Another simplification arises from the imposition of boundary conditions. By assuming that the Earth's mantle is a perfect insulator, the surface integral involving electromagnetic fluxes disappears. A similar assumption of stress-free boundary conditions allows the removal of the shear stress surface integral. This approach is convenient and will be followed here, although a no-slip boundary condition would be more realistic (see Roberts and Glatzmaier, 2000).

Equation [12] may thus be rearranged as follows:

$$\begin{aligned}Q_{cmb} &= \oint \mathbf{q}\cdot d\mathbf{S} \\ &= \int \rho h\, dV - \int \rho\frac{De}{Dt} dV \\ &\quad + \int \rho\mathbf{u}\cdot\nabla\psi\, dV - \int \nabla\cdot(P\mathbf{u})\, dV \qquad [13]\end{aligned}$$

where

$$\frac{\mathrm{D}}{\mathrm{D}t} = \frac{\partial}{\partial t} + \mathbf{u} \cdot \nabla \qquad [14]$$

and each integral is taken over the whole core and is time-averaged. The term in $\mathbf{u}. \nabla\psi$ arises because the long-term evolution of the gravitational potential depends only on $\mathbf{u}$. It will later prove useful to convert $\rho\mathbf{u} \cdot \nabla\psi$ to $\mathbf{u} \cdot \nabla P$ by using eqn [1]. The final term on the RHS is obtained by application of the divergence theorem and subsitution of $\mathbf{u}$ for $\mathbf{v}$ (see Gubbins *et al.*, 2003). Note that Q_{cmb} is the amount of heat being extracted from the core, and not the adiabatic heat flow.

Equation [13] shows that the long-term energy loss from the core may be estimated using integrals over the reference state, that is, the short-term convective velocity $\mathbf{v}$ is not important. Put another way, the short-term convective velocity is responsible for maintaining the reference state, but the energy available to drive the dynamo depends only on the long-term evolution of the core. In the absence of an IC, a further simplification arises: application of eqn [3] causes the last two terms on the RHS of eqn [13] to disappear and the integral in $\frac{\mathrm{D}e}{\mathrm{D}t}$ is replaced by one in $T\frac{\mathrm{D}s}{\mathrm{D}t}$ (see Gubbins *et al.*, 2003). However, in the case of a solidifying IC, there is a significant contribution to the gravitational energy from the redistribution of the light element (see below).

Thermal contraction and IC solidification both lead to a volume reduction in the core, which releases additional gravitational energy. Although it has been claimed (Hage and Muller, 1979) that this energy becomes available to drive the dynamo, Gubbins *et al.* (2003) showed that the actual contribution to the energy budget is very small.

Equation [13] is more directly applicable than eqn [12] to the Earth's core. In particular, it may be used to derive each of the terms given in eqn [11] in a relatively straightforward fashion. Each of these terms is derived in turn below; readers wishing to avoid the details will find all terms summarized in **Table 1**.

2.3.2.2 Core heat flow Q_{cmb}

From eqn [13], the heat flow across the CMB is given by

$$Q_{\mathrm{cmb}} = \oint \mathbf{q} \cdot \mathrm{d}\mathbf{S} \qquad [15]$$

The heat flux $\mathbf{q}$ in the case of a solidifying IC depends on the solute flux $\mathbf{i}$ (eqn [9]) as well as the temperature gradient in the boundary layer. However, assuming that no solute crosses the boundary, then $\mathbf{i} \cdot \mathrm{d}\mathbf{S} = 0$ and the CMB heat flow may be written as

$$Q_{\mathrm{cmb}} = -\oint k\nabla T \cdot \mathrm{d}\mathbf{S} \qquad [16]$$

2.3.2.3 Internal heating Q_{R}

Internal heat production may arise because of radiogenic elements or (less likely for the Earth) tidal dissipation (Greff-Lefftz and Legros, 1999) or core–mantle coupling (Touma and Wisdom, 2001). For the radiogenic case, it is reasonable to assume that heating is uniform since vigorous convection will homogenize the distribution of radiogenic elements. In such a case, the internal heating is given by

$$Q_{\mathrm{R}} = \int \rho h \, \mathrm{d}V \qquad [17]$$

where h is the volumetric heating rate. Note that this term does not include any contribution from Ohmic heating.

2.3.2.4 Heat of reaction Q_{H}

In the presence of a solidifying IC, the internal energy term in eqn [13] may be written with the help of eqn [3] as

$$\int \rho \frac{\mathrm{D}e}{\mathrm{D}t} \mathrm{d}V = \int \rho T \frac{\mathrm{D}s}{\mathrm{D}t} \mathrm{d}V + \int \frac{P}{\rho} \frac{\mathrm{D}\rho}{\mathrm{D}t} \mathrm{d}V + \int \rho\mu \frac{\mathrm{D}c}{\mathrm{D}t} \mathrm{d}V = \int \rho T \frac{\mathrm{D}s}{\mathrm{D}t} \mathrm{d}V - \int P \nabla \cdot \mathbf{u} \, \mathrm{d}V + \int \rho\mu \frac{\mathrm{D}c}{\mathrm{D}t} \mathrm{d}V \qquad [18]$$

where the second equality makes use of eqn [7].

The rate of change of internal energy depends on the rate of change of entropy (which depends on P, T, and c), pressure, and the concentration of the light element. The first term on the RHS may be written as

$$\rho T \frac{\mathrm{D}s}{\mathrm{D}t} = \rho T \left(\frac{\partial s}{\partial T}\right)_{P,c} \frac{\mathrm{D}T}{\mathrm{D}t} + \rho T \left(\frac{\partial s}{\partial P}\right)_{T,c} \frac{\mathrm{D}P}{\mathrm{D}t} + \rho T \left(\frac{\partial s}{\partial c}\right)_{P,T} \frac{\mathrm{D}c}{\mathrm{D}t} = \rho C_{\mathrm{p}} \frac{\mathrm{D}T}{\mathrm{D}t} - \alpha T \frac{\mathrm{D}P}{\mathrm{D}t} - \rho T \left(\frac{\partial \mu}{\partial T}\right)_{P,c} \frac{\mathrm{D}c}{\mathrm{D}t} \qquad [19]$$

where Maxwell's relation (eqn [4]) has been used. The specific heat capacity C_{p} is defined by

$$C_p = T\left(\frac{\partial s}{\partial T}\right)_{P,c} \quad [20]$$

and the thermal expansivity α is given by

$$\alpha = -\rho\left(\frac{\partial s}{\partial P}\right)_{T,c} = -\frac{1}{\rho}\left(\frac{\partial \rho}{\partial T}\right)_{P,c}. \quad [21]$$

The second term on the RHS of eqn [18] vanishes in conjunction with the last two terms on the RHS of eqn [13]. Combining the terms in eqn [18] that depend on c, we obtain

$$Q_H = \int \rho\left[\mu - T\left(\frac{\partial \mu}{\partial T}\right)_{P,c}\right]\frac{Dc}{Dt}dV = \int \rho R_H \frac{Dc}{Dt}dV \quad [22]$$

where R_H is the heat of reaction between iron and the light element. Here Q_H represents the change in internal energy due to chemical reactions; for an exothermic reaction, heat is absorbed at the ICB and released throughout the liquid core.

2.3.2.5 Secular cooling and pressure heating Q_s, Q_P

Making use of eqn [19], we can identify two terms in the internal energy budget not associated with compositional variations. The first is the secular cooling term:

$$Q_s = -\int \rho C_p \frac{DT}{Dt}dV \quad [23]$$

This term is simply the heat released as the core cools; it includes a small contraction term due to the Lagrangian derivative. Although the release of latent heat due to IC solidification can be included in this term (by modifying the specific heat), greater transparency is retained by including the latent heat as a separate term (see below).

The second term gives the heating that arises from a change in pressure, and is referred to as the pressure heating term by Gubbins *et al.* (2003). It is given by

$$Q_P = \int \alpha T P_T \frac{DT_c}{Dt}dV \quad [24]$$

where P_T is a numerical coefficient which relates the change of pressure at the ICB with time to the rate of core cooling:

$$\frac{DP}{Dt} = P_T \frac{dT_c}{dt} \quad [25]$$

where T_c is the core temperature at the CMB. A more extensive discussion of this small term is given in Gubbins *et al.* (2003).

2.3.2.6 Gravitational energy Q_g

The gravitational energy term arises due to the release of the light element at the ICB, and is thus really a compositional energy term (Braginsky and Roberts, 1995; Lister and Buffett, 1995). Considering only the density changes arising due to these compositional changes, the gravitational potential term in eqn [12] may be written (Gubbins *et al.*, 2004) as

$$Q_g = \int_\infty \rho \mathbf{v}\cdot\nabla\psi\, dV = \int_\infty \psi\left(\frac{\partial \rho}{\partial t}\right)_{P,T} dV = \int \rho\psi\alpha_c \frac{Dc}{Dt}dV \quad [26]$$

where the compositional expansion coefficient α_c is defined by eqn [6]. The gravitational energy contribution is proportional to the rate of expulsion of the light element, as parameterized by Dc/Dt.

2.3.2.7 Latent heat release Q_L

The latent heat released depends on the rate at which the IC solidifies:

$$Q_L = 4\pi r_i^2 L_H \rho_i \frac{dr_i}{dt} \quad [27]$$

where r_i is the IC radius, ρ_i is the density at the ICB, and L_H is the latent heat (assumed constant). The rate at which the ICB advances is determined by the rate at which the ICB temperature changes, and the relative slopes of the adiabat and melting curve (see Section 2.3.4). The rate at which T_i, the temperature at the ICB, changes may be directly related to the rate of change of the temperature T_c at the CMB, because both lie on the same adiabat. The quantity dr_i/dt may therefore be rewritten in terms of dT_c/dt (see Section 2.3.4).

2.3.2.8 Pressure effect on freezing Q_{PL}

If the pressure increases, the melting temperature of iron increases and thus the IC grows. A pressure change ΔP causes the IC radius to increase by a quantity Δr_i, where

$$\Delta r_i = \frac{T'_m}{T'_m - T'}\frac{\Delta P}{\rho_i g} \quad [28]$$

Here, T_m and T are the melting and adiabatic temperatures of the core at $r = r_i$, respectively, and T'_m and T' are their radial derivatives.

Making use of eqn [25], the additional latent heat released by this effect is given by

$$Q_{PL} = \frac{4\pi r_i^2 L_H T'_m}{T'_m - T'} \frac{P_T}{g} \frac{dT_c}{dt} \quad [29]$$

The effect can also be included by modifying the usual latent heat as follows:

$$L'_H = L_H \left(1 + P_T \frac{dT_m}{dP} \frac{T_c}{T_i}\right) \quad [30]$$

In practice, the increase in latent heat due to this effect is only 10–20% (Gubbins *et al.*, 2003).

2.3.2.9 Summary

Combining the results of Sections 2.3.2.2–2.3.2.8, the general energy budget for the core, eqn [13] may be rewritten as follows:

$$\begin{aligned} Q_{cmb} &= -\oint k\nabla T \cdot d\mathbf{S} = \int \rho h \, dV \\ &\quad - \int \rho C_P \frac{DT}{Dt} dV + \int \alpha T \frac{DP}{Dt} \\ &\quad + \int \rho R_H \frac{Dc}{Dt} dV + \int \rho \psi \alpha_c \frac{Dc}{Dt} dV \\ &\quad + 4\pi r_i^2 L'_H \rho_i \frac{dr_i}{dt} \\ &= Q_R + Q_s + Q_P + Q_H + Q_g + Q_L \end{aligned} \quad [31]$$

The utility of this equation is that it makes each of the mechanisms contributing toward the core energy budget obvious: the total amount of energy being extracted from the core depends on radioactive heat production within the core, secular cooling and contraction of the core, and chemical, gravitational, and latent heat release as the IC grows. With the exception of radioactive heating, all these terms turn out to be directly proportional to the rate of core cooling dT_c/dt (see Section 2.3.4). This equation also demonstrates that neither the adiabatic heat flow, Q_k, nor Ohmic heating, play any role in the global energy budget. Both these terms, however, do figure in the corresponding entropy balance, which is derived next.

2.3.3 Entropy Terms

As noted above, the Ohmic heating (dissipation) caused by the dynamo does not enter the global energy balance. However, dissipation is nonreversible and therefore a source of entropy. Thus, by considering the entropy budget of the core, criteria may be established for the operation (or failure) of the dynamo (see Section 2.4.2). In particular, an equation analogous to eqn [31] may be derived that describes the entropy budget of the core. In particular, both entropy sinks (heat diffusion, Ohmic dissipation, and molecular conduction) and sources (radioactive decay, secular cooling, latent heat release, pressure heating, chemical reactions, and compositional buoyancy) may be identified.

These entropy terms, derived below, are rates of entropy production. In general, each entropy term may be thought of as a heat flow, multiplied by some efficiency factor (<1) and divided by some characteristic operating temperature. Higher heat flows result in higher rates of entropy production.

The general equation corresponding to eqn [12] for the entropy terms is (Hewitt *et al.*, 1975; Gubbins *et al.*, 2004)

$$\rho \frac{Ds}{Dt} = -\frac{\nabla \cdot \mathbf{q}}{T} + \frac{\mu \nabla \cdot \mathbf{i}}{T} + \frac{\rho h}{T} + \frac{\Phi}{T} \quad [32]$$

where the heat flux $\mathbf{q}$ depends on the solute flux $\mathbf{i}$ (eqn [9]) and the entropy s depends on P, T, and c (eqn [3]). This equation summarizes the changes in entropy arising from both thermal and compositional effects. Here, Φ is the combined viscous and Ohmic dissipation. The former is assumed to be negligible and the volumetric Ohmic dissipation is given by

$$\Phi = \frac{\mathbf{J}^2}{\sigma} \approx \left(\frac{B^2}{\mu_0^2 \sigma l^2}\right) \quad [33]$$

where $\mathbf{J}$ is the electric current density, B and l are typical values for the magnetic field and the length-scale at which dissipation occurs (e.g., Labrosse, 2003), respectively, and σ is the electrical conductivity. It is because of the appearance of this term that the entropy balance may be used to determine whether or not a dynamo can operate.

Making use of eqns [9] and [10] and employing the divergence theorem, we have

$$\begin{aligned} \int \frac{\nabla \cdot \mathbf{q}}{T} dV &= \int \nabla \cdot \left(\frac{\mathbf{q}}{T}\right) dV + \int \mathbf{q} \cdot \frac{\nabla T}{T^2} dV \\ &= \frac{Q_{cmb}}{T_c} - \int k \left(\frac{\nabla T}{T}\right)^2 dV \\ &\quad + \int \frac{1}{\alpha_D T}\left(1 + \frac{\mu \alpha_D}{\beta T}\right)(-\alpha_D \nabla \mu \cdot \mathbf{i} - \mathbf{i}^2) \, dV \end{aligned} \quad [34]$$

The LHS of eqn [32] may be expanded using eqn [19] as before. The RHS may be further simplified by making use of eqn [8] and making the assumption that $\mu\alpha_D \ll \beta T$ (see Gubbins *et al.*, 2004). The resulting expression is

$$\int \frac{\rho C_{\mathrm{p}}}{T}\frac{\mathrm{D}T}{\mathrm{D}t}\mathrm{d}V - \int \alpha\frac{\mathrm{D}P}{\mathrm{D}t}\mathrm{d}V - \int \rho\frac{\partial\mu}{\partial T}\frac{\mathrm{D}c}{\mathrm{D}t}\mathrm{d}V = -\frac{Q_{\mathrm{cmb}}}{T_{\mathrm{c}}} + \int k\left(\frac{\nabla T}{T}\right)^2\mathrm{d}V + \int \frac{\mathbf{i}^2}{\alpha_D T}\mathrm{d}V + \int \frac{\rho h}{T}\mathrm{d}V + \int \frac{\Phi}{T}\mathrm{d}V \quad [35]$$

where the terms on the LHS represent contributions to the entropy budget from core cooling, contraction, and chemical reactions.

The term in $\mathbf{i}^2$ is the entropy of molecular conduction, E_α, which turns out be negligible (see below). The terms involving ρh and Φ are the entropy contibutions from internal heating and Ohmic (and viscous) dissipation, respectively. The term in $(\nabla T/T)^2$ is the entropy of thermal diffusion, E_k. This quantity depends on the temperature gradient and is an entropy sink; it reduces the entropy available to drive the dynamo and if large enough will ensure a conductive core and thus an absence of dynamo activity.

Since Q_{cmb} contains contributions from the different sources (eqn [31]), most of these source terms occur twice in eqn [35]. For instance, the secular cooling contribution to the overall entropy budget is

$$\int \left(\rho C_{\mathrm{p}}\frac{\mathrm{D}T}{\mathrm{D}t}\right)\left(\frac{1}{T} - \frac{1}{T_{\mathrm{c}}}\right)\mathrm{d}V \quad [36]$$

where the term in T_{c} is a consequence of the $Q_{\mathrm{cmb}}/T_{\mathrm{c}}$ term in eqn [35]. Similar equations may be derived for the contributions from latent heat, heat of solution, radioactive decay, and pressure heating. As before, the latent heat term may be incorporated into the secular cooling term.

The resulting expression for the entropy budget is as follows:

$$\int \rho C_{\mathrm{p}}\left(\frac{1}{T} - \frac{1}{T_{\mathrm{c}}}\right)\frac{\mathrm{D}T}{\mathrm{D}t}\mathrm{d}V - \int T\alpha\left(\frac{1}{T} - \frac{1}{T_{\mathrm{c}}}\right)\frac{\mathrm{D}T}{\mathrm{D}t}\mathrm{d}V - \frac{1}{T}\int \rho R_{\mathrm{H}}\frac{\mathrm{D}T}{\mathrm{D}t}\mathrm{d}V - \int \rho h\left(\frac{1}{T} - \frac{1}{T_{\mathrm{c}}}\right)\mathrm{d}V + \frac{Q_g}{T_{\mathrm{c}}} = \int k\left(\frac{\nabla T}{T}\right)^2\mathrm{d}V + \int \frac{\mathbf{i}^2}{\alpha_D T}\mathrm{d}V + \int \frac{\Phi}{T}\mathrm{d}V \quad [37]$$

which may be rewritten in an analogous form to eqn [31] as follows:

$$E_{\mathrm{s}} + E_P + E_{\mathrm{H}} + E_{\mathrm{R}} + E_g = E_k + E_\alpha + E_\Phi \quad [38]$$

Here, the latent heat terms have been incorporated into the secular cooling term E_{s}.

An important aspect of eqn [37] is the $((1/T_{\mathrm{c}})-(1/T))$ terms. These terms give the thermodynamic (Carnot) efficiency of the various processes driving the dynamo, and arise because the sources and sinks of buoyancy are operating at different temperatures.

Buoyancy sources which are distributed throughout the core volume have a lower efficiency than those which arise at the ICB, because the mean operating temperature of the former is smaller and thus the efficiency term $((1/T_{\mathrm{c}})-(1/T))$ is also smaller. Thus, the buoyancy forces due to radioactive decay or secular cooling are intrinsically less efficient at driving a dynamo compared to latent heat release.

An important exception is the term which arises due to compositional convection. This term only occurs once in eqn [37] and has the form Q_g/T_{c}. Its efficiency is thus intrinsically higher than those of the other energy sources. Thus, as we will see below, although compositional convection has only a moderate effect on the core's energy budget, its contribution to the core's entropy budget (and thus the operation of the dynamo) is very significant.

Table 1 summarizes the expressions derived above for the various entropy and energy terms.

2.3.4 Example Core Structure

The terms in energy and entropy derived above and summarized in **Table 1** are general in that they may be applied to any model of core structure. Some of the core properties required, such as density, compressibility, and gravitational acceleration, can be obtained from seismological observations (see **Figure 1**). Other properties, in particular the temperature structure, must be inferred from experimental studies or first-principles numerical calculations (see Section 2.4.3). Here a model core structure is derived which allows analytical expressions for the terms in **Table 1** to be obtained. These expressions give good agreement with numerical integrations carried out using the observed core density structure (see Nimmo *et al.*, 2004). A comparison of these results with those using different parameter choices is given in Section 2.4.3.9.

Given the large uncertainties associated with various core parameters, it is acceptable to make simplifying assumptions which allow the energy and entropy integrals to be treated analytically. The first major simplification is to assume spherical symmetry, although lateral variations in properties like the CMB heat flux may sometimes have important effects on the detailed behavior of the dynamo (e.g., Bloxham, 2000).

A second simplification is to assume that certain core properties, such as compressibility, thermal expansivity, and specific heat capacity, are constant. While only an approximation, this makes the integrals tractable and results in density and gravity profiles which are similar to those inferred (**Figure 1**). Here we describe the approach taken by Labrosse *et al.* (2001). Other authors, such as Buffett *et al.* (1996), Lister (2003), and Roberts *et al.* (2003) have adopted slightly different expressions.

The density within the Earth's core increases by about 35% from the CMB to the center of the planet (**Figure 1**). Following Labrosse *et al.* (2001), the varation of ρ with radial distance r from the center of the Earth is given approximately by

$$\rho(r) = \rho_{\text{cen}} \exp(-r^2/L^2) \quad [39]$$

where ρ_{cen} is the density at the center of the Earth and L is a lengthscale given by

$$L = \sqrt{\frac{3K_0\left(\text{In}\frac{\rho_{\text{cen}}}{\rho_0}+1\right)}{2\pi G\rho_0\rho_{\text{cen}}}} \quad [40]$$

Here, K_0 and ρ_0 are the compressibility and density at zero pressure, respectively, G is the universal gravitational constant, and $L \approx 7000$ km (see Section 2.4.3). From eqn [39], the mass of the core M_c is given by

$$M_c = \int_0^{r_c} \rho(r)\,dV = 4\pi\rho_{\text{cen}}\left[-\frac{L^2}{2}r\exp(-r^2/L^2) + \frac{L^3}{4}\sqrt{\pi}\,\text{erf}(r/L)\right]_0^{r_c} \quad [41]$$

where r_c is the core radius. Expanding erf (r/L) allows M_c to be written as

$$M_c = \frac{4}{3}\pi\rho_{\text{cen}}r_c^3 e^{-r_c^2/L^2}\left(1+\frac{2}{5}\frac{r_c^2}{L^2}+\cdots\right) \quad [42]$$

Similarly, the acceleration due to gravity g is given by

$$g(r) = \frac{4\pi}{3}G\rho_{\text{cen}}r\left(1-\frac{3r^2}{5L^2}\right) \quad [43]$$

This expression neglects the density jump $\Delta\rho$ across the ICB, introducing an error of order $r_i^4/L^4 \ll 1$, where r_i is the IC radius (Labrosse *et al.*, 2001). Note that the effect of this density jump is incorporated when considering compositional convection.

The adiabatic temperature T_a within the core is given by

$$T_a(r) = T_{\text{cen}}\exp(-r^2/D^2) \quad [44]$$

where T_{cen} is the temperature at the center of the Earth and D is another lengthscale given by

$$D = \sqrt{3C_p/2\pi\alpha\rho_{\text{cen}}G} \quad [45]$$

Here C_p is the specific heat capacity, α the thermal expansivity, and $D \approx 6000$ km (see Section 2.4.3). Note that eqn (44) assumes that the ratio α/C_p, or equivalently the ratio $\gamma/(v_p^2 - \frac{4}{3}v_s^2)$, is constant. This issue is discussed further below (Section 2.4.2).

The adiabatic profile (eqn [44]) allows the following useful simplification to be made (Gubbins *et al.*, 2003):

$$\frac{1}{T_a}\frac{DT_a}{Dt} = \frac{1}{T_c}\frac{dT_c}{dt} \quad [46]$$

where T_c is the temperature at the CMB. This approximation allows the term in $T^{-1}\,DT/Dt$ to be taken out of all the integrals given in **Table 1**, assuming that the temperature profile is adiabatic. Furthermore, the rate of IC growth dr_i/dt and the rate of change of concentration in the light element Dc/Dt may both be related to dT_c/dt as follows.

Equation [46] shows that the rate of cooling at the ICB is directly proportional to the rate of cooling at the CMB. In addition, **Figure 2** shows that the change in inner-core radius δr_i for a change in core temperature δT_c depends on the relative slopes of the adiabat and the melting curve. We may therefore write

$$\frac{dr_i}{dt} = \frac{1}{(dT_m/dP - dT/dP)}\frac{T_i}{\rho_i g}\frac{1}{T_c}\frac{dT_c}{dt} = C_r\frac{dT_c}{dt} \quad [47]$$

where the slopes of the adiabat and melting curve at the ICB are given by dT_a/dP and dT_m/dP, respectively, and T_i and ρ_i are the ICB temperature and density, respectively.

In a similar fashion, the rate of release of light material into the outer core may be written as

$$\frac{Dc}{Dt} = \frac{4\pi r_i^2\rho_i c}{M_{oc}}\frac{dr_i}{dt} = C_c C_r\frac{dT_c}{dt} \quad [48]$$

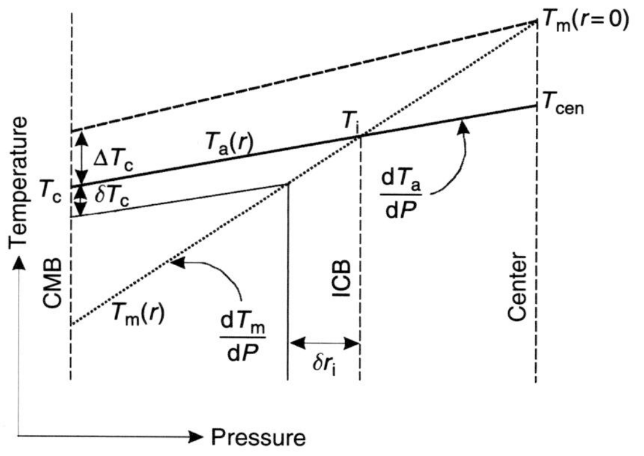

Figure 2 Schematic of melting and adiabatic temperature profiles. The IC (temperature T_i) is defined by the intersection of the adiabat T_a and the melting curve T_m. The corresponding temperature at the CMB, T_c, is obtained by following the adiabat. Changing this CMB temperature by a small amount δT_c results in a change in IC radius, δr_i, which depends on the relative slopes of the adiabat and melting curve, dT_a/dP and dT_m/dP, respectively. The temperature drop at the CMB since the onset of IC solidification is given by ΔT_c.

where M_{oc} is the mass of the outer core and $C_c = 4\pi r_i^2 \rho_i / M_{oc}$. Note that the rate of light element release is directly proportional to the core cooling rate.

It should further be noted that integrals involving Dc/Dt consist of two contributions: one denoting removal of light element from the IC at the ICB, and one involving the uniform redistribution of this material over the outer core. Assuming that the heat of reaction is constant, this conservation of species means that the energy term Q_H, but not the entropy term E_H, is identically equal to zero (see Gubbins *et al.*, 2004).

The pressure is given by

$$P(r) = P_c + \frac{4\pi G\rho_{cen}^2}{3}\left[\left(\frac{3r^2}{10} - \frac{L^2}{5}\right)\exp(-r^2/L^2)\right]_r^{r_c} \quad [49]$$

where P_c is pressure at the CMB.

The dependence of the core melting temperature on pressure and composition is often approximated using Lindemann's law (e.g., Buffett *et al.*, 1996; Labrosse *et al.*, 2001; Roberts *et al.*, 2003). For example, neglecting the compositional effects, one could write (Labrosse, 2003):

$$T_m(r) = T_{m0}\exp\left[-2\left(1 - \frac{1}{3\gamma}\right)\frac{r^2}{D^2}\right] \quad [50]$$

where T_{m0} is the melting temperature at zero pressure and γ is the Gruneisen parameter. Here we adopt a slightly different approach, following that of Stevenson *et al.* (1983), and parametrize the core melting temperature T_m as

$$T_m(P) = T_{m0}(1 + T_{m1}P + T_{m2}P^2) \quad [51]$$

where T_{m1} and T_{m2} are constants and T_{m0} incorporates the reduction in melting temperature due to the light element(s). In this work, it is assumed that $T_{m2} = 0$.

With the variation of ρ, g, and T within the core described by eqns [39], [43], and [44], respectively, analytical expressions for the entropy and energy terms are given below.

2.3.4.1 Specific heat (Q_s, E_s)

Table 1 and eqn [46] may be used to show that both Q_s and E_s involve a term $I_s = \int \rho T dV$ given (Labrosse *et al.*, 2001) as

$$I_s = 4\pi T_{cen}\rho_{cen}\left(-\frac{A^2 r_c}{2}e^{-r_c^2/A^2} + \frac{A^3\sqrt{\pi}}{4}\mathrm{erf}\,[r_c/A]\right) \quad [52]$$

where

$$A^2 = \left(\frac{1}{L^2} + \frac{1}{D^2}\right)^{-1} \quad [53]$$

It will be useful to expand I_s as follows:

$$I_s = \frac{4}{3}\pi T_{cen}\rho_{cen} r_c^3 e^{-r_c^2/A^2}\left(1 + \frac{2}{5}\frac{r_c^2}{A^2} + \cdots\right) \quad [54]$$

Strictly speaking, the adiabat is only followed in the outer core, but Labrosse (2003) argues that the integral may be taken over the entire core without introducing significant errors. The contributions Q_s and E_s may thus be written as

$$Q_s = -\frac{C_p}{T_c}\frac{dT_c}{dt}I_s, \quad E_s = \frac{C_p}{T_c}\left(M_c - \frac{I_s}{T_c}\right)\frac{dT_c}{dt} \quad [55]$$

where M_c is the mass of the core and T_c the core temperature at the CMB.

2.3.4.2 Radioactive heating (Q_R, E_R)

For radioactive heating, the term involved $I_T = \int (\rho/T)\, dV$ is more complicated and depends on the relative sizes of D and L. Although complete analytical solutions may be derived (see Nimmo *et al.*, 2004), it is more useful to carry out a series expansion, which yields

$$I_T = \frac{4\pi\rho_{cen}}{3T_{cen}}r_c^3\left(1 - \frac{3}{5}\frac{r_c^2}{B^2} + \cdots\right) \quad [56]$$

where

$$B^2 = \left(\frac{1}{L^2} - \frac{1}{D^2}\right)^{-1} \quad [57]$$

These equations allow the contributions Q_R and E_R to be determined:

$$Q_R = M_c h, \; E_R = \left(\frac{M_c}{T_c} - I_T\right) h \quad [58]$$

where h is the heat production per unit mass within the core.

2.3.4.3 Latent heat (Q_L, E_L)

The contributions Q_L and E_L given in **Table 1** may be written as:

$$Q_L = 4\pi r_i^2 L'_H \rho_i \frac{dr_i}{dt}, \; E_L = Q_L \frac{(T_i - T_c)}{T_c T_i} \quad [59]$$

where L'_H is the latent heat of fusion (incorporating the small pressure effect) and T_i is the temperature at the ICB.

2.3.4.4 Gravitational contribution (Q_g, E_g)

From [43], the potential $\psi(r)$ relative to zero potential at the CMB is given by

$$\psi(r) = \left[\frac{2}{3}\pi G \rho_{cen} r'^2 \left(1 - \frac{3r'^2}{10L^2}\right)\right]_{r_c}^{r} \quad [60]$$

Table 1 and eqn [48] give

$$Q_g = \left[\int_{oc} \rho\psi \, dV - M_{oc}\psi(r_i)\right] \alpha_c C_c C_r \frac{DT_c}{Dt} = \left[\int_{oc} \rho\psi \, dV - M_{oc}\psi(r_i)\right] \Delta\rho_c \frac{c}{\Delta c} \frac{4\pi r_i^2}{M_{oc}} \frac{dr_i}{dt} \quad [61]$$

Here, $\Delta\rho_c$ is the compositional density drop across the ICB, M_{oc} is the mass of the outer core ('oc' denotes the outer core), α_c is the compositional expansion coefficient (eqn [6]), $C_r dT_c/dt = dr_i/dt$ and C_c is defined after eqn [48]. If the IC completely excludes the light element(s), then $c = \Delta c$. Note that in this case Q_g depends only on the compositional density contrast $\Delta\rho_c$ and not on c, the actual concentration.

The integral $\int_{oc} \rho\psi \, dV$ is given by

$$\frac{8\pi^2 \rho_{cen}^2 G}{3} \left[\left(\frac{3}{20} r^5 - \frac{L^2}{8} r^3 - L^2 C^2 r\right) e^{-r^2/L^2} + \frac{C^2}{2} L^3 \sqrt{\pi} \, \text{erf}(r/L)\right]_{r_i}^{r_c} \quad [62]$$

where

$$C^2 = \frac{3L^2}{16} - \frac{r_c^2}{2}\left(1 - \frac{3r_c^2}{10L^2}\right) \quad [63]$$

The mass of the outer core M_{oc} may be obtained from [41] by changing the limits of integration. Equations [62] and [63] allow Q_g to be obtained; E_g is simply Q_g/T_c.

2.3.4.5 Entropy of heat of solution (E_H)

As discussed above, the quantity $Q_H = 0$. The contribution E_H in **Table 1** is given by

$$E_H = -R_H \left[\int_{oc} \frac{\rho}{T} dV - \frac{M_{oc}}{T_i}\right] C_c \frac{dr_i}{dt} \quad [64]$$

where R_H is the heat of reaction. The integral in this equation may be derived using the same approach as that for radioactive heating (eqns [56]–[58]). Note that this expression assumes that core contraction is negligible (see below).

2.3.4.6 Adiabatic contribution (Q_k, E_k)

The entropy change due to thermal diffusion is given by **Table 1**:

$$E_k = \int k \left(\frac{\nabla T}{T}\right)^2 dV \quad [65]$$

where k is the core thermal conductivity.

From [44], we have $\nabla T_a / T_a = -2r/D^2$; using [65], it can therefore be shown that for an adiabatic core

$$E_k = \frac{16\pi k r_c^5}{5D^4} \quad [66]$$

while Q_k at the CMB is simply given by

$$Q_k = 8\pi r_c^3 k T_c / D^2 \quad [67]$$

Note that both Q_k and E_k, unlike almost all the other terms in **Table 1**, are independent of the core cooling rate.

2.3.4.7 Other contributions

Gubbins *et al.* (2003, 2004) argued that the molecular diffusion term E_α and pressure heating term E_P in eqn [38] are negligible and may be ignored. The rate of change of the core temperature is governed by

$$\frac{DT_c}{Dt} = \frac{\partial T_c}{\partial t} + \mathbf{u} \cdot \nabla T_c \quad [68]$$

where **u** is the core contraction velocity. Gubbins *et al.* (2003) state that core contraction is negligible at the present day, in which case $DT_c/Dt = dT_c/dt$.

2.3.5 Summary

Table 1 gives the general expressions for each of the terms contributing to the core's energy (eqn [31]) and entropy (eqn [38]) budgets. Section 2.3.4 gives explicit expressions for these terms when the state of the core is described by a simple analytical model. Given a core cooling rate dT_c/dt, this allows the rates of heat transfer and entropy production for the core to be derived.

2.4 Present-Day Energy Budget

2.4.1 Introduction

Having derived approximate expressions for the individual terms in the energy and entropy equations, the magnitudes of these terms are estimated for the Earth in this section. In order to do so, various properties of the Earth's core are required, some of which are still highly uncertain. Section 2.4.2 derives some general conclusions regarding the present-day budgets which are independent of the specific property values assumed. Section 2.4.3 summarizes the likely values and discusses their uncertainties. Section 2.4.4 shows how these values may be used to establish the present-day global energy and entropy budget of the core, and compares these estimates with other constraints. For instance, the fact that a dynamo is operating at the present day places a constraint on the current CMB heat flow (eqn [71] and see below). This section draws heavily on previous works on the subject, as described below, in particular those of Buffett *et al.* (1996), Roberts *et al.* (2003), Labrosse (2003), Gubbins *et al.* (2003, 2004), Nimmo *et al.* (2004).

2.4.2 General Behavior

It is clear that the entropy budget within the core at the present day is capable of sustaining a dynamo. In this section, we review the constraints that this observation places on conditions in the core. In particular, we show how to calculate the rate at which the core must be cooling to sustain the dynamo, and the balance between thermal and compositional contributions to the dynamo. Considerations of core cooling inevitably lead to the question of the energy/entropy budget of the core through time; this topic is the subject of Section 2.5.

The entropy rate available to drive the dynamo is given by eqn [38] and, neglecting the small terms E_α and E_P, may be written as

$$\Delta E = E_R + E_s + E_L + E_H + E_g - E_k = E_R + \tilde{E}_T \frac{dT_c}{dt} - E_k \quad [69]$$

where E_s, E_L, E_H, and E_g depend on the core cooling rate dT_c/dt (see **Table 1**), E_R depends on the presence of radioactive elements in the core, and E_k depends on the adiabat at the CMB. $\tilde{E}_T$ is simply a convenient way of lumping together the terms which depend on core cooling rate (see Gubbins *et al.*, 2003), and is itself independent of dT_c/dt. This equation illustrates two important points. First, as expected, a higher cooling rate or a higher rate of radioactive heat production increases the entropy rate available to drive a dynamo. Second, a larger adiabatic contribution reduces the available entropy.

The equivalent energy balance (eqn [11]), neglecting Q_P and Q_H (=0), may be written

$$Q_{cmb} = Q_R + Q_s + Q_g + Q_L = Q_R + \tilde{Q}_T \frac{dT_c}{dt} \quad [70]$$

Again, $\tilde{Q}_T$ is simply a convenient way of lumping together the terms (Q_s, Q_g, and Q_L) that depend on the core cooling rate. By combining eqns [69] and [70], an expression may be obtained that gives the core heat flow required to sustain a dynamo characterized by a particular entropy production rate E_Φ:

$$Q_{cmb} = Q_R\left(1 - \frac{\tilde{Q}_T}{\tilde{E}_T}\frac{1}{T_R}\right) + \frac{\tilde{Q}_T}{\tilde{E}_T}(E_\Phi + E_k) \quad [71]$$

where T_R is the effective temperature such that $T_R = Q_R/E_R$. This equation encapsulates the basic energetics of the dynamo problem.

Equation [71] shows that larger values of adiabatic heat flow or Ohmic dissipation require a correspondingly higher CMB heat flow to drive the dynamo, as would be expected. In fact, in the absence of radiogenic heating, the CMB heat flow required is directly proportional to $E_k + E_\Phi$. Because the term $\left(1 - \frac{\tilde{Q}_T}{\tilde{E}_T}\frac{1}{T_R}\right)$ exceeds zero, a dynamo which is partially powered by radioactive decay will require a greater total CMB heat flow than the same dynamo powered without radioactivity. Alternatively, if the CMB heat flow stays constant, then an increase in the amount of radioactive heating reduces the entropy available to power the dynamo. These results are a consequence of the fact that radioactive heating has a lower thermodynamic efficiency than other methods of driving a dynamo (see Section 2.3.3).

Equation [71] also illustrates the fact that a dissipative dynamo can exist even if the CMB heat flow is subadiabatic (Loper, 1978). For instance, anticipating the results of Section 2.4.3, using **Table 4** we have $\tilde{Q}_T/\tilde{E}_T = 12\,200$ K for the present-day IC size. If the CMB heat flow is set to the adiabatic value ($Q_{cmb} = 4.9$ TW; $E_k = 162$ MW K^{-1}), application of [71] shows that the entropy available to drive the dynamo, $E_\Phi = 575$ MW K^{-1}. Only if the CMB heat flow fell below 2 TW, strongly subadiabatic, would the entropy available become negative. Thus, a subadiabatic CMB heat flow can sustain a dynamo, as long as an IC is present to drive compositional convection (e.g., Loper, 1978; Labrosse *et al.*, 1997). It should be noted that these results assume that the CMB heat flux does not vary in space; lateral variations in the heat flux may allow a dynamo to function even if the mean value of Q_{cmb} suggests the dynamo should fail (or vice versa).

In the absence of an IC, $\tilde{Q}_T \frac{dT_c}{dt} = Q_s$ and $\tilde{E}_T \frac{dT_c}{dt} = E_s$. Because the value of $\tilde{Q}_T/\tilde{E}_T$ increases when an IC is absent, eqn [71] shows that driving the same dynamo purely by thermal convection requires a higher CMB heat flow than when an IC is present. Again, this effect is a result of the higher efficiency of compositional convection (Section 2.3.3). Making use of eqns [43], [45], [54]–[56], [66], and [68], it can be shown that in the absence of an IC,

$$\frac{\tilde{Q}_T}{\tilde{E}_T} \approx \frac{Q_k}{E_k} = \frac{5}{2}\frac{D^2}{r_c^2} T_c \qquad [72]$$

and in a similar fashion (eqns [42], [44], and [56]–[58]) it may also be demonstrated that for $L < D$

$$T_R = T_c \frac{1 + \frac{2}{5}\frac{r_c^2}{L^2}}{\frac{2}{5}\frac{r_c^2}{D^2}} \qquad [73]$$

with a similar result for $L > D$.

In the absence of an IC, eqn [71] may therefore be rewritten as

$$Q_{cmb} = \frac{Q_R}{1 + \frac{5}{2}\frac{L^2}{r_c^2}} + Q_k\left(1 + \frac{E_\Phi}{E_k}\right) \qquad [74]$$

If $Q_R = 0$ and Ohmic heating is negligible ($E_\Phi = 0$), then eqn [74] shows that the heat flow at the CMB, Q_{cmb}, must exceed the adiabatic heat flow, Q_k, for a dynamo driven only by thermal convection to function. This result is well known (e.g., Nimmo and Stevenson, 2000), but the utility of eqn [74] is that it allows dynamo dissipation to be taken into account explicitly: a more strongly dissipative core requires a more superadiabatic CMB heat flow to operate.

Equation [74] also illustrates the effect of radiogenic heating in the absence of an IC. Because the term $(1 + (5/2)(L^2/r_c^2))$ is of order 10, the CMB heat flow required to maintain a thermally driven dynamo at a constant dissipation rate is not very sensitive to an increase in radiogenic element concentrations (see **Figure 3**). This is because the efficiencies of radiogenic heating and secular cooling are comparable (see Section 2.3.3).

The presence of radiogenic elements is more influential on the core cooling rate, and thus ultimately the age of the IC (e.g., Labrosse *et al.*, 2001). Rewriting eqn [70], we obtain a core cooling rate of

$$\frac{dT_c}{dt} = \frac{Q_{cmb} - Q_R}{\tilde{Q}_T} \qquad [75]$$

It is clear that the effect of the Q_R term is to reduce the rate of core cooling, and hence prolong the life of the IC. This is an issue we return to below.

2.4.3 Core Properties

Having made some general remarks, we now turn to the specific core properties we will assume for more detailed analysis. The temperature structure and

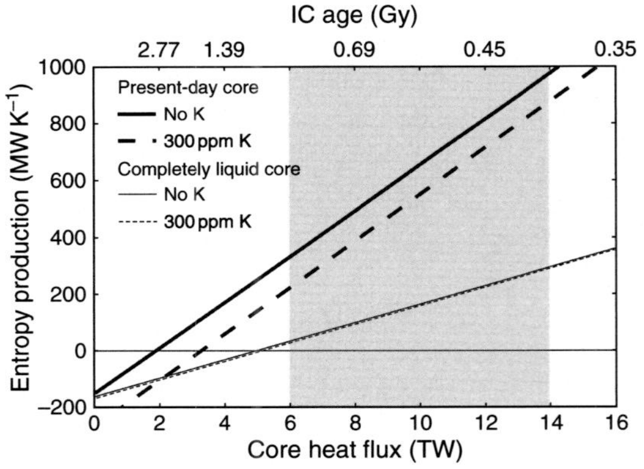

Figure 3 Rate of entropy production as a function of CMB heat flow. Calculations carried out using the expressions given in Section 2.3.4 and the parameter values given in **Table 2**. The IC ages given are only relevant to the case of a present-day IC containing no potassium, and are calculated assuming a constant heat flow. Shaded area denotes estimated present-day CMB heat flow (see text).

Table 2 Values of quantities assumed for core calculations

Symbol	Value	Units	Eqn	Symbol	Value	Units	Eqn
ρ_{cen}	12500	kg m^{-3}	[39]	r_i	1220	km	[43]
r_c	3480	km	[41]	P_c	139	GPa	[49]
K_0	500	GPa	[40]	D	5969	km	[45]
L	7272	km	[40]	T_i	5520	K	[59]
T_c	4100	K	[55]	T_{cen}	5756	K	[44]
α	1.35	10^{-5} K^{-1}	[45]	C_p	840	J kg^{-1} K^{-1}	[45]
L_H	750	kJ kg^{-1}	[59]	α_c	1.1		[6]
k	40	W m^{-1} K^{-1}	[65]	R_H	−27.7	MJ kg^{-1}	[64]
T_{m0}	2673	K	[51]	T_{m1}	3.25	10^{-12} Pa^{-1}	[51]
ρ_0	7900	kg m^{-3}	[40]				

The rationale for these values is given in Sections 2.2 and 2.4.3.
P_{icb} is the pressure at the ICB.

physical and compositional nature of the core are described in detail elsewhere in this treatise. The values adopted here are summarized in **Table 2**. Only a brief discussion of these values is given below; further discussion of the effects of uncertainties in these values on model results may be found. Note that in general the values are assumed to be constant throughout the core; as discussed below, such an assumption is incorrect, but unlikely to yield significant errors. The resulting theoretical profiles for density, temperature, and gravity are shown in **Figure 1** and may be compared with the seismologically inferred values.

2.4.3.1 Density and gravity

Figure 1 shows the seismologically inferred density and gravity profiles and compares them with those obtained using the theoretical expressions given in 2.3.4 and the values given in **Table 2**. There is good agreement; the only difference is that the simple density model (eqn [39]) neglects the increase in density of the IC. However, as noted above, this error has a negligible effect on the results.

2.4.3.2 Thermodynamic properties

From the point of view of the entropy and energy budgets, the most important thermodynamic properties are those influencing the temperature profile and the CMB heat flow, that is γ (or equivalently α and C_p) and k.

The thermal conductivity of iron at core conditions is obtained by using shock wave experiments and converting the measured electrical conductivity to thermal conductivity using the Wiedemann–Franz relationship (Stacey and Anderson, 2001). Although molecular dynamics simulations have been used to determine thermal conductivities in some systems (e.g., Recoules and Crocombette, 2005), doing so for iron has not yet been attempted. Shock experiments by Matassov (1977) led Stacey and Anderson (2001) to conclude that k at the CMB was 46 W m^{-1} K^{-1}. More recent shock experiments by Bi *et al.* (2002) give electrical resistivities 50% higher than those of Matassov (1977), suggesting a thermal conductivity closer to 30 W m^{-1} K^{-1}. A value of 40 ± 20 W m^{-1} K^{-1} spans the likely uncertainty and is in agreement with other recent choices for this parameter (see **Table 3**).

Thermodynamic arguments (e.g., Anderson, 1998) and molecular dynamics simulations (e.g. Alfe *et al.*, 2002b) both suggest that the value of Gruneisen's parameter γ is close to a constant value of 1.5 throughout the core. Since the seismic velocities vary significantly across the core, this result in turn implies that α is variable (here C_p is assumed constant). Labrosse (2003) derives an expression for α as a function of the core density, and argues that α varies from 1.25×10^{-5} K^{-1} at the centre to 1.7×10^{-5} K^{-1} at the CMB. Roberts *et al.* (2003) obtain a variation of 0.9–1.8 $\times 10^{-5}$ K^{-1} by assuming a depth-dependent γ. Labrosse (2003) also points out, however, that to the third-order accuracy required, the temperature profile (eqn [44]) is not affected by the variation in α and simply depends on the value of this parameter at the center of the Earth. We follow Labrosse (2003) and adopt a constant value for α of 1.35×10^{-5} K^{-1} to determine $T_a(r)$.

As Labrosse (2003) states, the temperature gradient and adiabatic heat flow at the CMB should be calculated using a larger value of α appropriate to CMB conditions. If $\gamma = 1.5$, then the values in **Table 2** give a temperature gradient at the CMB of

Table 3 Comparison of parameter values assumed by four different studies

Quantity	Units	Buffett	Roberts	Labrosse	This work
α	$\times 10^{-5}\ K^{-1}$	1.0	0.89–1.77	1.25–1.7	1.35
C_p	$J\ kg^{-1}\ K^{-1}$	800	819–850	850	840
γ		1.4	1.15–1.33	1.5	
L_H	$kJ\ kg^{-1}$	600	1560	660[a]	750
k	$W\ m^{-1}\ K^{-1}$	35	46	50	40
$\Delta\rho_c$	$kg\ m^{-3}$	400	255	500	575
α_c		0.93	1.0		1.1
T_i	K	4662	5100	5600	5520
T_c	K	3724	3949	4186	4100
ΔT_m	K	525	300		700
Q_k	TW	2.8	5.9	3.0	4.9

[a]Assuming a present-day IC temperature of 5600 K.
References are: Buffett *et al.* (1996), Roberts *et al.* (2003), Labrosse (2003), and this work.

$1\ K\ km^{-1}$, implying that the correct value of α to use in this situation is $1.9 \times 10^{-5}\ K^{-1}$. However, for simplicity, we prefer to retain a constant value of α throughout; doing so will result in an underestimate in Q_k and thus a conservatively low value of the CMB heat flow required to drive a dynamo (eqn [74]).

2.4.3.3 Temperature and melting

The temperature profile of the core is determined by the temperature at the ICB (see **Figure 1**). As explained in Section 2.2.3, we use the recent molecular dynamics simulations of Alfe *et al.* (2003), who find a drop in melting temperature ΔT_m due to the presence of light elements of 700 K. A linear fit to their results using eqn [51] results in $T_i = 5520\ K$ and melting parameters $T_{m0} = 2673\ K$ and $T_{m1} = 3.25 \times 10^{-12}\ K\ Pa^{-1}$. The adiabatic core temperature profile may be extrapolated from this point, with uncertainties arising due to uncertainties in α and C_p (see above).

2.4.3.4 Chemical properties

By far, the most important chemical property of the core is the compositional density drop $\Delta\rho_c$. Examination of eqn [62] demonstrates that it is this property, rather than the mass fraction of light element c, which actually determines the gravitational energy available. As noted in Section 2.2.1, the uncertainty in $\Delta\rho_c$ is probably a factor of 2, with a likely range of $400–800\ kg\ m^{-3}$. Other chemical properties of the core materials, such as the heat of reaction R_H and the material constants α_D and β, may be determined from molecular dynamics simulations (Gubbins *et al.*, 2004).

2.4.3.5 Ohmic dissipation

The entropy production rate E_Φ required to power the dynamo is a critical parameter because it ultimately determines how rapidly the core must cool to maintain such a dynamo (eqn [71]). It in turn depends directly on the Ohmic heating (eqn [33]), but this heating rate is currently very poorly constrained. The heating is likely to occur at lengthscales which are sufficiently small that they can neither be observed at the surface, nor resolved in numerical models (Roberts *et al.*, 2003). Moreover, the toroidal field, which is undetectable at the surface, may dominate the heating.

The Ohmic dissipation Q_Φ may be converted to an entropy production rate using $E_\Phi = Q_\Phi / T_D$ (Roberts *et al.*, 2003), where the characteristic temperature T_D is unknown but intermediate between T_i and T_c and is here assumed to be 5000 K. Theoretical lower bounds on the rate of Ohmic heating do exist, but are too low to be either realistic or useful; calculations based on simple assumed current geometries likewise produce unrealistically small values (e.g., 5 GW; Stacey, 1969). A more useful approach is to extrapolate from numerical dynamo simulations. Roberts *et al.* (2003) used the results of the Glatzmaier and Roberts (1996) simulation to infer that 1–2 TW are required to power the dynamo, equivalent to an entropy production rate of $200–400\ MW\ K^{-1}$. The dynamo model of Kuang and Bloxham (1997) gives an entropy production rate of $40\ MW\ K^{-1}$. Christensen and Tilgner (2004) gave a range of 0.2–0.5 TW, based on numerical and laboratory experiments, equivalent to $40–100\ MW\ K^{-1}$, and Buffett (2002) suggested 0.1–0.5 TW, equivalent to $20–100\ MW\ K^{-1}$. Labrosse (2003) argues for a range 350–700 MW K^{-1}. We shall regard the required Ohmic dissipation rate as currently unknown, but think it likely that entropy production rates in excess of 100 MW K^{-1} are sufficient to guarantee a geodynamo.

2.4.3.6 Internal heating

Section 2.4.2 shows that radioactive heat production (or, for that matter, tidal heating) in the core has a significant effect on the entropy and energy budgets. It has sometimes been suggested, for reasons explained below, that radioactive potassium is present in the core. This isotope, ^{40}K, has a half-life of 1.3 Gy, and makes up 0.012% of naturally occurring potassium at the present day.

Samples of the Earth's crust and upper mantle produce ratios of lithophile, refractory elements very similar to those found in primitive (chondritic) meteorites, and the solar photosphere. The constancy of these ratios allow the abundance of these elements (e.g., U and Th) in the bulk silicate Earth (BSE) to be established with some confidence (e.g., McDonough and Sun, 1995). However, it is also clear that Earth samples are generally depleted, relative to chondrites, in more volatile lithophile elements such as potassium. The amount of depletion appears to be roughly correlated with condensation temperature. A recent estimate of potassium concentration in the BSE yields a range of 120–300 ppm (Lassiter, 2004) and a K/U ratio range of 7000–12 000. Chondrites typically have K/U ratios an order of magnitude higher, indicating the extent of the terrestrial depletion.

One possible explanation for the depletion in volatile elements is that the Earth either never accreted them in the first place (due to the elevated temperatures in the protoplanetary disk), or lost them early in its history (e.g., due to late, large impacts). However, the lack of variation in potassium isotope ratios across measured solar system bodies implies that no fractionation of potassium occurred (Humayun and Clayton, 1995), which places severe limits on the classes of mechanisms which could induce volatile loss. An alternative possibility is that some of the potassium was sequestered into the Earth's core (e.g., Murthy and Hall, 1970; Lewis, 1971). Potassium would presumably not be the only element sequestered; other alkali metals (e.g., Na, Rb) should follow suit. Whether the inferred concentrations of these metals in the BSE are compatible with such sequestration is unclear, though it does seem likely that neither U nor Th resides in the core. Given the existing uncertainties, and the extent to which the BSE is depleted in potassium relative to the chondrites, it is hard to rule out ~100 ppm K in the core on cosmochemical grounds.

The hypothesis that the core might contain potassium was apparently not supported by low-pressure experiments (e.g., Chabot and Drake, 1999), which showed that K did not partition into Fe liquids. However, more recent studies overcame previously unrecognized experimental problems and demonstrated that K does in fact partition into Fe liquids, especially in the presence of S and/or Ni (Gessmann and Wood, 2002; Murthy *et al.*, 2003; Lee *et al.*, 2004). Further arguments regarding the presence or absence of potassium in the core may be found in Roberts *et al.* (2003) and McDonough (2007).

In view of the cosmochemical uncertainties surrounding the availability of potassium, the concentration of K in the core must currently be regarded as a free parameter. As discussed below, several groups (e.g., Labrosse *et al.*, 2001; Buffett, 2002; Nimmo *et al.*, 2004; Butler *et al.*, 2005) have recently advocated potassium in the core on the basis of theoretical calculations which are summarized below.

2.4.3.7 Present-day CMB heat flow

The rate of entropy production within the dynamo ultimately depends on the CMB heat flow, that is, the rate at which heat is extracted from the core. The CMB heat flow, in turn, is determined by the ability of the mantle to remove heat. Importantly, independent estimates on this cooling rate exist, based on our understanding of mantle behavior.

One approach to estimating the heat flow across the base of the mantle relies on the conduction of heat across the bottom boundary layer. As discussed in Section 2.2.3, the temperature at the bottom of this layer (the core) arises from extrapolating the temperature at the ICB outward along an adiabat, and is about 4100 K. The temperature at the top of the layer is obtained from extrapolating the mantle potential temperature inward along an adiabat, and is about 2700 K (Boehler, 2000). The thickness of the bottom boundary layer, based on seismological observations, is 100–200 km. For likely lower mantle thermal conductivities, the resulting conductive heat flow is probably in the range 9 ± 3 TW (Buffett, 2003). However, as pointed out by Buffett (2002), the presence of significant quantities of radioactive materials in the boundary layer at the base of the mantle would reduce the heat flux out of the core.

Early estimates of the CMB heat flow based on the inferred contribution from rising convective plumes (Davies, 1988; Sleep, 1990) are smaller by a factor of 2–4. However, these results are probably underestimates, both because the temperature contrast between plumes and the background mantle varies with depth (Bunge, 2005) and because not all plumes may reach the surface (Labrosse, 2002). A more recent seismological study of plumes by Montelli *et al.* (2004) yields a plume heat flow of 10 TW if the mean upwelling velocity is 1 cm yr^{-1}. Two recent theoretical studies of convection including, respectively, compositional layering and the post-perovskite phase transition result in CMB heat flows of ~13 TW (Zhong, 2006) and 7–17 TW (Hernlund *et al.*, 2005). These results are roughly consistent with the simple conductive heat flow estimate, and suggest that a range of 10 ± 4 TW is likely to encompass the real present-day CMB heat flow.

2.4.3.8 Geochemical constraints on IC age

As will become obvious, the age of the IC is a particularly important parameter. Obtaining observational constraints on this age is therefore of great significance. Several recent publications (Brandon *et al.*, 2003; Puchtel *et al.*, 2005) have suggested that (1) coupled ^{186}Os–^{187}Os isotope anomalies found in surface lavas carry a signal from the core (via material entrained across the CMB) and (2) these anomalies were generated as a result of IC formation, which began ~3.5 Gy BP. These results, if correct, are of great significance, but there is as yet little agreement that the current interpretation is the right one (e.g., Lassiter, 2006). First, recycled oceanic crust and/or sediments, rather than core material, could be responsible for the observations (Hauri and Hart, 1993; Baker and Jensen, 2004; Schersten *et al.*, 2004); though that conclusion has been disputed (Brandon *et al.*, 2003; Puchtel *et al.*, 2005). Second, a corresponding signal should also be observed in tungsten and lead isotope anomalies; such signals appear to be lacking in Hawaiian lavas (Schersten *et al.*, 2004; Lassiter, 2006), but may simply have been swamped by crustal contributions. Third, the partition coefficients for Re and Os under likely core conditions are unknown and thus currently have to be treated as free parameters. Nonetheless, this result is intriguing because it can potentially provide an observational constraint on the age of the IC, which can otherwise only be addressed with theoretical models.

2.4.3.9 Comparison with other models

Clearly, the choice of parameters presented above involves considerable uncertainties. One reason for adopting the analytical model presented in Section 2.3.4 is that the effect of varying individual parameters is relatively transparent. Estimated errors have been tabulated in several other recent works, notably Labrosse (2003), Roberts *et al.* (2003), and Nimmo *et al.* (2004). Rather than going through a similar exercise here, **Table 3** compares the parameter choices adopted by four different groups, to highlight those parameters which are particularly uncertain and those on which there is general agreement. Further discussion of the effects these uncertainties have on the model results may be found.

There is good general agreement on some parameters, in particular the specific heat capacity C_p, Gruneisen's parameter γ, and the compositional expansion coefficient α_c. Choices of the thermal conductivity k show more variability; together with the fact that some groups adopt a variable thermal expansivity, the resulting adiabatic heat flow at the CMB varies by a factor of 2. Although estimates of α_c do not vary significantly, the assumed compositional density contrast across the ICB $\Delta\rho_c$ also varies by a factor of roughly 2 (this is mainly because of the recent upward revision in the seismologically determined density contrast; see Section 2.2.1). Thus, the gravitational entropy production estimated by Labrosse (2003) will be almost double that of Roberts *et al.* (2003). Since the gravitational term is probably the dominant one in the overall entropy budget (see below), this kind of uncertainty can have potentially significant effects.

The assumed present-day temperature structure strongly affects the entropy estimates, because of terms having the form $(1/T)-(1/T_c)$. More recent estimates of T_i and T_c are higher than earlier estimates, mainly as a result of the recent molecular dynamics calculations (see Section 2.2.3). Estimates of the adiabatic temperature gradient have also increased in more recent works, and appear to be converging on a value of roughly 0.8 K km^{-1} at the CMB. The slope of the melting curve is, however, still uncertain: current estimates vary by roughly a factor of 2. Small uncertainties in the slopes of the adiabat and solidus lead to large variations in the growth history of the IC. This issue is discussed further, and tabulations of IC ages under different parameter choices are given, in Section 2.5.3 below.

2.4.4 Present-Day Behavior

Having summarized the likely parameter values, this section applies the results of Section 2.3.4 to calculate the present-day energy and entropy budget of the core for an assumed CMB heat flow of 9 TW. **Table 4** summarizes these results.

Figure 3 shows how the rate of entropy production available to drive a dynamo varies as a function of the heat flow out of the core, both for a set of core parameters appropriate to the present-day Earth, and for a situation in which the IC has not yet formed. As expected, higher core heat fluxes generate higher rates of entropy production; also, the same cooling rate generates more excess entropy when an IC exists than when thermal convection alone occurs.

As discussed above, when an IC is present, positive contributions to entropy production arise from core cooling, latent heat release, and gravitational energy; the adiabatic contribution is negative (eqn [70]). For a present-day, radionuclide-free core, a CMB heat flow of <2 TW results in a negative net entropy contribution and, therefore, no dynamo (**Figure 3**). This cooling rate would yield an IC roughly 2.8 Gy old, using the parameters adopted here. A higher core cooling rate generates a higher net entropy production rate; it also means that the IC must have formed more recently.

For a present-day estimated CMB heat flow of 6–14 TW, the net entropy production rate available to drive the dynamo is 300–1000 MW K^{-1}, sufficient to generate roughly 1.5–5 TW of Ohmic dissipation. Since most estimates of Ohmic heating are less than 2 TW (Section 2.4.3.5), it is clear that there is no difficulty in driving a dynamo at the present day. A heat flow of 6–14 TW also implies an IC age of about 0.4–1.2 Gy for the parameters given in **Table 2**.

Prior to the formation of an IC, the CMB heat flow had to exceed the adiabatic value (>4.9 TW) in order to maintain a dynamo for reasons discussed above (eqn [74]). For a dynamo requiring an entropy production rate of 200 MW K^{-1}, the core cooling rate had to be roughly 3 times as fast to maintain this rate before the onset of IC solidification. This again illustrates the rapid core cooling required if thermal convection is the only buoyancy source.

Figure 3 also shows that, as discussed above, a larger CMB heat flow is required for the same entropy production if radioactive heating is important in the present-day Earth. In this particular example, when there is 300 ppm potassium in the core, no dynamo can function if the CMB heat flow drops below 3.3 TW. However, the corresponding core cooling rate is 5 K Gy^{-1}, less than the potassium-free equivalent and permitting an IC to exist over the entire history of the Earth.

Prior to the existence of the IC, the effect of radioactive decay on the entropy production is small (**Figure 3**), because the prefactor for Q_R in eqn [74] is small. Again, however, the existence of radioactive potassium in the core has a significant effect on the cooling history of the core, and the growth of the IC. The role of potassium in controlling the age of the IC is discussed further in Section 2.5.3.

Table 4 Individual contributions to energy/entropy budgets for a present-day core with parameters given in **Table 2** and a CMB heat flow of 9 TW

	K = 0				K = 300 ppm			
	Q		E		Q		E	
	(TW)	(%)	(MW K^{-1})	(%)	(TW)	(%)	(MW K^{-1})	(%)
Q_S, E_S	2.2	25	73	8	1.7	19	56	7
Q_L, E_L	4.2	47	268	28	3.3	37	205	26
Q_g, E_g	2.5	28	618	64	1.9	21	474	59
Q_R, E_R	0	0	0	0	2.1	24	65	8
Q_k, E_k	4.9		−162		4.9		−162	
Q_H, E_H	0		−219		0		−168	
Q_{cmb}, ΔE	9.0		537		9.0		431	
dT_c/dt (K Gy^{-1})	−37				−30			
dr_i/dt (km Gy^{-1})	788				605			
IC age (My)	590				780			

Q and E refer to the energy and entropy contributions, respectively; ΔE is the entropy available to drive the dynamo, dT_c/dt is the core cooling rate and dr_i/dt the IC growth rate. Two cases are shown, one with no potassium (K) and one with 300 ppm potassium in the core. IC age is calculated assuming a constant CMB heat flow.

Table 4 summarizes the individual contributions to the overall entropy and energy budgets for the present-day core, both with and without potassium. In both cases, the CMB heat flow is specified to be 9 TW. It is clear that the different driving mechanisms are associated with different thermodynamic efficiencies. In particular, while the gravitational term only contributes roughly one quarter of the total heat budget when potassium is absent, it delivers two-thirds of the entropy budget. Similarly, the latent heat term delivers twice as much energy, but 4 times as much entropy, as the secular cooling term.

Table 4 also makes the role of radioactive heating clear. Its efficiency is comparable to that of secular cooling; thus, for the same total CMB heat flow, adding radioactive elements reduces ΔE, the entropy available to drive the dynamo. The presence of radioactivity also reduces the core cooling rate, and thus increases the age of the IC. However, since the radioactive contribution (2.1 TW) is small compared to the assumed CMB heat flow, the increase in IC age is relatively small.

In summary, **Figure 3** shows that the minimum CMB heat flow which could allow a dynamo to persist at the present day is about 2 TW. Under these circumstances, an IC could have persisted for about 3.5 Gy. However, such a small present-day CMB heat flow is incompatible with independent estimates of this quantity, which ranges from 6 to 14 TW. The latter range of heat flows allows a dynamo dissipating several terawatts via Ohmic heating to operate, but implies an IC of roughly 0.4–1.2 Gy. Radioactive heating can increase this age somewhat, but the present-day radioactive heat production is likely only a small fraction of the total energy budget, and thus the effects are modest. In practice, of course, both the core heat flux and the radiogenic heat production will vary with time; investigating the time evolution of the core and mantle is the subject of the next section.

2.5 Evolution of Energy Budget through Time

2.5.1 Introduction

From the above, it is evident that the energy and entropy budgets of the present-day core for a dynamo dissipating of order 1 TW of heat are consistent with the inferred CMB heat flow of 6–14 TW. Because a dynamo similar to the present-day one has apparently been operating for at least 3.5 Gy, constraints can also be applied to the energy and entropy budgets over this time period. Attempting to do so introduces additional, large uncertainties, in particular the fact that the evolution of the CMB heat flow through time is unknown. Nonetheless, this problem has been investigated by several groups, notably Yukutake (2000), Buffett (2002), Labrosse (2003), Nimmo *et al.* (2004), Nakagawa and Tackley (2004a), and Butler *et al.* (2005). Many of these groups have reached similar conclusions. In particular, the IC is commonly found to be a relatively young feature, with an age of ~1 Gy, and early core temperatures tend to be very high, an effect which is somewhat mitigated if radioactive potassium is present in the core.

Prior to IC solidification, the CMB heat flow over time depends on the adiabatic and Ohmic dissipation terms and is given by eqn [74]. Given a CMB heat flow, the core cooling rate, and hence the total core temperature drop prior to onset of IC solidification, is simple to determine (eqn [75]). Let the IC age be τ_{ic} and the age of the dynamo be τ_{d}. Then making the simplifying assumption that the heat capacity of the core is simply MC_{p}, the reduction in core temperature, T_{c}, over the interval from the onset of dynamo activity to the start of core solidification is given by

$$\Delta T_{\mathrm{c}}' \approx \left(T_{\mathrm{c}} \frac{8\pi r_{\mathrm{c}}^3 k}{D^2} \left[1 + \frac{E_{\Phi}}{E_k} \right] - \bar{Q}_{\mathrm{R}} \frac{1}{\frac{2r_{\mathrm{c}}^2}{5L^2} + 1} \right) \frac{[\tau_{\mathrm{d}} - \tau_{\mathrm{ic}}]}{MC_{\mathrm{p}}} \quad [76]$$

where $\bar{Q}_{\mathrm{R}}$ is the mean radiogenic heat flow over the interval of interest. Since the temperature at which IC solidification starts is fixed, eqn [76] allows the initial temperature of the core to be calculated. Higher Ohmic dissipation rates, later IC formation, and higher thermal conductivities all imply a larger core temperature drop, and a higher initial temperature (Labrosse *et al.*, 2001; Buffett, 2002). On the other hand, radioactive heating reduces the amount by which the temperature drops. Using the values of **Table 2** and assuming that $E_k \approx E_{\Phi}$, the core temperature drop over a 2 Gy time interval is roughly 400 K in the absence of radioactive heating. Such a high initial temperature for the early core likely implies widespread melting of the lowermost mantle (e.g., Boehler, 2000). Although the consequences of such melting are unclear, it is likely that advection of this melt will lead to high CMB heat flows and rapid initial cooling of the core and mantle. Whether such a scenario is compatible with the subsequent maintenance of a dynamo is unclear.

The advantage of eqn [76] is that it provides a robust and relatively transparent estimate of how much core cooling occurred prior to IC formation. However, in practice, the CMB heat flux is likely to have fluctuated with time, and the onset of IC solidification raises additional complications. Thus, a fuller picture of core and dynamo evolution requires more sophisticated approaches, which are discussed below.

2.5.2 Theoretical Models

The CMB heat flow ultimately controls the thermal evolution of the core, but its evolution through time is not observationally constrained. Accordingly, theoretical arguments have to be adopted. One such argument is to employ a numerical model of the Earth's thermal evolution, and predict the CMB heat flow through time. Models can either be relatively simple parametrized convection calculations (e.g., Yukutake, 2000; Buffett, 2002; Nimmo *et al.*, 2004), or more sophisticated (but two-dimensional (2-D)) approaches in which the full fluid-dynamical equations are solved (e.g., Nakagawa and Tackley, 2004a; Butler *et al.*, 2005). These models illustrate the important point that the evolution of the CMB heat flow cannot be decoupled from the thermal evolution of the Earth's mantle. The rate at which heat is transported across the CMB depends on the temperature gradient at the base of the mantle, which ultimately depends on the rate at which the mantle itself is cooling. Thus, a successful model must accurately reproduce the behavior of the mantle through time in order to calculate the evolution of the CMB heat flow.

None of the models currently in existence are likely to meet this criterion. First, the actual evolution of the mantle temperature through time is unknown, with estimated mantle potential temperatures at 3.5 Gy BP spanning the range from 1500°C to 1900°C (e.g., Abbott *et al.*, 1994; Grove and Parman, 2004). Second, the effect of mantle temperature on plate tectonic behavior is not well understood, with even the sign of the effect being disputed (e.g., Korenaga, 2003); similarly, the change in mean plate tectonic velocities over billions of years is unknown. Since it is plate tectonics that controls the cooling rate of the mantle, these are important issues. Third, it is not clear that any model incorporates all of the relevant physics. For instance, the proposed 'post-perovskite' phase near the base of the mantle (Murakami *et al.*, 2004; Oganov and Ono, 2004) likely has an appreciable effect on the evolution of the CMB heat flow (e.g., Nakagawa and Tackley, 2004b), as would dense chemical piles at the base of the mantle (McNamara and Zhong, 2004). The effect of a partially molten zone at the CMB (e.g., Williams and Garnero, 1996), either now or in the past, may also have important, but currently unquantified, effects on heat transfer.

An example of one such 2-D fluid-dynamical calculation is shown in **Figure 4**. This calculation is based on the code described in Xie and Tackley (2004) and includes a viscosity which depends on the mean mantle temperature, and which also increases by a factor of 100 over the depth of the mantle. The top panel shows the evolution of core and mantle temperature with time, both of which decrease due to the slow decay of radioactive elements in the mantle. The approximate present-day rate of core cooling is 150 K Gy^{-1}.

Because T_c never falls below 4500 K, an IC never forms in this particular example. The middle panel shows the evolution of the CMB heat flow with time. After an initial transient, the main trend is a slight decline, owing to the cooling of the mantle. However, strong fluctuations in the CMB heat flow occur on a timescale of $\sim$500 My, because mantle convection involves discrete thermals separating from the bottom boundary layer and ascending, leading to time-dependent behavior. The bottom panel shows the rate of entropy available to drive the dynamo, calculated using the methods and parameters given in Section 2.3. Because there is no IC contribution, the entropy production rate essentially mirrors the CMB heat flow and demonstrates similar fluctuations. The rate of entropy production is 100–200 MW K^{-1}, sufficient to drive a moderately dissipative dynamo for at least 3.5 Gy following the initial transient.

The advantage of the above approach is that arbitrarily complicated models, including phase changes, compositional variations, melting, and so on, can be included. The disadvantages are that such models are slow to run, which limits the parameter space that can be explored, and that it is not clear that all the relevant physical processes have been included. For instance, there is no agreement on whether or not plate tectonics was primarily responsible for transporting the Earth's heat during the Archean period (De Wit, 1998).

An alternative approach is to assume that the Ohmic dissipation in the core, and thus the rate of entropy production required to drive the dynamo E_Φ, has been constant, or varied in a prescribed manner,

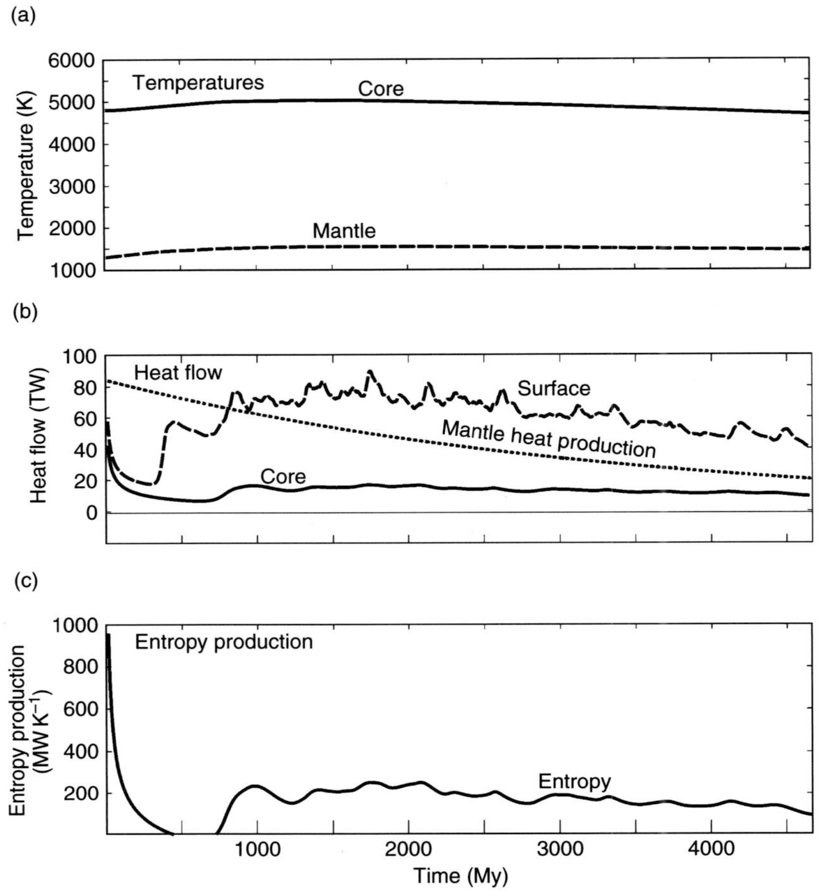

Figure 4 Example of 2-D numerical mantle thermal evolution calculations, based on the method described in Xie and Tackley (2004). (a) Evolution of mantle potential temperature and core temperature at the CMB as a function of time. (b) Evolution of surface heat flow, CMB heat flow, and radiogenic mantle heat production. (c) Rate of entropy production, calculated using the methods of Section 2.3.4 and the parameters given in Section 2.4. Model output courtesy of Paul Tackley.

through time (e.g., Buffett, 2002; Labrosse, 2003). Since the rate of entropy production depends on the core cooling rate (eqn [69]), specifying this rate of production specifies the core cooling rate and thus the CMB heat flow over time. This in turn allows the core temperature T_c to be integrated backward from the present day. A high rate of entropy production requires rapid cooling of the core and thus a young IC, and an early core which is very hot.

The CMB heat flow changes slowly, on timescales of several hundreds of millions of years (**Figure 4**). It can therefore be seen from eqn [71] that the onset of IC solidification will cause a large increase in the rate of entropy production available to drive the dynamo. This is because the heat flow Q_{cmb} is fixed in the short term, and the addition of latent heat and compositional driving forces (which are thermodynamically more efficient) reduce the value of $\tilde{Q}_T/\tilde{E}_T$. The specified rate of entropy production should therefore be appropriate to the situation just prior to the onset of IC solidification.

The advantage of this approach – specifying an entropy production rate and deducing the resulting temperature evolution – is that it is simple; for instance, no knowledge of the mantle is required. Thus, parameter space can be explored rapidly, and the sensitivity of the results to uncertainties in different parameters determined. On the other hand, this technique requires a large assumption – that dissipation within the core has stayed constant over time. Furthermore, as discussed above, the actual rate of dissipation involved in even the present-day dynamo is unknown.

Figure 5 is an example of the results obtained by specifying an entropy production rate. It starts from the present-day structure discussed in Section 2.4.3

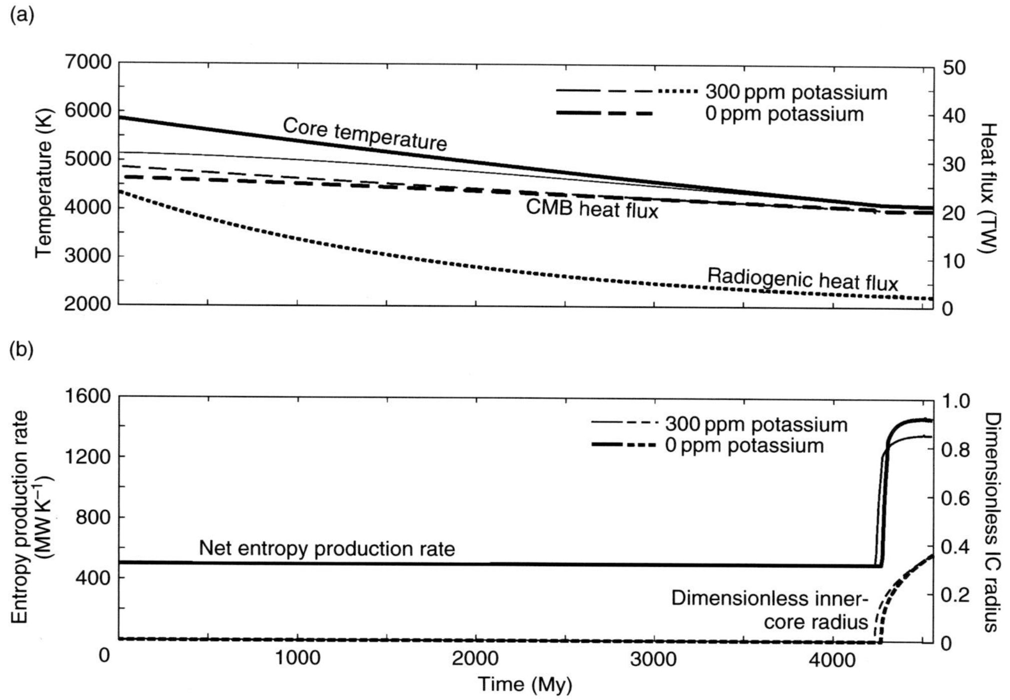

Figure 5 (a) Evolution of core temperature T_c (left-hand scale) and heat flow Q_{cmb} (right-hand scale) through time, assuming a constant rate of entropy production of 500 MW K^{-1} prior to IC solidification. During IC solidification, the heat flow is kept constant at 20 TW. Results were obtained by integrating backward in time using the present-day conditions and parameters specified in **Table 2**. Given a fixed heat flow or rate of entropy production, and a radioactive heat production rate, the change in core temperature with time may be calculated. Thick lines denote case with no core potassium; thin lines denote case with 300 ppm potassium. In the latter case, the radioactive heat production is given by the dotted line. (b) Evolution of rate of entropy production and growth of IC. Thick lines denote potassium-free case; thin lines show 300 ppm potassium in the core.

and integrates T_c backward in time such that the entropy production rate prior to IC formation stays constant at 500 MW K^{-1}. It is assumed that the core heat flow subsequent to IC formation remains constant, which is reasonable given the young IC ages obtained.

Figure 5(a) shows the evolution of T_c and CMB heat flow, integrated backward from the present. An entropy production rate prior to IC formation of 500 MW K^{-1} requires a CMB heat flow of 20 TW, and a present-day core cooling rate of 84 K Gy^{-1} in the absence of radioactive heating. This CMB heat flow is a significant fraction of the total global surface heat flow of 42 TW (Sclater *et al.*, 1980), and exceeds the likely value deduced from present-day observations (Section 2.4.3.7). These high heat fluxes suggest that a dynamo requiring an entropy production rate of 500 MW K^{-1} ($\approx$2.5 TW dissipation) is unrealistic.

Figure 5 shows two scenarios: one with no radioactive heating, the other with 300 ppm potassium in the core. In either case, the initial core temperatures are high, but the presence of potassium reduces the initial temperature, as expected from eqn [77], from 5863 to 5140 K. The magnitude of this effect is mainly due to the short half-life of potassium, which produces 15 times more heat at 3.5 Gy BP than at the present day.

Figure 5(b) shows the evolution of the IC radius and entropy production with time. The high core cooling rate means that the IC is a young feature (0.6 Gy old). As expected, the entropy production rate increases (by a factor of $\sim$3) when IC formation begins. While one might expect such an increase to have significant effects on the observed magnetic field, in practice the observations are sufficiently sparse that no such effects have been detected (Section 2.2.4). Because of the high CMB heat flow, the presence of potassium has little effect on the age of the IC, though it does affect the early core temperature as noted above.

As noted by Buffett (2002) and Labrosse (2003), and shown in eqn [76], the dissipation required to

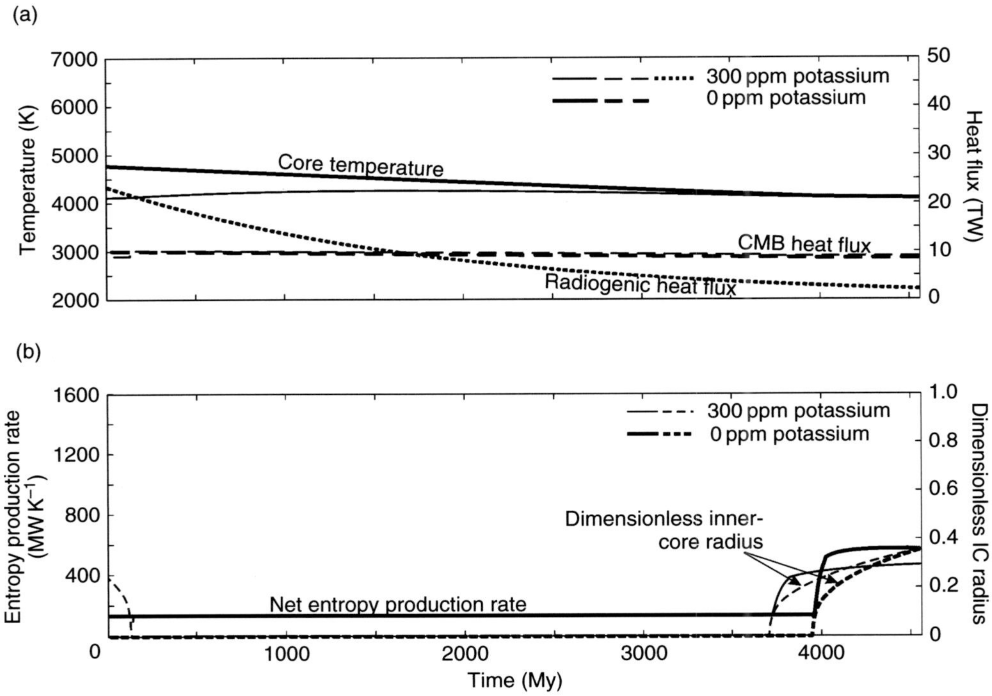

Figure 6 Same as for **Figure 5**, except that rate of entropy production prior to IC solidification is fixed at 135 MW K^{-1}, and CMB heat flow following onset of solidification is fixed at 9 TW.

drive the dynamo has a strong influence on the evolution of the core temperature. **Figure 6** shows an identical set of results to **Figure 5**, except assuming a constant (pre-IC) entropy production rate of 135 MW K^{-1} ($\sim$0.7 TW dissipation), appropriate to a more reasonable CMB heat flow of 9 TW. As one would expect, the lower heat flow results in a slower core cooling rate and an older IC (0.61–0.84 Gy, depending on the potassium concentration). The resulting changes in core temperatures are also smaller. In particular, if 300 ppm potassium is present, then the core initially heats up, because the radioactive decay alone is sufficient to account for the required entropy production. A corollary is that the IC may have been present, disappeared as the core heated up, and then began to resolidify (cf. Buffett, 2002). Such a scenario, however, is unlikely simply because the core is expected to have been initially hot due to the gravitational energy released during accretion of the Earth (see below).

There are two other interesting consequences of the lower entropy production rate assumed. First, the CMB heat flow stays almost constant over 4 Gy. While parametrized scalings (e.g., Nimmo *et al.*, 2004) suggest that the CMB heat flow should have declined over Earth history as the mantle temperature decreased, the numerical model shown in **Figure 4** demonstrates that the decrease may actually be rather small and thus a scenario invoking a constant CMB heat flow may not be unreasonable. Second, the presence of radioactive potassium has more of an effect on the IC age when the CMB heat flow, and total entropy production rate, are smaller.

Figure 7 summarizes these results. **Figure 7(a)** plots the initial core temperature as a function of core potassium and entropy production rate. It demonstrates that, as expected, higher rates of entropy production require more core cooling and thus higher initial temperatures. Adding potassium to the core counteracts this effect. These effects are essentially identical to those found by Buffett (2002) and Labrosse (2003). Unfortunately, as discussed below, the initial temperature of the core is sufficiently uncertain that only results indicating an initial temperature cooler than that of the present day can be excluded with any confidence.

The initial core temperature depends on both the manner in which the Earth accreted, and the process of core differentiation (Stevenson, 1989). The energy associated with the differentiation process may be calculated in a fairly straightforward manner, by comparing the gravitational potential energies of a

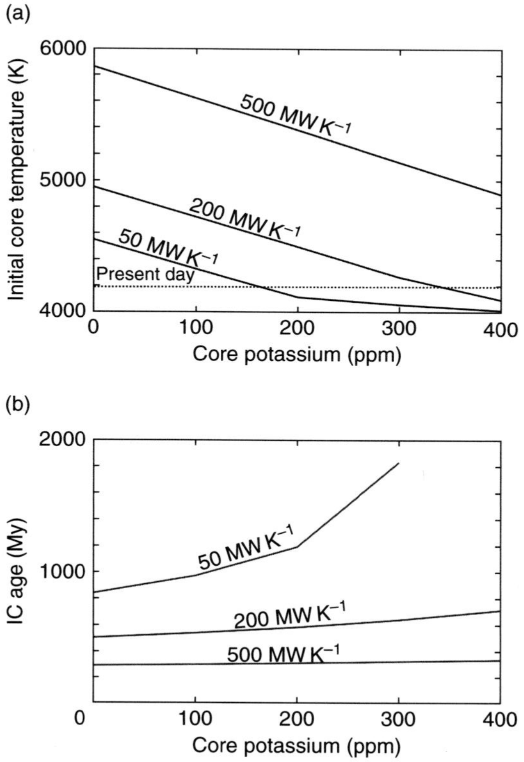

Figure 7 (a) Variation in initial core temperature (at 4.56 Gy BP) as a function of core potassium concentration and entropy production rate within the core. Calculations carried out as for **Figures 5** and **6**. (b) Variation in IC age as a function of core potassium concentration and entropy production rate.

uniform and layered body of the same total mass and radius (Flasar and Birch, 1973). For the case in which the core radius is half the planetary radius, and the core density is twice the mantle density, it may be shown that the change in energy ΔW_g associated with differentiation is given by

$$\Delta W_g = \frac{1}{15}\pi^2 r_e^5 \rho_m^2 G \qquad [77]$$

where ρ_m is the mantle density and is 8/9 of the mean planetary density, and r_e is the planetary radius. If this energy is distributed uniformly across the planet and is all converted into thermal energy, the resulting increase in temperature is roughly 1800 K. If all this energy were concentrated in the core, the increase would be roughly 8000 K. Such a large temperature increases implies an initially partially molten mantle, and suggests that scenarios involving an initial core colder than the present day are very unlikely.

2.5.3 Age of the IC

Figures 3, **5**, and **6** make it clear that the initiation of IC solidification significantly increases the entropy production available to drive the geodynamo, even though the consequences of this increase for the magnetic field intensity are unclear. If the IC age could be established observationally, it would provide an immediate constraint on the time-averaged CMB heat flow, and hence on the mechanisms available to drive the geodynamo. Establishing the age of the IC is therefore of great importance but, as outlined below, there is currently a significant disagreement between geochemical and geophysical arguments.

Over the interval since IC solidification began, the total amount of energy released W_t has come from four main sources: secular cooling of the whole core (W_s), gravitational energy (W_g), latent heat release (W_L), and perhaps some radioactive decay (W_R). These quantities may be calculated by time integration of the corresponding heat flow terms given in Section 2.3.2; for the similar set of equations investigated by Buffett *et al.* (1996), analytical expressions for these terms may also be derived.

Letting the IC age be τ_i and the mean CMB heat flow over this interval be $\bar{Q}_{cmb}$, we have

$$\tau_i = \frac{W_s + W_g + W_L + W_R}{\bar{Q}_{cmb}} = \frac{W_s + W_g + W_L + \frac{Q_R}{\lambda}\left(e^{\lambda\tau_i} - 1\right)}{\bar{Q}_{cmb}} \qquad [78]$$

where Q_R is the present-day radioactive heat production and λ is the decay constant. Note that this equation, unlike eqn [71], contains no information on whether a dynamo operates; in particular, if the mean CMB heat flow is too small, a dynamo will not occur.

The IC age depends on the total energy released since solidification began, and the CMB heat flow. Of the four energy terms, the first depends on the mean temperature drop ΔT in the core since solidification began ($W_s \approx M_c C_p \Delta T$); the others are independent of this quantity, and depend on the same variables as their time derivatives Q_g, Q_L, and Q_R (see Section 2.3.4). The mean temperature drop ΔT in turn depends on the behavior of the solidus and the adiabat, as discussed below.

Since the IC first began to solidify, the temperature at the ICB has dropped by an amount determined by the behavior of the melting curve and adiabat (see **Figure 2**). To achieve this growth in IC size requires a corresponding change in the temperature at the CMB, ΔT_c. Making the approximation that the solidus and adiabat are both linear (with respect to pressure) within the IC, then the change in CMB temperature since the onset of core solidification ΔT_c is given by (see **Figure 2**)

$$\Delta T_c = \frac{\Delta P}{f_{ad}}\left(\frac{dT_m}{dP} - \frac{dT_a}{dP}\right) = 22\,\mathrm{K}\left(\frac{\frac{dT_m}{dP} - \frac{dT_a}{dP}}{1\,\mathrm{K/GPa}}\right) \qquad [79]$$

where ΔP is the pressure difference between the present ICB and the center of the Earth, T_a and T_m are the adiabatic and melting temperature profiles, f_{ad} is a factor converting the core temperature at the CMB to that at the ICB, and the numerical values are obtained by using the parameter values given in **Table 2**. Note that ΔT_c depends on the difference in slopes and is thus sensitive to small variations in these slopes.

For the parameters given in **Table 2**, $f_{ad} = 1.35$, $\Delta P = 30$ GPa, and the difference in slopes at the ICB is 1.3 K GPa $^{-1}$. This indicates that the core has only cooled by approximately 30 K since the IC began to solidify. However, as noted above, the slopes of both the adiabat and the melting curve are subject to substantial uncertainties. In order to investigate these uncertainties, **Table 5** compares the values adopted for relevant quantities by the same four groups shown in **Table 3**. There is a considerable range in the value of ΔT_c obtained, from 31 to 146 K. The amount of heat released by core cooling W_s has a correspondingly large range: it dominates all other contributions in the Buffett *et al.* and Labrosse models, but is of lesser importance in the models of Roberts *et al.* and this work. Similarly, the release of latent heat W_L can range from minor (Buffett) to dominant (Roberts). The uncertainty in the total amount of energy released since IC solidification is uncertain by a factor of 2, ranging from 16 to 36 × 10^{28} J. This range will be adopted as representative of the likely uncertainties, and may be used to estimate the IC age. For a mean CMB heat flow range of 6–14 TW, eqn [78] shows that the resulting IC age in the absence of radiogenic heating is 0.37–1.90 Gy, much less than the age of the Earth. As discussed below, this result is of considerable importance, not least since it appears to contradict geochemical estimates of IC age.

A reduction in CMB heat flow would result in a reduction in Ohmic dissipation and a longer-lived IC. Just prior to the onset of dynamo formation, the CMB heat flow must have equalled or exceeded the adiabatic value, depending on the amount of Ohmic dissipation (eqn [74]). Assuming that the heat flow has not decreased since, the mean heat flow since IC formation cannot have been less than the adiabatic value. Taking this value to be 4.9 TW (likely too low an estimate) and using the total energy release range of 16–36 × 10^{28} J results in a maximum IC age range of 1.04–2.33 Gy from eqn [78]. This estimate is conservative and still results in an IC only half the age of the Earth at most. An even lower CMB heat flow would allow an older IC, but would not allow a dynamo to function prior to IC formation. Furthermore, such a low CMB heat flow is not consistent with the available present-day observational constraints (Section 2.4.3.7).

Equation [78] shows that another way to increase the age of the IC is to add radioactive heating. For instance, **Figure 6** shows that adding 300 ppm to the core increases the IC age from 0.61 to 0.84 Gy for a constant heat flow of 9 TW. **Figure 7(b)** shows how the IC age varies as a function of potassium concentration for different rates of Ohmic dissipation,

Table 5 Comparison of results obtained by different groups

Quantity	Units	Buffett	Roberts	Labrosse	This work
T_m ($r=0$)		5000	5379	5967	5800
$\frac{\partial T_m}{\partial P}\vert$cmb	$\times 10^{-8}$ K Pa^{-1}	1.0	1.25	1.7	0.87
$\frac{\partial T_a}{\partial r}\vert$cmb	K km^{-1}	0.54	0.88	0.75	0.80
ΔT_c	K	146	68	96	31
W_s	$\times 10^{28}$ J	23	8.7	18.2	5.3
W_g	$\times 10^{28}$ J	3.8	1.9	4.1	4.2
W_L	$\times 10^{28}$ J	5.6	15	7.0	6.7
W_{tot}	$\times 10^{28}$ J	36	26	29	16
IC age ($Q_{cmb} = 9$ TW)	Gy	1.26	0.92	0.64	0.57

References are Buffett *et al*. (1996), Roberts *et al*. (2003), and Labrosse (2003).
IC age assumes a constant heat flow of 9 TW.

calculated using the same method as that shown in **Figures 5** and **6**. As expected, lower rates of entropy production and the addition of potassium both result in longer-lived ICs. However, only at low dissipation rates does the addition of potassium have a significant effect, because at high dissipation rates the required CMB heat flow greatly exceeds that due to radioactive decay (2.1 TW at the present day for 300 ppm potassium). It is clear that the parameter values given in **Table 2** make an IC older than ~1 Gy unlikely for any significantly dissipative dynamo.

The effect of adopting different parameter values may be investigated by using eqn [78] and adopting the total energy release range of $16–36 \times 10^{28}$ J from **Table 5**; the results shown in **Figures 5** and **6** use the lower bound. In the absence of radioactive heating, a mean CMB heat flow of 9 TW results in an IC age of 0.57–1.26 Gy. Adding 300 ppm of potassium increases this range to 0.79–2.35 Gy. In the absence of potassium, an inner core age of 3.5 Gy implies a mean CMB heat flow of 1.5–3.3 TW. For the upper bound of 36×10^{28} J, a time-averaged heat flow of 8.5 TW with 300 ppm in the core results in an IC age equalling the age of the Earth; with 150 ppm in the core, this situation occurs for a heat flow of 6 TW. In both cases, the entropy production is a few hundred megawatts per kelvin (eqn [71]), likely sufficient to maintain a dynamo. Thus, an ancient IC can be obtained, but only with potassium and a time-averaged heat flow that does not significantly exceed the present-day value.

Investigations by Buffett (2002), Labrosse (2003), and Roberts *et al.* (2003) reach similar conclusions to those stated here. In the absence of radioactive heating, the IC is most likely young (~1 Gy); an ancient IC requires heat flows which are likely too low to maintain a dynamo. Radioactive heating in the core could extend the IC age to ≥3 Gy while permitting dynamo operation, but requires both significant concentrations of potassium and a relatively low value for the time-averaged CMB heat flow.

As discussed in Section 2.4.3.8, the possibility of an ancient IC is interesting because geochemical observations of Os isotope anomalies have been used to argue for ancient (~3.5 Gy) IC crystallization (e.g., Brandon *et al.*, 2003). If this interpretation of these anomalies is correct, the implications are profound; something is seriously wrong with the models tabulated in **Tables 3–5**. Possible errors might arise from incorrect core temperature models (e.g., a steeper melting curve would allow a more ancient IC; see **Figure 2**), a significantly overestimated core thermal conductivity, an underestimate of core radioactive heating, or an overestimate of the present-day CMB heat flow. Based on the discussion of Section 2.4.3, none of these errors currently appears very likely, but all should become the focus of future research if the geochemical arguments continue to stand up to scrutiny.

2.6 Summary and Conclusions

The results reviewed here are the outcome of three decades of intensive study of the energy and entropy budgets of the Earth's core. The theoretical description of these budgets has been understood for a long time, but only in recent years have experimental and computational constraints on many core parameters required (e.g., **Table 2**) become available. While most of these parameters are sufficiently well determined to allow relatively firm conclusions to be drawn, there are some lingering uncertainties (see Section 2.4.3).

First, and perhaps most important, the melting behavior of the iron–light element system is not yet sufficiently well understood. Although different computational groups are now converging on results which agree relatively well with experimental determinations, some discrepancies still remain. Similar uncertainties remain concerning the density contrast due to the light element, and in this case the uncertainties are exacerbated by differing interpretations of the seismological observations. Second, the present-day CMB heat flow, which is crucial in determining IC age and dynamo activity, is still uncertain by at least a factor of 2, and its value at earlier times is unknown. The mechanisms by which heat is transferred through the lower mantle are complicated, with the possibility of melting, phase or compositional boundaries, and local radioactive heating likely complicating the simple picture of conduction across a boundary layer. Third, it is not clear that current estimates of the thermal conductivity of the core, based on 30-year-old data, are reliable. Since this parameter has a major effect on the sustainability of a dynamo, more measurements are certainly needed. And finally, the amount of dissipation actually occurring within the present-day dynamo is unknown, but again has a major effect on the sustainability of the dynamo (eqn [71]).

Despite these uncertainties, there is actually a significant amount of agreement within the geophysical community concerning the present-day state and history of the core and dynamo. **Tables 3** and **5** show the extent to which different groups, working independently, have arrived at similar conclusions despite employing different parameter choices. For instance, there is general agreement that the single most important mechanism driving the dynamo at the present day is compositional convection. An important result is the likely age of the IC (**Table 5**), which depends only on the mean CMB heat flow and the total energy released since core solidification (eqn [78]). In the absence of core potassium, these different groups obtain an IC age range of 0.37–1.90 Gy for a time-averaged CMB heat flow equal to the inferred present-day value of 6–14 TW. A heat flow <6 TW would almost certainly cause the dynamo to fail prior to the onset of IC solidification (eqn [74]). Thus, a substantial disagreement currently exists between the geophysical results and geochemical observations which may indicate an IC 3.5 Gy old. Adding potassium to the core results in lower initial core temperatures, and also allows IC ages compatible with the geochemical estimates, but only if the CMB heat flow has remained low (e.g., <9 TW for 300 ppm potassium) for the whole of Earth history. Whether such a relatively constant heat flow is a likely outcome of the mantle's thermal evolution remains to be seen.

The age of the IC is clearly the most important outstanding question. Unless paleomagnetic observations can be used to identify its onset, which seems unlikely, the question is most likely to be settled by geochemistry. The role of potassium in the evolution of the core is also an important question; although better constraints from geochemistry are likely to help, this question is most likely to be settled by geophysics, in particular further investigation of the antineutrinos produced by radioactive decay in the Earth's interior (Araki *et al.*, 2005). Quantifying the CMB heat flow is also crucial; here a combination of higher-resolution seismology, better mineral physics constraints, and convection modeling is likely to yield dividends. In parallel with these advances, our improved understanding of the Earth's dynamo is likely to prove central in unraveling the histories of the dynamos on other terrestrial planets and satellites (Stevenson, 2003; Nimmo and Alfe, 2006).

Acknowledgment

This work is supported by NSF-EAR.

References

Abbott D, Burgess L, Longhi J, and Smith WHF (1994) An empirical thermal history of the Earth's upper mantle. *Journal of Geophysical Research* 99: 13835–13850.

Alfè D, Gillan MJ, and Price GD (1999) The melting curve of iron at the pressures of the Earth's core from *ab initio* calculations. *Nature* 401: 462–464.

Alfè D, Gillan MJ, and Price GD (2002a) *Ab initio* chemical potentials of solid and liquid solutions and the chemistry of the Earth's core. *Journal of Chemical Physics* 116: 7127–7136.

Alfè D, Price GD, and Gillan MJ (2002b) Iron under Earth's core conditions: Liquid-state thermodynamics and high-pressure melting curve from *ab intio* calculations. *Physical Review B* 65: 165118.

Alfè D, Gillan MJ, and Price GD (2003) Thermodynamics from first principles: Temperature and composition of the Earth's core. *Mineralogical Magazine* 67: 113–123.

Alfè D, Price GD, and Gillan MJ (2004) The melting curve of iron from quantum mechanics calculations. *Journal of Physics and Chemistry of Solids* 65: 1573–1580.

Anderson OL (1998) The Gruneisen parameter for iron at outer core conditions and the resulting conductive heat and power in the core. *Physics of the Earth and Planetary Interiors* 109: 179–197.

Araki T, Enomoto S, Furuno K, *et al.* (2005) Experimental investigation of geologically produced antineutrinos with KamLAND. *Nature* 436: 499–503.

Backus GE (1975) Gross thermodynamics of heat engines in deep interior of Earth. *Proceedings of the National Academy of Sciences of the United States of America* 72: 1555–1558.

Baker JA and Jensen KK (2004) Coupled Os-186-Os-187 enrichements in the Earth's mantle - Core–mantle interaction or recycling of ferromanganese crusts and nodules? *Earth and Planetary Science Letters* 220: 277–286.

Belonoshko AB, Ahuja R, and Johansson B (2000) Quasi-*ab-initio* molecular dynamic study of Fe melting. *Physical Review Letters* 84: 3638–3641.

Bi Y, Tan H, and Jin F (2002) Electrical conductivity of iron under shock compression up to 200 GPa. *Journal of Physics: Condensed Matter* 14: 10849–10854.

Bloxham J (2000) Sensitivity of the geomagnetic axial dipole to thermal core–mantle interactions. *Nature* 405: 63–65.

Boehler R (1993) Temperatures in the Earth's core from melting-point measurements of iron at high static pressures. *Nature* 363: 534–536.

Boehler R (2000) High-pressure experiments and the phase diagram of lower mantle and core materials. *Reviews of Geophysics* 38: 221–245.

Braginsky SI (1963) Structure of the F layer and reasons for convection in the Earth's core. *Dokland Akademii Nauk SSSR English Translation* 149: 1311–1314.

Braginsky SI (1999) Dynamics of the stably stratified ocean at the top of the core. *Physics of the Earth and Planetary Interiors* 111: 21–34.

Braginsky SI and Roberts PH (1995) Equations governing convection in Earth's core and the geodynamo. *Geophysical and Astrophysical Fluid Dynamics* 79: 1–97.

Braginsky SI and Roberts PH (2002) On the theory of convection in Earth's core. In: Ferriz-Mas A and Nunez M (eds.) *Advances in Nonlinear Dynamos*, pp. 60–82. London: Taylor & Francis.

Brandon AD, Walker RJ, Puchtel IS, Becker H, Humayun M, and Revillon S (2003) Os-186–Os-187 systematics of Gorgona Island komatiites: Implications for early growth of the inner core. *Earth and Planetary Science Letters* 206: 411–426.

Brown JM and McQueen RG (1986) Phase-transitions, Gruneisen parameter and elasticity for shocked iron

between 77 GPa and 400 GPa. *Journal of Geophysical Research* 91: 7485–7494.
Buffett BA (2002) Estimates of heat flow in the deep mantle based on the power requirements for the geodynamo. *Geophysical Research Letters* 29: 1566.
Buffett BA (2003) The thermal state of Earth's core. *Science* 299: 1675–1677.
Buffett BA, Huppert HE, Lister JR, and Woods AW (1996) On the thermal evolution of the Earth's core. *Journal of Geophysical Research* 101: 7989–8006.
Bullard EC (1950) The transfer of heat from the core of the Earth. *Monthly Notices of the Royal Astronomical Society* 6: 36–41.
Bunge HP (2005) Low plume excess temperature and high core heat flux inferred from non-adiabatic geotherms in internally-heated mantle circulation models. *Physics of the Earth and Planetary Interiors* 153: 3–10.
Busse FH (2000) Homogeneous dynamos in planetary cores and in the laboratory. *Annual Review of Fluid Mechanics* 32: 383–408.
Butler SL, Peltier WR, and Costin SO (2005) Numerical models of the Earth's thermal history: Effects of inner-core solidification and core potassium. *Physics of the Earth and Planetary Interiors* 152: 22–42.
Cao AM and Romanowicz B (2004) Constraints on density and shear velocity contrasts at the inner core boundary. *Geophysical Journal International* 157: 1146–1151.
Chabot NL and Drake MJ (1999) Potassium solubility in metal: The effects of composition at 15 kbar and 1900 degrees C on partitioning between iron alloys and silicate melts. *Earth and Planetary Science Letters* 172: 323–335.
Christensen UR and Olson P (2003) Secular variation in numerical geodynamo models with lateral variations of boundary heat flow. *Physics of the Earth and Planetary Interiors* 138: 39–54.
Christensen UR and Tilgner A (2004) Power requirements of the geodynamo from Ohmic losses in numerical and laboratory dynamos. *Nature* 429: 169–171.
Davies GF (1988) Ocean bathymetry and mantle convection. 1: Large-scale flow and hotspots. *Journal of Geophysical Research* 93: 10467–10480.
De Wit MJ (1998) On Archean granites, greenstones, cratons and tectonics: Does the evidence demand a verdict? *Precambrain Research* 91: 181–226.
Dunlop DJ and Yu Y (2004) Intensity and polarity of the geomagnetic field during Pre-cambrian time. In: Channell JET, Kent DV, Lowrie W, *et al.* (eds.) *Timescales of the Paleomagnetic Field, Geophysical Monograph*, vol. 145, pp. 85–100. Washington, DC: American Geophysical Union.
Dziewonski AM and Anderson DL (1981) Preliminary reference Earth model. *Physics of the Earth and Planetary Interiors* 25: 297–356.
Flasar FM and Birch F (1973) Energetics of core formation – A correction. *Journal of Geophysical Research* 78: 6101–6103.
Gessmann CK and Wood BJ (2002) Potassium in the Earth's core?. *Earth and Planetary Science Letters* 200: 63–78.
Glatzmaier GA (2002) Geodynamo simulations – How realistic are they? *Annual Review of Earth and Planatery Sciences* 30: 237–257.
Glatzmaier GA, Coe RS, Hongre L, and Roberts PH (1999) The role of the Earth's mantle in controlling the frequency of geomagnetic reversals. *Nature* 401: 885–890.
Glatzmaier GA and Roberts PH (1996) An anelastic evolutionary geodynamo simulation driven by compositional and thermal convection. *Physica D* 97: 81–94.
Greff-Lefftz M and Legros H (1999) Core rotational dynamics and geological events. *Science* 286: 1707–1709.
Grove TL and Parman SW (2004) Thermal evolution of the Earth as recorded by komatiites. *Earth and Planetary Science Letters* 219: 173–187.
Gubbins D (1977) Energetics of the Earth's core. *Journal of Geophysics* 43: 453–464.
Gubbins D, Alfe D, Masters G, Price GD, and Gillan MJ (2003) Can the Earth's dynamo run on heat alone? *Geophysical Journal International* 155: 609–622.
Gubbins D, Alfe D, Masters G, Price GD, and Gillan MJ (2004) Gross thermodynamics of two-component core convection. *Geophysical Journal International* 157: 1407–1414.
Gubbins D, Masters TG, and Jacobs JA (1979) Thermal evolution of the Earth's core. *Geophysical Journal of Royal Astronomical Society* 59: 57–99.
Hage H and Muller G (1979) Changes in dimensions, stresses and gravitational energy of the Earth due to crystallization at the inner core boundary under isochemical conditions. *Geophysical Journal of Royal Astronomical Society* 58: 495–508.
Hauri EH and Hart SR (1993) Re–Os isotope systematics of HIMU and EMII oceanic island basalts from the South Pacific ocean. *Earth and Planetary Science Letters* 114: 353–371.
Hernlund JW, Thomas C, and Tackley PJ (2005) A doubling of the post-perovskite phase boundary and structure of the Earth's lowermost mantle. *Nature* 434: 882–886.
Hewitt J, McKenzie DP, and Weiss NO (1975) Dissipative heating in convective flows. *Journal of Fluids Mechanics* 68: 721–738.
Humayun M and Clayton RM (1995) Potassium isotopic constraints on nebular processes. *Meteoritics* 30: 522–523.
Jacobs JA (1998) Variations in the intensity of the Earth's magnetic field. *Surveys in Geophysics* 19: 139–187.
Jackson A (2003) Intense equatorial flux spots on the surface of the Earth's core. *Nature* 424: 760–763.
Jephcoat A and Olson P (1987) Is the inner core of the Earth pure iron. *Nature* 325: 332–335.
Kono M and Roberts PH (2002) Recent geodynamo simulations and observations of the geomagnetic field. *Reviews of Geophysics* 40: 1013.
Koper KD and Dombrovskaya M (2005) Seismic properties of the inner core boundary from PKiKP/P amplitude ratios. *Earth and Planetary Science Letters* 237: 680–694.
Korenaga J (2003) Energetics of mantle convection and the fate of fossil heat. *Geophysical Research Letters* 30: 1437.
Kuang WL and Bloxham J (1997) An Earth-like numerical dynamo model. *Nature* 398: 371–374.
Labrosse S (2002) Hotspots, mantle plumes and core heat loss. *Earth and Planetary Science Letters* 199: 147–156.
Labrosse S (2003) Thermal and magnetic evolution of the Earth's core. *Physics of the Earth and Planetary Interiors* 140: 127–143.
Labrosse S and Macouin M (2003) The inner core and the geodynamo. *Comptes Rendus Geoscience* 335: 37–50.
Labrosse S, Poirier JP, and LeMouel JL (1997) On cooling of the Earth's core. *Physics of the Earth and Planetary Interiors* 99: 1–17.
Labrosse S, Poirier JP, and Le Mouel JL (2001) The age of the inner core. *Earth and Planetary Science Letters* 190: 111–123.
Laio A, Bernard S, Chiarotti GL, Scandolo S, and Tosatti E (2000) Physics of iron at Earth's core conditions. *Science* 287: 1027–1030.
Landau LD and Lifshitz EM (1959) *Course of Theoretical Physics, Vol. 6. Fluid Mechanics*. London: Pergamon.
Langel RA and Estes RH (1982) A geomagnetic field spectrum. *Geophysical Research Letters* 9: 250–230.
Lassiter JC (2004) Role of recycled oceanic crust in the potassium and argon budget of the Earth: Toward a resolution of the 'missing argon' problem. *Geochemistry Geophysics Geosystems* 5: Q11012.

Lassiter JC (2006) Constraints on the coupled thermal evolution of the Earth's core and mantle, the age of the inner core, and the origin of the $^{186}Os/^{188}Os$ 'core signal' in plume-derived lavas. *Earth and Planetary Science Letters* 250: 306–317.
Layer PW, Kroner A, and McWilliams M (1996) An Archean geomagnetic reversal in the Kaap Valley pluton, South Africa. *Science* 273: 943–946.
Lee KKM, Steinle-Neumann G, and Jeanloz R (2004) *Ab-initio* high-pressure alloying of iron and potassium: Implications for the Earth's core. *Geophysical Research Letters* 31: L11603.
Lewis JS (1971) Consequences of the presence of sulfur in the core of the Earth. *Earth and Planetary Science Letters* 11: 130–134.
Lister JR (2003) Expressions for the dissipation driven by convection in the Earth's core. *Physics of the Earth and Planetary Interiors* 140: 145–158.
Lister JR and Buffett BA (1995) The strength and efficiency of thermal and compositional convection in the geodynamo. *Physics of the Earth and Planetary Interiors* 91: 17–30.
Loper DE (1978) The gravitationally powered dynamo. *Geophysical Journal of Royal Astronomical Society* 54: 389–404.
Lowrie W and Kent DV (2004) Geomagnetic polarity timescales and reversal frequency regimes. In: Channell JET, Kent DV, Lowrie W, *et al.* (eds.) *Timescales of the Paleomagnetic Field, Geophysical Monograph*, vol. 145. Washington, DC: American Geophysical Union.
Ma YZ, Somayazulu M, Shen GY, Mao HK, Shu JF, and Hemley RJ (2004) *In situ* X-ray diffraction studies of iron to Earth-core conditions. *Physics of the Earth and Planetary Interiors* 143: 455–467.
Masters G and Gubbins D (2003) On the resolution of density within the Earth. *Physics of the Earth and Planetary Interiors* 140: 159–167.
Matassov G (1977) *The Electrical Conductivity of Iron-Silicon Alloys at High Pressures and the Earth's Core*. PhD Thesis, Lawrence Livermore Laboratory, University of California, CA.
McDonough W (2007) Core composition. In: Gubbins D and Herrero-Bervera E, (eds.) *Encyclopedia of Geomagnetism and Paleomagnetism*. Berlin: Springer.
McDonough W and Sun SS (1995) The composition of the Earth. *Chemical Geology* 120: 223–253.
McElhinny MW and Senanayake WE (1980) Paleomagnetic evidence for the existence of the geomagnetic field 3.5 Ga ago. *Journal of Geophysical Research* 85: 3523–3528.
McNamara AK and Zhong SJ (2004) Thermochemical structures within a spherical mantle: Superplumes or piles? *Journal of Geophysical Research* 109: B07402.
Merrill RT, McElhinny MW, and McFadden PL (1996) *The Magnetic Field of the Earth, Paleomagnetism, the Core and the Deep Mantle*. London: Academic.
Montelli R, Nolet G, Dahlen FA, Masters G, Engdahl ER, and Hung SH (2004) Finite-frequency tomography reveals a variety of plumes in the mantle. *Science* 303: 338–343.
Murakami M, Hirose K, Kawamura K, Sata N, and Ohishi Y (2004) Post-perovskite phase transition in $MgSi0_3$. *Science* 304: 855–858.
Murthy VR and Hall HT (1970) The chemical composition of the Earth's core: Possibility of sulfur in the core. *Physics of the Earth and Planetary Interiors* 2: 276–282.
Murthy VR, Van Westrenen W, and Fei YW (2003) Experimental evidence that potassium is a substantial radioactive heat source in planetary cores. *Nature* 323: 163–165.
Nakagawa T and Tackley PJ (2004a) Effects of thermo-chemical mantle convection on the thermal evolution of the Earth's core. *Earth and Planetary Science Letters* 220: 107–119.
Nakagawa T and Tackley PJ (2004b) Effects of a perovskite-post perovskite phase change near core–mantle boundary in compressible mantle convection. *Geophysical Research Letters* 31: 16611.
Nguyen JH and Holmes NC (2004) Melting of iron at the physical conditions of the Earth's core. *Nature* 427: 339–342.
Nimmo F and Alfe D (2006) Properties and evolution of the Earth's core and geodynamo. In: Sammonds PR and Thompson JMT, (eds.) *Advances in Science: Earth Science*. London: Imperial College Press.
Nimmo F, Price GD, Brodholt J, and Gubbins D (2004) The influence of potassium on core and geodynamo evolution. *Geophysical Journal International* 156: 363–376.
Nimmo F and Stevenson D (2000) The influence of plate tectonics on the thermal evolution and magnetic field of Mars. *Journal of Geophysical Research* 105: 11969–11980.
Oganov AR and Ono S (2004) Theoretical and experimental evidence for a post-perovskite phase of $MgSiO_3$ in Earth's D" layer. *Nature* 430: 445–448.
Poirier J-P (1994) Light elements in the Earth's core: A critical review. *Physics of the Earth and Planetary Interiors* 85: 319–337.
Prevot M, Derder ME, McWilliams M, and Thompson J (1990) Intensity of the Earth's magnetic field – Evidence for a Mesozoic dipole low. *Earth and Planetary Science Letters* 97: 129–139.
Puchtel IS, Brandon AD, Humayun M, and Walker RJ (2005) Evidence for the early differentiation of the core from Pt-Re-Os isotope systematics of 2.8 Ga komatiites. *Earth and Planetary Science Letters* 237: 118–134.
Recoules V and Crocombette JP (2005) *Ab initio* determination of electrical and thermal conductivity of liquid aluminum. *Physical Review B* 72: 104202.
Roberts PH and Glatzmaier GA (2000) Geodynamo theory and simulations. *Reviews of Modern Physics* 72: 1081–1123.
Roberts PH and Glatzmaier GA (2001) The geodynamo, past, present and future. *Geophysical and Astrophysical Fluid Dynamics* 94: 47–84.
Roberts PH, Jones CA, and Calderwood A (2003) Energy fluxes and Ohmic dissipation in the Earth's core. In: Jones CA, Soward AM, and Zhang K, (eds.) *Earth's core and Lower Mantle*. London, New York: Taylor & Francis.
Schersten A, Elliott T, Hawkesworth C, and Norman M (2004) Tungsten isotope evidence that mantle plumes contain no contribution from the Earth's core. *Nature* 427: 234–237.
Sclater JG, Jaupart G, and Galson D (1980) The heat flow through oceanic and continental crust and the heat loss of the Earth. *Reviews of Geophysics and Space Physics* 18: 269.
Selkin PA and Tauxe L (2000) Long-term variations in palaeointensity. *Philosophical Transactions of the Royal Society of London A* 358: 1065–1088.
Shearer P and Masters G (1990) The density and shear velocity contrast at the inner core boundary. *Geophysical Journal International* 102: 491–498.
Shen GY, Mao HK, Hemley RJ, Duffy TS, and Rivers ML (1998) Melting and crystal structure of iron at high pressures and temperatures. *Geophysical Research Letters* 25: 373–376.
Sleep NH (1990) Hot spots and mantle plumes: Some phenomenology. *Journal of Geophysical Research* 95: 6715–6736.
Stacey FD (1969) *Physics of the Earth*. New York: John Wiley and Sons.
Stacey FD and Anderson OL (2001) Electrical and thermal conductivities of Fe-Ni-Si alloy under core conditions. *Physics of the Earth and Planetary Interiors* 124: 153–162.
Stevenson DJ (1989) Formation and early evolution of the Earth. In: Peltier WR (ed.) *Mantle Convection Plate Tectonics and Global Dynamics*. New York: Gordon and Breach.

Stevenson DJ (2003) Planetary magnetic fields. *Earth and Planetary Science Letters* 208: 1–11.
Stevenson DJ, Spohn T, and Schubert G (1983) Magnetism and thermal evolution of the terrestrial planets. *Icarus* 54: 466–489.
Touma J and Wisdom J (2001) Nonlinear core–mantle coupling. *Astronomical Journal* 122: 1030–1050.
Valet JP (2003) Time variations in geomagnetic intensity. *Reviews of Geophysics* 41: 1004.
Verhoogen J (1961) Heat balance of the Earth's core. *Geophysical Journal of Royal Astronomical Society* 4: 276–281.
Williams Q and Garnero EJ (1996) Seismic evidence for partial melt at the base of the Earth's mantle. *Science* 273: 1528–1530.
Xie SX and Tackley PJ (2004) Evolution of U-Pb and Sm-Nd systems in numerical models of mantle convection and plate tectonics. *Journal of Geophysical Research* 109: B11204.
Yoo CS, Holmes NC, Ross M, Webb DJ, and Pike C (1993) Shock temperatures and melting of iron at Earth core conditions. *Physical Review Letters* 70: 3931–3934.
Yukutake T (2000) The inner core and the surface heat flow as clues to estimating the initial temperature of the Earth's core. *Physics of the Earth and Planetary Interiors* 121: 103–137.
Zhong S (2006) Constraints on thermochemical convection of the mantle from plume heat flux, plume excess temperature and upper mantle temperature. *Journal of Geophysical Research* 111: B04409.

3 Theory of the Geodynamo

P. H. Roberts, University of California, Los Angeles, CA, USA

3.1	**Introduction**	67
3.1.1	Dynamos and Self-Excited Dynamos	67
3.1.2	The Geodynamo Hypothesis	68
3.1.3	Plan of Chapter: Notation	69
3.2	**Basic Electrodynamics**	69
3.2.1	Pre-Maxwell Theory	69
3.2.2	Constitutive Relations	70
3.2.3	Induction Equation: Magnetic Reynolds Number	71
3.2.4	Electromagnetic Energy and Stress	72
3.2.5	The Perfect Conductor: Alfvén's Theorem	72
3.2.6	The Imperfect Conductor: Reconnection	74
3.2.7	The Low Conductivity Approximation	78
3.3	**Kinematic Dynamos**	78
3.3.1	The Dynamo Condition	78
3.3.2	EM Induction in Spherical Conductors	80
3.3.3	The Eigenvalue Problem for Steady Flows	82
3.3.4	Bounds on the Magnetic Reynolds Number	83
3.3.5	Antidynamo Theorems	84
3.4	**Laminar Dynamos**	86
3.4.1	Early Successful Models	86
3.4.2	One-Dimensional Models	86
3.4.3	Two-Dimensional Models	87
3.4.4	Three-Dimensional Models	88
3.5	**Turbulent Dynamos**	89
3.5.1	Induction by Cyclonic Turbulence	89
3.5.2	Mean Field Dynamos	91
3.5.3	Saturation: Intermediate Models	93
3.5.4	Application of MFE to the Core	94
3.6	**MHD Dynamos**	94
3.6.1	Basic Equations and Boundary Conditions	94
3.6.2	Classical Theory of Rotating Fluids	95
3.6.3	Coriolis Magnetohydrodynamics	98
3.7	**Final Remarks**	101
References		102

3.1 Introduction

3.1.1 Dynamos and Self-Excited Dynamos

'Dynamo' is the name given to any device that 'induces' electric current to flow, through the motion of an electrical conductor in an 'inducing' magnetic field. Dynamos were constructed first in the mid-nineteenth century but the first really significant technological advance was made by Werner von Siemens (1867). Siemens conceived of, constructed, and in 1866 publically demonstrated, the first self-excited dynamo, in which the inducing magnetic field was created by the induced currents; previously, it had been supplied by permanent magnets.

The physical principles governing dynamos are illustrated in **Figure 1**, where the homopolar dynamo is shown. Here D is a solid electrically-conducting disk rotating with angular velocity Ω about an axis of symmetry $A'A$, parallel to which we first suppose

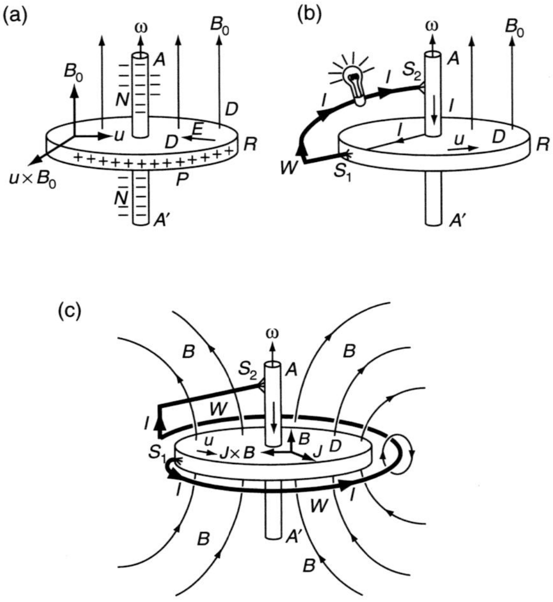

Figure 1 The homopolar dynamo (see text for explanation).

(see **Figure 1(a)**) that there is a uniform externally produced 'inducing field' $\mathbf{B}_0$, such as might be produced by permanent magnets. Through the velocity $\mathbf{u} = \boldsymbol{\Omega} \times \mathbf{x}$ of the conductor, an electromotive force (emf) $\mathbf{u} \times \mathbf{B}_0$ is induced that would, if there were a closed electric circuit, drive a current radially outward from A′A toward the rim R. (Here $\mathbf{x}$ is the position vector from an origin on A′A.) Instead, positive charges (P) build up on R, and negative charges (N) are set up on the electrically conducting axle. In this way an electric field **E** is created that cancels out the induced emf, as of course is necessary since no current flows.

Now let a conducting wire W join A′A to R, electrical contact being maintained by sliding contacts (or 'brushes') at S_1 and S_2 (see **Figure 1(b)**). Electric charges that in **Figure 1(a)** were trapped on the rim can now be 'drawn off' as an electric current I flowing along W and across D; the electric circuit has been completed and the emf has become motive! This induced current I can be directed to perform a useful task, such as powering the electric light bulb shown in **Figure 1(b)**, the concomitant expenditure of energy being provided by the energy source that maintains the rotation of D. It is this conversion of mechanical energy into electrical energy that is the essence of the dynamo, in this case a pre-Siemens, non-self-exciting dynamo.

Siemens' idea was to make use of the magnetic field that always accompanies the flow of electric current. Suppose that, instead of taking the shortest route between S_1 and S_2, the wire W is wound round, and in the plane of, D, just outside R and making almost one complete loop around R before being led to the two brushes (**Figure 1(c)**). The direction of the winding is such that the induced magnetic field **B** produced by the current I in the loop reinforces the applied magnetic field $\mathbf{B}_0$. In the plane of D, both fields are perpendicular to the disk, in the same sense A′ → A. Unlike $\mathbf{B}_0$ the magnetic field **B** is not uniform across D, so that the emf $\mathbf{u} \times \mathbf{B}$ it creates from the motion does not depend on $\mathbf{x}$ in the same way that $\mathbf{u} \times \mathbf{B}_0$ does, but it can perform the same function. If Ω exceeds some critical value Ω_c, the 'seed magnetic field' $\mathbf{B}_0$ can be removed and the dynamo will continue to operate, the current I being induced by the emf $\mathbf{u} \times \mathbf{B}$ alone. Thereafter, the device will create a magnetic field for as long as an angular velocity greater than Ω_c is maintained. The dynamo has become 'self-excited'.

If we persist in looking only at the electrodynamics of the dynamo, we would believe that, if Ω is increased beyond Ω_c, then **B** will grow without limit because the rotation of D would create magnetic energy faster than it is destroyed by the electrical resistance of the circuit. This absurdity is removed when we consider the mechanics of the device. A Lorentz force, $\mathbf{J} \times \mathbf{B}$ per unit volume, opposes the rotation $\mathbf{u} = \boldsymbol{\Omega} \times \mathbf{x}$ of the disk (see **Figure 1(c)**). In the marginal state, $\Omega = \Omega_c$, this creates a magnetic torque Γ_B about A′A that is equal and opposite to the applied torque Γ_A that keeps the disk in rotation. An increase in Γ_A makes Ω bigger, but this results in a larger **B** and a greater Γ_B. Soon a new balance between Γ_A and Γ_B is set up with a larger Ω_c.

3.1.2 The Geodynamo Hypothesis

The suggestion that the magnetic fields of the Earth and the Sun are produced by self-excited dynamos was first made by Joseph Larmor (1919). He visualized that the solid disk in **Figure 1** is replaced in the Earth and Sun by the movement of electrically conducting fluid. His idea gained ground as every other explanation of terrestrial and solar magnetism became increasingly implausible. Paleomagnetic studies found that there are no discernable differences between the normal and reversed polarity states of the geomagnetic field. This is particularly hard to reconcile with any other explanation of geomagnetism but is required of a self-excited dynamo because if $\mathbf{B}(\mathbf{x}, t)$ satisfies the governing equations, so does $-\mathbf{B}(\mathbf{x}, t)$. Once the seed field of the homopolar dynamo is removed, magnetic field of either sign can be sustained equally well. The idea that the main geomagnetic field is produced by a dynamo operating in the Earth's fluid core became known as 'the geodynamo hypothesis'.

It is clear from the homopolar dynamo that the geodynamo hypothesis demands answers to two questions. First, without seeking an origin for the fluid motion, can one find a motion that self-generates magnetic field? The answer to this 'kinematic dynamo' question is not obvious. The homopolar dynamo has a specially constructed lack of symmetry; the Earth's core is a nearly homogeneous electrical conductor. Are 'homogeneous dynamos' possible in such a body? Theoreticians were challenged either to prove the answer is 'No', or to create a working model. Nearly four decades elapsed before this challenge was met. During this time, and subsequently, it became increasingly well understood how structural asymmetry in the homopolar dynamo could be mimicked in a homogeneous dynamo by fluid motions that lack mirror symmetry.

Once the kinematic dynamo question had been answered in the affirmative, attention became increasingly focused on different questions: How is the driving motion $\mathbf{u}$ produced? How is it modified by the presence of the magnetic field $\mathbf{B}$ that it itself creates? These are the concerns of 'magnetohydrodynamic dynamo' theory, or 'MHD dynamo theory'. Obviously, the kinematic theory is 'contained' in the MHD theory and is an essential part of it.

3.1.3 Plan of Chapter: Notation

The objective of this chapter is to provide the basics of kinematic and MHD dynamo theory. In Section 3.2 the basic electromagnetic theory is briefly reviewed, and in Section 3.3 the kinematic theory is formulated. Dynamo models driven by laminar and turbulent flows are described in Sections 3.4 and 3.5. In Section 3.6 the ideas behind MHD dynamos are outlined. Discussion is everywhere slanted toward the geophysical application. SI units are used.

To simplify the presentation, some verbal abbreviations will usually be employed: 'velocity' = fluid velocity, 'density' = mass density, 'field' = magnetic field, 'current' = 'electric current density', 'conductor' = conductor of electricity, 'potential' = electric potential, 'core' = Earth's core, 'FOC' = fluid outer core, 'SIC' = solid inner core, 'CMB' = core–mantle boundary (radius r_c), 'ICB' = 'inner-core boundary' (radius r_i), 'emf' = electromotive force, and 'EM' = either electromagnetism or electromagnetic.

Script letters, $\mathcal{L}$, $\mathcal{T}$, $\mathcal{U}$, $\mathcal{B}$, $\mathcal{J}$, $\mathcal{E}, \ldots$, will indicate typical magnitudes of length, time, velocity, field, current, electric field, ... The angular velocity of the reference frame (approximately the angular velocity of the mantle) will be denoted by $\boldsymbol{\Omega}$, and will be assumed constant and in the z-direction. Sometimes cylindrical coordinates (s, ϕ, z) will be used and sometimes spherical coordinates (r, θ, ϕ). 'Zonal' will mean 'in the ϕ-direction' and 'meridional' will mean 'lying in meridional planes', that is, in the planes ϕ = constant. Except for the unit normal $\mathbf{n}$ to surfaces, unit vectors will be denoted by $\mathbf{1}_q$, where the q is the coordinate concerned; ∂_q will be short for $\partial/\partial q$, except that $\partial_t = \partial/\partial t$ is the (Eulerian) time derivative. Except in Section 3.5.1, $\bar{Q}$ will denote the average over ϕ of a quantity Q, and Q' will be short for $Q - \bar{Q}$. The ϕ-averages of the spherical or cylindrical components of a vector $\mathbf{Q}$ define $\bar{\mathbf{Q}}$. A similar notation will be used in Section 3.5.1, but there $\bar{Q}$ denotes the ensemble average of Q in a turbulent environment.

3.2 Basic Electrodynamics

3.2.1 Pre-Maxwell Theory

Magnetohydrodynamics is the study of the flow of electrically conducting fluids in the presence of magnetic fields. It has significant applications in technology and in the science of planets, stars, and galaxies. In this volume, its main focus is on its role in explaining the origin and properties of the geomagnetic field.

MHD is based on the equations governing fluid motion in the presence of magnetic fields and the equations governing EM fields in moving fluids. Nearly always, and invariably in geomagnetic studies, the nonrelativistic form of MHD is employed. This is appropriate when conditions are changing slowly, that is, when the characteristic velocity $\mathcal{U}$ of the conductor is small compared with the velocity of light c and when the time $\mathcal{L}/c$ taken by light to cross the system is small compared with all time scales $\mathcal{T}$ of interest. The EM fields are governed by the equations that were in use before Maxwell introduced displacement currents, and are therefore called the 'pre-Maxwell equations'. Unlike full Maxwell theory, in which the electric field $\mathbf{E}$ and the magnetic field $\mathbf{B}$ are on an equal footing, $\mathbf{E}$

plays a subsidiary role in pre-Maxwell theory. This is why the magnetic field is usually referred to simply as 'the field'. The aim of this subsection is to review pre-Maxwell theory briefly.

The pointwise form of the pre-Maxwell EM equations are, in SI units,

$$\nabla \times \mathbf{H} = \mathbf{J} \quad [1a]$$

$$\nabla \times \mathbf{E} = -\partial_t \mathbf{B} \quad [1b]$$

$$\nabla \cdot \mathbf{B} = 0 \quad [1c]$$

$$\nabla \cdot \mathbf{D} = \vartheta \quad [1d]$$

where $\mathbf{J}$ is the current density and ϑ is the volume charge density. Equations [1a] and [1b] are, respectively, Ampère's law and Faraday's law. The sources on the right-hand sides of [1a] and [1d] must satisfy charge conservation and, in pre-Maxwell theory, this requires

$$\nabla \cdot \mathbf{J} = 0 \quad [1e]$$

which is consistent with [1a]. If [1c] holds for any t, it holds for all t by [1b]. The CMB, and the ICB, is represented by a 'discontinuity surface', S, across which material properties change abruptly. The same integral laws that led to [1a]–[1e] imply that

$$[\mathbf{n} \times \mathbf{H}] = \mathbf{C} \quad \text{on} \quad \mathrm{S} \quad [2a]$$

$$[\mathbf{n} \times (\mathbf{E} + \mathbf{u} \times \mathbf{B})] = \mathbf{0} \quad \text{on} \quad \mathrm{S} \quad [2b]$$

$$[\mathbf{n} \cdot \mathbf{B}] = 0 \quad [2c]$$

$$[\mathbf{n} \cdot \mathbf{D}] = \Theta \quad \text{on} \quad \mathrm{S} \quad [2d]$$

$$[\mathbf{n} \cdot \mathbf{J}] = -\nabla_S \cdot \mathbf{C} \quad \text{on} \quad \mathrm{S} \quad [2e]$$

Here Θ is the surface charge density, $\mathbf{C}$ is the surface current density, ∇_S is the two-dimensional surface divergence, $\mathbf{n}$ is the unit vector to S, $\mathbf{u}$ is the velocity of a point P on S, and $[Q]$ is the difference in limiting values of a quantity Q at P, evaluated on opposite sides of S.

Geodynamo theory is mainly concerned with timescales much shorter than those over which the Earth evolves. It is then a good approximation to take $\mathbf{n} \cdot \mathbf{u} = 0$ on the CMB and ICB, and to assume that these surfaces are spherical, of constant radii r_c and r_i. Conditions [2] are simplified in Section 3.3.1 below.

3.2.2 Constitutive Relations

The pre-Maxwell equations must be supplemented by relations that define the physical nature of the medium in which they are applied. We shall suppose that the conductor is isotropic and homogeneous so that, at a point P within it and in a reference frame moving with the constant uniform velocity $\mathbf{u}_P = \mathbf{u}(\mathbf{x}_P)$ of P,

$$\mathbf{H}' = \mathbf{B}'/\mu_P \quad [3a]$$

$$\mathbf{D}' = \epsilon_P \mathbf{E}' \quad [3b]$$

$$\mathbf{J}' = \sigma_P \mathbf{E}' \quad [3c]$$

where μ, ϵ, and σ are the permeability, permittivity, and electrical conductivity of the medium. Except in the crust, the temperature within the Earth everywhere exceeds the Curie point, at which permanent ferromagnetism ceases to exist. We shall therefore assume that $\mu = \mu_0 = 4\pi \times 10^{-7}\,\mathrm{H\,m^{-1}}$, the vacuum permeability, but ϵ may differ from the vacuum permittivity ϵ_0.

To apply [3] it is necessary to relate $\mathbf{B}'$, $\mathbf{E}'$, $\mathbf{H}'$, $\mathbf{D}'$, $\mathbf{J}'$ to $\mathbf{B}$, $\mathbf{E}$, $\mathbf{H}$, $\mathbf{D}$, $\mathbf{J}$. To relate them at $\mathbf{x}_P$, the Lorentz transformation is approximated by its non-relativistic (Galilean) form, appropriate for $\mathcal{U} \ll c$:

$$\mathbf{x} = \mathbf{x}' + \mathbf{u}_P t' \quad [4a]$$

$$t = t' \quad [4b]$$

The EM fields and sources in the two reference frames are related by

$$\mathbf{B}' = \mathbf{B} \quad [5a]$$

$$\mathbf{E}' = \mathbf{E} + \mathbf{u}_P \times \mathbf{B} \quad [5b]$$

$$\mathbf{H}' = \mathbf{H} \quad [5c]$$

$$\mathbf{D}' = \mathbf{D} + \mathbf{u}_P \times \mathbf{H}/c^2 \quad [5d]$$

$$\mathbf{J}' = \mathbf{J} \quad [5e]$$

$$\vartheta' = \vartheta - \mathbf{u}_P \cdot \mathbf{J}/c^2 \quad [5f]$$

Since $\nabla' = \nabla$ and $\partial_t' = \partial_t + \mathbf{u}_P \cdot \nabla$ according to [4a] and [4b], it follows from [5] that [1] also hold in the primed frame of reference, that is, the pre-Maxwell equations are frame-indifferent. According to [5], only $\mathbf{E}$, $\mathbf{D}$, and ϑ are frame dependent. When translated back to the original frame, [3] are

$$\mathbf{H} = \mathbf{B}/\mu_0 \quad [6a]$$

$$\mathbf{D} = \epsilon \mathbf{E} + (\epsilon - \epsilon_0)\mathbf{u} \times \mathbf{B} \quad [6b]$$

$$\mathbf{J} = \sigma(\mathbf{E} + \mathbf{u} \times \mathbf{B}) \quad [6c]$$

The suffix P has been removed since these relations hold for every P. After its removal, $\mathbf{u}$, σ, and ϵ in [6] may be functions of $\mathbf{x}$.

Wherever **H** occurs, we shall now use [6a] to remove it, in favor of **B**. Equation [6c] is the generalization of Ohm's law to a moving conductor, and is centrally important. By [1a], [6a] and [6c]

$$\mu_0 \mathbf{J} = \nabla \times \mathbf{B} \quad [6d]$$

$$\mathbf{E} = -\mathbf{u} \times \mathbf{B} + \eta \nabla \times \mathbf{B} \quad [6e]$$

where $\eta = 1/\mu_0 \sigma$ is the 'magnetic diffusivity'. In estimating Earthlike values for the core, we shall take $\sigma = 4 \times 10^5\ \mathrm{S\,m^{-1}}$ so that $\eta = 2\ \mathrm{m^2\,s^{-1}}$. The form [6e] of Ohm's law provides a convenient way of removing **E** in favor of **B**; as mentioned earlier, **E** plays a subsidiary role in MHD as do **D** and ϑ. Nevertheless, it is worth observing that, by [1d], [6b], and [6c],

$$\vartheta = -\epsilon_0 \nabla \cdot (\mathbf{u} \times \mathbf{B}) + \mathbf{J} \cdot \nabla(\epsilon/\sigma) \quad [7]$$

Unlike the case of a stationary uniform conductor for which an initial ϑ flows to the boundaries in a time of order $\epsilon_0/\sigma = O(\eta/c^2)$ (i.e., instantaneously according to pre-Maxwell theory), the free charge density in a moving conductor is generally nonzero, and takes the value [7].

3.2.3 Induction Equation: Magnetic Reynolds Number

Eliminating **E** between [1b] and [6e], we see that **B** obeys the induction equation

$$\partial_t \mathbf{B} = \nabla \times (\mathbf{u} \times \mathbf{B}) - \nabla \times (\eta \nabla \times \mathbf{B}) \quad [8]$$

When η is constant (the usual assumption), [1c] shows that [8] can be rewritten as

$$\partial_t \mathbf{B} = \nabla \times (\mathbf{u} \times \mathbf{B}) + \eta \nabla^2 \mathbf{B} \quad [9]$$

Another useful form of the induction equation is

$$d_t \mathbf{B} = (\mathbf{B} \cdot \nabla)\mathbf{u} - \mathbf{B} \nabla \cdot \mathbf{u} + \eta \nabla^2 \mathbf{B} \quad [10a]$$

which follows from [9], [1c], and the vector identity

$$\nabla \times (\mathbf{u} \times \mathbf{B}) = (\mathbf{B} \cdot \nabla)\mathbf{u} - (\mathbf{u} \cdot \nabla)\mathbf{B} + \mathbf{u} \nabla \cdot \mathbf{B} - \mathbf{B} \nabla \cdot \mathbf{u} \quad [10b]$$

Here $d_t = \partial_t + \mathbf{u} \cdot \nabla$ is the motional or Lagrangian derivative, that is, the derivative following the motion of the conductor. Sometimes the term 'the dynamo equation' is applied to one or other of these three forms of the induction equation but this is a misnomer, since they apply in many contexts totally unrelated to dynamo theory.

MHD processes in rotating bodies such as the core are usually most readily studied using a reference frame rotating with the body. With the understanding that **u** is the velocity relative to the rotating frame, the same relations [6] hold and lead to the same induction equations.

Equation [9] exposes the evolution of **B** as a competition between EM induction (through $\nabla \times (\mathbf{u} \times \mathbf{B})$) and ohmic diffusion (through $\eta \nabla^2 \mathbf{B}$). The relative importance of these effects is quantified by

$$Rm = \mathcal{U}\mathcal{L}/\eta = \mu_0 \sigma \mathcal{U}\mathcal{L} \quad [11a]$$

In analogy with the familiar (kinetic) Reynolds number $Re = \mathcal{U}\mathcal{L}/\nu$ used in fluid mechanics to quantify the effects of the (kinematic) viscosity ν, Rm is called the 'magnetic Reynolds number'. It can also be written as the ratio

$$Rm = \mathcal{T}_\eta / \mathcal{T}_u \quad [11b]$$

of the 'electromagnetic diffusion time' $\mathcal{T}_\eta = \mathcal{L}^2/\eta$ and the 'fluid advection time' $\mathcal{T}_u = \mathcal{L}/\mathcal{U}$. The former quantifies the time in which a current system of scale $\mathcal{L}$ in a conductor will decay ohmically, unless some agency maintains it. For the Earth, $\mathcal{T}_\eta \approx 2 \times 10^5$ years, if we take $\mathcal{L} = r_c$. The Earth is known to have possessed a field for at least 3.4×10^9 years (Kono and Tanaka, 1995). Thus, the Earth's magnetic field is not a 'legacy of the past'; an agency is required to maintain the current system and, according to the geodynamo hypothesis, this is electromagnetic induction created by the term $\mathbf{u} \times \mathbf{B}$ in [9]. For the dynamo to be successful, this term must create new flux as rapidly as, or more rapidly than, it is destroyed by the $\eta \nabla^2 \mathbf{B}$ term. This implies that $\mathcal{T}_u < \mathcal{T}_\eta$ and $Rm > 1$. Most kinematic dynamo models require that $\mathcal{T}_\eta$ exceeds $\mathcal{T}_u$ by at least one, but more probably two, orders of magnitude.

The westward drift velocity on the CMB of recognizable features of the geomagnetic field is of order $3 \times 10^{-4}\ \mathrm{m\,s^{-1}}$. It is likely that part of this motion is due to diffusion of the features relative to the fluid (see Section 3.2.6 below). We shall estimate that $\mathcal{U} = 2 \times 10^{-4}\ \mathrm{m\,s^{-1}}$. Taking $\mathcal{L} = 10^6\ \mathrm{m}$ as a typical field scale, $\mathcal{T}_u \approx 2 \times 10^3$ years and $Rm = 100$.

Kinematic dynamo theory rests on the induction equation [9] and Gauss's law [1c]. It assumes that **u** is given, but it would make no sense to study physically meaningless flows. We shall restrict ourselves to 'eligible flows' that, for example, have no point sources or sinks, and that are everywhere and at all times bounded. Some models will involve discontinuous **u**, that is, 'tearing' of the fluid, but these models can be simply modified so that the tearing is replaced by a large but finite shear. For simplicity, it will usually be assumed that the fluid is incompressible:

$$\nabla \cdot \mathbf{u} = 0 \qquad [12]$$

Some results will apply to compressible fluids too.

3.2.4 Electromagnetic Energy and Stress

It follows from [1a] and [1b] that

$$\partial_t e^B + \nabla \cdot \mathbf{I}^B = \mathbf{E} \cdot \mathbf{J} \qquad [13a]$$

where e^B is the EM energy per unit volume and $\mathbf{I}^B$ is the EM energy flux (or Poynting vector):

$$e^B = B^2/2\mu_0 \qquad [13b]$$

$$\mathbf{I}^B = \mathbf{E} \times \mathbf{B}/\mu_0 \qquad [13c]$$

When integrated over a volume v bounded by a surface s, [13a] states that magnetic energy in v diminishes through its outward flow across s and by the rate at which the EM field transfers its energy to the material contents of v. By [6c] the latter has two parts:

$$-\mathbf{E} \cdot \mathbf{J} = -q^{\mathcal{J}} - \mathbf{u} \cdot \mathbf{L} \qquad [13d]$$

where

$$\mathbf{L} = \mathbf{J} \times \mathbf{B} \qquad [13e]$$

$$q^{\mathcal{J}} = \mathcal{J}^2/\sigma \geq 0 \qquad [13f]$$

Here $\mathbf{L}$, the 'Lorentz force', is the force per unit volume that the EM field exerts on the conductor; $q^{\mathcal{J}}$, which is necessarily non-negative, is the rate per unit volume at which EM energy is converted into heat through electrical resistance. This is called the 'Joule loss' or 'ohmic loss'. Summed over v it is

$$\Phi = \int_v \frac{\mathcal{J}^2}{\sigma}\, \mathrm{d}V \qquad [13g]$$

On combining [13a] and [13d] and integrating, we have

$$\frac{\mathrm{d}}{\mathrm{d}t}\int_v e^B\, \mathrm{d}V = -\Phi - \oint_s \mathbf{I}^B \cdot \mathbf{dS} - \int_v \mathbf{u} \cdot \mathbf{L}\, \mathrm{d}V \qquad [13h]$$

The final term is the rate of working of the Lorentz force. In an electric motor this is positive, representing conversion rate of EM energy into mechanical energy. In a dynamo the opposite is true: kinetic energy is transformed into EM energy, and the last term in [13h] is large enough to make good the integrated ohmic losses Φ.

The EM force and torque (about the origin of $\mathbf{x}$) on the contents of a volume v are

$$\mathbf{f} = \int_v \mathbf{L}\, \mathrm{d}V = \int_v \mathbf{J} \times \mathbf{B}\, \mathrm{d}V \qquad [14a]$$

$$\boldsymbol{\Gamma} = \int_v \mathbf{x} \times \mathbf{L}\, \mathrm{d}V = \int_v \mathbf{x} \times (\mathbf{J} \times \mathbf{B})\, \mathrm{d}V \qquad [14b]$$

By using [1a], [1c], and [13c], it is possible, and often useful, to express these as surface integrals. This is achieved by re-expressing $\mathbf{L}$ as the divergence of the EM stress tensor $\overleftrightarrow{\boldsymbol{\pi}}^B$:

$$L_i = \nabla_j \pi_{ij}^B \qquad [15a]$$

$$\pi_{ij}^B = \pi_{ji}^B = \frac{B_i B_j}{\mu_0} - \frac{B^2}{2\mu_0}\delta_{ij} \qquad [15b]$$

The first term on the right-hand side of [15b] represents tension in a field line (defined as a curve that is everywhere parallel to $\mathbf{B}$); the final term in [15b] represents an isotropic magnetic pressure, $P_{\mathrm{m}} = B^2/2\mu_0$, which is about 4 atm for a field of 1 T. These interpretations will be useful in Sections 3.2.5 and 3.2.6 below.

By using [15a], $\mathbf{f}$ and $\boldsymbol{\Gamma}$ may be replaced by surface integrals:

$$f_i = \oint_s \pi_{ij}^B\, \mathrm{d}S_j \qquad [15c]$$

$$\Gamma_i = \oint_s \epsilon_{ijk} x_j \pi_{kl}^B \mathrm{d}S_l \qquad [15d]$$

Equivalently,

$$\mathbf{f} = \oint_s \left[\frac{\mathbf{B}(\mathbf{B} \cdot \mathbf{dS})}{\mu_0} - \frac{B^2}{2\mu_0}\mathbf{dS}\right] \qquad [15e]$$

$$\Gamma = \oint_s \mathbf{x} \times \left[\frac{\mathbf{B}(\mathbf{B} \cdot \mathbf{dS})}{\mu_0} - \frac{B^2}{2\mu_0}\mathbf{dS}\right] \qquad [15f]$$

According to pre-Maxwell theory, the magnetic field alone determines the EM force on the conductor, the EM stresses, and the EM energy. The corresponding contributions made by the electric field are smaller by a factor of order $(\mathcal{U}/c)^2$.

3.2.5 The Perfect Conductor: Alfvén's Theorem

A 'perfect conductor' is one of infinite electrical conductivity σ, that is, $\eta = 0$. Since $\mathbf{J}$ must be finite even though $\sigma = \infty$, Ohm's law [6c] implies that

$$\mathbf{E} = -\mathbf{u} \times \mathbf{B} \quad (\text{if } \sigma = \infty) \qquad [16]$$

The induction equation can be written as

$$\partial_t \mathbf{B} = \nabla \times (\mathbf{u} \times \mathbf{B}) \quad (\text{if } \sigma = \infty) \qquad [17a]$$

$$d_t \mathbf{B} = \mathbf{B} \cdot \nabla \mathbf{u} - \mathbf{B} \nabla \cdot \mathbf{u} \quad (\text{if } \sigma = \infty) \qquad [17b]$$

In a perfect conductor, the magnetic Reynolds number, Rm, is infinite by [11a]. While $Rm \gg 1$ in many applications of MHD (including some considered here), $Rm = \infty$ is an idealization that is never achieved. It is however often helpful because of the following theorem:

***Alfvén's theorem:** Magnetic flux tubes move with a perfect conductor as though frozen to it.*

A flux tube is a 'bundle' of field lines. It is defined by a 'cross-section' Σ, the tube being the volume occupied by all the field lines intersecting Σ. The surface, M, of the tube is therefore a 'magnetic surface', that is, a surface composed of field lines and on which $\mathbf{n} \cdot \mathbf{B} = 0$. The net magnetic flux through Σ is

$$F = \int_\Sigma \mathbf{B} \cdot d\mathbf{S} \qquad [18]$$

and is called the 'strength' of the tube. It follows from [1c] that F is the same for every cross-section of the tube.

Alfvén's theorem can be proved by appealing to the kinematic result

$$d_t(d\mathbf{S}) = d\mathbf{S}(\nabla \cdot \mathbf{u}) - dS_j \nabla u_j \qquad [19]$$

which describes the change in area and orientation of an element $d\mathbf{S}$ of surface area as it is advected with the flow. Combining this with [17b], it is found that

$$d_t(\mathbf{B} \cdot d\mathbf{S}) = 0 \quad (\text{when} \quad \sigma = \infty) \qquad [20a]$$

It follows that

$$\mathbf{B} \cdot d\mathbf{S} = 0 \text{ at } t = 0 \text{ implies } \mathbf{B} \cdot d\mathbf{S} = 0 \text{ for all } t \qquad [20b]$$

so that, if M is initially a magnetic surface, it is always a magnetic surface. Thus a flux tube remains a flux tube for all time. Further, [20a] shows that the flux tube preserves its strength F as it moves.

According to Alfvén's theorem, field lines move with a perfect conductor as though frozen to it, and sometimes the theorem is stated in those terms. This statement is however a weaker form of the theorem that does not imply that F is conserved.

The following examples illustrate how the frozen flux picture, coupled with the concepts of stress and energy developed in Section 3.2.4, help in visualizing MHD processes:

- Conversion of kinetic energy into magnetic energy occurs in an incompressible fluid when motions stretch or bend flux tubes; conversely, shortening or straightening field lines creates kinetic energy from magnetic energy. The stretching of field lines stores magnetic energy in much the same way as a rubber band stores elastic energy when it is stretched. If a flux tube of cross-sectional area A_0 containing field B_0 is lengthened from L_0 to L, its cross-section will, according to [12], decrease in the same proportion ($A = A_0 L_0 / L$) and the field within it will increase by the same factor ($B = B_0 L / L_0$). The magnetic energy it contains, which is proportional to B^2, is therefore enhanced by a factor of $(L/L_0)^2$ from $(B_0^2/2\mu_0) L_0 A_0$ to $(B^2/2\mu_0) LA = (B^2/2\mu_0) L_0 A_0 = [(B_0^2/2\mu_0) L_0 A_0] (L/L_0)^2$. If $L = L_0 + \delta$ where $\delta \ll L_0$, the increase in magnetic energy is $(B_0^2/\mu_0) A_0 \delta$. This is the work done by the applied force in stretching the tube by δ in opposition to the magnetic tension $(B_0^2/\mu_0) A_0$ of the field lines.
- 'Alfvén waves' are the result of field line tension. The mass per unit length and tension of a flux tube of unit cross-section are ρ and B^2/μ_0 so that, as in a taut string, a wave can travel nondispersively in either direction along the flux tube with speed $\sqrt{[(B^2/\mu_0)/\rho]} = B/\sqrt{\mu_0 \rho}$. This defines the 'Alfvén velocity', $\mathbf{V}_A = \mathbf{B}/\sqrt{\mu_0 \rho}$. It is approximately $1\ \mathrm{cm\,s^{-1}}$ for $B = 1\ \mathrm{mT}$ and $\rho = 10^4\ \mathrm{kg\ m^{-3}}$; the Alfvénic timescale $\mathcal{T}_A = \mathcal{L}/\mathcal{V}_A$ is about a decade for $\mathcal{L} = r_c$. In a system such as the core that is rotating with an angular velocity Ω that is large compared with $\mathcal{V}_A/\mathcal{L}$, Alfvén waves lose much (but not all) of their significance, which is subsumed by *slow waves*, with velocity $\mathcal{V}_s = O(\mathcal{V}_A^2/\mathcal{L}\Omega) \ll \mathcal{V}_A$ and timescale $\mathcal{T}_s = O(\mathcal{L}^2\Omega/\mathcal{V}_A^2) \gg \mathcal{T}_A$; see Section 3.6.3.
- A zonal shearing motion $\zeta = \bar{u}_\phi / s$ can drag the lines of force of a meridional field $\bar{\mathbf{B}}_M$ out of the meridional planes, i.e., it can create a zonal component $\bar{B}_\phi$ in $\bar{\mathbf{B}}$ (see **Figure 2**). Dating from a time when the ζ was frequently denoted by ω, this is often called the *ω-effect*; in the astrophysical literature it is more often referred to as the 'Ω-effect'. Taking Alfvén's theorem literally, $\bar{u}_\phi$ will continue to wind the field lines around the symmetry axis for as long as the motion can be sustained, and $\bar{B}_\phi$ and e^B will increase monotonically. The magnetic stresses grow as the field lines become increasingly stretched in the ϕ-direction until eventually the agency that creates $\bar{u}_\phi$ can no longer maintain it, and growth of $\bar{B}_\phi$ ceases. This marks the end of the kinematic regime (in which fluid motion is prescribed *a priori* and the field grows) and the start of the MHD regime in which the field quenches its own further growth.

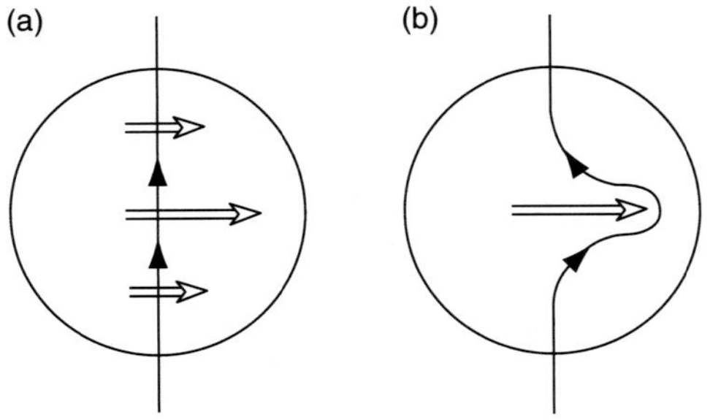

Figure 2 The ω-effect. (a) a field line in a meridian plane (single arrow) is sheared by a zonal flow (double arrows); (b) as a result of the shearing, the field line is bent, so creating a ϕ–component of **B**.

The intermittent character of MHD turbulence is reflected by the existence of flux tubes in which **B** is more intense than in their surroundings. Frequently, **B** is visualized as being discontinuous across the surface M of the tube, which is therefore a current sheet, by [2a]. Because of the field line tension, a closed flux tube will tend to collapse in much the same way as a stretched elastic band contracts when released. If there is no compensating stretching process, the collapse will continue until **B** has subsided to the level of the field in its environment. Similarly, a bent flux tube tends to straighten, returning some of its magnetic energy to the kinetic energy of the fluid frozen onto it; an example is given in the next subsection.

3.2.6 The Imperfect Conductor: Reconnection

Strictly speaking, a field line passes through every point in space, but only a finite number of lines can be shown in sketches of field line topology, where the crowding together of field lines indicates a more intense field. In these sketches, field lines may appear to disconnect and reconnect in a new topology as the field evolves. This phenomenon is called 'reconnection', the prior 'disconnection' being tacitly assumed.

According to Alfvén's theorem, field line topology is immutable. Fluid lying on a field line always remains on that field line; reconnection of field lines is impossible. This makes the dynamo problem meaningless; the magnetic flux trapped in the conductor can never be lost, it can only be rearranged. One cannot even ask how the conductor acquired its flux originally. Flux can be gained, retained, or lost only by its diffusion relative to the conductor, and diffusion happens only when σ and Rm are finite. Every successful dynamo relies on reconnection processes.

When $\eta \neq 0$, field lines diffuse relative to the conductor with a 'resistive drift speed' of order $\mathcal{U}_\eta = \eta/\mathcal{L}$, where $\mathcal{L}$ is a length characteristic of the field gradient ($\mathcal{L} \sim \mathcal{B}/\mu_0\mathcal{J}$). Several examples are given below.

When thinking about reconnection, it is perhaps helpful to regard Rm as a function of position. Wherever $Rm \gg 1$ (i.e., wherever $\mathcal{U} \gg \mathcal{U}_\eta$), Alfvén's theorem and the ideas of the last subsection are useful in picturing MHD processes. Reconnection is, however, likely to be significant wherever $\mathcal{U} \ll \mathcal{U}_\eta$ (i.e., $Rm \ll 1$). The following examples illustrate this idea and reconnection processes:

- The *ω-effect.* Besides the dynamical process described in the last subsection, there is a kinematic process that halts the production of zonal field by the ω-effect. As $\bar{B}_\phi$ is amplified by the shear ζ, diffusion increasingly acts to straighten the field lines depicted in **Figure 2(b)**. They move with the conductor less and less, drifting in the opposite direction with a velocity $\mathcal{U}_\eta$ of order $\varpi\eta$ relative to the conductor, where ϖ is the field line curvature. When $\bar{B}_\phi$ becomes of order $Rm\bar{\mathbf{B}}_M$ where $Rm = \zeta/\eta\varpi^2$, amplification of $\bar{B}_\phi$ ceases, the stretching process being offset by an equal but opposite resistive drift.

A similar example will be of interest in Section 3.3.5: Suppose that **B** is created by currents flowing in a spherical conductor. If the radial component of **u** vanishes, the collapse of the magnetic field cannot be prevented; resistive drift inexorably carries the field lines inwards.

- *The Sweet–Parker mechanism – SPM.* This, the simplest and most fundamental process of magnetic reconnection, is sketched in **Figure 3**. A steady flow, $\mathcal{U}$, of a weakly resistive fluid in the $\pm y$-directions, forces together two 'slabs' of oppositely directed field $\pm B$ (**Figure 3(a)**). The slabs are of length $\mathcal{L}$ and the field is approximately in the $\pm x$-directions. As the fluid is incompressible, **u** must have nonzero x-components, $\pm\mathcal{U}_e$, as indicated in **Figure 3(a)**. This figure shows (shaded) the formation of a reconnection region $\mathcal{R}$ which has such a small thickness ℓ that it resembles a current sheet. Because $\ell = O(\mathrm{Rm}^{-1/2})\mathcal{L} \ll \mathcal{L}$, ohmic diffusion can break the topological constraint of Alfvén's theorem and allow the field lines in $\mathcal{R}$ to reconnect (**Figure 3(c)**), even though field lines elsewhere are advected with the flow. The reconnected lines exit from $\mathcal{R}$ in the $\pm x$-directions. A field line labeled 'n' in **Figure 3(a)** is replaced by a field line '$n+1$' in **Figure 3(d)**. The entire process is then repeated.

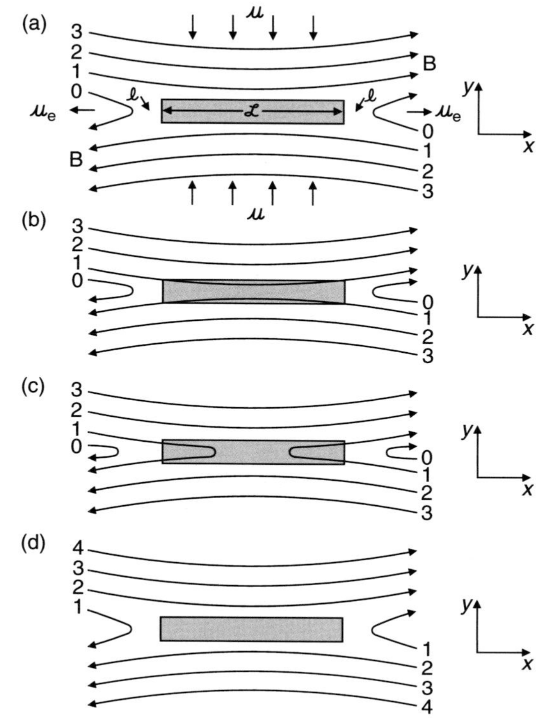

Figure 3 The Sweet–Parker reconnection machine (see text for explanation).

Many MHD phenomena appear to be explicable only through 'fast' reconnection, that is, reconnection on the ideal time scale $\mathcal{T}_u$. This is especially true of solar and astrophysical phenomena such as solar flares, but it is widely believed that reconnection in the Earth's core is slow, with a timescale of order $\mathcal{T}_\eta$; see Section 3.5.4. Between these two extremes are reconnection processes that are 'intermediate', that is, operate on a timescale that is short compared with $\mathcal{T}_\eta$ but long compared with $\mathcal{T}_u$. The SPM, which operates on the timescale $(\mathcal{T}_\eta \mathcal{T}_u)^{1/2}$ is of intermediate type.

The SPM is a kinematic reconnection machine but some dynamical aspects are clear, e.g., the curvature of the reconnected field lines in **Figure 3(c)** assists in ejecting them from $\mathcal{R}$ (Section 3.2.5).

Amongst other intermediate reconnection processes, those central to magnetic instabilities, such as the *tearing mode*, should be mentioned. This instability of sheared magnetic fields is most easily exhibited for the sheet pinch:

$$\mathbf{B}_0 = B_{0x}(z)\mathbf{1}_x + B_{0y}(z)\mathbf{1}_y \qquad [21a]$$

$$\mu_0 \mathbf{J}_0 = -B'_{0y}(z)\mathbf{1}_x + B'_{0x}(z)\mathbf{1}_y \qquad [21b]$$

In this model, the conductor is at rest and remains at rest since the Lorentz force $\mathbf{J}_0 \times \mathbf{B}_0$ is balanced by the pressure gradient $(B_{0x}^2 + B_{0y}^2)'/2\mu_0$. Since $\mathbf{u} \equiv \mathbf{0}$, the Alfvén velocity substitutes for $\mathcal{U}$ in defining the dynamic timescale $\mathcal{T}_A = \mathcal{L}/\mathcal{V}_A$, so that the Lundquist number,

$$\mathrm{Lu} = \mathcal{V}_A \mathcal{L}/\eta = \mathcal{T}_\eta/\mathcal{T}_A \qquad [22]$$

takes over from Rm. A perturbation of the equilibrium [21] proportional to $\exp(\mathrm{i}k_x x + \mathrm{i}k_y y + \lambda t)$ is envisaged. The instability arises at *resonant surfaces*, defined by $F = 0$, where $F = k_x B_{0x} + k_y B_{0y}$. Reconnection occurs in a thin boundary layer surrounding a resonant surface, in which the reconnected field lines can shorten, releasing magnetic energy to drive the instability. The timescale of reconnection, $\mathcal{T}_\eta^{3/5}\mathcal{T}_A^{2/5}$, is long compared with $\mathcal{T}_A$ but short compared with $\mathcal{T}_\eta$; see Furth *et al.* (1963). In a rapidly rotating system such as the core in which $\mathcal{V}_A/\mathcal{L} \ll \Omega$, tearing is dramatically slowed by Coriolis forces. Its dynamic timescale becomes the slow timescale defined in Section 3.2.5 and the Elsasser number, $\Lambda = \mathcal{V}_A^2/\Omega\eta = \mathcal{T}_\eta/\mathcal{T}_s$ takes on the role of Lu; see Kuang and Roberts (1990).

- *Flux expulsion.* A simple illustration of this phenomenon is provided by a solid sphere of radius a, embedded in a solid conductor with which it is in perfect electrical contact. A uniform magnetic field $\mathbf{B}_0$ is applied, and the sphere is set into rotation with angular velocity ω about an axis perpendicular to $\mathbf{B}_0$. The appropriate magnetic Reynolds number is $Rm = \omega a^2/\eta$ and is assumed to be large. To an observer on the surface S of the sphere and rotating with it, the applied field seems to be oscillatory and, since $Rm \gg 1$, it ultimately penetrates only a short distance, of order $\delta_m = Rm^{-1/2}a = (\eta/\omega)^{1/2}$, into the sphere. This phenomenon is well known in the EM of solid conductors and is called the 'skin effect', because the induced currents are confined to a thin 'skin' on S. These currents create a dipolar magnetic field, $\mathbf{b}$, that almost completely excludes $\mathbf{B}_0$ from the interior of the sphere $\mathbf{n} \cdot (\mathbf{B}_0 + \mathbf{b}) \approx 0$ on S. During the time the system evolves into this steady state, the field lines within the sphere gradually diffuse out of it, and reconnect in the skin and in the vicinity of the sphere. This process, called 'flux expulsion', is sketched in **Figure 4**.

Field lines tend to be similarly expelled from circulating flows in fluid conductors and to be crowded together in EM boundary layers at the edges of the circulations, so forming flux ropes and

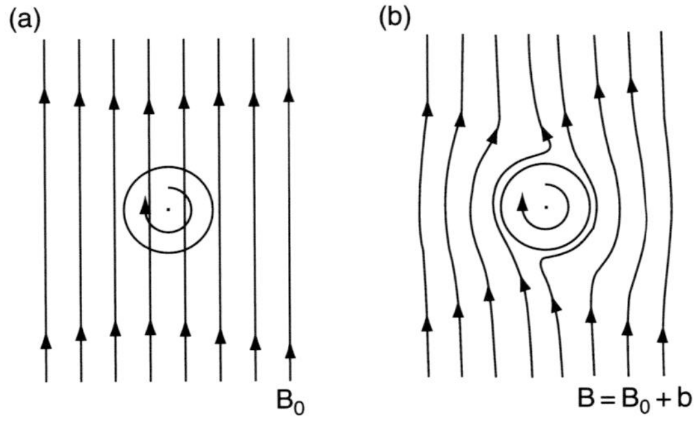

Figure 4 Flux expulsion. (a) A conducting sphere embedded in a similarly conducting, stationary medium lies in a uniform magnetic field $\mathbf{B}_0$ and is set in rotation. (b) The field is ultimately almost completely expelled from the sphere.

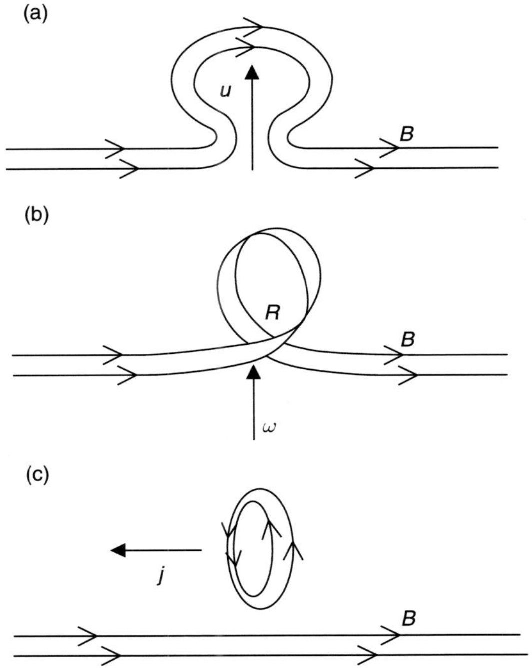

Figure 5 The alpha-effect mechanism. (a) A flux rope bent by the velocity **u** of a cyclonic eddy, (b) is twisted by the vorticity $\boldsymbol{\omega}$ of the eddy, creating large field gradients near R where the flux loop detaches as indicated in (c).

flux sheets. It might seem at first sight that this can only be detrimental to dynamo action but this is not necessarily the case. Consider induction by a steady flow. An X-type stagnation point may exist between one circulation and its neighbor. (An example is given in Section 3.4.3 below.) The expelled flux surrounds this point and, since the streamlines separate there, the field lines almost frozen to them are 'exponentially stretched'. This is an efficient way of transforming kinetic energy into magnetic energy. Even more efficient are flows with chaotic streamlines, that is, flows in which a finite volume of space is filled by only a finite number of streamlines (i.e., the streamlines do not lie on stream surfaces but fill a volume). Exponential stretching can be even more efficient when the chaos includes time, as in a turbulent flow.

- *Parker's* α*-effect mechanism.* Let us use Alfvén's theorem to visualize what happens when a helical eddy meets a flux rope. Helical motions are those in which the velocity **u** is correlated to the vorticity $\boldsymbol{\omega}$ so that the 'helicity',

$$H = \mathbf{u} \cdot \boldsymbol{\omega} \qquad [23a]$$

$$\boldsymbol{\omega} = \nabla \times \mathbf{u} \qquad [23b]$$

is nonzero. A helical eddy may be pictured as a screw motion in which a blob of fluid twists around the direction in which it moves. Although the inductive effects of **u** and $\boldsymbol{\omega}$ act simultaneously, it is convenient to consider them successively. The 'vertical' motion u creates an Ω-shaped indentation on the flux rope, as sketched in **Figure 5(a)**. The twisting motion associated with $\boldsymbol{\omega}$ turns this Ω out of the plane of the paper (**Figure 5(b)**). The field gradients are large near the point marked 'R' in **Figure 5(b)**. Here diffusion can rapidly detach the Ω as an independent flux loop that, in the idealization depicted here, lies in the plane perpendicular to the flux rope.

Helicity is the simplest manifestation of a lack of mirror-symmetry, and the destruction of this symmetry is often called 'symmetry breaking'. Helicity is a pseudoscalar, that is, a scalar that changes sign under a reflection of coordinates, either in a plane (e.g., $z \to -z$) or through the origin ($\mathbf{x} \to -\mathbf{x}$). It arises naturally in a rotating, convecting system. Perhaps a compressible fluid provides the simplest example of this: Imagine that the sequence shown in **Figure 5** concerns a cold, sinking 'blob' of fluid at high 'southern' latitudes, the 'top' of each panel being nearer than the bottom to the center of the fluid body. As the eddy shown sinks, it compresses under the increasing pressure of its environment but attempts to conserve its angular momentum relative to the inertial frame. This translates, in the reference frame rotating with the fluid body, to an increase in the angular velocity ω of the eddy about the rotation axis, as indicated in panel b. The sense of helicity is the same for a hot, buoyant blob expanding as it rises through the fluid. This example also indicates a preference for 'right-handed' helicity, $H > 0$ in the Southern Hemisphere of a rotating, convecting body and 'left-handed' helicity, $H < 0$, in its Northern Hemisphere. It is clear from [23a] that $|H| \leq |\mathbf{u}||\boldsymbol{\omega}| = O(\mathcal{U}^2/\mathcal{L})$; a 'maximally helical flow' is one in which $|H| = |\mathbf{u}||\boldsymbol{\omega}|$.

• *The stretch-twist-fold dynamo – STF.* Obtain an elastic (rubber) band, loosely corresponding to the flux tube in **Figure 6(a)**, and use its tension to roughly simulate the tension of the field lines in the flux tube. The tension in the band may be systematically increased by a three-step, stretch-twist-fold process:

(1) 'stretch' the rubber band to double its length;
(2) 'twist' it into a figure-8; and
(3) 'fold' the loops of the 8 on top of one another to form two linked loops of the same size as the original band.

These steps are illustrated in Figures 6(a)–6(c), in which panel (a) depicts the initial flux tube, panel (b) sketches it when it is stretched and twisted, and panel (c) shows it after it is subsequently folded. The energy within, and the tension of, the flux tube are increased by a factor of 4 in step (1). Steps (2)–(3) can, in principle, be repeated over and over again, the sense of twist in step (2) being always the same. This gives the band the sense of 'handedness' or helicity of a screw motion, and the broken reflection symmetry that is crucial to the success of this dynamo. The tension in the band (analogous to magnetic tension) increases progressively as steps (1)–(3) are repeated, and at some stage the reader's hands will tire, making it impossible to increase the tension further. This marks the end of the kinematic regime (in which fluid motion is prescribed *a priori* and the field grows) and the start of the dynamic regime (in which the field quenches its own further growth). If the reader lets go of the band at any stage, it will immediately relax back to its initial state. This illustrates the crucial importance of magnetic reconnection as a means of 'locking-in' the amplified magnetic field. Said another way, magnetic reconnection provides the crucial element of irreversibility in the dynamo. The process of reconnection in the flux tube occurs near the point marked 'R' in panel 'c', where the field gradients are greatest. The result of the reconnection, two separate rings, is shown in panel (d).

The STF dynamo, originally proposed by Vainshtein and Zeldovich (1972), has become the paradigm for 'fast dynamos', which are defined as processes of field amplification that operate on timescales independent of $\mathcal{T}_\eta$ in systems where *Rm* is large (see Section 3.3.3 below and Childress and Gilbert (1995)). Clearly, a fast dynamo necessarily requires fast reconnection, in order to lock in the dynamo-generated field as fast as it is produced.

• *Alfvén's twisted kink dynamo – ATK.* Take a piece of taut rope and start twisting its ends, as in **Figure 7(a)**. As the result of an instability, it will develop kinks like those often seen on the cord joining a telephone handset to its cradle (**Figure 7(b)**). A magnetic flux rope twisted in the same way will also develop a kink and, because of the large field gradients near the crossing point R in the figure, a tube will detach (**Figure 7(c)**) and move away from the flux rope. Simultaneously, the torsion of the rope is released, but increases again as the twisting continues until further loops detach. The flux rope has become a machine for generating flux tubes. A sufficiently large twisting rate creates a fast dynamo (Alfvén, 1950).

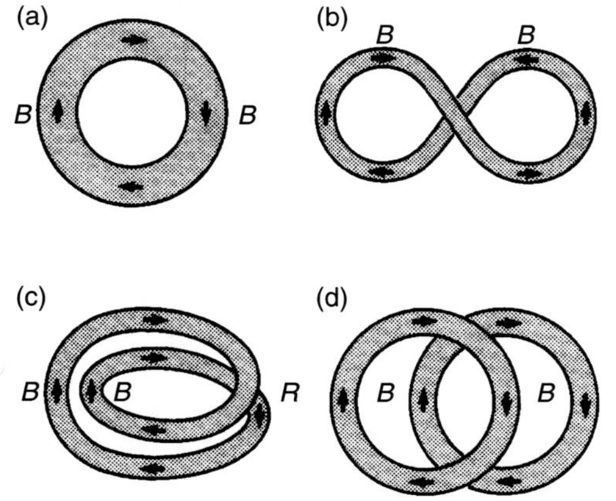

Figure 6 Stretch-twist-fold. (a) A closed flux tube is (b) stretched, twisted, and (c) folded over onto itself, so creating large field gradients near *R* where one loop detaches from the other, as indicated in (d).

When $\eta \neq 0$, the boundary conditions [2] simplify. From a physical standpoint, it is clear that an infinitely thin concentration of current, even if it could be set up initially, would instantly diffuse into a layer of finite thickness, in which $\mathbf{J}$ (though possibly large) is finite. This layer replaces the surface current, that is, $\mathbf{C} \equiv \mathbf{0}$ instantaneously. Thus, when $\eta \neq 0$, the conditions [2] reduce to

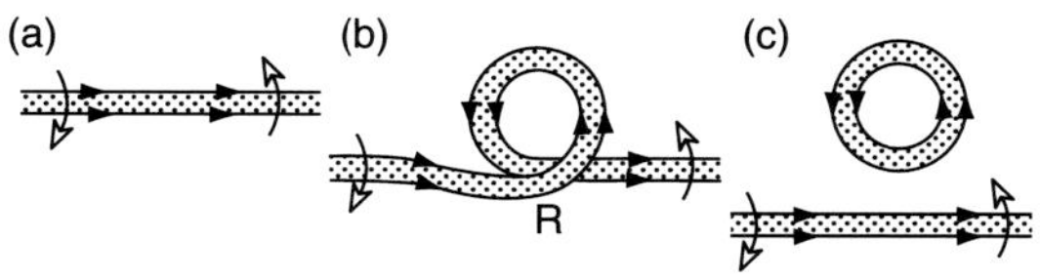

Figure 7 Alfvén's twisted kink dynamo. (a) A flux rope is twisted and (b) becomes unstable to the formation of kinks. The large field gradients near *R* allows the kink to detach from the rope, as indicated in (c).

$$[\mathbf{n} \times \mathbf{B}] = 0 \quad \text{on S} \qquad [24a]$$

$$[\mathbf{n} \times \mathbf{E}] = 0 \quad \text{on S} \qquad [24b]$$

$$[\mathbf{n} \cdot \mathbf{B}] = 0 \quad \text{on S} \qquad [24c]$$

$$[\mathbf{n} \cdot \mathbf{J}] = 0 \quad \text{on S} \qquad [24d]$$

$$[\mathbf{n} \cdot \mathbf{D}] = \Theta \quad \text{on S} \qquad [24e]$$

Equations [24a] and [24b] imply

$$\left[\mathbf{n} \cdot \mathbf{I}^B\right] = 0 \text{ on S} \qquad [24f]$$

Condition [24d] is superfluous since, according to [1a] and [6a], it is satisfied when [24a] holds. Similarly [24c] implies that $[\mathbf{n} \cdot \partial_t \mathbf{B}] = 0$, so that one (linear combination) of the two scalar conditions [24b] is automatically obeyed by [24c]. Thus, only four of the six scalar conditions [24a]–[24d] are independent. The usual choice is

$$[\mathbf{B}] = \mathbf{0} \quad \text{on S} \qquad [25]$$

together with one condition from [24b]. According to [1c] and [24a], $\left[(\mathbf{n} \cdot \nabla)(\mathbf{n} \cdot \mathbf{B})\right] = 0$ but, because $[\sigma] \neq 0$ in general, $\left[(\mathbf{n} \cdot \nabla)(\mathbf{n} \times \mathbf{B})\right] \neq \mathbf{0}$ and therefore $[\mathbf{n} \times \mathbf{J}] \neq \mathbf{0}$ on S.

3.2.7 The Low Conductivity Approximation

Alfvén's theorem and the ideas just presented in Sections 3.2.5 and 3.2.6 are useful when $Rm \gg 1$. We now consider the opposite extreme, the 'low conductivity approximation' in which $Rm \ll 1$. Since then $\mathcal{T}_\eta \ll \mathcal{T}_u$, the EM field adjusts almost instantaneously to changes in $\mathbf{u}$ so that [1a] and [6e] give, to leading order in Rm,

$$\mathbf{E} = -\nabla\psi \qquad [26a]$$

$$\sigma^{-1}\mathbf{J} = \eta\nabla \times \mathbf{B} = -\nabla\psi + \mathbf{u} \times \mathbf{B} \qquad [26b]$$

just as though $\mathbf{B}$ were time independent. By [26b], the Lorentz force [13e] is

$$\mathbf{L} \equiv \mathbf{J} \times \mathbf{B} = \sigma\mathbf{B} \times \nabla\psi - \sigma B^2 \mathbf{u}_\perp \qquad [26c]$$

where $\mathbf{u}_\perp = \mathbf{u} - (\mathbf{u} \cdot \mathbf{B})\mathbf{B}/B^2$ is the component of $\mathbf{u}$ perpendicular to $\mathbf{B}$. This demonstrates that the field opposes the motion in part through an anisotropic 'magnetic friction', $-\sigma B^2 \mathbf{u}_\perp/\rho$ per unit mass. Comparison with the fluid acceleration $\partial_t \mathbf{u}$ shows the importance of $\mathcal{T}_\mathrm{m} = \rho/\sigma B^2 = \eta/V_\mathrm{A}^2$, which is often called the 'magnetic damping time', not to be confused with the magnetic diffusion time $\mathcal{T}_\eta = \mathcal{L}^2/\eta$. Their ratio $\mathcal{T}_\eta/\mathcal{T}_\mathrm{m}$ is Lu^2. The magnetic damping time is significant in MHD turbulence theory for liquid metals such as the Earth's core (Section 3.5.4). It is independent of $\mathcal{L}$ and, for $\eta = 2\ \mathrm{m^2\,s^{-1}}$ and $V_\mathrm{A} = 1\ \mathrm{cm\,s^{-1}}$, it is the shortest timescale relevant to core dynamics: $\mathcal{T}_\mathrm{m} \approx 6\ \mathrm{h}$.

The integrated rate of working of the leading order Lorentz force [26c] is

$$\int_v \mathbf{u} \cdot \mathbf{L}\, \mathrm{d}V = -\int_v \mathbf{J} \cdot (\mathbf{u} \times \mathbf{B})\, \mathrm{d}V$$
$$= -\Phi - \int_s \psi\mathbf{J} \cdot \mathrm{d}\mathbf{S} \qquad [26d]$$

which is $-\Phi$ when no current flows across s. The Lorentz force then merely satisfies the ohmic demands of v. In the approximation [26b], the magnetic energy density e^B does not change since, by [26a], $\nabla \cdot \mathbf{I}^B = -\mathbf{E} \cdot \nabla \times \mathbf{B}/\mu_0 = -\mathbf{E} \cdot \mathbf{J}$. The field, $\mathbf{b} = \mathbf{B} - \mathbf{B}_0$, induced by motions in v is $O(Rm\ \mathbf{B}_0)$, that is, tiny compared with the field $\mathbf{B}_0$ created by currents flowing outside v. Nevertheless, the currents $\mathbf{J}$ $(=\mathbf{j})$ induced in v, which are $O(\sigma\mathcal{U}B_0)$, create an $\mathbf{L}$ of order $\sigma B_0^2 \mathcal{U}$ that may significantly alter $\mathbf{u}$. If $Ha \gg 1$, $\mathbf{L}$ dominates the viscous force $\rho\nu\nabla^2\mathbf{u} = O(\rho\nu\mathcal{U}/\mathcal{L}^2)$; if $I \gg 1$, it dominates the inertial force $\rho\mathbf{u} \cdot \nabla\mathbf{u} = O(\rho\mathcal{U}^2/\mathcal{L})$. Here Ha is the Hartmann number and I is the interaction parameter:

$$Ha = \mathcal{BL}\left(\frac{\sigma}{\rho\nu}\right)^{1/2} = \frac{\mathcal{V}_A\mathcal{L}}{\sqrt{\nu\eta}} \qquad [27a]$$

$$I = \frac{\sigma\mathcal{B}^2\mathcal{L}}{\rho\mathcal{U}} = \frac{\mathcal{V}_A^2\mathcal{L}}{\eta\mathcal{U}} \qquad [27b]$$

Approximations [26] are commonly applied to laboratory experiments involving liquid metals for which EM induction processes are usually insignificant. It is perhaps surprising though true that, even when the low conductivity approximation is appropriate, it is sometimes helpful to use the concept of frozen-in fields when picturing the evolution of $\mathbf{B}$. This is because, even when $Rm \ll 1$, there is usually a slight tendency for the field to evolve as it would in a perfect conductor. When [26] hold, the EM field depends only parametrically on t but an improved solution may be obtainable as an expansion in Rm, the leading term obeying [26a]. Of course, steady-state solutions of the induction equation for any Rm must satisfy [26b] (see Section 3.3.3).

3.3 Kinematic Dynamos

3.3.1 The Dynamo Condition

The STF and ATK models of Section 3.2.6 are heuristic and remote from geophysical needs. This subsection derives more explicitly the condition of self-excitation. We do this in a framework relevant to the Earth: a conducting fluid V surrounded by an insulator $\hat{V}$. In fact, the mantle is such a poor electrical conductor that little is lost by assuming (as is usually done) that $\hat{V}$ consists not only of the exterior of Earth but also of the entire mantle and crust. Quantities in $\hat{V}$ will be distinguished from those in V by a superimposed hat. Since $\hat{\sigma}=0$, it follows from [19]–[1c] and [6a] that

$$\nabla \times \hat{\mathbf{B}} = \mathbf{0} \quad [28a]$$

$$\nabla \cdot \hat{\mathbf{B}} = 0 \quad [28b]$$

$$\nabla \times \hat{\mathbf{E}} = -\partial_t \hat{\mathbf{B}} \quad [28c]$$

The first two of these equations imply that $\hat{\mathrm{B}}$ is a potential field:

$$\hat{\mathbf{B}} = -\nabla \hat{V} \quad [29a]$$

where

$$\nabla^2 \hat{V} = 0 \quad [29b]$$

In the case of the Earth, [29b] led Gauss to his well-known representation of $\hat{\mathbf{B}}$ at and above the Earth's surface $r=r_e$:

$$\hat{V} = r_e \sum_{n=1}^{\infty} \sum_{m=0}^{n} \left[\left(\frac{r_e}{r}\right)^{n+1} \left(g_n^m \cos m\phi + h_n^m \sin m\phi\right) + \left(\frac{r}{r_e}\right)^{n} \left(q_n^m \cos m\phi + s_n^m \sin m\phi\right) \right] P_n^m(\theta) \quad [29c]$$

where $P_n^m(\theta)$ is the Legendre function (usually Schmidt normalized now, though not in Gauss's day). Because we assume that the mantle is insulating, we adopt Gauss's expansion everywhere above the CMB, that is, for all $r \geq r_c$.

The coefficients g, h, q, and s in [29c] are time dependent; the g and h coefficients describe the field created within V and decreasing with distance from it, while the q and s coefficients correspond to fields created by external sources that, according to [29c], are 'at infinity'. The 'dynamo condition' expresses the absence of sources outside V:

Dynamo condition: form 1: $\quad q_n^m \equiv 0, \quad s_n^m \equiv 0 \quad$ [30]

When this is enforced, the magnetic field is produced totally by sources within V, and [29c] gives

$$\hat{V} = r_c \sum_{n=1}^{\infty} \left(\frac{r_c}{r}\right)^{n+1} Y_n(\theta, \phi) \quad [31a]$$

$$\hat{B}_r = \sum_{n=1}^{\infty} (n+1) \left(\frac{r_c}{r}\right)^{n+2} Y_n(\theta, \phi) \quad [31b]$$

etc., where $Y_n(\theta, \phi)$ is the surface harmonic

$$Y_n(\theta, \phi) = \left(\frac{r_e}{r_c}\right)^{n+2} \sum_{m=0}^{n} \left(g_n^m \cos m\phi + h_n^m \sin m\phi\right) P_n^m(\theta) \quad [31c]$$

Equation [31b] shows that the field becomes increasingly dipolar with increasing r, and an alternative way of expressing the dynamo condition is

Dynamo condition: form 2:

$$\hat{\mathbf{B}} = O(r^{-3}) \text{ for } r \to \infty \quad [32]$$

The three $n=1$ terms in [31a] define the Earth's centered dipole moment $\mathbf{m}$. This is customarily expressed in terms of $\mathbf{H}$ rather than $\mathbf{B}$ so that $V_{\text{dipole}} = \mu_0 \mathbf{m} \cdot \mathbf{r}/4\pi r^3$ (the 4π being there because SI are rationalized units). Since $P_1(\theta)=\cos\theta$ and $P_1^1(\theta)=\sin\theta$, it therefore follows that $\mathbf{m} = (4\pi r_e^3/\mu_0)\left[g_1^1 \mathbf{1}_x + h_1^1 \mathbf{1}_y + g_1^0 \mathbf{1}_z\right]$, where $4\pi r_e^3/\mu_0 \approx 2.580 \times 10^{27}\ \mathrm{A\,m^{-1}\,T^{-1}}$; the z-component of $\mathbf{m}$ defines the 'axial dipole', and the x- and y-components the 'equatorial dipole'. Because $g_1^0 < 0$ currently, the angle between $\mathbf{m}$ and Oz, commonly called the 'tilt' of the dipole axis, is large: $\cos^{-1}(m_z/m) \approx 170°$, but more usually it is expressed as $\cos^{-1}(-m_z/m) \approx 10°$. The five $n=2$ terms in [31a] represent the quadrupolar part of $\hat{\mathbf{B}}$.

It is worth noting that, according to [29a] and [31], the magnetic energy stored outside the core is

$$\hat{M} = \frac{1}{2\mu_0} \int_{\hat{V}} \hat{B}^2\, \mathrm{d}V = \frac{1}{2\mu_0} \int_{\text{CMB}} \hat{V} \hat{B}_r\, \mathrm{d}S = \frac{2\pi r_c^3}{\mu_0} \sum_{n=1}^{\infty} \frac{M_n}{2n+1} \quad [33a]$$

where

$$M_n = (n+1) \left(\frac{r_e}{r_c}\right)^{2n+4} \sum_{m=0}^{n} \left[(g_n^m)^2 + (h_n^m)^2 \right] \quad [33b]$$

is the (Mauersberger–Lowes) field spectrum at the CMB, which is B^2 for the nth harmonic averaged over the CMB. This can be computed from observation for $n \lesssim 13$, and is remarkably constant for $n \geq 3$. If other planetary dynamos share this property, it provides a valuable way of estimating the radii of their conducting cores from flybys (Hide, 1978) (see also Glatzmaier and Roberts (1996b), where further references are given). The constancy of M_n cannot hold for all n, as this would give infinite $\hat{M}$, by [33a].

Since $\mathbf{B} = \hat{\mathbf{B}}$ on the CMB by [25], the dynamo conditions [30] and [32] restrict the solutions of the induction equation on S and therefore in V. They are rather inconvenient conditions as they stand; a simpler, third form will be derived in Section 3.3.2.

In Section 3.2.6 it was shown that, in addition to [25], one scalar condition on $\mathbf{E}$ must be obeyed at a discontinuity surface such as the CMB. It is important to show that this condition does not restrict solutions of the induction equation.

To see this, introduce scalar and vector potentials Ψ and $\mathbf{A}$ to satisfy [1b] and [1c]

$$\mathbf{B} = \nabla \times \mathbf{A} \qquad [34a]$$

$$\mathbf{E} = -\nabla\Psi - \partial_t \mathbf{A} \qquad [34b]$$

$$\nabla \cdot \mathbf{A} = 0 \qquad [34c]$$

Without essential loss of generality, we may assume for simplicity that $\hat{\vartheta} = 0$ and $\hat{\epsilon} = \epsilon_0$, so that, by [1d], $\nabla \cdot \hat{\mathbf{E}} = 0$, which implies that $\hat{\Psi}$ satisfies Laplace's equation. Condition [24c] is satisfied by $[\mathbf{n} \times \mathbf{A}] = \mathbf{0}$ and therefore, according to [24b], $\mathbf{n} \times \mathbf{E}$ determines a unique $\hat{\Psi}$ on S and (by solving $\nabla^2\hat{\Psi} = 0$) a unique $\hat{\Psi}$ throughout $\hat{V}$ that vanishes for $r \to \infty$. The resulting $\mathbf{n} \cdot \hat{\mathbf{D}}$ will generally differ from $\mathbf{n} \cdot \mathbf{D}$ on S but their difference does no more than determine Θ by [24e]. It does not constrain the solution of the induction equation in V any way whatever. Because magnetic charges do not exist, a similar argument applied to $\mathbf{B}$ has a radically different outcome. Although Laplace's equation and an arbitrary $\mathbf{n} \cdot \mathbf{B}$ on S determine the magnetic potential $\hat{V}$ uniquely, the resulting $[\mathbf{n} \times \mathbf{B}]$ would be nonzero in general, so contradicting [24a]. Only [25] restricts $\mathbf{B}$ on S and therefore in V.

On the ICB, condition [24b] on $\mathbf{n} \times \mathbf{E}$ is not superf luous, and it is in general necessary to supplement [25] with one independent condition from [24b]. When however, as is often the case, the SIC is modeled as a solid of the same conductivity as the fluid, electrical properties are continuous across S and, if in addition it is assumed that $[\mathbf{u}] = \mathbf{0}$ (the no-slip condition), then $[\mathbf{B}] = \mathbf{0}$ is enough to imply $[\mathbf{E}] = \mathbf{0}$ and $[\mathbf{J}] = \mathbf{0}$. One is then free to treat the whole core, SIC + FOC, as a single domain in which

$$\mathbf{u} = \boldsymbol{\Omega}_i \times \mathbf{r} \quad \text{for} \quad r \le r_i \qquad [35a]$$

In kinematic theory, the angular velocity $\boldsymbol{\Omega}_i(t)$ is specified at the same time as $\mathbf{u}$ is assigned to the FOC. In MHD theory, the torque $\boldsymbol{\Gamma}_i$ on the ICB determines $\boldsymbol{\Omega}_i$ from Euler's equation

$$\overleftrightarrow{\mathbf{I}} \cdot (\partial_t \boldsymbol{\Omega}_i + \boldsymbol{\Omega} \times \boldsymbol{\Omega}_i) = \boldsymbol{\Gamma}_i \qquad [35b]$$

where $\overleftrightarrow{\mathbf{I}}$ is the moment of inertia tensor. Plausibly, $\boldsymbol{\Gamma}_i$ is dominated by the gravitational torque from the mantle and the magnetic torque [15f] from the FOC. (Some modelers treat the SIC as an insulator. Then, as for the CMB, only [25] is needed.)

3.3.2 EM Induction in Spherical Conductors

Nearly all investigations of dynamos in the Earth, planets, and stars make the simplifying assumption that these bodies are spherical. It may then be convenient to satisfy [1c] automatically by separating $\mathbf{B}$ into 'toroidal' and 'poloidal' parts:

$$\mathbf{B} = \mathbf{B}_\mathrm{T} + \mathbf{B}_\mathrm{P} = \nabla \times (T\mathbf{r}) + \nabla \times \nabla \times (S\mathbf{r}) \qquad [36a]$$

where $T(r, \theta, \phi, t)$ is the 'toroidal scalar' and $S(r, \theta, \phi, t)$ is the 'poloidal scalar'. Obviously, according to [1b] and [35a], the toroidal field $\mathbf{B}_T = \nabla \times (T\mathbf{r})$ is created solely by poloidal currents. A particularly pleasant feature is the reverse: the poloidal field $\mathbf{B}_\mathrm{P} = \nabla \times \nabla \times (S\mathbf{r})$ is created solely by toroidal currents, since

$$\mu_0 \mathbf{J} = \nabla \times \mathbf{B} = \nabla \times ([-\nabla^2 S]\mathbf{r}) + \nabla \times \nabla \times (T\mathbf{r}) \qquad [36b]$$

A second curl shows that $\nabla^2\mathbf{B} = \nabla \times ([\nabla^2 T]\mathbf{r}) + \nabla \times \nabla \times ([\nabla^2 S]\mathbf{r})$.

In spherical polar components, $\mathbf{B}_\mathrm{T}$ and $\mathbf{B}_\mathrm{P}$ are

$$\mathbf{B}_\mathrm{T} = -\mathbf{r} \times \nabla T = \frac{1}{\sin\theta}\frac{\partial T}{\partial \phi}\mathbf{1}_\theta - \frac{\partial T}{\partial \theta}\mathbf{1}_\phi \qquad [36c]$$

$$\begin{aligned}\mathbf{B}_\mathrm{P} &= \nabla\left[\frac{\partial(rS)}{\partial r}\right] - \mathbf{r}\nabla^2 S \\ &= \frac{L^2 S}{r}\mathbf{1}_r + \frac{1}{r}\frac{\partial}{\partial\theta}\left[\frac{\partial(rS)}{\partial r}\right]\mathbf{1}_\theta + \frac{1}{r\sin\theta}\frac{\partial}{\partial\phi}\left[\frac{\partial(rS)}{\partial r}\right]\mathbf{1}_\phi\end{aligned} \qquad [36d]$$

where

$$\begin{aligned}L^2 S &= r\frac{\partial^2(rS)}{\partial r^2} - r^2\nabla^2 S \\ &= -\left\{\frac{1}{\sin\theta}\frac{\partial}{\partial\theta}\left[\sin\theta\frac{\partial S}{\partial\theta}\right] + \frac{1}{\sin^2\theta}\frac{\partial^2 S}{\partial\phi^2}\right\}\end{aligned} \qquad [36e]$$

The salient characteristic of [36c] is that toroidal fields have no radial components.

The simplest examples of toroidal and poloidal fields are the axisymmetric fields $\bar{\mathbf{B}}_\phi$ and $\bar{\mathbf{B}}_\mathrm{M}$ defined by $T = \bar{T}(r, \theta, t)$ and $S = \bar{S}(r, \theta, t)$. It may be seen from [36c] and [36d] that $\bar{\mathbf{B}}_\phi$ is 'zonal' (has only a ϕ-component) and $\bar{\mathbf{B}}_\mathrm{M}$ is meridional (has no ϕ-component):

$$\bar{\mathbf{B}} = \bar{\mathbf{B}}_\phi + \bar{\mathbf{B}}_M = \bar{B}_\phi \mathbf{1}_\phi + \nabla \times (\bar{A}_\phi \mathbf{1}_\phi) \qquad [37]$$

where $\bar{A}_\phi = -\partial_\theta \bar{S}$ and $\bar{B}_\phi = -\partial_\theta \bar{T}$.

The usefulness of the representation [36] depends partly on the ease with which T and S can be 'projected' from a given $\mathbf{B}$ by solving

$$L^2 T = r(\nabla \times \mathbf{B})_r = \mu_0 r \mathcal{J}_r \qquad [38a]$$

$$L^2 S = rB_r \qquad [38b]$$

Since $L^2 S = f(r)$ has no solutions, fields having monopolar components (if they existed) could not be represented by [36] but, since magnetic monopoles do not exist, [38a] and [38b] are easily solved through their spherical harmonic components, using the fact that

$$L^2 Y_n = n(n+1) Y_n \qquad [38c]$$

For example, by [38b],

$$B_r = \sum_{n=1}^{\infty} B_{r,n}(r, t) Y_n(\theta, \phi) \qquad [38d]$$

implies

$$S = r \sum_{n=1}^{\infty} \frac{B_{r,n}(r, t)}{n(n+1)} Y_n(\theta, \phi) \qquad [38e]$$

T can be similarly extracted from [38a].

By using this projection method, partial differential equations governing $T_n(r, t)$ and $S_n(r, t)$, equivalent to the induction equation, can be derived. A further attraction is the ease with which the dynamo condition can be applied. In $\hat{V}$, [36b] implies

$$\hat{T} = 0 \qquad [39a]$$

and

$$\nabla^2 \hat{S} = 0 \qquad [39b]$$

Equations [36c] and [36d] then show that

$$\hat{\mathbf{B}} = -\nabla \hat{V}, \quad \text{where } \hat{V} = -\partial_r (r\hat{S}) \qquad [39c]$$

As in [29], $\hat{S}$ satisfies Laplace's equation and, as in [31b], the dynamo condition shows that

$$\hat{S} = \sum_{n=1}^{\infty} \hat{S}_n(r, t) Y_n(\theta, \phi) \qquad [39d]$$

where

$$\hat{S}_n(r, t) = \hat{S}_n(r_c, t)\left(\frac{r_c}{r}\right)^{n+1} \qquad [39e]$$

This, together with [39d], means that, everywhere in $\hat{V}$ and in particular on the CMB,

$$\hat{T}_n = 0 \qquad [39f]$$

$$\frac{\partial \hat{S}_n}{\partial r} + \frac{(n+1)\hat{S}_n}{r} = 0 \qquad [39g]$$

According to [36c] and [36d], [25]is equivalent to

$$[T_n] = 0, \quad [S_n] = 0, \quad [\partial S_n/\partial r] = 0, \quad \text{on S} \qquad [39h]$$

implying, by [39f] and [39g],

$$\text{Dynamo condition: form 3:} \quad T_n = 0, \quad \frac{\partial S_n}{\partial r} + \frac{(n+1)S_n}{r} = 0, \quad \text{at } r = r_c \qquad [40]$$

These are boundary conditions on the solutions for $\mathbf{B}$ in V. Using these, the induction equation can be solved without reference to $\hat{\mathbf{B}}$. From the resulting $\mathbf{B}$ one can, but only if desired (perhaps for presentational purposes), extract $S_n(r_c, t)$ and determine the field

$$\hat{\mathbf{B}} = -\nabla \hat{V} \qquad [41a]$$

where

$$\hat{V} = \sum_{0}^{\infty} n S_n(r_c, t)\left(\frac{r_c}{r}\right)^{n+1} Y_n(\theta, \phi) \qquad [41b]$$

that satisfies the dynamo condition and the continuity requirement [25].

The simplicity of the third form of the dynamo condition partly explains the current popularity of spectral methods in solving both kinematic and MHD dynamo problems. To illustrate this simplicity, we derive the 'decay modes' for a spherical conductor such as the Earth's core. These are solutions of the induction equation [9] when $\mathbf{u} \equiv \mathbf{0}$, so that

$$\partial_t T = \eta \nabla^2 T \qquad [42a]$$

$$\partial_t S = \eta \nabla^2 S \qquad [42b]$$

The solutions that are nonsingular at $r=0$ are proportional to

$$T_n(r) = j_n(k_n^{\mathrm{T}} r) Y_n(\theta, \phi) \exp(-\eta k_n^{\mathrm{T}2} t) \qquad [42c]$$

$$S_n(r) = j_n(k_n^{S} r) Y_n(\theta, \phi) \exp(-\eta k_n^{S2} t) \qquad [42d]$$

where j_n is the spherical Bessel function of order n. They obey [40] provided

$$j_n(k_n^T r_c) = 0 \qquad [42e]$$

$$j_{n-1}(k_n^S r_c) = 0 \qquad [42f]$$

These equations are satisfied by an infinite set of positive admissible k_n^T and k_n^S, the smallest of which we denote by k_{n1}^T and k_{n1}^S. The e-folding times over which T_n and S_n disappear through ohmic diffusion are therefore $(\eta k_{n1}^{T2})^{-1}$ and $(\eta k_{n1}^{S2})^{-1}$. The longest of these times is given by the smallest number in the combined k_{n1}^T and k_{n1}^S sets, which belongs to the poloidal dipole mode for which $k_{S1} = \pi/r_c$, we denote this by $k_{\min}$. In units of r_c^2/η, the e-folding times of the first five dipole modes are approximately 0.101321, 0.025330, 0.011258, 0.006333, 0.004053; for the quadrupole modes and for the toroidal $n=1$ modes, they are 0.049528, 0.016756, 0.008410, 0.005054, 0.003372, and for the toroidal $n=2$ modes they are 0.030105, 0.012089, 0.006585, 0.004154, 0.002863. Other cases can be obtained from table 10.6 of Abramowitz and Stegun (1964).

The decay modes provide the basis for what is called a 'variational inequality', which will be useful in Section 3.3.4 below. This exploits the fact that the modes form a complete orthogonal set, in terms of which an arbitrary magnetic field **B** satisfying [25] and the dynamo condition, but not necessarily the induction equation, can be expanded. Since every other mode in that sum decays more rapidly than the longest-lived poloidal $n=1$ mode, it follows that mode provides a lower bound on the ohmic losses:

$$\int_V (\nabla \times \mathbf{B})^2 dV \geq k_{\min}^2 \int_{V_\infty} B^2 dV \qquad [43a]$$

where V_∞ denotes all space $(V + \hat{V})$. (For any bounded nonspherical container, a variational inequality of the form [43a] holds, but with a different $k_{\min}$.) Inequality [43a] can be rewritten in terms of the total magnetic energy M of the dynamo and its ohmic dissipation Φ:

$$\Phi \geq 2\eta k_{\min}^2 M \qquad [43b]$$

where

$$M \equiv \int_{V_\infty} e^B dV = \frac{1}{2\mu_0} \int_{V_\infty} B^2 dV \qquad [43c]$$

$$\Phi \equiv \int_V q^J dV = \frac{\eta}{\mu_0} \int_V (\nabla \times \mathbf{B})^2 dV \qquad [43d]$$

The 'efficiency' of a dynamo, if defined as

$$E = k_{\min}^2 \int_{V_\infty} B^2 dV \Big/ \int_V (\nabla \times \mathbf{B})^2 dV \qquad [44a]$$

cannot exceed 1 (see Gubbins *et al.* (2000a)).

The value of $k_{\min}$ provides an improved estimate of $\mathcal{T}_\eta = (\eta k_{\min}^2)^{-1} = r_c^2/\pi^2\eta$, or about 10^4 years, an order of magnitude smaller than our previous estimate r_c^2/η. Nevertheless, it refers only to the longest-lived mode, not to the entire **B** spectrum. An appropriately smaller $\mathcal{T}_\eta$, more representative of the ohmic losses, is

$$T_\eta = E(\eta k_{\min}^2)^{-1} = \int_{V_\infty} B^2 dV/\eta \int_V (\nabla \times \mathbf{B})^2 dV \qquad [44b]$$

Both E and $\mathcal{T}_\eta$ can be assessed from any geodynamo simulation, but this is not possible for the Earth since $\mathbf{B}_T$ is unknown and only the $n \lesssim 13$ harmonics of $\mathbf{B}_P$ are available.

A few remarks about nonspherical V are apposite. Many large-scale bodies are significantly flattened by centrifugal forces (or distorted by tidal forces), but an oblate spheroid (or a triaxial ellipsoid) may be an acceptable representation of V. Standard analytic methods can then determine the solution $\hat{V}$ of Laplace's equation that vanishes at infinity, and a dynamo condition similar to [40] can be derived (see Walker and Barenghi (1994)).

Experiments with liquid metals, at values of Rm large enough in some cases for regeneration, are discussed in Chapter **. When V is a cylinder of finite length, it is hard to solve [29] analytically and a dynamo condition of the form [40] does not exist. One expedient, that is also sometimes used in constructing galactic dynamos, is to introduce a sphere S_0 surrounding V and to solve Laplace's equation for $\hat{V}$ numerically in the domain D between S_0 and S, subjecting the solution to [40] on S_0. It is not possible to treat D as an extension of V in which $\eta = \infty$, since $\nabla^2 \hat{\mathbf{B}} = \mathbf{0}$ in D does not imply $\hat{\mathbf{J}} = \mathbf{0}$.

3.3.3 The Eigenvalue Problem for Steady Flows

In studying kinematic dynamo action, it is often convenient to keep the form of **u** fixed but to vary its amplitude. This is most easily done by writing the induction equation in nondimensional form by the transformation $\mathbf{r} \rightarrow \mathcal{L}\mathbf{r}$ and $t \rightarrow (\mathcal{L}/\mathcal{U})t$. Then [9] and [1c] are

$$\partial_t \mathbf{B} = \nabla \times (\mathbf{u} \times \mathbf{B}) + \mathrm{Rm}^{-1} \nabla^2 \mathbf{B} \qquad [45a]$$

where

$$\nabla \cdot \mathbf{B} = 0 \qquad [45b]$$

Time-dependent **u** will be briefly considered at the end of Section 3.4.4, but we now focus on the easier case $\partial_t \mathbf{u} = \mathbf{0}$ for which normal mode solutions of [45] exist of the form

$$\mathbf{B} \propto e^{\lambda t} \qquad [46]$$

Equations [45] become

$$\lambda \mathbf{B} = \nabla \times (\mathbf{u} \times \mathbf{B}) + \mathrm{Rm}^{-1} \nabla^2 \mathbf{B} \qquad [47a]$$

where

$$\nabla \cdot \mathbf{B} = 0 \qquad [47b]$$

The boundary conditions and the equations governing $\hat{\mathbf{B}}$ are unaltered. Together they define an eigenvalue problem of a type commonly encountered in fluid stability theory. The eigenvalue is the growth rate λ, and Rm is the 'control parameter'.

For bounded volumes V, the eigenfunctions, $\mathbf{B}_\alpha$, form a complete set, in terms of which an arbitrary solution of [45] can be expanded:

$$\mathbf{B} = \sum_\alpha \mathbf{B}_\alpha(\mathbf{x}) \exp(\lambda_\alpha t) \qquad [48a]$$

and the spectrum of λ_α is discrete, with limit point at $-\infty$. The eigenvalue problem is not self-adjoint in general. The growth rates λ_α may be complex, although they then occur in conjugate complex pairs, so that the corresponding eigenfunctions can ensure the reality of the sum [48a]. Greatest interest attaches to the λ_α (or λ_αs) having the greatest real part, because these dominate the solution for large t; for presentational simplicity we suppose that only one such eigenvalue exists and denote it by λ_{max}.

The eigenvalues are continuous functions of the control parameter. As Rm varies, mode crossing may occur, and a complex eigenvalue pair may become real or conversely. If

$$\mathrm{Re}(\lambda_{max}) \geq 0 \qquad [48b]$$

the corresponding term in [48a] will persist for all time, showing that the flow $Rm\,\mathbf{u}$ is dynamo. In fact, if $\mathrm{Re}(\lambda_{max}) > 0$, the field grows without bound, a physical absurdity that arises from the linearity of the kinematic dynamo problem. It is removed by the nonlinear Lorentz force in the MHD dynamo problem. If [48b] does not hold, all terms in [48a] disappear as $t \to \infty$, and the motion $Rm\,\mathbf{u}$ is not a dynamo for this Rm.

As in fluid stability theory, the 'marginal' or 'critical' mode is of special interest. This is defined by the smallest value Rm_c of Rm for which

$$\mathrm{Re}(\lambda_{max}) = 0 \qquad [48c]$$

The critical mode may not exist, in which case the postulated $\mathbf{u}$ is not a dynamo for 'any' Rm. When the critical mode exists, it is of one of two types:

$$\begin{cases} \mathrm{Im}(\lambda_{max}) = 0, & \text{called a DC mode} \\ \mathrm{Im}\,(\lambda_{max}) \neq 0, & \text{called an AC mode} \end{cases} \qquad [48d]$$

Here 'AC' and 'DC' follow everyday usage of these abbreviations: AC for alternating current; DC for direct current. The DC case is a 'direct bifurcation'; the AC case is often called a 'Hopf bifurcation' (although such bifurcations had earlier been studied by Eddington, who called the bifurcation 'overstability').

It will be shown in Section 3.3.4 that, $\mathrm{Re}(\lambda_{max})$ is bounded above so that, in the original dimensional variables, $\mathrm{Re}(\lambda_{max})$ is at most of order $1/\mathcal{T}_u$, which is independent of η. It might be supposed that a limit of this order is necessarily attained as $Rm \to \infty$, but this is not so. As the dimensional $\mathbf{u}$ is increased, flux expulsion as described in Section 3.2.6 becomes more effective and regeneration is increasingly confined to current sheets of diminishing thickness. As a result, the dimensional $\mathrm{Re}(\lambda_{max})$, after reaching a maximum, may decrease. Three types of kinematic dynamo are usually distinguished: If the dimensionless $\mathrm{Re}(\lambda_{max})$ tends to a 'positive' limit as $\mathrm{Rm} \to \infty$, the dynamo is 'fast'. In this limit, the dimensional $\mathrm{Re}(\lambda_{max})$ is independent of η and of order $\mathcal{T}_u^{-1}$. If the dimensional $\mathrm{Re}(\lambda_{max})$ is of order $\mathcal{T}_\eta^{-1}$ (though positive), the dynamo is 'slow'; if it is positive and lies between $\mathcal{T}_\eta^{-1}$ and $\mathcal{T}_u^{-1}$, it is an 'intermediate dynamo'. Plausibly, the geodynamo is slow (Section 3.5.4).

If $\mathbf{u}(-\mathbf{x}) = -\mathbf{u}(\mathbf{x})$, the MHD equations admit solutions of two distinct parities: either $\mathbf{B}(-\mathbf{x}) \equiv \mathbf{B}(\mathbf{x})$ or $\mathbf{B}(-\mathbf{x}) \equiv -\mathbf{B}(\mathbf{x})$. But it is not generally true that, if $\mathbf{u}(\mathbf{x})$ regenerates field, so will $-\mathbf{u}(\mathbf{x})$; in general $\lambda(-\mathrm{Rm}) \neq \lambda(\mathrm{Rm})$. A striking example of this is the homopolar dynamo of Section 3.1.1, where reversing $\mathbf{u}$ destroys regeneration and actually hastens the demise of $\mathbf{B}$. The eigenvalue problem for λ is not self-adjoint. (The adjoint dynamo problem was first derived by Roberts (1960). It has the same eigenvalues as [47] but different eigenfunctions. See Gibson and Roberts (1966), Proctor (1977b), Kono and Roberts (1991), and Sarson and Gubbins (1996)).

When the dynamo is of DC type, the critical mode can be found by setting $\lambda = 0$ in [47a]. This transfers the eigenvalue from λ to Rm_c:

$$\mathbf{0} = \mathrm{Rm_c}\boldsymbol{\nabla} \times (\mathbf{u} \times \mathbf{B}) + \nabla^2\mathbf{B} \qquad [49]$$

3.3.4 Bounds on the Magnetic Reynolds Number

The discussion following [11b] indicates that dynamo action is possible only if Rm is big enough. Lower bounds on Rm have been derived by Backus (1958), Childress (1969b), and Roberts (1967). We derive only the first of these here, in dimensional units.

The Backus and Childress bounds follow from the evolution equation for the 'total' magnetic energy M defined in [43c]. Applying [13h] to V and $\hat{V}$ and adding, using [25f] and the dynamo condition [32], we have

$$\frac{\mathrm{d}M}{\mathrm{d}t} = -\Phi - \int_v \mathbf{u}\cdot(\mathbf{J}\times\mathbf{B})\,\mathrm{d}V \qquad [50]$$

The Backus bound requires that $\mathbf{u}=0$ on S. Then, by [15] and the divergence theorem,

$$\int_V \mathbf{u}\cdot(\mathbf{J}\times\mathbf{B})\,\mathrm{d}V = \int_V u_i\nabla_j\pi^B_{ij}\,\mathrm{d}V = \int_V \pi^B_{ij}\nabla_j u_i\,\mathrm{d}V = \frac{1}{\mu_0}\int_V d_{ij}B_iB_j\,\mathrm{d}V \qquad [51a]$$

where $d_{ij} = \frac{1}{2}\left(\nabla_i u_j + \nabla_j u_i - \boldsymbol{\nabla}\cdot\mathbf{u}\,\delta_{ij}\right)$ reduces to the rate of strain tensor when the fluid is incompressible. Evidently,

$$\int_V d_{ij}B_iB_j\,\mathrm{d}V \le \zeta_{\max}\int_V \mathbf{B}^2\mathrm{d}V \le 2\mu_0 M\zeta_{\max} \qquad [51b]$$

where $\zeta_{\max}$ is the largest absolute value taken by any eigenvalue of d_{ij} anywhere in V. It now follows from [50], [51a], and [51b] and the variational inequality [43b] that

$$\frac{\mathrm{d}M}{\mathrm{d}t} \le 2\left(\zeta_{\max} - \eta k^2_{\min}\right)M \qquad [51c]$$

In terms of the magnetic Reynolds number $Rm_{\mathrm{B}} = \zeta_{\max}/\eta k^2_{\min}$, [51c] shows that a necessary condition for dynamo action is $Rm_{\mathrm{B}}>1$. It also demonstrates that the maximum growth rate of $\mathbf{B}$ is $\zeta_{\max} - \eta k^2_{\min}$. The result [51c] was first established by Backus (1958) for the incompressible case [12] (see also Proctor (1977a)).

The Childress bound

$$\frac{\mathrm{d}M}{\mathrm{d}t} \le 2k_{\min}(u_{\max} - \eta k_{\min})M \qquad [52]$$

is also obtained from the energy equation without assuming that $\mathbf{u}\equiv\mathbf{0}$ on S. In terms of the magnetic Reynolds number $Rm_{\mathrm{C}} = u_{\max}/\eta k_{\min}$, [52] shows that a necessary condition for dynamo action is $Rm_{\mathrm{C}}>1$. It also demonstrates that the maximum growth rate of $\mathbf{B}$ is $k_{\min}(u_{\max} - \eta k_{\min})$. In any specific situation, the bound Rm_{C} may be lowered by rotating the reference frame to reduce $u_{\max}$ since, as noted in Section 3.2.3, the induction equation is invariant under rotation of frame.

It was recognized in Section 3.2.6 that, if $u_r\equiv 0$ in a sphere (i.e., if $\mathbf{u}_P\equiv\mathbf{0}$), $\mathbf{B}$ collapses as the field lines diffuse inexorably inwards with a velocity of order $\mathcal{U}_\eta = \eta/r_c$. This can be countered by a sufficiently large u_r and, the bigger the u_r, the larger the ratio $\mathcal{B}_{\mathrm{P}}/\mathcal{B}_{\mathrm{T}}$ for the $\mathbf{B}_{\mathrm{P}}$ created from $\mathbf{B}_{\mathrm{T}}$. It is therefore plausible that, to maintain the dynamo for the specified $\mathcal{B}_{\mathrm{P}}/\mathcal{B}_{\mathrm{T}}$, u_r must exceed $O([\mathcal{B}_{\mathrm{P}}/\mathcal{B}_{\mathrm{T}}]\mathcal{U}_\eta)$. This idea was made precise by Busse (1975), who showed that, if M_{P} and M_{T} are the magnetic energies of the poloidal and toroidal fields, a necessary condition for dynamo action is

$$M_{\mathrm{P}} \equiv \frac{1}{2\mu_0}\int_{V_\infty} B^2_{\mathrm{P}}\,\mathrm{d}V \le \frac{1}{2}Rm_r^2 M_{\mathrm{T}}, \qquad M_{\mathrm{T}} \equiv \frac{1}{2\mu_0}\int_V B^2_{\mathrm{T}}\,\mathrm{d}V \qquad [53a]$$

and $Rm_r = (|ru_r|)_{\max}/\eta$ (see also Roberts (1987)).

Proctor (2004) derived a similar but stronger result. He showed that a necessary condition for dynamo action is

$$Rm_{\mathrm{P}} \ge \left(\frac{1}{2}Rm_{\mathrm{T}} + \sqrt{2}\right)^{-1} \qquad [53b]$$

where $Rm_{\mathrm{P}} = u_{\mathrm{P}_{\max}}r_c/\eta$ and $Rm_{\mathrm{T}} = u_{\mathrm{T}_{\max}}r_c/\eta$ are the magnetic Reynolds numbers based on the maxima of the poloidal and toroidal flow speeds.

3.3.5 Antidynamo Theorems

The bounds of the last subsection provide precise necessary conditions for dynamos to exist, but these conditions are not sufficient. This became important in 1934 when Cowling announced an unexpected and negative result that was the first of so many other 'antidynamo theorems' (ADTs) that many believed it would only be a matter of time before a general ADT was proved:

- ***Cowling's theorem:*** *Axisymmetric magnetic fields cannot be maintained by a dynamo.*

By [9], a $\mathbf{u}$ that has ϕ-dependent components necessarily creates from an axisymmetric $\mathbf{B}$, a field with ϕ-dependent components. The theorem therefore pre-supposes that $\mathbf{u}(=\bar{\mathbf{u}})$ is axisymmetric. It should be particularly noticed that Cowling's theorem does not rule out dynamos driven by axisymmetric motions, but forbids such dynamos from maintaining a field that has an axisymmetric part (see Section 3.4.3).

In the steady state envisaged by Cowling (1934), in which [26a] and [26b] hold, the axisymmetry of $\bar{\Psi}$ means that $\bar{E}_\phi = 0$, so that

$$\eta\boldsymbol{\nabla}\times\bar{\mathbf{B}}_{\mathrm{M}} = \bar{\mathbf{u}}_{\mathrm{M}}\times\bar{\mathbf{B}}_{\mathrm{M}} \qquad [54]$$

Cowling reversed the dynamo problem: instead of seeking $\mathbf{B}$ for given $\mathbf{u}$, he sought $\mathbf{u}$ for given $\mathbf{B}$. He noticed that an axisymmetric field satisfying the dynamo condition possesses at least one singular latitude circle C (in V or on S) where $\bar{\mathbf{B}}_M = \mathbf{0}$. In general $\nabla \times \bar{\mathbf{B}}_M \neq \mathbf{0}$ on C (or if it vanishes then $\bar{\mathbf{B}}_M$ vanishes to higher order) so that [54] requires $\bar{\mathbf{u}}_M = \infty$; in the terminology of Section 3.2.3, the flow is ineligible. There is only one escape: $\bar{\mathbf{B}}_M \equiv \mathbf{0}$, i.e., $\bar{\mathbf{B}} = \bar{B}\mathbf{1}_\phi$, but this too can be ruled out by geometrical reasoning (Roberts, 1967).

Cowling's argument applied to both compressible and incompressible fluid motions but it had a lacuna (Roberts, 1967). Backus and Chandrasekhar (1956), and later Braginsky (1965a), provided more robust proofs that applied to incompressible, but not necessarily steady, fields and motions. Ivers and James (1984) gave a proof that is valid for both compressible and incompressible unsteady motions. These results generalize the theorem beyond what Cowling had in mind.

Braginsky's (1965a) proof, which applies to unsteady $\mathbf{B}$ and incompressible $\mathbf{u}$, is useful here. It starts by extracting from [9] the equations governing the functions $\bar{A}_\phi(s, z, t)$ and $\bar{B}_\phi(s, z, t)$ defining the general axisymmetric field [37]:

$$\partial_t \bar{A}_\phi + s^{-1}\bar{\mathbf{u}}_M \cdot \nabla(s\bar{A}_\phi) = \eta\Delta\bar{A}_\phi \qquad [55a]$$

$$\partial_t \bar{B}_\phi + s\bar{\mathbf{u}}_M \cdot \nabla(s^{-1}\bar{B}_\phi) = \eta\Delta\bar{B}_\phi + s\bar{\mathbf{B}}_M \cdot \nabla\zeta \qquad [55b]$$

where $\Delta = \nabla^2 - s^{-2}$ and ζ is the zonal shear $\bar{u}_\phi/s$. The last term in [55b] is a source that can create zonal field from meridional field by the ω-effect (Sections 3.2.5 and 3.2.6). There is however no corresponding source in [55a] to generate meridional field from the zonal field; the left-hand side of [55a] merely represents the advection of $\bar{\mathbf{B}}_M$ and not its creation. Braginsky's proof consists in showing that $\bar{A}_\phi \to 0$ as $t \to \infty$, where upon the last term in [55b] vanishes also, so that $\bar{B}_\phi$, having lost its source, disappears too. It is interesting to realize that the magnetic energy of an initially specified field may at first grow enormously through the ω-effect, and yet the entire field must eventually become arbitrarily small. It is true quite generally in dynamo theory, that a demonstration that a flow initially amplifies a field does not establish that it is a dynamo. This is a consequence of the non-self-adjointness of the dynamo problem (see Section 3.3.3).

• The Cartesian analog of Cowling's theorem is due to Cowling (1957):

Two-dimensional field theorem: *Two-dimensional magnetic fields cannot be maintained by a dynamo.*

Here 'two-dimensional' (2D) means that all three components are independent of one Cartesian coordinate; it does not mean that any particular component is identically zero. A 3D velocity $\mathbf{u}$ necessarily induces a 3D field from a 2D field. The theorem therefore presupposes that $\mathbf{u}$ is 2D. It should be particularly noticed that the theorem does not rule out dynamos driven by 2D motions, but forbids such dynamos from maintaining a field that has a 2D part (see Section 3.4.3).

• Perhaps the ADT next in significance to Cowling's theorem is the

Toroidal velocity theorem: *The toroidal motion of an incompressible fluid in a spherical conductor cannot maintain a magnetic field by dynamo action.*

Following the discussion in Section 3.2.6, this ADT is hardly a surprise! Its proof has points of similarity with Braginsky's proof of Cowling's theorem, with S and T taking over the roles of $\bar{A}_\phi$ and $\bar{B}_\phi$, respectively. It is more convenient to work with $\mathcal{Q} \equiv \mathbf{r} \cdot \mathbf{B} (= L^2 S)$ and T rather than S and T. The induction equation [9] implies that

$$\partial_t \mathcal{Q} + \mathbf{u} \cdot \nabla \mathcal{Q} = \eta\nabla^2 \mathcal{Q} + \mathbf{B} \cdot \nabla(\mathbf{r} \cdot \mathbf{u}) \qquad [56a]$$

The final term in [56a] is the only source that can maintain $\mathcal{Q}$, and it is absent if the motion is purely toroidal $(\mathbf{r} \cdot \mathbf{u} \equiv \mathbf{0})$. Then $\mathcal{Q}$, and therefore S and $\mathbf{B}_P$, disappear in a time of order $\mathcal{T}_\eta$. When neither poloidal field nor poloidal flow is present, the induction equation reduces to

$$\partial_t T + \mathbf{u} \cdot \nabla T = \eta\nabla^2 T \qquad [56b]$$

This resembles the heat conduction equation with T being temperature. There is no source of T within V and $T = 0$ on S. Therefore $\mathbf{B}_T$ disappears in a time of order $\mathcal{T}_\eta$.

The toroidal velocity ADT is due to Elsasser (1947) and Bullard and Gellman (1954) (see also Cowling (1957)). There is no analogous poloidal velocity ADT; Love and Gubbins (1996a) have constructed a dynamo maintained by a purely poloidal flow.

• The Cartesian analog of the toroidal velocity theorem (Zeldovich, 1956) is

Planar velocity theorem: *Motions in a plane layer that everywhere lack a component perpendicular to the boundaries cannot maintain a magnetic field by dynamo action.*

Bachtiar *et al.* (2006) recently pointed out that the shape of V is important: in a sphere, planar motions can maintain a dynamo.

• Kaiser *et al.* (1994) established a result that also follows from Proctor's theorem [53b]:

Toroidal field theorem: *A purely toroidal field cannot be maintained by a dynamo.*

This is also called the 'invisible dynamo theorem' since it shows that a dynamo necessarily signals its existence magnetically outside the conductor. (A dynamo generally produces an external electric field since charges on S must be present to prevent the currents from leaving V.)

For other ADTs, see Hide (1979), Hide and Palmer (1982), Ivers and James (1981, 1986, 1988a,b), Namikawa and Matsushita (1970), and Zeldovich and Ruzmaikin (1980).

3.4 Laminar Dynamos

3.4.1 Early Successful Models

The difficulty faced by early dynamo theorists is apparent from the last subsection: It was necessary either to prove a general ADT or to construct an explicit working model. As the computers then available were incapable of deriving convincing numerical solutions of the partial differential equations posed by the induction equation [9], the explicit model could emerge only from mathematical analysis, perhaps supplemented by sufficiently undemanding numerical work. At first there seemed to be little hope of finding simple, exact solutions of [9] satisfying the dynamo condition, and thoughts naturally turned to approximate procedures, but a convincingly converged result required expansion in a small parameter. The bounds of Section 3.3.4 show that Rm is certainly not that parameter! Eventually, two successful models were produced (Backus, 1958; Herzenberg, 1958). These provided the first, and much needed, mathematical demonstrations that homogeneous dynamos exist, so establishing that a search for a general ADT would be fruitless. The Herzenberg flow **u** is steady. It inspired the first homogeneous dynamo experiments (Lowes and Wilkinson, 1963, 1968) and further theoretical studies (see for example Gibson (1968) and, most recently, Brandenburg *et al.* (1998)).

A step closer to geophysical reality was taken by Braginsky (1965a), who based an asymptotic model on a simple idea: if the magnetic Reynolds number $Rm = \bar{u}_\phi r_c/\eta$ based on the zonal core motion $\bar{u}_\phi$ is large, even a small deviation $\mathbf{u}'$ from axisymmetry may suffice to produce (from the $\bar{B}_\phi$ created by the ω-effect) the $\bar{\mathbf{B}}_M$ necessary to defeat Cowling's theorem. To achieve this, **u** and **B** are expanded in powers of $Rm^{-1/2}$, the leading terms being $\bar{u}_\phi$ and $\bar{B}_\phi$. Then $\mathbf{u}'$ of order $Rm^{-1/2}\bar{u}_\phi$ induces from $\bar{B}_\phi$ an asymmetric $\mathbf{B}'$ of order $Rm^{-1/2}\bar{B}_\phi$ that in turn generates an emf $\bar{\mathcal{E}}_\phi = \overline{(\mathbf{u}' \times \mathbf{B}')}_\phi$ of order $Rm^{-1}\bar{B}_\phi$. This is a source that must be added to the right-hand side of [55a]. It may maintain a meridional field $\bar{\mathbf{B}}_M$ of order $Rm^{-1}\bar{B}_\phi$, which is precisely the strength necessary to create $\bar{B}_\phi$ by the ω-effect. This regenerative loop, $\bar{B}_\phi \leftrightarrow \bar{\mathbf{B}}_M$, parallels the one that maintains a turbulent dynamo in Section 3.5.2 below. To leading order,

$$\bar{\mathcal{E}}_\phi = \alpha \bar{B}_\phi \mathbf{1}_\phi \qquad [57]$$

where α is $O(Rm^{-1})$ and can be calculated explicitly from $\mathbf{u}'$ and $\bar{u}_\phi$, though the recipe is rather complicated. Soward (1972) showed how α is linked to the helicity of **u**. After the source [57] has been inserted into the right-hand side of [55a], Cowling's theorem is no longer a threat; self-excited $\bar{\mathbf{B}}$ may exist and may be found by solving [57a] and [57b] for $\bar{\mathbf{B}}$ numerically, a task that is far less challenging than integrating the 3D induction equation [9] for **B**. Braginsky (1964a) successfully solved three such axisymmetric models. One of these even contains a dynamical element: the angular velocity Ω_i of the inner core was determined from the magnetic torque $\Gamma_{z,i}$ by solving the equation of motion for the SIC (see [35b]).

Two features of Braginsky's asymptotic analysis are of special geophysical interest. First, since $\mathbf{B}'/\bar{\mathbf{B}}_M = O(Rm^{1/2}) \gg 1$, it might seem at first sight that his expansion is geophysically irrelevant. Because $Rm \gg 1$ however, the largest terms in his expansion of $\mathbf{B}'$ are frozen to the fluid and do not escape from the core. In fact, $\widehat{\mathbf{B}}' = O(Rm^{-1}\mathbf{B}') = O(Rm^{-1/2}\widehat{\mathbf{B}}_M)$, consistent with the small tilt of the centered dipole. Second, as is clear from his generalization (Braginsky, 1965b), $\mathbf{u}'$ and $\mathbf{B}'$ should be attributed mainly to large-scale, longitudinally moving planetary waves that ride on the axisymmetric state $\bar{u}_\phi$ and $\bar{B}_\phi$, creating as they do so the α that sustains $\bar{\mathbf{B}}_M$. This attractive idea gives a new significance to the westward drift of the observed geomagnetic field.

3.4.2 One-Dimensional Models

The existence of simple solutions of [9] satisfying the dynamo condition was overlooked until 1973 when Ponomarenko devised a 1D model, that is, one in which **u** depends solely on one coordinate s and is independent of ϕ, z (and t), so that [36] simplifies to

$$\mathbf{B} = \mathbf{B}_1(s)\exp(\imath m\phi + \imath kz + \lambda t) \qquad [58a]$$

Cowling's theorem demands that $m \neq 0$ and the 2D field ADT that $k \neq 0$.

In the model of Ponomarenko (1973), the conductor fills all space but is stationary outside a cylinder of radius a. The dynamo condition requires that $\mathbf{B}_1(s)$ is exponentially small for $s \rightarrow \infty$. Within C, **u** is a 'solid-body' motion, that is, a motion that C can execute even if solid. The velocity of C is helical, a combination of a uniform velocity U along the axis Oz and a rotation about Oz with uniform angular velocity ω:

$$\mathbf{u} = \begin{cases} s\omega \mathbf{1}_\phi + U\mathbf{1}_z, & \text{if } s < a \\ \mathbf{0}, & \text{if } s > a \end{cases} \qquad [58b]$$

see **Figure 8**. If $U = 0$, the motion has no component in the z-direction and the dynamo fails by the planar velocity ADT. Clearly, there is an infinity of possible models, distinguished by the value of $p = a\Omega/U$ where $-\infty < p < \infty$. For $p \neq 0$, the motion [58b] is non-mirror-symmetric, the positive helicity of a model with $p > 0$ being mirrored by the negative helicity of the model of opposite p.

The magnetic Reynolds number is defined as $Rm = u_{\max}a/\eta$, where $u_{\max} = \sqrt{U^2 + \omega^2 a^2}$. The eigenvalue problem for $\lambda(m, k)$ only requires that ordinary differential equations for $\mathbf{B}_1(s)$ be solved, and that can be done analytically, leading to a dispersion relationship for λ that can be solved even on a pocket calculator! As described in Section 3.3.3, the smallest value of Rm for which $\mathrm{Re}(\lambda) = 0$ is minimized over m and k to give Rm_c and, since $\mathrm{Im}(\lambda_c) \neq 0$, the dynamo is of AC type. The minimum of Rm_c over $|p|$ gives the most efficient generator, and $Rm_c \approx 17.7$ for this model.

The Ponomarenko dynamo spawned a number of similar 'screw dynamos': Léorat (1995), Lupian and Shukurov (1992), Lupian *et al.* (1996), Ruzmaikin *et al.* (1988, 1989), Sokoloff *et al.* (1989), Solovyev (1985a, 1985b, 1987). Because of the infinite shear on $s = a$, the Ponomarenko dynamo is fast (Gilbert, 1988). Screw dynamos, in which the infinite shear is smoothed out in a narrow layer, are also efficient generators but, in the terminology of Section 3.3.3, they are 'intermediate'. A Ponomarenko-type dynamo of finite z-length (Gailitis, 1990) developed into a homogeneous dynamo successfully demonstrated in laboratory experiments; *see* Chapter 11.

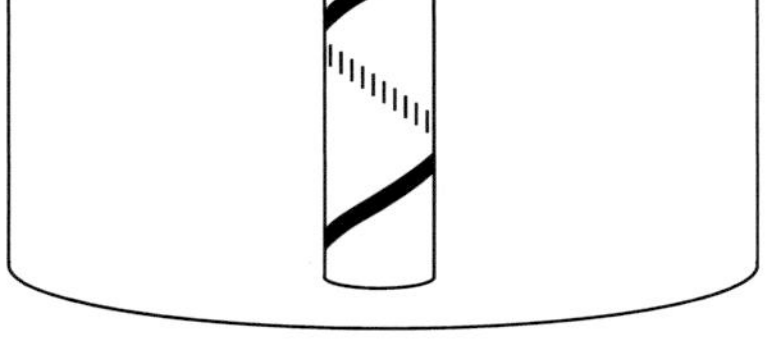

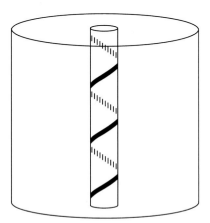

Figure 8 The Ponomarenko dynamo. The outer cylinder represents a stationary conductor that fills all space, apart from the central cylinder that moves in the spiral indicated.

3.4.3 Two-Dimensional Models

As noted in Section 3.3.5, the 2D field ADT does not exclude models driven by 2D motions, that is, flows independent of one Cartesian coordinate, z. For these [46] simplifies to

$$\mathbf{B} = \mathbf{B}_2(x, y)\exp(\imath kz + \lambda t) \qquad [59]$$

The 2D field ADT merely requires that $k \neq 0$.

The earliest 2D model was the spatially periodic dynamo of G. O. Roberts (1972), in which

$$\mathbf{u} = \mathbf{1}_x \sin y + \mathbf{1}_y \sin x + \mathbf{1}_z(\cos x - \cos y) \qquad [60a]$$

The projection onto an xy-plane of this flow is sketched in **Figure 9(a)**. There is, in addition to the xy-components of motion shown, a z-component into (out of) the paper for each clockwise-turning (counterclockwise-turning) 'cell'. Although the model predates Ponomarenko's, each cell resembles a Ponomarenko dynamo, adjacent cells moving in opposite z-directions but having the same positive helicity H.

From the real part of [59], the xy-average of **B** at $t = 0$ is

$$\bar{\mathbf{B}} = B_0(\mathbf{1}_x \cos kz + \mathbf{1}_y \sin kz), \quad \text{where } k > 0 \qquad [60b]$$

It is exponentially stretched where it crosses the dividing streamlines at the stagnation points shown in **Figure 9(a)**. This makes the motion [60a] an efficient generator (Section 3.2.6). Strictly it is an intermediate dynamo although it would be fast, but for a logarithmic factor (Soward, 1987). It is the basis of successful laboratory experiments demonstrating homogeneous dynamo action (*see* Chapter 11).

The way in which the flow [60a] generates $\bar{\mathbf{B}}$ is sketched in **Figure 9(b)**, where three constant-z 'decks' are shown spanning one-quarter of a z-wavelength. On the 'lower' deck, the flow has an 'upward' component in the shaded patches and the mean field (in the direction indicated) is 'lifted upwards' and twisted by the 'vertical' vorticity so

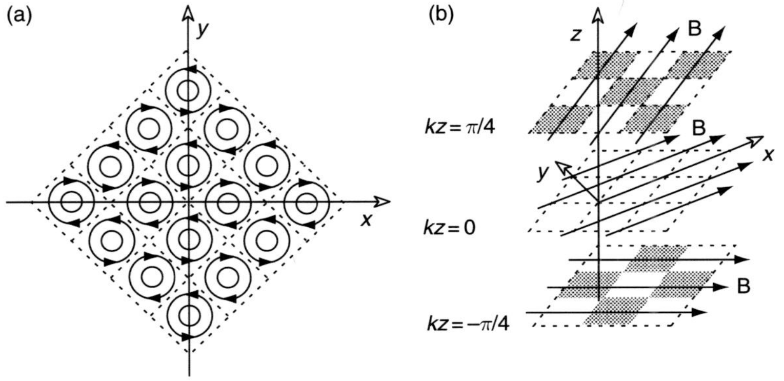

Figure 9 The G.O. Roberts dynamo. (a) The *xy*-projection of the periodic pattern of motion that fills all space. The flow also has a component out of (into) the plane of the paper in the counter-clockwise-moving (clockwise-moving) cells; (b) the effect of the motion on the mean field (see text).

that it is roughly in the direction Ox of the mean field on the center deck. On the upper deck ($z = \pi/4k$), the flow has a 'downward' component in the unshaded patches and the mean field (in the direction indicated) is 'depressed' to the center deck and twisted by the vorticity so that it too is roughly in the x-direction of $\bar{\mathbf{B}}$. The same process reinforces $\bar{\mathbf{B}}$ on every z-section of the flow. The mean field [60b] is helical, in the sense that $\nabla \times \bar{\mathbf{B}} = -k\bar{\mathbf{B}}$; the current helicity $H_J = \bar{\mathbf{B}} \cdot \nabla \times \bar{\mathbf{B}}$ is negative, as is the magnetic helicity $H_B = \bar{\mathbf{A}} \cdot \nabla \times \bar{\mathbf{A}} = \bar{\mathbf{A}} \cdot \bar{\mathbf{B}}$.

In addition to [60a], Roberts (1972) investigated three other dynamos driven by spatially periodic flows (see also Roberts (1969) and Childress (1969a, 1970)). Since then there have been several other studies of such systems, most recently by Tilgner and Busse (1995), Zheligovsky and Galloway (1998), and Demircan and Seehafer (2002).

The term '2D' can be generalized to cover motions 'independent of one coordinate', either z as in the Cartesian 2D models described above, or ϕ as in the spherical 2D dynamos described next. As noted in Section 3.3.5, Cowling's ADT does not exclude models driven by axisymmetric motions $\bar{\mathbf{u}}$. For these [46] simplifies to

$$\mathbf{B} = \mathbf{B}'(r, \theta)\exp(\imath m\phi + \lambda t) \qquad [61]$$

Cowling's ADT merely requires that $m \neq 0$. When $\mathrm{Im}(\lambda_c) \neq 0$, the field [61] moves longitudinally as a 'dynamo wave', regenerating itself as it goes (see also Section 3.5.2 below).

Although spherical 2D models were first proposed more than four decades ago, perhaps the first convincing model was that of Roberts (1971b), which was solved numerically. It is driven by an adaptation of a Cartesian periodic motion that forces it into a sphere (see also Childress (1970)). A simpler flow was employed by Gubbins (1973). Three models devised by Dudley and James (1989) deserve special notice. They explore symmetries with simple $\bar{\mathbf{u}}$ and have proved useful in interpreting laboratory experiments (*see* Chapter 11).

Gailitis (1993, 1995) investigated several spherical 2D models, in which the conductor fills all space but is stationary outside a sphere. Although electric currents now leak into $\hat{V}$, the dynamo condition still imposes [32]. Gailitis made an interesting comparison between two similar flows, one having helicity and one without it. Both regenerated field but Rm_c was much smaller for the helical model. He concluded that 'helicity is favorable but not indispensible' for dynamo action.

3.4.4 Three-Dimensional Models

When $\mathbf{u}$ depends on all three coordinates, separable solutions of the form [58a], [59], or [61] do not exist, so that [46] is the only simplification in solving [9]. We classify these '3D models' as cylindrical, Cartesian, or spherical.

The cylindrical 3D class consists of the models of Lortz (1968). As for the Ponomarenko model, these operate in a conductor filling all space. They capitalize on the effectiveness of helicity in field generation, and they can be solved analytically. They will not be described here. Their theory has been developed most recently by Eltayeb and Loper (1988), Lortz (1990), and Soward (1990).

The Cartesian 3D class is exemplified by Beltrami motions, defined as those in which the flow $\mathbf{u}$ and the vorticity $\boldsymbol{\omega} = \nabla \times \mathbf{u}$ are everywhere parallel, so that the helicity is maximal (i.e., $|H| = |\mathbf{u}|\,|\boldsymbol{\omega}|$; see Section 3.2.6). The G. O. Roberts dynamo [60a] is a 2D Beltrami flow. A famous 3D example is the Arnold–Beltrami–Childress or ABC flow, defined by

$$\mathbf{u} = \mathbf{1}_x(C\sin z + B\cos y) + \mathbf{1}_y(A\sin x + C\cos z) + \mathbf{1}_z(B\sin y + A\cos x) \quad [62]$$

Depending on the choice of the constants A, B, and C, this flow possesses chaotic streamlines (see Section 3.2.6) and may be a fast dynamo (see Galloway and Frisch (1986)).

The earliest example of the 3D spherical class was that of Herzenberg (1958). The first successful computational models were those of Pekeris *et al.* (1973) and Kumar and Roberts (1975). The latter aimed at comparing 3D models with corresponding 2D models derived from Braginsky's asymptotic analysis (Section 3.4.1). Four independent parameters define this model, one for the zonal flow, one for the meridional flow, and two for the asymmetric (poloidal) flow. This 4D parameter space has been thoroughly investigated in papers by Gubbins *et al.* (2000a, 2000b), Gubbins and Gibbons (2002), Holme (1997), Love and Gubbins (1996a, 1996b), Sarson and Gubbins (1996). They found few regenerating solutions, prompting Gubbins *et al.* (2000a) to opine that "Dynamo action might, therefore, be quite unusual, at least for a large-scale flow." The particular case in which the zonal flow parameter is zero, led to the successful poloidal field model referred to in Section 3.3.5 (see Love and Gubbins (1996a)).

Two further issues about spherical kinematic models should be mentioned. The first concerns the paradoxical result of Bullard and Gubbins (1977): it appeared that, contrary to intuition, Rm_c can be reduced by bringing to rest the outermost layer of fluid (see also Liao *et al.* (2005), Sarson and Gubbins (1996), Serebrianya (1988)). This enhancement of dynamo action is welcome news for those who believe on other grounds that the uppermost layers of the FOC are stably stratified (see, e.g., Braginsky (1984) and Moffatt and Loper (1994)). The reduction of Rm_c is attributed by Hutcheson and Gubbins (1994) to the diffusion (into the stagnant layer) of toroidal field that would otherwise concentrate more intensely near the boundary and produce a larger ohmic loss. Some laboratory dynamos also make use of stagnant surroundings to make field regeneration easier; *see* Chapter 11.

The second issue concerns a case in which even [46] does not apply: dynamos driven by time-dependent, periodic $\mathbf{u}$, such as those produced by precessional forcing. Then $\mathbf{B}$ oscillates within an envelope that grows or diminishes in amplitude, corresponding respectively to regeneration or nonregeneration; Floquet theory can decide which. (See Willis and Gubbins (2004), who study a model in which $\mathbf{u}$ has the same spatial form as the Kumar–Roberts dynamo but oscillates sinusoidally in time.) The success of the Backus (1958) dynamo depended on the choice of a time-dependent $\mathbf{u}$ that is periodic but far from sinusoidal.

3.5 Turbulent Dynamos

3.5.1 Induction by Cyclonic Turbulence

The importance of helicity in promoting dynamo action was first realized by Parker (1955) in the context of solar magnetism. He argued that Coriolis forces make solar turbulence 'cyclonic'. He visualized that, through the process idealized in Section 3.2.6, each turbulent eddy generates a flux loop in the plane perpendicular to the inducing field (represented by the flux rope in **Figure 5**). He pointed out that a large-scale mean field $\mathbf{b}$ is then produced in that plane through the diffusive merging of each loop with those created by neighboring cyclonic eddies. Similarly in spherical geometry, a mean zonal field $\bar{B}_\phi$ induces a mean meridional field $\bar{\mathbf{B}}_\mathrm{M}$, leaving the ω-effect to complete the regenerative cycle $\bar{B}_\phi \leftrightarrow \bar{\mathbf{B}}_\mathrm{M}$ and so defeat Cowling's theorem.

In cyclonic turbulence, the helicity is positive, and each induced flux loop encircles its parent flux rope in the left-handed sense so that the associated electric current $\mathbf{j}$ is antiparallel to the flux rope, and the sum $\mathbf{B}(=\mathbf{B}_0 + \mathbf{b})$ of the induced and inducing fields has negative current helicity $H_\mathrm{J} = \mu_0\mathbf{J}\cdot\mathbf{B}$. Positive H_J is produced by anticyclonic turbulence. This heuristic description of Parker's idea relies on Alfvén's theorem, and this is appropriate if the magnetic Reynolds number Rm' of the turbulence is large. To show more formally that helicity acts in a similar way for all Rm', we represent turbulent fields traditionally as $\mathbf{u} = \bar{\mathbf{u}} + \mathbf{u}'$, where $\bar{\mathbf{u}}$ is the turbulent (ensemble) average of $\mathbf{u}$ and $\mathbf{u}'$ is the fluctuating remnant. This notation will apply in this subsection even though it conflicts with that of Sections 3.3 and 3.4, where $\bar{\mathbf{u}}$ and $\mathbf{u}'$ were the axisymmetric and asymmetric parts of $\mathbf{u}$. No confusion should arise; in fact, some parallels with Section 3.4 may emerge more clearly. In the following qualitative arguments, $\bar{\mathcal{L}} \sim r_\mathrm{c}$, $\bar{\mathcal{T}}_u = \bar{\mathcal{L}}/\bar{\mathcal{U}}$, $\bar{\mathcal{T}}_\eta = \bar{\mathcal{L}}^2/\eta$, and $\bar{\mathcal{U}}$ are characteristic of $\bar{\mathbf{u}}$, while $\mathcal{L}'$, $\mathcal{T}'_u(=\mathcal{L}'/\mathcal{U}')$, $\mathcal{T}'_\eta = \mathcal{L}'^2/\eta$, and $\mathcal{U}' = \sqrt{\mathbf{u}'^2}$ are characteristic of $\mathbf{u}'$; the corresponding magnetic Reynolds numbers are $\overline{Rm} = \bar{\mathcal{U}}\bar{\mathcal{L}}/\eta$ and $Rm' = \mathcal{U}'\mathcal{L}'/\eta$.

Substitution of $\mathbf{u} = \bar{\mathbf{u}} + \mathbf{u}'$ and $\mathbf{B} = \bar{\mathbf{B}} + \mathbf{B}'$ into [6e] and averaging leads to

$$\eta\nabla \times \bar{\mathbf{B}} = \bar{\mathbf{E}} + \bar{\mathbf{u}} \times \bar{\mathbf{B}} + \bar{\mathcal{E}} \qquad [63a]$$

where

$$\mathcal{E} = \mathbf{u}' \times \mathbf{B}' \qquad [63b]$$

Attention focuses on the new term $\bar{\mathcal{E}}$ in the Ohms law [63a] for the mean field $\bar{\mathbf{B}}$. Out of the recognition of the existence and importance of this term, a new subject was born, 'mean field electrodynamics' or 'MFE'. To evaluate $\bar{\mathcal{E}}$, it is necessary to find, or at least evaluate approximately, the solution of the equation governing $\mathbf{B}'$. This equation is derived from [6e] and [63a] by subtraction:

$$\eta\nabla \times \mathbf{B}' = \mathbf{E}' + \mathbf{u}' \times \bar{\mathbf{B}} + \bar{\mathbf{u}} \times \mathbf{B}' + \mathcal{E}' \qquad [63c]$$

(The possibility that a dynamo operates with $\mathbf{B}'$ alone, that is, from solutions of [63c] with $\mathbf{B} \equiv \mathbf{0}$, is irrelevant to MFE and is not considered here.)

Consider induction by homogeneous turbulence, that is, turbulence whose statistical properties are the same for all points $\mathbf{x}$ so that, for example, the correlation between the velocity at two different points depends only on t and their separation $\mathbf{x}$, that is,

$$\mathcal{Q}_{ij}(\mathbf{x}, t) = \overline{u_i(\mathbf{x}_1, t)u_j(\mathbf{x}_1 + \mathbf{x}, t)} = \mathcal{Q}_{ji}(-\mathbf{x}, t) \quad [64]$$

is independent of $\mathbf{x}_1$; similarly, $\bar{\mathbf{u}}$ is constant, assumed zero by choice of reference frame. Suppose that $Rm' \ll 1$. Two simplifications are immediate. First, the emf $\mathbf{u}' \times \bar{\mathbf{B}}$ produces a field at most of order $Rm'\bar{\mathbf{B}}$ and this in turn creates a contribution to $\mathcal{E}'$ of order $Rm'\mathbf{u}' \times \bar{\mathbf{B}}$ which is negligible compared with $\mathbf{u}' \times \bar{\mathbf{B}}$ in [63c]. We may therefore set $\mathcal{E}' = \mathbf{0}$, a step usually called 'first-order smoothing' and here justified. Second, we recall from Section 3.2.6 that, when $Rm' \ll 1$, $\nabla \times \mathbf{E}' = \mathbf{0}$ to leading order. By [10b], [12], and [63c], we therefore have

$$\eta\nabla^2\mathbf{B}' = \mathbf{u}' \cdot \nabla\bar{\mathbf{B}} - \bar{\mathbf{B}} \cdot \nabla\mathbf{u}' \qquad [65]$$

The sources of $\mathbf{B}'$ on the right-hand side of [65] make two contributions to $\bar{\mathcal{E}}$:

- The first source creates $\mathbf{B}'$ of order $\mathcal{L}'Rm'\nabla\bar{\mathbf{B}}$. This generates an $O(\eta Rm'^2\nabla\bar{\mathbf{B}})$ part of $\bar{\mathcal{E}}$ that may be written as $\bar{\mathcal{E}}_i^{(s)} = -\beta_{ijk}\nabla_j\bar{B}_k$, where β_{ijk} depends only on the symmetric part $\mathcal{Q}_{ij}^{(s)} = \frac{1}{2}(\mathcal{Q}_{ij} + \mathcal{Q}_{ji})$ of $\mathcal{Q}_{ij}$. If the turbulence is isotropic, that is, has statistical properties independent of direction, $\beta_{ijk} = \beta\epsilon_{ijk}$ with $\beta > 0$. Then $\bar{\mathcal{E}} = -\beta\nabla \times \bar{\mathbf{B}}$ and, if this term is transferred to the left-hand side of [62a], it is immediately seen that its effect on $\bar{\mathbf{B}}$ is purely diffusive, just as though the molecular diffusivity η were enhanced by a turbulent diffusivity β of order $\mathcal{U}'^2\mathcal{T}'_\eta$.
- The second source in [65] creates $\mathbf{B}'$ of order $Rm'\bar{\mathbf{B}}$. This generates an $O(\mathcal{U}'Rm'\bar{\mathbf{B}})$ part of $\bar{\mathcal{E}}$ that may be written as $\bar{\mathcal{E}}_i^{(a)} = \alpha_{ij}\bar{B}_k$ where α_{ij} depends only on the antisymmetric part $\mathcal{Q}_{ij}^{(a)} = \frac{1}{2}(\mathcal{Q}_{ij} - \mathcal{Q}_{ji})$ of $\mathcal{Q}_{ij}$. The classic theory of turbulence supposes mirror symmetry, which means that its statistical properties are unchanged by the coordinate reflection $\mathbf{x} \to -\mathbf{x}$, so that $\mathcal{Q}_{ij}^{(a)} = 0$, by [64]. Cyclonic turbulence lacks mirror symmetry and $\mathcal{Q}_{ij}^{(a)} \neq 0$. The simplest case is 'pseudoisotropic turbulence', which is statistically independent of direction, but possesses mean helicity $\bar{h} = \overline{\mathbf{u}' \cdot \boldsymbol{\omega}'}$. Then $\alpha_{ij} = \alpha\delta_{ij}$ and $\bar{\mathcal{E}} = \alpha\bar{\mathbf{B}}$, where $\alpha = O(\bar{h}\mathcal{T}'_\eta)$. In agreement with the earlier heuristic arguments, α and $\bar{h}$ usually have opposite signs.

Summarizing the discussion so far, we have found that

$$\bar{\mathcal{E}}_i = \alpha_{ij}\bar{B}_j - \beta_{ijk}\nabla_j\bar{B}_k \qquad [66a]$$

or in the isotropic case

$$\bar{\mathcal{E}} = \alpha\bar{\mathbf{B}} - \beta\nabla \times \bar{\mathbf{B}} \qquad [66b]$$

The creation of the large-scale emf $\alpha\bar{\mathbf{B}}$ parallel to $\bar{\mathbf{B}}$ is called the α-effect, for no better reason than the choice made by Steenbeck *et al.* (1966) for the constant of proportionality. The α-effect is a classic example of 'inverse cascade' or what is sometimes called an 'up-the-spectrum process', where the small scales are not merely an energy sink for the large scales.

The case $Rm' \ll 1$ is the easiest to analyze but the least interesting since $\bar{\mathcal{E}}$ has little effect on $\bar{\mathbf{B}}$; for example, the turbulent diffusivity β is of order $\mathcal{U}'^2 T'_\eta = Rm'^2_{\mathrm{m}}\eta \ll \eta$. A more interesting and more challenging case is $Rm' \gg 1$. It is more interesting because $\bar{\mathcal{E}}$ in [63a] now plays a very significant role in the electrodynamics. It is more challenging because there is no simple route to [66]; indeed, [66] is an oversimplification and first-order smoothing is unjustified. On the assumption that [66] is nevertheless an acceptable approximation, it is often argued that, because of fast reconnection (Section 3.2.3), α and β are independent of η. Replacing therefore $\mathcal{T}'_\eta$ by $\mathcal{T}'_u$ in the small Rm'

estimates for α and β, we obtain $\alpha = O(\bar{h}\mathcal{T}'_u)$ and $\beta = O(\mathcal{U}'^2\mathcal{T}'_u) = O(\mathcal{U}'\mathcal{L}') \gg \eta$. The diffusive timescale of $\bar{\mathbf{B}}$ is $\bar{\mathcal{L}}^2/\beta$ rather than $\bar{\mathcal{L}}^2/\eta$. In the case of the Sun, $\bar{\mathcal{L}}^2/\eta$ may be billions of years but the turbulence is so violent that $\bar{\mathcal{L}}^2/\beta$ may be the decadal timescale of magnetic activity, the time during which the Sun cleanses itself of old flux and makes new reversed flux. The dynamo efficiency, E, defined by [44a], is plausibly reduced by a factor of order η/β by the ohmic losses of the turbulence. In order of magnitude, the ratio $\alpha\bar{\mathbf{B}}/\beta\nabla \times \bar{\mathbf{B}}$ of the two sources on the right-hand sides of [66] is $\alpha\bar{\mathcal{L}}/\beta = O(\bar{h}\bar{\mathcal{L}}/\mathcal{U}'^2)$. Because $\mathcal{L}' \ll \bar{\mathcal{L}}$, the first dominates the second if $\bar{h}$ takes its maximum possible value of order $\mathcal{U}'^2/\mathcal{L}'$. Even a helicity much smaller than this maximum can create a significant α-effect when $Rm' \gg 1$.

The form of [66b] of $\bar{\boldsymbol{\mathcal{E}}}$ relies on the (pseudo) isotropy of the turbulence and ignores significant questions, such as how $Q_{ij}^{(a)}$ can be created and maintained. Although the α-effect and helicity arise naturally in rotating convective turbulence, Coriolis and buoyancy forces, together with the Lorentz forces from the $\bar{\mathbf{B}}$ created by the dynamo, make the turbulence nonisotropic, possibly strongly so. Generally [66a] is more realistic than [66b], but modelers have often preferred the simpler choice because it contains the essence, while involving fewer *ad hoc* parameters. (But see, e.g., Krause and Rädler (1980), Roberts (1972), Tilgner (2004)).

This discussion of MFE merely touches the fringes of what has become a massive subject. Further details can be found in Krause and Rädler (1980), Moffatt (1978), and Rüdiger and Hollerbach (2004).

3.5.2 Mean Field Dynamos

The impact of MFE and especially the α-effect has been enormous, particularly in astrophysics. In exploring a tiny fraction of what has been done, we shall revert to the notation of Section 3.4 in which $\bar{\mathbf{u}}$ and $\mathbf{u}'$ are the axisymmetric and asymmetric parts of $\mathbf{u}$, but where now $\mathbf{u}(=\bar{\mathbf{u}}+\mathbf{u}')$ is the statistical mean of the velocity; similarly for $\mathbf{B}$. We adopt the simpler [66b] of the two expressions for the emf. This, combined with [63a] and the ensemble averages of [1b] and [1c], gives the simplest form of the 'MF induction equation':

$$\partial_t\mathbf{B} = \nabla \times (\alpha\mathbf{B} + \mathbf{u} \times \mathbf{B}) + \bar{\eta}\nabla^2\mathbf{B} \qquad [67]$$

where $\bar{\eta} = \eta + \beta$. Self-sustaining solutions satisfying the dynamo conditions are called 'mean field dynamos' or 'MF dynamos'. These have been proffered as models of planetary, stellar, and galactic magnetism.

The additional source $\alpha\mathbf{B}$ in [67] means that Cowling's theorem (Section 3.3.5) does not prevent $\mathbf{B}$ from having an axisymmetric part $\bar{\mathbf{B}}$. We concentrate mainly on this part, using the representation [37] and obtaining, in place of [55],

$$\partial_t\bar{A}_\phi + s^{-1}\bar{\mathbf{u}}_{\mathrm{M}} \cdot \nabla(s\bar{A}_\phi) = \bar{\eta}\Delta\bar{A}_\phi + \alpha\bar{B}_\phi \qquad [68a]$$

$$\partial_t\bar{B}_\phi + s\bar{\mathbf{u}}_{\mathrm{M}} \cdot \nabla(s^{-1}\bar{B}_\phi) = \bar{\eta}\Delta\bar{B}_\phi + s\bar{\mathbf{B}}_{\mathrm{M}} \cdot \nabla\zeta - \alpha\Delta\bar{A}_\phi \qquad [68b]$$

As already noted in Section 3.2.6, dynamical considerations favor the equatorial symmetry

$$\alpha(-z) = -\alpha(z), \quad \bar{u}_s(-z) = \bar{u}_s(z)$$
$$\bar{u}_\phi(-z) = \bar{u}_\phi(z), \quad \bar{u}_z(-z) = -\bar{u}_z(z) \qquad [69a]$$

applying for each s. In this case there are two distinct types of solutions:

$$\text{dipole family:} \quad \bar{B}_s(-z) = -\bar{B}_s(z),$$
$$\bar{B}_\phi(-z) = -\bar{B}_\phi(z), \quad \bar{B}_z(-z) = \bar{B}_z(z) \qquad [69b]$$

$$\text{quadrupole family:} \quad \bar{B}_s(-z) = \bar{B}_s(z),$$
$$\bar{B}_\phi(-z) = \bar{B}_\phi(z), \quad \bar{B}_z(-z) = -\bar{B}_z(z) \qquad [69c]$$

These imply $\bar{A}_\phi(-z) = \bar{A}_\phi(z)$ for the dipolar family and $\bar{A}_\phi(-z) = -\bar{A}_\phi(z)$ for the quadrupolar family. When [69a] does not hold, the solutions must generally involve both symmetry types.

The concept of equatorial symmetry applies to general $\mathbf{B}$ after the overbars have been removed from [69] and the MHD equations again allow solutions of a single parity provided $\mathbf{u}(-\mathbf{x}) = -\mathbf{u}(\mathbf{x})$. To avoid confusion, however, the words 'dipolar' and 'quadrupolar' are usually then replaced by 'antisymmetric' and 'symmetric', respectively, since the equatorial dipole actually belongs to the symmetric family. There has been considerable interest in the equatorial symmetry of the paleofield (see Merrill *et al.* (1996)).

Both the final term in [68a] and at least one of the last two terms in [68b] are needed to maintain $\bar{\mathbf{B}}$. We first suppose that $\bar{\mathbf{u}}_{\mathrm{M}}$ is negligible or zero. There are three possibilities:

- If $\alpha \gg \bar{u}_\phi$, the α-effect provides the dominating source in [68b] and creates from $\bar{A}_\phi$ a zonal field of strength $\bar{B}_\phi = O(Rm_\alpha\bar{A}_\alpha/\mathcal{L})$, where $Rm_\alpha = \alpha\mathcal{L}/\bar{\eta}$ is the α-effect Reynolds number; the α-source in [68a] produces $\bar{A}_\phi = O(Rm_\alpha\bar{B}_\phi\mathcal{L})$ from $\bar{B}_\phi$. Clearly regeneration requires that Rm_α exceeds some critical value $Rm_{\alpha c}$. Such a dynamo, that operates through the 'product' of two α-effects, is called an α^2-dynamo.

- If $\bar{u}_\phi \gg \alpha$, the ω-effect provides the dominating source in [68b] and creates from $\bar{A}_\phi$ a zonal field of strength $\bar{B}_\phi = O(Rm_{\boldsymbol{\omega}} \bar{A}_\phi \mathcal{L})$, where $Rm_{\boldsymbol{\omega}} = \zeta \mathcal{L}^2/\bar{\eta}$ is the ω-effect Reynolds number. The α-source in [68a] produces $\bar{A}_\phi = O(Rm_\alpha \bar{B}_\phi \mathcal{L})$ from $\bar{B}_\phi$, as before. Clearly, regeneration requires that the 'dynamo number', $D = Rm_\alpha Rm_{\boldsymbol{\omega}} = \alpha\zeta\mathcal{L}^3/\bar{\eta}^2$, exceeds some critical value D_c. Such a dynamo, that operates through the 'product' of α- and ω-effects, is called an $\alpha\omega$-dynamo.
- If $\alpha \sim \bar{u}_\phi$, the two sources in [68b] are equally effective and must be retained, giving an $\alpha^2\omega$-dynamo. This less-studied case will not be considered here.

Many solutions of each type have been generated. The α^2-dynamos satisfy [67] with $\mathbf{u} \equiv \mathbf{0}$, and are usually of DC type. This is so for the very simple spherical solutions that exists when α ($\neq 0$) is a constant. These can be readily found by the technique of Section 3.3.2, which applied to [67] gives

$$\partial_t T = \bar{\eta}\nabla^2 T - \alpha \nabla^2 S \qquad [70a]$$

$$\partial_t S = \bar{\eta}\nabla^2 S + \alpha T \qquad [70b]$$

Marginal solutions are of DC type and satisfy (in nondimensional variables) $\nabla^2 T + Rm_{\alpha c}^2 T = 0$, where $Rm_\alpha = \alpha r_c/\bar{\eta}$. Instead of [42c] and [42d], we have

$$T = Rm_{\alpha c} j_n(Rm_{\alpha c} r) Y_n(\theta, \phi) \qquad [70c]$$

$$S = [j_n(Rm_{\alpha c} r) - C_n r^n] Y_n(\theta, \phi) \qquad [70d]$$

If $C_n = Rm_{\alpha c} j'_n(Rm_{\alpha c})/(2n+1)$ and $j_n(Rm_{\alpha c}) = 0$, the dynamo condition [40] is satisfied. Again $n = 1$ is the critical mode and $Rm_{\alpha c} = 4.493$. This constant-α model does not satisfy the symmetry condition [69a].

The $\alpha\omega$-dynamos are usually of AC type, and involve a particularly interesting phenomenon: the 'dynamo wave'. Parker, who invented the $\alpha\omega$-dynamo, also exhibited dynamo waves for the first time using a simple planar model in which the α-effect creates $B_x(y, t)\mathbf{1}_x$ from $B_z(y, t)\mathbf{1}_z$ and a shearing motion $\zeta x \mathbf{1}_z$ does the reverse; α and ζ are constants. Equation [67] gives

$$\partial_t B_x = \bar{\eta}\partial_y^2 B_x + \alpha \partial_y B_z \qquad [71a]$$

$$\partial_t B_z = \bar{\eta}\partial_y^2 B_z + \zeta B_x \qquad [71b]$$

These equations admit solutions of the form

$$\mathbf{B} \propto \exp[\imath(ky - \omega t)] \qquad [71c]$$

provided that

$$(\imath\omega - \bar{\eta}k^2)^2 = \imath\alpha\zeta k \qquad [71d]$$

Only one root of this equation can be regenerative, namely

$$\omega = -\imath\bar{\eta}k^2 - \left(\frac{1}{2}|\alpha\zeta k|\right)^{1/2}[\mathrm{sgn}(\alpha\zeta k) - \imath] \qquad [71e]$$

This gives growing $\mathbf{B}$ if $\mathrm{Im}(\omega) > 0$, that is, if $D > D_c = 2$, where $D = |\alpha\zeta k^{-3}|/\bar{\eta}^2$ is the dynamo number. In the critical state $D = D_c$, the frequency $\omega[= -\bar{\eta}k^2\,\mathrm{sgn}\,(\alpha\zeta k)]$ is nonzero, so the dynamo is of AC type. Magnetic activity progresses as a wave moving with phase velocity $-\bar{\eta}|k|\mathrm{sgn}(\alpha\zeta)\mathbf{1}_y$. As in Section 3.2.6, the drift speed $\mathcal{U}_\eta$ of $\mathbf{B}$ relative to the conductor is of order $\eta/\mathcal{L}$, where $\mathcal{L}$ is the wavelength.

Parker (1957) used dynamo waves to explain the solar activity cycle, which progresses from the poles of the Sun to its equator. The timescale is consistent with its 22-year period provided β is sufficiently large, so that $\bar{\mathcal{T}}_\eta$ is sufficiently small (see Section 3.5.1). Parker visualized that (x, y, z) in the planar model correspond to (r, θ, ϕ), in that order, for spherical $\alpha\omega$-dynamos. The planar model then indicates that, if $\zeta < 0$ and if $\alpha > 0$ in the Northern Hemisphere, a dynamo wave originating near the north pole will progress toward the equator. The symmetry conditions [69a] show that one starting near the south pole will also move toward the equator. Subsequently, numerous integrations of [67] in spherical geometry confirmed this (see, e.g., P.H. Roberts (1972). More recently, as mentioned in Section 3.4.4, Gubbins and Gibbons (2002) found similar solutions in a 3D kinematic model. As Section 3.2.6 has shown, $\alpha > 0$ ($\alpha < 0$) in the Northern (Southern) Hemisphere is plausible, but advances in helioseismology did not support $\zeta < 0$, and this led to the abandonment of Parker's model of the solar cycle in the simple form in which he originally conceived it. Dynamo waves may nevertheless be significant for the geodynamo (Section 3.6.3). In the 3D MHD geodynamo simulation of Glatzmaier and Roberts (1996a), waves of magnetic activity progress westward through the core even though the fluid itself, together with the SIC, moves predominantly eastward. It is tempting to believe that these are dynamo waves, responsible for maintaining the $\bar{\mathcal{E}}_\phi$ that the dynamo requires, the relative motion of field and conductor being the resistive drift $\eta/\mathcal{L}$ described in Section 3.2.6. This might explain both the observed geomagnetic westward drift of magnetic features across the CMB and the seismically inferred (if controversial) eastward motion of the SIC. This discussion of the α^2- and $\alpha\omega$-models has so far supposed that $\bar{\mathbf{u}}_M \equiv \mathbf{0}$. We now consider $\alpha\omega m$-dynamos in which $\bar{\mathbf{u}}_M \neq \mathbf{0}$. It was observed in Section 3.3.5 that $\bar{\mathbf{u}}_M$ advects $\bar{A}_\phi$ and $\bar{B}_\phi$ but does not enable them to defeat Cowling's theorem. Nevertheless, $\bar{\mathbf{u}}_M$ can change the character of MF dynamos by carrying $\bar{B}_\phi$ to regions where the α-source is particularly effective, and by advecting $\bar{A}_\phi$ to places where the ω-effect is especially potent, or it can do the reverse. This can reduce or increase D_c considerably and can turn the dynamo from AC to DC type. This is what happens to the Braginsky models described in Section 3.4.1. These do not create α at a point through helical turbulence in the neighborhood of that point. Instead, they produce α through the helicity of planetary waves of global scale. Nevertheless, they are MF dynamos, where the M refers to an average over ϕ. As shown in [56], their α-effect is confined to the production of $\bar{A}_\phi$, so they are $\alpha\omega$-dynamos or more precisely $\alpha\omega m$-dynamos, since $\bar{\mathbf{u}}_M$ is large enough and of the right sign to reduce D_c significantly and to create a steady $\bar{\mathbf{B}}$.

Some vestiges of Cowling's theorem infect models based on [66b]. As pointed out in Section 3.3.5, every regenerative $\bar{\mathbf{B}}$ has at least one singular latitude circle C, in V or on S, where $\bar{\mathbf{B}}_M = \mathbf{0}$ and only ineligible, infinite $\bar{\mathbf{u}}$ can maintain $\bar{\mathbf{B}}$. Cowling's argument also excludes MF dynamos based on [66b] that have a singular C on any zero-α surface, such as the equatorial plane for the symmetry [69a]. This means that not only $\bar{\mathbf{B}}$ of the quadrupolar family [69c] but also $\bar{\mathbf{B}}$ of the dipole family [69b] must possess at least two singular C, one in each hemisphere. This tends to make each hemisphere act as a dynamo, independently of the other, so that reversing the sign of $\mathbf{B}$ in one hemisphere almost creates an acceptable solution of the opposite symmetry. This means that $Rm_{\alpha c}$ for α^2-dynamos, or D_c for $\alpha\omega$-dynamos, is numerically nearly the same for one symmetry as for the other (see also Proctor (1977b)). The more general MF dynamos based on [66a] can in principle maintain the $\bar{\mathcal{E}}_\phi(-z) = -\bar{\mathcal{E}}_\phi(z)$ symmetry of the dipole family without excluding C from the equatorial plane.

The reversals produced by the AC models are on far too short a timescale to be relevant to the polarity reversals of the geomagnetic field, which are much more likely to have a dynamic than a kinematic cause. Some of the DC models create $\bar{\mathbf{B}}$ remarkably similar to those produced by 3D MHD geodynamo simulations. Perhaps this is not too surprising since (even though this tends to be hidden beneath its greater complexity) an MHD model must still create an $\bar{\mathcal{E}}_\phi$ large enough to sustain the axisymmetric part of its field.

3.5.3 Saturation: Intermediate Models

Nonlinear generalizations of MF dynamos provide comparatively simple ways of exploring dynamical issues. By assigning α, Cowling's theorem is defeated and the collapse of $\bar{\mathbf{B}}$ is prevented, leaving the investigator free to study nonlinear processes such as 'saturation', in which the Lorentz force equilibrates $\bar{\mathbf{B}}$ at some statistically constant amplitude.

The only mechanism that can equilibrate an α^2-dynamo is a $\mathbf{B}$-dependent α. The heuristic discussion of Parker's α-effect in Section 3.2.6 makes it clear that the larger the $\mathbf{B}$ in the flux tube, the more it will resist deformation by a cyclonic eddy. This motivates the commonly used but *ad hoc* Ansatz

$$\alpha = \frac{\alpha_0(\mathbf{x})}{1 + (\mathbf{B}/B_0)^2} \quad (B_0 = \text{constant}) \qquad [72]$$

Additionally, it is reasonable to include a dependence of α on the current helicity H_J, in recognition of the decreased angle through which the Ω in **Figure 5(a)** is turned by ω when $\mathbf{B}$ is increased (Roberts, 1971a). The question of whether the full $\mathbf{B}$ or its statistical mean should feature in [72] is of considerable interest in solar physics and astrophysics but is probably less significant for the geodynamo (see Section 3.5.4).

Nonlinear $\alpha\omega$-models can saturate by a different route: $\mathbf{B}$-dependence of the ω-effect. The energy source powering these dynamos is usually assumed to be pole-equator temperature differences produced by convection. In a rotating system this creates a thermal wind, $\bar{u}_T(s, z)\mathbf{1}_\phi$ (see Section 3.6.2). In addition, Lorentz and viscous forces make two other contributions to the ω-effect: a magnetic wind $\bar{u}_M(s, z)\mathbf{1}_\phi$ of order $\bar{B}_\phi^2/2\Omega\mu_0\rho r_c$ that tends to suppress the zonal differential rotation, and a geostrophic wind $\bar{u}_G(s)\mathbf{1}_\phi$ that depends only on distance s from the polar axis (and t) (see Section 3.6.2). In total,

$$\bar{u}_\phi = \bar{u}_T(s, z) + \bar{u}_M(s, z) + \bar{u}_G(s) \qquad [73]$$

Because the α-effect and thermal wind are specified, these 2D, nonlinear α^2- and $\alpha\omega$-models cannot properly be called 'MHD dynamos' but, as they take an important step in that direction, they are often termed 'intermediate dynamos'. No attempt will be made here to review the, by now, extensive literature on intermediate dynamo models; interested readers will find Soward (1991) and Rüdiger and Hollerbach (2004) informative. Despite significant progress, perplexing issues remain unresolved, a state of affairs that is unlikely to improve quickly now that fully 3D MHD geodynamo simulations have become commonplace.

3.5.4 Application of MFE to the Core

In analogy with the name traditionally given to $Pr = \nu/\kappa$, the diffusivity ratios

$$Pm = \nu/\eta \qquad [74a]$$

$$P\kappa = \kappa/\eta \qquad [74b]$$

$$P\delta = \delta/\eta \qquad [74c]$$

where ν is the kinematic viscosity, κ is the thermal diffusivity, and δ is the compositional diffusivity, are often called 'Prandtl numbers'. Since the FOC is a liquid metal, Pm and $P\kappa$ are small, of order 10^{-6}–10^{-5}, and $P\delta$ is even smaller, of order 10^{-10}–10^{-9}. In most laboratory experiments with liquid metals, $Rm \ll 1$ so that EM induction is weak (Section 3.2.6). For the FOC, $Rm = O(10^2)$, so that the kinetic Reynolds number $Re = \mathcal{U}\mathcal{L}/\nu$ and the Péclet number $Pe = \mathcal{U}\mathcal{L}/\kappa$ are of order 10^7–10^8. It can hardly be doubted therefore that the FOC is highly turbulent and that the molecular transport of momentum, heat, and composition are, except on tiny dissipation scales, negligible compared with their turbulent transport. Because η is so large compared with ν, κ, and δ, it is far from obvious that the turbulent transport of magnetic flux similarly dwarfs its molecular diffusion. Plausibly, the magnetic energy spectrum tails off with increasing wave number k as $k^{-11/3}$, that is, much more rapidly than the $k^{-5/3}$ expected for the kinetic and internal energy spectra. Therefore, considering Rm to be a function of $\mathcal{L}$, the turbulence in the range $Rm' \lesssim 1$ rides on a comparatively smooth background field $\bar{\mathbf{B}}$.

A rough assessment of the importance of the turbulent transport of large-scale magnetic flux in the Earth's core follows from [63a] and [63b], which imply that η is increased to $\bar{\eta} = \eta + \beta$ and $\bar{\mathcal{T}}_\eta$ is correspondingly reduced, as argued above. No characteristic of the observed geomagnetic field demands this. For example, paleomagnetism proffers no compelling evidence that the duration of a polarity reversal is significantly shorter than $\bar{\mathcal{L}}^2/\pi^2\eta$. This suggests that, unlike the Sun, β is not large compared with η, though conceivably they are comparable. By omitting $\bar{\mathcal{E}}$ from [63a], one is asserting that the geodynamo is driven by motions and waves of global scale, and that MFE is irrelevant. Even if this attitude is simplistic, omitting turbulent α- and β-effects from core electrodynamics is arguably far less significant than many other geophysical uncertainties that beset the theory, and is much less serious than the inadequate way that the transport of momentum, heat and composition by the turbulence is currently handled.

3.6 MHD Dynamos

3.6.1 Basic Equations and Boundary Conditions

The aim of this section is to survey the basics of MHD dynamo theory. The addition of the equations of fluid dynamics is a major step demanding more than a greater computational committment; it requires also a deeper insight when interpreting the results of numerical work. The discussion will focus on convective dynamos in which $\mathbf{u}$ obeys

$$\partial_t \mathbf{u} + \mathbf{u}\cdot\nabla\mathbf{u} + 2\boldsymbol{\Omega}\times\mathbf{u} = -\nabla\Pi - \gamma T\mathbf{g} + \mathbf{J}\times\mathbf{B}/\rho + \nu\nabla^2\mathbf{u} \qquad [75]$$

This is the usual Navier–Stokes equation for a rotating Boussinesq fluid with an added Lorentz force. In the Boussinesq approximation ρ is uniform, so that mass conservation reduces to the condition [12] of incompressibility. Momentum conservation [75] is expressed in the reference frame that co-rotates with the mantle, at a constant angular velocity Ω. The resulting Coriolis acceleration, $2\boldsymbol{\Omega}\times\mathbf{u}$, appears on the left-hand side of [75]. The centrifugal acceleration, $\boldsymbol{\Omega}\times(\boldsymbol{\Omega}\times\mathbf{x})$, has been combined with the kinetic pressure, P, to form the 'reduced pressure', $\Pi = P/\rho - \frac{1}{2}(\boldsymbol{\Omega}\times\mathbf{x})^2$. The last two terms in [75] represent the Lzorentz and viscous forces per unit mass. The buoyancy force per unit mass is $-\gamma T\mathbf{g}$, where T is the temperature, γ is the thermal expansion coefficient, and $\mathbf{g}$ is the gravitational acceleration. Compositional buoyancy is ignored. Most MHD geodynamo simulations are based on [12] and [75], although a geophysically more realistic approach was initiated by Glatzmaier and Roberts (1996a), whose simulations used the anelastic theory of Braginsky and Roberts (1995).

The temperature is governed by an energy equation of the form

$$\partial_t T + \mathbf{u}\cdot\nabla T = \kappa\nabla^2 T + q/\rho C_p \qquad [76]$$

where q comprises three heat sources: $q = q^R + q^\nu + q^{\mathcal{J}}$. Here q^R arises from internal sources, if any, such as dissolved radioactivity and q^ν $(=\rho\nu d_{ij}d_{ij})$ is the viscous dissipation rate per unit volume and C_p is the specific heat at constant pressure. Usually, and here, q^ν and $q^{\mathcal{J}}$ are omitted from q, and q^R is required to drive convection only when the SIC is ignored.

The conditions for **B** derived in Sections 3.2.1 and 3.3.2 are supplemented by boundary conditions on **u** and T. Often selected are

$$\left.\begin{aligned}\mathbf{u} &= \mathbf{0} \\ T &= T_c\end{aligned}\right\}\ \text{on the CMB} \qquad \begin{aligned}&[77a]\\&[77b]\end{aligned}$$

$$\left.\begin{aligned}\mathbf{u} &= \mathbf{0} \\ T &= T_i\end{aligned}\right\}\ \text{on the ICB} \qquad \begin{aligned}&[77c]\\&[77d]\end{aligned}$$

where T_c and T_i $(>T_c)$ are constants; $T_i - T_c$ will be denoted by ΔT. Because of its small moments of inertia, the SIC responds readily to torques exerted by the mantle and FOC, but for simplicity $\mathbf{\Omega}_i = \mathbf{0}$ has been assumed in [77c]. When viscous forces are ignored, the differential order of [75] is reduced, and it is possible to satisfy only one scalar condition on **u** at the boundaries; [77a] and [77c] are then replaced by

$$\mathbf{n}\cdot\mathbf{u} = 0, \quad \text{on the CMB and ICB for } \nu = 0 \qquad [77e]$$

Two nondimensional parameters are commonly used in describing convection, the 'Rayleigh number', Ra and the (thermal) 'Prandtl number', Pr:

$$Ra = g\beta\gamma\mathcal{L}^4/\nu\kappa \qquad [78a]$$

$$Pr = \nu/\kappa \qquad [78b]$$

Here β is typical of the applied temperature gradient, for example, $\beta = \Delta T/(r_c - r_i)$ or $\beta = q^R r_c/\rho C_p\kappa$. The Rayleigh number is a measure of how effective the buoyancy force is in overcoming the diffusive processes that oppose convection.

A popular way to gain insight into convective processes, popular because the theory is relatively tractable, is to study the onset of convection. A motionless 'conductive state' is defined in which heat is carried across the system by thermal conduction; the linear stability of this state is then analyzed. It is found that when the 'control parameter', Ra, reaches some 'critical' or 'marginal' value, Ra_c, the conduction solution becomes convectively unstable. The eigenfunction corresponding to Ra_c gives the structure of the marginal state. In the absence of Coriolis and Lorentz forces, Ra_c is typically of order 10^3 and the marginal mode is a pattern of overturning convection cells, each having about the same horizontal scale as the depth of the fluid.

While studies of this type provide some insight into convective flows, they clearly have limited value. They predict that, when $Ra > Ra_c$, the convective motions increase without limit, though in reality the nonlinearities $\mathbf{u}\cdot\nabla\mathbf{u}$ and $\mathbf{J}\times\mathbf{B}/\rho$ in [75], prevents this and cause the solution to 'saturate'. The enhanced amplitudes and modified structure of the convective motions as Ra is increased beyond Ra_c are topics beyond the scope of this chapter. We shall use the results of linear theory to obtain clues about how Coriolis and Lorentz forces affect thermal convection.

3.6.2 Classical Theory of Rotating Fluids

In this subsection we shall suppose that $\mathbf{B} \equiv \mathbf{0}$; at first we exclude buoyancy too $(\mathbf{g} \equiv \mathbf{0})$. We give a rudimentary account of relevant concepts in the theory of rotating fluids. For a more complete treatment, see the classic text of Greenspan (1968).

Two dimensionless numbers quantify the importance of viscosity and inertia relative to the Coriolis force, the 'Ekman number', E, and the 'Rossby number', Ro:

$$E = \nu/\Omega\mathcal{L}^2 \qquad [79a]$$

$$Ro = \mathcal{U}/\Omega\mathcal{L} \qquad [79b]$$

The molecular viscosity, ν_M, of the FOC is uncertain but is commonly estimated to be about $10^{-6}\,\mathrm{m^2\,s^{-1}}$. Then taking $\mathcal{L} = 2\times10^6$ m, we obtain $E \sim 10^{-15}$. Such a small value suggests that large-scale momentum is transported more effectively by small turbulent eddies than by molecular diffusion, and that E should be estimated using a larger, turbulent viscosity, ν_T. A popular paradigm (see, e.g., Braginsky and Meytlis (1990)) assumes that $\nu_T \sim \eta$, but even then E does not exceed 10^{-9}. Taking $\mathcal{U} = 2\times10^{-4}$ m s^{-1}, we have $Ro \sim 10^{-5}$. From these estimates, we see that the FOC is a 'rapidly rotating fluid', if defined as one for which

$$E \ll 1 \qquad [79c]$$

$$Ro \ll 1 \qquad [79d]$$

For the geodynamo to exist, $Rm = \mathcal{U}\,\mathcal{L}/\eta$ must be of order 10^2 and, since $Re = Ro/Pm$ and $Pm \ll 1$, it follows that $Re \gg 1$. Because $E = Ro/Re$, [79c] is a consequence of [79d]. The smallness of Ro suggests that we may discard one of the inertial terms, $\mathbf{u}\cdot\nabla\mathbf{u}$, from [75]. Looking ahead to Section 3.6.3, we shall recognize that the main nonlinear feedback equilibrating the geodynamo is the Lorentz force and not the inertial force.

The remaining inertial term, $\partial_t\mathbf{u}$, in [75] is responsible for 'inertial waves'. These are most easily studied by abbreviating [75] to

$$\partial_t\mathbf{u} + 2\boldsymbol{\Omega}\times\mathbf{u} = -\nabla\Pi \qquad [80]$$

by assuming that $\mathbf{u}$ is proportional to $\exp(\imath\omega t)$, and by solving this equation and [12] subject to [77e]. The waves are dispersive; the spectrum of possible ω is discrete, with infinitely many possible eigenvalues ω, all in the range $|\boldsymbol{\omega}| \le 2\Omega$ (see Greenspan (1968)). The closer $|\omega|$ is to zero, the denser the packing of eigenvalues and the more 2D the eigenfunctions are with respect to $\boldsymbol{\Omega} = \Omega\mathbf{1}_z$. The extreme case, $\omega = 0$, gives a single, infinitely degenerate eigenfunction called the 'geostrophic mode'. For this mode, $\partial_t\mathbf{u} = \mathbf{0}$ and [12] and [80] give

$$2\boldsymbol{\Omega}\cdot\nabla\mathbf{u} = \mathbf{0} \qquad [81a]$$

that is,

$$\mathbf{u} = \mathbf{u}(x,\ y,\ t) \qquad [81b]$$

Stated in words:

Proudman–Taylor theorem: *The slow steady motion of a rotating inviscid fluid is 2D with respect to the rotation axis.*

Since the CMB and ICB are assumed spherical, [77e] and [81b] imply

$$\mathbf{u} = \bar{\mathbf{u}}_G(s,t)\mathbf{1}_\phi \qquad [81c]$$

The function $\bar{\mathbf{u}}_G(s,\ t)$ is arbitrary; this is the infinite degeneracy referred to above. The 'geostrophic flow' $\bar{\mathbf{u}}_G$ is axisymmetric, zonal, and constant on 'geostrophic cylinders', $\mathcal{C}(s)$. These are cylinders of constant radius s. **Figure 10** shows a typical geostrophic cylinder and also a particularly significant one, $\mathcal{C}_i = \mathcal{C}(r_i)$, that touches the SIC on its equator and is therefore called the 'tangent cylinder'. The arbitrariness of $\bar{\mathbf{u}}_G$ expresses the fact that the Coriolis force created by $\bar{\mathbf{u}}_G$ is ineffective, because $2\boldsymbol{\Omega}\times\bar{\mathbf{u}}_G = \nabla\Upsilon$, where $\Upsilon = -2\Omega\int \bar{u}_G(s,t)\,ds$. Thus, the Coriolis force can be absorbed into $\nabla\Pi$. This means that, for $\nu = 0$, each geostrophic cylinder

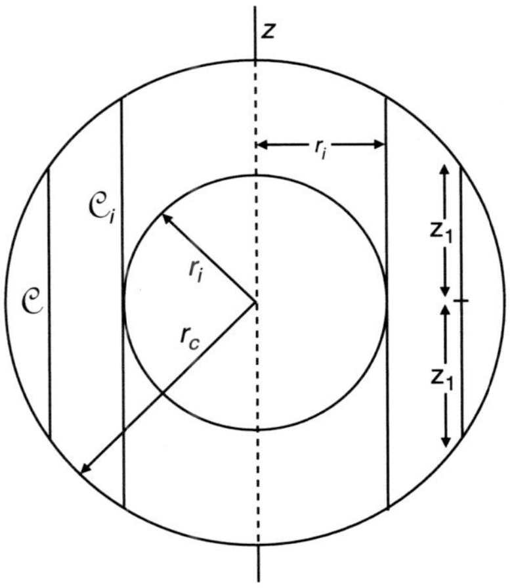

Figure 10 Geostrophic cylinders. A meridional section of the core is shown. One geostrophic cylinder, $\mathcal{C}(s)$, of radius $s > r_i$, appears in cross-section as two lines parallel to Oz of length $2z_1$, where $z_1 = \sqrt{r_i^2 - s^2}$. The tangent cylinder, $\mathcal{C}_i$, is also shown; this is the geostrophic cylinder of radius r_i that touches the SIC on its equator.

can turn about its axis Oz without disturbing, or being disturbed by, its neighbors.

Consider next the dynamical effect of a buoyancy force produced by an axisymmetric temperature distribution $\bar{T}$. The steady linearized inviscid form of [81] is then

$$2\boldsymbol{\Omega}\times\bar{\mathbf{u}} = -\nabla\bar{\Pi} - \gamma\bar{T}\mathbf{g} \qquad [82a]$$

Since $\nabla\times\mathbf{g} = \mathbf{0}$, this implies, in place of [81a]

$$2\boldsymbol{\Omega}\cdot\nabla\bar{\mathbf{u}} = -\gamma\mathbf{g}\times\nabla\bar{T} \qquad [82b]$$

Because $\mathbf{g}(= -g\mathbf{1}_r)$ is radial, this reduces to $\partial_z\bar{u}_\phi = (g\gamma/2\Omega r)\partial_\theta\bar{T}$, which integrates to

$$\bar{\mathbf{u}} = [\bar{u}_T(s,\ z) + \bar{u}_G(s)]\mathbf{1}_\phi \qquad [82c]$$

The flow $\bar{u}_T$ is called the 'thermal wind'. Its magnitude is $O(g\gamma\delta T/\Omega)$ and, if this is comparable with the assumed characteristic velocity $\mathcal{U} = 2\times10^{-4}$ m s^{-1}, the pole-equator temperature difference δT is of order 10^{-4} K for $\gamma \sim 10^{-5}$ K^{-1}. This may be regarded as typical of the temperature differences between rising and falling convecting streams in the core though, more precisely, when the compressibility of the core is properly allowed for, it is typical of the excess or deficit of the temperature relative to the adiabat.

There is clearly some arbitrariness in [82c], since all or any part of $\bar{u}_G$ can be absorbed into $\bar{u}_T$. A convenient way of removing this arbitrariness is to introduce a 'geostrophic average'. The geostrophic average of a scalar field $Q(\mathbf{x}, t)$ is denoted here by angle brackets and is defined for $s > R_i$ by:

$$\langle Q \rangle(s,\ t) = \frac{1}{A(s)} \int_{C(s)} Q\,(\mathbf{x},\ t)\,\mathrm{d}S = \frac{1}{2z_1} \int_{-z_1}^{z_1} \bar{Q}\,(s, z, t)\mathrm{d}z \qquad [83]$$

Here $\pm z_1(s) = \pm\sqrt{r_c^2 - s^2}$ are the z-coordinates of the latitude circles where the geostrophic cylinder $C(s)$ meets the CMB, $A = 4\pi s z_1(s)$ is the area of the cylinder, and $\mathrm{d}S = s\,\mathrm{d}\phi\,\mathrm{d}z$ (see **Figure 10**). The 'ageostrophic part' of $Q(\mathbf{x}, t)$ is what is left over after the geostrophic part has been subtracted: $\tilde{Q}\,(\mathbf{x}, t) = Q\,(\mathbf{x}, t) - \langle Q \rangle(s, t)$. The geostrophic part of a vector field $\mathbf{Q}(\mathbf{x}, t)$ is defined from its zonal component by $\langle \mathbf{Q} \rangle(s, t) = \langle Q_\phi \rangle(s, t)\mathbf{1}_\phi$. These definitions apply only when $s > r_i$. Inside the TC, there is a geostrophic average for the fluid to the north of the SIC and another for the fluid to the south, but we shall often avoid this complication by ignoring the SIC entirely. The separation [82c] is made unique by defining $\bar{\mathbf{u}}_G = \langle \bar{\mathbf{u}} \rangle$. It is particularly significant because the z-component, $\rho s \mathcal{U}_\phi$, of the angular momentum density of $\mathbf{u}$ is carried by $\langle \mathbf{u} \rangle$ alone (Greenspan, 1968).

The thermal wind is illustrative of a more general situation: the response of a rotating fluid to forcing. Ignoring the $\mathbf{u} \cdot \nabla \mathbf{u}$ part of the inertial acceleration, [75] has the form

$$\partial_t \mathbf{u} + 2\mathbf{\Omega} \times \mathbf{u} = -\nabla \Pi + \mathbf{F} \qquad [84a]$$

where $\mathbf{F}(\mathbf{x}, t)$ is the forcing and the response $\mathbf{u}$ is sought. There are interesting cases in which the forcing has a high frequency. For example, a precessionally driven flow has the same diurnal timescale as the inertial waves. Also, diurnal frequencies may arise from core turbulence. But, on the timescale $\mathcal{T}$ of large-scale convection, the modified Rossby number, $1/\Omega\mathcal{T}$, is small. The response of $\mathbf{u}$ to $\mathbf{F}$ is then a combination of 'free' inertial waves (the solution to the homogeneous part [80] of [84]), and a particular solution that varies slowly, on the same timescale as $\mathbf{F}$. This is the part of $\mathbf{u}$ of greatest interest; the free inertial waves are of lesser significance and can be 'filtered out' by discarding the $\partial_t \mathbf{u}$ term in [83a], so dispensing with the inertial force entirely.

Discarding $\partial_t \mathbf{u}$ is tempting but dangerous. The geostrophic average of [84a] is

$$\partial_t \bar{u}_G + 2\mathbf{\Omega}\langle \bar{u}_s \rangle = \langle F_\phi \rangle \qquad [84b]$$

By definition, the second term is proportional to the mass flux across, and out of, $\mathcal{C}$. By [77e], the mass flux

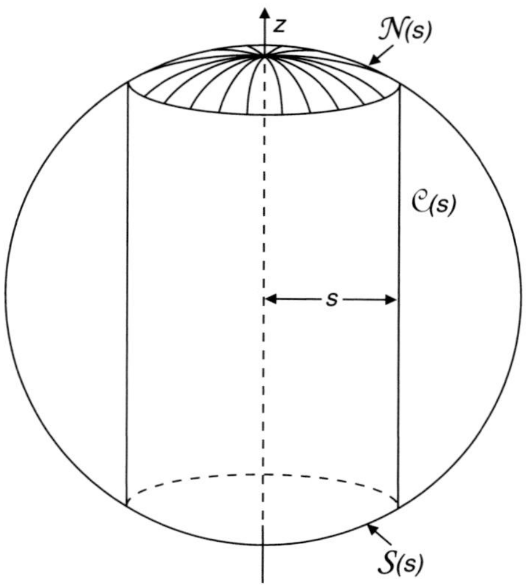

Figure 11 Spherical caps, $\mathcal{N}(s)$ and $\mathcal{S}(s)$, on the CMB provide the ends of the geostrophic cylinder $\mathcal{C}(s)$. The volume enclosed by $S = \mathcal{C}(s) + \mathcal{N}(s) + \mathcal{S}(s)$ is denoted by $\mathcal{V}(s)$ in the text. (for simplicity, the existence of the SIC is ignored in the figure).

through the spherical caps, $\mathcal{N}(s)$ and $\mathcal{S}(s)$ is zero (see **Figure 11**). Therefore, denoting by $\mathcal{V}(s)$ the volume surrounded by the surface $S = \mathcal{C}(s) + \mathcal{N}(s) + \mathcal{S}(s)$, we have

$$\langle \bar{u}_s \rangle \equiv \frac{1}{A(s)} \int_{C(s)} \mathbf{u} \cdot \mathrm{d}\mathbf{S} = \frac{1}{A(s)} \oint_S \mathbf{u} \cdot \mathrm{d}\mathbf{S} = \frac{1}{A(s)} \int_{\mathcal{V}(s)} \nabla \cdot \mathbf{u}\,\mathrm{d}V = 0 \qquad [84c]$$

by [12]. Equation [83b] therefore simplifies to

$$\partial_t \bar{u}_G = \langle F_\phi \rangle \qquad [84d]$$

If we discard the inertial term $\partial_t \mathbf{u}$ in [84a], we also remove $\partial_t \bar{\mathbf{u}}_G$ in [84d] and obtain a contradiction unless $\langle F_\phi \rangle = 0$. When $\mathbf{F}$ is the buoyancy force $-\gamma T\mathbf{g}$ alone, $\langle F_\phi \rangle = 0$ because $\mathbf{g}$ has no ϕ-component. If $\mathbf{F}$ contains the Lorentz force, however, it is not necessarily true that $\langle F_\phi \rangle = 0$, a point to which we return in Section 3.6.3. Meanwhile we see that one way of evading this difficulty, while at the same time removing the free inertial waves, is to retain only the geostrophic part of $\partial_t \mathbf{u}$, replacing [84a] by

$$\partial_t \bar{\mathbf{u}}_G + 2\mathbf{\Omega} \times \mathbf{u} = -\nabla \Pi + \mathbf{F} \qquad [84e]$$

Another way is to restore viscous effects. The smallness [79c] of E encourages an asymptotic approach to determining core flow. Conceptually, the core is divided into boundary layers, in which viscosity is significant, and the remaining 'mainstream' in which viscosity does not act at leading order, as in [84e].

The most important boundary layer in rotating fluids is the Ekman layer, the thickness of which is of order $\delta_\nu = (\nu/\Omega)^{1/2} = E^{1/2}\mathcal{L} \sim 0.1$ m (see Greenspan (1968)). The main task of an Ekman layer is to reconcile mainstream flows with the no-slip conditions [77a] and [77b]. For example, the tangential component $\mathbf{n} \times \bar{\mathbf{u}}(s, z_1)$ of the mainstream velocity obtained from [84e] will not in general obey [77a] at the latitude circle (s, z_1) on the CMB, and an Ekman layer forms that smoothly brings $\mathbf{n} \times \bar{\mathbf{u}}(s, z_1)$ to zero. In the process, the normal component $\mathbf{n} \cdot \mathbf{u}$ acquires a small additional part, of order $E^{1/2}\mathbf{n} \times \bar{\mathbf{u}}(s, z_1)|$. This vanishes on the CMB, but is nonzero on the edge of the boundary layer. This process, called 'Ekman pumping' (if the flow is away from the boundary) or 'Ekman suction' (if it is toward the boundary), creates a meridional mainstream flow $\bar{\mathbf{u}}_M(s)$ of order $E^{1/2}\bar{u}_\phi$.

Ekman pumping (or suction) provides an alternative way of overcoming the difficulty encountered when $\langle F_\phi \rangle \neq 0$. The mass flux across, and out of, $\mathcal{C}(s)$, can be balanced by an inward Ekman pumping from the boundary layers on $\mathcal{N}(s)$ and $\mathcal{S}(s)$. Pumping through the caps vitiates [84c] and leads to a modification of [84e]:

$$\partial_t \bar{\mathbf{u}}_G + E^{1/2}\Omega \bar{\mathbf{u}}_G + 2\mathbf{\Omega} \times \mathbf{u} = -\nabla\Pi + \mathbf{F} \qquad [84f]$$

The obstacle to setting $\partial_t \mathbf{u} = \mathbf{0}$ has disappeared; [84f] gives $\bar{u}_G = E^{-1/2}\langle F_\phi \rangle/\Omega$. Equation [84f] also determines the most important viscous timescale of a rapidly rotating system, the 'spin-up timescale' $\mathcal{T}_{su}$. The angular momentum of the mantle about Oz is viscously exchanged with that of the FOC on this timescale. By comparing the first two terms in [84f], it may be seen that $\mathcal{T}_{su} = O(E^{-1/2}\Omega^{-1}) = O(\mathcal{L}/(\nu\Omega)^{1/2}) \sim 10^5$ years (When the SIC is included, complicated shear layers surround the tangent cylinder whenever the SIC does not corotate with the mantle. For a brief discussion of these 'Stewartson layers' and references to recent work, see Rüdiger and Hollerbach (2004).)

Viscosity also plays a crucial role in thermal convection. We shall consider only the case [79c] of small E and for simplicity, suppose that $Pr > 1$. We again ignore the SIC and suppose that convection is driven by heat sources q^R. The flow in the marginal state, $Ra = Ra_c$, is asymmetric, consisting of a 'cartridge belt' of 2D cells, often called 'Taylor cells', parallel to the axis of rotation and regularly spaced round that axis (Busse, 1970; Jones *et al.*, 2000; Roberts, 1968). Adjacent cells spin around their axes in a sequence of cyclonic and anticyclonic vortices, their vorticity being respectively parallel and antiparallel to $\mathbf{\Omega}$. The name 'Taylor cell' is a useful reminder of the Proudman–Taylor theorem, which the flow is trying to obey by being as 2D as possible, consistent with allowing convection to occur at all. We have already seen that, if $\nu = 0$, small amplitude motions must be geostrophic. But geostrophic motions have no radial components to carry heat outwards. Convection can occur only if viscous forces are large enough to 'break the rotational constraint' of the theorem. Viscous forces are most effective at small $\mathcal{L}$ but such motions viscously dissipate energy rapidly and therefore require a strong buoyancy force to maintain them. This explains why, when $E \ll 1$, the width of the cells in the marginal state is small, of order $E^{1/3}r_c$, and why Ra_c is large, of order $E^{-4/3}$.

3.6.3 Coriolis Magnetohydrodynamics

This section focuses on MHD, but only in the form relevant to core dynamics where Coriolis forces are strong and drastically transform classical MHD, so much so that the subject deserves its own name and acronym. The obvious choice is 'rotating MHD', but 'RMHD' is an abbreviation that is often used for either 'relativistic MHD' or 'reduced MHD'. Here it will be called 'Coriolis magnetohydrodynamics' or 'CMHD'.

The Alfvén number (aka the 'magnetic Mach number') is defined by

$$Al = \mathcal{U}/\mathcal{V}_A \qquad [85a]$$

With $\mathcal{U} = 2 \times 10^{-4}$ m s^{-1} and $\mathcal{V}_A = 1$ cm s^{-1} as before, $Al \approx 0.02$. Even though this is not very small, we write for simplicity

$$Al \ll 1 \qquad [85b]$$

for it is Al^2 ($\approx 4 \times 10^{-4}$) that is significant. Not only is it the ratio of kinetic and magnetic energy densities but also it quantifies the importance of the inertial force $\rho\mathbf{u} \cdot \nabla\mathbf{u} = O(\rho\mathcal{U}^2/\mathcal{L})$ relative to the Lorentz force $\mathbf{J} \times \mathbf{B} = O(\mathcal{B}^2/\mu_0\mathcal{L})$. The smallness of Al^2 indicates that, for all $\mathcal{L}$, it is Lorentz force that equilibrates the system; the inertial term $\rho\mathbf{u} \cdot \nabla\mathbf{u}$ in [75] is mostly ignored in what follows.

The Lorentz force can equilibrate the system in two main ways, the first of which has already been noted in Section 3.2.7: a balance between Lorentz and viscous forces requires

Weak field regime:

$$\mathcal{B} = \sqrt{\mu_0 \rho \nu \eta}/\mathcal{L} \qquad [86a]$$

or

$$Ha = O(1) \quad [86b]$$

(Here Ha is the Hartmann number defined in [27a].) Alternatively, equilibration may occur when the Lorentz and Coriolis forces are comparable:

Strong field regime:

$$\mathcal{B} = \sqrt{\mu_0 \rho \boldsymbol{\Omega} \eta} \quad [86c]$$

or

$$\Lambda = O(1) \quad [86d]$$

Here Λ is the Elsasser number, a parameter that is independent of both $\mathcal{L}$ and $\mathcal{U}$:

$$\Lambda = \frac{\sigma \mathcal{B}^2}{\Omega \rho} = \frac{\mathcal{V}_A^2}{\Omega \eta} \quad [86e]$$

The adjectives 'weak' and 'strong' are appropriate since $\mathcal{B}_{\text{weak}}/\mathcal{B}_{\text{strong}} = O(E^{1/2}) \ll 1$.

Magnetic parameters analogous to [79a] and [79b] are the magnetic Ekman number, E_{m}, and magnetic Rossby number, Ro_{m}:

$$E_{\text{m}} = \eta/\Omega\mathcal{L}^2 \quad [87a]$$

$$Ro_{\text{m}} = \mathcal{V}_A/\Omega\mathcal{L} \quad [87b]$$

When $\eta = 2\,\text{m}^2\text{s}^{-1}$, $\mathcal{L} = 10^6\,\text{m}$ and $\mathcal{V}_A = 1\,\text{cm}\,\text{s}^{-1}$ as before, $E_{\text{m}} \approx 3 \times 10^{-7}$, $Ro_{\text{m}} \approx 2 \times 10^{-3}$, so that

$$E_{\text{m}} \ll 1 \quad [87c]$$

$$Ro_{\text{m}} \ll 1 \quad [87d]$$

the first of which follows from [79d] since $E_{\text{m}} = Ro/Rm$ and $Rm \gg 1$. Since $Al^2 = Pm^{-1}(Rm/Ha)^2$, the magnetic energy in a weak field dynamo operating in a liquid metal is necessarily much less than the kinetic energy of the motions that sustain it. According to [87c], the reverse is true for a strong field dynamo, since $Al^2 = E_{\text{m}}(Rm^2/\Lambda)$.

Weak field dynamos exist, but apparently the geodynamo is not one of them since $\Lambda = 1$ gives $\mathcal{B}_{\text{strong}} \approx 1\,\text{mT}$, whereas, even if $\nu = \nu_{\text{T}} \approx \eta$, [86a] gives $\mathcal{B}_{\text{weak}} \approx 10^{-7}\,\text{T}$, that is, $\mathcal{B}_{\text{strong}}$ is typical of the observed field strength while $\mathcal{B}_{\text{weak}}$ is very much weaker. Equilibration in the strong field regime is partly due to the magnetic wind (Section 3.5.3), which reduces the potency of the ω-effect, a fact confirmed by several fully 3D MHD geodynamo simulations. Ignoring the inertial force as before, [75] gives, in place of [82b],

$$2\boldsymbol{\Omega} \cdot \nabla \bar{\mathbf{u}} = -\gamma \mathbf{g} \times \nabla \bar{T} - \rho^{-1} \nabla \times \left(\overline{\mathbf{J} \times \mathbf{B}}\right) \quad [88]$$

the ϕ-component of which integrates to give [73].

The magnetic field removes the geostrophic degeneracy of the inertial waves discussed in Section 3.6.2. The geostrophic cylinders $\mathcal{C}(s)$ shown in **Figure 10** are no longer free to turn about Oz independently; each is threaded to its neighbors by the s-component of $\mathbf{B}$. Thus $\bar{\mathbf{u}}_{\text{G}}$ is no longer arbitrary but is controlled by B_s. It was seen in Section 3.6.2 that the Coriolis force associated with geostrophic motions can be absorbed into the pressure gradient. Since this otherwise dominating force is then effectively removed, the remaining forces become influential, including the inertial term $\rho \partial_t \mathbf{u}$. Torsional waves therefore resemble Alfvén waves and have the same timescale, $\mathcal{T}_{\text{A}s} = r_{\text{c}}/\mathcal{V}_{\text{A}s}$, where $V_{\text{A}s}$ is now based on the rms strength of B_s on the geostrophic cylinder $\mathcal{C}(s)$:

$$\begin{aligned} \langle B_s^2 \rangle (s,t) &= \frac{1}{A(s)} \int_{\mathcal{C}(s)} B_s^2(s, \phi, z, t)\, \text{d}S \\ &= \mu_0 \rho V_{\text{A}s}^2(s,t) \end{aligned} \quad [89]$$

Even if $\mathcal{V}_{\text{A}s}$ is as small as $1\,\text{cm}\,\text{s}^{-1}$, $\mathcal{T}_{\text{A}s}$ is less than a decade. This may be compared with the timescale of the ageostrophic waves which (see below) is of order 10^3 years. It is also short in comparison with the timescales, $\mathcal{T}_{\text{su}}$ and $\mathcal{T}_\eta$, of the diffusive processes. This means that, in a first approximation, we may ignore the time dependence of $\langle B_s^2 \rangle$ and all diffusive effects.

For $\nu = \eta = 0$, torsional waves are governed by the induction equation [17a] for a perfect conductor and by [84d] with $\mathbf{F}$ as the Lorentz force:

$$\rho \partial_t \bar{u}_{\text{G}} = \langle (\mathbf{J} \times \mathbf{B})_\phi \rangle \quad [90]$$

In a steady state, $\langle (\mathbf{J} \times \mathbf{B})_\phi \rangle = 0$, that is,

$$\int_{\mathcal{C}(s)} (\mathbf{J} \times \mathbf{B})_\phi\, \text{d}S = 0 \quad [91]$$

This important result (Taylor, 1963) is called Taylor's condition or Taylor's constraint; it can also be derived from the s-component of [88] and [12]. A field obeying [91] is called a Taylor state.

The torque exerted by the Lorentz force on the interior $\mathcal{V}(s)$ of $\mathcal{C}(s)$ is

$$\begin{aligned} \Gamma_z(s) &= \int_{\mathcal{V}(s)} s(\mathbf{J} \times \mathbf{B})_\phi\, \text{d}V \\ &= \int_0^s \left\langle (\mathbf{J} \times \mathbf{B})_\phi \right\rangle s A(s)\, \text{d}s \end{aligned} \quad [92a]$$

According to [15f], this may also be written as

$$\Gamma_z(s) = \frac{1}{\mu_0} \oint_S s B_\phi (\mathbf{B} \cdot \text{d}\mathbf{S}) \quad [92b]$$

where S is the complete boundary of $\mathcal{V}(s)$. In the axisymmetric case, $\bar{B}_\phi = 0$ on the spherical caps of S, so that the magnetic torques exerted by these caps on $\mathcal{V}(s)$ is zero, leaving only the magnetic torque from $\mathrm{C}(s)$. Therefore [91] implies

$$\int_{-z_1}^{z_1} \bar{B}_\phi \frac{\partial \bar{A}_\phi}{\partial z} \mathrm{d}z = 0 \qquad [92c]$$

This is the form of Taylor's condition used in intermediate $\alpha\omega$-dynamo theory (Section 3.5.3). If [91] does not hold initially, a torsional wave is launched in which the geostrophic cylinders oscillate about a Taylor state, $\bar{u}_G^T(s)$. Denoting by $v(s,t)$ the departure, $\bar{u}_G(s,t) - \bar{u}_G^T(s)$, of the geostrophic motion from this Taylor state, Braginsky (1970) showed that the associated shear $\zeta = v/s$ satisfies the 'torsional wave equation'

$$\frac{\partial^2 \zeta}{\partial t^2} = \frac{1}{s^2 A} \frac{\partial}{\partial s}\left[V_{A_s}^2 s^2 A \frac{\partial \zeta}{\partial s}\right] \qquad [93a]$$

The waves transport the z-component of the angular momentum density to and fro across the core, but do not change its integral:

$$m_{\mathrm{FOC}} = \int_{\mathrm{FOC}} \rho s u_\phi \, \mathrm{d}V = \rho \int_0^{r_c} s \bar{u}_G^T A \, \mathrm{d}s \qquad [93b]$$

In this simplest application of [93a], the SIC is ignored. The ordinary differential equation obtained by substituting $\zeta\ (s,t) = \mathcal{Z}(s)\ \exp\ (\iota\omega t)$ into [93a] has regular singularities both at $s=0$ and at $s=r_c$, where $A=0$. The implicit requirement that ζ be bounded at both these points transforms [93a] into an eigenvalue problem for the torsional wave frequencies, ω. Similar conclusions hold if the SIC is included, and more generally still if coupling between the FOC, SIC, and mantle is introduced: the sum, $m_{\mathrm{FOC}} + m_{\mathrm{SIC}} + m_{\mathrm{mantle}}$, of their angular momenta is conserved. A torsional wave, initiated by a turbulent fluctuation in the FOC, then creates an increase (or decrease) in the length of day that correlates with an increase (or decrease) in $m_{\mathrm{FOC}} + m_{\mathrm{SIC}}$. By estimating $\bar{u}_G(s,t)$ from $\hat{\mathbf{B}}$ on the CMB and comparing with the known variation in length of day, Jault *et al.* (1988) inferred that such a correlation exists (see also Jackson *et al.* (1993)).

If viscous and ohmic dissipation is restored, it is plausible that any departure of u_G from a Taylor state will die out, but this can happen only on the same long diffusive timescales $\mathcal{T}_{su}$ and $\mathcal{T}_\eta$ as $\mathbf{B}$ evolves on; the target $\bar{u}_G^T(s)$ changes as rapidly as $|v|$ declines. Nevertheless, it is reasonable to suppose that $\bar{u}_G(s)$ will asymptotically approach an evolving Taylor state and remain in a Taylor state thereafter. This explains why Taylor's condition and Taylor states occupy such an important place in MHD dynamo theory. In what is sometimes called the 'Malkus–Proctor scenario' (Malkus and Proctor, 1975), solutions of the MHD equations continuously satisfying [91] are sought. Intermediate models of this type have been constructed but so far no fully 3D MHD dynamo.

Knowledge of $\hat{\mathbf{B}}$ on the CMB tells little about $\mathbf{B}_\mathrm{P}$ and nothing about $\mathbf{B}_\mathrm{T}$ in the core. The torsional wave potentially provides badly needed information about $\mathbf{B}$ via $\langle B_s^2 \rangle$, a quantity to which not only $\mathbf{B}_\mathrm{P}$ but also $\mathbf{B}_\mathrm{T}$ contributes. This thought has motivated several models of torsional waves, for example, Braginsky (1970) and Zatman and Bloxham (1997). Extrapolation of $\mathbf{B}_\mathrm{P}$ to the CMB and comparisons with geodynamo simulations indicate that $\mathcal{T}_{\mathrm{A}s}$ is about a decade, but there are indications that it may be nearly 10 times larger. A 60-year peridocity in both the geomagnetic field and the length of day has been claimed and denied intermittently for the last 50 years, but recently modern methods of data analysis appear to have put its existence beyond reasonable doubt (Roberts *et al.*, 2007). The torsional wave provides the natural explanation. Braginsky (1994) proposed a type of dynamo in which $\mathbf{B}$ is not in a Taylor state and in which, except in boundary layers, $\bar{\mathbf{B}}_\mathrm{P}$ is almost in the z-direction, so prompting its name: 'model-$\mathcal{Z}$.' Characteristic of this dynamo is a large u_G and a small $\langle B_s^2 \rangle$ consistent with a large $\mathcal{T}_{\mathrm{A}s}$. Intermediate models of this type have been constructed but so far no fully 3D MHD dynamo.

Consider next the ageostrophic waves. If $Ro_\mathrm{m} \ll 1$, these are of one of two types: inertial waves (slightly modified by the Lorentz force) and slow waves (slightly modified by inertia). The only wave in the system of Alfvén type is geostrophic, the torsional wave. The timescale of the slow waves is $\mathcal{T}_s = \Omega\mathcal{L}^2/\mathcal{V}_\mathrm{A}^2$ and is of order 10^3 years. This is comparable with the timescale of the secular variation of the main geomagnetic field. The corresponding velocity, $\mathcal{V}_s = \mathcal{L}/\mathcal{T}_s = \mathcal{V}_\mathrm{A}^2/\Omega\mathcal{L}$, is of order $10^{-4}\ \mathrm{m\,s^{-1}}$, which is roughly the speed at which discernable magnetic features at the CMB drift westward. The time dependence of the slow waves is governed by the left-hand side, $\partial_t\mathbf{B}$, of the induction equation [17a]; the inertial term $\partial_t\mathbf{u}$ in [75] plays essentially no role. For this reason the waves are sometimes called 'MC waves', to emphasize that they are governed by the Magnetic and Coriolis forces alone (and of course the pressure gradient).

MC waves, like inertial waves, are dispersive. They are characteristically large-scale planetary waves, that is, they are asymmetric. Therefore, if they have finite amplitude, they are potentially able to create an emf $\bar{\varepsilon}_\phi = \left(\overline{\mathbf{u}' \times \mathbf{B}'}\right)_\phi$ that can defeat Cowling's theorem. But then they feed energy to $\mathbf{B}_{\mathrm{M}}$, and to maintain themselves they must draw energy from buoyancy. This provides a strong motivation for studying 'MAC waves', which are MC waves modified by the addition of the Archimedean (buoyancy) force. Braginsky proposed this acronym and provided the first model, which was planar (Braginsky, 1964b), and the second model, which was spherical (Braginsky, 1967). In a perfect conductor, $\mathbf{u}'$ and $\mathbf{B}'$ tend to be parallel and, not surprisingly, the phase relationship between $\mathbf{u}'$ and $\mathbf{B}'$ in MAC waves is unfavorable for creating the emf $\left(\overline{\mathbf{u}' \times \mathbf{B}'}\right)_\phi$. To obtain a nonzero $\bar{\mathcal{E}}_\phi$, it is necessary to include diffusion. The proper exploitation of the idea of a 'MAC wave dynamo', therefore requires that the theory of diffusive, finite amplitude MAC waves be advanced, a task scarcely less challenging than that of studying the MHD dynamo in full!

Magnetoconvection theory gives a hint why $\Lambda = O(1)$ is the relevant dynamical balance for the core. Consider again the rotating convecting sphere discussed in Section 3.6.2, but now suppose that a strong magnetic field $\mathbf{B}$ is present. The viscous force is no longer necessary to break the rotational constraint; the magnetic field can achieve this unaided. In contrast to Section 3.6.2, where the length scale $\mathcal{L}$ of the marginal state was very small in order to enhance the viscous forces, the scale of the convection cells in the marginal state when $\Lambda = O(1)$ is the same as that of the container: $\mathcal{L} = O(r_c)$. Cells of this scale are much less dissipative and therefore demand a much smaller buoyant energy input. It is found that $Ra_c = O(E^{-1})$, instead of the $Ra_c = O(E^{-4/3})$ of Section 3.6.2, that is, the magnetic field allows the system to convect more readily. The result $Ra_c = O(E^{-1})$ may be written as $\widetilde{Ra}_c = O(1)$, where $\widetilde{Ra} = g\gamma\beta r_c^2/\Omega\kappa = Ra.\,E$ is the 'modified Rayleigh number', which is often used in preference to Ra in CMHD convection studies. The fact that $\widetilde{Ra}_c$, Λ, and the cell size are all independent of ν emphasizes how completely the Lorentz forces have taken over from viscous forces in breaking the Proudman–Taylor constraint. It appears that $\Lambda = O(1)$ may give the field strengths that are most effective in breaking the constraint when $E \ll 1$. If so, this may explain why the geomagnetic field has varied so little in strength over geological time: it is tied to the rotation of the Earth through [86c]. It is often said that the fact that the magnetic compass needle points approximately north proves that the Coriolis force dominates core dynamics, but this is an oversimplification. The Coriolis force has a preferred direction, $\mathbf{\Omega}$. To counter the rotational constraint, the magnetic field configures itself so that the Lorentz force shares the same preferred direction (and magnitude).

This example, like others in this section, is intended to be suggestive rather than definitive. It ignores important questions and in particular, "Is the dynamo–generated $\mathbf{B}$ dynamically stable?" Instabilities driven by the Lorentz force have been extensively studied in plasma physics, because they threaten the generation of magnetic fusion energy. The instabilities are of two types: ideal and resistive. The latter depend on the reconnection of field lines; the former do not. An example of a resistive instability is the tearing mode (see Section 3.2.6). In a nonrotating system, ideal instabilities grow on the dynamic timescale $\mathcal{T}_{\mathrm{A}}$, but the timescale of a resistive instability is usually intermediate between $\mathcal{T}_{\mathrm{A}}$ and $\mathcal{T}_\eta$. Magnetic instabilites in the core are greatly affected by the Coriolis force, which lengthens their dynamic timescale from $\mathcal{T}_{\mathrm{A}}$ to $\mathcal{T}_s$. They have been extensively studied by Fearn and his associates (see Fearn (1998) where earlier references are given). Obviously magnetic instabilities cannot drive the geodynamo, but the larger the Λ, the greater the menace of instability. Perhaps there is a limit to Λ that, if exceeded by the dynamo, causes a magnetic instability that leads to a radically different field configuration? This is one explanation of polarity reversals (Fearn, 1998). Another is that the meridional flow $\bar{\mathbf{u}}_{\mathrm{M}}$ becomes temporarily weaker, putting the dynamo into an AC state (see Section 3.5.2), from which it may emerge with its axial dipole in either direction with equal probability (Glatzmaier and Roberts, 1995; Sarson *et al.*, 1998). Sarson and Jones (1999) argue that the disruption of $\bar{\mathbf{u}}_{\mathrm{M}}$ is due to buoyancy surges.

3.7 Final Remarks

This chapter has focused on basic electromagnetic theory as applied to electrically conducting fluids, and on the associated kinematic dynamo problem in which the fluid motion is specified and a self-exciting magnetic field is sought. These topics are contained within the larger MHD dynamo problem in which an energy source for the fluid motions is specified and both the fluid motion and a self-excited magnetic field are sought. Fundamentals of convective MHD dynamos have been described in Section 3.6; more details will be given in Chapter 5.

Hopefully the reader of this chapter will realize that kinematic theory has brought important concepts and significant understanding to field generation by homogeneous dynamos, that is, dynamos that operate in simply connected, homogeneous masses of electrically conducting fluids. Among its successes, recognition of the importance of broken symmetry, including helicity, stands out. The kinematic theory has also provided motivation for laboratory experiments that study electromagnetic induction and dynamo action in liquid metals.

Geodynamo theory is now so dominated by MHD simulations that the reader may wonder whether kinematic theory has any useful further role to play. One may argue however that the purpose of such simulations is not merely to produce numerical analogs of the geodynamo. It is also to decipher their message. This daunting objective requires a good understanding of kinematic theory. Unfortunately, it also demands proper treatment of turbulence, dominated by Coriolis, Lorentz, and buoyancy forces. The question of how to incorporate the turbulent transport of momentum, heat, and composition into geodynamo theory is the most severe challenge faced by geodynamo theory today. Large increases in ν, κ, and δ are currently necessary to ensure that numerical simulations are adequately resolved. They can be, and often are, advertised as turbulent diffusivities representing the effect of the small scales (i.e., the 'subgrid scales' that cannot be resolved in numerical simulations) on the large (resolved) scales. But it is questionable whether scalar turbulent diffusivities can adequately describe transport by a turbulence that, because of the dominance of the Coriolis and Lorentz forces, must surely be highly anisotropic. In view of these major obstacles, it is truly remarkable how successful numerical simulations of the geodynamo have been; the main features of the observed geomagnetic field have all been replicated, leading some commentators to assert that the geodynamo problem has been solved. This success is sometimes called the 'geodynamo paradox'. Its resolution is currently the main target of geodynamo theory. Further discussion of core turbulence will be found in Chapter 6.

References

Abramowitz M and Stegun IA (eds.) (1964) *Handbook of Mathematical Functions with Formulas, Graphs and Mathematical Tables*. Washington, DC: National Bureau of Standards.

Alfvén H (1950) Discussion of the origin of the terrestrial and solar magnetic fields. *Tellus* 2: 74–82.

Bachtiar AA, Ivers DJ, and James RW (2006) Planar velocity dynamos in a sphere. *Proceedings of the Royal Society of London A* 462: 2439–2456.

Backus G (1958) A class of self-sustaining dissipative spherical dynamos. *Annals of Physics (NY)* 4: 372–447.

Backus G and Chandrasekhar S (1956) On Cowling's theorem on the impossibility of self-maintained axisymmetric dynamos. *Proceedings of the National Academy of Sciences Washington* 42: 105–108.

Braginsky S (1994) The nonlinear dynamo and model-*Z*. In: Proctor M, and Gilbert A (eds.) *Lectures on Solar and Planetary Dynamos*, pp. 267–304. Cambridge, UK: Cambridge University Press.

Braginsky SI (1964a) Kinematic models of the Earth's hydromagnetic dynamo. *Geomagnetism and Aeronomy* 4: 572–583.

Braginsky SI (1964b) Magnetohydrodynamics of the Earth's core. *Geomagnetism and Aeronomy* 4: 698–712.

Braginsky SI (1965a) Self-excitation of a magnetic field during the motion of a highly conducting fluid. *Proceedings of the National Academy of Sciences Washington* 20: 726–735.

Braginsky SI (1965b) Theory of the hydromagnetic dynamo. *Soviet Physics JETP* 20: 1462–1471.

Braginsky SI (1967) Magnetic waves in the Earth's core. *Geomagnetism and Aeronomy* 7: 851–859.

Braginsky SI (1970) Torsional magnetohydrodynamic vibrations in the Earth's core and variations in day length. *Geomagnetism and Aeronomy* 10: 1–8.

Braginsky SI (1984) Short-period geomagnetic secular variation. *Geophysical and Astrophysical Fluid Dynamics* 30: 1–78.

Braginsky SI and Meytlis VP (1990) Local turbulence in the Earth's core. *Geophysical and Astrophysical Fluid Dynamics* 55: 71–87.

Braginsky SI and Roberts PH (1995) Equations governing convection in the Earth's core and the geodynamo. *Geophysical and Astrophysical Fluid Dynamics* 79: 1–97.

Brandenburg A, Moss D, and Soward AM (1998) New results for the Herzenberg dynamo: Steady and oscillatory solutions. *Proceedings of the Royal Society of London A* 454: 1283–1300, erratum p. 3275.

Bullard EC and Gellman H (1954) Homogeneous dynamos and terrestrial magnetism. *Philosophical Transactions of the Royal Society of London A* 247: 213–255.

Bullard EC and Gubbins D (1977) Generation of magnetic fields of global scale. *Geophysical and Astrophysical Fluid Dynamics* 8: 43–56.

Busse FH (1970) Thermal instabilities in rapidly rotating systems. *Journal of Fluid Mechanics* 44: 441–460.

Busse FH (1975) A necessary condition for the geodynamo. *Journal of Geophysical Research* 80: 278–280.

Childress S (1969a) A class of solutions of the magnetohydrodynamic dynamo problem. In: Runcorn SK (ed.) *The Application of Modern Physics to the Earth and Planetary Interiors*, pp. 629–648. New York: Wiley.

Childress S (1969b) Théorie magnetohydrodynamique de l'effet dynamo. Technical report, Departement Méchanique de la Faculté des Sciences, Université de Paris.

Childress S (1970) New solutions of the kinematic dynamo problem. *Journal of Mathematical Physics* 11: 3063–3076.

Childress S and Gilbert AD (1995) *Stretch, Twist, Fold: The Fast Dynamo*. Heidelberg: Springer.
Cowling TG (1934) The magnetic field of sunspots. *Monthly Notices of the Royal Astronomical Society* 94: 39–48.
Cowling TG (1957) The dynamo maintenance of steady magnetic fields. *Quarterly Journal of Mechanics and Applied Mathematics* 10: 129–136.
Demircan A and Seehafer N (2002) Dynamo in asymmetric square convection. *Geophysical and Astrophysical Fluid Dynamics* 96: 461–479.
Dudley ML and James RW (1989) Time-dependent kinematic dynamos with stationary flows. *Proceedings of the Royal Society of London A* 425: 407–429.
Elsasser WM (1947) Induction effects in terrestial magnetism. Part III. Electric modes. *Physical Review* 72: 831–833.
Eltayeb IA and Loper DE (1988) On steady kinematic helical dynamos. *Geophysical and Astrophysical Fluid Dynamics* 44: 259–269.
Fearn DR (1998) Hydromagnetic flows in planetary cores. *Reports on Progress in Physics* 61: 175–235.
Furth HP, Killeen J, and Rosenbluth MN (1963) Finite resistive instabilities of a sheet pinch 6: 459–484.
Gailitis A (1990) The helical MHD dynamo. In: Moffatt HK and Tsinober A (eds.) *Topological Fluid Mechanics*, pp. 147–156. Cambridge, UK: Cambridge University Press.
Gailitis AK (1993) Magnetic field generation by axisymmetric flows of conducting liquids in a spherical stationary conductor cavity. *Magnetohydrodynamics* 29: 107–115.
Gailitis AK (1995) Magnetic field generation by the axisymmetric conducting fluid flow in a spherical cavity of a stationary conductor. 2. *Magnetohydrodynamics* 31: 38–42.
Galloway D and Frisch U (1986) Dynamo action in a family of flows with chaotic streamlines. *Geophysical and Astrophysical Fluid Dynamics* 36: 53–83.
Gibson RD (1968) The Herzenberg dynamo. II. Quart, *Journal of Mechanics and Applied Mathematics* 21: 257–267.
Gibson RD and Roberts PH (1966) Some comments on the theory of homogenous dynamos. In: Hindmarsh W, Lowes FJ, Roberts PH, and Runcorn S (eds.) *Magnetism and the Cosmos*, pp. 108–120. Edinburgh: Oliver and Boyd.
Gilbert AD (1988) Fast dynamo action in the Ponomarenko dynamo. *Geophysical and Astrophysical Fluid Dynamics* 44: 241–258.
Glatzmaier GA and Roberts PH (1995) A three-dimensional self-consistent computer simulation of a geomagnetic field reversal. *Nature* 377: 203–209.
Glatzmaier GA and Roberts PH (1996a) An anelastic evolutionary geodynamo simulation driven by compositional and thermal convection. *Physica D* 97: 81–94.
Glatzmaier GA and Roberts PH (1996b) On the magnetic sounding of planetary interiors. *Physics of the Earth and Planetary Interiors* 98: 207–220.
Greenspan H (1968) *The Theory of Rotating Fluids*. Cambridge, UK: Cambridge University Press.
Gubbins D (1973) Numerical solution of the kinematic dynamo problem. *Philosophical Transactions of the Royal Society of London A* 274: 493–521.
Gubbins D, Barber CN, Gibbons S, and Love JJ (2000a) Kinematic dynamo action in a sphere. I. Effects of differential rotation and meridional circulation on solutions with axial dipole symmetry. *Proceedings of the Royal Society of London A* 456: 1333–1353.
Gubbins D, Barber CN, Gibbons S, and Love JJ (2000b) Kinematic dynamo action in a sphere. II. Symmetry selection. *Proceedings of the Royal Society of London A* 456: 1669–1683.
Gubbins D and Gibbons S (2002) Three dimensional dynamo waves in a sphere. *Geophysical and Astrophysical Fluid Dynamics* 96: 481–498.
Herzenberg A (1958) Geomagnetic dynamos. *Philosophical Transactions of the Royal Society of London A* 250: 543–583.
Hide R (1978) How to locate the electrically conducting fluid core of a planet from external magnetic observation. *Nature* 271: 640–641.
Hide R (1979) Dynamo theorems. *Geophysical and Astrophysical Fluid Dynamics* 14: 183–186.
Hide R and Palmer TN (1982) Generalisation of Cowling's theorem. *Geophysical and Astrophysical Fluid Dynamics* 19: 301–309.
Holme R (1997) Three-dimensional kinematic dynamos with equatorial symmetry: Application to the magnetic fields of Uranus and Neptune. *Physics of the Earth and Planetary Interiors* 102: 105–122.
Hutcheson KA and Gubbins D (1994) Kinematic magnetic field morphology at the core mantle boundary. *Geophysical Journal International* 116: 304–320.
Ivers DJ and James RW (1981) On the maintenance of magnetic fields by compressible flows and the Nernst–Ettingshausen effect. *Geophysical and Astrophysical Fluid Dynamics* 16: 319–323.
Ivers DJ and James RW (1984) Axisymmetric antidynamo theorems in non-uniform compressible fluids. *Philosophical Transactions of the Royal Society of London A* 312: 179–218.
Ivers DJ and James RW (1986) Extension of the Namikawa–Matsushita anti-dynamo theorem to toroidal fields. *Geophysical and Astrophysical Fluid Dynamics* 36: 317–324.
Ivers DJ and James RW (1988a) An anti-dynamo theorem for partly symmetric flows. *Geophysical and Astrophysical Fluid Dynamics* 44: 271–278.
Ivers DJ and James RW (1988b) Anti–dynamo theorems for non–radial flows. *Geophysical and Astrophysical Fluid Dynamics* 40: 147–163.
Jackson A, Bloxham J, and Gubbins D (1993) Time-dependent flow at the core surface and conservation of angular momentum in the coupled core-mantle system. In: Le Mouël J-L, Smylie D, and Herring T (eds.) *Geophysical Monograph 72: Dynamics of the Earth's Deep Interior and Earth Rotation*, pp. 97–107. Washington, DC: AGU.
Jault D, Gire G, and Mouël J-LL (1988) Westward drift, core motions and exchanges of angular momentum between core and mantle. *Nature* 333: 353–356.
Jones C, Soward A, and Mussa A (2000) The onset of thermal convection in a rapidly rotating sphere. *Journal of Fluid Mechanics* 405: 157–179.
Kaiser R, Schmitt BJ, and Busse FH (1994) On the invisible dynamo. *Geophysical and Astrophysical Fluid Dynamics* 77: 93–109.
Kono M and Roberts PH (1991) Small amplitude solutions of the dynamo problem: 1. The adjoint system amd its solution. *Journal of Geomagnetics and Geoelectricity* 43: 839–862.
Kono M and Tanaka H (1995) Intensity of the geomagnetic field in geological time: A statistical study. In: Yukutake T (ed.) *The Earth's Central Part: Its Structure and Dynamics*, pp. 75–94. Tokyo, Japan: Terra Scientific Publishing Co.
Krause F and Rädler K-H (1980) *Mean-field Magnetohydrodynamics and Dynamo Theory*. Berlin: Akademic Press.
Kuang W and Roberts PH (1990) Resistive instabilities in rapidly rotating fluids: Linear theory of the tearing mode. *Geophysical and Astrophysical Fluid Dynamics* 55: 199–239.
Kumar S and Roberts PH (1975) A three-dimensional kinematic dynamo. *Proceedings of the Royal Society of London A* 414: 235–258.
Larmor J (1919) How could a rotating body such as the Sun become a magnet? 159–160.
Léorat J (1995) Linear dynamo simulations with time-dependent helical flows. *Magnetohydrodynamics* 31: 367–373.

Liao X, Zhang K, and Gubbins D (2005) A multi-layered kinematic dynamo model: Implications of a stratified upper layer in the Earth's core. *Geophysical and Astrophysical Fluid Dynamics* 99: 377–395.

Lortz D (1968) Exact solutions of the hydromagnetic dynamo problem. *Plasma Physics* 10: 967–972.

Lortz D (1990) Mathematical problems in dynamo theory. In: Brinkmann W, Fabian AC, and Giovannelli F (eds.) *Physical Processes in Hot Cosmic Plasmas*, pp. 221–234. Dordrecht: Kluwer.

Love JJ and Gubbins D (1996a) Dynamos driven by poloidal flow exist. *Geophysical Journal International* 23: 857–860.

Love JJ and Gubbins D (1996b) Optimized kinematic dynamos. *Geophysical Journal International* 124: 787–800.

Lowes FJ and Wilkinson I (1963) Geomagnetic dynamo: A laboratory model. *Nature* 198: 1158–1160.

Lowes FJ and Wilkinson I (1968) Geomagnetic dynamo: An improved laboratory model. *Nature* 219: 717–718.

Lupian EA and Shukurov A (1992) The screw dynamo in realistic flows. *Magnetohydrodynamics* 28: 234–240.

Lupian EA, Shukurov AM, and Sokolov DD (1996) Dynamo action in swirling jets. *Magnetohydrodynamics* 32: 101–104.

Malkus WVR and Proctor MRE (1975) The macrodynamics of α–effect dynamos in rotating fluid systems. *Journal of Fluid Mechanics* 67: 417–443.

Merrill R, McElhinny M, and McFadden P (1996) *The Magnetic Field of the Earth. Paleomagnetism, the Core and the Deep Mantle,* 2nd edn. xii + 527 pp. San Diego CA: Academic Press.

Moffatt HK (1978) *Magnetic Field Generation in Electrically Conducting Fluids*, 264 pp. Cambridge UK: Cambridge University Press.

Moffatt HK and Loper DE (1994) Hydromagnetics of the Earth's core, I. The rise of a buoyant blob. *Geophysical Journal International* 117: 394–402.

Namikawa T and Matsushita S (1970) Kinematic dynamos problem. *Geophysical Journal of the Royal Astronomical Society* 19: 319–415.

Parker EN (1955) Hydromagnetic dynamo models 121: 293–314.

Parker EN (1957) The solar hydromagnetic dynamo. *Proceedings of the National Academy of Sciences Washington* 43: 8–14.

Pekeris CL, Accad Y, and Schkoller B (1973) Kinematic dynamos and the Earth's magnetic field. *Philosophical Transactions of the Royal Society of London A* 275: 425–461.

Ponomarenko YB (1973) On the theory of the hydromagnetic dynamo. *Journal of Applied Mechanics and Technical Physics* 14: 775–778.

Proctor MRE (1977a) On Backus' necessary condition for dynamo action in a conducting sphere. *Geophysical and Astrophysical Fluid Dynamics* 9: 89–93.

Proctor MRE (1977b) The role of mean circulation in parity selection by planetary magnetic fields. *Geophysical and Astrophysical Fluid Dynamics* 8: 311–324.

Proctor MRE (2004) An extension of the toroidal theorem. *Geophysical and Astrophysical Fluid Dynamics* 98: 235–240.

Roberts GO (1969) Dynamo waves. In: Runcorn SK (ed.) *The Application of Modern Physics to the Earth and Planetary Interiors*, pp. 602–628. New York: Wiley.

Roberts GO (1972) Dynamo action of fluid motions with two-dimensional periodicity. *Philosophical Transactions of the Royal Society of London A* 271: 411–454.

Roberts PH (1960) Characteristic value problems posed by differential equations arising in hydrodynamics and hydromagnetics. *Journal of Mathematical Analysis and Applications* 1: 195–214.

Roberts PH (1967) *An Introduction to Magnetohydrodynamics*, pp. 264. London: Longmans, Green and Co.

Roberts PH (1968) On the thermal instability of a rotating-fluid sphere containing heat sources. *Philosophical Transations of the Royal Society of London A* 263: 93–117.

Roberts PH (1971a) Dynamo theory. In: Reid WH (ed.) *Lectures in Mathematics 14: Mathematical Problems in the Geophysical Sciences. 2. Inverse Problems, Dynamo Theory and Tides*, pp. 129–206. Providence, RI: American Mathematical Society.

Roberts PH (1971b) Dynamo theory of geomagnetism. In: Zmuda AJ (ed.) *World Magnetic Survey 1957–1969*, pp. 123–131. Paris: IUGG.

Roberts PH (1972) Kinematic dynamo models. *Philosophical Transactions of the Royal Society of London A* 271: 663–697.

Roberts PH (1987) Dynamo theory. In: Nicolis C and Nicolis G (eds.) *NATO ASI Series 192: Irreversible Phenomena and Dynamical Systems Analysis in Geosciences,* pp. 73–133. Dordrecht: Reidel.

Roberts PH, Yu ZJ, and Russell CT (2007) On the 60–year signal from the core. *Geophysical and Astrophysical Fluid Dynamics* 101: 11–35.

Rüdiger G and Hollerbach R (2004) The Magnetic Universe, Geophysical and Astrophysical Dynamo Theory, 332 pp. Weinheim: Wiley-VCH.

Ruzmaikin AA, Sokoloff DD, and Shukurov AM (1988) A hydromagnetic screw dynamo. *Journal of Fluid Mechanics* 197: 37–56.

Ruzmaikin AA, Sokoloff DD, Solovev AA, and Shukurov AM (1989) Couette–Poiseuille flow as a screw dynamo. *Magnetohydrodynamics* 25: 6–11.

Sarson GR and Gubbins D (1996) Three-dimensional kinematic dynamos dominated by strong differential rotation. *Journal of Fluid Mechanics* 306: 223–265.

Sarson GR, Jones C, and Longbottom A (1998) Convection driven geodynamo models of varying Ekman number. *Geophysical and Astrophysical Fluid Dynamics* 88: 225–259.

Sarson GR and Jones CA (1999) A convection driven geodynamo reversal model. *Physics of the Earth and Planetary Interiors* 111: 3–20.

Serebrianya PM (1988) Kinematic stationary geodynamo models with separated toroidal and poloidal motions. *Geophysical and Astrophysical Fluid Dynamics* 44: 141–164.

Siemens W (1867) Über die Uwandlung von Arbeitskraft in electrischen Strom ohne Anwendlung permanenter Magnete. Monats. König. Preuss. Akad. Wiss: Berlin, 55–58.

Sokoloff DD, Shukurov AM, and Shumkina TS (1989) The second approximation in the screw dynamo problem. *Magnetohydrodynamics* 25: 1–6.

Solovyev AA (1985a) Couette spiral flows of conducting fluids consistent with magnetic field generation. *Physics of Solid Earth* 12: 40–47.

Solovyev AA (1985b) Magnetic field generation by the axially-symmetric flow of conducting fluid. *Physics of Solid Earth* (4): 101–103.

Solovyev AA (1987) Excitation of a magnetic field by the movement of a conductive liquid in the presence of large values of the magnetic Reynolds number. *Physics of Solid Earth* 23: 420–423.

Soward AM (1972) A kinematic theory of large magnetic Reynolds number dynamos. *Proceedings of the Royal Society of London A* 275: 611–651.

Soward AM (1987) Fast dynamo action in a steady flow. *Journal of Fluid Mechanics* 180: 267–295.

Soward AM (1990) A unified approach to a class of slow dynamos. *Geophysical and Astrophysical Fluid Dynamics* 53: 81–107.

Soward AM (1991) The Earth's dynamo. *Geophysical and Astrophysical Fluid Dynamics* 62: 191–209.

Steenbeck M, Krause F, and Rädler K-H (1966) Berechnung der mittleren Lorentz–Feldstärke $\overline{\mathbf{v} \times \mathbf{B}}$ für ein elektrisch leitendendes Medium in turbulenter, durch Coriolis–Kräfte

beeinflußter Bewegung. Zeitschrift Für Naturforschung. 21a: 369–376.
Taylor J (1963) The magnetohydrodynamics of a rotating fluid and the earth's dynamo problem. *Proceedings of the Royal Society of London A* 274: 274–283.
Tilgner A (2004) Small scale kinematic dynamos: Beyond the α–effect. *Geophysical and Astrophysical Fluid Dynamics* 98: 225–234.
Tilgner A and Busse FH (1995) Subharmonic dynamo action of fluid motions with two-dimensional periodicity. *Proceedings of the Royal Society of London A* 448: 237–244.
Vainshtein SI and Zeldovich YB (1972) Origin of magnetic fields in astrophysics. *Soviet Physics Uspekhi* 15: 159–172.
Walker MR and Barenghi CF (1994) High resolution numerical dynamos in the limit of a thin disk galaxy. *Geophysical and Astrophysical Fluid Dynamics* 76: 265–281.
Willis AP and Gubbins D (2004) Kinematic dynamo action in a sohere: Effects of periodic time dependent flows on solutions with axial dipole symmetry. *Geophysical and Astrophysical Fluid Dynamics* 98: 537–554.
Zatman S and Bloxham J (1997) Torsional oscillations and the magnetic field within the Earth's core. *Nature* 388: 760–763.
Zeldovich YB (1956) The magnetic field in the two-dimensional motion of a conducting fluid. *Soviet Physics JETP* 31: 154–156.
Zeldovich YB and Ruzmaikin AA (1980) Magnetic field in a conducting fluid in two dimensional motion. *Soviet Physics JETP* 51: 493–497.
Zheligovsky V and Galloway D (1998) Dynamo action in Christopherson hexagonal flow. *Geophysical and Astrophysical Fluid Dynamics* 88: 277–293.

4 Large-Scale Flow in the Core

R. Holme, University of Liverpool, Liverpool, UK

4.1	**Introduction**	107
4.2	**Surface Core Flow from Observed Secular Variation**	108
4.2.1	Westward Drift	108
4.2.2	The Frozen-Flux Hypothesis	109
4.2.3	Poloidal–Toroidal Decomposition	110
4.2.4	Boundary Conditions	110
4.2.5	Nonuniqueness	111
4.2.6	The Large-Scale Approximation	111
4.2.7	Steady Flows	112
4.2.8	Toroidal Flows	112
4.2.9	Tangentially Geostrophic Flows	113
4.2.10	Comparison of Solutions Constructed with Different Nonuniqueness-Reducing Assumptions	114
4.2.11	Other Flow Modeling Assumptions	115
4.2.11.1	Flow in a drifting reference frame	115
4.2.11.2	Helical flow	115
4.2.11.3	Columnar flow	115
4.2.12	Limitations in Resolution of Nonuniqueness	115
4.2.13	The Horizontal Field Components	116
4.2.14	Effects of Diffusion	117
4.3	**Higher-Resolution Flows from Detailed Models of SV from Satellite Observations**	118
4.4	**Angular Momentum – LOD Variation, and Correlation with Core Angular Momentum**	121
4.5	**Torsional Oscillations as a Probe of Core Structure**	122
4.5.1	Probing the Magnetic Field Interior to the Core	122
4.5.2	Coupling of Torsional Oscillations to the Outer and Inner Core	123
4.5.3	Geomagnetic Jerks, and Links to Earth's Rotation	124
4.6	**Wave Motion as an Explanation for Secular Variation**	124
4.7	**Polar Vortices**	125
4.8	**Modeled Core Flow and the Dynamics of the Core**	127
References		128

4.1 Introduction

Timescales of processes in the Earth tend in general to be very short (e.g., the rupture associated with an earthquake) or very long (millions of years or longer) (e.g., the movement of the tectonic plates). The geomagnetic field provides the primary example of intermediate timescales. The internal magnetic field, with origin in the Earth's fluid core, varies on timescales from yearly or less (the so-called geomagnetic jerks) through decadal variation, secular change over centuries, to more significant variations (excursions and reversals) on timescales of millenia. These changes are named the secular variation. This chapter is concerned with variations on timescales of years to centuries, and what can be deduced about the interior of the Earth and processes in the core from such changes. This subject has a long history: the first attempt to infer motion in the Earth's interior from geomagnetic secular variation was by Halley (1692). Modern analysis interprets the secular variation as being primarily due to advection of magnetic field at the top of the core, and so uses magnetic observations to constrain flow there.

In Section 4.2, we begin by outlining how neglect of diffusion in the core allows the simplification of the magnetic induction equation to a form that can be used to solve for flow. This solution is highly nonunique: we review the approaches that have been taken to deal with this nonuniqueness, and discuss to what extent they influence the resulting models of flow. We also discuss some attempts to allow for diffusional effects. In Section 4.3, we consider recent fine-scale flow models which have attempted to take advantage of new, detailed models of the magnetic field and (particularly) secular variation from satellite data. Effective modeling of flow from such models is proving difficult.

In Section 4.4, we consider the link between flow modeling and an important independent geophysical observable, the variation in length of day (LOD). Decadal variations in LOD almost certainly originate from the exchange of angular momentum between the fluid core and solid Earth, and simple models of whole-core flow constrained by surface flow modeling provide strong support for this. The existence of this correlation, and the underlying theory of torsional oscillations, provides an opportunity to probe physical processes not just at the surface of the core, but also at depth: this is described in Section 4.5. There is increasing evidence that torsional processes are also responsible for the most rapid feature seen in the internal secular variation, the so-called geomagnetic jerks (Section 4.5.3), promising a probe of core processes on timescales as little as a year or even less.

Large-scale flow provides one paradigm for secular variation, but another, hydromagnetic waves, is considered briefly in Section 4.6. In Section 4.7, we focus in detail on a particular small-scale structure in the flow, in the polar regions north and south of the inner core, the so-called polar vortices. Finally, in Section 4.8, we discuss briefly attempts to unify the kinematics of flow modeling with more dynamical approaches, for further understanding of core processes.

4.2 Surface Core Flow from Observed Secular Variation

4.2.1 Westward Drift

The time variation (secular variation) of the magnetic field was first recognized in the late seventeenth century. In attempting to provide a theory to explain it, Halley (1692) noticed that a large part of the secular variation could be explained in terms of a 'westward drift' of the field. When maps of the field are produced over the Earth's surface at different epochs, there is a clear westward movement of features in the field. This apparent motion is seen particularly clearly if lines of zero declination (so-called agonic lines) are plotted at different epochs (e.g., Langel, 1987). Halley estimated that a full revolution of the field would take about 700 years, giving a rotation rate of just over $0.5° \text{ yr}^{-1}$. To explain the drift, he posited a model of the interior of the Earth consisting of concentric shells of magnetic material rotating at varying rates with respect to the Earth's surface. While much of his theory was unavoidably flawed (the interior of the Earth is too hot to support permanent magnetization at depths greater than about 100 km, and there are no large void spaces within the Earth, and in particular, no people living in them awaiting religious conversion!), the fundamental insight of the magnetic secular variation resulting from large-scale differential motion between different components of the Earth remains fully valid. In many ways, his theory has remarkable similarities to our current understanding. The Earth does consist of different layers, one of which (the fluid outer core) gives rise to the secular variation, of which a large part can indeed be represented by westward motions at the surface of the core.

As understanding of the Earth's interior improved, it became clear that Halley's theory was inadequate both observationally and theoretically, not least because the thermal state of the Earth does not allow permanent magnetization at depth. With the advent of dynamo theory, and the understanding that the magnetic field originates from fluid motions in the molten iron core, the first 'modern' theory of westward drift was proposed by Bullard *et al.* (1950). They argued that cooling of the Earth would lead to a pattern of largescale convection in the core. As a parcel of fluid rose, it would move further from the Earth's rotation axis. Therefore, to conserve angular momentum, the angular velocity of the parcel would reduce. Similarly, a sinking fluid parcel would reduce its moment arm, and so need to increase its angular velocity. Bullard *et al.* suggested that this would lead to a net eastward flow with respect to the solid Earth at the base of the core near the inner-core boundary (ICB), and westward flow at the core surface, which would then carry the magnetic field in a westerly direction, giving rise to westward drift. While the basic physical ideas of this theory are very attractive, modeling of rotating convection has shown that the interaction of convection and rotation is much more complicated, and this pleasingly simple explanation for the drift is not physically viable.

Further, westward drift is an oversimplification of the magnetic secular variation. Estimates of westward drift obtained vary depending on the precise definition: should it involve a fit to the whole field, just the equatorial dipole, the secular variation, or some combination? The value of Bullard *et al.* (1950) of $0.2°\ yr^{-1}$ has entered the geomagnetism consciousness as a standard, but estimates have varied from 0.8° eastward per year, up to 0.733° westward (for a summary, see Langel, 1987). What this range of values demonstrates is that the picture of westward drift is much too simplistic. Even in the eighteenth century, it was realized that some features of the field drift northwards rather than westward, and once Gauss had developed a method for measuring magnetic intensity, the observed decay of the dipole could clearly not be explained by drift alone. Further, westward drift is much more clearly visible in Europe and North America than in Asia and the Pacific Hemisphere. Yukutake and Tachinaka (1969) argued that other features in the field did not move at all, and proposed a division of the field into drifting and standing components, while Yukutake (1979) further suggested that the equatorial dipole field could be separated into two components, drifting in opposite directions, a process reminiscent of wave motion.

With the advent of detailed time-dependent models of the magnetic field at the core–mantle boundary (CMB), the idea of simple westward drift has been superseded, although as we shall see, westward motion is still a major component of most models of core-surface flow. It is these models that are the primary focus of this chapter.

4.2.2 The Frozen-Flux Hypothesis

We turn therefore to a more detailed mathematical description of the generation of the secular variation. The evolution of the magnetic field in the core is governed by the magnetic induction equation

$$\frac{\partial \mathbf{B}}{\partial t} = \nabla \wedge (\mathbf{u} \wedge \mathbf{B}) + \eta \nabla^2 \mathbf{B} \qquad [1]$$

where $\mathbf{B}$ is the magnetic field, $\mathbf{u}$ the fluid velocity and $\eta = 1/(\mu_0 \sigma)$ is the magnetic diffusivity, defined in terms of the permeability of free space μ_0 and the electrical conductivity σ of the core fluid. For the derivation of this equation, see the discussion of secular variation by Jackson. Using this kinematic description of the system, we will derive and apply methods to determine the flow at the top of the core from time-varying models of the magnetic field.

For a full description of the flow in the core, we must also consider the dynamics, governed by the Navier–Stokes equations (in the Boussinesq approximation)

$$\rho_0 \left(\frac{\partial \mathbf{u}}{\partial t} + \mathbf{u} \cdot \nabla \mathbf{u} + 2\boldsymbol{\Omega} \wedge \mathbf{u} \right) = -\nabla p + \rho' \mathbf{g} + \mathbf{J} \wedge \mathbf{B} + \rho_0 \nu \nabla^2 \mathbf{u} \qquad [2]$$

where ρ_0 is the hydrostatic density, ρ' the departure of density from the hydrostatic state, $\boldsymbol{\Omega}$ is the Earth's rotation vector, p is the nonhydrostatic pressure, $\mathbf{g}$ is the acceleration due to gravity, and $\mathbf{J} = (1/\mu_0)\nabla \wedge \mathbf{B}$ is the current density. Although the dynamics are not a primary focus of this chapter, the dynamical equation will be used in a simplistic way to provide additional constraints on flow. For more details on the full dynamical system *see* Chapters 3 and 5.

The primary tool for observational constraint of core fluid motions is the radial component of the induction equation at the CMB. The CMB is approximated as spherical, and because it is a material boundary the radial component of the flow there is zero. Then, the radial component of eqn [1] can be written as

$$\dot{B}_r = -\nabla_H \cdot (\mathbf{u}_H B_r) + \frac{\eta}{r} \nabla^2 (r B_r) \qquad [3]$$

where $\mathbf{u}_H$ are the flow components in the horizontal directions, and $\nabla_H \cdot$ is the horizontal part of the divergence operator. In other words, the rate of change of the magnetic field at the CMB is the sum of the contributions from the advection of the radial field (the first term) and its diffusion (the second term). A simple scaling analysis of this equation allows estimation of the relative importance of advection to diffusion in the generation of the secular variation. Typical scales in the core might be $U \simeq 5 \times 10^{-4}\ \mathrm{m\,s^{-1}}$, $L \simeq 1000\ \mathrm{km}$, $\eta \simeq 1\ \mathrm{m^2\,s^{-1}}$, giving $R_\mathrm{m} \approx 500$, suggesting that advection dominates diffusion. If we neglect diffusion, then eqn [3] becomes

$$\dot{B}_r + \nabla_H \cdot (\mathbf{u} B_r) = 0 \qquad [4]$$

This equation is the starting point for modeling of flow at the core surface, as originally proposed by Roberts and Scott (1965). It is named the frozen-flux induction equation, because it can be shown that the magnetic field is carried along passively with the flow and so provides a tracer of that flow. The justification for dropping diffusion is not particularly strong: Bloxham and Jackson (1991) among others discuss many of the factors that might make this assumption fail. Perhaps most obviously, the estimate of length scale may be inadequate: it is motivated by possible horizontal length scales of the magnetic field. It is likely that the radial length scale for magnetic field variation at the CMB may be shorter, due to the formation of hydromagnetic boundary layers, strongly increasing the importance of the diffusion term. Unfortunately, the radial dependence of the magnetic field cannot be constrained by observations. Even neglect of the diffusion term in the horizontal directions may be inappropriate, a point to which we will return later. Nevertheless, eqn [4] provides the primary tool for the modeling of core-surface flow. Given a model of the radial field B_r, and either a model of secular variation $\dot{B}_r$, or direct estimates at the Earth's surface (e.g., from yearly differences of magnetic observatory annual means) it is possible to model (or at least constrain) the surface flow **u**.

4.2.3 Poloidal–Toroidal Decomposition

For a spherical geometry, it proves useful to consider both the magnetic field and flow in terms of a poloidal–toroidal decomposition (*see*, e.g., Gubbins and Roberts (1987); *see* Chapter 3). Because of Gauss's theorem for magnetic field (and the absence of magnetic monopoles), and the incompressibility condition for flow from the Boussinesq approximation, it follows that both the field and flow are divergence free

$$\nabla \cdot \mathbf{B} = 0 \qquad [5]$$

$$\nabla \cdot \mathbf{u} = 0 \qquad [6]$$

As a result, we may write field and flow as

$$\mathbf{B} = \nabla \wedge (T\mathbf{r}) + \nabla \wedge \nabla \wedge (S\mathbf{r}) \qquad [7]$$

$$\mathbf{u} = \nabla \wedge (t\mathbf{r}) + \nabla \wedge \nabla \wedge (s\mathbf{r}) \qquad [8]$$

where T, t and S, s are toroidal and poloidal scalars, respectively, and **r** is the position vector. This leads to expressions for the flow in spherical coordinates (r, θ, ϕ) (radius, colatitude, longitude) of

$$\mathbf{u}_T = \left(0, \frac{1}{\sin\theta}\frac{\partial t}{\partial \phi}, -\frac{\partial t}{\partial \theta}\right) \qquad [9]$$

$$\mathbf{u}_S = \left(\frac{L^2 s}{r}, \frac{1}{r}\frac{\partial (rs)}{\partial r \partial \theta}, \frac{1}{r\sin\theta}\frac{\partial (rs)}{\partial r \partial \phi}\right) \qquad [10]$$

and similar expressions for the field, where

$$L^2 = -\left\{\frac{1}{\sin\theta}\frac{\partial}{\partial\theta}\left(\sin\theta\frac{\partial}{\partial\theta}\right) + \frac{1}{\sin^2\theta}\frac{\partial^2}{\partial\phi^2}\right\} \qquad [11]$$

is the angular momentum operator from quantum mechanics. (Note that this expansion for the poloidal–toroidal decomposition is by no means unique: often, e.g., the unit vector $\hat{\mathbf{r}}$ replaces the position vector **r** in eqns [7] and [8].) We must consider the boundary conditions for both field and flow. As already noted, defining an impenetrable, spherical CMB gives zero radial velocity there, leading to an alternative form for the flow decomposition. From eqn [10], zero radial flow requires that $s = 0$, but $\partial s/\partial r$ is nonzero. Writing $\mathcal{T} = t$, $\mathcal{S} = (1/r)(\partial(rs)/\partial r)$, we obtain an expression for the horizontal flow

$$\mathbf{u}_H = \nabla \wedge (\mathcal{T}\mathbf{r}) + \nabla_H r\mathcal{S} \qquad [12]$$

To solve for the flow, each of the scalars is further expanded on a basis of spherical harmonics, most commonly real, Schmidt-normalized harmonics, defining

$$\mathcal{T} = \sum_{l=1}^{\infty}\sum_{m=0}^{l} P_l^m(\cos\theta)\left[\mathcal{T}_l^{mc}\cos m\phi + \mathcal{T}_l^{ms}\sin m\phi\right]$$
$$\mathcal{S} = \sum_{l=1}^{\infty}\sum_{m=0}^{l} P_l^m(\cos\theta)\left[\mathcal{S}_l^{mc}\cos m\phi + \mathcal{S}_l^{ms}\sin m\phi\right] \qquad [13]$$

where $P_l^m(\cos\theta)$ are associated Legendre functions of degree l and order m (see, e.g., Langel (1987)). These expansions are then substituted into eqn [4], along with expressions for field and secular variation derived from spherical harmonic models of the potential magnetic field at the Earth's surface. Spherical harmonic orthogonality is used to derive a linear system of equations for $\{\mathcal{T}_l^m, \mathcal{S}_l^m\}$. Solution for this set of flow coefficients provides a solution for the flow (e.g., Whaler, 1986).

4.2.4 Boundary Conditions

In order to estimate the flow near the surface of the core, we must consider the boundary conditions in detail. One boundary condition has already been discussed, which is that no penetration of the CMB requires zero radial velocity. However, as the CMB is a rigid boundary, considerations of hydrodynamics also require that the horizontal components of velocity there are zero. Clearly, therefore, determining the flow precisely at the CMB is trivial, but not very informative – it is zero there! When we speak of obtaining the flow at the top of the core, we are generally referring to the flow at the top of the 'free stream' – the region just below the hydromagnetic boundary layers at the CMB. The jump in the radial component of the field across this layer is constrained by a scaling analysis of the condition $\nabla \cdot \mathbf{B} = 0$, which gives

$$\frac{\partial B_r}{\partial r} \sim \nabla_H \cdot \mathbf{B}_H \qquad [14]$$

As the radial length scale of the boundary layer is much smaller than the horizontal length scale of the whole core, it follows that the change in B_r is small. Therefore, we may adopt the radial field at the CMB as appropriate for the radial field at the top of the free stream. Analysis of the perpendicular field components is less clear, particularly concerning the toroidal magnetic field (eqn [7]), but as we shall not use these perpendicular components here, we will pass over these problems.

4.2.5 Nonuniqueness

Equation [4] provides the fundamental tool with which it is possible to determine a surface core flow from observations. Its use was first attempted by Kahle *et al.* (1967), but problems soon became apparent. An ambiguity in part of the flow was noticed by Roberts and Scott (1965), and later quantified by Backus (1968). Backus wrote the product of the flow and radial field in the form

$$\mathbf{u}B_r = \nabla_H\Phi - \hat{\mathbf{r}} \wedge \nabla_H\Psi \qquad [15]$$

Substituting this decomposition into eqn [4] leads to

$$\dot{B}_r + \nabla_H^2\Phi = 0 \qquad [16]$$

eliminating the scalar Ψ. Thus, the part of the flow corresponding to this scalar does not contribute to secular variation and is not constrained by field observations except for some weak continuity conditions at null-flux curves (lines of $B_r = 0$). (In other words, Ψ defines the 'null space' of the flow.) This problem is also clear from a very simple argument: even given the assumed zero radial velocity, there remain two components of flow to be determined (e.g., northwards and eastwards) over the surface of the core. Equation [4] provides only one equation with which to do this, so with fewer equations than unknowns, nonuniqueness should be no surprise. From expanding eqn [4] as

$$\dot{B}_r + B_r\nabla_H \cdot \mathbf{u} + \mathbf{u} \cdot \nabla_H B_r = 0 \qquad [17]$$

Backus (1968) observed that a component of the flow is only determined independently at points on the core surface on the null-flux curves, and even at these locations, only that component of the flow perpendicular to the curve.

Clearly, a second equation for the flow (at least) is required: while this does not guarantee a formally unique solution, clearly two equations for two unknown flow components are better than just one. Such an equation can be obtained by considering various dynamical approximations; we discuss the most frequently used below, but begin with a 'hidden' assumption of at least equal importance.

4.2.6 The Large-Scale Approximation

The most important condition required for flow calculation does not provide an additional equation, and is often overlooked, because it is so ingrained in the methods of solution. The so-called large-scale approximation is the assumption that the flows that generate the observed SV are large scale at the surface of the core. It would always be possible to fit observed SV exactly by taking a flow of arbitrarily small scale. However, such small scales are not required to explain the observations, and therefore cannot be reliably recovered from them. Two approaches have been used to apply this condition. The simpler is to limit the small scales of the flow by truncation of the spherical harmonic series in eqn [14] (e.g., Voorhies, 1986): by not parametrizing small scales, they are required to be zero. This approach requires no additional physical assumptions about the flow, but is arbitrary and mathematically unattractive, similar to applying Fourier methods without any tapering, leading to possible 'ringing'. A more common approach is to apply some form of regularization or 'damping'. One physically motivated form of damping is to minimize the mean-square velocity in the flow integrated over the area of the CMB,

$$\int_{\mathrm{CMB}} (v_\theta^2 + v_\phi^2)\,\mathrm{d}S \qquad [18]$$

limiting its kinetic energy (Madden and Le Mouël, 1982). Problems from the implementation of this norm (essentially due to the lack of convergence of the flow and its implications for other physical constraints) led Bloxham (1988) to propose what has become known as the strong norm, which can be written as

$$\int_{\mathrm{CMB}} \left((\nabla_H^2 v_\theta)^2 + (\nabla_H^2 v_\phi)^2\right)\mathrm{d}S \qquad [19]$$

Physical interpretation of this norm can be attempted, but it is probably best to regard it merely as penalizing small-scale structure (roughness) more severely than condition [18]. Expanding the flow into its toroidal and poloidal parts following eqn [14], damping condition [18] requires minimization of

$$\sum_{l,m}\left(\frac{l(l+1)}{2l+1}\right)\left(\mathcal{T}_l^{m2}+\mathcal{S}_l^{m2}\right) \qquad [20]$$

while damping condition [19] minimizes to

$$\sum_{l,m}\left(\frac{l^3(l+1)^3}{2l+1}\right)\left(\mathcal{T}_l^{m2}+\mathcal{S}_l^{m2}\right) \qquad [21]$$

Hence, the strong norm increases the penalty on higher-degree flow components.

We emphasize that all of the more physical (and 'higher profile') approximations that follow are useful only because of the inherent assumption that the flow is large scale.

4.2.7 Steady Flows

The first *a priori* assumption adopted was to model the flow as steady (Gubbins, 1982) – that is, $\mathcal{T}$, $\mathcal{S}$ in eqn [13] are independent of time. This was motivated partially by computational convenience – considering a flow which is steady rather than time dependent greatly reduces the number of free parameters which must be considered. However, more physical justification is also possible. The steady part of the secular variation accounts for a large fraction of its power, which can be related to a steady flow (consider eqn [4]). Further, many features of the field at the CMB change approximately uniformly with time: for example, the close-to-uniform westward progression of flux patches in the equatorial region under the Atlantic Hemisphere suggests that they are being carried along by a broadly steady flow.

Steady flows have been determined in two ways: first, a simple approach of fitting estimates of the secular variation at a range of epochs, given the magnetic field at those epochs; second, and more consistently, the nonlinear problem posed by advecting an initial field model over the period of interest using a steady flow (e.g., Bloxham, 1992). Both methods produce similar-looking flows that can explain a large fraction of the secular variation, particularly its steady component – variance reductions of over 98% being perfectly possible (Bloxham, 1992).

Resolution of the flow nonuniqueness relies on sufficient change of the field over the time of analysis to allow for view of the flow in the null space. This requirement was given in a determinant condition by Voorhies and Backus (1985), and yields a minimum time over which data must be considered for the flow to be formally unique. However, whatever the mathematical conditions imposed, the important factor is whether the assumption is physically sensible: is the flow really steady (or sufficiently slowly varying) over the time period over which it is solved for? Bloxham (1992) demonstrates that even with very high variance reductions, such flows do not well explain fine-scale detail in the SV, particularly the 60-year oscillation in the axial dipole and other more rapid variations in field coefficients. Voorhies (1993) argues that such small variations are well below the error budget imposed by unmodeled physical processes relating to core flow (such as the influence of variations in mantle conductivity, core conductivity, and core asphericity), but nonetheless, these features of the SV appear to be robust, and have the potential to provide much information about the dynamics of the top of the core – see, for example, the description of geomagnetic jerks in Section 4.5.3. In particular, time variations can be correlated against variations in Earth rotation, and as described in Section 4.4, this correlation provides perhaps the only independent confirmation that core-surface flow modeling is meaningful. In this respect, assuming a simple steady flow is unsatisfactory, despite its success in explaining the observed secular variation in a very compact way.

4.2.8 Toroidal Flows

A toroidal motion is one in which upwelling of the flow is suppressed, so that there is no radial motion of the fluid, giving $u_r = 0$ not only at the CMB, but also within the free stream. Various workers, particularly Braginsky (1999) and references therein, have suggested that the top of the core could be stably stratified, which might justify this assumption. Therefore, the radial derivative of the radial component of the flow is zero. The continuity equation gives

$$\frac{1}{r}\frac{\partial(ru_r)}{\partial r}+\nabla_H\cdot\mathbf{u}_H=0 \qquad [22]$$

giving the condition for no upwelling as

$$\nabla_H\cdot\mathbf{u}_H=0 \qquad [23]$$

implying no horizontal divergence in the surface flow. Substituting from eqn [12] gives

$$\nabla_H^2 r\mathcal{S}=0 \qquad [24]$$

and hence the condition that $\mathcal{S}=0$ over the core surface, requiring no poloidal component to the motion: hence the name toroidal flow.

The additional constraint on the flow provided by eqn [23] reduces the formal nonuniqueness in the flow determination, but does not eliminate it. Combining eqn [23] with eqn [17] leads to

$$\dot{B}_r + \mathbf{u}\cdot\nabla_H B_r = 0 \qquad [25]$$

Therefore, flow along contours of B_r (in the direction in which $\nabla_H B_r = 0$) generates no secular variation, and so is unconstrained by observations. A particularly important such contour is the geomagnetic equator, meaning that under the toroidal flow assumption, the net flow around the equator is not constrained. As already discussed, Backus (1968) demonstrated that flow can always be determined uniquely perpendicular to null-flux contours; the toroidal flow assumption extends this result to include all contours of B_r. With additional physical assumptions, a toroidal flow can be considered as being determined fully uniquely: Lloyd and Gubbins (1990) demonstrated that, in the case of an completely insulating mantle, a toroidal flow is unique if calculated using both radial and horizontal components of the induction equation. However, even if the mantle is only a weak conductor, the horizontal components of the induction equation are no longer useful, and so this result is not applied in current flow modeling.

Whaler (1980) provided observational support for the toroidal flow assumption from a local analysis of the geomagnetic field. She observed, following Roberts and Scott (1965), that at local extrema in B_r, where $\nabla_H B_r = \mathbf{0}$, eqn [25] requires

$$\dot{B}_r = 0 \qquad [26]$$

She tested this condition against geomagnetic field models (choosing available international geomagnetic reference field (IGRF) models) and found evidence that this condition holds within observational uncertainties. However, because the field is downward continued from surface observations, no such test can be conclusive, because only 'patch-averages' of the field can be determined from surface observations: the formal error on a point estimate of the field at the CMB is infinite (e.g., Backus *et al.*, 1996).

Toroidal flows have been calculated by several authors (e.g., Voorhies, 1984; Whaler, 1986; Bloxham, 1992). They fit the secular variation less well than flows with both toroidal and poloidal components, but this is not surprising as they possess half the number of free parameters for a given spherical harmonic truncation.

4.2.9 Tangentially Geostrophic Flows

An alternative dynamical assumption, and the one that has been used most frequently in flow modeling for several decades, is that of tangential geostrophy, independently proposed by Hills (1979) and Le Mouël (1984). An assumption is made that the dynamics of the flow at the top of the core is dominated by a balance between pressure gradients, rotation (via the Coriolis force), and buoyancy. Because the mantle is approximately an electrical insulator (at least compared with the core), the Lorentz force is assumed to be small. Justification for this assumption is well-summarized by Bloxham and Jackson (1991). In this approximation, the dynamical eqn [2] reduces to

$$2\rho(\boldsymbol{\Omega}\wedge\mathbf{u}) = -\nabla p + \rho'\mathbf{g} \qquad [27]$$

Curling this equation eliminates the pressure term, giving the 'thermal wind equation'

$$2\rho(\boldsymbol{\Omega}.\nabla)\mathbf{u} = \mathbf{g}\wedge\nabla\rho' \qquad [28]$$

Assuming that gravity is fully radial, the radial component of eqn [28] implies

$$\nabla_H\cdot(\mathbf{u}\cos\theta) = 0 \qquad [29]$$

(compare with the equivalent expression for toroidal flow, eqn [23]). As for toroidal flow, this assumption reduces, but does not eliminate, the formal nonuniqueness in flow determination. Substituting eqn [29] into the frozen-flux induction, eqn [4] yields

$$\dot{B}_r + \cos\theta\,\mathbf{u}\cdot\nabla_H.(B_r/\cos\theta) = 0 \qquad [30]$$

If follows that flow is undetermined along contours of $B_r/\cos\theta$. However, this remaining nonuniqueness is less severe than that for toroidal flow, as realized by Backus and Le Mouël (1986). Many of these contours intersect the geographic equator at points where $B_r = 0$. The flow along such contours is not undetermined: it is only closed contours of $B_r/\cos\theta$ around which the flow is ambiguous. These depend on the morphology of the main field, and so differ slightly from epoch to epoch, but typically make up approximately 40% of the core surface. Such areas have been named 'ambiguous patches'. The tangential geostrophic approximation must also fail close to the geographic equator. In this region, the radial component of the Coriolis term is identically zero, and so the thermal wind balance [28] must break down (Backus and Le Mouël, 1986). Numerically, eqn [29] still applies a constraint on flow at the equator, giving $u_\theta = 0$ there, equatorial flow is entirely east–west, but this constraint is not physically well-posed.

Two methods have been used to implement the tangentially geostrophic approximation. First, the flow can be expanded in terms of allowed geostrophic functions, the nonorthogonal geostrophic basis (e.g., Le Mouël *et al.*, 1985; Backus and Le Mouël, 1986; Gire and Le Mouël, 1990; Jackson, 1997). This conditions the flow to be precisely tangentially geostrophic. An alternative approach is to damp nongeostrophic components of the flow numerically, for example, by minimizing the norm

$$\int_{\text{CMB}} (\nabla_H \cdot (\mathbf{u}\cos\theta))^2 \mathrm{d}\Omega \qquad [31]$$

(e.g., Le Mouël *et al.*, 1985; Voorhies, 1993; Holme, 1998; Pais *et al.*, 2004). If this integral is identically equal to zero, then the flows are constrained to be identically tangentially geostrophic. An advantage of this second approach, pursued in detail by Pais *et al.* (2004), is that it is possible to study the effects of small departures from geostrophy. Such a study seems appropriate, because the tangential geostrophic assumption is derived depending on other terms in the force balance being small, not zero, and so it is important to check that forcing the flow to be exactly tangentially geostrophic does not impose artifacts on the solution.

4.2.10 Comparison of Solutions Constructed with Different Nonuniqueness-Reducing Assumptions

Bloxham and Jackson (1992) present a comparison of flows derived by several groups with each of the above assumptions (steady, toroidal, and tangentially geostrophic). Here we present flow of each type derived in the same way from the same field models, so as to focus on the different physical features of the resulting flows, rather than potential methodological differences. We consider the field and secular variation around 1980, calculated from the model GUFM of Jackson *et al.* (2000). This is a continuous, time-dependent model of the Earth's magnetic field, regularized to be smooth at the CMB. For more details of this model. We present toroidal and geostrophic flows calculated for 1980 (the epoch of the Magsat mission, at which the model is expected to be best constrained by the data), and a steady flow, using field and secular variation in 1970, 1980, and 1990, a sufficiently long interval to ensure formal uniqueness. Each flow is regularized with the strong norm (eqn [19]), and with damping parameters chosen to produce similar flow strength. The flows are presented in **Figure 1**. We see some similarities between the flows, but also some differences. The three models have similar maximum and root mean square (rms) flow speeds. All three flows show some evidence of large-scale symmetry between the North and South Hemispheres, and are dominated by toroidal components, particularly zonal toroidal modes (Hulot *et al.*, 1990; Bloxham, 1992). There is a strong westward flow under the Atlantic Hemisphere in all three cases. Under the Pacific, flow is generally weaker, due to weaker secular variation there; but the tangentially geostrophic flow shows in places a strong eastward flow. This flow is not required to explain the observations, but arises from the combination of the

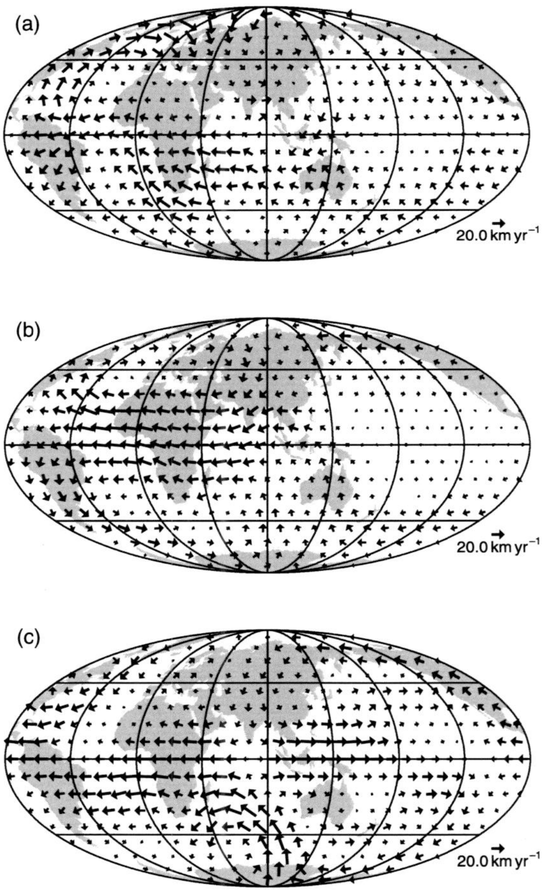

Figure 1 (a) Steady, (b) toroidal, and (c) tangentially geostrophic flows for 1980 calculated from the field model GUFM (Jackson *et al.*, 2000). Maximum/rms flow speeds in km yr^{-1} are (a) 31.0/12.2, (b) 29.6/12.6, and (c) 31.2/14.3.

large-scale and tangential geostrophy assumptions. Furthermore, it is strongest at the equator, where the latter assumption does not necessarily hold, and so should be viewed with suspicion. All three flows show some evidence of flow circulation in the Southern Hemisphere under Africa and the south Atlantic (christened the southern gyre by Voorhies), and weaker common circulation under North America. However, the differences in detail between the flow models demonstrate how any conclusions drawn from a particular model depend on the validity of the underlying physical assumptions used to derive it.

4.2.11 Other Flow Modeling Assumptions

More recently, other assumptions have been tested to further constrain flow modeling. In Section 4.2.14, we discuss briefly modeling with diffusion, but within the framework of frozen flux, three additional methods have been proposed.

4.2.11.1 *Flow in a drifting reference frame*

Voorhies and Backus (1985) suggested examination of flows that were steady in a drifting reference frame – a steady flow pattern rotating around the Earth's rotation axis. This could be conceived of as a crude solid body rotation of the whole outer core, of the core near-surface, or as a phase propagation of wave motion at the top of the core. Such flows provide a parsimonious fit to the observed secular variation (Davis and Whaler, 1997; Holme and Whaler, 2001), in the sense that addition of one further free parameter to a steady flow, the drift rate, results in a large reduction in the misfit to the secular variation, largely by a better fit to variations in the axial dipole field. Allowing a time-dependent drift does not greatly further reduce secular variation misfit, but the variations in drift rate show good correlation with decadal variations in LOD – see Section 4.4.

4.2.11.2 *Helical flow*

Helical flow is one of two new types of *a priori* information introduced by Amit and Olson (2004). Dynamos are known to operate well in the presence of helicity (giving rise to the so-called α-effect), and therefore helicity is a sensible choice of a physical quantity with which to constrain flow structure. Amit and Olson (2004) assume that

$$\nabla_H \cdot \mathbf{u}_H = \mp k_0 \zeta \qquad [32]$$

where k_0 is a constant to be chosen, and ζ is the radial component of the vorticity $\nabla \wedge \mathbf{u}$. Using the toroidal–poloidal decomposition, the condition on the flow components is

$$\nabla_H^2 \mathcal{S} = \mp k_0 \nabla_H^2 \mathcal{T} \qquad [33]$$

The negative sign applies in the Northern Hemisphere, the positive sign in the Southern Hemisphere. Note that k_0 is considered constant, for which there is some evidence from some dynamo simulations. Amit and Olson (2004) show that a formulation which combines any amount of helical flow (any nonzero constant value of k_0) with the tangential geostrophic assumption produces a flow solution that is formally unique everywhere.

4.2.11.3 *Columnar flow*

The second assumption provided by Amit and Olson (2004) is columnar flow, which assumes that the flow internal to the dynamo is organized dominantly in columns; then, because these columns must make contact with the slanted spherical surface of the CMB, the surface flow is constrained such that

$$\nabla_H \cdot \mathbf{u}_H = \frac{2 \tan \theta}{c} u_\theta \qquad [34]$$

where c is the core radius. The columnar assumption is closely related to the tangential geostrophic approximation – eqn [29] can be rewritten as

$$\nabla_H \cdot \mathbf{u}_H = \frac{\tan \theta}{c} u_\theta \qquad [35]$$

differing only by a factor of 2 from eqn [34]. Some observational evidence for columnar flow is provided by very detailed models of the core-surface field constrained using maximum entropy techniques (Jackson, 2003).

4.2.12 Limitations in Resolution of Nonuniqueness

Considerable effort has been expended on investigating the reduction or elimination of formal nonuniqueness by the various dynamical flow assumptions. This has led to a perception that, for example, constraining a flow to be tangentially geostrophic yields a unique result which can be applied to examine other physical problems. Such confidence is misplaced, as even in the case where a formally unique flow can be obtained, in practice the flow is obtained by a fit to noisy 'data'. As in all data modeling problems, the 'data' (here the radial field secular variation at the Earth's surface) are not fit exactly, but instead to within a given tolerance, generally to about one standard deviation defined by the expected error. The field models (and especially the secular variation models) used are uncertain because of sparse, unevenly distributed, noisy data, and contributions to the field from sources other than the core (for the field itself, particularly from the long-wavelength lithospheric field, and for the secular variation, from the field external to the Earth, particularly from induced components in the conducting mantle related to the solar cycle). Direct estimates of magnetic secular variation from obsevatory data (e.g., Whaler, 1986) have also been used; these allow better quantification of noise, but the problem of uncertainty of degree of fit remains. For convenience, the further uncertainty in the field models is usually ignored; the consequences of this are considered formally by Jackson (1995). Numerically, a model fit provides a unique solution (the flow which fits the data to within a given error tolerance, subject to additional physical information, and smoothness defined by the chosen from of damping – the large-scale approximation); however, there are an infinite number of flows which fit the data to such a given tolerance, while allowing perhaps only slightly greater complexity (a slight relaxation of the damping). Thus, there is an infinite space of flows which explain the observed secular variation. This is particularly the case as the errors on the field coefficients are formally unbounded (or rather, any rigorous formal bound is so weak as to be practically useless (Backus, 1988)), due to the need to downward continue the field from the Earth's surface to the CMB, an operation which is mathematically unstable. An example of the pitfalls of overestimating the uniqueness of the flow was given by Holme (1998) in a study of electromagnetic core–mantle coupling. While forward modeling of the electromagnetic torque of the core on the mantle from a calculated field model displays little similarity with that torque required to explain observed variations in LOD, inverse models calculated to explain both the secular variation and the torque do so with little additional flow complexity, showing that this core–mantle coupling mechanism cannot be ruled out on observational (flow modeling) grounds alone. Pais *et al.* (2004) took such ideas further, by exploring the effects of relaxing the tangential geostrophic constraint, rather than applying it exactly. Physically this is sensible: while we might believe that tangential geostrophy is a good approximation to the physics at the top of the core – for example, that the influence of the Lorentz force is weaker than the Coriolis force and pressure gradients – this is far from stating that the Lorentz force must be exactly zero. A small Lorentz force could give rise to a large change in the physical behavior of the flow.

We should therefore be very careful in interpreting the flows modeled, because in the Earth the nonuniqueness-reducing assumptions will not apply exactly – essentially, the flows obtained are only as good as the physics used as a constraint. This is particularly of concern when flow models are further analyzed to determine other physical properties, for example, in deriving torsional oscillation structure, or constraining core–mantle coupling.

4.2.13 The Horizontal Field Components

The analysis thus far has focused entirely on the radial component of the induction equation; what can be determined from the perpendicular components of the equation? Sadly, the answer appears to be not very much. The horizontal field components include contributions from the toroidal field, which, in the assumption that the mantle is an insulator, is totally contained within the core. The mantle is not a perfect insulator, and in principle it should be possible to obtain as estimate of the toroidal field at the top of the core by the measurement of global-scale electric currents at the Earth's surface – for example, in abandoned telecommunications cables (Lanzerotti *et al.*, 1985). However, the effects are very small, and very difficult to distinguish from other sources of electric currents, especially oceanic motional induction (see, e.g., Shimizu *et al.*, 1998; Shimizu and Utada, 2004). This leaves the toroidal parts of the horizontal field, as given in eqn [12], essentially unknown. At the top of the free stream, it is likely that the toroidal field is at least as large as the poloidal field in the horizontal directions.

Nevertheless, Bloxham and Jackson (1991) attempted to use the horizontal components of the poloidal part of the magnetic field to map the vertical shear of the flow. They generated maps of the shear, but these were contested by Jault and Le Mouël (1991). Further investigation of the effects unearthed several unfortunate surprises: in particular, in inverting for shear from synthetic flows with shears of opposite signs, the same shear sign was obtained in both (as reported in Whaler and Davis (1997)). Therefore, the determination of shear at the top of the core does not appear to be robust, and will not be discussed further. Gubbins (1996) has further considered the influence of shear in deriving methodology for including diffusion in flow modeling.

4.2.14 Effects of Diffusion

The analysis developed above, particularly questions of flow uniqueness, has relied fundamentally on the neglect of magnetic diffusion, assuming that secular variation is dominated by advection, in particular on decadal timescales. However, this is by no means certain. In particular, eqn [3] allows an alternative self-consistent scaling interpretation: rather than the dominant term balance being the match of the secular variation against advection, the largest terms could instead be the advection and diffusion, almost cancelling, with the secular variation a result of small departures from this equilibrium. Support for such a state was provided by analysis of kinematic dynamo output by Gubbins and Kelly (1996): they argued that a steady flow will evolve to just such a balance, with secular variation tending to zero over time. Love (1999) argued this point with vigor, suggesting that because we expect the dynamo process to be close to steady on decadal timescales, this is precisely the type of state we should expect in the core. If this physical picture truly holds, then clearly flow models neglecting diffusion are in serious error.

Some studies suggest that this problem may be less severe than at first glance. Voorhies (1993) argued for just such a state, and solved for a steady flow and parametrized steady diffusion. Interestingly, he found that the form of the steady flow did not change greatly with the additional allowance for diffusion, even at quite low magnetic Reynolds numbers (when diffusion is in principle dominant over advection). This result is supported by the study of Rau *et al.* (2000), who conducted synthetic experiments based on the output of dynamo codes. They took the synthetic secular variation generated by the dynamo model, solved for a flow based on this secular variation, and compared their solution with the true flow at the dynamo surface. Despite the diffusive contribution to the flow being of the same order of magnitude as the advective contribution, the modeled flow recovered well many features of the true flow.

Many workers have continued to ignore diffusion in modeling for core flow, perhaps encouraged by the result of Rau *et al.* (2000), or perhaps motivated at least in part by convenience: the flow inversion is already a severely underdetermined inverse problem, and adding additional parametrization for diffusion only makes this problem worse. The method of Voorhies (1993) relied on the steady flow assumption, together with steady diffusion, which makes the number of parameters tractable. However, it does not allow for time-dependent flow. Voorhies argues that part of the secular variation which must be explained by nonsteady flows is small, particularly in comparison with errors resulting from neglect of other effects (such as diffusion, or a noninsulating mantle). Bloxham (1992) made a similar point. Nevertheless, direct comparison with observations from magnetic observatories shows that a steady flow is unable to explain clear features in the observed secular variation (see, e.g., the results of Bloxham *et al.* (2002), who solve for a particularly detailed steady flow, which is nevertheless unable to explain notable features of the secular variation). Further, these parts of the secular variation are among the most interesting, and provide the one independent test that the flow models are in any way meaningful – see Section 4.4. Therefore, it is unfortunate to ignore them. Gubbins (1996) has provided a methodology for solving for both time-dependent flow and diffusion, but this has not as yet been implemented.

Alternative treatments of diffusion have been proposed. Olson *et al.* (2002) appeal to mean-field analysis, which has been useful in dynamo theory, to consider the different scales on which diffusion can act. They separate both field and flow into large and small scales

$$B_r = B + b \qquad [36]$$

$$\mathbf{u}_H = \mathbf{U} + \mathbf{u} \qquad [37]$$

yielding after some analysis a relation matching the upwelling of the small-scale flow to the diffusion of the small-scale field

$$B(\nabla_H \cdot \mathbf{u}) = \frac{\eta}{r}\nabla^2(rb) \qquad [38]$$

where these smaller-scale magnetic field features are in advective–diffusive equilibrium in a frame moving with the large-scale velocity (not dissimilar to the concept of steady flow in a drifting frame described above). The diffusive term still depends on unknown radial derivatives of the magnetic field at or within the free stream, but Olson *et al.* (2002) argue that, at least away from the equator, horizontal diffusion should be dominant, supporting this contention with examples from dynamo simulations.

Usually, the effects of diffusion are considered at small scales, but Holme and Olsen (2006) instead considered the influence of diffusion at large scales, particularly in its influence on the observed secular variation. Decomposing the field into its spherical harmonic constituents, the secular variation from diffusion might be expected to increase with harmonic degree, a simple scaling argument giving perhaps

$$\dot{B}_r(l) \sim l^2 B_r(l) \qquad [39]$$

But the strength of the observed secular variation also increases with degree as a function of the field strength, approximately

$$\dot{\mathrm{B}}_r(l) \sim l^{3/2} \mathrm{B}_r(l) \qquad [40]$$

Thus, the observed secular variation power increases almost as rapidly with degree as the simple scaling argument for diffusion would predict, leaving the predictive fractional diffusive power in the observed SV only a weak function of harmonic degree. Holme and Olsen (2006) obtain crude estimates on the possible magnitude of diffusion at each harmonic degree using the well-defined magnetic free-decay modes (e.g., Gubbins and Roberts, 1987) to match the observed field. With this analysis, the largest contribution of diffusion to the observed secular variation occurs at low degrees (large scales) rather than the usually assumed high degrees (small scales). In particular, the strength of the axial dipole coefficient of the field is so dominant at the Earth's surface (where the geomagnetic field is observed) that the action of diffusion on this field component may be the most important diffusional element in the observed secular variation. Interestingly, some tangentially geostrophic flows, for example, the flows of Jackson (1997) and even more clearly flows of the same author reported by Whaler and Davis (1997), underfit the decay of the axial dipole by 5–10 nT yr^{-1}. However, applying diffusional estimates as uncertainties on flow determination from the frozen-flux assumption gives surprisingly little change in flow from conventional error estimates, perhaps because the flow is unable to easily explain such secular variation, and so does not.

4.3 Higher-Resolution Flows from Detailed Models of SV from Satellite Observations

Even within the constraints of nonuniqueness and unmodeled diffusion, a model of the core-surface flow can be no better than the models of the radial field and secular variation used to calculate it. Recent models of the secular variation in particular have a considerable improvement in resolution arising from data from recent satellite missions: the Danish Ørsted satellite (Neubert *et al.*, 2001), the German CHAMP satellite (Reigber *et al.*, 2002), and the Argentine SAC-C mission. These have provided effectively continuous data coverage from early 1999 to the time of writing (2007), with global three-component vector measurements of the magnetic field allowing construction of high-quality models of the global magnetic field. For more details of these missions. Only one earlier mission, the Magsat satellite which flew for 6 months in 1979/1980 (Langel *et al.*, 1982) provided data quality and coverage even approaching the current period. These data allow models of secular variation to be constrained to smaller scales: for example, the CHAOS model of Olsen *et al.* (2006), which shows coherent signal in secular variation to at least spherical harmonic degree 15.

Hulot *et al.* (2002) made the first attempt to use these higher-resolution models to calculate detailed flow models. They estimated a 20-year time-averaged secular variation from the difference between the field in 1980 and 2000, and used this to model the mean flow for the intervening period, centered on 1990, presented in **Figure 2(a)**. This flow model included finer-scale features than previous models, in particular many small vortices in mid-latitudes. Such features are characteristic of some geodynamo simulations (e.g., Olson *et al.*, 1999). In contrast, Holme and Olsen (2006) calculated flows based on the secular variation from 4 years of data from the satellite missions. **Figure 3** presents possible flows calculated with both the toroidal and tangentially geostrophic assumptions for these epochs. Fine-scale detail is seen which varies greatly in detail between the two flow approximations, and even more compared with the flow of Hulot *et al.* (2002), particularly lacking the many mid-latitude vortices. The differences between the flows are a consequence of their construction. The vortices observed by Hulot *et al.* (2002) dominantly flow along closed contours around field maxima and minima; this corresponds

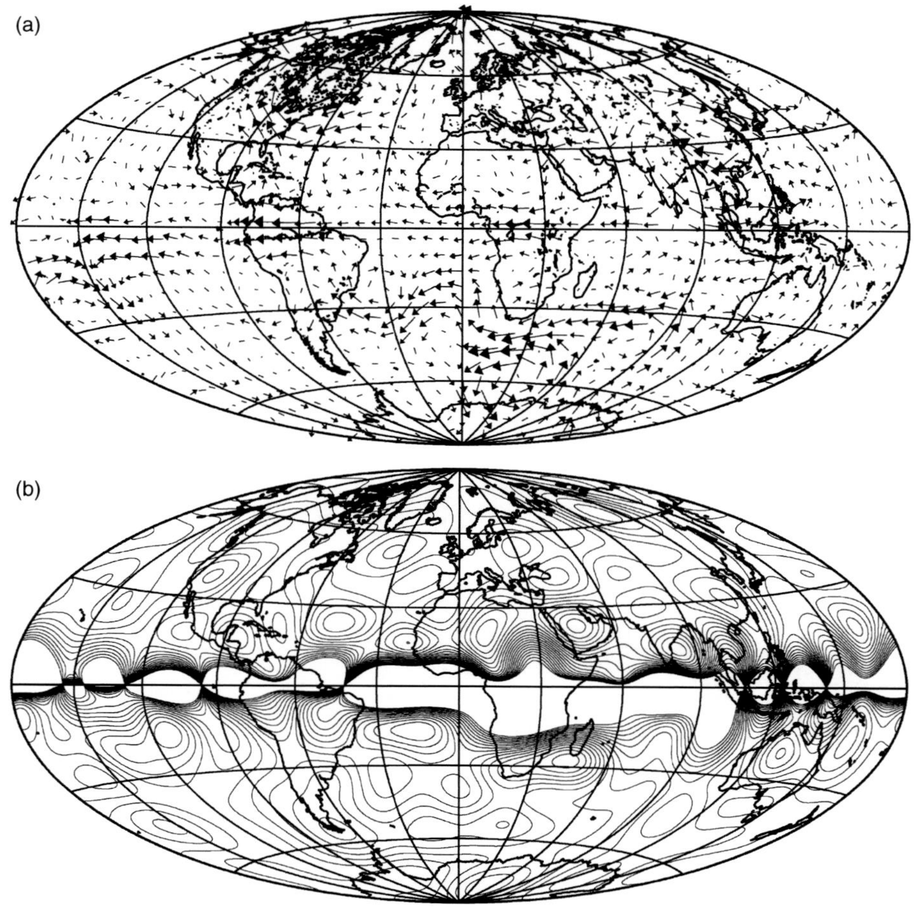

Figure 2 Flow of Hulot *et al.* (2002), plotted above the contours of the geostrophic nonuniqueness. Many of the small vortices which can be identified in (a) coincide with closed contours of the nonuniqueness in (b). Adapted by permission from Macmillan Publishers Ltd: *Nature* (Hulot G, Eymin C, Langlais B, Mandea M, and Olsen N (2002) Small-scale structure of the geodynamo inferred from Oersted and MAGSAT satellite data. 416 (6881): 620–623), copyright (2002).

closely to the nonuniqueness described by eqn [30], and shown in the **Figure 2(b)**, and so these flows produce no secular variation. Such flows might be expected precisely from the balance between advection and diffusion described in Section 4.2.14, and predicted by, for example, Gubbins and Kelly (1996). The vortex flows suggested by Hulot *et al.* (2002) are not required by the observations, but neither can they be ruled out by them.

Previous flow models had shown broad agreement between different methods and authors (see, e.g., the survey provided by Bloxham and Jackson (1991) and the flows shown in **Figure 1**); now, clearly, this agreement is breaking down. The reason for the emergence of this problem (compared with earlier, larger-scale flows) can be understood quite simply. The nonuniqueness of eqn [30] is for flows along closed contours of $B_r/\cos\theta$. These contours are themselves small scale, and only now, with attempts to fit higher-resolution secular variation, is the inversion able to generate flows of fine-enough scale to match them. The flows of Holme and Olsen (2006) do not show obvious evidence of the resulting vortices; however, this by no means proves that such features are not present in the real core flow. Their flows are damped so as to minimize the overall magnitude of the flow. Therefore, the nonunique flow around these closed contours will be chosen so as to minimize the global strength of the flow. The flow models are equally influenced by the presence of the nonuniqueness, if less obviously than the flows of Hulot *et al.* (2002).

An even greater problem for resolution of small-scale flow may be the limits to field resolution. Even

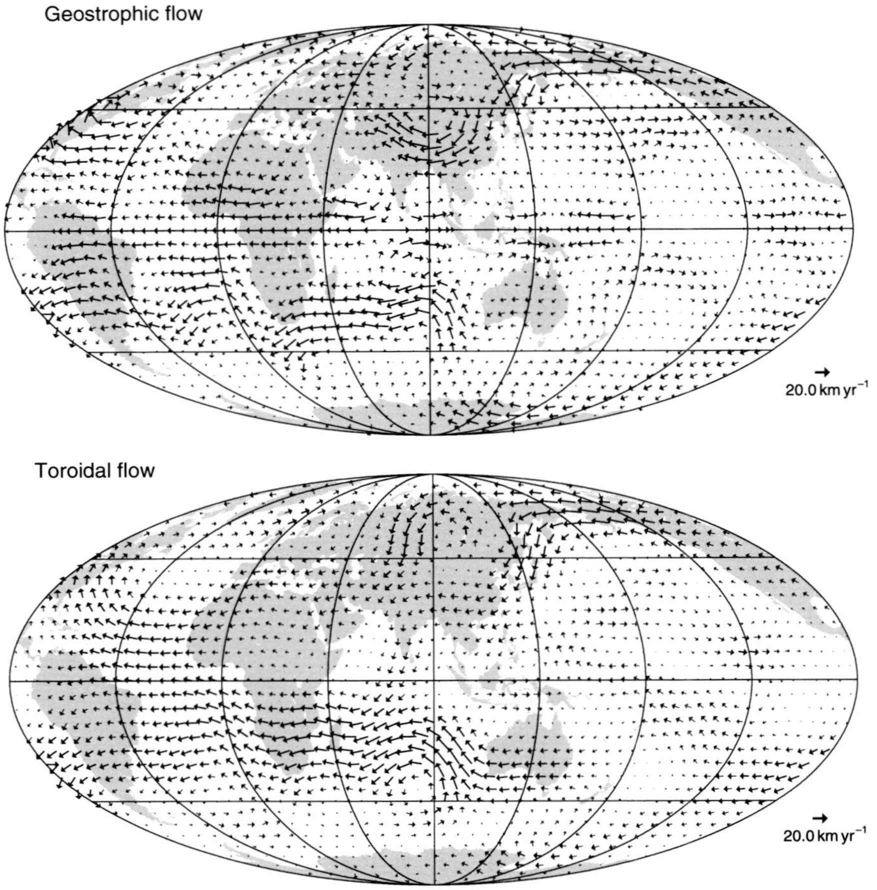

Figure 3 A toroidal and a tangentially geostrophic flow fit from a model derived from high-quality magnetic satellite data for the period 1999–2004 (Holme and Olsen, 2006). Note circulation foci which might be evidence for columnar flow (North and South Hemisphere vortices at matching latitudes and opposite rotational sense), although these features are not robust to variations in inversion parameters. Maximum/rms flow speeds in km yr^{-1} are 35.0/15.2 and 36.4/13.6, respectively. Adapted from Holme R and Olsen N (2006) Core surface flow modelling from high-resolution secular variation. *Geophysical Journal International* 166 (2): 518–528.

with more detailed secular variation estimates, the new field models cannot resolve details in the core field smaller than spherical harmonic degree 13, because of the presence of the lithospheric field which screens out these scales. Eymin and Hulot (2005) demonstrated that flow models were strongly compromised by missing knowledge of the main field, supporting the earlier contention of Hulot *et al.* (1992). The problem arises because field and flow interact with each other at all scales: therefore, the modeled large-scale flow will generate large-scale secular variation from the unknown small-scale field. Eymin and Hulot (2005) consider a suite of field models with varied 'unknown' small-scale field, and conclude that this lack of knowledge critically impairs modeling of the true flow from modeled secular variation.

Rau *et al.* (2000) arrived at similar conclusions when analyzing the output of dynamo simulations. They found that neglect of diffusion did not critically damage the integrity of the flows obtained, but that lack of knowledge of the smaller-scale features in the field was much more damaging. A filter on the field chosen to mimic the effects of the crustal field took a fine-scale field with no clear large-scale westward drift, and produced a larger-scale field with apparent large westward drift. In effect, they argue (at least for their model) that a wave mechanism (see Section 4.6 below) is made to appear as if it were generated by large-scale flow.

4.4 Angular Momentum – LOD Variation, and Correlation with Core Angular Momentum

Core-surface flows can be calculated using the methodology described in Section 4.2, but without independent verification they cannot be confirmed as meaningful. Fortunately, there is another piece of geophysical information against which at least part of the flow model can be tested: the variation in Earth's LOD. The rotation period of the Earth is not a uniform 86 400 s, but varies on timescales of days to millennia, due to the action of external torques, and exchanges of angular momentum between different parts of the Earth system. Over the longest timescales, the tidal drag of the Moon and Sun on the rotating Earth is known to produce a secular slowing of rotation (increase in LOD), currently of order 1.4 ms per century. On yearly and subyearly timescales, the angular momentum variation predicted from models of the atmospheric global circulation explains most of the observed LOD variation, and a further part of the signal can also be predicted from oceanic general circulation models (see, e.g., Marcus *et al.* (1998)) and hydrology (Chen, 2005). However, there are further variations of a couple of milliseconds in LOD over several decades that are not easily explained by surface processes. It was realized as early as the 1950s that the missing angular momentum was likely being taken up by the core: there was a good correlation between variations in declination at some magnetic observatories and LOD, and in particular Vestine (1953) demonstrated a good correlation between bulk core rotation predicted from westward drift of the magnetic field and the LOD signal. The observation was put on a much stronger footing by Jault *et al.* (1988), based on the concept of Braginsky (1970) that angular momentum in the core on decadal timescales should be carried by torsional oscillations, rotation of fluid on concentric cylinders coaxial with the Earth's rotation axis. Jault *et al.* (1988) realized that if such motions were important, they would be manifest in variations in surface flow patterns (see **Figure 4**). Therefore, using surface flow models the angular momentum of the whole core could be approximated, and compared with observed LOD. Perhaps surprisingly,for uniform flow on cylinders, only two surface flow harmonics are required to calculate the change in angular momentum of the core, the zonal toroidal harmonics of degree 1 and 3. The angular momentum (in the direction of the rotation axis) is given by (Jault *et al.*, 1988)

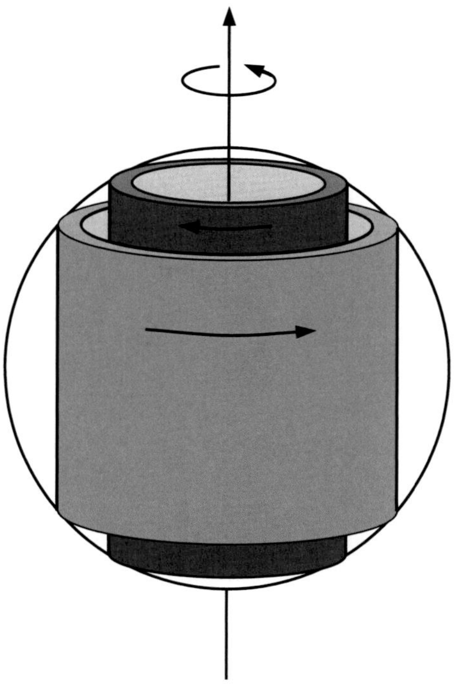

Figure 4 Schematic diagram of the structure of torsional oscillations. The vertical axis coincides with the Earth's rotation axis, and core flow moves on uniform cylinders. The expression in the surface flow in the Northern and Southern Hemispheres should be the same. Figure courtesy of Dr. Stephen Zatman.

$$\mathcal{J}_z = \frac{8\pi c^4 \rho}{15}\left(\mathcal{T}_1^0 + \frac{12}{7}\mathcal{T}_3^0\right) \qquad [41]$$

where c and ρ are the radius and mean density of the core, respectively. Conservation of angular momentum between the solid Earth and core leads to a prediction for changes in observed LOD δT (measured in milliseconds) of

$$\delta T = 1.138\left(\delta \mathcal{T}_1^0 + \frac{12}{7}\delta \mathcal{T}_3^0\right) \qquad [42]$$

where the toroidal flow coefficients are measured in units of kilometers per year. Using this formulation, Jault *et al.* (1988) obtained a promising correlation between observed and predicted variations in LOD for the previous 40 years determined from a sequence of tangentially geostrophic flow models, especially the rate of change; convincing support followed from the work of Jackson *et al.* (1993), who provided an excellent correlation between predicted core angular momentum changes and observed LOD variations for the past century. Prior to this period, the correlation is poorer, in particular with an apparent phase shift, but the general pattern is still provoking. Pais and Hulot (2000) argue that the misfit of ΔLOD prior to 1900 is not significant compared with model uncertainty. The results of Jault *et al.* (1988), Jackson *et al.* (1993), and some other flow calculations are presented in **Figure 5**, taken from Ponsar *et al.* (2003). The way in which the correlation with LOD changes is achieved is also interesting. Jackson *et al.* (1993) show that the two relevant toroidal harmonics $\mathcal{T}_1^0$ and $\mathcal{T}_3^0$ are approximately anticorrelated for much of the period. The flow giving rise to the LOD signal is a small part of the overall flow, and the secular variation it generates a small part of the total secular variation, making the observed correlation with the variation in LOD even more impressive.

Other assumptions for flow dynamics also produce a good correlation with variations in LOD. Toroidal and even unconstrained motions (flow constrained to fit the secular variation without further dynamical assumptions) show evidence of the correlation, although weaker than for geostrophic flow. This could be regarded as support for the assumption that flow at the top of the core is approximately tangentially geostrophic. However, estimating core angular momentum from a toroidal flow is more subject to uncertainty from nonuniqueness, because part of the toroidal nonuniqueness involves flow along contours of B_r which encircle the Earth, and can make strong contributions to the calculated angular momentum. (Note that uncertainty for geostrophic flows arising from the failure of the approximation near the equator is not significant: although the moment arm from the rotation axis is maximum, the amout of fluid in the cylinder touching the equator is small because it is short, and so little angular momentum is carried related to surface flows at this latitude.) Drifting flows with time-dependent drift (Holme and Whaler, 2001) show a correlation with ΔLOD that is almost as good as that from tangentially geostrophic flows, and arguably better prior to 1900, as shown in **Figure 5**. A yet better correlation prior to 1900 is probably given by the original Vestine (1953) analysis of westward drift! The greater success of simpler flow parametrization from 1870 to 1900 may be due to the field and secular variation models: over this period a major improvement in data quality and quantity was achieved, particularly the establishment of a global network of magnetic observatories. The improvement in data allows increasing resolution (to smaller scales) of secular variation in the field models, which a fit by a flow model attempts to explain by core processes. It may be that more parsimonious representations of flow are less influenced by this change.

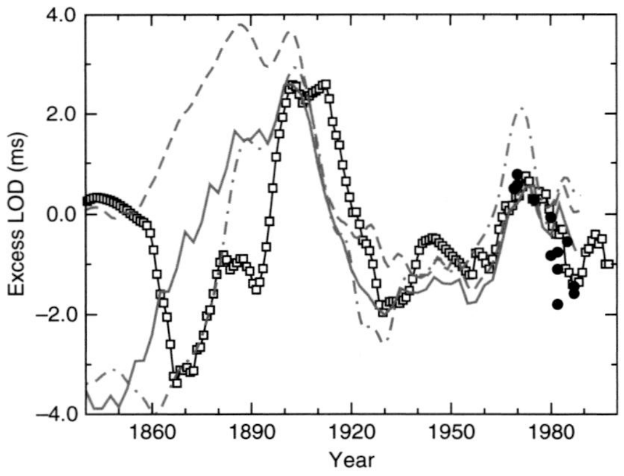

Figure 5 Comparison of observed variation in length of day (squares, from McCarthy and Babcock, 1986) and prediction from core-flow models: Jault *et al.* (1988) (circles), Pais and Hulot (2000) (dashed line), Jackson *et al.* (1993) (solid line), and Holme and Whaler (2001) (dot-dashed line). From Ponsar S, Dehant V, Holme R, Jault D, Pais A, and van Hoolst T(2003) The core and fluctuations in the Earth's rotation. In: Dehant V, Creager KC, Karato S-I, Zatman S (eds.) *Geodynamics Series, Vol. 31: Earth's Core: Dynamics, Structure, Rotation*, pp. 251–261. Washington, DC: AGU.

4.5 Torsional Oscillations as a Probe of Core Structure

In the previous section, torsional motions have been introduced as an important contributor to the secular variation, even the amplitude of the resulting secular variation signal is small. Further consideration of torsional oscillations provides a direct probe of processes within the core, rather than just at its surface.

4.5.1 Probing the Magnetic Field Interior to the Core

If torsional oscillations are excited, and assuming that core–mantle coupling is sufficiently weak so that they can oscillate freely, then the characteristics of such oscillations can constrain magnetic field structure in the core. The coupling force between the cylinders arises primarily from their linkage by the magnetic field, in particular its cylindrical radial component B_s. The work of Zatman and Bloxham (1997, 1998) was pioneering in this respect. They examined the structure of a time-dependent flow model, and identified at least one and possibly two separate torsional waves within the structure of the oscillations, shown in **Figure 6**. Both the frequency and decay rate of these oscillations provide useful constraints, suggesting a value for B_s from 0.2 mT rising to over 1 mT at the ICB in models including boundary friction at the CMB and ICB. Note, however, that caution is necessary: it would be possible to construct a range of possible flow models from the observations which provide a sufficient fit to decadal LOD variation, but differ in their fine-scale temporal structure. Because this analysis is of a model of a flow generated from a model of secular variation constrained by field observations, there is much scope for uncertainties to arise depending on the precise nature of the models chosen at each stage.

4.5.2 Coupling of Torsional Oscillations to the Outer and Inner Core

Using the framework of torsional oscillations, observed variations in LOD can also be used directly to probe core–mantle interaction. Drawing on a detailed theory giving the normal modes of oscillation including the inner core, coupled electromagnetically to the outer

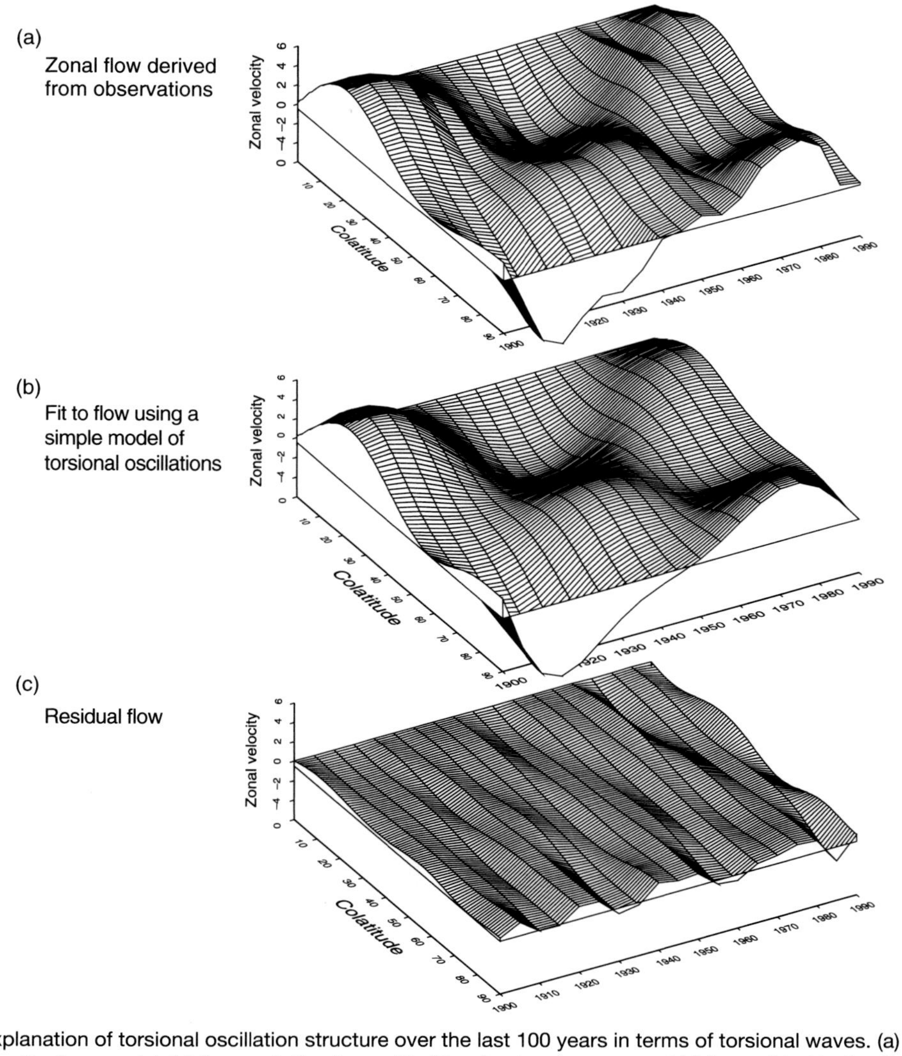

Figure 6 Explanation of torsional oscillation structure over the last 100 years in terms of torsional waves. (a) The waves calculated from the flow model, (b) the prediction from a fit of two torsional waves, and (c) the residual to the fit. Reprinted by permission from Macmillan Publishers Ltd: *Nature* (Zatman S and Bloxham J (1997) Torsional oscillations and the magnetic field within the Earth's core. 388 (6644): 760–763), copyright (1997).

core and gravitationally to the mantle (Mound and Buffett, 2003, 2005; Buffett and Mound, 2005), Mound and Buffett (2006) focus on a variation with approximate 6-year peroid detected by Abarco del Rio *et al.* (2000). They argue that the 6-year oscillation is a natural consequence of the coupling of the modes with motion of the inner core (gravitationally coupled to the mantle). By matching their model to the LOD time series, they are able to constrain the strength of gravitational coupling between the inner core and the mantle, and place a lower bound on the viscosity of the inner core.

4.5.3 Geomagnetic Jerks, and Links to Earth's Rotation

Geomagnetic jerks are sharp changes in the rate of change of geomagnetic secular variation (in other words, close to a discontinuity in the second time derivative of the geomagnetic field) at points on the Earth's surface, first identified by Courtillot and Le Mouël (1984) (e.g., Mandea *et al.* (2000) and references therein). They are seen particularly clearly in the eastward (*Y*) field component from magnetic observatory records in Europe, although some jerks are seen globally. As an important detail in the secular variation, it is sensible to interpret their generation in terms of core-surface flow. Bloxham *et al.* (2002) have provided a particularly attractive interpretation. They modeled the background (almost) steady secular variation with a steady (if complex) flow, and showed that for the 1969 geomagnetic jerk, the magnetic field variation could be well-fit by flows consisting only of odd-degree, zonal toroidal, harmonics: the surface expression of torsional oscillations. Unfortunately, the mechanism does not explain some other jerks as effectively (e.g., Southern Hemisphere jerk features in 1971). However, direct examination of the variation in LOD provides additional evidence for a link between geomagnetic jerks and Earth rotation (and so perhaps torsional oscillations in the core). Holme and de Viron (2005) identify features in the decadal rate of change in LOD at or slightly before epochs of identified jerks (**Figure 7**). These features are consistent with impulsive torques on the core, producing discontinuities in the rate of change of torsional rotation, again consistent with torsional motions in the core being associated with geomagnetic jerks. To date, a mechanism which can generate such changes has not been identified, but the possibility remains to probe core processes on a very short timescale (a year or less).

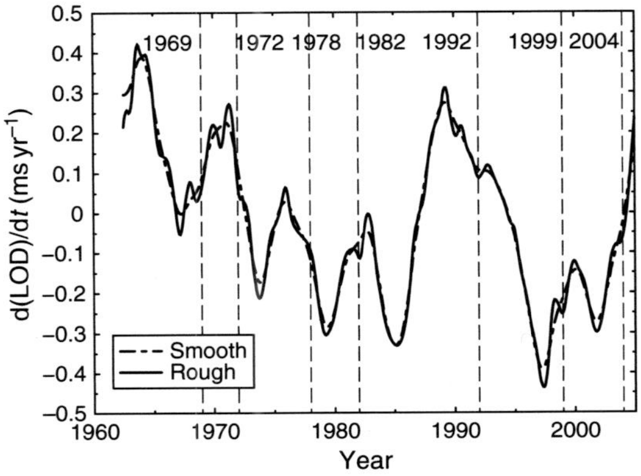

Figure 7 Rate of change of two smooth fits to LOD on decadal timescales. The LOD time series is corrected for modeled variations in atmospheric angular momentum, and a 1-year running average is applied to remove much remaining signal on annual, biannual, and terannual timescales (presumably resulting from oceanographic and hydrological sources as well as misfit atmospheric angular momentum). The smoothed data are fit using penalized least-squares splines, and differentiated. The difference between the different fits to the curve shows small 'double-couple'-like features at or just before the reported time of geomagnetic jerks (given by the vertical dashed lines), consistent with a sharp discontinuity in rate of change of LOD. Figure extended from Holme and de Viron (2005) to include possible jerk in approximately 2004 located in magnetic satellite data (Olsen and Mandea, 2007).

4.6 Wave Motion as an Explanation for Secular Variation

The analysis presented so far has assumed that the secular variation is primarily a result of large-scale flow at the top of the core. The correlation between predicted core angular momentum and observed variations in LOD suggests that at least part of the flow can be represented in this way. However, for much of the secular variation, there is another alternative, originating from a suggestion by Hide (1966) of an alternative mechanism to explain westward drift. Whereas Bullard *et al.* (1950) had argued for a origin of westward drift in large-scale flow, Hide instead suggested that it could arise as a result of wave motion. Magnetic waves called Alfven waves were known to be supported in a fluid penetrated by a magnetic field. Adding the effect of rotation, additional families of diffusionless magnetohydrodynamic waves were shown to exist, some with periods of order days (similar to inertial waves), but others, named magnetic Rossby waves or planetary waves, with much longer periods, for reasonable estimates of the magnetic field strength perhaps 300 years. Initial analysis based on propagation in a thin shell (similar to analysis for the atmosphere) unfortunately suggested that these waves would propagate eastward, not westward! Hide provided an intuitive argument, later confirmed by more detailed analysis, as to why in the core (a thick shell) such waves would propagate westward instead. Further examination has suggested that westward motion will dominate when the toroidal field is stronger than the poloidal field, as is thought to be the case within the core (Hide and Roberts, 1979).

There are a number of tantalizing observations that suggest that a wave origin for much of the secular variation may be appropriate. When dividing the observed surface field into its standing and drifting components, Yukutake (1979) further suggested that the equatorial dipole field could be separated into two components, drifting in opposite directions; such a process is reminiscent of wave motion, consistent with Hide's ideas. Holme and Whaler (2001) parametrized the large-scale flow into a steady flow in a drifting frame, and derived optimal solutions with equal and opposite drift rate, which if added together (the solutions are linear) give rise to a standing wave. Finlay and Jackson (2003) have provided a detailed analysis of the variation of the nonaxisymmetric part of the field as revealed by the GUFM model (Jackson *et al.*, 2000). They revealed evidence of uniform westward motion of field near the core equator over all longitudes (although much less clear in the Pacific Hemisphere) suggesting that the difference in secular variation between the Atlantic and Pacific Hemispheres could have more to do with the morphology and strength of the background field than with any significant difference in physical processes in the two areas. Hydromagnetic waves provide one possible explanation for this signal.

Wave theory can also explain some level of complexity in the secular variation very concisely, as the local 'drift rate' of any feature can very with location, as the wave velocity is a function of the background field strength, which will vary over the core's surface. Many features of the core field could be explained by a wave-based interpretation: for example, the westward motion of equatorial flux patches under the Atlantic Hemisphere has been suggested to be due to trapped barotropic and baroclinic modes in a thin, stably stratified equatorial band (Bergman, 1993). Some numerical dynamo results (e.g., Rau *et al.*, 2000) also show the motion of fine-scale field structure in ways that could be interpreted in terms of waves, but which if viewed through the 'crustal filter' (in other words, only the long-wavelength parts) appear to be large-scale westward drift, not dissimilar from that we see in models from the historical record.

How is a wave explanation for secular variation consistent with induction through flow (and specifically with eqn [4])? If a part of SV can be explained through a wave process, then this equation must still be justified, but the flows which satisfy it may be small scale, in particular too small scale to be resolved by the limit we have observationally on the observed secular variation (and particularly the observed field). Evidence from the correlation with LOD suggests that at least one feature of the flow, the torsional oscillations, are large scale, although they themselves are now interpreted in terms of wave processes! While interpretations of secular variation in terms of large-scale flow are currently dominant, it is certainly not impossible that wave processes have an important role. If this is identified, it opens up new avenues for constraining the physical processes in the core: for example, the nature of the propagation is likely to be a function of the structure of the unseen toroidal field.

4.7 Polar Vortices

If, as has been argued for torsional oscillations, models of core-surface flow reflect processes deeper within the core, then it might be expected that flow near the poles should differ in structure from flow elsewhere. The theory of torsional oscillations relies on the Taylor–Proudman theorem, and that motions are uniform on cylinders throughout the core. However, inside the 'tangent cylinder' defined by the region of the core along the rotation axis above and below the solid inner core, the Taylor cylinders cannot extend from top to bottom of the core. The possible expression in surface flow of the region is small: the two caps at North and South Poles make up less than 7% of the total core-surface area, but analysis of the differences between flow in these caps and flow over the rest of the core surface offers the tantalzing prospect of probing the different dynamics in the different regions of the core. Flow in the tangent cylinder is particularly of interest, as it might reflect rotation of the inner core, controversially claimed to have been observed seismically (Song and Richards, 1996; Souriau and Poupinet, 2000; Zhang *et al.*, 2005). Olson and Aurnou (1999) first identified the possible presence of a polar vortex from examination of the UFM field model of

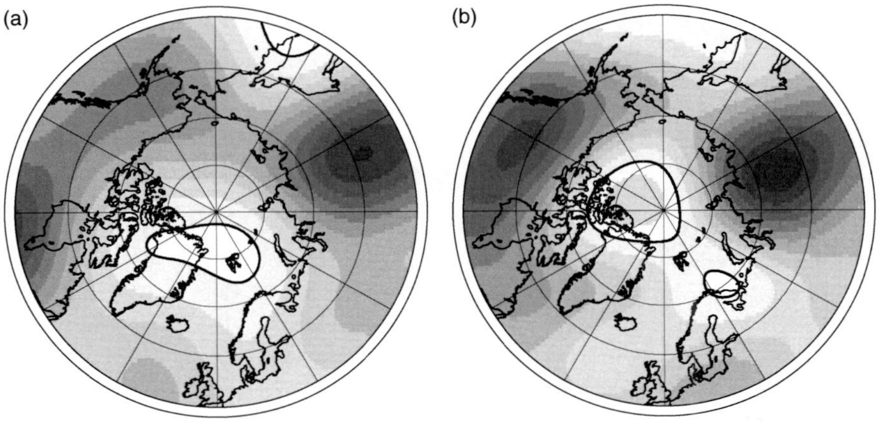

Figure 8 Evolution of the radial field of the GUFM model in the polar region ((a) 1870; (b) 1990) following Olson and Aurnou (1999). The movement of the null-flux curve (included as a thick contour) shows the direction of fluid motion of the vortex.

Bloxham and Jackson (1992) over more than 100 years. They identified a rotation of field features with time (**Figure 8**), and using a local (rather than global) flow modeling algorithm, a finite difference scheme regularized by horizontal diffusion, derived an average westward drift rate within the tangent cylinder of nearly a degree per year. They were aware of the inherent nonuniqueness of their flow model, and solved for a flow that was cylindrically symmetric about the rotation axis, which they argued would yield a unique flow field.

Further support for polar vortices has come from the analysis of field models from recent satellite data. Hulot *et al.* (2002) identified vortices in the flow not only in the north polar region, but also in the south polar region. **Figure 9** shows the calculated axially averaged flow from the model of Hulot *et al.* (2002), based on the difference in field between 1980 and 2000. This profile shows a remarkably uniform profile in axisymmetric velocity outside the tangent cylinders, but with vortex-like profiles within them. Further support is provided by results of Holme and Olsen (2006) for both tangentially geostrophic and toroidal flows for the period 1999–2004. However, Eymin and Hulot (2005) cast some doubt on the robustness of these conclusions, arguing that the polar vortices determined are poorly resolved, echoing other earlier (unpublished) criticism. The root of this controversy is (once again) nonuniqueness of the flow determination. Tangentially geostrophic flows are nonunique to flows along contours of $B_r/\cos\theta$ (eqn [30]). The contours are shown for the north polar region in **Figure 10**. Because they encircle the pole, the net circulation of flow around the North Pole could take any value, positive or negative.

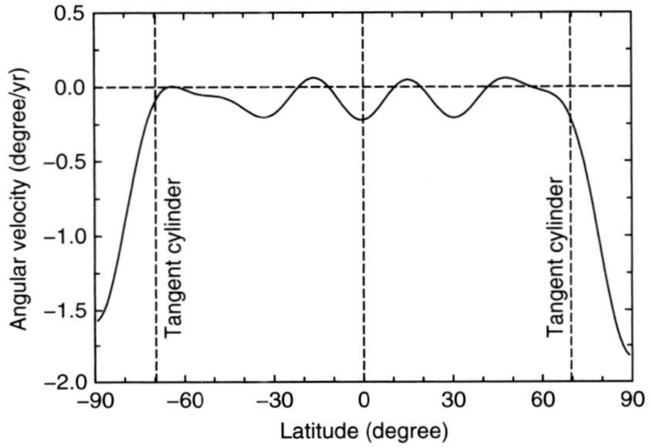

Figure 9 Axisymmetric flows from the model of Hulot *et al.* (2002) suggesting evidence for polar vortices in both North and South Hemispheres. Reprinted by permission from Macmillan Publishers Ltd: *Nature* (Hulot G, Eymin C, Langlais B, Mandea M, and Olsen N (2002) Small-scale structure of the geodynamo inferred from Oersted and MAGSAT Satellite data. 416 (6881): 620–623), copyright (2002).

Olson and Aurnou (1999) tried to bypass these dificulties by solving for an axially symmetric component of the flow, which they argued can then be determined uniquely. Exact axial symmetry excludes all of the flows around the contours of geostrophic degeneracy, because they themselves are not axially symmetric. However, such a claim of uniqueness is only valid if the physical assumption is truly valid, which at best is difficult to prove. Nevertheless, if the flow near the poles is large scale (or has a strong large-scale component), then the results of Hulot *et al.* (2002) and Holme and Olsen (2006) may be robust, because such large-scale flow cannot closely follow the contours of nonuniqueness, because small-scale structure cannot be matched.

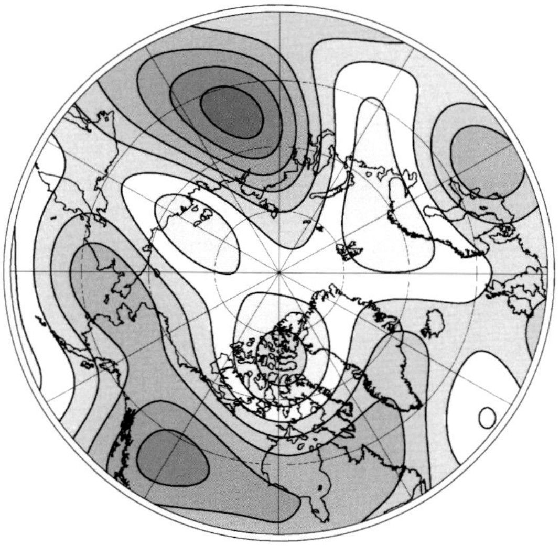

Figure 10 Contours of tangentially geostrophic degeneracy in the region of the North Pole for 2000 from the model CO2003 (Holme and Olsen, 2006). Note that the North Pole is contained within a local maximum of such contours, and so the net flow around the North Pole can take any value. Adapted from Holme R and Olsen N (2006) Core surface from modelling from high-resolution secular variation. *Geophysical Journal International* 166(2): 518–528.

Examination of dynamo simulations may provide some insight. In dynamo simulations of Sreenivasan and Jones (2005), flow occurs around a local field maximum, and so would be poorly resolved in a flow determination from secular variation, as flows around such contours are close to the null-space of the inversion. However, this field area itself migrates around the pole. Thus the flow models may reflect both these processes, with the Olson and Aurnou (1999) observation primarily sensitive to the long-term drift of the field maximum, and the models from satellite data more sensitive to possible flow around this maximum.

Given an observed polar vortex, what further can be deduced? Voorhies (1999) argued that from observation of the vortex, it was possible to constrain a rate of inner-core rotation, obtaining a maximum bound of $1.5^\circ \text{ yr}^{-1}$. However, the simulations of Sreenivasan and Jones (2005) yield a polar vortex with a stationary inner core, gravitationally coupled to the mantle, so it is clear that observational measurements of geomagnetic secular variation cannot constrain possible inner-core motion until the physics is much better understood. As with many questions as to the core flow, more rigorous statements await a better understanding of the flow regime at the top of the core, leading to better physical constraints on core motion.

4.8 Modeled Core Flow and the Dynamics of the Core

Except in very general terms, this review has avoided discussions of flow dynamics. The calculation of the flow models has been kinematic (from the induction equation), and while dynamical information has been included to reduce the nonuniqueness of this determination, dynamical implications of the flow models have not been considered in detail. This is in part due to the uncertainties in the models: there are large differences between calculated flow models, and it is difficult to argue objectively which model to accept. Nonetheless, it is important to consider what further inferences can be made, and how the flow models can be fitted into the dynamical framework, particularly with the ever-growing number of numerical dynamo models.

One method is to examine the range of behavior which comes from relaxing nonuniqueness assumptions, and whether this can provide any dynamical information. To this end, Pais *et al.* (2004) considered the relaxation of the tangential geostrophy approximation, modeling the required departure from equilibrium as being due to the magnetic Lorentz force. The equation for the current density which they solved is essentially identical to the flow modeling eqn [4] here, and so subject to the same issues of nonuniqueness. Nonetheless, it is possible to seek current distributions that minimize, for example, the energy dissipation required; they found that substantial departures from tangential geostrophy required an electrical dissipation inconsistent with the energy budget of the core. This methodology may be applicable more broadly, to test whether the Lorentz forces required to support a particular flow are physically possible.

It is of particular interest to consider what might control the long-term structure of the flow, and in particular, whether any features are likely to demonstrate coherence on long timescales. For this to occur, the most likely mechanism is thermal core–mantle coupling (Bloxham and Gubbins, 1987), with the thermal regime at the base of the mantle having a strong influence on the dynamo, and so the flow patterns observed. Assuming that Lorentz forces are weak near the CMB, the thermal wind equation [28] can in principle be used to constrain the buoyancy, and so the temperature field. However, in order to do this, information is required on the flow shear. To construct the temperature field, either a thin-layer approximation must be used for flow at the top of the core (as applied by Kohler and Stevenson (1990) looking at topographic rather than thermal coupling), or the shear must be determind (Bloxham and Jackson, 1990), which as already mentioned is problematic. Instead, progress is being made through comparison with the results of dynamo codes (e.g., Amit and Olson, 2006), although as always using such methods, a significant caveat remains when interpreting the results as to the usefulness of simulations running in a physical regime far from that of the Earth.

What overall conclusions can be drawn with regards to the nature of large-scale flow in the core? If we wish to be able to make fully rigorous statements, then unfortunately there are very few. The still uncertain role of diffusion, the uncertainty as to the validity of the assumptions made to reduce nonuniqueness, and the screening influence of the lithospheric field, mean that almost any conclusion has to come with a caveat. Nonetheless, if we are willing to accept some of the assumptions as reasonable, we may draw some overall conclusions. The correlation of predicted core angular momentum from flow models with observed variations in LOD suggest that at the very least, the large-scale time-varying flow is being well recovered by our flow modeling. The steady component is less clear, but given results of testing output from dynamo simulations, it is likely that at least the gross features of flow are being successfully recovered. Root mean square flow speed has not increased greatly with the higher-resolution models calculated from satellite data (compare the values in **Figures** 1 and 3) and so a general rms flow speed of not more than 20 km yr^{-1} seems reasonable. The flow is probably well-defined under the Atlantic Hemisphere, where secular variation is strong and regular, but less so under the Pacific Hemisphere, where the secular variation is weaker, and models with different *a priori* information display strong differences. The success of the tangentially geostrophic assumption in fitting observed and modeled secular variation, and its comparitively strong theoretical underpinnings, means that it remains the most attractive form of *a priori* information for flow modeling. In this case, thermal winds seem a probable driving mechanism for surface flows, at least over periods of centuries or so. Finally, some finer features of core flow are likely to be robust: in particular, the polar vortices are likely to be present, and should provide a constraint on the dynamics in the tangent cylinder, although constraining inner-core rotation is likely to be overambitious.

How may we improve on this rather vague state of affairs? We have looked briefly at studies which have used insights from numerical dynamo modeling. The logical extension of such studies is to combine observational and computational methods directly – applying data assimilation to dynamo modeling. In this way, the dynamo models are constrained to fit the observations as far as possible. Such work is in its infancy (e.g., Kuang *et al* (in press)), but only in this way is there any real hope of resolving the issues of flow ambiguity described in this chapter, and of making real progress in direct constraint of dynamical processes from observed secular variation and inferred core flow.

References

Abarco del Rio R, Gambis D, and Salstein DA (2000) Interannual signals in length of day and atmospheric angular momentum. *Annales de Geophysique* 18: 347–364.

Amit H and Olson P (2004) Helical core flow from geomagnetic secular variation. *Physics of the Earth and Planetary Interiors* 147: 1–25.

Amit H and Olson P (2006) Time-average and time-dependent parts of core flow. *Physics of the Earth and Planetary Interiors* 155: 120–139.

Backus G, Parker R, and Constable C (1996) *Foundations of Geomagnetism*. Cambridge, UK: Cambridge University Press.

Backus GE (1968) Kinematics of geomagnetic secular variation in a perfectly conducting core. *Philosophical Transactions of the Royal Society of London A* 263: 239–266.

Backus GE (1988) Bayesian inference in geomagnetism. *Geophysical Journal* 92: 125–142.

Backus GE and Le Mouël J-L (1986) The region on the core–mantle boundary where a geostrophic velocity field can be determined from frozen flux magnetic data. *Geophysical Journal of the Royal Astronomical Society* 85: 617–628.

Bergman MI (1993) Magnetic rossby waves in a stably stratified layer near the top of the Earth's outer core. *Geophysical and Astrophysical Fluid Dynamics* 68: 151–175.

Bloxham J (1988) The determination of fluid flow at the core surface from geomagnetic observations. In: Vlaar NJ, Nolet G, Wortel MJR, and Cloetingh SAPL (eds.) *Mathematical Geophysics, A Survey of Recent Developments in Seismology and Geodynamics.* Dordrecht, The Netherlands: Reidel.

Bloxham J (1992) The steady part of the secular variation of the Earth's magnetic field. *Journal of Geophysical Research* 97: 19565–19579.

Bloxham J and Gubbins D (1987) Thermal core–mantle interactions. *Nature* 325: 511–513.

Bloxham J and Jackson A (1990) Lateral temperature variations at the core–mantle boundary deduced from the magnetic field. *Geophysical Research Letters* 17: 1997–2000.

Bloxham J and Jackson A (1991) Fluid flow near the surface of the Earth's outer core. *Reviews of Geophysics* 29: 97–120.

Bloxham J and Jackson A (1992) Time-dependent mapping of the magnetic field at the core–mantle boundary. *Journal of Geophysical Research* 97: 19537–19563.

Bloxham J, Zatman S, and Dumberry M (2002) The origin of geomagnetic jerks. *Nature* 420: 65–68.
Braginsky SI (1970) Torsional magnetohydrodynamic vibrations in the Earth's core and variations in day length. *Geomagnetism, and Aeronomy (English translation)* 10: 1–8.
Braginsky SI (1999) Dynamics of the stably stratified ocean at the top of the core. *Physics of the Earth and Planetary Interiors* 111: 21–24.
Buffett BA and Mound JE (2005) A Green's function for the excitation of torsional oscillations in the earth's core. *Journal of Geophysical Research Letters* 110: B08105.
Bullard EC, Freeman C, Gellman H, and Nixon J (1950) The westward drift of the Earth's magnetic field. *Philosophical Transactions of the Royal Society of London A* 243: 61–92.
Chen J (2005) Global mass balance and the length-of-day variation. *Journal of Geophysical Research* 110: B08404 (doi:10.1029/2004JB003474).
Courtillot V and Le Mouël J-L (1984) Geomagnetic secular variation impulses. *Nature* 311: 709–716.
Davis RG and Whaler KA (1997) The 1969 geomagnetic impulse and spin-up of the Earth's liquid core. *Physics of the Earth and Planetary Interiors* 103: 181–194.
Eymin C and Hulot G (2005) On core surface flows inferred from magnetic satellite data. *Physics of the Earth and Planetary Interiors* 152: 200–220.
Finlay CC and Jackson A (2003) Equatorially dominated magnetic field change at the surface of Earth's core. *Science* 300: 2084–2086.
Gire C and Le Mouël J-L (1990) Tangentially geostrophic flow at the core–mantle boundary compatible with the observed geomagnetic secular variation: The large-scale component of the flow. *Physics of the Earth and Planetary Interiors* 59: 259–287.
Gubbins D (1982) Finding core motions from magnetic observations. *Philosophical Transactions of the Royal Society of London A* 306: 247–254.
Gubbins D (1996) A formalism for the inversion of geomagnetic data for core motions with diffusion. *Physics of the Earth and Planetary Interiors* 98: 193–206.
Gubbins D and Kelly P (1996) A difficulty with using the frozen flux hypothesis to find steady core motions. *Geophysical Research Letters* 23: 1825–1828.
Gubbins D and Roberts PH (1987) Magnetohydrodynamics of the Earth's core. In: Jocobs JA (ed.) *Geomagnetism*, vol. 2, ch. 1. San Diego, CA: Academic Press.
Halley E (1692) On the cause of the change in the variation of the magnetic needle; with an hypothesis of the structure of the internal parts of the earth. *Philosophical Transactions of the Royal Society of London* 17: 470–478.
Hide R (1966) Free hydromagnetic oscillations of the Earth's core and the theory of the geomagnetic secular variation. *Philosophical Transactions of the Royal Society of London A* 259: 615–650.
Hide R and Roberts PH (1979) How strong is the magnetic field in the Earth's liquid core? *Physics of the Earth and Planetary Interiors* 20: 124–126.
Hills RG (1979) *Convection in the Earth's Mantle Due to Viscous Shear at the Core–Mantle Interface and Due to Large-Scale Buoyancy.* PhD Thesis, New Mexico State University, Las Cruces.
Holme R (1998) Electromagnetic core–mantle coupling. I: Explaining decadal variations in the Earth's length of day. *Geophysical Journal International* 132: 167–180.
Holme R and de Viron O (2005) Geomagnetic jerks and a high-resolution length-of-day profile for core studies. *Geophysical Journal International* 160: 435–439.
Holme R and Olsen N (2006) Core surface flow modelling from high-resolution secular variation. *Geophysical Journal International* 166: 518–528.
Holme R and Whaler KA (2001) Steady core flow in an azimuthally drifting frame. *Geophysical Journal International* 145: 560–569.
Hulot G, Eymin C, Langlais B, Mandea M, and Olsen N (2002) Small-scale structure of the geodynamo inferred from Oersted and MAGSAT satellite data. *Nature* 416: 620–623.
Hulot G, Le Mouël JL, and Jault D (1990) The flow at the core–mantle boundary – Symmetry properties. *Journal of Geomagnetism and Geoelectricity* 42: 857–874.
Hulot G, Le Mouël JL, and Wahr JA (1992) Taking into account truncation problems and geomagnetic model accuracy in assessing computed flows at the core mantle boundary. *Geophysical Journal International* 108: 224–246.
Jackson A (1995) An approach to estimation problems containing uncertain parameters. *Physics of the Earth and Planetary Interiors* 90: 145–156.
Jackson A (1997) Time dependency of geostrophic core surface motions. *Physics of the Earth and Planetary Interiors* 103: 293–311.
Jackson A (2003) Intense equatorial flux splots on the surface of the earth's core. *Nature* 424: 760–763.
Jackson A, Bloxham J, and Gubbins D (1993) Time-dependent flow at the core surface and conservation of angular momentum in the coupled core–mantle system. In: Le Mouël J-L, Smylie DE, and Herring T (eds.) *Dynamics of the Earth's Deep Interior and Earth Rotation*, pp. 97–107. Washington, DC: AGU/IUGG.
Jackson A, Jonkers ART, and Walker MR (2000) Four centuries of geomagnetic secular variation from historical records. *Philosophical Transactions of the Royal Society of London A* 358: 957–990.
Jault D, Gire C, and Le Mouël JL (1988) Westward drift, core motions and exchanges of angular momentum between core and mantle. *Nature* 333: 353–356.
Jault D and Le Mouël JL (1991) Physical properties at the top of the core and core surface motions. *Physics of the Earth and Planetary Interiors* 68: 76–84.
Kahle AB, Vestine EH, and Ball RH (1967) Estimated surface motions of the Earth's core. *Journal of Geophysical Research* 72: 1095–1108.
Kohler MD and Stevenson DJ (1990) Modelling core fluid motions and the drift of magnetic field patterns at the CMB by use of topography obtained by seismic inversion. *Geophysical Research Letters* 17: 1473–1476.
Kuang W, Tangborn A, and Sabaka T (in press) A stable model mapping geomagnetic data to geodynamo solution: Towards geomagnetic data assimiliation. *Physics of the Earth and Planetary Interiors.*
Langel RA (1987) The main field. In: Jacobs JA (ed.) *Geomagnetism*, vol. 1, ch. 4. San Diego, CA: Academic Press.
Langel RA, Ousley G, Berbert J, Murphy J, and Settle M (1982) The MAGSAT mission. *Geophysical Research Letters* 9: 243–245.
Lanzerotti LJ, Medford LV, Maclennan CG, Thomson DJ, Meloni A, and Gregori GP (1985) Measurements of the large-scale direct-current Earth potential and possible implications for the geomagnetic dynamo. *Science* 229: 47–49.
Le Mouël J-L (1984) Outer core geostrophic flow and secular variation of Earth's magnetic field. *Nature* 311: 734–735.
Le Mouël J-L, Gire C, and Madden T (1985) Motions at the core surface in the geostrophic approximation. *Physics of the Earth and Planetary Interiors* 39: 270–287.
Lloyd D and Gubbins D (1990) Toroidal fluid motion at the top of the Earth's core. *Geophysical Journal International* 100: 455–467.
Love JJ (1999) A critique of frozen-flux inverse modelling of a nearly steady geodynamo. *Geophysical Journal International* 138: 353–365.
Madden T and Le Mouël J-L (1982) The recent secular variation and the motions at the core surface. *Philosophical Transactions of the Royal Society of London* 306: 271–280.

Mandea M, Bellanger E, and Le Mouël J-L (2000) A geomagnetic jerk for the end of the 20th century? *Earth and Planetary Science Letters* 183: 369–373.

Marcus SL, Chao Y, Dickey JO, and Gregout P (1998) Detection and modelling of nontidal oceanic effects on Earth's rotation rate. *Science* 281: 1656–1659.

McCarthy DD and Babcock AK (1986) The length of day since 1656. *Physics of the Earth and Planetary Interiors* 44: 281–292.

Mound JE and Buffett BA (2003) Interannual oscillations in length of day: Implications for the structure of the mantle and core. *Journal of Geophysical Research* 108: 2334.

Mound JE and Buffett BA (2005) Mechanisms of core–mantle angular momentum exchange and the observed spectral properties of torsional oscillations. *Journal of Geophysical Research* 110: B08103.

Mound JE and Buffett BA (2006) Detection of a gravitational oscillation in length-of-day. *Earth and Planetary Science Letters* 243: 383–389.

Neubert T, Mandea M, Hulot G, *et al.* (2001) Ørsted satellite captures high-precision geomagnetic field data. *EOS Transactions of the American Geophysical Union* 82: 81–88.

Olsen N, Lühr H, Sabaka TJ, *et al.* (2006) CHAOS - A model of the Earth's magnetic field derived from CHAMP, Ørsted, and SAC-C magnetic satellite data. *Geophysical Journal International* 166: 67–75.

Olsen N and Mandea M (2007) Investigation of a secular variation impulse using satellite data: The 2003 geomagnetic jerk. *Earth and Planetary Science Letters* 255: 94–105.

Olson P and Aurnou J (1999) A polar vortex in the Earth's core. *Nature* 402: 170–173.

Olson P, Christensen U, and Glatzmaier GA (1999) Numerical modeling of the geodynamo: Mechanisms of field generation and equilibration. *Journal of Geophysical Research* 104: 10383–10404.

Olson P, Sumita I, and Aurnou J (2002) Diffusive magnetic images of upwelling patterns in the core. *Journal of Geophysical Research* 107: 2348.

Pais A and Hulot G (2000) Length of day decade variations, torsional oscillations and inner core superrotation: Evidence from recovered core surface zonal flows. *Physics of the Earth and Planetary Interiors* 118: 291–316.

Pais MA, Oliveria O, and Nogueira F (2004) Nonuniqueness of inverted core–mantle boundary flows and deviations from tangential geostrophy. *Journal of Geophysical Research* 109: B08105 (doi: 10.1029/2004JB003012).

Ponsar S, Dehant V, Holme R, Jault D, Pais A, and van Hoolst T (2003) The core and fluctuations in the Earth's rotation. In: Dehant V, Creager KC, Karato S-I, and Zatman S (eds.), *Geodynamics Series Vol. 31: Earth's Core: Dynamics, Structure, Rotation*, pp. 251–261. Washington, DC: AGU.

Rau S, Christensen U, Jackson A, and Wicht J (2000) Core flow inversion tested with numerical dynamo models. *Geophysical Journal International* 141: 485–497.

Reigber C, Lühr H, and Schwintzer P (2002) CHAMP mission status. *Advances in Space Research* 30: 129–134.

Roberts PH and Scott S (1965) On analysis of the secular variation. 1: A hydromagnetic constraint: Theory. *Journal of Geomagnetism and Geoelectricity* 17: 137–151.

Shimizu H, Koyama T, and Utada H (1998) An observational constraint on the strength of the toroidal magnetic field at the CMB by time variation of submarine cable voltages. *Geophysical Research Letters* 25: 4023–4026.

Shimizu H and Utada H (2004) The feasibility of using decadal changes in the geoelectric field to probe Earth's core. *Physics of the Earth and Planetary Interiors* 142: 297–319.

Song XD and Richards PG (1996) Seismological evidence for differential rotation of the Earth's inner core. *Nature* 382: 221–224.

Souriau A and Poupinet G (2000) Inner core rotation: A test at the worldwide scale. *Physics of the Earth and Planetary Interiors* 118: 13–17.

Sreenivasan B and Jones CA (2005) Structure and dynamics of the polar vortex in the Earth's core. *Geophysical Research Letters* 32: L20301 (doi: 10.1029/2005GL023841).

Vestine EH (1953) On variations of the geomagnetic field, fluid motions and the rate of Earth's rotation. *Journal of Geophysical Research* 58: 127–145.

Voorhies CV (1984). *Magnetic Location of Earth's Core–Mantle Boundary and Estimates of the Adjacent Fluid Motion*. PhD Thesis, University of Colorado, Boulder.

Voorhies CV (1986) Steady flow at the top of Earth's core derived from geomagnetic field models. *Journal of Geophysical Research* 91: 12444–12466.

Voorhies CV (1993) Geomagnetic estimates of steady surficial core flow and flux diffusion: Unexpected geodynamo experiments. In: Le Mouël J-L, Smylie DE, and Herring T (eds.) *Dynamics of the Earth's Deep Interior and Earth Rotation*, pp. 113–125. Washington, DC: AGU/IUGG.

Voorhies CV (1999) Inner core rotation from geomagnetic westward drift and a stationary spherical vortex in Earth's core. *Physics of the Earth and Planetary Interiors* 112: 111–123.

Voorhies CV and Backus GE (1985) Steady flows at the top of the core from geomagnetic-field models – The steady motions theorem. *Geophysical and Astrophysical Fluid Dynamics* 32: 163–173.

Whaler KA (1980) Does the whole of the Earth's core convect? *Nature* 287: 528–530.

Whaler KA (1986) Geomagnetic evidence for fluid upwelling at the core–mantle boundary. *Geophysical Journal of the Royal Astronomical Society* 86: 563–588.

Whaler KA and Davis RG (1997) Probing the Earth's core with geomagnetism. In: Crossley DJ (ed.) *Earth's Deep Interior*, pp. 114–166. Amsterdam, The Netherlands: Gordon and Breach.

Yukutake T (1979) Review of the geomagnetic secular variations on the historical time scale. *Physics of the Earth and Planetary Interiors* 20: 83–95.

Yukutake T and Tachinaka H (1969) Separation of the Earth's magnetic field into drifting and standing parts. *Bulletin of the Earthquake Research Institute* 47: 65.

Zatman S and Bloxham J (1997) Torsional oscillations and the magnetic field within the Earth's core. *Nature* 388: 760–763.

Zatman S and Bloxham J (1998) A one-dimensional map of B_S from torsional oscillations of the Earth's core. In: Gurnis M, Wysession ME, Knittle E, and Buffett BA (eds.) *The Core–Mantle Boundary Region*, pp. 183–196. Washington, DC: AGU.

Zhang J, Song X, Li Y, Richards PG, Sun X, and Waldhauser F (2005) Inner core differential motion confirmed by earthquake doublets. *Science* 309: 1357–1360.

5 Thermal and Compositional Convection in the Outer Core

C. A. Jones, University of Leeds, Leeds, UK

5.1	**Introduction to Core Convection**	133
5.1.1	The Need for a Dynamo	133
5.1.2	Velocities in the Core	134
5.1.3	The Heat Flux and Core Cooling	135
5.1.4	The Dynamics of Core Convection	137
5.2	**Equations Governing Convection**	139
5.2.1	Compressible and Anelastic Equations	139
5.2.2	Adiabatic Reference State	142
5.2.3	Equations for the Convective Perturbations	143
5.2.4	Boundary Conditions	144
5.2.5	The Boussinesq Limit	145
5.2.6	Formulation of the Equations for Numerical Solution	146
5.3	**Convection in the Absence of Rotation and Magnetic Field**	147
5.3.1	Thermal Convection	147
5.3.2	Compositional Convection	149
5.4	**Effect of Rotation on Convection**	150
5.4.1	The Proudman–Taylor Theorem	150
5.4.2	The Onset of Instability	152
5.4.3	The Onset of Instability in the Rapidly Rotating Limit	154
5.4.4	The Ekman Boundary Layers	156
5.4.5	The Busse Annulus	158
5.4.5.1	Linear properties of the annulus model	158
5.4.5.2	Weakly nonlinear theory and the annulus model	159
5.4.5.3	Zonal flows and multiple jets	160
5.4.6	The Quasi-Geostrophic Approximation	161
5.4.7	Thermal Wind	161
5.4.8	Scaling Laws and Heat Transport in Nonlinear Rapidly Rotating Convection	162
5.4.8.1	The inertial theory of rapidly rotating convection	162
5.4.8.2	Heat transport in rapidly rotating convection	164
5.5	**Convection with and without Rotation in the Presence of Magnetic Field**	165
5.5.1	Flux Expulsion and Flux Rope Formation	165
5.5.2	Linear Theory of Magnetoconvection in Plane Geometry	166
5.5.2.1	Nonrotating magnetoconvection	166
5.5.2.2	Onset of plane-layer rotating magnetoconvection	167
5.5.2.3	Waves in the core	168
5.5.2.4	Small-scale dynamics and the Braginsky–Meytlis theory of plate-like motion	169
5.5.2.5	Boundary layers in rotating, magnetic fluids	169
5.5.3	Onset of Rotating Magnetoconvection in Spherical Geometry	170
5.5.4	Magnetic Instabilities	171
5.5.5	Taylor's Constraint	172
5.5.6	Numerical Simulations of Nonlinear Convection-Driven Dynamos	173
5.5.6.1	Convection outside the tangent cylinder: picture from numerical simulations	175
5.5.6.2	Convection inside the tangent cylinder: picture from numerical simulations	176

5.5.7 Scaling Laws and Dynamo Simulations 177
5.6 Heterogeneous Boundary Conditions and Stable Layers Near the CMB 178
5.7 Conclusions and Future Developments 180
References 181

Nomenclature

c_p	specific heat at constant pressure ($\mathrm{J\,kg^{-1}\,K^{-1}}$)
d	$r_{cmb} - r_{icb}$ (m)
g	acceleration due to gravity ($\mathrm{m\,s^{-2}}$)
h^{ξ}	heat of reaction ($\mathrm{J\,kg^{-1}}$)
ℓ	typical width of convection rolls (m)
$\mathbf{j}$	current density ($\mathrm{A\,m^{-2}}$)
k	thermal conductivity ($\mathrm{W\,m^{-1}\,K^{-1}}$)
$\mathbf{k}$	vector wave number ($\mathrm{m^{-1}}$)
k^{ξ}	mass diffusion coefficient ($\mathrm{kg\,m^{-1}\,s^{-1}}$)
m	azimuthal wave number (exp $im\phi$)
p	pressure (fluctuating pressure)
q	Roberts number (κ/η)
r	distance from Earth's center (m)
r_{cmb}	radius of CMB (m)
r_{icb}	radius of ICB (m)
t	time (s)
$\mathbf{u}$	velocity ($\mathrm{m\,s^{-1}}$)
A	magnetic potential (T m)
$\mathbf{B}$	magnetic field (T)
B_*	typical core field strength (T)
$\mathcal{C}$	Coriolis operator in toroidal-poloidal expansion equations ($\mathrm{m^{-1}}$)
D	dissipation number
$\mathbf{E}$	electric field ($\mathrm{V\,m^{-1}}$)
E, ε	Ekman number
F_{conv}	the convective part of the heat flux ($\mathrm{W\,m^{-2}}$)
$\mathbf{F}^{v}$	viscous force per unit mass ($\mathrm{N\,kg^{-1}}$)
H	internal heat source ($\mathrm{W\,kg^{-1}}$)
I_{ad}	heat flux per unit area conducted down the adiabat ($\mathrm{W\,m^{-2}}$)
I_{cmb}	heat flux per unit area passing through CMB ($\mathrm{W\,m^{-2}}$)
L	height of Busse annulus (m)
L_H	latent heat per unit mass ($\mathrm{J\,kg^{-1}}$)
$\mathcal{L}^2$	r^2 times the angular part of the Laplacian operator
N	Nusselt number based on the adiabatic temperature gradient
Nu	Nusselt number based on the superadiabatic temperature gradient
Pm	magnetic Prandtl number (ν/η)
Pr	Prandtl number (ν/κ)
Q	Chandrasekhar number ($B_0^2/\mu\rho\nu\eta$)
Q_j	ohmic dissipation ($\mathrm{W\,m^{-3}}$)
Q_v	viscous dissipation ($\mathrm{W\,m^{-3}}$)
Ra	Rayleigh number $g\alpha\Delta T_s d^3/\kappa\nu$
Ra^*	nondiffusive Rayleigh number $g\alpha\Delta T_s/\Omega d^2$
Ra_Q	convective flux Rayleigh number $Ra(Nu-1)$
R_m	magnetic Reynolds number $U_* d/\eta$
Ro	Rossby number $U_*/\Omega d$
$\mathcal{R}$	scaled Rayleigh number $\varepsilon^{4/3}Ra$
S, (S_c)	specific entropy (entropy perturbation) ($\mathrm{J\,K^{-1}\,kg^{-1}}$)
$\mathcal{S}$	poloidal scalar for velocity ($\mathrm{m^3\,s^{-1}}$)
T	temperature (K)
T_a	temperature of adiabatic state (K)
T_c	temperature perturbation due to convection (K)
T'	$T_c - T_0$, T_0 being the static temperature distribution (K)
T_0'	$\partial T_0/\partial r$ ($\mathrm{K\,m^{-1}}$)
T_*'	typical temperature perturbation in the core (K)
$\mathcal{T}$	toroidal scalar for velocity ($\mathrm{m^2\,s^{-1}}$)
U_*	typical core velocity ($\mathrm{m\,s^{-1}}$)
$\mathcal{U}$	toroidal scalar for magnetic field (T m)
α	coefficient of thermal expansion ($\mathrm{K^{-1}}$)
α^{ξ}	adiabatic compositional expansion coefficient
β	rotation parameter in the annulus model $4\chi\Omega D^3/L\nu$
γ	Grüneisen parameter
δ	thermal boundary layer thickness (m)
δ_B	magnetic dissipation length scale (m)
ζ	vorticity ($\mathrm{s^{-1}}$)
η	magnetic diffusivity ($\mathrm{m^2\,s^{-1}}$)
κ	thermal diffusivity ($\mathrm{m^2\,s^{-1}}$)
κ^{ξ}	mass diffusivity ($\mathrm{m^2\,s^{-1}}$)
μ	magnetic permeability ($\mathrm{H\,m^{-1}}$)
$\mathcal{V}$	poloidal scalar for magnetic field ($\mathrm{T\,m^2}$)
ν_T	turbulent kinematic viscosity ($\mathrm{m^2\,s^{-1}}$)

ξ	mass fraction of light element	Δ_2	dimensionless quantity derived from melting point curve
ρ, ρ_a	density of reference state ($kg\,m^{-3}$)	ΔT_a	adiabatic ICB-CMB temperature difference (K)
ρ_c	density perturbation ($kg\,m^{-3}$)	ΔT_s	superadiabatic ICB-CMB temperature difference (K)
σ	electrical conductivity ($S\,m^{-1}$)	$\Delta\xi$	jump of ξ across the ICB
τ_{diss}	magnetic dissipation time (energy/dissipation rate) (s)	Λ	Elsasser number $B_0^2/\mu\rho\eta\Omega$
χ	endwall slope angle in the annulus model	$\boldsymbol{\Omega}$	rotation rate of mantle (assumed constant) (s^{-1})
ψ	stream function ($m^2\,s^{-1}$)	$\boldsymbol{\Omega}_{IC}$	rotation rate of inner core (s^{-1})
ω_A	buoyancy frequency (internal wave frequency) (s^{-1})		
ω_C	inertial wave frequency (s^{-1})		
ω_M	Alfvén frequency (torsional wave frequency) (s^{-1})		

5.1 Introduction to Core Convection

The Earth's liquid outer core is stirred by thermal and compositional convection. In thermal convection the buoyancy is due to temperature fluctuations, while in compositional convection it is produced by light material released at the inner-core boundary. Convection in the core is strongly supercritical, and is greatly affected by rotation and magnetic field. In this chapter, we focus on the issues of how fast the convective motions are, what type of flows occur, and how this relates to the heat and mass transport. Core convection is believed to be a primary source of dynamo action, and so it is responsible for the Earth's magnetic field. We therefore also consider how core convection affects the magnetic field generated.

5.1.1 The Need for a Dynamo

The basic structure of the Earth's interior has been deduced from seismic observations. It is known from these data that there is a high-density core surrounded by a lower density mantle, with the transition occurring at the core–mantle boundary (CMB), which is located 3480 km from the center of the Earth. The high-density core, which is believed to be composed mainly of iron, is divided into a fluid outer core and a solid inner core, with the inner-core boundary (ICB) having a radius of 1220 km. There is a density jump across the ICB of roughly 500 $kg\,m^{-3}$, the density of the inner core itself being around 12 000 $kg\,m^{-3}$. The fluid nature of the outer core is deduced from its inability to support transverse seismic waves (S-waves), which can only propagate in a solid.

The basic structure of the Earth's interior was mainly worked out in the first half of the twentieth century. It had been known much earlier that the Earth's magnetic field varies slowly with time; indeed, this secular variation was known to Henry Gellibrand in 1635, and Edmond Halley noticed that the field appeared to drift westwards. He suggested that the Earth might have a layered concentric shell structure, with the magnetic field being carried along with one of the interior shells. Joseph Larmor (1920) proposed that the magnetic fields of the Earth and the Sun might be generated by a dynamo, driven by motion in an electrically conducting fluid in the core, which is the basis of the modern geodynamo theory of the Earth's magnetic field. The physical principles behind the dynamo theory are the laws of electromagnetism (Maxwell's equations) and the laws of fluid motion (the Navier–Stokes equations). These equations were well known in Larmor's time, but finding useful solutions of them has proved difficult, though it is now possible to find numerical solutions using powerful computers.

The iron in the outer core has an electrical conductivity of around $4 \times 10^5\,S\,m^{-1}$, though there is some uncertainty because the exact composition of the outer core is not known, and the very high pressure adds a further difficulty to estimating this (and other physical properties of the core) accurately. It is not difficult to show that with this conductivity, the geomagnetic field would decay on a 20 000 year timescale (see, e.g., Moffatt (1978)) if the fluid in the core is not actively driven. Core material has electrical resistance which leads to ohmic dissipation, giving rise to heating. Unless fluid motion is continuously driven, the current systems that maintain the field in the core decay away. The amount of ohmic heating is a measure of the power input required to maintain the dynamo. Since the geomagnetic field induced magnetism in rocks laid down 3.5 Ga (see, e.g., Kono and Roberts (2002)), soon after the Earth's formation, it is clear that a long-term energy source is needed to maintain the Earth's field. Many alternative theories to the dynamo mechanism have been proposed for maintaining the geomagnetic field, but none has stood the test of time. Gauss (1839) showed that the nature of the field is such that the main field must have an internal rather than an external origin. Chemical inhomogeneity (the battery effect) and thermo-electricity have been considered as possible sources, but while they can generate very small fields, they cannot plausibly provide the substantial energy needed to maintain the observed field.

While the dynamo origin of the geomagnetic field has been largely accepted, there is much less agreement on the nature of the energy source driving it. Thermal and compositional convection have been explored in considerable detail as possible energy sources, but precession and tidal forcing have also been considered seriously (Malkus, 1994). The gravitational torques on the nonspherical Earth make the polar axis precess around the ecliptic pole approximately once every 26 000 years. The liquid outer core of the Earth will therefore be forced at its boundary by this motion, and consequently fluid motion is induced inside the outer core. The rotation axis of the fluid outer core will lag a little behind the rotation axis of the mantle, and the resulting torque will extract energy from the Earth's rotation, eventually bringing the rotation axis into its minimum energy configuration parallel to the ecliptic pole. Oceanic and solid Earth tides cause the Earth to gradually rotate more slowly, and stresses at the CMB transmit this angular momentum loss down to the fluid outer core by generating fluid motion there. These precessional and tidal energy sources for core motion ultimately derive from the rotational energy of the Earth. There is somewhat less energy in the Earth's rotation than there is in the gravitational and thermal energy that drives convection, but sufficient rotational energy exists to maintain a dynamo provided that the conversion of kinetic energy to magnetic energy is reasonably efficient. We do not yet know whether the fluid motions induced by precession or tides can drive a dynamo, but this is being investigated (Tilgner, 2005). The difficulty is that the forcing provided by these mechanisms has the diurnal frequency, while the dynamo operates on a much longer timescale. Here we adopt the hypothesis that the dynamo is driven by convection, which provides an adequate power source on a suitable timescale.

5.1.2 Velocities in the Core

Direct evidence for core motions comes from the temporal behavior of the geomagnetic field, the secular variation. To a first approximation, fluid carries magnetic field with it as it moves. The change in the field is therefore due partly to the velocity just under the CMB. Magnetic diffusion allows field to move through the fluid, but even if this is neglected, we cannot specify the velocity solely from magnetic field observations (Roberts and Scott, 1965). For example, if we monitor the radial component of the field B_r at the CMB, there will be no change in the field if the flow is parallel to lines of constant B_r, so this component of the field is invisible to secular variation measurements. Additional assumptions have to be made in order to specify the core flow. Two common assumptions are either that the flow is tangentially geostrophic, or that it is toroidal, just below the CMB. Tangential geostrophy implies that $\nabla_{\mathrm{H}} \cdot \cos\theta\, \mathbf{u} = 0$, while toroidal flow implies $\nabla_{\mathrm{H}} \cdot \mathbf{u} = 0$. Here θ is the co-latitude and $\mathbf{u}$ is the fluid velocity just below the CMB. ∇_{H} means the horizontal part of the divergence is taken, so for toroidal flow

$$\frac{1}{r\sin\theta}\frac{\partial(\sin\theta u_\theta)}{\partial\theta} + \frac{1}{r\sin\theta}\frac{\partial u_\phi}{\partial\phi} = 0$$

(for further details, see Holme and Olsen (2006)). When either of these assumptions are made, it is possible to reconstruct the flow field (Bloxham and Jackson, 1991). **Figure 1**, which is based on data from the Orsted and CHAMP magnetic satellite observations, is reproduced from Holme and Olsen (2006).

The figure shows that some features are universal even if different assumptions are made, and the typical velocity of 15 km yr^{-1}, which is 5×10^{-4} m s^{-1}, is a measure of the faster moving flows that are transporting magnetic field. Of course, this is just the flow beneath the CMB, and the flow at larger depths may be different. Also, small-scale flows are invisible

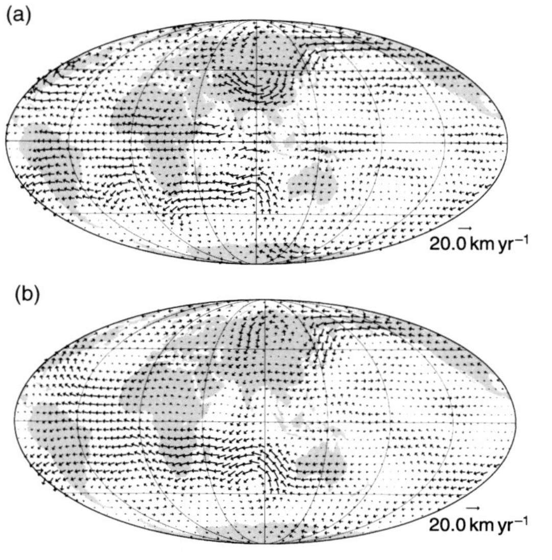

Figure 1 Core flows inferred from secular variation observations using the (a) geostrophic flow assumption and (b) toroidal flow assumption. Reproduced from Holme R Olsen N (2006) Core surface flow modelling from high-resolution secular variation. *Geophysical Journal International* 166: 518–528.

because only large-scale field structures can be seen. Small-scale structures are obscured by crustal magnetism. The flow speeds may be somewhat larger if there is a significant small-scale contribution. Nevertheless, flow speeds of this order are what is predicted from our understanding of core convection, and we will adopt the point of view that convective velocities of the order of $5 \times 10^{-4}\,\mathrm{m\,s^{-1}}$ are present in the core.

5.1.3 The Heat Flux and Core Cooling

It is likely that a great deal of heat was released when the iron differentiated from the mantle to the core of the Earth. The Earth has therefore been cooling down ever since its birth. The current rate of loss of heat from the Earth is around 44 TW. The fraction of this 44 TW that comes from the core is controversial, with estimates ranging from 3 TW (Sleep, 1990) to 15 TW (Roberts *et al.*, 2003). Since the temperature increases as we approach the Earth's center, heat can be transmitted outwards by thermal conduction. If all the heat flux coming from the core can be so transmitted, there will be no thermal convection. Thermal convection occurs when there is more heat to be transmitted than can be carried by conduction. As we shall see below, the criterion for the onset of core convection is essentially that the actual temperature gradient is steeper than the adiabatic temperature gradient. The key question is therefore whether the required heat flux can be carried down the adiabatic temperature gradient entirely by conduction. The adiabatic temperature gradient can be estimated reasonably well (see, e.g., Anufriev *et al.* (2005)), but the thermal conductivity k is still rather uncertain with estimates being in the range $k \approx 25\text{–}50\,\mathrm{W\,m^{-1}\,K^{-1}}$. With $k = 46\,\mathrm{W\,m^{-1}\,K^{-1}}$, Roberts *et al.* (2003) estimated the heat flux conducted down the adiabat at 0.3 TW near the ICB and 6 TW near the CMB. Models suggest that a large fraction of the core heat flux comes from latent heat released at the ICB as the inner core grows. If this is the case, even with a low estimate of CMB heat flux there is more than 0.3 TW to be carried outward near the ICB, and convection must occur there. Near the CMB the situation is much less clear; high estimates of CMB heat flux give convection throughout the core, but if the low estimates are correct then there will be convectively stable regions in the core where the heat is being extracted solely by conduction down the adiabat.

If the core is cooling, the solid inner core must be slowly growing in size as the iron gradually freezes out. It might seem surprising that the core freezes first near the center, where it is hottest, rather than at the CMB where it is coolest. This is due to the rapid increase in pressure as we move toward the center of the core, pressure raising the melting point temperature. Because $(\mathrm{d}T/\mathrm{d}p)_{\mathrm{liq}} > (\mathrm{d}T/\mathrm{d}p)_{\mathrm{ad}}$, where the subscript 'liq' denotes variation along the melting point curve and the subscript 'ad' denotes variation along the adiabat, freezing first occurred at the center of the Earth as the core gradually cooled. As the inner core grows, light material is released, the iron content of the inner core being higher than that of the outer core. The exact nature of this light material is uncertain, but *ab initio* quantum calculations of the properties of materials at core temperatures and pressures suggest that a mixture of several different light elements, oxygen, sulfur, and silicon, form the light component (Alfè *et al.*, 2003). This light material is buoyant, and drives compositional convection. The rate of growth of the inner core is therefore directly related to the buoyancy flux of light material driving compositional convection. The density jump across the ICB can be measured directly by seismic observations. Recent estimates (Masters and Gubbins, 2003) give the density of the inner core to be about

5% greater than that of the outer core at the ICB. Some of this is due to the fact that iron (unlike water) contracts as it freezes, so there would be a density jump even if there was no light material; after this jump is subtracted off, the remaining part of the density jump relates the compositional buoyancy flux to the rate of growth of the inner core. Knowledge of the melting point curve allows us to relate the rate of growth of the inner core to the rate of cooling of the outer core, which in turn relates to the heat flux passing through the CMB. The latent heat released at the ICB is also related to the rate of growth of the inner core.

While the basic physics governing inner-core growth, compositional buoyancy flux, latent heat release, and the core cooling rate is understood, unfortunately the key physical properties of materials at very high pressure and temperature are not well known. They are measured by a combination of experiments and theory, but different authors give estimates which can vary by a factor 2 or more, which means that the estimated rate of growth of the inner core can vary considerably. A further complication is the uncertainty over whether radioactive elements occur in the core. They are believed to be responsible for most of the heat coming out of the mantle, but it is uncertain whether radioactive elements such as potassium (^{40}K) are present in the core material or not (*see* Chapter 2). ^{40}K is strongly depleted in the mantle from the value expected from the primordial composition, but this could be because it evaporated into space when the Earth formed. Alternatively, it could have alloyed with elements such as nickel, and hence entered the core; our knowledge of geochemistry is not currently sufficient to be certain which scenario occurred. If ^{40}K and other radioactive elements are present in significant quantities, the whole heat budget of the core is changed dramatically. The core might not be cooling at all, with the radioactive heat released balancing the heat coming out at the core mantle boundary. The inner core would then not be growing, so there would be no compositional convection and no latent heat released at the ICB. The dynamo would then have to be driven entirely by thermal convection, or some other source such as precession. If compositional convection is occurring, with a thermal conductivity of 40 W m^{-1} K^{-1}, it is likely that the rate of release of buoyant material at the ICB is about 10^{5} kg s^{-1}, corresponding to an inner core age of only 1.2 billion years. The rate of release of gravitational energy is then about 0.5 TW and the latent heat release is between 3 TW and 6 TW. Reducing the thermal conductivity of the core reduces the amount of heat conducted down the adiabat, and hence slows down the rate of growth of the inner core. The age estimate also depends on the amount of latent heat released per unit mass at the ICB.

As mentioned above, there is no agreement about the amount of heat passing through the core–mantle boundary. Physically, this is determined by mantle convection. From the point of view of the mantle, the core provides a uniform temperature bottom boundary. There appears to be a thermal boundary layer at the base of the mantle, the D″ layer, through which heat is conducted into the mantle, but in the absence of reliable information about mantle convection, we have to accept that there is a wide range of plausible estimates from 3 TW to 10–15 TW for the CMB heat flux, and we need to consider what happens in both high and low CMB heat flux models. In the case of a high CMB heat flux, that is, one greater than the heat flux conducted down the adiabat near the CMB (about 6 TW on current estimates of the thermal conductivity), the situation is straightforward because convection can occur throughout the core. If the heat flux passing through the CMB is less than that conducted down the adiabat, the situation is more complicated. If compositional convection is occurring, and this is sufficiently effective to maintain the temperature gradient close to adiabatic, then the temperature gradient will actually be slightly subadiabatic and there is a negative convective heat flux (Loper, 1978), that is the convection pumps heat back into the core, so that the total convected and conducted flux near the CMB equals the flux passing through the CMB. Alternatively, compositional convection may lead to an accumulation of light material just below the CMB. In this case, the temperature gradient would depart very strongly from adiabatic until the conducted flux fell to the CMB value. Then the very strong stable stratification would allow only oscillatory radial motion. Braginsky (1993) has termed this the 'inverted ocean', since it would consist of a sea of relatively light material floating up against the CMB. The dynamo would then have to be driven in the deeper part of the core, below the level where the actual CMB flux equals the flux carried by the adiabat. These arguments suppose that the conducted and convected heat flux is approximately spherically symmetric. This may well not be the case. If the heat flux passing into the mantle is strongly inhomogeneous, there could be regions of the CMB where the heat flux is greater than adiabatic, so there would be strong convection beneath these features, but in other regions the heat flux might be subadiabatic, and consequently relatively little convection in the core beneath these regions. The total CMB heat flux might then be at the low end of the estimates, but still allow some strong upwellings in the core just below the CMB, and hence possible dynamo activity at all depths in the core. In Section 5.6 below we therefore discuss convection with a heterogeneous CMB boundary condition to explore this further.

5.1.4 The Dynamics of Core Convection

It is clear from the above discussion that while we do not know the exact amounts of heat flux and compositional flux that are being convected in the core, it is very likely that these processes are occurring. Thermal convection is essentially a process by which hot parcels of fluid move outward, while cold parcels of fluid move inward. Natural questions are what is the temperature difference between the warm and cool fluid parcels, and how fast are they moving? The heat flux transported constrains the product of these two quantities, but does not on its own specify either the velocity or the temperature perturbations individually. For example, suppose thermal convection at a radius $r_0 = (r_{icb} + r_{cmb})/2$ convects 1 TW of heat. Suppose also that exactly one-half of the fluid at $r = r_0$ is moving radially outward at speed u_r, and has temperature T'_* greater than the remaining fluid, which moves radially inward at speed u_r. If we take the density as the mean density 11 340 kg m^{-3} and the specific heat as 860 J kg^{-1} K^{-1}, then to transport one terawatt of heat requires $T'_* u_r = 3.7 \times 10^{-8}$ m s^{-1} K. From the arguments in Section 5.1.2, the typical core velocity can be estimated from the typical speed with which flux patches move, say 5×10^{-4} m s^{-1}. Then $T'_* \sim 7 \times 10^{-5}$ K, a surprisingly small value. However, it might be that radial motion is a factor ten less, in which case we could transport the same amount of heat by increasing T'_* by a factor 10. Of course, this example assumed that the hot fluid was perfectly correlated with rising fluid. This might not be the case, in which case a slightly higher T'_* would be required to transport the required heat flux at a given velocity. Actually, as we shall discuss later, simulations suggest that there does generally seem to be a fairly high correlation between the rising fluid and the horizontal temperature perturbations. Now that very small horizontal temperature perturbations have been proposed for thermal convection in the core, one might ask whether such small temperatures can actually drive motions at this speed. As we shall see below, this is indeed possible, and core speeds of the order of 5×10^{-4} m s^{-1} emerge naturally from convection models. This is of course strong evidence for the existence of core convection. We can repeat these estimates for compositional convection. Again we find that very small density perturbations of the order of 10^{-9} of the actual density are capable of delivering the required mass fluxes.

Our very small estimate for T'_* raises an issue that it is important to understand. The actual temperature difference between the ICB and the CMB is over 1000 K. So how can the temperature differences driving the convection be so much less? The answer is that in a convecting region the core is almost adiabatically stratified, and it is only the difference between the actual temperature gradient and the adiabatic gradient which drives the convection. This difference is called the superadiabatic temperature gradient, and it is more than a million times smaller than the adiabatic gradient. So in a compressible fluid such as the Earth's core, we call ΔT_s the superadiabatic temperature difference across the core. T'_* is of the same order as ΔT_s, which is much less than 1000 K. This is illustrated in **Figure 2**.

Convection in the core will be influenced by rotation and magnetic field. Since the velocities are so

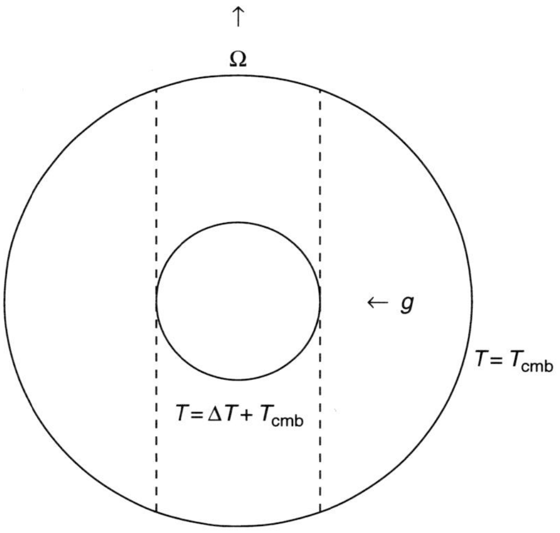

Figure 2 Geometry of convection between spherical shells. The dashed lines mark the position of the tangent cylinder (TC). For a Boussinesq liquid, ΔT is the total temperature difference across the outer core, but for compressible convection $\Delta T = \Delta T_{ad} + \Delta T_s$, and the superadiabatic temperature difference ΔT_s is much less than the adiabatic temperature difference ΔT_{ad}.

slow, the dimensionless Rossby number $Ro = U_*/d\Omega$, taking $d = r_{cmb} - r_{icb}$ as a typical length scale, is very small. It is therefore natural to work in the rotating frame and introduce Coriolis and centrifugal forces. The centrifugal force is small compared with gravity and it is usually ignored in core convection. Note that for constant density, the centrifugal force is the gradient of a potential, $\nabla \rho \Omega^2 s^2/2$, so it can be exactly balanced by the pressure. Only when the very small variations due to density are taken into account does centrifugal force contribute, and then it is dominated by the corresponding terms involving the gravitational acceleration, g. It is therefore through the Coriolis force that rotation makes an impact. In crude terms, we might expect this to happen when the magnitude of the Coriolis force $|2\rho\mathbf{\Omega} \times \mathbf{u}|$ exceeds the magnitude of the inertial acceleration $|\rho(\partial \mathbf{u}/\partial t + \mathbf{u} \cdot \nabla \mathbf{u})|$. If we estimate the gradient as $1/d$, we see that Coriolis force dominates acceleration provided the Rossby number Ro is small, and the velocity changes on timescales longer than the rotation rate. If we accept the estimate of $|\mathbf{u}| \sim 5 \times 10^{-4}\,\mathrm{m\,s^{-1}}$ suggested above, then $Ro \sim 3 \times 10^{-6}$, so that from this crude estimate Coriolis force completely dominates inertia in the core, and so we expect rotation to be very influential in the core. On the timescales we are interested in for core convection, which are much longer than a day, we therefore expect inertia to be small, so that the core is always in equilibrium, and changes occur in the core by evolving slowly through a sequence of equilibrium states. We need to be careful, however, because there are a number of ways in which this estimate can be radically altered. If the fluid motion is geostrophic, this means that the Coriolis force is a potential force, and so it can be balanced entirely by pressure. Flows which are geostrophic are then controlled by the smaller terms in the equation of motion after the pressure force and the Coriolis force have cancelled each other out. So for geostrophic flows, inertial terms can be more important than our crude estimate implies. The condition for geostrophy is that

$$\nabla \times (\rho\mathbf{\Omega} \times \mathbf{u}) = 0$$

Motions in the Earth's atmosphere and oceans are mainly geostrophic, and this is likely to be true in the core too. However, because the Rossby number is so small in the core, any significant departure from exact geostrophy will lead to a large Coriolis force that cannot be pressure balanced, which will normally be balanced by buoyancy and magnetic forces. An example of geostrophic core motion is the torsional oscillation, where the fluid moves azimuthally only and the velocity is constant on cylinders. For this motion, which has a typical timescale of 60 years, the balance is between the magnetic Lorentz force and inertia, since the large Coriolis and pressure forces exactly cancel at all times. There is another way in which inertia can be important for slow core flows; the length scale for the gradient in $\mathbf{u} \cdot \nabla \mathbf{u}$ may be much less than the core size, making inertia relatively more important. We shall return to this important issue of the effect of inertia in rotating convection below.

If viscosity and variations of density are ignored, the condition for geostrophy reduces to $\partial \mathbf{u}/\partial z = 0$. The requirement that slow steady motions with negligible ageostrophic forces (such as buoyancy and magnetic field) must be geostrophic, and hence satisfy this equation, is known as the Proudman–Taylor theorem, predicted by Proudman (1916) and verified experimentally by Taylor (1923). Core motions driven by convection are indeed slow, and are steady on timescales corresponding to the rotation timescale, so we expect the velocity generally to be independent of z, that is, the motion is columnar, with whole columns of fluid moving together. Within the geometry of the Earth's core, there is then going to be an important distinction between the behavior inside and outside the tangent cylinder (see **Figure 2**). The tangent cylinder (TC) is the imaginary cylinder whose axis is parallel to the rotation axis and which touches the inner core. Outside the tangent cylinder, columns of fluid can extend from one hemisphere right through to the other hemisphere. Inside the tangent cylinder, fluid columns can only reach from the CMB down to the ICB. Furthermore, a column of fluid moving outside the tangent cylinder will have to change its length (and hence induce ageostrophic motion) quite slowly as it changes its distance from the rotation axis. A column of fluid moving from outside to inside the tangent cylinder is cut in two and undergoes a very rapid change in length as it crosses the tangent cylinder. Strongly ageostrophic motion is required to do this, so it is not very likely to happen; although there is no physical barrier at the tangent cylinder, the dynamics of rapidly rotating fluids makes it unlikely that there is much fluid transport across the TC.

Convection inside and outside the tangent cylinder is likely to be rather different in character, for geometrical reasons. Outside the TC, columnar motions can transport heat effectively outward in the s-direction (using cylindrical polar coordinates s, ϕ, z with z parallel to the rotation axis). A relatively small amount of ageostrophic motion in the z-direction is required to transport the heat parallel to the columns. Inside the TC, convection faces more difficulties, because to get heat from the ICB to the CMB inside the TC requires mainly motion in the z-direction, which must vary with z because there can be no penetration through the ICB or CMB. Even though there is no penetration through the ICB and CMB, there can be a thin boundary layer, the Ekman layer, inside which there is some horizontally diverging flow. This leads to Ekman suction, that is a small flow in the radial direction into the boundary layer (see section 5.4.4 below). This Ekman suction may affect the flow in the core, but it is too small to carry much heat outwards. Not surprisingly, convection inside the TC is harder to get going, and will rely more strongly on ageostrophic forces such as the Lorentz force to control it. Nevertheless, at Rayleigh numbers well above critical, convection inside the TC can be as vigorous as that outside the TC, because the buoyancy forces are strong enough to maintain the ageostrophic part of the flow. Indeed, the heat transport inside the TC can even exceed that outside the TC (Tilgner and Busse, 1997; Busse and Simitev, 2006).

Magnetic fields influence core flow to some extent everywhere in the core. In a dynamo generated field this is inevitable, because the induction equation is linear in the magnetic field $\mathbf{B}$. For a dynamo, there must be growing modes which can potentially increase the field without limit. Magnetic field saturation occurs by the field altering the dynamo properties of the flow so that magnetic growth no longer occurs. However, the ability to generate magnetic field from a flow is a fairly subtle process, and small changes in the flow may be sufficient to prevent further field growth. We can try estimating the magnitude of the Lorentz force $|\mathbf{j} \times \mathbf{B}|$ and comparing it with the crude estimate of Coriolis force. From geomagnetic observations, the radial field at the CMB is typically 0.5 mT, and if we assume the field inside the core is slightly larger at 1 mT, then the current density $|\mathbf{j}| = |\boldsymbol{\nabla} \times \mathbf{B}|/\mu \sim |\mathbf{B}|/\mu d$ is around $4 \times 10^{-4}\,\mathrm{A\,m^{-2}}$ giving $|\mathbf{j} \times \mathbf{B}| \sim 4 \times 10^{-7}\,\mathrm{N\,m^{-3}}$. The comparable estimate for $|2\rho\boldsymbol{\Omega} \times \mathbf{u}|$ is $8 \times 10^{-4}\,\mathrm{N\,m^{-3}}$, considerably larger. However, it would be wrong to conclude that the magnetic field has no dynamical influence in the core. First, the field in the core may be considerably larger than the observed radial field at the CMB. There may be 'invisible' toroidal components which contribute to the core field but never emerge at the CMB. Even if 1 mT is a fair estimate of the field, we must remember that the field will be important in the dynamics even if it balances only the ageostrophic part of the Coriolis force (that part not balanced by pressure) which may be considerably smaller than our $8 \times 10^{-4}\,\mathrm{N\,m^{-3}}$ estimate. Finally, the scale of variation of the magnetic field might be considerably smaller than d. Indeed, a crude balance between the generation term and the diffusion term in the induction equation suggests that the length scale for evaluating the current should be $dR_\mathrm{m}^{-1/2}$, roughly a factor 20 less than d. Nevertheless, as we shall see below, although the magnetic field is important in the dynamics of core convection, it is mainly effective only over length scales significantly less than that of the whole core. Since convection can be controlled by what happens in thin boundary layers, this can be sufficient to strongly influence the nature of convection. In particular, it is believed that the small-scale dissipation which balances the convective energy input comes primarily from magnetic (ohmic) dissipation.

5.2 Equations Governing Convection

5.2.1 Compressible and Anelastic Equations

Much of our information about core convection has come from numerical simulations based on the governing equations. The full equations are quite complicated, and a wide range of simplifications have been used to study particular aspects of convection. In this section we consider some of the most commonly used approximations, and the assumptions on which they are based. Since magnetic fields can influence convection, we include them in our discussion. In consequence, these equations can be used for dynamo simulations as well as for convection calculations. We shall assume that we are dealing with a two-component fluid with a light component, sometimes called a binary fluid. The mass fraction of a small volume of fluid that is light is denoted by ξ. This is already a simplification of the situation in the core, as the material properties of the core are better described by a mixture of light components (Alfè *et al.*, 2003). The heavy component is usually identified with iron (mass fraction $1 - \xi$) and the light component with sulfur, though we are really lumping together all the different light elements in the core into one component. The density of the inner core near the ICB is greater than that of the outer core near the ICB, and this difference is greater than that due to the density change on solidification. So the value of ξ is lower in the inner core, and excess ξ is released at the ICB during the freezing process. This gives rise to buoyant compositional convection. The variable ξ is usually called the composition. We also allow for variations of temperature to drive thermal convection.

The momentum conservation equation can be written as

$$\rho \frac{\mathrm{D}\mathbf{u}}{\mathrm{D}t} + 2\rho\boldsymbol{\Omega} \times \mathbf{u} = -\nabla p + \rho\mathbf{g} + \mathbf{j} \times \mathbf{B} + \rho\mathbf{F}^{v}$$
$$\frac{\mathrm{D}}{\mathrm{D}t} = \frac{\partial}{\partial t} + \mathbf{u}\cdot\nabla \qquad [1]$$

Here the density ρ is the averaged density of both fluid components, $\mathbf{u}$ is the fluid velocity, and t the time. $\boldsymbol{\Omega}$ is the Earth's rotation vector, so the second term on the left is the Coriolis force. Strictly speaking, it is an acceleration times a density, but it is convenient to consider it as a force in the rotating frame.

A list of numerical values for the parameters used in core convection is given in **Table 1**, taken mostly from Anufriev *et al.* (2005). Note that there is uncertainty about many of these numbers. For example, while the value given for the kinematic viscosity is that usually adopted, Palmer and Smylie (2005) argue for a value five orders of magnitude larger, giving an Ekman number also five orders of magnitude greater than that quoted in **Table 1**. Some quantities such as the rotation rate are known to very high accuracy, but others are little more than informed guesses. The centrifugal force has not been included explicitly as it is added into the gravitational acceleration term. In core convection it is usually omitted altogether, so gravity is radial. Note that in some experiments, centrifugal force dominates over gravity, so in these cases $\mathbf{g}$ is perpendicular to the rotation axis. The oblateness of the core due to rotation may be important in studies of precession and tidal forcing, but is usually omitted for simplicity in convection studies.

The pressure is denoted by p, and the magnetic term $\mathbf{j} \times \mathbf{B}$ is usually called the Lorentz force, sometimes the Laplace force. $\mathbf{j}$ is the current density (units A m^{-2}) and $\mathbf{B}$ the magnetic field, measured in tesla (T). Older papers often used gauss as the unit of magnetic field (1 gauss $= 10^{-4}$ tesla). Most estimates of the core field strength lie in the range 1–10 mT. We include in [1] the viscous force $\rho\mathbf{F}^{\nu}$. All estimates of the viscosity in the core suggest that viscous forces are very weak when acting over length scales the size of the core, so it is strange that we include them in our discussion. The reason is that it has not yet proved possible to devise stable numerical schemes with no viscosity. Schemes which do not include viscosity explicitly invariably contain numerical diffusion which provides an effective viscosity. Viscous

Table 1 Definitions and numerical values of core quantities

r_{cmb}	Outer core radius	3.48×10^6 m
r_{icb}	Inner core radius	1.22×10^6 m
d	Gap width $r_{\mathrm{cmb}} - r_{\mathrm{icb}}$	2.26×10^6 m
ρ_0	Mean core density	11340 kg m^{-3}
g	Gravitational acceleration	7.8 m s^{-2}
Ω	Earth rotation rate	7.29×10^{-5} s^{-1}
α	Thermal expansion coefficient	1.5×10^{-5} K^{-1}
c_p	Specific heat at constant pressure	860 J kg^{-1} K^{-1}
L_H	Latent heat energy per unit mass	$\sim 10^6$ J kg^{-1}
$\Delta\xi$	Jump in ξ across the ICB	$\sim$0.05
γ	Grüneisen parameter	1.5
η	Magnetic diffusivity	$\sim$2 m^2 s^{-1}
κ	Thermal diffusivity	$\sim 5 \times 10^{-6}$ m^2 s^{-1}
κ_{T}	Turbulent thermal diffusivity	$\sim\eta = 2$ m^2 s^{-1}
ν	Kinematic viscosity	$\sim 5 \times 10^{-7}$ m^2 s^{-1}
ν_{T}	Turbulent kinematic viscosity	$\sim\eta = 2$ m^2 s^{-1}
U_*	Typical core velocity	$\sim 5 \times 10^{-4}$ m s^{-1}
T'_*	Typical superadiabatic temperature perturbation	$\sim 10^{-4}$ K
Ro	Rossby number $U_*/\Omega d$	$\sim 3 \times 10^{-6}$
R_{m}	Magnetic Reynolds number $U_* d/\eta$	$\sim$500
E	Ekman number $\nu/\Omega d^2$	$\sim 10^{-15}$ (laminar) $\sim 5 \times 10^{-9}$ (turbulent)

Where the quantity varies over the core, a typical value is given.

forces act to smooth out very sharp variations in the flow, particularly in boundary layers. There have been attempts to solve the equations with no viscosity (e.g., Walker *et al.* 1998), but none as yet has been particularly successful. The most fruitful approach has been to include viscosity, but to reduce its magnitude as much as possible using high-resolution numerical schemes. In a compressible flow, the viscosity is often written as

$$F_i^v = \partial_j \left[\bar{\rho} \nu_{\mathrm{T}} \left(\partial_j u_i + \partial_i u_j - \frac{2}{3} \delta_{ij} \nabla \cdot \mathbf{u} \right) \right] \qquad [2]$$

where ν_{T} is the turbulent kinematic viscosity (see the article by Loper).

The mass conservation equation is

$$\frac{\partial \rho}{\partial t} = -\nabla \cdot (\rho \mathbf{u}) \qquad [3]$$

The time-dependent perturbations in density are small, and the scale of time variations in the core is long. In consequence, the term $\partial \rho / \partial t$ in this equation is extremely small compared with the other terms. It is helpful to introduce the convective turnover time, $t_{\mathrm{conv}} = d / U_*$, where $d = r_{\mathrm{cmb}} - r_{\mathrm{icb}}$ and U_* is a typical speed of convection. The turnover time is then the time an element of convecting fluid takes to travel a distance *d*. Following the arguments given in the introduction, the typical density perturbation driving core convection is only $10^{-9}\rho$, so the magnitude of the term $\partial \rho / \partial t$ is $10^{-9} \rho / t_{\mathrm{conv}}$, whereas the magnitude of the other terms is ρ / t_{conv}, estimating the divergence as $1/d$. The term $\partial \rho / \partial t$ is therefore a factor 10^{-9} smaller than the other terms in [3]. This argument is of course very crude, but the factor 10^9 is so huge that the errors incurred by our rough approximations are negligible in comparison. It can also be shown that neglecting $\partial \rho / \partial t$ is equivalent to assuming the sound travel time is small compared to the convective turnover time. We therefore immediately replace [3] by

$$\nabla \cdot (\rho \mathbf{u}) = 0 \qquad [4]$$

The approximation of neglecting $\partial \rho / \partial t$ in the continuity equation is known as the anelastic approximation.

The density across the core varies by about 20% only, and many authors ignore this variation. This is part of the Boussinesq approximation, which will be systematically derived from the full equations below. Crudely speaking, the Boussinesq approximation neglects variations of density everywhere except in the buoyancy term. Density perturbations are retained in the buoyancy term because although they are a very small fraction of the total density, they are multiplied by *g*, which is a very large acceleration compared to U_* / t_{conv}.

The electromagnetic equations governing **j** and **B** are (e.g., Moffatt, 1978; Davidson, 2001)

$$\frac{\partial \mathbf{B}}{\partial t} = \nabla \times (\mathbf{u} \times \mathbf{B}) + \nabla \times (\eta \nabla \times \mathbf{B}) \qquad [5]$$

$$\mu \mathbf{j} = \nabla \times \mathbf{B} \qquad [6a]$$

$$\nabla \cdot \mathbf{B} = 0 \qquad [6b]$$

Here μ is the magnetic permeability. Because the core is hot, well above the Curie point, the magnetic permeability is essentially that of free space, $\mu = 4\pi \times 10^{-7}\,\mathrm{H\,m^{-1}}$. The magnetic diffusivity η is defined by $\eta = 1/(\mu\sigma)$, where σ is the electrical conductivity measured in Siemens per meter, or equivalently in some papers $\mathrm{ohm^{-1}\,m^{-1}}$, $(\Omega^{-1}\,\mathrm{m^{-1}})$. The units of η are $\mathrm{m^2\,s^{-1}}$. Equation [5] is known as the induction equation. It is derived from Maxwell's equations, together with Ohm's law in a moving medium

$$\mathbf{j}/\sigma = \mathbf{E} + \mathbf{u} \times \mathbf{B} \qquad [7]$$

The composition equation, or mass transport equation, is

$$\rho \frac{\mathrm{D}\xi}{\mathrm{D}t} = \nabla \cdot (k^{\xi} \nabla \xi) \qquad [8]$$

The coefficient k^{ξ} is the mass diffusion coefficient. It is very small in the core, and mass diffusion will be negligible in the core unless it is strongly enhanced by turbulence (see Chapter 6). In simulations, a much larger eddy diffusion value of k^{ξ} is usually assumed. The heat transport equation is

$$\rho T \frac{\mathrm{D}S}{\mathrm{D}t} = \nabla \cdot (k \nabla T) + H\rho \qquad [9]$$

Here *T* is the temperature and *S* is the entropy. *k* is the thermal conductivity, which again in simulations is often enhanced to crudely model turbulent diffusion. *H* is the rate of release of heat per unit mass. One source for this could be radioactivity in the core. As we see below, this term must also include the heating from viscous and ohmic dissipation.

To complete the equations, we need an equation of state giving the density in terms of pressure, temperature, and composition, and an equation for the entropy, also in terms of pressure, temperature, and composition,

$$\rho = \rho(p, T, \xi) \qquad [10a]$$

$$S = S(p, T, \xi) \qquad [10b]$$

Equations [1]–[10] could be used, in principle, to solve numerically for convection inside the core. In practice they are almost always simplified much further before numerical solution is attempted. The perturbations of density that drive the convection are a tiny fraction of the full density, so errors of even one part in a million for the density would overwhelm the true density perturbations. Any practical scheme must separate out the perturbations in the thermodynamic variables from the reference state variables, a procedure sometimes called thermodynamic linearization.

5.2.2 Adiabatic Reference State

We must first specify the reference state about which we linearize the thermodynamic quantities. We write

$$p = p_a + p_c \qquad [11a]$$

$$\rho = \rho_a + \rho_c \qquad [11b]$$

$$T = T_a + T_c \qquad [11c]$$

$$\xi = \xi_a + \xi_c \qquad [11d]$$

$$S = S_a + S_c \qquad [11e]$$

where the subscript a refers to the reference state values and c to the small perturbations associated with the convection. A number of different choices of reference state are possible. When Boussinesq convection experiments are performed, the natural reference state is the conduction state. Thus for convection in a horizontal layer of liquid, where the only source of heat is from the boundary planes, the reference state has $T_a = T_0 - cz$ which is a solution of $\nabla^2 T_a = 0$. The equation of state is taken as $\rho_a = \rho_0(1 - \alpha T)$, α being the expansion coefficient, and the reference state pressure is the solution of the hydrostatic equation $\mathrm{d}p_a/\mathrm{d}z = -g\rho_a$, which is $p_a = p_0 - p_0 g(1-\alpha T_0)z - 0.5\rho_0 g\alpha cz^2$. In a spherical Boussinesq system with no heat sources, the equivalent reference state temperature is a solution of $\nabla^2 T_a = 0$ in spherical polar coordinates, $T_a = T_0 + c/r$, which corresponds to a uniform heat flux of $4\pi kc$ passing from the ICB to the CMB.

The core is believed to be in a well-mixed state and close to adiabatic stratification. In consequence, a reference state in which the composition is uniform, ξ_a independent of position, and uniform entropy, S_a independent of position, is adopted (e.g., Braginsky and Roberts, 1995). The argument in favor of the well mixed, near adiabatic state, is that if it were violated, the very large forces would lead to a rapid adjustment back to such a state. Also, these large forces would lead to rapid changes in the secular variation which are not seen (geomagnetic jerks (*see* Chapter 4) are sudden changes in the rate of change of secular variation, not sudden jumps in the field itself). We note that constant entropy in the core does not imply constant temperature, and indeed defines the reference state temperature gradient. This will therefore not in general satisfy Fourier's heat conduction law. This leads to effective additional heat sources in the core as detailed below.

We now need some thermodynamic relations, to express the change in entropy in terms of other variables,

$$\mathrm{d}S = \left(\frac{\partial S}{\partial T}\right)_{p,\xi} \mathrm{d}T + \left(\frac{\partial S}{\partial p}\right)_{T,\xi} \mathrm{d}p + \left(\frac{\partial S}{\partial \xi}\right)_{p,T} \mathrm{d}\xi$$
$$= \frac{c_p}{T}\mathrm{d}T - \frac{\alpha}{\rho}\mathrm{d}p + \frac{h^\xi}{T}\mathrm{d}\xi \qquad [12]$$

see, e.g., Braginsky and Roberts (1995, eq. (D6)). The first of these relations expresses the chain rule, using the fact that the entropy is a function of the three variables p, T, and ξ. c_p is the specific heat at constant pressure $c_p = T(\partial S/\partial T)_{p,\xi}$. The relation $(\partial S/\partial p)_{T,\xi} = (1/\rho^2)(\partial\rho/\partial T)_{p,\xi}$ is one of Maxwell's thermodynamic relations, and $\alpha = -\rho^{-1}(\partial\rho/\partial T)_{p,\xi}$ is the coefficient of thermal expansion. h^ξ is known as the heat of reaction, the heat released (or absorbed) when the composition changes.

Since the entropy and composition are assumed constant in the reference state, it follows that

$$\frac{\mathrm{d}p_a}{\mathrm{d}r} = -g\rho_a \qquad [13a]$$

$$\frac{d}{T_a}\frac{\mathrm{d}T_a}{\mathrm{d}r} = -D \qquad [13b]$$

$$\frac{d}{\rho_a}\frac{\mathrm{d}\rho_a}{\mathrm{d}r} = -\frac{D}{\gamma} \qquad [13c]$$

$$D = \frac{g\alpha d}{c_p} \qquad [13d]$$

where the dimensionless parameter $D \approx 0.3$ in the core is called the dissipation parameter (see, e.g., Schubert *et al.* (2001)), and γ is a dimensionless parameter called the Grüneisen parameter. As we see below, D also measures the importance of ohmic and viscous dissipation in a planetary core. The equation for $\mathrm{d}p_a/\mathrm{d}r$ is simply the hydrostatic equation, and since $\mathrm{d}S$ and $\mathrm{d}\xi$

are zero for the chosen reference state, the equation for dT_a/dr follows from eqn [12]. The equation for dp_a/dr then follows from the definition of the Grüneisen parameter, $\gamma = (\rho/T)(\partial T/\partial \rho)_{S,\xi}$. The numerical value of γ is discussed in Alfè *et al.* (2002).

The reference state equations can be solved either using the values from a reference Earth model such as PREM (Dziewonski and Anderson, 1981) constructed from seismic data, or using a simple model such as that of Nimmo, described in Chapter 2. The temperature has to be specified at either the ICB or CMB. In principle, the melting point of iron determines the ICB temperature, but in practice there is uncertainty about how much the melting point is depressed by the chemical impurity; Nimmo takes $T_{\rm icb} = 5\,520$ K. The models suggest a temperature drop across the outer core of approximately 1300 K.

5.2.3 Equations for the Convective Perturbations

We can now formulate the anelastic equations for convection in the Earth's core. The treatment here follows that of Anufriev *et al.* (2005), though here we ignore any perturbations in gravity induced by the perturbations ρ_c, though these are formally the same order as the other terms. The convective density perturbation ρ_c can be written in three parts in terms of the entropy, pressure, and composition perturbations:

$$\rho_c = S_c\left(\frac{\partial\rho}{\partial S}\right)_{p,\xi} + p_c\left(\frac{\partial\rho}{\partial p}\right)_{S,\xi} + \xi_c\left(\frac{\partial\rho}{\partial \xi}\right)_{p,S} = -\frac{\rho_a\alpha T_a}{c_p}S_c - \frac{p_c}{g\rho_a}\frac{d\rho_a}{dr} - \rho_a\alpha^\xi\xi_c \qquad [14]$$

Here $\alpha^\xi = -\rho^{-1}(\partial\rho/\partial\xi)_{p,S}$ is the adiabatic compositional expansion coefficient; because the light element has very low density compared to iron, it is nearly unity. We can now write the momentum equation in a form close to that of Braginsky and Roberts (1995):

$$\frac{D\mathbf{u}}{Dt} + 2\boldsymbol{\Omega}\times\mathbf{u} = -\nabla\left(\frac{p_c}{\rho_a}\right) + 1_r g\left(\frac{\alpha T_a}{c_p}S_c + \alpha^\xi\xi_c\right) + \frac{\mathbf{j}\times\mathbf{B}}{\rho_a} + \mathbf{F}^v \qquad [15]$$

This form of the equation of motion makes it convenient to eliminate the pressure perturbation by taking the curl. We then need the equations for the composition and entropy perturbation. The composition equation is

$$\frac{\partial\xi_c}{\partial t} + \mathbf{u}\cdot\nabla\xi_c = \frac{1}{\rho_a}\nabla\cdot\rho_a\kappa^\xi\nabla\xi_c - \dot{\xi}_a \qquad [16]$$

where $\kappa^\xi = k^\xi/\rho_a$ is mass diffusivity. The term $\dot{\xi}_a$ arises because light material is continually being introduced at the ICB, and there is no way of getting rid of it. We assume that in time the material becomes well mixed, so that from the point of view of the convective composition component, the effect is equivalent to having a uniform sink of composition which averaged over the core exactly balances the input of ξ from the ICB.

The entropy equation is

$$\rho_a T_a\frac{DS_c}{Dt} = \nabla\cdot\rho_a T_a\kappa_T\nabla S_c + \nabla\cdot k\nabla T_a - \rho_a T_a\dot{S}_a + \rho_a H \qquad [17]$$

Here κ_T is the turbulent diffusivity of entropy. The reasoning behind this expression is that the turbulence makes elements of fluid move isotropically carrying their entropy, which then dissolve into the ambient fluid releasing that entropy. Braginsky and Meytlis (1990) pointed out that such elements will probably not move isotropically, so this form of diffusion might only be a crude model of the actual turbulent diffusion process. The term $\nabla\cdot k\nabla T_a$ is the heat flux deficit (Anufriev *et al.*, 2005) which comes from the heat being conducted along the adiabat. This is a large term, and since this conducted heat flux increases rapidly from the ICB to the CMB, the heat flux transported by the convection has to fall rapidly, possibly even going negative as the CMB is approached. This term is therefore a strong heat sink. The core is likely to be cooling (unless radioactivity is very significant), and so $\rho_a T_a\dot{S}_a$ is negative, so core cooling is equivalent to a uniform heat source. The final term

$$\rho_a H = Q_v + Q_j + \rho_a H^{\rm R} \qquad [18]$$

gives the heat sources, Q_v being the viscous dissipation, Q_j being the ohmic dissipation and $H^{\rm R}$ the heat released by radioactivity.

Anufriev *et al.* (2005) point out that for a liquid, where $\alpha T << 1$, the entropy is dominated by the temperature term, that is, if in [12] we replace dS by S_c, dT by T_c, etc., then the terms involving p_c and ξ_c are small compared to the term involving T_c. So we have the anelastic liquid approximation (ALA):

$$S_c = \frac{c_p}{T_a}T_c \qquad [19]$$

In the core, $\alpha T \sim 6 \times 10^{-2}$, so the errors arising from the ALA will be small. The advantage of the ALA is that the boundary conditions are more naturally formulated in terms of the temperature, not the entropy. Note that for a perfect gas, $\alpha T = 1$, and the ALA is not appropriate.

5.2.4 Boundary Conditions

The mechanical boundary conditions in the core are that there is no flow through the ICB and CMB, and there is no slip at these boundaries. The growth rate of the inner core is so slow that it can be ignored on the convective timescale. If the inner core is rotating relative to the mantle (whose rotation defines $\mathbf{\Omega}$), then separating the core flow into radial and tangential components, $\mathbf{u} = u_r \mathbf{1}_r + \mathbf{u}_t$,

$$u_r = 0, \quad \mathbf{u}_t(\mathbf{r}) = \mathbf{\Omega}_{\mathrm{IC}} \times \mathbf{r}, \text{ at } r = r_{\mathrm{icb}}, \quad \text{and } \mathbf{u} = 0, \text{ at } r = r_{\mathrm{cmb}} \qquad [20]$$

where $\mathbf{\Omega}_{\mathrm{IC}}$ is the rotation vector of the inner core relative to the mantle. Kuang and Bloxham (1997) suggested that since the Ekman layer is so thin (see Section 5.4.4) it should be ignored, and a stress-free condition applied instead (see also Section 5.5.6).

The electromagnetic boundary conditions at the ICB are that **B** is continuous across the ICB and the continuity of tangential **E** implies that $\eta \mathbf{j}_t$ is also continuous. Denoting the jump in a quantity across the boundary by square brackets [ ],

$$[\mathbf{B}] = 0, \quad [j_r] = 0, \quad [\eta \mathbf{j}_t] = 0, \quad \text{at} \quad r = r_{\mathrm{icb}} \qquad [21a]$$

where r denotes the radial component and t the tangential component. Note that if the electrical conductivity of the inner and outer cores is different, tangential current is not continuous. Assuming the electrical conductivity of the mantle is much less than that of the core, there is negligible current in the mantle, and the core field has to match to a potential field at the CMB. This implies

$$[\mathbf{B}] = 0, \qquad j_r = 0 \qquad [21b]$$

This assumption does, of course, rule out any electromagnetic coupling between core and mantle, which might be important for length of day models.

We also need a condition on the temperature and the composition at both the ICB and CMB. At the ICB, there are two contributions to the input heat flux per unit area into the outer core, the heat conducted out of the inner core, and the latent heat, released as a consequence of inner core freezing. If $r_{\mathrm{icb}}(\theta, \phi, t)$ is the location of the ICB, the rate of release of latent heat energy there is $L_{\mathrm{H}} \rho \dot{r}_{\mathrm{icb}}$ per unit area (*see* Chapter 2), and the rate of release of light material is $\rho \Delta \xi \dot{r}_{\mathrm{icb}}$ also per unit area. Here L_{H} is the latent heat energy per unit mass and $\Delta \xi$ is the jump in ξ across the ICB; numerical values are given in **Table 1**.

The location of the ICB, $r = r_{\mathrm{icb}}$, is determined by the melting point of iron which is a function of p and ξ. Braginsky and Roberts (1995) and Roberts *et al.* (2003) show that its rate of change at the ICB is governed by

$$\dot{S}_a = -c_p \Delta_2 \dot{r}_{\mathrm{icb}} / r_{\mathrm{icb}} \qquad [22]$$

$\Delta_2 \sim 0.04$ being a parameter dependent on the variation of the melting point with pressure (see Roberts *et al.* (2003, equation [3.16]) and Chapter 2). Assuming that the heat conducted out of the inner core balances the heat conducted along the adiabat near the base of the outer core, and using the ALA [19], the rate of release of latent heat at the ICB balances the heat flux conducted in the thermal boundary layer:

$$\frac{L_{\mathrm{H}} \rho_a r_{\mathrm{icb}}}{\Delta_2 c_p T_a} \left(T_a \dot{S}_a + c_p \frac{\partial T_c}{\partial t} \right) = \rho_a c_p \kappa_{\mathrm{T}} \frac{\partial T_c}{\partial r} \quad \text{on} \quad r = r_{\mathrm{icb}} \qquad [23]$$

which is the thermal boundary condition on the ICB at $r = r_{\mathrm{icb}}$. In general, there will be a jump in the temperature gradient at the ICB, and the thermal conductivity may be different across the ICB also. Nevertheless, Buffett *et al.* (1992) and Labrosse *et al.* (2001) point out that only small errors arise in assuming the inner core is adiabatically stratified, so the assumption that the conducted heat flux is the same on both sides of the ICB is reasonable. Note that $\dot{S} = \dot{S}_a + c_p \dot{T}_c / T_a$, and both terms are of comparable magnitude, as the very large S_a changes only very slowly on the core evolution timescale, while the much smaller $c_p T_c / T_a$ changes more rapidly on the convective turnover timescale. Similarly, the rate of release of light material at the ICB gives

$$\frac{\Delta \xi \rho_a r_{\mathrm{icb}}}{\Delta_2 c_p T_a} \left(T_a \dot{S}_a + c_p \frac{\partial T_c}{\partial t} \right) = \rho_a \kappa_{\mathrm{T}} \frac{\partial \xi_c}{\partial r} \quad \text{on} \quad r = r_{\mathrm{icb}} \qquad [24]$$

which is the boundary condition on ξ at the ICB. Although both [23] and [24] contain time derivatives of T_c, Glatzmaier and Roberts (1996) found them to be numerically stable, though most work on core convection uses simplified boundary conditions, such as $T_c = 0$ or $\partial T_c / \partial r$ constant at the ICB.

On the CMB, the heat flux is determined by mantle convection. It may be possible to get information about the distribution of the heat flux through the CMB, I_{cmb}, from seismic measurements (see Section 5.6 and, e.g., Olson (2003)), but in principle it should be determined by mantle convection calculations. Assuming that this total heat flux is greater than the heat conducted down the adiabat at the CMB, if we adopt as a simple model a uniform distribution of CMB heat flux:

$$\rho_a c_p \kappa_T \frac{\partial T_c}{\partial r} = -I_{cmb} + I_{ad}(r_{cmb}) \quad \text{on} \quad r = r_{cmb} \qquad [25]$$

where I_{cmb} is the heat flux per unit area passing through the CMB, and $I_{ad}(r_{cmb})$ is the heat flux conducted along the adiabat near the CMB. There has been a substantial amount of work on models with a nonuniform heat flux across the CMB, recently reviewed by Olson (2003). Since there can be no flux of light material across the CMB,

$$\frac{\partial \xi_c}{\partial r} = 0 \text{ on } r = r_{cmb} \qquad [26]$$

This completes the anelastic boundary conditions. Possible choices for the magnetic boundary conditions are either to assume an insulating inner core and mantle, or to give a more accurate representation of the field inside the core by solving the magnetic diffusion problem in the solid inner core.

5.2.5 The Boussinesq Limit

The Boussinesq limit is found by letting the dissipation number $D = g\alpha d/c_p$ go to zero. It can therefore be thought of as a 'thin layer' approximation, since it is equivalent to $d << c_p/g\alpha$. As noted, it is only marginally satisfied in the core, but nevertheless most studies of core convection are based on the Boussinesq equations. There are several reasons for this. First, Boussinesq convection is a well-studied problem, and so a wealth of intuition about it has built up, so that Boussinesq studies are built on a firm foundation. Second, the Boussinesq equations are significantly simpler to solve numerically, and have fewer free parameters than the full anelastic equations. Since the convection equations are numerically demanding, particularly in three-dimensional (3-D) calculations, this simplicity is a major advantage. A third reason is that much of our understanding has come from the study of laboratory experiments, which are all in the $\nabla \cdot \mathbf{u} = 0$ regime. So simulations designed to make contact with this work are naturally Boussinesq.

We start with the anelastic mass conservation (eqn [4]). Combining with [13c] we get

$$\nabla \cdot (\rho_a \mathbf{u}) = \rho_a \nabla \cdot \mathbf{u} - \frac{u_r D \rho_a}{\gamma d} = 0 \qquad [27]$$

so in the limit $D \to 0$.

$$\nabla \cdot \mathbf{u} = 0 \qquad [28]$$

The entropy equation can be reduced to a temperature equation by substituting in the anelastic liquid approximation [19]. A large number of terms arise that involve ∇T_a and ∇c_p, but in the limit $D \to 0$ they all vanish because of [13b] and [13c]. Then [17] becomes

$$\rho_a c_p \frac{DT_c}{Dt} = \rho_a c_p \kappa_T \nabla^2 T_c + \nabla \cdot k \nabla T_a - \rho_a T_a \dot{S}_a + \rho_a H \qquad [29]$$

In the Boussinesq limit, the expression for H, [18], simplifies because the viscous and ohmic dissipation are proportional to D, and so drop out; see Anufriev *et al.* (2005) for details. The basic reason for this is that in the Boussinesq limit the thermal energy is much larger than the kinetic energy or the magnetic energy. The heat transport equation then becomes a temperature equation, because the dominant form of energy is the temperature-dependent internal heat energy. This is rather a drawback for the Boussinesq approximation, because the ohmic dissipation might be a significant player in the overall heat budget. In consequence, the last three terms on the right-hand side of [29] can be bundled into a single space-dependent heat source term:

$$\hat{H} = \frac{\nabla \cdot k \nabla T_a}{\rho_a c_p} - \frac{T_a \dot{S}_a}{c_p} + \frac{H}{c_p} \qquad [30]$$

to get

$$\frac{DT_c}{Dt} = \kappa_T \nabla^2 T_c + \hat{H} \qquad [31]$$

which is the Boussinesq form of the heat transport equation. Equation [30] shows that the effective heat source for Boussinesq models is considerably more complicated than is sometimes appreciated. The first term on the right-hand side is the heat flux deficit term arising from the fact that more and more heat is conducted down the adiabat as we move outwards from the ICB to the CMB. From the point of view of the convection this is a large, almost uniform, heat sink. The second term on the right-hand side is the core cooling term, which is a source term, but probably not as large as the sink term. The third term is radioactive heating, and is a source that is currently completely unknown. It could be zero or it could dominate the other two sources. Note that if it is small, $\hat{H}$ will be a net heat sink, and convection is then driven by the heat input at the ICB.

The composition equation [16] becomes

$$\frac{\partial \xi_c}{\partial t} + \mathbf{u} \cdot \nabla \xi_c = \kappa^\xi \nabla^2 \xi_c - \dot{\xi}_a \qquad [32]$$

The Boussinesq momentum equation is not very different from [15], the entropy being replaced by the temperature according to [19], and the viscous term simplifies to give

$$\frac{\mathrm{D}\mathbf{u}}{\mathrm{D}t} + 2\boldsymbol{\Omega} \times \mathbf{u} = -\nabla\left(\frac{p_c}{\rho_a}\right) + \mathbf{1}_r g(\alpha T_c + \alpha^\xi \xi_c) + \frac{\mathbf{j} \times \mathbf{B}}{\rho_a} + \nu_T \nabla^2 \mathbf{u} \qquad [33]$$

Of course, ρ_a is now a constant.

Equations [28] and [31]–[33] constitute the Boussinesq equations for convection in the core. The boundary conditions are not affected by taking the Boussinesq limit, so the thermal and compositional equations [23]–[26] can be applied, though in practice much simpler conditions are usually adopted. Another frequent simplification is to bundle together the composition and temperature variables by multiplying [31] by α, [32] by α^ξ and adding, defining the co-density $C = \alpha T_c + \alpha^\xi \xi_c$. This device only works if both composition and temperature have the same diffusion coefficient and the same boundary conditions, see Section 5.3.2.

Both the equations and the boundary conditions contain inhomogeneous forcing terms. In some codes, these equations are handled directly, but sometimes the inhomogeneous terms are eliminated. This procedure is detailed in Anufriev *et al.* (2005); here we illustrate with the composition equation. We define $\xi_c = \xi_0 + \xi'$, and choose the basic distribution ξ_0 to be a function of r only and to satisfy

$$\kappa^\xi \nabla^2 \xi_0 - \dot{\xi}_a = 0 \qquad [34]$$

with boundary conditions

$$\frac{\mathrm{d}\xi_0}{\mathrm{d}r} = \frac{\Delta \xi r_{\mathrm{icb}}}{\Delta_2 c_p T_a \kappa_{\mathrm{T}}} T_a \dot{S}_a \quad \text{on} \quad r = r_{\mathrm{icb}}$$
$$\frac{\mathrm{d}\xi_0}{\mathrm{d}r} = 0 \quad \text{on} \quad r = r_{\mathrm{cmb}} \qquad [35]$$

Now ξ' satisfies

$$\frac{\partial \xi'}{\partial t} + \mathbf{u} \cdot \nabla \xi' = -\frac{\mathrm{d}\xi_0}{\mathrm{d}r} u_r + \kappa^\xi \nabla^2 \xi' \qquad [36]$$

Similarly we define $T_c = T_0 + T'$ and $T_0' = \partial T_0 / \partial r$. Because they are spherically symmetric and constant in time, these basic state contributions do not affect the momentum equation significantly, because the extra terms can be balanced by a pressure gradient. The homogeneous Boussinesq equations are then [28] with

$$\frac{\mathrm{D}\mathbf{u}}{\mathrm{D}t} + 2\boldsymbol{\Omega} \times \mathbf{u} = -\nabla\left(\frac{p'}{\rho_a}\right) + \mathbf{1}_r g(\alpha T' + \alpha^\xi \xi') + \frac{\mathbf{j} \times \mathbf{B}}{\rho_a} + \nu_{\mathrm{T}} \nabla^2 \mathbf{u} \qquad [37]$$

$$\frac{\mathrm{D}T'}{\mathrm{D}t} = -\frac{\mathrm{d}T_0}{\mathrm{d}r} u_r + \kappa_{\mathrm{T}} \nabla^2 T' \qquad [38]$$

$$\frac{\mathrm{D}\xi'}{\mathrm{D}t} = -\frac{\mathrm{d}\xi_0}{\mathrm{d}r} u_r + \kappa^\xi \nabla^2 \xi' \qquad [39]$$

with the same mechanical and electromagnetic boundary conditions as in the inhomogeneous Boussinesq equations, but with thermal and compositional boundary conditions

$$\left.\begin{aligned} \frac{\partial T'}{\partial r} &= \frac{L_{\mathrm{H}} r_{\mathrm{icb}}}{\Delta_2 c_p \kappa_{\mathrm{T}} T_a} \frac{\partial T'}{\partial t} \\ \frac{\partial \xi'}{\partial r} &= \frac{\Delta \xi r_{\mathrm{icb}}}{\Delta_2 \kappa_T T_a} \frac{\partial T'}{\partial t} \end{aligned}\right\} \text{on} \quad r = r_{\mathrm{icb}} \qquad [40]$$

$$\frac{\partial T'}{\partial r} = 0, \quad \frac{\partial \xi'}{\partial r} = 0 \quad \text{on} \quad r = r_{\mathrm{cmb}} \qquad [41]$$

Since only derivatives of T' and ξ' are specified, they are arbitrary up to a constant. This degeneracy can be removed by specifying that the average of T' and ξ' is zero over the ICB. This completes the specification of the homogeneous Boussinesq equations, but it should not be forgotten that T_0 and ξ_0 must be added on to reconstruct the actual temperature and composition perturbations. The homogeneous Boussinesq equations can be linearized in a natural way, both in the nonmagnetic and magnetic cases (in which case the induction equation is needed), to give an eigenvalue problem for the critical Rayleigh number and wave frequency at the onset of convection.

5.2.6 Formulation of the Equations for Numerical Solution

It is useful to see how the full anelastic equations could be approached (see e.g. Glatzmaier (1984). One method (many others are possible) would be to take

$\mathbf{u}$, $\mathbf{B}$, ξ_c, and S_c as the primary variables which are advanced at every timestep using [15], [5], [16], and [17]. Because of eqns [4] and [6b], there are only two independent scalar fields in $\mathbf{u}$ and $\mathbf{B}$, but this can be taken into account by writing $\mathbf{B}$ in terms of toroidal and poloidal scalar fields $\mathbf{B} = \nabla \times \mathcal{U}\mathbf{r} + \nabla \times \nabla \times \mathcal{V}\mathbf{r}$, with a similar expansion for $\rho\mathbf{u}$. This expansion means that [4] and [6b] are automatically exactly satisfied. Two equations for $\mathcal{U}$ and $\mathcal{V}$ can be conveniently obtained by taking the radial component and the radial component of the curl of [5], and similarly for ρ times [15]. It is then the toroidal and poloidal scalar fields for the magnetic field and momentum density that are time-stepped forward, along with S_c and ξ_c. At each timestep, the pressure p_c can be found by taking the divergence of ρ times [15] and solving the resulting time-independent Poisson equation for p_c. Equation [14] and the equation of state then give the temperature and density perturbations. At this stage every term on the right-hand sides of [15], [5], [16], and [17] is known at the current timestep, so we can advance to the next timestep. An appropriate set of boundary conditions would of course be needed to complete the numerical solution.

5.3 Convection in the Absence of Rotation and Magnetic Field

5.3.1 Thermal Convection

The thermal diffusivity in the core is estimated at $\kappa \sim 5 \times 10^{-6}\mathrm{m}^2\,\mathrm{s}^{-1}$, with the kinematic viscosity a factor 10 smaller at $\nu \sim 5 \times 10^{-7}\mathrm{m}^2\,\mathrm{s}^{-1}$ and the compositional diffusion coefficient another factor of 10^3 smaller. In consequence, the dimensionless Rayleigh number $g\alpha\Delta T d^3/\kappa\nu$ is very large in the core, much greater than the critical value for the onset of convection in the absence of rotation, which is typically $Ra \sim 10^3$. The onset of convection is discussed in considerable detail both for the case of a plane layer and a spherical shell, in chapters 2 and 6 of Chandrasekhar (1961). As we shall see, convection in the core is strongly influenced by rotation and magnetic field, and much useful information about the onset of convection in the presence of these constraints is given in the same book. It is nevertheless of interest to examine what form convection in the core might take if these constraints are ignored. One advantage of this approach is that unconstrained Boussinesq plane layer convection has been extensively studied both theoretically and experimentally up to quite high Rayleigh numbers, so that far more is known about the strongly nonlinear behavior in this case. Much less is known about the corresponding problems when rotation and magnetic field are important.

Experiments typically measure the heat flux in terms of the Nusselt number

$$Nu = \frac{Fd}{\rho c_p \kappa \Delta T} \qquad [42]$$

which is the ratio of the total convective plus conductive heat flux per unit area, F, to the heat flux that would occur in the absence of convection, $\rho c_p \kappa \Delta T/d$. When comparing with the Earth's core, there is a potential for confusion here, because there are two possible definitions for the conducted heat flux. If this is taken as the heat conducted down the adiabat, then the Nusselt number in the Earth's core turns out to be of order unity. We denote the Nusselt number defined in this way, with $\Delta T = \Delta T_a$ the adiabatic temperature drop across the core (approximately 1300 K) as $N = Fd/\rho c_p \kappa \Delta T_a$. However, for comparison with Boussinesq convection models, $\Delta T = \Delta T_s$, the superadiabatic temperature drop across the core. This is much smaller, probably less than 10^{-3} K. We denote the Nusselt number defined in this way as $Nu = Fd/\rho c_p \kappa \Delta T_s$, and its typical value in the core is about 10^6. Note that no direct numerical simulation of rotating spherical convection has ever achieved Nu values this large, and there is no prospect of ever doing so.

At Rayleigh numbers just above critical, weakly nonlinear theory (described in more detail in Section 5.4) can be used to predict the Nusselt number in terms of the Rayleigh number. Weakly nonlinear theory is valid when the pattern of nonlinear convection is still close to that given by the linear theory of the onset of convection. It is no longer valid when convection is turbulent. It predicts that $Nu - 1 \propto Ra - Ra_c$, Ra_c being the critical Rayleigh number for the onset of convection. Initially as Ra is increased above critical this is what is found experimentally, but if the Rayleigh number is further increased it is found that the Nusselt number starts to grow more slowly with Ra than this. Experiments at very large Ra usually find $Nu \sim Ra^n$, and in the range of Ra up to 10^{17}, which is about the largest value that has been tested experimentally (Niemela *et al.*, 2000), n is typically about 0.3. As we see below, theories exist for both $n = 1/3$ and $n = 2/7$, but it is conjectured that at $Ra \gg 10^{17}$, n may rise to 0.5 (see, e.g., Grossmann and Lohse, 2000). Note that this experimental value of $Ra \sim 10^{17}$ is much larger than can be obtained with 3-D numerical experiments, $Ra \sim 10^7$ (note that for rapidly rotating convection, where the critical Rayleigh number is much larger than 10^3, considerably higher Ra can be achieved) though 2-D simulations can reach significantly larger Ra, up to 10^{12} (Rogers *et al.*, 2003). The Rayleigh number in the Earth's core is probably around 10^{23} (Christensen and Aubert, 2006).

Since we cannot measure the temperature difference ΔT_s directly, it is more appropriate to work in terms of the convective heat flux per unit area, F_{conv}. Even this is not well known, but a total of 2TW near the CMB (giving $0.013\,\mathrm{W\,m^{-2}}$ for the flux per unit area) for the 'convective' heat flux (ignoring that carried by conduction down the adiabat) is a reasonable order of magnitude estimate. Note that the actual convective heat flux varies considerably across the core (see Section 5.2) and may even be negative on average at the CMB, but nevertheless we adopt $0.013\,\mathrm{W\,m^{-2}}$ for the purpose of estimation here. Since

$$Ra = \frac{g\alpha\Delta T_s d^3}{\kappa\nu}, \qquad F_{conv} = (Nu - 1)\frac{\rho c_p \kappa \Delta T_s}{d} \qquad [43]$$

eliminating ΔT_s gives

$$Ra(Nu - 1) = \frac{g\alpha F_{conv} d^4}{\rho c_p \kappa^2 \nu} \qquad [44]$$

A simple early estimate for Nu in plane layer Boussinesq convection at large Ra was given by Priestley (1959) and Kraichnan (1962) (see Siggia (1994) for a more complete discussion than that given here). They suppose thermal boundary layers of thickness δ occur at the top and bottom boundaries, across which most of the temperature drop takes place. Since the temperature drop across each boundary is approximately $\Delta T_s/2$, and all the heat is transported by conduction in the boundary layer,

$$Nu \approx \frac{0.5d}{\delta} \qquad [45]$$

They further assume that the Rayleigh number for the boundary layer is close to critical. This gives

$$\frac{g\alpha\Delta T_s \delta^3}{2\kappa\nu} \approx 1000 \qquad [46]$$

assuming that the critical Rayleigh number is around 10^3, which linear theory calculations suggest is reasonable. Eliminating δ between [45] and [46] gives $Nu \propto Ra^{1/3}$. Goldstein *et al.* (1990) found

$$Nu \approx 0.066 Ra^{1/3} \qquad [47]$$

from experimental data, where the coefficient 0.066 is not far from the $16\,000^{-1/3} \approx 0.04$ expected from [45] and [46].

The interior velocity and temperature perturbations are obtained by balancing inertia and buoyancy, the mixing-length balance, together with a simple estimate of the convective heat flux. The mixing length balance can be thought of as an element of fluid being accelerated by buoyancy for a turnover time d/U_*, where U_* is a typical velocity. The acceleration is therefore

$$U_*^2/d \approx g\alpha T'_* \qquad [48]$$

where T'_* is the typical temperature perturbation in the fluid interior. The convective heat flux is carried by blobs of hot fluid moving upwards, with cold fluid descending, so

$$F_{conv} \approx \rho c_p U_* T'_* \qquad [49]$$

This gives the Deardorff velocity and temperature scale (Deardorff 1970):

$$U_* = \left(\frac{g\alpha d F_{conv}}{\rho c_p}\right)^{1/3} \qquad [50a]$$

$$T'_* = \left(\frac{F_{conv}}{\rho c_p}\right)^{2/3}\left(\frac{1}{g\alpha d}\right)^{1/3} \qquad [50b]$$

Note that the Deardorff velocity and temperature are independent of the diffusion coefficients. Experiments indicate that U_* is close to the rms velocity in the interior of the fluid; we therefore do not need a numerical scaling factor in [50a].

Putting in the standard values for the core for g, α, d, and c_p from **Table 1**, together with our estimate of $F_{conv} = 0.013\,\mathrm{W\,m^{-2}}$ gives the Deardorff velocity $U_* = 7 \times 10^{-3}\,\mathrm{m\,s^{-1}}$. The faster moving magnetic features at the CMB move with a speed of about $5 \times 10^{-4}\,\mathrm{m\,s^{-1}}$, about 10 times slower than this, so our velocity estimate has turned out 10 times too big. The overestimate is due to neglecting the inhibiting forces of rotation and magnetic field. While our estimate is clearly too large, it is nevertheless of interest to pursue this nonrotating nonmagnetic plane layer model a little further. Using [44] and [47] with the **Table 1** standard values and our estimate for F_{conv}, we obtain $Ra \sim 10^{23}$ and $Nu \sim 3 \times 10^6$. $\Delta T_s \sim 2 \times 10^{-4}$ which is substantially larger than the Deardorff temperature, as it must be for model consistency as the Deardorff temperature is the internal temperature perturbation, but the main temperature drop is across the boundary layer. Note that $Nu\Delta T_s \sim 600$ K, the same order of magnitude as $\Delta T_a \sim 1300$ K, consistent with N being of order unity. The thermal boundary layer thickness $\delta = 0.4$ m only. We have, of course, no way of telling whether the core boundaries are sufficiently smooth for such a thin boundary layer to be physically plausible.

For completeness, we should note that in a number of experiments, $Nu \sim Ra^{2/7}$ gives a better fit than the $Ra^{1/3}$ law (Castaing *et al.*, 1989; Siggia, 1994). Rather than assuming a boundary layer marginally unstable to convection, they observe thin sheet like plumes of typical temperature ΔT_s and of similar width δ as the boundary layer thickness. These plumes form an intermediate layer which carries the heat into the interior. These plumes have the Deardorff velocity, and viscosity balances buoyancy in them so

$$g\alpha\Delta T_s \sim \frac{\nu U_*}{\delta^2} \qquad [51]$$

This equation defines the boundary layer thickness in the Castaing *et al.* picture, rather than [46], and together with [50a] and [45] leads to

$$Nu \sim Ra^{2/7} Pr^{-1/7} \qquad [52]$$

While this theory is reasonable up to experimental values of $Ra \sim 10^{17}$, it is unlikely to hold for larger Ra as the viscous control of the plumes will then break down.

Kraichnan (1962) pointed out that at very large Ra, even a thin boundary layer like this might be shear unstable, and hence turbulent. If so, this undermines the assumption of a quiescent boundary layer in which heat transfer is by conduction. Indeed, the arguments leading to [51] require a laminar mixing layer, while the basis of [46] is that it is a laminar layer which thickens due to thermal diffusion until it becomes unstable due to local convection. With our estimates, the local Reynolds number of the boundary layer is $U_*\delta/\nu \sim 5000$, probably large enough to expect a turbulent boundary layer. The boundary layer then starts to change its character, becoming passive rather than active. This means that the temperature drop across the boundary layer becomes the same as the Deardorff temperature, so that it is the interior that controls the heat flux, not the boundary layer. F_{conv} then becomes independent of the diffusivities, since the Deardorff velocity and temperature are independent of diffusivities. We get

$$\Delta T_s \sim T'_* \qquad [53]$$

and using [48] and [49] gives

$$Nu \approx \frac{U_* d}{\kappa} \sim (RaPr)^{1/2} \qquad [54]$$

A more precise treatment of the boundary layer (e.g., Siggia, 1994) suggests

$$Nu \approx \frac{U_* d}{\kappa} \sim (RaPr)^{1/2}/(\ln Ra)^{3/2} \qquad [55]$$

Note, however, that experiments even at $Ra \sim 10^{17}$ show no sign of the $Ra^{1/2}$ dependence, the exponent typically being $n \approx 0.3$ (Niemela *et al.*, 2000). As we expect that rotation and magnetic field will inhibit heat transport, and for nonmagnetic nonrotating convection the interior is already beginning to take over control of the heat transport at Earth core Rayleigh numbers, it is likely that the interior will be in control in the rotating magnetic Earth's core.

5.3.2 Compositional Convection

It was suggested by Braginsky (1963) that the gravitational energy released by the growth of the inner core might help to power the geodynamo. The density jump at the ICB suggests that the outer core must contain less dense elements than iron, so that the freezing process allows buoyant material to float upwards, supplementing the buoyancy from thermally driven convection (Fearn, 1998). Most analyses of the energy balance in the core suggest that compositional convection does a similar amount of work as thermal convection, though the exact ratio is model dependent (*see* Chapter 2). Unlike heat, the light material cannot escape from the outer core. There are two possibilities; the usual assumption is that the light material mixes in the outer core, so that the density of the inner core is a slowly decreasing function of time, the term $\dot{\xi}_a$ in [16]. Because this rate of change of density is very slow compared to the flow turnover time, this is equivalent to having a uniform sink of buoyancy distributed throughout the core. A second possibility is that the light material forms a sediment at the top of the core (Buffett *et al.*, 2000), near the CMB. If this is the case, there is a release of heavy material (compositionally closer to pure iron) near the CMB, which would give an additional convective driving, increasing the amount of gravitational power to drive the geodynamo.

Compositional convection can be seen in the laboratory using a saturated solution of ammonium chloride (e.g., Chen *et al.*, 1994). If the container is strongly cooled from below (placing it on dry ice, for example), ammonium chloride comes out of solution releasing relatively fresh water which is less dense than its surroundings, and so rises. When the experiment is performed, the ammonium chloride does not crystallize as a solid block on the bottom of the container, but instead forms a dendritic structure with a 'mushy zone' consisting partly of crystallized ammonium chloride and partly of liquid. The lower-density solution released in the neighborhood of the crystallization process rises rapidly in thin 'chimneys' formed inside the mushy zone. Typically, the chimneys are narrow and spaced quite far apart, so there is a strong asymmetry between the narrow upwelling plumes coming out of the chimneys, and the slowly sinking fluid being sucked back into the mushy zone (Aussillous *et al.*, 2006). There is also evidence that mushy zones form when liquid iron freezes in metallurgical processes.

These observations have led to the suggestion that there may be a mushy zone at the ICB (Fearn *et al.*, 1981) though this is controversial (Morse, 2002). There may even be chimneys of light material emerging from the mushy zone. Loper *et al.* (2003) have explored the dynamics of such plumes, though it is uncertain whether such plumes can remain stable and maintain their identity (Classen *et al.*, 1999), or whether the action of rotation and magnetic field might make the plumes mix into the core fluid.

While the physical origins of thermal and compositional convection are different, it is not so clear that the dynamics of the two phenomena are very different, and indeed modelers often lump compositional and thermal convection together, the co-density C as mentioned in Section 5.2.5. Admittedly, the composition and temperature have different diffusion coefficients, but both are so low that small-scale turbulence is probably responsible for mixing both heat and composition. The main difference is in the boundary conditions at the CMB; the light material cannot escape the core, eqn [26], but heat can, [25]. If fluid cannot easily pass across the tangent cylinder (see also Sections 5.5.6.1 and 5.5.6.2), light material released at the ICB might have difficulty escaping outside the TC, and hence give rise to a higher concentration inside the TC. This has been proposed as a possible explanation for the anticyclonic polar vortices (Olson and Aurnou, 1999; see also Section 5.5.6.2).

Anufriev *et al.* (2005) argued that the natural model of compositional convection consisting of a source at the ICB and a uniformly distributed sink of equal strength distributed throughout the core may be a reasonable model for the thermal convection too, because the heat flux deficit term arising from conduction down the adiabat is so large, eqn [17].

5.4 Effect of Rotation on Convection

In this section we consider how rotation affects convection in the absence of magnetic field. As we shall see later, magnetic fields do affect rotating convection very significantly, but it is necessary first to consider how rotation alone can affect convection before the additional complications of the magnetic field are considered. We will use two coordinate systems, cylindrical polars (s, ϕ, z), z being parallel to the rotation axis, and spherical polar coordinates (r, θ, ϕ), θ being the co-latitude. Cylindrical coordinates turn out to be more natural for understanding the dynamics of rotating fluids, but spherical polar coordinates are preferred for numerical simulations in spherical shells.

5.4.1 The Proudman–Taylor Theorem

Slow steady flows in homogeneous rotating fluids have the remarkable property that the velocity is independent of the rotation axis. This result, known as the Proudman–Taylor theorem, can be demonstrated in the laboratory by towing a small obstacle along the bottom of a rotating tank; the whole column of fluid above the obstacle moves along with it as though the fluid column was a rigid body. This surprising result can best be understood in terms of the vorticity equation. Taking coordinates with $\boldsymbol{\Omega} = \Omega \mathbf{1}_z$, the curl of the Boussinesq momentum equation [37] with the magnetic field and composition set to zero gives

$$\frac{\partial \boldsymbol{\zeta}}{\partial t} + \mathbf{u} \cdot \nabla \boldsymbol{\zeta} - (2\boldsymbol{\Omega} + \boldsymbol{\zeta}) \cdot \nabla \mathbf{u} = \nabla \times g\alpha T' \mathbf{1}_r + \nu \nabla^2 \boldsymbol{\zeta} \quad [56]$$

$\boldsymbol{\zeta}$ being the vorticity. For simplicity we drop the subscript on ν_{T}; in this section the viscosity can be thought of either as the molecular value when discussing low *Ra* convection, or as a turbulent value for very high *Ra* convection. Taking U_* as the typical magnitude of the velocity, slow flows are those for which the Rossby number

$$Ro = \frac{U_*}{\Omega d} << 1 \quad [57]$$

The Rossby number in the Earth's core is about 3×10^{-6}.

To understand the Proudman–Taylor theorem it is helpful to first consider flows for which the length scale of variation in all directions (including the z-direction) is of order d. Then the vorticity has magnitude U_*/d, and small Rossby number implies that $|\mathbf{u}\cdot\nabla\boldsymbol{\zeta} - \boldsymbol{\zeta}\cdot\nabla\mathbf{u}| << |2\Omega\partial\mathbf{u}/\partial z|$. Since $\boldsymbol{\Omega}$ is constant, eqn [56] can be thought of as an equation for $2\boldsymbol{\Omega} + \boldsymbol{\zeta}$. In oceanography and atmospheric sciences, the part $2\boldsymbol{\Omega}$ is called the planetary vorticity, because it is due to the rotation of the Earth rather than the behavior of the fluid. We therefore assume that the planetary vorticity is much greater than the fluid vorticity when we make this small Rossby number approximation. Given the slow flows, this will only break down for vortices with radii less than a few tens of meters in the core. The steady flow approximation requires that $|\partial\boldsymbol{\zeta}/\partial t| << |2\Omega\partial\mathbf{u}/\partial z|$, which will generally be the case provided that the timescale over which the velocity varies is much longer than a day. Since the convective turnover time d/U_* is several hundred years, such short timescales are unlikely to be of much significance in the dynamics of convection. The dominant term on the left-hand-side of [56] is therefore the term $2\Omega\partial\mathbf{u}/\partial z$. Of the terms on the right-hand side, viscosity is small except on very short length scales. The dimensionless ratio of viscous to Coriolis force is measured by the Ekman number $E = \nu/\Omega d^2 \sim 10^{-15}$ in the core. Buoyancy forces (and Lorentz forces) are significant in the core, but do not occur in homogeneous nonmagnetic fluids. The Proudman–Taylor theorem then arises because there is no term on the right-hand side to balance $2\Omega\partial\mathbf{u}/\partial z$. In consequence, the z-derivative of $\mathbf{u}$ must be zero, and fluid has to move in columns. If the time-dependent vorticity term is restored, but the forces on the right-hand side are ignored, we obtain

$$\frac{\partial\boldsymbol{\zeta}}{\partial t} - 2\boldsymbol{\Omega}\cdot\nabla\mathbf{u} = 0 \qquad [58]$$

Taking the curl of this and eliminating $\boldsymbol{\zeta}$ using [58], we derive the inertial wave equation

$$\frac{\partial^2}{\partial t^2}\nabla^2\mathbf{u} = -4\Omega^2\frac{\partial^2}{\partial z^2}\mathbf{u} \qquad [59]$$

This shows that if a z-dependent flow is set up initially, the fluid responds by oscillating on a time scale of days. In the absence of any dissipation mechanism, these oscillations would go on ringing forever. In the core, however, these fast oscillations would excite rapid changes in the magnetic field, which would lead to a rapid damping though ohmic dissipation. Inertial waves may well exist in the core, but they are likely to have a small amplitude only and we will ignore any such oscillations. There are, however, a class of inertial waves which have much smaller frequencies, namely those with a short wavelength ℓ in the plane perpendicular to z. These waves have frequency $2\Omega\ell/d$, and are considered in more detail below.

Another way of thinking of the Proudman–Taylor theorem is that in z-independent motion the pressure force is in balance with the Coriolis acceleration. This is geostrophic motion. While the buoyancy force and the magnetic field can induce some z-dependence, many numerical models and laboratory experiments indicate that the motion is mostly in the form of almost 2-D columnar rolls. This is not to say, however, that the ageostrophic parts of the motion are unimportant.

The spherical geometry means that convection inevitably involves some z-dependent motion. The only purely geostrophic motion in a sphere is azimuthal flow $\mathbf{u} = u_\phi(s)\mathbf{1}_\phi$ which cannot transport any heat radially. If the CMB is bumpy, the geostrophic contours are those on which the length of fluid columns is preserved. In general, therefore, flow along geostrophic contours might have a weak radial dependence and hence permit convective heat transport (Bell and Soward, 1996; Westerburg and Busse, 2003). In a perfect sphere, any column moving in the s-direction changes its length. To avoid excessive changes of length of fluid columns, which inevitably involve z-dependent motion, tall thin columns are preferred. These can transport heat while minimizing the departure from geostrophy. If there is no magnetic field, and the velocity is small enough for the nonlinear terms to be negligible, no balance between pressure, Coriolis force, and buoyancy is possible. Viscosity must be included for convection to occur at all. Viscous forces are very small in the core, so columns that are viscously controlled are very thin indeed. In the presence of magnetic field this problem is alleviated, because an inviscid balance is possible by replacing the viscous force with Lorentz force.

There is one further consequence of the difficulty fluid columns have changing their length in rapidly rotating fluid, which is that there is unlikely to be much fluid motion across the tangent cylinder. Inside the tangent cylinder the inner core cuts the fluid columns in half. Any motion in the s-direction across the tangent cylinder will therefore involve strong

z-dependence, and so will need correspondingly strong thermal or magnetic forcing to occur. Another factor tending to divide the regions inside and outside the tangent cylinder into two distinct parts is that columns inside the tangent cylinder increase their length as they move outward in s, but columns outside the tangent cylinder decrease in length as they move outward in s. As we see below, this means that the Rossby waves propagate eastward (that is prograde, in the same direction as the rotation) outside the tangent cylinder but westward (retrograde) inside the tangent cylinder. This discontinuity in the phase speed of the waves is another reason why convection in the two regions has to be considered separately.

5.4.2 The Onset of Instability

We are therefore led to study two types of linear problem for the onset of convection in a rotating sphere or spherical shell. The first is the nonmagnetic viscous problem, and the second is the equivalent magnetic problem in which a simple imposed magnetic field is specified. The nonmagnetic problem can be solved either numerically at finite (possibly small) E or asymptotically in the limit $E \to 0$. The magnetic problem can be solved numerically, or asymptotically in the weak field limit (see Section 5.5). We start by describing the nonmagnetic problem.

Numerical solutions were obtained by Gilman (1975), Zhang and Busse (1987), Zhang (1992), Jones *et al.* (2000), and Dormy *et al.* (2004), with later workers generally reaching lower E, due to the increase in computer speeds. An accurate method for obtaining solutions is to expand the velocity in toroidal and poloidal scalars

$$\mathbf{u} = \nabla \times \mathcal{T}\mathbf{r} + \nabla \times \nabla \times \mathcal{S}\mathbf{r} \qquad [60]$$

which automatically satisfies the continuity equation, and reduces the velocity field to two independent scalars. The equations are nondimensionalized taking the unit of time as r_{cmb}^2/ν, the viscous time, the unit of length as the sphere radius r_{cmb}, and the unit of temperature as $-Pr\, r_{\text{cmb}} \mathrm{d}T_0/\mathrm{d}r$ where T_0 is the basic state temperature. The discussion below is in terms of temperature, but this could be replaced by composition. In this case, the boundary conditions adopted should be those appropriate for compositional convection (see [40] and [41]) but otherwise nothing is significantly changed. Note that the length scale used for nondimensionalization in this problem is r_{cmb} rather than the gap width d. This affects the definition of Ekman number and Rayleigh number, so care must be exercised when comparing results from these stability problems with numerical dynamo simulations, which often use d as the length scale.

Two cases are commonly studied: a uniformly heated sphere where $\mathrm{d}T_0/\mathrm{d}r$ is proportional to r, appropriate for a strongly radioactive heat source, and differential heating where $\mathrm{d}T_0/\mathrm{d}r$ is proportional to $1/r^2$, suitable for a model where all the heat (or composition flux) is input at the ICB. In the differential heating model the total input heat flux at the ICB equals the heat flux coming out of the CMB, so no account is taken of the variable amount of heat conducted down the adiabat. The gravity field is $-g\mathbf{r}$.

The linearized vorticity equation [56], which omits the advection of vorticity and the stretching of fluid vorticity (but not the much larger stretching of planetary vorticity), can be written as

$$\frac{\partial \boldsymbol{\zeta}}{\partial t} - 2\boldsymbol{\Omega} \cdot \nabla \mathbf{u} = \nabla \times g\alpha T' \mathbf{1}_r + \nu \nabla^2 \boldsymbol{\zeta} \qquad [61]$$

and the scalar equations used are the radial components of [61] and its curl:

$$\mathcal{E}\left(\frac{\partial}{\partial t} - \nabla^2\right)\mathcal{L}^2\mathcal{T} - \frac{\partial \mathcal{T}}{\partial \phi} + \mathcal{CS} = 0 \qquad [62]$$

$$\mathcal{E}\left(\frac{\partial}{\partial t} - \nabla^2\right)\mathcal{L}^2\nabla^2\mathcal{S} - \frac{\partial}{\partial \phi}\nabla^2\mathcal{S} - \mathcal{CT} + \mathcal{E}Ra\mathcal{L}^2 T' = 0 \qquad [63]$$

where (r, θ, ϕ) are spherical polar coordinates, the temperature perturbation is T', and

$$\begin{aligned} \mathcal{L}^2 &= -\frac{1}{\sin\theta}\frac{\partial}{\partial\theta}\left(\sin\theta\frac{\partial}{\partial\theta}\right) - \frac{1}{\sin^2\theta}\frac{\partial^2}{\partial\phi^2} \\ \mathcal{C} &= \mathbf{1}_z \cdot \nabla - \frac{1}{2}(\mathcal{L}^2\mathbf{1}_z \cdot \nabla + \mathbf{1}_z \cdot \nabla\mathcal{L}^2) \end{aligned} \qquad [64]$$

The temperature equation [38] becomes

$$Pr\frac{\partial T'}{\partial t} = \mathcal{Q}\mathcal{L}^2\mathcal{S} + \nabla^2 T' \qquad [65]$$

where $\mathcal{Q} = 1$ for internal heating and $\mathcal{Q} = 1/r^3$ for differential heating. The dimensionless parameters are the Ekman number, the Rayleigh number, and the Prandtl number:

$$\mathcal{E} = \frac{\nu}{2\Omega r_{\text{cmb}}^2}, \qquad Ra = -\frac{g\alpha r_{\text{cmb}}^5}{\kappa\nu}\frac{\mathrm{d}T_0}{\mathrm{d}r}, \qquad Pr = \frac{\nu}{\kappa} \qquad [66]$$

Note that many nonlinear convection and dynamo codes use the same basic formulation of the equations as this, with a further toroidal and poloidal expansion of the magnetic field, and the radial components of induction equation and its curl giving the two scalar equations for the field. Note also the factor 2 in this definition of the Ekman number. Unfortunately, there is no consistency in the literature over the definition of the Ekman number. In dynamo simulations, $E = \nu/\Omega d^2$ is often used. This gives $\mathcal{E} = E(1 - r_{\rm icb}/r_{\rm cmb})^2/2$ so at radius ratio 0.35, $\mathcal{E} = 0.21E$. Similarly, care is needed with Ra; although the same symbol is used in many papers, the definitions are frequently different, so suitable conversion factors must be worked out if numerical results from one paper are used in another.

The quantities $\mathcal{S}$, $\mathcal{T}$, and T' are then expanded in spherical harmonics, taking account of the symmetry about the equator. Busse (1970) showed that solutions with u_r, u_ϕ, and ω_z symmetric about the equator, and u_θ and u_z antisymmetric about the equator are preferred, in that they have the lowest critical Rayleigh number. The expansion for $\mathcal{S}$ is then

$$\mathcal{S} = \sum_{l=0}^{L} \mathcal{S}_l(r) P^m_{2l+m}(\cos\theta)\exp \mathrm{i}(m\phi - \omega t) \qquad [67]$$

the other expansions being found in Jones *et al.* (2000). The radial dependence can be dealt with for example either by expanding $\mathcal{S}_l(r)$ in Chebyshev polynomials and using collocation to derive the equations for the coefficients, or by finite difference methods. In either case, a key point is that there is no coupling between modes with different azimuthal wave number m. The operators $\mathcal{L}^2$ and ∇^2 are particularly convenient in this representation, as $\mathcal{L}^2 P^m_{2l+m} = -(2l+m+1)(2l+m)P^m_{2l+m}$ so they do not couple spherical harmonics, and the operator $\mathcal{C}$ only couples adjacent spherical harmonics, that is, $\mathcal{C}P^m_{2l+m}$ can be written as a simple linear combination of P^m_{2l+m+1} and P^m_{2l+m-1}. In consequence the matrix equations arising from the expansion or discretization of the radial structure can be written in banded form, with each m value treated separately and only the blocks corresponding to $l-1$, l, and $l+1$ having nonzero elements. The eigenvalue ω is the frequency and the value of the Rayleigh number Ra_c which makes ω real, the neutral mode condition, can then be found simply using a variety of different methods, for example inverse iteration.

The Rayleigh number Ra_c is then minimized over m, to determine m_c, ω_c, and Ra_c as a function of $\mathcal{E}$, Pr, and the radius ratio $r_{\rm icb}/r_{\rm cmb}$. Calculations using a variety of boundary conditions, either stress free or rigid, and with fixed temperature and fixed heat flux, have been performed. Values of $\mathcal{E}$ down to 10^{-7} can be reached without excessive computational resources, which is sufficiently small to get excellent agreement with the asymptotic methods described below.

The results show that as $\mathcal{E}$ is reduced, the azimuthal wave number increases with $m_c \sim \mathcal{E}^{-1/3}$, the Rayleigh number increases as $Ra_c \sim \mathcal{E}^{-4/3}$ and frequency increases as $\omega_c \sim \mathcal{E}^{-2/3}$. This means that at small $\mathcal{E}$ the horizontal wavelength gets small, and in the limit of zero viscosity the critical Rayleigh number goes to infinity, showing that no convection can happen without viscosity in this problem. Numerical simulations at four different Ekman numbers are shown in **Figure 3**, which is for the case of no internal heating, so convection onsets first close to the tangent cylinder. The roll structures seen in these numerical calculations are also found in laboratory experiments (Busse and Carrigan, 1976) where the gravity is replaced by centrifugal acceleration arising from the very rapid rotation. In these experiments, the sphere has to be cooled from within. With $\mathcal{E} \sim 10^{-15}$ in the core, these results suggest $m \sim 3 \times 10^4$, giving a typical roll diameter of about 300 m only (there is a cyclonic and an anticyclonic roll in each wavelength). Tall thin columns stretching all the way to the boundaries are seen in experiments, but it seems unlikely structures with such a large aspect ratio exist in the core. As we see in Sections 5.4.8.1 and 5.5.3 below, magnetic fields and nonlinear inertial effects can thicken the

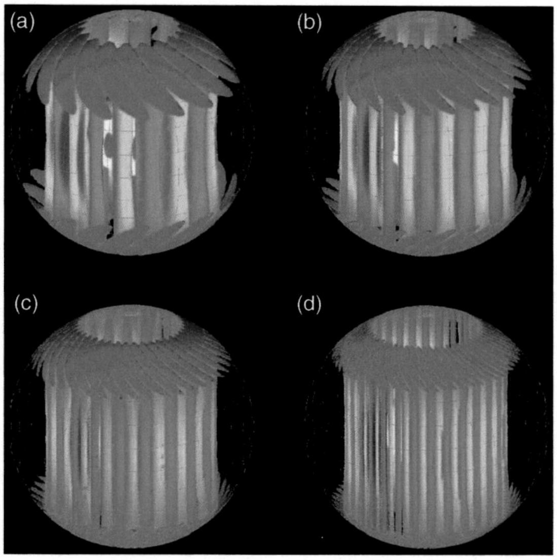

Figure 3 Contours of axial vorticity, colored according to the temperature perturbation, at the onset of convection in a spherical shell, radius ratio 0.35: (a) $\mathcal{E} = 3 \times 10^{-5}$, (b) $\mathcal{E} = 10^{-5}$, (c) $\mathcal{E} = 3 \times 10^{-6}$, (d) $\mathcal{E} = 10^{-6}$. This figure appeared in Dr E. Dormy's PhD. thesis.

columns, and time-dependent motions may break the columns up. Having said that, it is possible that tall thin columns with a large (but less than 10^4) aspect ratio are present in the core. Dynamo simulations, especially those in the most Earth-like regimes, often show columnar convection. Interestingly, convection columns are particularly good at generating dipolar magnetic fields.

5.4.3 The Onset of Instability in the Rapidly Rotating Limit

The asymptotic $\mathcal{E} \to 0$ theory of convection was developed by Roberts (1968) and Busse (1970), but the distinction between the local theory and the global theory of the onset of convection was elucidated more recently (Jones *et al.*, 2000; Dormy *et al.*, 2004). We use cylindrical polar coordinates (s, ϕ, z) and use an axial poloidal and toroidal decomposition:

$$\mathbf{u} = \nabla \times \psi \mathbf{1}_z + \nabla \times \nabla \times \Phi \mathbf{1}_z \qquad [68]$$

Since all the components of velocity are of similar order of magnitude (the boundary conditions ensure u_r and u_z are the same order but $1/s\ \partial/\partial\phi$ and $\partial/\partial s$ are $O(\mathcal{E}^{-1/3})$), Φ is $O(\mathcal{E}^{1/3})$ smaller than ψ and so to leading order

$$\mathbf{u} = \nabla \times \psi \mathbf{1}_z + u_z \mathbf{1}_z \qquad [69]$$

Inserting this equation into the z-components of the linearized [61] and its curl, the temperature equation [38], we obtain

$$\mathcal{E}\left(\frac{\partial}{\partial t} - \nabla^2\right)\nabla_{\mathrm{H}}^2 \psi + \frac{\partial u_z}{\partial z} = \mathcal{E} Ra \frac{\partial T'}{\partial \phi} \qquad [70]$$

$$\mathcal{E}\left(\frac{\partial}{\partial t} - \nabla^2\right)\nabla^2 u_z - \frac{\partial}{\partial_z}(\nabla_{\mathrm{H}}^2 \psi) = \mathcal{E} Ra\left[z\nabla_{\mathrm{H}}^2 T' - \frac{1}{s}\frac{\partial}{\partial s}\left(s^2 \frac{\partial T'}{\partial z}\right)\right] \qquad [71]$$

$$\left(Pr\frac{\partial}{\partial t} - \nabla^2\right) T' = \mathcal{Q}\left(\frac{\partial \psi}{\partial \phi} + z u_z + s\frac{\partial^2 \Phi}{\partial s \partial z}\right) \qquad [72a]$$

$$\nabla_{\mathrm{H}}^2 \Phi = -u_z \qquad [72b]$$

$$\nabla_{\mathrm{H}}^2 = \frac{1}{s}\frac{\partial}{\partial s} s \frac{\partial}{\partial s} + \frac{1}{s^2}\frac{\partial^2}{\partial \theta^2} \qquad [72c]$$

with Q defined in [65].

The local theory starts by seeking solutions with dependence $\exp \mathrm{i}(\tilde{k}s + \tilde{m}\phi - \tilde{\omega}t)$, and assuming that the wavelengths perpendicular to the z-direction are $O(\mathcal{E}^{1/3})$, and the Rayleigh number and frequency follow the scalings mentioned above. We write

$$Ra = \mathcal{E}^{-4/3}\mathcal{R}, \quad \tilde{\omega} = \mathcal{E}^{-2/3}\omega, \quad \tilde{m} = \mathcal{E}^{-1/3} m$$
$$\tilde{k} = \mathcal{E}^{-1/3} k, \quad a^2 = \frac{m^2}{s^2} + k^2 \qquad [73]$$

Some terms drop out, and the leading order equation for u_z then becomes a second-order equation for z, known as the Roberts–Busse equation,

$$\mathcal{F}\frac{\mathrm{d}}{\mathrm{d}z}\frac{1}{\mathcal{F}}\frac{\mathrm{d}u_z}{\mathrm{d}z} + \left[(a^2 - \mathrm{i}\omega)\mathcal{G} - \frac{\mathrm{i}m}{a^2 - \mathrm{i}Pr\omega}\mathcal{F}\frac{\mathrm{d}}{\mathrm{d}z}\left(\frac{\mathcal{R}\mathcal{Q}z}{\mathcal{F}}\right)\right] u_z = 0 \qquad [74]$$

following the notation of Dormy *et al.* (2004), where

$$\mathcal{F} = \frac{m^2 \mathcal{R}\mathcal{Q}}{a^2 - \mathrm{i}Pr\omega} - a^2(a^2 - \mathrm{i}\omega)$$
$$\mathcal{G} = \frac{(m^2 + a^2 z^2)\mathcal{R}\mathcal{Q}}{a^2 - \mathrm{i}Prw} - a^2(a^2 - \mathrm{i}\omega) \qquad [75]$$

The stress-free boundary conditions are

$$\frac{\mathrm{d}u_z}{\mathrm{d}z} - \frac{\mathrm{i}a^2}{m}(a^2 - \mathrm{i}\omega) z u_z = 0 \text{ at } z = \pm(1 - s^2)^{1/2} \qquad [76]$$

and for no-slip conditions

$$\frac{\mathrm{d}u_z}{\mathrm{d}z} - \frac{\mathrm{i}a^2}{m}(a^2 - \mathrm{i}\omega) z u_z = \mathcal{E}^{1/6}\sqrt{\frac{r}{2z}}\frac{m^2 + a^2 z^2}{m^2}\mathcal{F} u_z \text{ at}$$
$$z = \pm(1 - s^2)^{1/2} \qquad [77]$$

from which it is clear that in the low $\mathcal{E}$ limit the distinction between stress-free and no-slip boundary conditions is not crucial. Physically, this is because the rolls are so thin that most of the dissipation is occurring in the bulk of the fluid and not in the boundary layer. Note that if these thin rolls excite a large-scale zonal flow, the bulk dissipation for that zonal flow will be very small, so then the dissipation in the boundary layer is crucial in determining the strength of the zonal flow. Just because the boundary conditions do not affect the linear rolls very much does not imply that boundary conditions are unimportant in nonlinear rotating convection.

The Roberts–Busse equation [74] can be solved numerically subject to the boundary conditions [76] or [77] as an eigenvalue problem, using any standard ODE eigenvalue numerical method. The real and imaginary parts determine the complex frequency ω for any given Rayleigh number $\mathcal{R}$. The imaginary part of the frequency is the exponential growth rate of the linear mode. The output of the eigenvalue solver is then the dispersion relation

$$\omega = \omega(s, k, m, \mathcal{R}) \qquad [78]$$

The difference between the original local theory of Roberts and Busse and the newer global theory lies entirely in how this dispersion relation is treated. The same equation (and hence the same dispersion relation) is used in both approaches. The local theory derives the equations for the five unknowns ω_L, s_L, k_L, m_l and $\mathcal{R}_L$ from the five imaginary part conditions

$$\Im\{\omega\}=0,\quad \Im\left\{\left(\frac{\partial\omega}{\partial m}\right)_{\mathrm{L}}\right\}=0,\quad \Im\left\{\left(\frac{\partial\omega}{\partial \mathcal{R}}\right)_{\mathrm{L}}\right\}=0$$
$$\Im\left\{\left(\frac{\partial\omega}{\partial k}\right)_{\mathrm{L}}\right\}=0,\quad \Im\left\{\left(\frac{\partial\omega}{\partial s}\right)_{\mathrm{L}}\right\}=0 \qquad [79]$$

which are that the scaled growth rate (the imaginary part of ω) should be zero and that the growth rate should be a local maximum of $\mathcal{R}$, m, k, and s. s_{L} is then the location of the local convective instability, that is, where the rolls first become unstable. For internal heating the critical s_{L} is typically about $0.5r_{\mathrm{cmb}}$. The differential heating case does not have an internal local maximum for the growth rate (equivalently a minimum critical Rayleigh number) but is maximized on the ICB.

In the differentially heated case, the local theory prediction for the critical Rayleigh number agrees with the fully 2-D spherical coordinate results, in the sense that as $\mathcal{E}$ is reduced the results steadily approach the asymptotic results. However, in the internally heated case, where s_{L} is in the interior rather than at a boundary, there is no such agreement. In either case, the local solution must be part of a WKBJ expansion

$$u_z(s,\phi,z) \sim \mathcal{W}(x)u_z(z)\exp\left(\mathrm{i}\int \tilde{k}\,\mathrm{d}s\right)\exp \mathrm{i}(\tilde{m}\phi-\tilde{\omega}t) \qquad [80]$$

where the variable x is given by

$$x = (s - s_c)/\mathcal{E}^{1/6} \text{ for internal heating} \qquad [81a]$$

$$x = (s - s_{\mathrm{icb}})/\mathcal{E}^{2/9} \text{ for differential heating} \qquad [81b]$$

In the internally heated case, s_c is determined by a procedure described below. The rapid variation on the $O(\mathcal{E}^{1/3})$ length scale is taken care of by the factor $\exp\left(\mathrm{i}\int \tilde{k}\,\mathrm{d}s\right)$, and the z-dependence comes from the Roberts–Busse equation. The equation for the amplitude of the envelope, $\mathcal{W}(x)$, is derived by inserting [80] into the governing equations, and with the differential heating case scaling for x we obtain

$$-\frac{1}{2}\left(\frac{\partial^2\omega}{\partial k^2}\right)_{\mathrm{L}}\frac{\mathrm{d}^2\mathcal{W}}{\mathrm{d}x^2}+\left(\frac{\partial\omega}{\partial s}\right)_{\mathrm{L}}x\mathcal{W}$$
$$+\left[\left(\frac{\partial\omega}{\partial\mathcal{R}}\right)_{\mathrm{L}}\mathcal{R}_1-\omega_1+\left(\frac{\partial\omega}{\partial m}\right)_{\mathrm{L}}m_1\right]\mathcal{W}=0 \qquad [82]$$

where

$$\mathcal{R}=\mathcal{R}_{\mathrm{L}}+\mathcal{E}^{2/9}\mathcal{R}_1,\qquad m=m_L+\mathcal{E}^{2/9}m_1$$
$$\omega=\omega_L+\mathcal{E}^{2/9}\omega_1 \qquad [83]$$

The Airy equation [82] can be solved subject to the boundary conditions

$$\mathcal{W}=0 \text{ at } x=0(s=s_{\mathrm{icb}}) \text{ and } \mathcal{W}\to 0 \text{ at } x\to\infty \qquad [84]$$

The Airy function solution therefore gives the leading order asymptotic behavior in the radial direction, together with the z-dependence from the local Roberts–Busse equation. The leading order critical Rayleigh number and frequency are therefore $\mathcal{R}_{\mathrm{L}}$ and ω_{L}, and the first-order corrections come from the solution of [82] which is an eigenvalue problem, the real and imaginary parts giving $\mathcal{R}_1$ and $\omega_1 - m_1\partial\omega/\partial m$. How ω_1 and m_1 are found individually is described in Dormy *et al.* (2004). In the asymptotic limit of small $\mathcal{E}$, the s-structure of the solution takes the form of a succession of convective rolls, each with a radial and azimuthal length scale $O(\mathcal{E}^{1/3})$, beginning close to the tangent cylinder and of gradually diminishing amplitude, on a longer length scale $O(\mathcal{E}^{2/9})$, as we move outwards. However, because the difference between $O(\mathcal{E}^{1/3})$ and $O(\mathcal{E}^{2/9})$ is so small, extremely small ε is required before more than one roll is seen in the radial direction. In **Figure 4**, reproduced from Dormy *et al.* (2004), we see the planform of the rolls at onset for $\mathcal{E}=10^{-7}$, $Pr=1$. In the top row of **Figure 4**, numerical solution of the full equations has been used, while in the bottom row the asymptotic theory was used. There is no great difference, suggesting that the asymptotic method works very well at $E=10^{-7}$ or less. The variation with radius ratio at onset was studied by Al-Shamali *et al.* (2004), using the numerical method rather than the asymptotic method.

The natural way to deal with the internally heated case would be to solve [82] subject to the boundary conditions

$$\mathcal{W}\to 0 \text{ at } x\to\pm\infty \qquad [85]$$

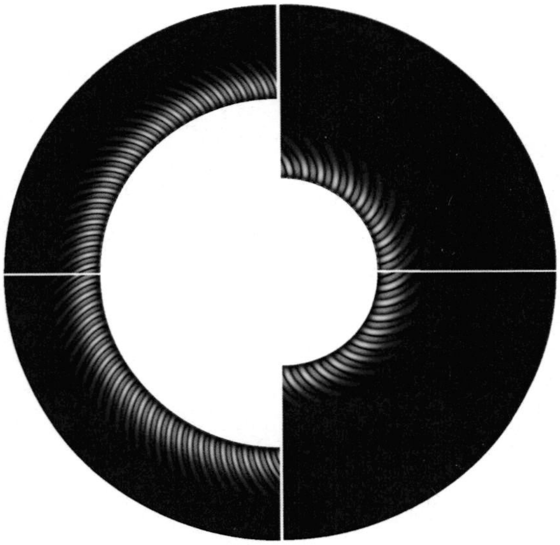

Figure 4 Comparison between equatorial cross-sections of the axial vorticity in the full numerics for $\mathcal{E}=10^{-7}$ (top row) and the asymptotic eigenfunction ψ for the same Ekman number (bottom row). Internal heating with large aspect ratio r_{icb}/r_{cmb} on the left, differential heating with aspect ratio 0.35 on the right. Reproduced from Dormy E, Soward AM, Jones CA, Jault D, and Cardin P (2004) The onset of thermal convection in rotating spherical shells. *Journal of Fluid Mechanics* 501: 43–70.

Unfortunately, no solution of [82] satisfying [85] exists, because of the term proportional to $(\partial\omega/\partial s)_L$. This means that when the onset of convection occurs away from the boundary, the local solution cannot be embedded in a consistent WKBJ solution. The solution to this difficulty is quite radical (Jones *et al.*, 2000) and consists of abandoning the s_L found by local theory, and instead seeking a point in the complex s-plane, s_c, at which the real part

$$\Re\{\partial\omega/\partial s\} = 0 \qquad [86]$$

as well as the imaginary part. We therefore solve [79] together with [86], so with s complex we now have six equations for six unknowns. The coefficient of the term proportional to x, which is the term which makes it impossible to satisfy the boundary conditions [85], is now zero. This is why the scaling for x changes to [81a], and then the amplitude equation becomes (with scaling $\mathcal{E}^{1/3}$ instead of $\mathcal{E}^{2/9}$ in [83])

$$-\frac{1}{2}\left(\frac{\partial^2\omega}{\partial k^2}\right)_c\frac{d^2\mathcal{W}}{dx^2}+\frac{1}{2}\left(\frac{\partial^2\omega}{\partial s^2}\right)_c x^2\mathcal{W}$$
$$+\left[\left(\frac{\partial\omega}{\partial\mathcal{R}}\right)_c\mathcal{R}_1-\omega_1+\left(\frac{\partial w}{\partial m}\right)_c m_1\right]\mathcal{W}=0 \qquad [87]$$

which now has a satisfactory solution in parabolic cylinder functions satisfying [85].

Since the equations used to find the critical Rayleigh number and frequency are different from the local theory (six equations, not five), the results are different. The critical Rayleigh number is significantly higher (typical 25% higher at $Pr\sim 1$ or greater, but with a much larger difference at low Prandtl number) for this global theory than for local theory. The way in which the solutions for complex $s=s_c$ are converted into real eigenfunctions is detailed in Jones *et al.* (2000). In physical terms this surprising result can be understood in terms of phase mixing. Disturbances that try to grow at the local critical Rayleigh number are sheared apart by the variation of ω with s. Because of the increasing slope of the boundaries with s in spherical geometry, waves at larger s travel with faster phase speed. It is therefore more difficult to get a disturbance inside an envelope where each part has to travel at the same speed, when each part wants to travel at a different speed. At low Prandtl number, the disturbance occupies a much greater range of s, and the problem of 'orchestration' is more difficult, which is why the global critical Rayleigh number has to be much greater than the local critical Rayleigh number.

5.4.4 The Ekman Boundary Layers

In eqns [76] and [77] above, we noted that the boundary condition in the Roberts–Busse equation was affected by whether the boundary was no slip or stress free. This difference comes about because of the presence of Ekman layers near the boundary. Ekman layers are the thin boundary layers of thickness $O((\Omega/\nu)^{1/2})$ over which the velocity changes from its interior value to the value at the rigid boundary, usually taken to be zero. In this respect, the boundary layer is similar to the thin layers in nonrotating flow, for example, over aircraft wings. There are, however, some important differences. Ekman layers have an unusual 'suction' property. At the boundary itself, the velocity is zero, while at the 'edge' of the boundary layer the velocity parallel to the wall merges into its interior value. Somewhat surprisingly, at the edge of the boundary layer there is a nonzero component of velocity perpendicular to the wall, called the Ekman suction. This velocity can be either outward or inward, depending on the sign of the interior fluid vorticity. If the horizontal length scale of this external vorticity is d, we can define an

Ekman number $E=\nu/\Omega d^2$, and then this suction velocity is $O(E^{1/2})$ times the parallel velocity components. This might seem negligibly small, but in fact it can have quite a significant effect on the dynamics.

The Ekman suction affects the spin-up (or spin-down) time of rotating fluid, that is, the time taken for rotating fluid initially at rest to achieve rigid body rotation, subsequent to a sudden change in the container rotation speed. Naively, one might expect the spin-up time to be $O(d^2/\nu) \sim 10^{11}$ years for the core, this being the typical time for momentum to diffuse through the fluid by internal friction. Actually, the time is $O(E^{1/2})$ times this value, or about 5000 years, because all the fluid in the sphere is pumped into the boundary layer, where it acquires the boundary velocity much more rapidly. Recent experiments (Brito *et al.*, 2004) have measured the spin-up time and find good agreement with the times predicted by Ekman layer theory. They also find that if low Prandtl number fluid is convecting, the spin-up time is significantly shortened, suggesting that the convective turbulence is affecting the Ekman layer. The usual theory assumes a laminar Ekman layer.

The Ekman layer is studied by assuming E is small, and considering the thin layer in which viscous forces, pressure forces, and Coriolis forces are in balance:

$$2\mathbf{\Omega} \times \mathbf{u} = \frac{1}{\rho}\nabla p + \nu\nabla^2\mathbf{u} \qquad [88]$$

Because the layer is thin, the gradient of $\mathbf{u}$ in the direction perpendicular to the boundary is much larger than gradients in other directions, and these can therefore be ignored by comparison. Greenspan (1968) shows that the Ekman suction is

$$\mathbf{u}\cdot\mathbf{1}_{n|z=\pm H} = \mp\frac{1}{2}\left(\frac{\nu}{\Omega}\right)^{1/2}\mathbf{1}_n\cdot\nabla \times \left[\frac{1}{(\mathbf{1}_n\cdot\mathbf{1}_z)^{1/2}}(\mathbf{1}_n\times\mathbf{u}+\mathbf{u})\right] \qquad [89]$$

where $\mathbf{1}_n$ is the unit vector normal to the wall pointing out of the fluid. Here $\mathbf{u}\cdot\mathbf{1}_{n|z=H}$ denotes the asymptotic value of the normal velocity at the Northern Hemisphere boundary ($-H$ for Southern Hemisphere boundary) as we move out of the boundary layer into the interior.

The simplest case is when the boundary is normal to the rotation axis, as near the poles. For the south pole, $\mathbf{1}_n=-\mathbf{1}_z$, when [89] reduces to $u_z=(\nu/\Omega)^{1/2}\zeta_z/2$, so that the suction is proportional to the z-component of the interior vorticity just above the boundary layer. Anticyclonic vorticity gives suction into the boundary at either pole, cyclonic (in the same sense as the rotation) vorticity to motion away from the boundary. We take the z-component of the curl and double curl of [88] to get

$$-2\Omega\frac{\partial u_z}{\partial z} = \nu\frac{\partial^2\zeta_z}{\partial z^2} \qquad [90a]$$

$$2\Omega\frac{\partial\zeta_z}{\partial z} = \nu\frac{\partial^4 u_z}{\partial z^4} \qquad [90b]$$

which can be combined to give

$$\frac{\partial u_z}{\partial z} = -4E^2d^4\frac{\partial^5 u_z}{\partial z^5} \qquad [91]$$

We can add on H to z without changing these formulas, thus bringing the wall to the level $z=0$. The solution then consists of a constant plus four exponential terms corresponding to the four complex fourth roots of -4, $(\pm 1 \ \pm \mathrm{i})$. The two with positive real part correspond to solutions which grow as we move out of the boundary layer, and are therefore unacceptable. We are left with

$$u_z = u_z^i + \exp\left(-\frac{z}{d\sqrt{E}}\right)\left(A\cos\frac{z}{d\sqrt{E}} + B\sin\frac{z}{d\sqrt{E}}\right) \qquad [92a]$$

$$\zeta_z = \zeta_z^i + \exp\left(-\frac{z}{d\sqrt{E}}\right) \times \left(\frac{A+B}{d\sqrt{E}}\cos\frac{z}{d\sqrt{E}} + \frac{B-A}{d\sqrt{E}}\sin\frac{z}{d\sqrt{E}}\right) \qquad [92b]$$

and the boundary conditions are that $u_z=0$, $\partial u_z/\partial z=0$, and $\zeta=0$ at the wall, $z=0$. The last two follow from the fact that $u_x=u_y=0$ at the wall, so horizontal derivatives are zero, and $\nabla\cdot\mathbf{u}=0$. Inserting these into [92] gives $A=B=-u_z^i$, $\zeta_z^i = 2E^{-1/2}u_z^i/d$, consistent with [89]. Since the A and B terms in [92] vanish as we move up into the interior, the constants of integration ζ_z^i and u_z^i have a natural interpretation as the interior vorticity just after leaving the boundary layer and the Ekman suction, respectively. The Ekman layer in the core is less than 1 m thick, and so any roughness at the CMB or ICB may lead to a thickening of the layer.

The effect of Ekman suction on the onset of convection was investigated by Zhang and Jones (1993). They found that at high Prandtl numbers it is stabilizing, but at low Prandtl numbers it can be destabilizing. At low E the effect of Ekman suction on the tall thin columns is small, and then the most important effect of the Ekman suction is on the zonal flow.

5.4.5 The Busse Annulus

The discussion in Sections 5.4.1–5.4.3 showed that the sloping boundaries are crucial when considering the onset of rotation in convecting spherical geometry. Busse (1970) pointed out that a simplified model for the convection in a sphere is to consider a cylindrical annulus with sloping top and bottom boundaries (see **Figure 5**). The gravity is assumed to act in the $-\mathbf{i}_s$ direction.

If we now take the limit in which the angle of slope of the annulus $\chi << 1$, then the motion is almost geostrophic. The z-component of the velocity is now small compared to the other two components, and so we can write

$$\mathbf{u} = \nabla \times \psi(s, \phi)\mathbf{1}_z + u_z\mathbf{1}_z \qquad [93]$$

because the horizontal part of the flow is much larger and 2-D flow with zero divergence must be the curl of a scalar. There is a distinction between [69] and [93] because in [69] we retained the z-dependence of ψ, whereas in the annulus model this disappears because of the small slope assumption. The annulus is also assumed to be thin, so that the curvature terms can be neglected, which means that the ϕ-coordinate is replaced by a Cartesian x-coordinate, that is, $(1/s)\partial/\partial\phi$ becomes $\partial/\partial x$, and s is replaced by $-y$.

Our fundamental equation is [56], the vorticity equation. We assume small Rossby number so the planetary vorticity dominates the fluid vorticity, so the term $\boldsymbol{\zeta}\cdot\nabla\mathbf{u}$ is omitted. The z-component of the vorticity equation [56] is then

$$\frac{\partial \zeta}{\partial t} + \mathbf{u}\cdot\nabla\zeta - 2\Omega\frac{\partial u_z}{\partial z} = -g\alpha\frac{\partial T'}{\partial x} + \nu\nabla^2\zeta \qquad [94]$$

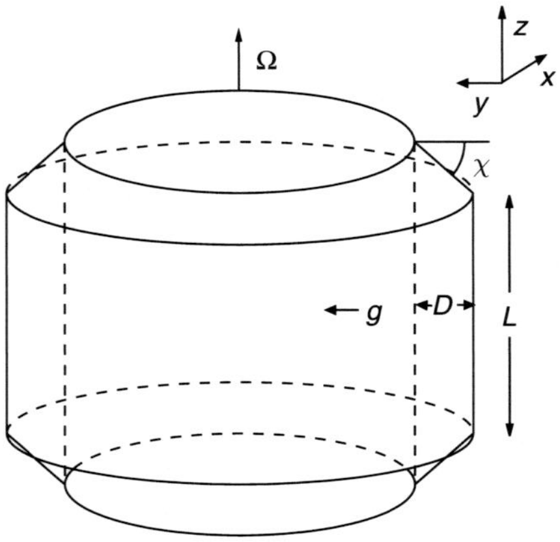

Figure 5 Geometry of the Busse annulus. The angle χ is assumed small, as is the gap width to depth ratio, D/L. The x-direction corresponds to the ϕ-direction, and the y-direction to the $-s$-direction in spherical geometry.

Although the z-component of velocity u_z is small, $O(\chi)$, compared to the other components in [93], the rotation is large, so that the stretching of the planetary vorticity is important. Integrating over z and applying the boundary conditions $u_z = \pm\chi u_y$ on $z = \pm L/2$ gives

$$\frac{\partial \zeta}{\partial t} + \frac{\partial(\psi, \omega)}{\partial(x, y)} - \frac{4\chi\Omega}{L}\frac{\partial\psi}{\partial x} = -g\alpha\frac{\partial\theta}{\partial x} + \nu\nabla^2\zeta \qquad [95]$$

Note that in assuming this boundary condition we have discarded any Ekman layer contribution from the endwalls. We are therefore assuming implicitly that they are stress free. Nondimensionalizing on the gap between the cylinders D, timescale D^2/ν, and temperature $\nu\Delta T/\kappa$, ΔT being the temperature drop between the cylindrical boundaries, we get

$$\frac{\partial \zeta}{\partial t} + \frac{\partial(\psi, \zeta)}{\partial(x, y)} - \beta\frac{\partial\psi}{\partial x} = -Ra\frac{\partial T'}{\partial x} + \nabla^2\zeta \qquad [96a]$$

$$\zeta = \nabla^2\psi \qquad [96b]$$

The z-vorticity can be written in terms of the stream function ψ from [93]. The temperature equation [38] becomes

$$Pr\left[\frac{\partial T'}{\partial t} + \frac{\partial(\psi, T')}{\partial(x, y)}\right] = -\frac{\partial\psi}{\partial x} + \nabla^2 T' \qquad [97]$$

where

$$Ra = \frac{g\alpha\Delta T D^3}{\kappa\nu}, \qquad \beta = \frac{4\chi\Omega D^3}{L\nu} \qquad [98]$$

The rotation enters these equations explicitly only through the β-parameter; apart from this term the equations are identical to those of 2-D Rayleigh–Bénard convection. Rotation has been invoked implicitly, though to justify the two dimensionality of the flow. Equation [96a] is commonly used in the atmospheric sciences community where it is known as the β-plane equation.

5.4.5.1 Linear properties of the annulus model

If we linearize these equations, by dropping the Jacobian terms in [96] and [97], and adopt the convenient no penetration, stress-free, constant temperature boundary conditions at the side walls

$$\psi = \frac{\partial^2\psi}{\partial y^2} = 0, \quad T' = 0, \quad \text{on } y = \pm\frac{1}{2} \qquad [99]$$

then there is a simple solution

$$\psi = \exp i(kx - \omega t)\cos\pi y, \qquad T' = \frac{-ik}{k^2 + \pi^2 - i\omega Pr}\psi \qquad [100]$$

provided

$$\omega = \frac{\beta k}{(1 + Pr)(\pi^2 + k^2)} \quad [101a]$$

$$Ra = \frac{(\pi^2 + k^2)^3}{k^2} + \frac{Pr^2 \beta^2}{(\pi^2 + k^2)(1 + Pr)^2} \quad [101b]$$

The equation for ω is the dispersion for thermal Rossby waves, and the expression for Ra is the critical Rayleigh number for the onset of convection. If $\beta = 0$ the onset of convection is steady, but the presence of the rotation means that convection onsets in the form of traveling waves. The critical wave number k for the onset of instability is found by setting

$$\frac{\mathrm{d}Ra}{\mathrm{d}k} = 0 \quad [102]$$

Note also that if $\beta = 0$ we recover Rayleigh's famous result for the onset of convection $Ra_c = 27\pi^4/4$ with $k = \pi/\sqrt{2}$.

The thermal Rossby waves travel with phase speed c_s and group velocity c_g

$$c_s = \frac{\omega}{k} = \frac{\beta}{(1 + Pr)(\pi^2 + k^2)} \quad [103a]$$

$$c_g = \frac{\mathrm{d}\omega}{\mathrm{d}k} = \frac{\beta(\pi^2 - k^2)}{(1 + Pr)(\pi^2 + k^2)^2} \quad [103b]$$

If β is positive, the phase speed is positive, so the waves propagate eastward. This will be the case outside the tangent cylinder. Inside the TC, the integration over z leads to a negative β, so waves travel westward there. From [101b] we see that rotation delays the onset of convection, as we expect. The rapid rotation limit $\beta \to \infty$ is the most instructive; we find

$$k = \frac{\beta^{1/3} Pr^{1/3}}{2^{1/6}(1 + Pr)^{1/3}} \quad [104a]$$

$$\omega = \frac{\beta^{2/3} 2^{1/6}}{Pr^{1/3}(1 + Pr)^{2/3}} \quad [104b]$$

$$Ra = \frac{3\beta^{4/3} Pr^{4/3}}{2^{2/3}(1 + Pr)^{4/3}} \quad [104c]$$

Since β is essentially $1/E$ multiplied by factors to account for the particular geometry of the annulus, it is comforting to note that these are the same scalings found for Ra, ω, and k in Sections 5.4.2 and 5.4.3 in the rapid rotation limit. It is also interesting to note that although the phase speed is always eastward, the group velocity is westward for rapid rotation. Note also that it follows from the dispersion relation that the larger wavelengths have a higher phase speed than the smaller 'tall thin column' modes.

5.4.5.2 Weakly nonlinear theory and the annulus model

Equations [96a], [96b], and [97] can also be solved in the nonlinear regime to investigate rapidly rotating convection at large Ra. One method is to solve these 2-D equations numerically using the pseudo-spectral method, and this is described in Section 5.4.5.3. It is also possible to use weakly nonlinear theory to investigate nonlinear behavior in rotating (and magnetic) systems at Rayleigh numbers just above critical.

The fundamental idea of weakly nonlinear theory is that just above critical the spatial form of the convection is still approximately that given by linear theory. It can be shown that for Rayleigh numbers sufficiently close to critical this will always be true, but in some situations weakly nonlinear theory still gives a reasonable picture of behavior well above critical while in other situations it breaks down quite quickly. The great advantage of weakly nonlinear theory is that it avoids the solution of multidimensional partial differential equations. This means that although the theory may be quite hard to formulate, once that task is done the parameter space can be covered much more thoroughly than is possible for direct numerical simulations (DNS). Weakly nonlinear theory is most useful for suggesting possible behavior and developing understanding of rotating convection. Because its domain of validity cannot be determined *a priori*, results using it do need to be checked against DNS simulations.

Weakly nonlinear theory for fluids problems was developed in the late 1950s, but the theory for the annulus model was given by Busse and Or (1986). The velocity is expanded in a power series

$$\psi = \epsilon\psi_1 + \epsilon^2\psi_2 + \epsilon^3\psi_3 + \cdots \quad [105]$$

with a similar expansion for the temperature perturbation

$$T' = \epsilon T'_1 + \epsilon^2 T'_2 + \epsilon^3 T'_3 + \cdots \quad [106]$$

Here ϵ is small parameter, and the Rayleigh number and the frequency must also be expanded in ϵ:

$$\begin{aligned} Ra &= Ra_c + \epsilon R_1 + \epsilon^2 R_2 + \cdots \\ \omega &= \omega_c + \epsilon\omega_1 + \epsilon^2\omega_2 + \cdots \end{aligned} \quad [107]$$

We insert these expansions into eqns [96a], [96b], and [97] and equate like powers of ϵ. At first order in ϵ there are no contributions from the nonlinear terms, and we just recover the linear theory. The solution then determines Ra_c and ω_c just as above, and ψ_1, T'_1 are the standard linear eigenfunctions. The nonlinear terms start to appear at $O(\epsilon^2)$. In the annulus problem these nonlinear terms can be balanced by second-order terms in ψ and T', for example, the nonlinear terms in the temperature equation give rise to a term proportional to $\epsilon^2 \sin \pi y \cos \pi y$, which leads to a T'_2 term also proportional to $\sin 2\pi y$. However, at the next order, $O(\epsilon^3)$, some nonlinear terms have the same x- and y-dependence as the linear solution, $\exp ikx \cos \pi y$. These are called resonant terms, and they cannot be allowed, because no finite terms in ψ and T' can balance them. We must arrange that these resonant terms have a zero coefficient, and this gives equations which determine R_2 and ω_2. The absence of resonant terms at $O(\epsilon^2)$ in this problem, which is symmetric about $y=0$, means that $R_1 = \omega_1 = 0$. This procedure can be extended to higher orders, and at third order a mean flow appears, that is, a flow u_x which does not average to zero in the x-direction. These flows are called zonal flows and have been much studied recently because large zonal flows are seen in the atmospheres of giant planets.

The idea of weakly nonlinear theory can be extended to the case of the rapidly rotating annulus (Abdulrahman *et al.*, 2000). Now we have two small parameters, the amplitude ϵ and the small horizontal wavelength $O(\beta^{-1/3})$. The behavior depends on the the relative size of these small quantities, but the most interesting case is when $\epsilon \sim \beta^{-1/3}$. It is then possible to derive nonlinear partial differential equations in y and t only. In this system, the x-dependence is given by a wavy mode with the linear critical wave number, but the y-dependence can vary from its onset form. This makes it possible to analyze the bifurcations that occur as the Rayleigh number is increased. Interestingly, all these bifurcations occur in a range of Rayleigh number for which Ra/Ra_c is only $O(\beta^{-2/3})$ above critical, so that the transition from steady rolls to chaotic convection takes place when the Rayleigh number is only slightly supercritical. Because the curvature of the endwalls is neglected in the annulus model, the system is symmetric about the midplane in y, that is, if the sidewalls are at $y = \pm d/2$, there is a symmetry about $y=0$. The zonal flow driven by the convection initially has this symmetry, and has an approximately parabolic form. However, as the Rayleigh number is increased, a symmetry-breaking bifurcation occurs, and the zonal flow becomes asymmetric about $y=0$ (Or and Busse, 1987; Abdulrahman *et al.*, 2000). With no curvature, there is nothing to distinguish the two asymmetric zonal flow patterns corresponding to eastward or westward flow at the outer boundary, but if small curvature is introduced, then the zonal flow which is eastward on the outer boundary is preferred, corresponding to a prograde equatorial zonal flow in spherical shell geometry.

5.4.5.3 Zonal flows and multiple jets

The 2-D nonlinear equations [96a], [96b], and [97] can be integrated forward in time very efficiently, using fast Fourier transforms in Cartesian geometry with a pseudo-spectral method in which ψ and T' have expansions of the form

$$\psi = \sum_{l=-(N_x-1)}^{N_x-1} \sum_{m=1}^{N_y} \psi_{lm} e^{ilx(2\pi/L_x)} \sin m\pi y \qquad [108]$$

This means that values of $\beta \sim 10^6$ (low E) can be achieved (see Brummell and Hart, 1993; Jones *et al.*, 2003). This makes the annulus model a powerful tool for the study of zonal flows. Zonal flow strength can be significantly affected by Ekman suction, which we omitted when we set $u_z = \pm \chi u_y$ on $z = \pm L/2$. If we now include the Ekman suction term [89], remembering that χ is small, there is now a boundary term when we integrate [94] over z, and [96] becomes

$$\frac{\partial \zeta}{\partial t} + \frac{\partial(\psi, \zeta)}{\partial(x, y)} - \beta \frac{\partial \psi}{\partial x} = -Ra \frac{\partial T'}{\partial x} - C|\beta|^{1/2} \zeta + \nabla^2 \zeta \qquad [109]$$

details being given in Jones *et al.* (2003). The extra term proportional to the geometrical constant of order unity, C, represents a damping due to the Ekman boundary layer. In experiments, this term can often be larger than the internal friction $\nabla^2 \zeta$, for the zonal flow.

Strong zonal flow can affect the convection which sustains it. Indeed, the shear can suppress convection, so that the zonal flow can bite the hand that feeds it. This leads to a range of interesting dynamical behavior, for example, convection may become localized in space or in time, so that convection does not occur continuously, but only in bursts when the zonal flow is particularly weak (Busse, 2002; Rotvig and Jones, 2006).

In the Earth's core, zonal flows can be strongly affected by magnetic fields (Aubert, 2005), but the formation of zonal flow patterns is an important feature of rotating convecting flows. Simulations show that these east–west flows can build up to a very large amplitude, so that the kinetic energy in the jets can be much larger than the kinetic energy of the nonaxisymmetric rolls which are transporting the heat. An issue which has attracted much attention recently is the formation of multiple jets. The giant planets, particularly Jupiter, have a zonal flow pattern which has an alternating pattern of eastward and westward jets as the latitude varies. Numerical simulations of spherical convection at moderate Rayleigh and Ekman number show a much simpler pattern, with a prograde (eastward) flow near the equator and a retrograde flow near the poles. The Sun has a differential rotation pattern of this form. In terms of the annulus model, the zonal flow comes from the $l=0$ mode in [108], and a multiple jet solution is one on which this $l=0$ mode is dominated not by $m=1$ but by a higher m value. The boundary conditions have some effect on the appearance of multiple jets. If there is a no-slip boundary on the sloping boundaries, an Ekman layer is created which leads to Ekman suction and hence damping in the interior. This reduces the magnitude of the zonal flows, which makes multiple jets more common. Increasing β to very high values also favors the formation of multiple jets. However, in the core, the pattern of the zonal flow is determined by the magnetic field, so we will not pursue these interesting nonmagnetic problems further here.

5.4.6 The Quasi-Geostrophic Approximation

The Busse annulus model is a 2-D model which captures many of the essential features of rapidly rotating convection in spherical geometry found in experiments (*see* Chapter 11): convection occurs in the form of tall thin columns, it onsets as thermal Rossby waves propagating prograde, and large zonal flows develop which may have a multiple jet structure. The annulus model can only be rigorously derived when the slope of the boundaries is small, and it is limited because it does not take into account the strong variation of the boundary slope χ that occurs in spherical geometry (see **Figure 5**). The key property of the annulus model is that the axial vorticity is z-independent, thus reducing the problem from three to two dimensions. Both the linear theory of convection and experiments suggest that the axial vorticity does not vary strongly with z even in spherical geometry (Gillet and Jones, 2006). There is some variation with z but it is numerically small, suggesting that assuming axial vorticity is z-independent may be a good approximation, even though it cannot be rigorously derived in any limit. If this is done, it follows that u_s and u_ϕ are also z-independent, and that u_z is only linearly dependent on z. This is the essence of the quasi-geostrophic approximation (QGA) for rapidly rotating fluids. Detailed discussions of this approximation are given in Aubert *et al.* (2003), and numerical results using it in Morin and Dormy (2004) and Gillet and Jones (2006). Another way of viewing the QGA is that the terms omitted by assuming χ is small are not very important even when χ is $O(1)$, so that the QGA behavior is qualitatively similar to that of full spherical convection provided the rotation is very large. Note that one of these omitted terms is the z-component of gravity. In the experiments, where gravity is replaced by centrifugal acceleration (*see* Chapter 11), there is no z-component of gravity, so the QGA may be particularly appropriate. The QGA will certainly break down if the Rossby number becomes $O(1)$ or larger, because then the convection will no longer be columnar. The QGA gives useful results which can be compared with experiments (Gillet and Jones, 2006) and has the great advantage that because it is relatively easier to solve 2-D problems, the heat transport and the zonal flow can be explored over a wide range of parameter space. The zonal flow equation is found by taking the ϕ average of the ϕ-component of [37], remembering that the z average has already been taken, so we are effectively averaging over cylinders of radius s:

$$\left(\frac{\partial}{\partial t} - \nu\left[\nabla_{\mathrm{H}}^2 - \frac{1}{s^2} - \frac{E^{-1/2} r_{\mathrm{cmb}}^{1/2}}{d H^{3/2}}\right]\right)\bar{u}_\phi = -\overline{\mathbf{u}\cdot\nabla u_\phi} \qquad [110]$$

where H is the height of the CMB above the equatorial plane, $H = (r_{\mathrm{cmb}}^2 - s^2)^{1/2}$. On the left-hand side, the viscous force splits into two parts, the bulk friction and the Ekman suction, and in a steady state, these must balance the driving by the Reynolds stresses. Frequently, the driving is quite strong and the viscosity is small, so a large zonal flow is needed to bring this equation into balance. However, even quite weak Lorentz forces will upset this balance, so we can expect zonal flows to be radically different in the magnetic case. The QGA cannot be used (at least in its present form) for convection inside the tangent cylinder because the axial vorticity there does depend quite strongly on z; indeed it can change sign as z varies.

5.4.7 Thermal Wind

Perhaps the least satisfactory feature of the QGA is that it is necessary to assume the temperature has no variation perpendicular to gravity. This may not be true in the core; it certainly is not true in the Earth's atmosphere where the poles are much colder than the equator. The ϕ-component of [56], omitting viscosity, time dependence, and small Rossby number, gives

$$2\Omega\frac{\partial u_\phi}{\partial z} = \frac{g\alpha}{r}\frac{\partial T'}{\partial \theta} \quad [111]$$

This is the thermal wind equation. Note that now we have omitted the Reynolds stress term on the right-hand side of [110] as well as internal friction and Ekman suction, so this is a very different balance from that envisaged in [109] and [110], but it may be more realistic in the core. In the Northern Hemisphere of the Earth's atmosphere, where cold arctic air comes close to warm equatorial air, there is a strong positive $\partial T'/\partial\theta$ and hence the zonal east–west flow increases rapidly upwards from [111]. At ground level, the atmosphere is constrained to rotate at the same speed as the Earth, but at great heights the resulting jetstream significantly shortens aeroplane travel times going from west to east. In the core, there is evidence from the secular variation that there are anticyclonic vortices near the poles in the Earth's core (Olson and Aurnou, 1999). If the origin of these vortices is a thermal wind (Aurnou *et al.*, 2003; Sreenivasan and Jones, 2005), then the polar regions inside the core must be warmer (less dense) than the equatorial regions. Order of magnitude estimates show that even very small latitudinal temperature variations, of the order of 10^{-3} K, can give rise to the observed anticyclonic vortices. As mentioned in Section 5.4.1, the motion across the tangent cylinder is likely to be small in the core, so a buildup of light material and hot material released from the inner core may accumulate inside the tangent cylinder.

Since the thermal wind produces a variation of zonal flow with z, not only may the flow be anticyclonic near the CMB, it may be cyclonic (eastward) near the ICB. This could lead to a super-rotation of the inner core, that is, an inner core rotating faster than the mantle. There is some seismological evidence that the inner core is rotating faster than the mantle, by about 0.3° per year (Song, 1996; Collier and Helffrich, 2001; Zhang *et al.*, 2005), but it is controversial (Souriau *et al.*, 2003). If the inner core and the mantle are both slightly nonaxisymmetric, gravitational coupling (Buffett and Glatzmaier, 2000) may lock the inner core to the mantle, allowing only an oscillation between the two rather than a continuous relative rotation.

5.4.8 Scaling Laws and Heat Transport in Nonlinear Rapidly Rotating Convection

The weakly nonlinear theory of Section 5.4.5.2 can be used to predict behavior just above critical, but when the flow becomes turbulent we would not expect a theory based on convection having the same pattern as it has near onset to be appropriate. In particular, at low Prandtl number the flow speed and the heat transport grow only slowly with $Ra/Ra_c - 1$ when Ra is very close to Ra_c, but simulations indicate that much more rapid growth of heat transport and flow speed occurs when $Ra/Ra_c - 1$ is $O(1)$ (Plaut and Busse, 2002; Gillet and Jones, 2006). A turbulent 2-D flow is one where columnar eddies grow and decay continually throughout the fluid. They typically last a turnover time, which means their lifetime is about ℓ/U_*, where ℓ is the typical length scale transverse to the rolls ($\ell << d$, where d is the integral length scale, either the radius of the sphere or the gap width as appropriate) and U_* is the typical fluid velocity. Near onset, we expect $\ell \sim dE^{1/3}$, but in a strongly nonlinear regime we might expect that ℓ becomes independent of ν as a turbulent cascade develops. The usual situation in turbulence is that the very small value of ν only affects the very small length scale at which the viscous dissipation occurs, and the only effect of reducing ν further is to extend the cascade so that the dissipation length becomes even smaller, without significantly affecting the behavior at the larger length scales. Note that here we are only discussing rapidly rotating convection, by which we mean that the velocity is small enough so the Rossby number is small, and so the convection remains columnar. As we increase the Rayleigh number, the typical velocity will increase, so we must also increase the rotation rate in order to remain in the rapidly rotating regime. If the Rayleigh number is increased at fixed rotation rate, eventually the Rossby number will become order unity and a different regime is entered. Because the Earth's core has low Rossby number, we focus here only on the rapidly rotating regime, and do not discuss what happens at intermediate Rossby number.

5.4.8.1 The inertial theory of rapidly rotating convection

A scaling independent of viscosity was proposed by Ingersoll and Pollard (1982) and developed by Aubert *et al.* (2001). The key idea is that there is a three-term balance in the axial vorticity (eqn [56]) between vorticity advection, vortex stretching, and vorticity generation by buoyancy. We cannot give exact results without detailed numerical simulation, but here we simply try to identify the dominant terms in the equations and hence to see how quantities such as the typical velocity scale. In this spirit, we denote the typical vorticity by ζ_* and the typical temperature perturbation by T'_* to get

$$\zeta_* \sim \frac{U_*}{\ell}, \quad U_*\zeta_*/\ell \sim \Omega U_*/d \sim g\alpha T'_*/\ell \qquad [112]$$

We have omitted the factor 2 in the vortex stretching term, because it can be absorbed in the definition of the typical length scales, and we are only interested in the scalings. Note however that it is d not ℓ that appears in the vortex stretching term, because the axial velocity only varies on the long length integral scale in columnar convection. The balance of inertial advection against buoyancy in nonrotating convection, [48], would be the same as in [112] if ℓ is taken as the mixing length. In rapidly rotating convection the vorticity constraint impedes movement in the s-direction (the Proudman–Taylor constraint) because of the sloping endwalls, so we can view [112] as a mixing length balance but with the mixing length much reduced from d to ℓ. Another way of viewing this is that as a hot fluid column attempts to move outward in the s-direction, the Coriolis force turns it sideways in a distance ℓ.

The temperature perturbation is determined by the convective heat flux per unit area as in [49]

$$F_{\text{conv}} \sim \rho c_p U_* T'_* \qquad [113]$$

ρ being the fluid density, c_p the specific heat at constant pressure, and F_{conv} a typical value of the convective heat flux per unit area. To complete the theory we need a relation giving the convective heat flux in terms of the superadiabatic temperature difference between the ICB and the CMB. We discuss this below, but for geophysical (and astrophysical) applications, usually more is known about the heat flux than about temperature differences, so [112] and [113] are often all that is needed. Eliminating T'_* and U_* between them gives

$$\ell \sim \left(\frac{g\alpha F_{\text{conv}} d^3}{\Omega^3 \rho c_p}\right)^{1/5} \qquad [114]$$

and then we get

$$U_* \sim \left(\frac{d}{\Omega}\right)^{1/5} \left(\frac{g\alpha F_{\text{conv}}}{\rho c_p}\right)^{2/5} \qquad [115a]$$

$$g\alpha T'_* \sim \left(\frac{\Omega}{d}\right)^{1/5} \left(\frac{g\alpha F_{\text{conv}}}{\rho c_p}\right)^{3/5} \qquad [115b]$$

It is interesting to compare these with the Deardorff velocity and temperature scalings [50]. The exponents in the scalings with F_{conv} are different, but not that different! In the inertial theory we have $U_* \sim F_{\text{conv}}^{0.4}$, whereas for the Deardorff scaling $U_* \sim F_{\text{conv}}^{0.5}$. The small difference arises because of the weak power of 1/5 in [114]. This length scale ℓ is sometimes called the Rhines scale (Rhines, 1975) and it arises from the balance of vorticity advection and vortex stretching. For the inertial theory to be valid, this length scale must be larger than the thickness of the rolls at onset $\ell_c \sim d^{1/3}(\nu/\Omega)^{1/3}$. Of course, at sufficiently small ν the Rhines length will be larger than onset length, but E has to be very small before this is achieved. In the core, if we take the usual estimate of $E \approx 10^{-15}$ and $d = 2.26 \times 10^6$ m, then ℓ_c is only 20 m (but numerical factors from simulations can increase this to 300 m), but the Rhines scale $(U_* d/\Omega)^{1/2} \approx 4$ km, assuming a typical velocity of 5×10^{-4} m s^{-1}.

In simulations and experiments it is quite difficult to achieve such a low viscosity, and then an alternative viscous scaling

$$\Omega U_*/d \sim g\alpha T'_*/\ell_c, \quad \ell_c \sim d\left(\frac{E(1+P)}{P}\right)^{1/3} \qquad [116]$$

together with [113] may give better results (Gillet and Jones, 2006).

The convective heat flux may be estimated in terms of the Nusselt number

$$F_{\text{conv}} \sim (Nu - 1)\rho c_p \kappa \Delta T_s / d \qquad [117]$$

exact in plane layer geometry but only an estimate in spherical geometry. Here ΔT_s is the superadiabatic temperature gradient across the layer. If we define the Rayleigh number as $Ra = g\alpha\Delta T_s d^3/\kappa\nu$, the Ekman number as $E = \nu/\Omega d^2$, and define the convective flux Rayleigh number as $Ra_Q = Ra(Nu - 1)$, then [114] and [115] can be expressed in dimensionless form as

$$\frac{\ell}{d} \sim E^{3/5} Pr^{-2/5} Ra_Q^{1/5} \qquad [118]$$

$$\frac{U_* d}{\kappa} \sim (EPr)^{1/5} Ra_Q^{2/5} \qquad [119]$$

$$Ra\frac{T'_*}{\Delta T_s} \sim (EPr)^{-1/5} Ra_Q^{3/5} \qquad [120]$$

5.4.8.2 Heat transport in rapidly rotating convection

We noted above that both the inertial scaling and the viscous scaling are incomplete, because the Nusselt number–Rayleigh number relation is undetermined. In nonrotating convection, this relationship is determined by the formation of boundary layers, the interior outside these boundary layers being almost isothermal, because convection is very efficient there. In rapidly rotating convection the formation of boundary layers is delayed, because the vorticity constraint makes convection less efficient. As the Rayleigh number is raised, eventually boundary layers form and we expect that the *Nu–Ra* relationship will become like that in nonrotating convection. The experiments of Sumita and Olson (2000), where very high Rayleigh numbers were achieved, showed this behavior. For moderate *Ra*, $Nu - 1$ is almost proportional to $Ra/Ra_c - 1$ but $\mathrm{d}\log(Nu - 1)/\mathrm{d}\log(Ra/Ra_c - 1)$ dropped down to 0.41 at the highest Rayleigh numbers they could achieve.

A natural extension to the scaling laws in the absence of boundary layers is to balance nonlinear advection of heat with transport down the mean gradient,

$$\overline{(\mathbf{u}\cdot\nabla)T'} \sim u_s\frac{\mathrm{d}\bar{T}}{\mathrm{d}s} \quad \text{or} \quad T'_* \sim \frac{\ell}{d}\Delta T_s \qquad [121]$$

Fluid elements carry their temperature until they break after a distance ℓ and merge into their surroundings, which have temperature differing by $\ell\Delta T_s/d$.

With the inertial scaling formulas [114], [115a], [115b], and [117], this gives

$$\begin{aligned} Nu - 1 &\sim \frac{(g\alpha\Delta T_s)^{3/2} d^{1/2}}{\Omega^2\kappa} \quad \text{or} \\ Nu - 1 &\sim E^2 Ra^{3/2} Pr^{-1/2} \quad \text{or} \\ Nu - 1 &\sim E^{4/5} Pr^{-1/5} Ra_Q^{3/5} \end{aligned} \qquad [122]$$

The viscous scaling [116] gives

$$\begin{aligned} Nu - 1 &\sim \frac{g\alpha\Delta T_s d^{1/3}}{\Omega^{4/3}}\frac{(\nu+\kappa)^{1/3}}{\kappa} \sim Ra\left(\frac{1+Pr}{Pr}\right)^{1/3} E^{4/3} \\ &\sim \left(\frac{1+Pr}{Pr}\right)^{1/6} E^{2/3} Ra_Q^{1/2} \end{aligned} \qquad [123]$$

Both these formulas, which apply only when thermal boundary layers are negligible, show that the heat transport is strongly reduced at low *E*, the effect of the vorticity constraint blocking the convection, but $Nu-1$ increases much more rapidly with *Ra* than in nonrotating convection. Eventually, boundary layers develop and [121] is replaced by a formula in which ΔT_s is not the total superadiabatic temperature drop across the layer but the drop across the interior region outside the boundary layers (see Gillet and Jones (2006) for details).

The inertial scaling suggests that viscosity is unimportant in determining the velocity and heat transport. Christensen (2002) made a numerical study of convection in a rotating spherical shell and concluded that the results could be better represented in terms of a modified Rayleigh number and Nusselt number

$$Ra^* = \frac{g\alpha\Delta T_s}{\Omega^2 d}, \qquad Nu^* = \frac{F_{\mathrm{conv}}}{\rho c_p \Delta T_s \Omega d} \qquad [124a]$$

$$Ra_Q^* = Nu^* Ra^* \qquad [124b]$$

This has the advantage that it is independent of all diffusivities. Note that Ra^* is very small in the core, unlike the usual definition which is very large in the core. Now assuming that $Nu \gg 1$, [119] and [122] can be written as

$$\frac{U_*}{\Omega d} = Ro \sim (Ra_Q^*)^{2/5}, \qquad Nu^* \sim (Ra_Q^*)^{3/5} \qquad [125]$$

showing explicitly that the velocities and the convective heat transport are completely independent of diffusion. Christensen (2002) has compared these formulas with the output from numerical simulations. He found that they fit fairly well to the data. When there is no internal heating, the total convective flux across any spherical surface of radius *r* is constant, so $F_{\mathrm{conv}} \sim 1/r^2$. At large *Ra* the distinction between *Nu* and $Nu - 1$ is unimportant, and he defines

$$Ra_Q^* = \frac{r_{\mathrm{icb}} g\alpha F}{r_{\mathrm{cmb}} \rho c_p \Omega^3 d^2} \qquad [126]$$

where *F* is the heat flux per unit area on the outer boundary. The Rossby number is measured by taking the typical velocity as the root mean square poloidal velocity averaged over the outer core. The poloidal velocity is a measure of the convective velocity, as it excludes the zonal flow, which can be much larger. He then finds a good fit for the relations

$$Ro = 0.54(Ra_Q^*)^{2/5} \qquad [127a]$$

$$Nu^* = 0.077(Ra_Q^*)^{5/9} \qquad [127b]$$

The typical velocity therefore agrees well with the inertial scaling, and the power of 5/9 is between the 1/2 of [123] and the 3/5 of [122]. These results are encouraging, though some caution is needed in applying them. Gillet and Jones (2006) find that the viscous scaling gives rather similar power laws to the inertial scaling, and at moderate *Ra* these may give a better fit to simulation data. Also, at fixed E and large *Ra* thermal boundary layers become important and [127b] will break down. However, the core may well be in a regime where E is so small that diffusion is unimportant, and *Ra* is large but not so large that thermal boundary layers determine the heat transport.

5.5 Convection with and without Rotation in the Presence of Magnetic Field

The Earth's core is made up of electrically conducting fluid permeated by magnetic field. These magnetic fields affect convection in many different ways. To simplify the different effects, theoreticians have studied simplified models which highlight the behavior in various circumstances. From the intuition developed by these studies, we hope to be able to understand the interplay of the many different forces at work in core convection. We first distinguish between the dynamo process, which addresses the question of how the field is generated, described in Chapter 3, and the back-reaction that the created field has on the convection. In this section, we do not attempt to describe the dynamo process, but instead assume that a field has been generated by a dynamo, and we study the interaction of this given field with the convection. The study of convection in the presence of an imposed field is usually known as magnetoconvection, whereas the full problem, including generation of field from only a small seed field, is called the hydromagnetic dynamo problem. Magnetoconvection studies therefore have an imposed field. This is usually done by specifying that there is a nonzero magnetic field at the boundaries. Sometimes this field is fixed at the boundaries, sometimes there is a specified amount of flux passing through the boundary which can be moved around but not destroyed. While it is convenient to split the problem into magnetoconvection and dynamo action, there are of course connections between these two problems. The dynamo generated field saturates when the back-reaction of the created field alters the flow pattern into something that no longer generates large quantities of field, so that the field strength is determined by magnetoconvection processes.

The issues we concentrate on here are as follows: (1) If we start with a uniform field, how is the field altered by the flow? (2) If we impose a uniform magnetic field, how is the pattern of convection at onset changed? (3) What magnetic field structures emerge in a fully convecting conducting fluid? The first two questions are essentially linear problems, and so analytic progress is possible. The third question is nonlinear, and so much difficult; we have to rely on a relatively small number of computer simulations.

While much of our understanding is based on the effect of a simple large-scale field on convection, it should be remembered that the field inside the core is not necessarily like this. Indeed, dynamo simulations usually show a rather complex field pattern inside the core, not just the simple dipolar type field seen at the Earth's surface. Indeed, a much more complex pattern emerges even when the external potential field is extrapolated down to the CMB. In consequence, although our intuition is based on the effect of uniform fields, this may not always give us a reliable picture, so we should not be too surprised if dynamo simulations give a rather different picture from that predicted by magnetoconvection.

5.5.1 Flux Expulsion and Flux Rope Formation

We start by assuming that an initially uniform magnetic field permeates a fluid layer. At time $t=0$ a specified velocity field is switched on, and the magnetic field is moved around. What happens to the field? This is called the kinematic flux expulsion problem (Weiss, 1966). It is kinematic because the flow is assumed inexorable, unaffected by the field itself. It is flux expulsion because that is what happens; the flux is expelled from regions where the flow is vigorous into regions where very little shear occurs. The simplest case is that of 2-D roll flow, in a square box of height and width d,

$$\begin{aligned}\mathbf{u} &= \nabla \times \psi \mathbf{1}_z \\ &= \left(-U \sin\frac{\pi x}{d}\cos\frac{\pi y}{d},\ U\cos\frac{\pi x}{d}\sin\frac{\pi y}{d},\ 0\right)\end{aligned}$$

with a magnetic field

$$\mathbf{B} = \nabla \times A\mathbf{1}_z = \left(\frac{\partial A}{\partial y}, -\frac{\partial A}{\partial x}, 0\right),$$

$$A = 0 \text{ on } x = 0, \quad A = B_0 d \text{ on } x = d$$

The initial field is a uniform vertical field, $A = B_0 x$. Note that the boundary conditions on A ensure that the total amount of vertical flux is conserved, so there is no possibility of the field disappearing. The equation for A is the induction equation, which can be written in two dimensions as

$$\frac{\partial A}{\partial t} + \frac{\partial(\psi, A)}{\partial(x, y)} = R_{\mathrm{m}}^{-1}\nabla^2 A$$

Here the magnetic Reynolds number $R_{\mathrm{m}} = Ud/\eta$, where η is the magnetic diffusivity. If R_{m} is small, the magnetic field is not disturbed much from its initial state, as the diffusion term dominates and the initial field satisfies $\nabla^2 A = 0$. If R_{m} is large, flux expulsion occurs, and a typical solution is shown in **Figure 6**, based on Galloway and Weiss (1981), which shows how flux ropes are formed. The figure shows behavior in a square box at a sequence of times in units of the turnover time d/U with $R_{\mathrm{m}} = 1000$.

The concentrated fields that develop at $x = 0$ and $x = d$ are called flux sheets. At large R_{m} their thickness scales as $dR_{\mathrm{m}}^{-1/2}$ (Galloway *et al.*, 1977). This represents a balance of advection and diffusion in the flux sheet, advection into the sheet being at a rate Ud and diffusion out being at a rate $\eta/\delta_{\mathrm{B}}^2$, δ_{B} being the sheet thickness. In three dimensions, an axisymmetric convective flow produces a rope at the center of the cell rather than a flux sheet, but the thickness is still approximately $dR_{\mathrm{m}}^{-1/2}$, though logarithmic factors can enter (Galloway *et al.*, 1977). The magnetic Reynolds number in the core is in the range $10^2 - 10^3$, so we expect flux expulsion to be significant. There is evidence from geomagnetic field observations that flux patches exist, most notably under Alaska and Siberia in the present field. Unfortunately, the width of the patches is rather uncertain, because crustal magnetism prevents us seeing small-scale structures, and it may be that the patches are thinner but more intense than is apparent from the necessarily smoothed out published CMB fields. This is certainly suggested by geodynamo simulations. On the surface of the Sun, the intense turbulence just below the photosphere sweeps the flux into sunspots, which are relatively quiescent regions. Since the convection is reduced in the spots, they appear cooler, and hence dark by comparison with the rest of the solar surface.

5.5.2 Linear Theory of Magnetoconvection in Plane Geometry

We now consider the classic problem of the onset of magnetoconvection in a plane layer of electrically conducting fluid confined between horizontal boundaries at $z = \pm d/2$. The imposed field is uniform, but the convection generally produces a small induced field which can significantly affect the convection. Chandrasekhar (1961) describes in great detail the case where the imposed magnetic field and the rotation axis are parallel to gravity. Many other cases have been studied. Convection near the equator in the presence of a strong azimuthal field might be best modeled by having gravity, rotation, and the magnetic field mutually perpendicular (Roberts and Jones, 2000; Jones and Roberts, 2000).

5.5.2.1 Nonrotating magnetoconvection

We start with the nonrotating case, to isolate the effects of magnetic field alone. If the imposed magnetic field is horizontal, then the convection onsets in the form of 2-D rolls with their axes aligned with the magnetic field. This resembles the alignment of convection rolls with the rotation axis in rotating convection. The magnetic alignment can be simply understood, because if $\mathbf{B}_0$ is the imposed field, and $\mathbf{b}$ is the induced field created by the motion, [5] implies

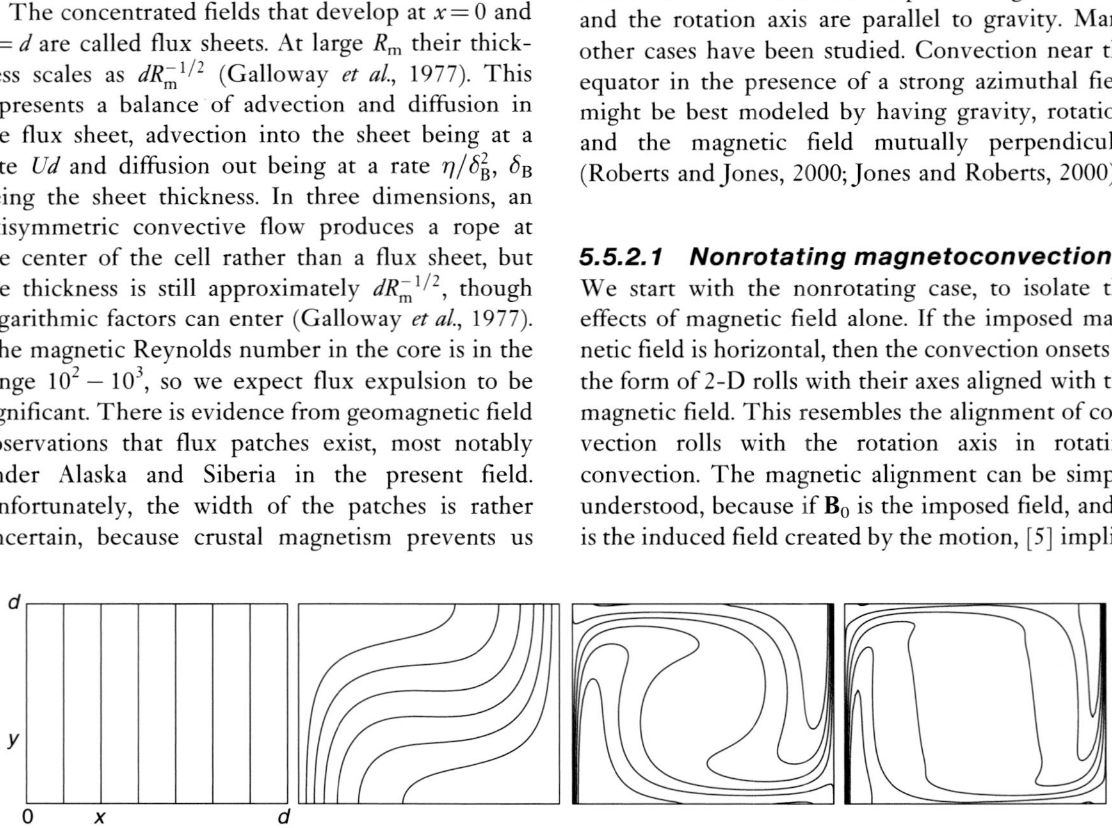

Figure 6 Flux expulsion. The field is initially vertical, and the turnover time is d/U. Four plots shown are at times 0, 0.5, 2.0, and 20.0 turnover times from left to right. The flux is expelled into ropes near the sides of the box.

$$\frac{\partial \mathbf{b}}{\partial t} = \mathbf{B}_0 \cdot \nabla \mathbf{u} + \eta \nabla^2 \mathbf{b} \qquad [128]$$

so if the flow does not vary in the field direction $(\mathbf{B}_0 \cdot \nabla)\mathbf{u} = 0$, and there is no source term and hence $\mathbf{b} = 0$. There is then no Lorentz force, and so convection proceeds as though there was no magnetic field, and the critical Rayleigh number at onset is the same as in the absence of magnetic field.

In contrast, if the field is vertical, then the magnetic field restrains the convection, as indeed occurs in a sunspot. Because of the horizontal boundaries, rising hot fluid has to turn over and so cannot be z-independent, so $(\mathbf{B}_0 \cdot \nabla)\mathbf{u}$ is nonzero, and there is an induced field. This gives rise to a perturbed current $\mathbf{j} = \nabla \times \mathbf{b}/\mu$ and hence a Lorentz force $\mathbf{j} \times \mathbf{B}_0$ in the linearized equation of motion:

$$\frac{\partial \mathbf{u}}{\partial t} = -\nabla\left(\frac{p'}{\rho}\right) + \mathbf{1}_z g\alpha T' + \frac{\mathbf{j} \times \mathbf{B}_0}{\rho} + \nu \nabla^2 \mathbf{u} \qquad [129]$$

The linearized temperature equation (we assume compositional convection behaves similarly) is

$$\frac{\partial T'}{\partial t} = -T'_0 u_z + \kappa \nabla^2 T' \qquad [130]$$

In addition to the velocity components being zero at the boundaries, we also need additional boundary conditions for the magnetic field. The simplest case is for a perfect conductor, as then $b_z = 0$ on the horizontal boundaries, and the horizontal components of the current are zero, which since the current is divergence free implies that $\mathrm{d}j_z/\mathrm{d}z = 0$ on the boundaries (Chandrasekhar, 1961). Together with a condition on the temperature, $T' = 0$ on the boundaries being the usual choice, this provides a convenient set of conditions. Somewhat more realistic for the core are insulating boundary conditions, that is, the material outside the core is assumed to be an insulator. This is a sensible condition for the CMB (though some iron could have leaked into cracks in the mantle, giving a finite conductivity). Then there is no current out of the fluid, so $j_z = 0$, and the horizontal components of $\mathbf{b}$ must match onto an external potential field. In the case where the field on the boundary is naturally decomposed into Fourier modes $\sim \exp(ik_x x + ik_y y)$, this is straightforward to apply, because the external potential field has the form $\exp(-az + ik_x x + ik_y y)$, where $a = \pm\left(k_x^2 + k_y^2\right)^{1/2}$, the plus sign for the field above the layer, the minus sign for the field below. From $\nabla \cdot \mathbf{b} = 0$ it follows that $\mathrm{d}b_z/\mathrm{d}z = \mp a b_z$ on the top and bottom boundaries, respectively. A similar trick works for spherical geometry, but the cases of rectangular or cylindrical containers are much harder, as there is then no simple solution for the external field, which must therefore be calculated numerically.

The Lorentz force opposes the convection, so that the onset of convection is delayed, that is, the critical Rayleigh number is increased. Details are given in Chandrasekhar (1961). The dimensionless parameter that measures the importance of magnetic field is now called the Chandrasekhar number

$$Q = \frac{B_0^2}{\mu \rho \nu \eta} \qquad [131]$$

If this parameter is large, the onset of convection is delayed until $Ra \sim Q^{1/2}$, and when convection occurs it takes the form of tall thin cells, which minimizes the bending of the field lines. The Lorentz force for a general nonuniform field is

$$\frac{1}{\mu}(\nabla \times \mathbf{B}) \times \mathbf{B} = \frac{1}{\mu}(\mathbf{B} \cdot \nabla)\mathbf{B} - \frac{1}{2\mu}\nabla \mathbf{B}^2 \qquad [132]$$

the first term being the curvature and the second the magnetic pressure. The magnetic pressure is not so important in Boussinesq convection, since it merely adds to the fluid pressure, and has no curl. It is the curvature force that impedes convection by opposing the convection rolls so that stronger buoyancy forces are needed to maintain the convection. If the field is oblique to gravity, the horizontal component lines up the convection rolls with the field, while the vertical component raises the critical Rayleigh number.

5.5.2.2 Onset of plane-layer rotating magnetoconvection

Magnetic field on its own either opposes convection, or is neutral to it in the case of horizontal field. However, in the presence of rotation it is possible for the magnetic field to break the Proudman–Taylor constraint and hence lower the critical Rayleigh number. The linearized equation of motion is now

$$\frac{\partial \mathbf{u}}{\partial t} + 2\boldsymbol{\Omega} \times \mathbf{u} = -\nabla\left(\frac{p'}{\rho}\right) + \mathbf{1}_z g\alpha T' + \frac{\mathbf{j} \times \mathbf{B}_0}{\rho} + \nu \nabla^2 \mathbf{u} \qquad [133]$$

where again $\mathbf{B}_0$ is a uniform magnetic field and $\mathbf{j}$ is the small current induced by the small velocity. The case where the gravity, rotation, and applied field are all vertical (appropriate for the polar regions in the core) was studied by Chandrasekhar (1961). The rapid rotation limit was examined by Eltayeb (1972). The case where the rotation, applied field, and gravity are mutually perpendicular was studied by Roberts and Jones (2000) and Jones and Roberts (2000). When the field and rotation are in the same direction, not surprisingly the convection rolls have their axes parallel to this direction. In the case where there is an angle between the rotation and magnetic field vectors, it is possible for the rolls to line up with either the rotation or the magnetic field, or in some circumstances to be oblique to both (see Roberts and Jones (2000) for details).

It is most useful to consider the case of rapid rotation, that is, small E, with the field gradually increasing in strength. A key parameter is the Elsasser number

$$\Lambda = \frac{B_0^2}{\mu\rho\eta\Omega} \quad [134]$$

At the onset of convection, the induction equation gives a relation between the typical perturbed field b_* and the velocity U_*:

$$\frac{B_0 U_*}{d} \sim \frac{\eta b_*}{d^2} \quad [135]$$

which means that the typical value of the Coriolis force and the Lorentz force balances when $\Lambda \sim 1$. Note that this is not necessarily the case far from onset.

In the polar regions, where gravity, rotation, and applied field are all approximately parallel, a non-magnetic mode onsets first until the Elsasser number reaches $\Lambda_c = 2^{1/3}\pi^{2/3}E^{1/3}$ in the case of stress-free, thermally conducting and electrically insulating boundaries. Note that at small E this is a very weak field, because $E^{1/3}$ is small. For $\Lambda < \Lambda_c$ these non-magnetic modes have the form of tall thin columns with wavelength $O(E^{1/3})$ perpendicular to the rotation axis. However, as Λ is increased past Λ_c, there is a sudden transition to a magnetic mode of convection, which has horizontal wavelength the size of the layer depth. If Λ is increased further, the horizontal wavelength starts to decrease again, as in the non-rotating case, and the critical Rayleigh number starts to rise. There is therefore an optimum field strength which minimizes the Rayleigh number and has critical wavelengths of order d. The case of a vertical field is slightly exceptional, because with more general field orientations the optimum Rayleigh number occurs when the Elsasser number is of order unity (Eltayeb, 1972).

5.5.2.3 Waves in the core

Equations [128]–[130] can be used to study wave motion in the core as well as convection. We ignore the boundary conditions, assume a uniform field, and look for solutions of the form $\exp i(\mathbf{k}\cdot\mathbf{x} - \omega t)$ which represent traveling waves. Substituting this form into the equations gives the dispersion relation between $\mathbf{k}$ and ω. If the temperature gradient is stabilizing (subadiabatic), then real values of ω are found if diffusion is neglected. If diffusion is retained, ω is complex with a negative imaginary part corresponding to damped oscillations. If the temperature gradient is superadiabatic, growing waves (instabilities) are found provided the diffusion is not too large.

The full dispersion relation is complicated, but it can be simplified when, as in the Earth's core, different types of waves have very different frequencies (Fearn *et al.*, 1988). It also helps to ignore diffusion in the first analysis. The fastest waves are inertial waves which balance the Coriolis and inertial accelerations as in [59], and the inertial wave frequency is given by

$$\omega_C = 2(\mathbf{\Omega}\cdot\mathbf{k})/|\mathbf{k}| \quad [136]$$

The typical period is therefore of the order of a day, though for columnar modes with $\mathbf{k}$ almost perpendicular to $\mathbf{\Omega}$ they are slower. These waves have not yet been observed in the core, but they are expected to be driven by tidal forcing. Alfvén waves result from a balance of inertia and Lorentz force in the equation of motion, when combined with the induction equation. It is useful to think of the magnetic field lines as stretched strings, with magnetic curvature forces, [132], providing a restoring force whenever the field lines are bent. These waves have frequency

$$\omega_M = (\mathbf{B_0}\cdot\mathbf{k})/(\mu\rho)^{1/2} \quad [137]$$

and they travel at the Alfvén speed, $\mathbf{B}_0/(\mu\rho)^{1/2}$, which for a moderate 1 mT core field is around 10^{-2} m s^{-1}, giving around 60 years for the wave to travel round the core. An important class of Alfvén waves are those corresponding to azimuthal motion constant on cylinders, which are called torsional oscillations (Braginsky, 1967). These are believed to be important in the core and are probably responsible for decadal length of day variations (Jault *et al.*, 1988; Jackson, 1997; Jault, 2003), and possibly shorter period elements of the geomagnetic secular variation, such as the so-called geomagnetic jerks.

Another timescale comes from the temperature gradient, from the balance of buoyancy and inertia in the equation of motion, combined with the temperature equation [130] with $\kappa = 0$. The frequency of internal gravity waves is $\omega_A = (g\alpha T'_0)^{1/2} k_H/|\mathbf{k}|$, where T'_0 is the subadiabatic temperature gradient, and k_H is the component of $\mathbf{k}$ perpendicular to gravity. In a convectively unstable region, $T'_0 < 0$, the temperature gradient is superadiabatic. Then ω_A is imaginary, which corresponds to an exponentially growing unstable mode, with $|\omega_A|$ being the growth rate. With an estimate of 10^{-4} K for a typical superadiabatic temperature perturbation, $-T'_0 \sim 10^{-4}/d$, giving a typical growth rate of about 5 months.

When all the nondiffusive terms in the equations are present, the dispersion relation gives a fast inertial wave and a slow wave in which only the time derivative terms in the induction and temperature equations are important, inertia being negligible. These slow waves are known as MAC waves and have frequency

$$\omega_{MAC} = \omega_{MC}(1 + \omega_A^2/\omega_M^2)^{1/2}$$
$$\omega_{MC} = \frac{\omega_M^2}{\omega_C} \sim \frac{B_0^2}{\mu\rho\Omega d^2} \quad [138]$$

In the superadiabatic case, when $|\omega_A| > |\omega_M|$, ω_{MAC} is imaginary, corresponding to a growing convective mode. The rate of growth is slow, since $\tau_{MC} = 2\pi/\omega_{MC}$ is of the order of some thousands of years. It is comparable to the magnetic diffusion time, since the ratio $\tau_{diff}/\tau_{MC} = B_0^2/\mu\rho\Omega\eta = \Lambda$ and the Elsasser number Λ has a value of $O(1)$ in the core. The MAC growth rate will be a little larger than $1/\tau_{MC}$ because $|\omega_A| > \omega_M$, but these slow growth times are consistent with the time taken for the typical convective velocity to take fluid across the core, 200 years, so the dynamical picture does seem to be self-consistent.

5.5.2.4 Small-scale dynamics and the Braginsky–Meytlis theory of plate-like motion

Although for general wave vectors $\mathbf{k}$ the MAC wave timescale is slow, it is much faster if $\mathbf{k}$ and $\boldsymbol{\Omega}$ are perpendicular. This is the case for columnar motion independent of z, the coordinate parallel to the rotation axis. For these motions, ω_{MC} is singular, and we have to restore inertia, leading to the much faster torsional wave frequency. It is also possible for $\mathbf{k}$ to be perpendicular to both $\boldsymbol{\Omega}$ and $\mathbf{B}_0$. This type of motion does not bend the field lines, and so has no Alfvén wave behavior. For these motions with $\mathbf{k}$ in the direction $\boldsymbol{\Omega} \times \mathbf{B}_0$ neither magnetic field nor rotation can impede the growth of convective instability. In a Boussinesq fluid, waves have $\mathbf{k} \cdot \mathbf{u} = 0$, so such waves have motion only along 'plates', planes containing the rotation vector and the magnetic field vector. If there is an unstable temperature gradient in a direction parallel to these planes, instability will grow on the much shorter ω_A timescale. Large-scale motion along plates may be blocked by the boundary curvature, and therefore 'feel' the rotation, but small-scale convection in the core will be preferentially in the form of plate-like motions aligned with the rotation and magnetic field. Since small-scale turbulence in the core is driven primarily by convective instability rather than turbulent cascade, Braginsky and Meytlis (1990) argued that this would mean that turbulence is likely to be highly anisotropic. This is a cause for concern, because most dynamo simulations assume an isotropic eddy diffusion. Introducing an anisotropic diffusion adds considerably to the computational cost of a dynamo code, though some work has been done (e.g., Matsushima, 2005).

A promising approach to the problem of how to include small-scale behavior into dynamo models is subgrid scale modeling. A number of approaches have been tried (Buffett, 2003), but the similarity method is perhaps currently the most successful (Chen and Glatzmaier, 2005). The idea is to filter simulation runs at two different filter scales in order to estimate the effect of the smallest scales present in the run. Then the similarity assumption is made, which comes down to assuming that the unresolved scales behave similarly to the smallest resolved scales. This allows an extrapolation to be made that takes into account the unresolved scales; details are given in Chen and Glatzmaier (2005). The advantage of this method is that if the smallest resolved scales are anisotropic, then the correction from the unresolved scales will also be anisotropic, so Braginsky–Meytlis ideas could be captured by this approach. On the down side, if there are new modes entering the problem at the subgrid scale level, they will not be detected by the similarity method, and so any effect they have will be missed in the simulation. The application of subgrid scale methods to core convection is very recent, and it will be interesting to see how successful they are.

5.5.2.5 Boundary layers in rotating, magnetic fluids

The presence of magnetic field affects the Ekman boundary layer structure discussed in Section 5.4.4.

The simplest problem is to consider the boundary layer just above a plane wall $z=0$ with rotation and an imposed magnetic field B_0 in the z-direction. The horizontal flow in the boundary layer then induces horizontal field there. We seek a steady solution, $\partial \mathbf{B}/\partial t=0$, so by Maxwell's equations $\nabla \times \mathbf{E}=0$, and we assume no externally imposed electric field, so $\mathbf{E}=0$. Then from Ohm's law,

$$\mathbf{j} = \frac{1}{\mu\eta}\mathbf{u} \times \mathbf{B} \qquad [139]$$

There is some velocity normal to the wall in an Ekman layer (see Section 5.4.4) but it is small, $O(E^{1/2})$, and so to leading order the horizontal components of $\mathbf{j}$ are

$$\mathbf{j} = \frac{u_y B_0}{\mu\eta}\mathbf{1}_x - \frac{u_x B_0}{\mu\eta}\mathbf{1}_y \qquad [140]$$

The curl of the steady momentum equation is

$$-2\Omega\frac{\partial \mathbf{u}}{\partial z} = \frac{1}{\rho}B_0\frac{\partial \mathbf{j}}{\partial z} + \nu\frac{\partial^2 \boldsymbol{\zeta}}{\partial z^2} \qquad [141]$$

since the term involving $(\mathbf{j}\cdot\nabla)\mathbf{B}$ is negligible, because the boundary layer is thin (formally this requires the boundary layer to be thinner than $R_{\mathrm{m}}^{-1/2}$ where R_{m} is the magnetic Reynolds number based on the large horizontal length scale). The x- and y-components of [141] give

$$-2\Omega\frac{\partial}{\partial z}(u_x + \mathrm{i}u_y) = \frac{-iB_0^2}{\rho\mu\eta}\frac{\partial}{\partial z}(u_x + \mathrm{i}u_y) + \mathrm{i}\nu\frac{\partial^3}{\partial z^3}(u_x + \mathrm{i}u_y) \qquad [142]$$

Seeking solutions of the form $u_x+\mathrm{i}u_y \sim \exp \lambda z$ gives a zero root (allowing matching to the interior flow) and roots satisfying

$$\lambda^2 = \frac{B_0^2}{\rho\mu\eta\nu} + \frac{2\mathrm{i}\Omega}{\nu} \qquad [143]$$

When $B_0=0$ this gives an Ekman layer of thickness $(\nu/\Omega)^{1/2}$, and when $\Omega=0$ it gives a Hartmann layer of thickness $(\rho\mu\eta\nu)^{1/2}/B_0$. These layers are of comparable thickness when the Elsasser number (see [134]) is order unity. In this case the layer thickness is determined by [143] and the layer is called an Ekman–Hartmann layer.

The stability of Ekman–Hartmann layers has been analyzed (see, e.g., Desjardins *et al.* (1999)). An analysis of Ekman–Hartmann layers in spherical geometry was given by Loper (1970).

5.5.3 Onset of Rotating Magnetoconvection in Spherical Geometry

The onset of instability in the presence of a magnetic field is studied by techniques similar to those described in Section 5.4.2. Most work has been done for magnetic fields of the form $\mathbf{B}=B_0 s\mathbf{1}_\phi$, the so-called Malkus field (Malkus, 1967; Fearn 1979a, 1979b; Jones *et al.*, 2003), and the case of a uniform field parallel to the spin axis (Sakuraba, 2002). In addition to expanding the velocity in toroidal and poloidal scalars, [60], we also expand the magnetic field as

$$\mathbf{b} = \nabla \times \mathcal{U}\mathbf{r} + \nabla \times \nabla \times \mathcal{V}\mathbf{r} \qquad [144]$$

As usual, the dynamical equations are treated by taking the r components of the curl and double curl of the momentum equation, giving equations for the toroidal and poloidal components of $\mathbf{u}$. We take the radial components of the induction equation and its curl to get equations for $\mathcal{U}$ and $\mathcal{V}$. The temperature equation completes the set. Details are given in Zhang (1995). Although at first sight this formulation seems to involve a great many curl operations, the resulting equations are surprisingly straightforward, particularly if expansion in spherical harmonics P_l^m is adopted, so that the horizontal part of the ∇^2 operator reduces to multiplication by $-l(l+1)$. The Malkus form of the field is particularly attractive because it only couples adjacent spherical harmonics together, as does the Coriolis term. This means that the equation for P_l^m only involves the coefficients in front of P_{l-1}^m, P_l^m, and P_{l+1}^m. In consequence, the matrices for the linear eigenproblem have a banded structure, which makes it much easier to find their eigenvalues. If a more general form of the field is adopted, all the different l components are coupled, so the eigenvalues of a dense matrix must be found. Of course, if a nonaxisymmetric field were considered the problem would be even worse, because then modes with different m are coupled. Solving such a problem would be as difficult as the full nonlinear dynamo, and so it is not surprising that little is known about magnetoconvection in fields with complex geometries.

The effect of the magnetic field is measured by the Elsasser number, Λ, see [134]. If Λ is very small, the magnetic field has little effect, and the convection onsets in the usual manner for rotating convection, that is, columnar modes are preferred. The first noticeable effect is when Λ is $O(E^{1/3})$. This is a rather small value, and core values of Λ must certainly be greater than this. For values in the weak field range, $O(E^{1/3}) < \Lambda < O(1)$, the convection is still columnar, but the critical value of m is significantly reduced, indeed the preferred value drops from $O(E^{-1/3})$ down to $O(1)$ in this range. So we still have columnar convection, but the rolls are now much fatter due to the presence of the magnetic field. In this range the critical Rayleigh number typically first increases with Λ, then reaches a maximum after which it falls significantly (Fearn, 1979b). This shows that magnetic field can break the Proudman–Taylor constraint and hence be favorable for convection. The fattening of the rolls due to magnetic field may be important in core convection. The expected roll thickness without magnetic field is extremely thin due to the very low value of the Ekman number. The actual value in the core is difficult to estimate; magnetoconvection calculations such as these suggest that there might be comparatively few (less than a dozen) but this is a linear theory valid only at onset, and low E nonlinear dynamo simulations suggest considerably more rolls than this.

Another feature of magnetoconvection is that the frequency at onset drops rapidly as Λ increases (Fearn, 1979b), that is, the phase speed of propagation is much slower in the presence of magnetic field. It is also possible for the waves to propagate westward, rather than eastward, as generally seems to be the case in the core. The westward drift is not, though, a universal feature of the secular variation, as the Pacific hemisphere has no strong westward tendency, whereas the Atlantic hemisphere has. The phase frequency actually drops to the inverse thermal diffusion time in magnetoconvection for the main thermal mode of convection (Fearn, 1979b), but there are other modes where the phase speed is the inverse magnetic diffusion time (Zhang and Jones, 1996). At small $q = \kappa/\eta$ these timescales are very different. Quite surprisingly, the Malkus field can be unstable for negative Rayleigh number (Roberts and Loper, 1979). The instability is driven by magnetic field energy being converted into fluid motion, and it is these magnetic modes which have a frequency corresponding to the inverse magnetic diffusion time.

If magnetic field is increased into the $\Lambda \sim O(1)$ regime, the critical Rayleigh number typically has a minimum, and then starts to increase when Λ becomes large. There are however exceptions to this behavior when the applied field is more complicated: thus, if the applied toroidal field is antisymmetric about the equator (as may well be the case in the core), at large Λ the critical Rayleigh number continues to decrease as Λ increases (Zhang and Jones, 1994), and Ra_c goes negative because convection is then driven by magnetic instability (see Section 5.5.4 below) rather than by thermal or compositional buoyancy, so these unstable modes have the character of the magnetic modes mentioned above.

In the weak field case, it is possible to use the same asymptotic techniques as discussed in Section 5.4.3 to find solutions at very low E (Jones *et al.*, 2003). This also provides a check on the numerical codes. As in nonmagnetic rotating convection, the corrections to the critical Rayleigh number due to the formation of boundary layers are small, but they can be calculated.

In the case where the applied magnetic field is axial and uniform, Sakuraba (2002) found broadly similar behavior to that found in the Malkus field, though here the broadening of the columns only occurs at $\Lambda \sim O(1)$ rather than at $\Lambda \sim O(E^{1/3})$, probably because in this model magnetic field and rotation are parallel. In addition to the almost geostrophic columnar modes, another polar mode of convection was found, which has the opposite equatorial parity to the geostrophic mode, that is, the axial vorticity is antisymmetric about the equatorial plane while the axial velocity is symmetric. This mode tends to be strongest in the polar regions (see also Section 5.5.6.2 below).

5.5.4 Magnetic Instabilities

It is well known that magnetic instabilities make it very difficult to confine plasma in fusion devices, and we might also expect that magnetic instabilities are significant in the core. A valuable review on magnetic instabilities is given in Fearn (1998). Magnetic instabilities require a nonuniform field, so the simple uniform field models usually employed for convection models are not subject to them, though as noted above, even the very simple Malkus field can be destabilized by the addition of a stable thermal gradient. The simplest models with magnetic instability are those where the basic field is assumed to be azimuthal. The two cases which have received most attention are (1) $\mathbf{B} = B(s)\mathbf{1}_\phi$ and (2) $\mathbf{B} = B(s, z)\mathbf{1}_\phi$. The first of these is often studied in cylindrical geometry, using various profiles $B(s)$, for example, Fearn (1988). The advantage of this is that the linearized stability problem has coefficients which are functions of s only, so that disturbances of the form $\exp i(k_z z + m\phi - \omega t)$ can be assumed, leading to a simple 1-D eigenvalue problem in s for the complex eigenvalue ω. The disadvantage of this simple model is that the toroidal field of the Earth is most likely to be antisymmetric about the equator, and so must be a function of z. These models become magnetically unstable at lower values of the Elsasser number than models in which B is only a function of s, typically $\Lambda \sim O(10)$ rather than $O(100)$ (see Zhang and Fearn (1994, 1995)). These instabilities can interact with thermal convection, and as noted above can have a significant effect in magnetoconvection models (Zhang and Jones (1994)). Some magnetic instabilities can function in the absence of diffusion, and these are known as ideal instabilities, while the so-called resistive instabilities require diffusion and typically have a slower growth rate. In the core, where the magnetic diffusion time is similar to the inverse ideal MAC wave frequency, the distinction between these modes is not so clear-cut. It is also possible for differential rotation to provide a source of instability, and as one might expect differential rotation can interact with magnetic instabilities to affect the critical Elsasser numbers for onset (Fearn, 1998).

5.5.5 Taylor's Constraint

In Section 5.5.2.3 we noted that torsional waves in the core have a frequency of around 60 years for a moderate 1 mT core poloidal field. This is much faster than the rate of evolution of the large-scale field, so that if we average over this longer timescale we expect the forces that excite geostrophic torsional oscillations to be in equilibrium. Assuming that the velocities in the core are of order those given by the secular variation, the Reynolds stresses are comparatively small in the core, and we ignore them here. Then we integrate the ϕ-component of [37] over cylinders of radius s. The Coriolis term reduces to an integral of $2\Omega u_s$ over the cylinder, which is zero because in a Boussinesq fluid ($\nabla \cdot \mathbf{u} = 0$) there can be no net flow through these cylinders. The ϕ-component of the pressure also integrates to zero, and the buoyancy has no ϕ-component, so we get

$$\frac{\partial}{\partial t}\int_{C(s)} \rho u_\phi \, \mathrm{d}S = \int_{C(s)} \mathbf{j} \times \mathbf{B} \cdot \mathbf{1}_\phi \, \mathrm{d}S + \rho\nu \int_{C(s)} \mathbf{1}_\phi \cdot \nabla^2 \mathbf{u} \, \mathrm{d}S \qquad [145]$$

The viscous term is usually dominated by the contribution from the Ekman boundary layers shown in [110], and is small in the core compared to the Lorentz force term. On the dynamo timescale we expect the time-dependent term to average to zero, so we expect Taylor's constraint (Taylor, 1963)

$$\int_{C(s)} \mathbf{j} \times \mathbf{B} \cdot \mathbf{1}_\phi \, \mathrm{d}S = 0 \qquad [146]$$

to be satisfied. The essential point is that this constraint depends on the form of the magnetic field. It is not satisfied by $\mathbf{B}$ being small, but rather by positive and negative parts of the integrand exactly canceling each other out. The magnetic fields arising from the linear solutions of the dynamo equations or the induced fields from magnetoconvection problems will not in general satisfy Taylor's constraint. A state in which Taylor's constraint is not satisfied is called an Ekman state (see, e.g., Fearn (1998)) and in such a state the magnetic field strength has to be small, controlled by the Ekman suction. Malkus and Proctor (1974) suggested that a geostrophic flow would develop which alters the induction process until a field satisfying Taylor's constraint is generated; this type of field is called a Taylor state. The Malkus–Proctor scenario was shown to occur in the context of α-effect dynamos by Soward and Jones (1983), with Taylor states emerging. However, for a plane layer magnetoconvection problem, Jones and Roberts (1990) were able to show that no Taylor state emerged as the Rayleigh number was increased so the Malkus–Proctor scenario is not necessarily universal, while Hollerbach *et al.* (1992) found some evidence of an approach to an inviscid state, but no true Taylor state emerged. In numerical dynamo simulations, it is difficult to obtain solutions at low E, so that the viscous term is often still significant in [145]. The measure of whether a Taylor state has been achieved is called the Taylorization (Anufriev *et al.* (1995)):

$$\mathrm{Tay} = \frac{\int_{C(s)} \mathbf{j} \times \mathbf{B} \cdot \mathbf{1}_\phi \mathrm{d}S}{\int_{C(s)} |\mathbf{j} \times \mathrm{B} \cdot \mathbf{1}_\phi| \mathrm{d}S} \qquad [147]$$

In an Ekman state, Tay is of order unity, but when a Taylor state is achieved, Tay is small. Strong evidence that Taylorization occurs in a convectively driven plane layer dynamo was given by Rotvig and Jones (2002) and recent work by Aubert (2005) reports that at $E = 10^{-4}$ the cylindrical average of $\mathbf{j} \times \mathbf{B} \cdot \mathbf{1}_\phi$ was only one-seventh of its maximum value suggesting that spherical dynamo simulations are approaching a Taylor state too.

As mentioned in Section 5.5.2.3, torsional oscillations are believed to occur in the core. These can be thought of as oscillations about the Taylor state, but it is not currently known what is exciting them. In dynamo models, Reynolds stresses (see Eqn [110]) are significant, and these give rise to some torsional oscillations (Dumberry and Bloxham, 2003), so that at any instant in time, a dynamo model will not exactly obey Taylor's condition, though it should do if averaged over time. In the core, Reynolds stresses are thought to be small, but Dumberry and Bloxham (2003) argue that they may nevertheless play a role in exciting torsional oscillations.

5.5.6 Numerical Simulations of Nonlinear Convection-Driven Dynamos

The next natural step after considering linear magnetoconvection is nonlinear magnetoconvection. However, the equations that need to be solved for this problem are the same as for numerical simulations of the convection-driven dynamos, namely [5], [6], [28], [37], and [38]. Usually, compositional effects are assumed to give similar results to thermal effects, so ξ' is set to zero in [37]. Most recent work has concentrated on the full dynamo problem rather than the nonlinear magnetoconvection problem. Indeed, the only essential difference is that in the magnetoconvection problem an imposed magnetic field is added at the boundaries, so that it is not necessary to have a configuration which has growing dynamo modes. However, it transpires that in convecting rotating spherical shells, dynamo action is surprisingly easy to achieve, so the incentive for doing magnetoconvection calculations is correspondingly reduced. The ease with which rotating spherical convection models produce dynamo action came as somewhat of a surprise, given that many simple flows do not give dynamo action even at large R_{m} (*see* Chapter 3). However, spherical geometry is not the easiest configuration in which to study either magnetoconvection or dynamo action, and so the Busse annulus geometry has also been used for understanding nonlinear convection in the presence of a magnetic field (see, e.g., Kurt *et al.* (2004) and references contained therein).

The four basic parameters needed to define a convective dynamo are the Rayleigh number, the Ekman number, the Prandtl number, and the magnetic Prandtl number ν/η. The parameter range in which it is feasible to run convective dynamo simulations is restricted. Currently, simulations are restricted to $E > 10^{-6}$, $\kappa/\eta > 0.05$, and $R/R_{\mathrm{c}} < 100$ (R_{c} being the critical value in the absence of magnetic field), though these values are being improved all the time due mainly to faster computers, but also partly to improved numerical methods. The first successful self-consistent dynamo models were designed for solar and stellar convection (Gilman and Miller, 1981; Gilman, 1983; Glatzmaier, 1984). Zhang and Busse 1989 found steady drifting roll solutions which sustained a magnetic field. The first fully time-dependent low Ekman number dynamos made use of hyperdiffusion (Glatzmaier and Roberts, 1995, 1997) or had very restricted resolution in ϕ (Jones *et al.*, 1995), but with increasing computer speed these drawbacks are no longer necessary. Jones (2000) reviewed models up to that date, but many more recent papers have appeared. The models have been surprisingly successful at reproducing many features of the observed field. They are frequently dipole dominated, they can give occasional reversals rather reminiscent of the behavior of the actual Earth, the field strength produced is the right order of magnitude and the secular variation is broadly similar to geomagnetic secular variation. Of course, varying the input parameters gives different results, but it is nevertheless rather encouraging that it is possible to make contact with geophysical observations from first-principles solutions of the fundamental equations. Although there is much still to discover about the relation between the mathematical solutions of the equations and the geophysical observations, it does appear that the fundamental physics behind the models is correct.

A particularly simple convective dynamo, at a rather modest set of parameter values, has been adopted as a benchmark (Christensen *et al.*, 2001) against which to test dynamo codes. This is necessary, as the dynamo computer programs are very complex, allowing many possibilities for coding error. A snapshot of the field and flow produced by a typical dynamo code is shown in **Figures 7(a)** and **7(b)**. Details of the boundary conditions and definitions of the parameters are in Sreenivasan and Jones (2006a). A modified Rayleigh number, often used in dynamo simulations and defined as $R = g\alpha\Delta T_{\mathrm{s}}d/\eta\Omega$, is used. These simulations are for a fairly modest value of $R = 750$.

The flow pattern is columnar, not very different from that shown in **Figure 3** which is for the linear onset of nonmagnetic convection. **Figures 7(a)** and **7(b)** are for $E = 10^{-4}$; reducing E gives more, thinner columns. The pattern drifts westward. The magnetic field seems to be primarily generated in the convection columns (Olson *et al.*, 1999), and indeed the strongest field is found near the tangent cylinder,

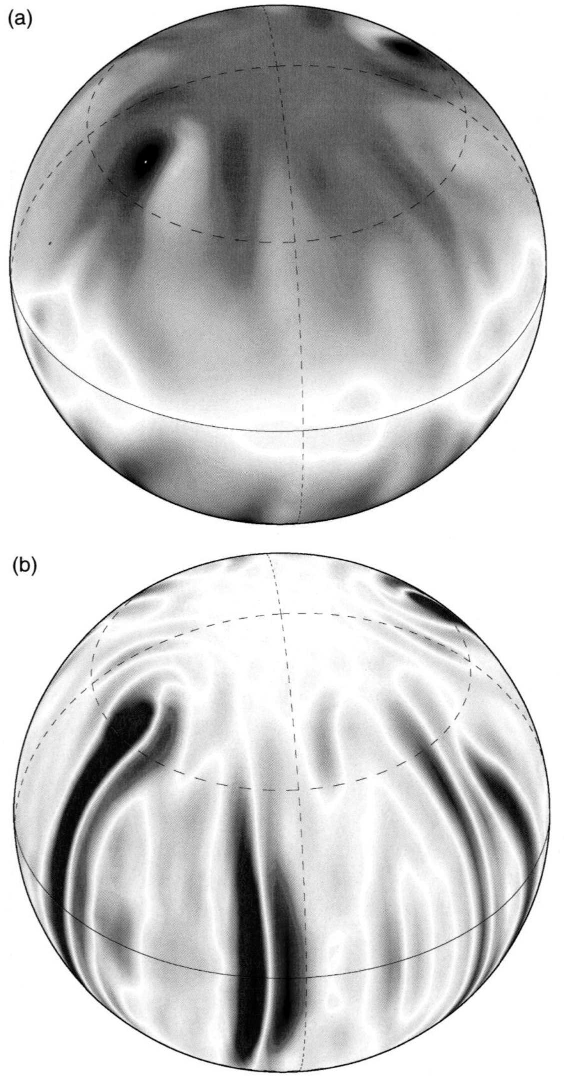

Figure 7 Contour plots from a dynamo simulation with $R = 750$, $E = 10^{-4}$, $Pr = Pm = 1$. (a) Shaded contours of B_r at the CMB. (b) Shaded contours of u_r at $r = 0.8r_{cmb}$. Reproduced from Sreenivasan and Jones (2006a) The role of inertia in the evolution of spherical dynamos. *Geophysical Journal International* 164: 467–476.

coming out of the tops of the rolls. Magnetic field near the equator is concentrated in the anticyclonic rolls (Kageyama and Sato, 1997), which then expands the anticyclonic rolls leading to westward flow near the CMB and eastward flow near the ICB (Sakuraba and Kono, 1999). Models in which the driving is from the boundaries only (with no internal heat source) (e.g., Christensen *et al.*, 1999) differ in a systematic way from models with a uniform heat source (e.g., Grote and Busse, 2000; Busse, 2002). Models with no internal heating are typically much more dipolar, while with internal heating quadrupolar dynamos are almost as common as dipolar ones. Generally, when the convective columns are concentrated close to the tangent cylinder, dipolar dynamos are more common, whereas with distributed heating, convective columns can occur well away from the tangent cylinder and the resulting dynamos are much less dipole dominated.

Another interesting issue is the importance of inertia in dynamo models. At the modest values of E possible, if the Prandtl number is small, inertia plays a significant role in the dynamo process (Simitev and Busse, 2005; Sreenivasan and Jones, 2006a; Christensen and Aubert, 2006). The dynamo-generated field is then typically weaker, much more complex and no longer dipole dominated. Velocity estimates from the secular variation suggest inertia is not important at large scales in the core, and the dynamo simulations also suggest that the dipolar structure of the Earth is more easily reproduced in the low inertia regime. Having said that, it should be noted that the low inertia, strongly dipole-dominated models show no sign of reversing. To get reversals, it is generally necessary to increase the Rayleigh number (and not to have E too small) to produce a more strongly time-dependent and less dipole-dominated field. Although dynamo models give rise to reversals (Glatzmaier and Roberts, 1995; Sarson and Jones 1999; Kutzner and Christensen, 2002), there is no general agreement about how they are achieved. Sarson and Jones (1999) noted that the reversal typically takes place first in one hemisphere, suggesting the growth of a time-dependent quadrupolar mode, cancelling out the dipolar field in one hemisphere, strengthening it in the other. The field can then reverse in one hemisphere, and this reversed flux can lead to a complete reversal. They also suggested that meridional circulation might be important in controlling the reversal process.

As we learn more about convective dynamos, models which might be suitable for simulating the magnetic fields found on other planets are being developed (Jones, 2003; Stevenson, 2003). There is considerable variety in the types of magnetic fields found on other planets. For example, the field on Mercury is particularly weak, while the fields on Uranus and Neptune are not dipolar dominated but appear to be a mixture of dipolar and quadrupolar fields with no strongly preferred parity (Holme and Bloxham, 1996; Stanley and Bloxham, 2006). Developing planetary dynamo models is an area likely to develop rapidly over the next few years.

5.5.6.1 Convection outside the tangent cylinder: picture from numerical simulations

The inner core appears to have an important effect on the dynamics of convection, and the behavior of convection outside the tangent cylinder (TC) is rather different from that inside the TC. As mentioned in the introduction, the large amount of vortex stretching required makes it difficult for fluid to cross the TC, though magnetic field may help to allow some transport. Dynamo simulations suggest that the pattern of convection outside the TC is not strongly influenced by magnetic field, the columnar pattern found in nonmagnetic simulations persisting, and the rolls have the same parity, that is, axial vorticity symmetric and axial velocity antisymmetric about the equator (see **Figure 7(b)**). The magnetic field slows down the drift rate, as expected from magnetoconvection calculations, but the roll width is not increased as much as would be expected from linear magnetoconvection theory. Indeed, the roll thickness seems to go down with E according to the $E^{1/3}$ scaling.

In **Figures 8(a)** and **8(b)**, which are outputs from a dynamo simulation run by Dr B. Sreenivasan, the Ekman number $E = 3 \times 10^{-6}$, and comparing **Figure 8(b)** and **7(b)** we can see that the columnar structures are indeed much thinner in the lower E case, despite the fact that both have fields with Elsasser number of order unity. Possibly, this is due to the field being quite nonuniform. It is known that in nonlinear magnetoconvection without rotation that once convection starts, flux expulsion occurs from the convecting region and the heat transport is then relatively unimpeded by the field. Nevertheless, we should remember that in the simulations it is not possible to get to the very thin roll regime, and that on small scales when the local magnetic Reynolds number gets small we expect the magnetic field to be smooth, so it may be that magnetic field does thicken the rolls at very low Ekman number.

As the Rayleigh number is increased, the velocities increase and the flow becomes more chaotic. The Earth's core has low $q = Pm/Pr$, so it is of interest to see what happens to the convection pattern at low q. This is a hard regime to reach, because the magnetic Reynolds number has to be greater than about 50 to get dynamo action, but low q then implies very high Péclet number $U_* d/\kappa$, which makes it difficult to get resolved simulations. Nevertheless at sufficiently low E (typically about 10^{-6}), it is possible to get dynamos with $q \sim 0.1$ (Christensen and Aubert, 2006) and see also **Figures 8(a)** and **8(b)**, which are for $Pr = 1$, $Pm = 0.1$. Note that although the Rayleigh number here is 50 times critical, because of the low E, the magnetic Reynolds number $R_{\mathrm{m}} \approx 125$, low for the core. This indicates that we need higher Ra, but unfortunately this is very expensive computationally. Note that these low E dynamos are remarkably

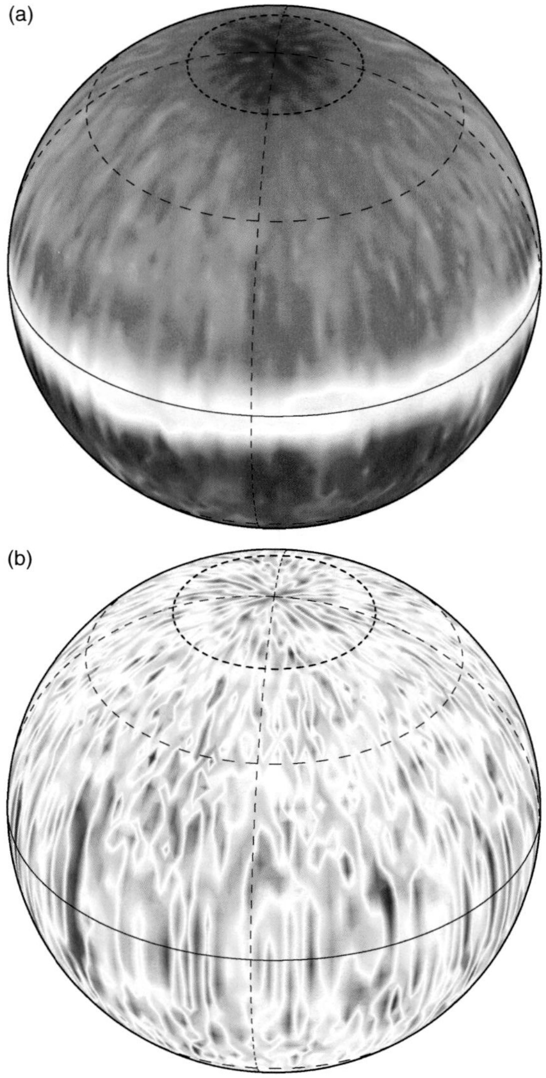

Figure 8 Contour plots from a dynamo simulation with $R = 50R_c, E = 3 \times 10^{-6}, Pr = 1, Pm = 0.1$. (a) Shaded contours of B_r at the CMB. (b) Shaded contours of u_r at $r = 0.8r_{\mathrm{cmb}}$. These figures were supplied by Binod Sreenivasan.

dipole dominated (**Figure 8(a)**), even more so than the field shown in **Figure 7(a)**. Possibly higher *Ra* would introduce a more chaotic flow, which might help to reduce this strong dipole dominance, which prevents reversals occurring.

Figures 8(a) and **8(b)** show that at small q the flow structure is finer than the field structure, as we expect in the core. It is unfortunate, though, that this does not happen until very low E is reached, because the very long CPU time required to perform the simulations in this regime makes it very difficult to explore the parameter space. If the value of Pr is reduced, then inertial effects become important, while if Pr is increased, the roll thickness becomes even smaller, as predicted by linear nonmagnetic theory. The actual value of Pr is probably not very relevant in the core, because both thermal and viscous diffusion are so small it can only operate on very short length scales on dynamo timescales.

The mechanism by which the convective rolls generate magnetic field by dynamo action in simulations was discussed by Olson *et al.* (1999). The toroidal field is mostly generated by twisting of the poloidal field, as in α^2 dynamos and unlike $\alpha - \omega$ dynamos, where toroidal field is generated by the stretching out of poloidal field by differential rotation. Low Ekman number dynamo simulations driven by rolls appear not to have very large differential rotation. In the quasi-geostrophic approximation, it is assumed that u_z varies linearly with z, but in the numerical simulations u_z changes more rapidly near the equator than at the boundary. Olson *et al.* (1999) point out that this variation is significant for dynamo action, so it is not clear that quasi-geostrophic flows will necessarily have the same dynamo properties as fully 3-D simulations give.

5.5.6.2 Convection inside the tangent cylinder: picture from numerical simulations

The linear theory of nonmagnetic convection in a sphere (Busse, 1970) suggests that the onset occurs at substantially high Rayleigh number inside the TC than outside the TC. This result is verified by numerical simulations (e.g., Dormy *et al.* 2004). Whereas the s-component of gravity predominates in convection outside the TC, it is the z-component that is most significant inside the TC. Indeed, of the standard idealized geometry convection models, the one most relevant to convection inside the TC is the plane horizontal parallel layer with vertical rotation and gravity (Sreenivasan and Jones, 2006b), discussed in Section 5.5.2.2. In this model, the first mode to onset has vertical velocity symmetric about the midplane and vertical vorticity antisymmetric, so the axial vorticity will not be even approximately constant for convection inside the TC.

Magnetic field appears to affect the onset of convection much more radically inside the TC than outside it (Sreenivasan and Jones, 2005, 2006b). In particular, nonmagnetic convection has the tall thin column configuration both inside and outside the TC, whereas magnetic field has only a limited impact outside the TC; inside the TC, the convection columns are much thicker with magnetic field, and magnetic field enhances the vigor of convection inside the TC (Olson and Glatzmaier, 1996). In the plane horizontal layer geometry, the effect of magnetic field is not gradual, but onsets at a specific value of the Elsasser number as noted in Section 5.5.2.2. In a recent study of behavior inside the TC in dynamo simulations, Sreenivasan and Jones (2005, 2006b) found anticyclonic vortices just below the CMB when inertial effects are small, and cyclonic vortices when they are large. These vortices are associated with the large-scale magnetic mode of convection, and their horizontal scale can be predicted surprisingly accurately using the Chandrasekhar linear theory. Since there is evidence from secular variation that there are anticyclonic polar vortices (Olson and Aurnou, 1999), this supports the view that the geodynamo is in a low inertia regime. They also found that the polar vortices are not actually centered on the pole but are in fact typically 10° off axis, with the vortex patches drifting slowly westward round the pole.

The magnetic mode of convection in a plane layer has a somewhat different physical character from conventional nonmagnetic, nonrotating convection. In the simple case, hot fluid rises directly because of its buoyancy. In rotating magnetic mode convection, hot fluid rotates about the convection column axis, because of the local thermal wind that results from the horizontal temperature gradient between the hot interior and relatively cool surroundings. This local rotation, or vertical vorticity, winds up the field lines to produce a current and a corresponding Lorentz force. This Lorentz force breaks the Proudman–Taylor constraint, and allows a u_z that varies with z. It is then possible for fluid to rise in the interior.

There is still much that is uncertain about convection inside the TC. Simulations up to 10 times critical generally have stronger convection outside the TC, because the delay in onset inside the TC is still having an effect. At very large Ra, the large-scale magnetic mode may eventually give rise to convection inside the TC more vigorous than outside it. There is also some uncertainty about how the field inside the TC is created. On the Earth, there is evidence that the poles have weaker field than the region where the TC intersects the CMB (about latitude 70°). This may be due either to weak convection failing to generate much field or to vigorous convection expelling flux from the polar regions. The reversed flux patch seen inside the North Pole in field models of the current geomagnetic field (see, e.g., Olson and Aurnou, 1999) is not often seen in geodynamo simulations. These often produce more reversed flux patches near the equator, whereas the Earth seems to prefer high latitude reversed flux patches.

5.5.7 Scaling Laws and Dynamo Simulations

In order to overcome the problem of the inaccessibility of the geodynamo parameter regime, there has been recent work on how the more important outputs from geodynamo simulations scale with the input parameters. If an asymptotic regime in which the role of the small diffusion coefficients can be identified, it may be possible to extrapolate to the very small values that occur in the Earth. Two approaches have been tried; Starchenko and Jones (2002) used results from plane layer models and the general understanding of rotating magnetoconvection to estimate typical velocities and magnetic field strengths expected from very low E, low q dynamos. An alternative approach (Christensen and Aubert, 2006) is to analyze the data from a large number of dynamo simulations and see whether asymptotic trends are evident in the data. In practice, these two approaches are not so different, as theoretical ideas inevitably affect the way the dynamo simulation data are analyzed.

Starchenko and Jones started by assuming that at very low E the Coriolis and buoyancy forces would be in balance, and that the magnetic field would bring the horizontal length scale ℓ appearing in [112] to a fixed ratio with $d = r_{cmb} - r_{icb}$. The idea here is that as E is reduced, the magnetic field prevents the roll-width reducing as $E^{1/3}$. Then [112] is replaced simply by

$$\Omega U_*/d \sim g\alpha T'_*/\ell \qquad [148]$$

Then eliminating the temperature perturbation using [113],

$$U_* \sim \left(\frac{g\alpha F_{conv}}{\rho c_p \Omega}\right)^{1/2}\left(\frac{d}{\ell}\right)^{1/2} \qquad [149a]$$

$$g\alpha T'_* \sim \Omega^{1/2}\left(\frac{g\alpha F_{conv}}{\rho c_p}\right)^{1/2}\left(\frac{\ell}{d}\right)^{1/2} \qquad [149b]$$

If the ratio d/ℓ tends to a fixed limit at small E due to magnetic field, as postulated by Starchenko and Jones, U_* is proportional to $F_{conv}^{1/2}$. Interestingly, Christensen and Aubert (2006) find that the exponent of $2/5$ gives a better fit to their data than the exponent of $1/2$ in [149]. Of course, the difference is not that great, but it may be connected with the observation made in Section 5.5.6.1 that in dynamo simulations there is not much evidence of the magnetic field controlling the roll-width, as suggested by Starchenko and Jones (2002). It remains possible, though, that this control will start to occur at E less than 10^{-6}. Christensen and Aubert (2006) suggest replacing [149a] by

$$U_* \sim (\Omega d)\left(\frac{g\alpha F_{conv}}{\rho c_p \Omega^3 d^2}\right)^{2/5} \qquad [150]$$

which is the same formula as in the nonmagnetic case, [115a]. They note that the best fit with their simulation data can be expressed in dimensionless variables

$$Ro = \frac{U_*}{\Omega d}, \qquad Ra^*_Q = \frac{g\alpha Q_{adv}}{4\pi r_{icb} r_{cmb} \rho c_p \Omega^3 d^2} \qquad [151]$$

by

$$Ro = 0.83 Ra^{*\,0.41}_Q \qquad [152]$$

very close to [150]. Here Q_{adv} is the total convective heat flux over a spherical surface of radius r, assumed constant with r throughout the core. Note that with this assumption, the convective heat flux F_{conv} in [148]–[150] is best taken as

$$F_{conv} = \frac{Q_{adv}}{4\pi r_{icb} r_{cmb}} \qquad [153]$$

The Rayleigh number Ra^*_Q introduced by Christensen and Aubert is essentially the Ra^*_Q defined in [126]. Equation [152] therefore expresses the independence of the velocity scaling on any diffusion coefficient. The emergence of [152], with no dependence on any Prandtl numbers, expresses the fact that the simulations do appear to be approaching an asymptotic regime which is independent of diffusion, a very encouraging result.

Christensen and Aubert (2006) also go on to estimate the typical magnetic field strength in dynamo simulations. The rate of working of the buoyancy forces is $g\alpha F/c_p$, and assuming that the dissipation is primarily ohmic, this must balance the rate of ohmic dissipation, so

$$\eta\mu \mathbf{j}^2 \sim \frac{g\alpha F_{\text{conv}}}{c_p} \qquad [154]$$

To convert this into a formula for $\mathbf{B}^2$ we need to know the magnetic dissipation time, that is,

$$\tau_{\text{diss}} = \frac{\int_V \mathbf{B}^2/2\mu \, \mathrm{d}v}{\int_V \eta\mu \mathbf{j}^2 \, \mathrm{d}v} \qquad [155]$$

or equivalently the magnetic dissipation length scale

$$\delta_B = \left(\frac{\tau_{\text{diss}}}{\eta}\right)^{1/2} \qquad [156]$$

Christensen and Tilgner (2004) argue that $\delta_B \sim dR_{\text{m}}^{-1/2}$, again on the basis of an analysis of numerical simulations. This is consistent with non-rotating flux expulsion arguments (Galloway *et al.* (1977)). Then [154] becomes

$$\eta\mu \mathbf{j}^2 \sim \eta \frac{(\nabla \times \mathbf{B})^2}{\mu} \sim \frac{\eta \mathbf{B}^2}{\mu \delta_B^2} \sim \frac{g\alpha F}{c_p} \qquad [157]$$

giving

$$B_* \sim \left(\frac{g\alpha F_{\text{conv}}\mu d}{U_* c_p}\right)^{1/2} \qquad [158]$$

with the Starchenko and Jones scaling [149] this gives

$$B_* \sim \mu^{1/2}\left(\frac{g\alpha F_{\text{conv}}\rho\Omega\ell d}{c_p}\right)^{1/4} \qquad [159]$$

and with the Christensen and Aubert scaling [150] we get

$$B_* \sim \mu^{1/2} d^{2/5} \rho^{1/5} \Omega^{1/10} \left(\frac{g\alpha F}{c_p}\right)^{3/10} \qquad [160]$$

Note that the small difference in the velocity scalings has led to a significant difference in the scaling of B_* with Ω, because the Christensen and Aubert scaling has a remarkably weak scaling of B_* with Ω. Indeed, their best fit of the simulation data is

$$Lo = 0.87 f_{\text{ohm}}^{1/2} Ra_Q^{*\,0.33} \qquad [161]$$

where $Lo = B_*/(\mu\rho)^{1/2}\Omega d$ and f_{ohm} is the fraction of the total dissipation (ohmic + viscous) which is ohmic. This gives

$$B_* \approx 0.9\mu^{1/2}\rho^{1/6}\left(\frac{g\alpha Q_{adv} d}{4\pi r_{\text{icb}} r_{\text{cmb}}}\right)^{1/3} \qquad [162]$$

so that B_* is completely independent of the rotation.

When compositional convection is present, in the above formulas the thermal buoyancy flux $g\alpha Q_{\text{conv}}$ is replaced by the compositional buoyancy flux gQ_{buoy}, where Q_{buoy} is the mass of light material released per unit time.

The scalings [152] and [161] can be used to find the ratio of kinetic to magnetic energy, which is close to $Ra_Q^{*\,0.15}$. Since Ra_Q^* is very small in the core, this implies that magnetic energy is indeed larger than kinetic energy in the core, but the weak power shows that it is not surprising that in simulations the two forms of energy have similar magnitudes. Also, although ohmic dissipation is predicted to dominate over viscous dissipation at very low E, simulations have great difficulty getting down to this regime (at $Pr = 1$), and generally at $Pr = 1$ they show comparable ohmic and viscous dissipations.

5.6 Heterogeneous Boundary Conditions and Stable Layers Near the CMB

Because the departure from an adiabatic temperature gradient is so small in the core, the core–mantle boundary will be maintained at an almost uniform temperature. However, it is believed that in the mantle, cold descending slabs may reach close to the CMB, and the locations where this happens may have an anomalously high CMB heat flux, while in regions where the mantle has been relatively static the temperature gradients, and hence the heat flux, may be lower. It is therefore likely that there are substantial lateral variations of heat flux at the CMB (e.g., Schubert *et al.*, 2001). If so, this lateral heterogeneity may well have an impact on core convection; for a recent review, see Olson (2003). It is now possible to measure the velocity of shear waves near the CMB, using seismic tomography (e.g., Kuo *et al.*, 2000). The simplest interpretation is that high velocity zones are associated with colder material and hence with higher heat flux. A natural assumption is then that the heat flux is simply proportional to the seismic velocity variation. If this is correct, then the lowest heat fluxes are found under the central Pacific and the eastern Atlantic and Africa, while high heat flux is found under Australia and the gulf of Mexico. This distribution is often modeled as a spherical harmonic P_2^2 pattern of variation.

As mentioned in the introduction, and discussed in Chapter 2, there is no general agreement about the total heat flux coming out of the core. If it is as large as 12 TW or above, then the heat flux inhomogeneities are only a small perturbation on the total heat flux, but if it is at the low end of the estimates, say 3 TW, then the core could be convectively stable below the low heat flux regions, but unstable under the high heat flux regions. Thus, we would expect the effect of lateral inhomogeneity to be strongest for low overall heat flux models.

One possible effect of laterally inhomogeneous heat flux would be to lock core convection to the mantle, that is, prevent any systematic westward or eastward drift. This could in turn lock the dynamo to the mantle. This could explain the preferred paths for the direction of the dipole field during reversals. There seems to be a preference for the virtual geomagnetic pole to track through the Americas or through east Asia (Laj *et al.*, 1991) but the results are controversial (see, e.g., the review of Dormy *et al.* (2000) for a discussion). Another possible effect is that long-term changes in mantle convection might affect the reversal rate. It is well established that during certain eras the geomagnetic field has not reversed, an extreme example being the Cretaceous superchron, during which the field maintained the same polarity for about 35 million years. The current reversal rate is about 4.5 reversals per million years. It is difficult to explain why the reversal rate should vary on such a long timescale, because the magnetic diffusion time and the turnover time are much shorter. It is tempting to identify these long timescales with mantle convection timescales.

It has also been noticed that downwelling fluid tends to concentrate radial magnetic flux into patches (Bloxham and Gubbins, 1987). Downwelling fluid occurs where the CMB temperature is coolest, and the outward temperature gradient is largest there (e.g., see figure 10 of Sarson *et al.* 1997) so the heat flux is a local maximum. So according to this argument, there should be long-term local maxima in the radial flux where the CMB heat flux is strongest, under Australia and the gulf of Mexico, while upwelling fluid giving weaker field should be associated with relatively low, or negative, CMB heat flux. There is some evidence at least for a long-term excess field under Australia, and relatively low field in the central Pacific (see Olson (2003) for a review of the paleomagnetic evidence).

A number of authors have studied the effect of lateral inhomogeneous boundary conditions on both magnetic and nonmagnetic fluids. It is helpful to divide the work into those studies where the fluid is subadiabatically stratified, that is stable to convection, and those where it is superadiabatically stratified.

Lister (2003) has considered the effect of lateral variations in the thermal boundary conditions at the horizontal boundary above stably stratified fluid. If the stably stratified layer is unstirred, the subadiabatic temperature gradient will be of the same order as the adiabatic gradient itself, which implies a very strong stable stratification. Lister finds that a vertical magnetic field makes a very substantial difference to the resulting flows. With no magnetic field, even a 5% lateral variation in heat flux could lead to a horizontal temperature difference of $T'_* \sim 1\,\mathrm{K}$, which would drive a thermal wind of order $g\alpha T'_*/2\Omega \sim 1\,\mathrm{m\,s^{-1}}$ through a layer of depth 200 km, which is far too big to be consistent with secular variation observations. Note that because of the strong stable stratification, $T'_* \sim 1\,\mathrm{K}$ is much larger than the usual estimate of the temperature perturbation in a convecting core. However, with a magnetic field, the estimate is reduced by a factor which scales with $E^{1/4}$. Even if a turbulent value of $E \sim 10^{-8}$ is adopted, this gives a huge reduction in flow speed, and with a radial field of order 0.5 mT he finds a velocity of $7\,\mathrm{mm\,s^{-1}}$ in a layer 35 km deep. With a molecular Ekman number, the velocity drops to $0.3\,\mathrm{mm\,s^{-1}}$ which is compatible with the secular variation observations. Given the uncertainties in the estimates of the strength of the stable stratification, it seems that stably stratified layers below the CMB cannot be ruled out on the grounds that they would produce too strong a thermal wind.

Studies of the influence of heterogeneous boundaries on rotating convection have been made by Zhang and Gubbins (1996), and by Gibbons and Gubbins (2000). In the latter paper, the core is unstably stratified, and a steady-state convection pattern is found, locked to the inhomogeneous boundary condition. The downwelling regions generally coincide with regions of enhanced heat flux, but at low Ekman number there is a tendency for maximum downwelling to occur to the east of the peak heat flux. Sarson *et al.* (1997), using a truncated dynamo model, also found locked solutions, this time with a steady dynamo, and again the downwelling broadly coincides with enhanced heat flux, this time associated with concentrated field patches, as predicted by

Bloxham and Gubbins (1987). However, Sarson *et al.* (1997) found a tendency for the downwelling to be slightly west of the maximum heat flux. Glatzmaier *et al.* (1999) worked at a higher (and more realistic) value of the Rayleigh number, and found that the solution was not locked, but time dependent. A similar result was found earlier in a magnetoconvection calculation by Olson and Glatzmaier (1996). They found that though some locking occurs near CMB, at deeper levels the flow and field drifted just as they do with spherically symmetric CMB conditions. However, even though the Glatzmaier *et al.* (1999) solutions were not locked, a heterogeneous heat flux distribution gave enhanced field where the heat flux is maximal when averaged over time, so there is clearly a tendency for strong magnetic flux patches to form over regions of strong heat flux even when the solution is not completely locked. Kutzner and Christensen (2004) used a reversing numerical dynamo model with an imposed heterogeneous CMB heat flux to study whether reversal paths were systematically correlated with the heat flux distribution. They found that although individual reversal paths had considerable variation, there was a tendency for reversal paths to cluster around the regions of high heat flux on the CMB.

In summary, at the present time it seems that inhomogeneous thermal boundary conditions, with the heat flux varying by an $O(1)$ amount, can certainly affect the flow near the CMB, and this is of course where the secular variation field and inferred flow is measured. However, it is less clear that the whole convection pattern and dynamo are also locked to the boundary inhomogeneities. It is certainly possible for some parameter regimes, but unfortunately not those that the geodynamo is most likely to be in.

5.7 Conclusions and Future Developments

While uncertainties about the physical properties of material at high pressure mean that we cannot be absolutely certain that convection occurs in the core, the evidence strongly suggests that both thermal and compositional convection are present at least in some parts of the core. It is also likely that convection is driving the geodynamo, though we cannot rule out the possibility that tides and precession might make a contribution in the Earth and other planets. The form that core convection takes is more uncertain. The linear theory of rotating, magnetic convection at onset is now fairly well understood, but core convection is strongly supercritical. Numerical simulations have carried our understanding into the nonlinear regime, and a picture that is consistent with our knowledge about the linear problem is beginning to emerge. However, numerical simulations cannot at present reach the true parameter regime for the core, and there is no prospect of doing so using our current techniques. Nevertheless, the combination of numerical simulations, asymptotic analysis, and physical intuition is a powerful one, and rapid progress is being made.

The most fruitful approach to date has been through geodynamo simulations, using the constraint that the pattern of convection must be consistent with the known facts of the secular variation and the paleomagnetic data. Recent analyses of how the properties of convection scale with the parameters have been particularly encouraging. This has given strong support to the notion that at least outside the tangent cylinder columnar convection dominates. This type of motion not only is expected on theoretical grounds, it also generates fields consistent with those of the Earth in a robust manner, without the need for any special tuning of the parameters. The size of the rolls is however still controversial. Some believe that the large-scale flux patches seen in geomagnetic data are directly connected with the convection roll size, others believe that the roll size is much thinner and that the observed flux patches are the result of organized larger-scale flows.

Generally, our lack of understanding of small-scale convection is a matter for some concern. The regime where $\eta >> \kappa$ is a difficult one, and simulations always use a hugely enhanced value of κ to avoid it. This means that the small-scale motions which must be present to transport the heat at very large Péclet number are not present in the simulations, and we do not know what effect they have. The current assumption, which is made for computational reasons rather than for physical reasons, is that these small-scale motions act to produce an enhanced isotropic eddy diffusion. This assumption may not be correct.

Even more controversial is whether the whole core is convecting or whether the regions near the CMB are at least partly stably stratified. This can be condensed down to the question of how much heat is flowing through the CMB. One view is that the heat flux is relatively low, say of the order of 3 TW, while the other view is that it is of the order of 7 TW or larger. In favor of the low heat flux theory are the observations that the thermal core–mantle coupling seems to be significant, and the (tentative) conclusion that there is no difficulty driving a dynamo, giving fields as strong as the Earth's, with only 3 TW of heat going through the CMB. Opposing this, the high CMB heat flux adherents argue that the observational evidence suggests that the dynamo has been working for 3.5 Gy, before inner-core formation, because current estimates of the age of the inner core are only of the order of 1 Gy. They argue that 3 TW is insufficient to drive a dynamo with no inner core. They also point out that the only way found so far to make a dynamo simulation reverse is to drive it strongly, so that while a nonreversing dynamo can work with low heat flux (and an inner core), it may be more difficult to make a low heat flux reversing dynamo.

At present, none of these arguments is entirely compelling, though hopefully more evidence will accumulate as our understanding of core convection and the dynamo process improves. Possibly the evidence for the effect of heterogeneous heat flux at the CMB will strengthen, and it will become clear that this requires a partially stably stratified upper core. Possibly, estimates of the age of the inner core will change, due to improvements in our knowledge of the thermal conductivity of the core. The development of *ab initio* quantum calculations is helping to improve our knowledge of material properties at very high pressure, and this could provide a key input to our understanding of core processes. Further developments in dynamo theory may clarify the issue of whether reversing dynamos with low heat flux can exist.

Another exciting development is the growth of experimental programs studying the dynamics of rotating magnetic fluids. Like the simulations, these cannot reach core conditions either, but they do extend the parameter range in useful directions, particularly low magnetic Prandtl number. In combination with numerical experiments they provide a very useful weapon in the fight to gain understanding of the fundamental processes. Crucial understanding of how small-scale and large-scale flows interact can be gained from laboratory experiments, while direct numerical simulations have great difficulty with this issue, because of the very large resolution required to study it.

Finally, the geophysical data available are continually improving. New satellites are providing more information about the core field and its secular variation, though masking by the crustal field does limit what is possible. Seismic measurements give more information about the inner core, its structure and rotation, which are directly connected with flow patterns in the core. Length of day measurements and the torsional oscillations are also a potentially important source of information about the core. The detection of the short period oscillations of the inner core about its mean position (the Slichter modes) are currently controversial, but nevertheless could provide very valuable evidence about the state of the core, as indeed can nutation and precession data. In the longer term, there is even the possibility of that the detection of Earth's core neutrinos (Araki *et al.*, 2005) could give us direct window into the core. In an age when the furthest reaches of the universe are being explored, it is perhaps surprising that what is going on beneath our feet is still a mystery. Solving this mystery is an outstanding scientific challenge, and one that is being actively addressed by scientists from many different fields. We must hope that this simultaneous attack on the problem from such a large number of fronts will eventually prove decisive.

References

Abdulrahman A, Jones CA, Proctor MRE, and Julien K (2000) Large wavenumber convection in the rotating annulus. *Geophysical and Astrophysical Fluid Dynamics* 93: 227–252.

Alfè D, Gillan MJ, and Price GD (2003) Thermodynamics from first principles: Temperature and composition of the Earth's core. *Mineralogical Magazine* 67: 113–123.

Alfè D, Price GD, and Gillan MJ (2002) Iron under Earth's core conditions: Liquid-state thermodynamics and high-pressure melting curve from *ab initio* calculations. *Physical Review B* 65: Art. No. 165118.

Al-Shamali FM, Heimpel MH, and Aurnou JM (2004) Varying the spherical shell geometry in rotating thermal convection. *Geophysical and Astrophysical Fluid Dynamics* 98: 153–169.

Anufriev AP, Cupal I, and Hejda P (1995) The weak Taylor state in an alpha–omega dynamo. *Geophysical and Astrophysical Fluid Dynamics* 79: 125–145.

Anufriev AP, Jones CA, and Soward AM (2005) The Boussinesq and anelastic liquid approximations for convection in the Earth's core. *Physics of the Earth and Planetary Interiors* 152: 163–190.

Araki T, Enomoto S, Furuno K, *et al.* (2005) Experimental investigation of geologically produced antineutrinos with KamLAND. *Nature* 436: 499–503.

Aubert J (2005) Steady zonal flows in spherical shell dynamos. *Journal of Fluid Mechanics* 542: 53–67.

Aubert J, Brito D, Nataf HC, Cardin P, and Masson JP (2001) A systematic experimental study of rapidly rotating spherical convection in water and liquid gallium. *Physics of the Earth and Planetary Interiors* 128: 51–74.

Aubert J, Gillet N, and Cardin P (2003) Quasigeostrophic models of convection in rotating spherical shells. *Geochemistry Geophysics Geosystems* 4: Art. No. 1052.

Aurnou J, Andreadis S, Zhu L, and Olson P (2003) Experiments on convection in the Earth's core tangent cylinder. *Earth and Planetary Science Letters* 212: 119–134.

Aussillous P, Sederman AJ, Gladden LF, Huppert HE, and Worster MG (2006) Magnetic resonance imaging of structure and convection in solidifying mushy layers. *Journal of Fluid Mechanics* 552: 99–125.

Bell PI and Soward AM (1996) The influence of surface topography on rotating convection. *Journal of Fluid Mechanics* 313: 147–180.

Bloxham J and Gubbins D (1987) Thermal core–mantle interactions. *Nature* 325: 511–513.

Bloxham J and Jackson A (1991) Fluid flow near the surface of the Earth's outer core. *Reviews of Geophysics* 29: 97–120.

Braginsky SI (1963) Structure of the F layer and reasons for convection in the Earth's core. *Soviet Physics – Doklady* 149: 8–10.

Braginsky SI (1967) Magnetic waves in the Earth's core. *Geomagnetism and Aeronomy* 7: 851–859.

Braginsky SI (1993) MAC-oscillations of the hidden ocean of the core. *Journal of Geomagnetism and Geoelectricity* 45: 1517–1538.

Braginsky SI and Meytlis VP (1990) Local turbulence in the Earth's core. *Geophysical and Astrophysical Fluid Dynamics* 55: 71–87.

Braginsky SI and Roberts PH (1995) Equations governing convection in the Earth's core and the geodynamo. *Geophysical and Astrophysical Fluid Dynamics* 79: 1–97.

Brito D, Aurnou J, and Cardin P (2004) Turbulent viscosity measurements relevant to planetary core–mantle dynamics. *Physics of the Earth and Planetary Interiors* 141: 3–8.

Brummell NH and Hart JE (1993) High Rayleigh number beta convection. *Geophysical and Astrophysical Fluid Dynamics* 68: 85–114.

Buffett BA (2003) A comparison of subgrid-scale models for large-eddy simulations of convection in the Earth's core. *Geophysical Journal International* 153: 753–765.

Buffett BA, Garnero EJ, and Jeanloz R (2000) Sediments at the top of the Earth's Core. *Science* 290: 1338–1342.

Buffett BA and Glatzmaier GA (2000) Gravitational braking of inner-core rotation in geodynamo simulations. *Geophysical Research Letters* 27: 3125–3128.

Buffett BA, Huppert HE, Lister J, and Woods AW (1992) Analytical model for solidification of the Earth's core. *Nature* 356: 329–331.

Busse FH (1970) Thermal instabilities in rapidly rotating systems. *Journal of Fluid Mechanics* 44: 441–460.

Busse FH (2002) Convective flows in rapidly rotating spheres and their dynamo action. *Physics of Fluids* 14: 1301–1314.

Busse FH and Carrigan CR (1976) Laboratory simulation of thermal convection in rotating planets and stars. *Science* 191: 81–83.

Busse FH and Or AC (1986) Convection in a rotating cylindrical annulus: Thermal Rossby waves. *Journal of Fluid Mechanics* 166: 173–187.

Busse FH and Simitev RD (2006) Parameter dependences of convection-driven dynamos in rotating spherical fluid shells. *Geophysical and Astrophysical Fluid Dynamics* 100: 341–361.

Castaing B, Gunaratne G, Heslot, *et al.* (1989) Scaling of hard thermal turbulence in Rayleigh–Bénard convection. *Journal of Fluid Mechanics* 204: 1–30.

Chandrasekhar S (1961) *Hydrodynamic and Hydromagnetic Stability*. Oxford: Clarendon Press.

Chen FL, Lu JW, and Yang TL (1994) Convective instability in ammonium chloride solution directionally solidified from below. *Journal of Fluid Mechanics* 276: 163–187.

Chen Q and Glatzmaier GA (2005) Large eddy simulations of two-dimensional turbulent convection in a density-stratified fluid. *Geophysical and Astrophysical Fluid Dynamics* 99: 355–375.

Christensen UR (2002) Zonal flow driven by strongly supercritical convection in rotating spherical shells. *Journal of Fluid Mechanics* 470: 115–133.

Christensen UR and Aubert J (2006) Scaling properties of convection-driven dynamos in rotating spherical shells and application to planetary magnetic fields. *Geophysical Journal International* 166: 97–114.

Christensen UR, Aubert J, Cardin P, *et al.* (2001) A numerical dynamo benchmark. *Physics of the Earth and Planetary Interiors* 128: 25–34.

Christensen U, Olson P, and Glatzmaier GA (1999) Numerical modelling of the geodynamo: A systematic parameter study. *Geophysical Journal International* 138: 393–409.

Christensen UR and Tilgner A (2004) Power requirement of the geodynamo from ohmic losses in numerical and laboratory dynamos. *Nature* 429: 169–171.

Classen S, Heimpel M, and Christensen U (1999) Blob instability in rotating compositional convection. *Geophysical Research Letters* 26: 135–138.

Collier JD and Helffrich G (2001) Estimate of inner core rotation rate from United Kingdom regional seismic network data and consequences for inner core dynamical behaviour. *Earth and Planetary Science Letters* 193: 523–537.

Davidson PA (2001) *An Introduction to Magnetohydrodynamics*. Cambridge: Cambridge University Press.

Deardorff JW (1970) Convective velocity and temperature scales for unstable planetary boundary layer and for Rayleigh convection. *Journal of Atmospheric Sciences* 27: 1211.

Desjardins B, Dormy E, and Grenier E (1999) Stability of mixed Ekman–Hartmann boundary layers. *Nonlinearity* 12: 181–199.

Dormy E (1997) *Modélisation numérique de la dynamo terrestre*. PhD Thesis, IPGP.

Dormy E, Soward AM, Jones CA, Jault D, and Cardin P (2004) The onset of thermal convection in rotating spherical shells. *Journal of Fluid Mechanics* 501: 43–70.

Dormy E, Valet J-P, and Courtillot V (2000) Numerical models of the geodynamo and observational constraints. *Geochemistry Geophysics Geosystems* 1: Art. No. 200GC000062.

Dumberry M and Bloxham J (2003) Torque balance, Taylor's constraint and torsional oscillations in a numerical model of the geodynamo. *Physics of the Earth and Planetary Interiors* 140: 29–51.

Dziewonski AM and Anderson DL (1981) Preliminary reference Earth model. *Physics of the Earth and Planetary Interiors* 25: 297–356.

Eltayeb IA (1972) Hydromagnetic convection in a rapidly rotating fluid layer. *Proceedings of the Royal Society Series A* 326: 229–254.

Fearn DR (1979a) Thermally driven hydromagnetic convection in a rapidly rotating sphere. *Proceedings of the Royal Society Series A* 369: 227–242.

Fearn DR (1979b) Thermal and magnetic instabilities in a rapidly rotating fluid sphere. *Geophysical and Astrophysical Fluid Dynamics* 14: 103–126.

Fearn DR (1988) Hydromagnetic waves in a differentially rotating annulus. 4. Insulating boundaries. *Geophysical and Astrophysical Fluid Dynamics* 44: 55–75.

Fearn DR (1998) Hydromagnetic flow in planetary cores. *Reports on Progress in Physics* 61: 175–235.

Fearn DR, Loper D, and Roberts PH (1981) Structure of the Earth's inner core. *Nature* 292: 232–233.

Fearn DR, Roberts PH, and Soward AM (1988) Convection, stability and the dynamo. In: Galdi GP and Straughan B (eds.) *Energy, Stability and Convection*, pp. 64–324. New York: Longmans.

Galloway DJ, Proctor MRE, and Weiss NO (1977) Formation of intense magnetic fields near the surface of the Sun. *Nature* 266: 686–689.
Galloway DJ and Weiss NO (1981) Convection and magnetic fields in stars. *Astrophysical Journal* 243: 945–953.
Gauss CF (1839) Allgemeine Theories des Erdmagnetismus. In: Gauss CF and Weber W (eds.) *Resultate aus den Beobachtungen des magnetischen Vereins. im Jahre 1838*, pp. 1–57. Göttingen: Dieteriche Buchhandlung.
Gibbons SJ and Gubbins D (2000) Convection in the Earth's core driven by lateral variations in the core–mantle boundary heat flux. *Geophysical Journal International* 142: 631–642.
Gillet N and Jones CA (2006) The quasi-geostrophic model for rapidly rotating spherical convection outside the tangent cylinder. *Journal of Fluid Mechanics* 554: 343–369.
Gilman PA (1975) Linear simulations of Boussinesq convection in a deep rotating spherical shell. *Journal of Atmospheric Sciences* 32: 1331–1352.
Gilman PA (1983) Dynamically consistent nonlinear dynamos driven by convection in a rotating spherical shell. 2. Dynamos with cycles and strong feedbacks. *Astrophysical Journal Supplement* 53: 243–268.
Gilman PA and Miller J (1981) Dynamically consistent nonlinear dynamos driven by convection in a rotating spherical shell. *Astrophysical Journal Supplement* 46: 211–238.
Glatzmaier GA (1984) Numerical simulations of stellar convective dynamos. 1. The model and the method. *Journal of Comparative Physiology* 55: 461–484.
Glatzmaier GA, Coe RS, Hongre L, and Roberts PH (1999) The role of the Earth's mantle in controlling the frequency of geomagnetic reversals. *Nature* 401: 885–890.
Glatzmaier GA and Roberts PH (1995) A 3-dimensional self-consistent computer simulation of a geomagnetic field reversal. *Nature* 377: 203–209.
Glatzmaier GA and Roberts PH (1996) An anelastic evolutionary geodynamo simulation driven by compositional and thermal convection. *Physica D* 97: 81–94.
Glatzmaier GA and Roberts PH (1997) Simulating the geodynamo. *Contemporary Physics* 38: 269–288.
Goldstein RJ, Chiang HD, and See DL (1990) High-Rayleigh number convection in a horizontal enclosure. *Journal of Fluid Mechanics* 213: 111–126.
Greenspan H (1968) *The Theory of Rotating Fluids*. Cambridge: Cambridge University Press.
Grossmann S and Lohse D (2000) Scaling in thermal convection: A unifying theory. *Journal of Fluid Mechanics* 407: 27–56.
Grote E and Busse FH (2000) Hemispherical dynamos generated by convection in rotating spherical shells. *Physical Review E* 62: 4457–4460.
Hollerbach R, Barenghi CF, and Jones CA (1992) Taylor's constraint in a spherical $\alpha\omega$-dynamo. *Geophysical and Astrophysical Fluid Dynamics* 67: 3–25.
Holme R and Bloxham J (1996) The magnetic fields of Uranus and Neptune: Methods and models. *Journal of Geophysical Research – Planets* 101: 2177–2200.
Holme R and Olsen N (2006) Core surface flow modelling from high-resolution secular variation. *Geophysical Journal International* 166: 518–528.
Ingersoll AP and Pollard D (1982) Motions in the interiors and atmospheres of Jupiter and Saturn: Scale analysis, anelastic equations, barotropic stability criterion. *Icarus* 52: 62–80.
Jackson A (1997) Time-dependency of tangentially geostrophic core surface motions. *Physics of the Earth and Planetary Interiors* 103: 293–311.
Jault D (2003) Electromagnetic and topographic coupling, and LOD variations. In: Jones CA, Soward AM, and Zhang K (eds.) *Earth's Core and Lower Mantle*, pp. 56–76. London: Taylor and Francis.
Jault D, Gire C, and Lemouel JL (1988) Westward drift, core motions and exchanges of angular-momentum between core and mantle. *Nature* 333: 353–356.
Jones CA (2000) Convection driven geodynamo models. *Philosophical Transactions of the Royal Society of London A* 358: 873–897.
Jones CA (2003) Dynamos in planets. In: Thompson MJ and Christensen-Dalsgaard JC (eds.) *Stellar Astrophysical Fluid Dynamics*, pp. 159–176. Cambridge: Cambridge University Press.
Jones CA, Longbottom AW, and Hollerbach R (1995) A self-consistent convection driven geodynamo model, using a mean field approximation. *Physics of the Earth and Planetary Interiors* 92: 119–141.
Jones CA, Mussa AI, and Worland SJ (2003) Magnetoconvection in a rapidly rotating sphere: The weak-field case. *Proceedings of the Royal Society of London A* 459: 773–797.
Jones CA and Roberts PH (1990) Magnetoconvection in rapidly rotating Boussinesq and compressible fluids. *Geophysical and Astrophysical Fluid Dynamics* 55: 263–308.
Jones CA and Roberts PH (2000) The onset of magnetoconvection at large Prandtl number in a rotating layer. II. Small magnetic diffusion. *Geophysical and Astrophysical Fluid Dynamics* 93: 173–226.
Jones CA, Rotvig J, and Abdulrahman A (2003) Multiple jets and zonal flow on Jupiter. *Geophysical Research Letters* 30: Art. No. 1731.
Jones CA, Soward AM, and Mussa AI (2000) The onset of convection in a rapidly rotating sphere. *Journal of Fluid Mechanics* 405: 157–179.
Kageyama A and Sato T (1997) Velocity and magnetic field structures in a magnetohydrodynamic dynamo. *Physics of Plasmas* 4: 1569–1575.
Kono M and Roberts PH (2002) Recent geodynamo simulations and observations of the geomagnetic field. *Reviews of Geophysics* 40: Art. No. 1013.
Kraichnan RH (1962) Turbulent thermal convection at arbitrary Prandtl number. *Physics of Fluids* 5: 1374–1389.
Kuang W and Bloxham J (1997) An Earth-like numerical dynamo model. *Nature* 389: 371–374.
Kuo B-Y, Garnero EJ, and Lay T (2000) Tomographic inversion of S-SKS times for shear velocity heterogeneity in D″: Degree 12 and hybrid models. *Journal of Geophysical Research* 105: 28138–28157.
Kurt E, Busse FH, and Pesch W (2004) Hydromagnetic convection in a rotating annulus with an azimuthal magnetic field. *Theoretical and Computational Fluid Dynamics* 18: 251–263.
Kutzner C and Christensen UR (2002) From stable dipolar towards reversing numerical dynamos. *Physics of the Earth and Planetary Interiors* 131: 29–45.
Kutzner C and Christensen UR (2004) Simulated geomagnetic reversals and preferred virtual geomagnetic pole paths. *Geophysical Journal International* 157: 1105–1118.
Labrosse S, Poirier JP, and LeMouel JL (2001) The age of the inner core. *Earth and Planetary Science Letters* 190: 111–123.
Laj C, Mazaud A, Weeks R, Fuller M, and Herrero-Bervera E (1991) Geomagnetic reversal paths. *Nature* 351: 447.
Larmor J (1920) How could a Rotating Body such as the Sun become a Magnet? Report of 87th meeting of the British Association for the Advancement of Science, September, 1919. Published by John Murray.
Lister JR (2003) Thermal winds forced by inhomogeneous boundary conditions in rotating, stratified hydromagnetic fluid. *Journal of Fluid Mechanics* 505: 163–178.

Loper DF (1970) General solution for the linearized Ekman–Hartmann layer on a spherical boundary. *Physics of Fluids* 13: 2995–2998.

Loper DF (1978) Some thermal consequences of a gravitationally powered dynamo. *Journal of Geophysical Research* 83: 5961–5970.

Loper DE, Chulliat A, and Shimizu H (2003) Buoyancy-driven perturbations in a rapidly rotating, electrically conducting fluid. Part 1. Flow and magnetic field. *Geophysical and Astrophysical Fluid Dynamics* 97: 429–469.

Malkus WVR (1967) Hydromagnetic planetary waves. *Journal of Fluid Mechanics* 28: 793–802.

Malkus WVR (1994) Energy Sources for Planetary Dynamos. In: Proctor MRE and Gilbert AD (eds.) *Lectures on Solar and Planetary Dynamos*, pp. 161–179. Cambridge: Cambridge University Press.

Malkus WVR and Proctor MRE (1974) The macrodynamics of α-effect dynamos in rotating fluids. *Journal of Fluid Mechanics* 67: 417–443.

Masters G and Gubbins D (2003) On the resolution of density within the Earth. *Physics of the Earth and Planetary Interiors* 140: 159–167.

Matsushima M (2005) A scale-similarity model for the subgrid-scale flux with application to MHD turbulence in the Earth's core. *Physics of the Earth and Planetary Interiors* 153: 74–82.

Moffatt HK (1978) *Magnetic Field Generation in Electrically Conducting Fluids*. Cambridge: Cambridge University Press.

Morin V and Dormy E (2004) Time dependent β convection in rapidly rotating spherical shells. *Physics of Fluids* 16: 1603–1609.

Morse SA (2002) No mushy zones in the Earth's core. *Geochimica et cosmochimica Acta* 66: 2155–2165.

Niemela JJ, Skrbek L, Sreenivasan KR, and Donnelly RJ (2000) Turbulent convection at very high Rayleigh numbers. *Nature* 404: 837–840.

Olson P (2003) Thermal interaction of the core and the mantle. In: Jones CA, Soward AM, and Zhang K (eds.) *Earth's Core and Lower Mantle*, pp. 1–38. London: Taylor and Francis.

Olson P and Aurnou J (1999) A polar vortex in the Earth's core. *Nature* 402: 170–173.

Olson P, Christensen UR, and Glatzmaier GA (1999) Numerical modeling of the geodynamo: Mechanisms of field generation and equilibration. *Journal of Geophysical Research* 104: 10383–10404.

Olson P and Glatzmaier GA (1996) Magnetoconvection and thermal coupling of the Earth's core and mantle. *Philosophical Transactions of the Royal Society of London A* 354: 1413–1424.

Or AC and Busse FH (1987) Convection in a rotating cylindrical annulus. Part 2. Transitions to asymmetric and vacillating flow. *Journal of Fluid Mechanics* 174: 313–326.

Palmer A and Smylie DE (2005) VLBI observations of free core nutations and viscosity at the top of the core. *Physics of the Earth and Planetary Interiors* 148: 285–301.

Plaut E and Busse FH (2002) Low Prandtl number convection in a rotating cylindrical annulus. *Journal of Fluid Mechanics* 464: 345–363.

Priestley CHB (1959) *Turbulent Transfer in the Lower Atmosphere*. Chicago: University of Chicago Press.

Proudman J (1916) On the motion of solids in a liquid possessing vorticity. *Proceedings of the Royal Society Series A* 92: 408–424.

Rhines PB (1975) Wave and turbulence on a beta-plane. *Journal of Fluid Mechanics* 122: 417–443.

Roberts PH (1968) On thermal instability of a rotating fluid sphere containing heat sources. *Philosophical Transactions of the Royal Society of London A* 263: 93–117.

Roberts PH and Jones CA (2000) The onset of magnetoconvection at large Prandtl number in a rotating layer. I. Finite magnetic diffusion. *Geophysical and Astrophysical Fluid Dynamics* 92: 289–325.

Roberts PH, Jones CA, and Calderwood A (2003) Energy fluxes and ohmic dissipation in the Earth's core. In: Jones CA, Soward AM, and Zhang K (eds.) *Earth's Core and Lower Mantle*, pp. 100–129. London: Taylor and Francis.

Roberts PH and Loper DE (1979) Diffusive instability of some simple steady magnetohydrodynamic flows. *Journal of Fluid Mechanics* 90: 641–668.

Roberts PH and Scott S (1965) On the analysis of the secular variation. 1. A hydromagnetic constraint: Theory. *Journal of Geomagnetism and Geoelectricity* 17: 137–151.

Rogers TM, Glatzmaier GA, and Woosley SE (2003) Simulations of two-dimensional turbulent convection in a density-stratified fluid. *Physical Review E* 67: Art. No. 026315.

Rotvig J and Jones CA (2002) Rotating convection-driven dynamos at low Ekman number. *Physical Review E* 66: 056308-1–15.

Rotvig J and Jones CA (2006) Multiple jets and bursting in the rapidly rotating convecting two-dimensional annulus model with nearly plane-parallel boundaries. *Journal of Fluid Mechanics* 567: 117–140.

Sakuraba A (2002) Linear magnetoconvection in rotating fluid spheres permeated by a uniform axial magnetic field. *Geophysical and Astrophysical Fluid Dynamics* 96: 291–318.

Sakuraba A and Kono M (1999) Effect of the inner core on the numerical solution of the magnetohydrodynamic dynamo. *Physics of the Earth and Planetary Interiors* 111: 105–121.

Sarson GR and Jones CA (1999) A convection driven geodynamo reversal model. *Physics of the Earth and Planetary Interiors* 111: 3–20.

Sarson GR, Jones CA, and Longbottom AW (1997) The influence of boundary region heterogeneities on the geodynamo. *Physics of the Earth and Planetary Interiors* 101: 13–32.

Schubert G, Turcotte DL, and Olson P (2001) *Mantle Convection in the Earth and Planets*. Cambridge: Cambridge University Press.

Siggia ED (1994) High Rayleigh number convection. *Annual Review of Fluid Mechanics* 26: 137–168.

Simitev R and Busse FH (2005) Prandtl-number dependence of convection-driven dynamos in rotating spherical fluid shells. *Journal of Fluid Mechanics* 532: 365–388.

Sleep NH (1990) Hot spots and mantle plumes: Some phenomenology. *Journal of Geophysical Research* 95: 6715–6736.

Song XD (1996) Seismological evidence for differential rotation of the Earth's inner core. *Nature* 382: 221–224.

Souriau A, Garcia R, and Poupinet G (2003) The seismological picture of the inner core: Structure and rotation. *Comptes Rendu Geoscience* 335: 51–63.

Soward AM and Jones CA (1983) α^2-dynamos and Taylor's constraint. *Geophysical and Astrophysical Fluid Dynamics* 27: 87–122.

Sreenivasan B and Jones CA (2005) Structure and dynamics of the polar vortex in the Earth's core. *Geophysical Research Letters* 32: Art. No. L20301.

Sreenivasan B and Jones CA (2006a) The role of inertia in the evolution of spherical dynamos. *Geophysical Journal International* 164: 467–476.

Sreenivasan B and Jones CA (2006b) Azimuthal winds, convection and dynamo action in the polar region of planetary cores. *Geophysical and Astrophysical Fluid Dynamics* 100(4–5): 319–339.

Stanley S and Bloxham J (2006) Numerical dynamo models of Uranus' and Neptune's magnetic fields. *Icarus* 184: 556–572.

Starchenko S and Jones CA (2002) Typical velocities and magnetic field strengths in planetary interiors. *Icarus* 157: 426–435.

Stevenson DJ (2003) Planetary magnetic fields. *Earth and Planetary Science Letters* 208: 1–11.
Sumita I and Olson P (2000) Laboratory experiments on High Rayleigh number thermal convection in a rapidly rotating hemispherical shell. *Physics of the Earth and Planetary Interiors* 117: 153–170.
Taylor GI (1923) The motion of a sphere in a rotating liquid. *Proceedings of the Royal Society Series A* 102: 180–189.
Taylor JB (1963) The magneto-hydrodynamics of a rotating fluid and the Earth's dynamo problem. *Proceedings of the Royal Society Series A* 274: 274–283.
Tilgner A (2005) Precession driven dynamos. *Physics of Fluids* 17: Art. No. 034104.
Tilgner A and Busse FH (1997) Finite-amplitude convection in rotating spherical fluid shells. *Journal of Fluid Mechanics* 332: 359–376.
Walker MR, Barenghi CF, and Jones CA (1998) A note on dynamo action at asymptotically small Ekman number. *Geophysical and Astrophysical Fluid Dynamics* 88: 261–275.
Weiss NO (1966) The expulsion of magnetic flux by eddies. *Proceedings of the Royal Society Series A* 293: 310.
Westerburg M and Busse FH (2003) Centrifugally driven convection in the rotating cylindrical annulus with modulated boundaries. *Nonlinear Processes in Geophysics* 10: 275–280.
Zhang K (1992) Spiralling columnar convection in rapidly rotating spherical shells. *Journal of Fluid Mechanics* 236: 535–556.
Zhang K (1995) Spherical shell rotating convection in the presence of a toroidal magnetic field. *Proceedings of the Royal Society Series A* 448: 245–268.
Zhang K and Busse FH (1987) On the onset of convection in rotating spherical shells. *Geophysical and Astrophysical Fluid Dynamics* 39: 119–147.
Zhang K and Busse FH (1989) Convection driven magnetohydrodynamic dynamos in rotating spherical shells. *Geophysical and Astrophysical Fluid Dynamics* 49: 97–116.
Zhang K and Fearn DR (1994) Hydromagnetic waves in rapidly rotating spherical shells generated by magnetic toroidal decay modes. *Geophysical and Astrophysical Fluid Dynamics* 77: 133–157.
Zhang K and Fearn DR (1995) Hydromagnetic waves in rapidly rotating spherical shells generated by poloidal decay modes. *Geophysical and Astrophysical Fluid Dynamics* 81: 193–209.
Zhang K and Gubbins D (1996) Convection in a rotating spherical fluid shell with an inhomogeneous temperature boundary condition at finite Prandtl number. *Physics of Fluids* 8: 1141–1158.
Zhang K and Jones CA (1993) The influence of Ekman boundary layers on rotating convection. *Geophysical and Astrophysical Fluid Dynamics* 71: 145–162.
Zhang K and Jones CA (1994) Convective motions in the Earth's fluid core. *Geophysical Research Letters* 21: 1939–1942.
Zhang K and Jones CA (1996) On small Roberts number magnetoconvection in rapidly rotating systems. *Proceedings of the Royal Society Series A* 452: 981–995.
Zhang J, Song XD, Li YC, Richards PG, Sun XL, and Waldhauser F (2005) Inner core differential motion confirmed by earthquake waveform doublets. *Science* 309: 1357–1360.

6 Turbulence and Small-Scale Dynamics in the Core

D. E. Loper, Florida State University, Tallahassee, FL, USA

6.1	**Introduction**	187
6.1.1	What Is Turbulence?	187
6.1.2	Why Does Turbulence Occur?	187
6.1.3	Forces and Fluxes Affecting Core Turbulence	188
6.1.4	Features of Core Turbulence	189
6.1.5	Dynamic Regions in the Outer Core	189
6.1.5.1	The plume region	189
6.1.5.2	The well-mixed interior region	190
6.1.5.3	A stable region beneath the CMB?	190
6.2	**Governing Equations**	190
6.2.1	Reference State	190
6.2.2	Convective Equations	193
6.3	**Parameters and Scaling**	194
6.3.1	Momentum and Magnetic Diffusion Equations	194
6.3.2	Composition and Specific Entropy	196
6.4	**Scaling and Structure of Plumes**	196
6.5	**Dynamics of the Plume Region**	197
6.5.1	Plume Flux	198
6.5.2	Relative Magnitudes of Forces and Fluxes	199
6.5.3	Plume Dynamo Action	199
6.5.4	Stratification in Downwelling Regions	200
6.6	**Cascades and Transfers of Energy in Core Turbulence**	200
6.7	**Approaches to Parametrization of Turbulence**	201
6.7.1	The Need for Parametrization	201
6.7.2	Diffusive Parametrizations	202
6.7.3	Alternative Parametrizations	202
6.8	**Unresolved Issues and Future Directions**	202
References		205

6.1 Introduction

6.1.1 What Is Turbulence?

Turbulence is fluid motion that is chaotic in time and space. It is characterized by vortices and eddies on many scales. Fluid motions that are not turbulent are said to be laminar. It is difficult to define turbulence more precisely, as it takes many forms depending on circumstances. The classic definition of turbulence, arising in engineering and applied to flow in or around smooth bodies, is that instability of laminar motion which occurs when the Reynolds number exceeds approximately 2300. However, it is not unreasonable to characterize flow in Earth's mantle, for example, as turbulent, even though it occurs at a Reynolds number of order 10^{-20}. It is clear from this example that the Reynolds number is not always the best measure of the occurrence of turbulence. As explained below, it is not an appropriate measure for Earth's core; more appropriate measures of turbulence in Earth's core are the Lorentz and magnetic Reynolds numbers, quantifying the effects of rotation and the magnetic field.

6.1.2 Why Does Turbulence Occur?

One reason why turbulence occurs can be understood with reference to the momentum equation for an incompressible fluid:

$$\partial \mathbf{u}/\partial t + (\mathbf{u}\cdot\nabla)\mathbf{u} = -\nabla\Pi + \mathbf{F} \quad [1]$$

where $\mathbf{u}$ is the fluid velocity, Π is the dynamic pressure, and $\mathbf{F}$ represents the forces per unit mass acting on the fluid. In classic Newtonian fluid dynamics, only the viscous force due to molecular viscosity contributes to $\mathbf{F}$:

$$\mathbf{F} = \mathbf{F}_{\mathrm{vm}} = \nu\nabla^2\mathbf{u} \qquad [2]$$

where ν is the kinematic viscosity. The ratio of inertial to viscous forces is quantified by the Reynolds number

$$R_{\mathrm{e}} = UL/\nu \qquad [3]$$

where U is a typical speed of fluid motion and L is a typical lengthscale. (If the flow structure of interest is characterized by several lengths, L is normally the smallest of these.) When the viscous force is small in comparison with inertial effects, the Reynolds number is large. Setting $\mathbf{F} = \mathbf{0}$ and assuming the flow to be steady, the momentum equation [1] simplifies to

$$(\mathbf{u}\cdot\nabla)\mathbf{u} = -\nabla\Pi \qquad [4]$$

This equation describes a special balance; the inertial term on the left-hand side is a full vector, with three scalar degrees of freedom, whereas the pressure term on the right is a gradient, having only one. If the balance described by [4] prevails, a large-scale steady laminar flow is possible. However, when this balance is perturbed by a velocity field of arbitrary structure, the inertial force can no longer be balanced by the pressure gradient. With $\mathbf{F}$ negligibly small, a steady force balance is not possible; the unsteady term in [1] must make up the difference. But this implies that the flow becomes unsteady, that is, unstable, leading to turbulence. The time-derivative term generates smaller scales of motion, and eventually scales are produced having $R_{\mathrm{e}} = O(1)$, and $\mathbf{F}_{\mathrm{vm}}$ restores the balance. In this example, the cascade of energy to smaller scales is driven by the nonlinear inertia term.

Classic homogeneous isotropic turbulence is described by equation [1] with $\mathbf{F}$ given by [2], together with the equation of conservation of mass:

$$\nabla\cdot\mathbf{u} = 0 \qquad [5]$$

Its behavior is characterized by a single parameter, the Reynolds number, given by [3].This problem has been the subject of countless studies, and much of our intuition regarding turbulence is based on this problem. This is somewhat unfortunate in the present context, as turbulence in Earth's core bears little resemblance to classic homogeneous turbulence, and it is doubtful that fundamental concepts of homogeneity (independence of position), isotropy (independence of direction), and an inertial range (in which kinetic energy cascades, without dissipative loss, to progressively smaller spatial scales) apply to core turbulence. As explained in the following subsection, the forces affecting flow in the core are such that core motions and turbulence are both nonhomogeneous and anisotropic. In particular, whereas there are no preferred directions in isotropic turbulence, at least three are relevant to the dynamics of Earth's core. The existence of an inertial range relies on the dominant force balance in the momentum equation involving only conservative forces. As we shall see, this is not the case in Earth's core; the Lorentz force is dominant and dissipative.

6.1.3 Forces and Fluxes Affecting Core Turbulence

Turbulent motions take differing forms and have differing mathematical representations depending on the forces contributing to $\mathbf{F}$. Those relevant to Earth's fluid outer core include the Coriolis force, the Lorentz force, and buoyancy forces, due both to the ambient (background) stratification and to local density differences between plumes and their surroundings. (In this chapter, 'plume' means a small-scale parcel of buoyant material.)

Additionally the form of turbulence depends on the nature of the forcing for fluid motion. In general, motions are driven by movement of boundaries (forced convection) or by internal density differences (natural convection), with the former common in engineering and the latter in geophysics and astrophysics. This distinction is not sharp; in some circumstances, natural convection leads to large-scale flows (e.g., the Jet Stream in the atmosphere and the Gulf Stream in the ocean) which can be unstable and lead to turbulent motions that are similar to those driven by forced convection.

Motions in Earth's fluid outer core are due ultimately to radioactive heating and secular cooling. More specifically, motions are driven by sources of buoyancy at the inner core boundary (ICB) and core–mantle boundary (CMB); in the absence of these forcings, conduction of heat would cause the core fluid to evolve to a thermally stably stratified state. The sources at the ICB include buoyant material segregated into the outer core by the progressive solidification and growth of the inner core, latent heat released by that solidification process, and heat sources (secular cooling, Ohmic heating, and possibly radioactivity) arising within the inner core. The sources at the CMB are much less certain. It is not known with any certainty whether or how much material is being transferred between the core and mantle. Such transfers will be ignored in what follows. Thermal buoyancy capable of driving convection is generated at the CMB only if the rate of transfer of heat from core to mantle exceeds the rate that heat is conducted down the adiabat in the outer core.

6.1.4 Features of Core Turbulence

While relatively few definitive statements can be made about core turbulence, there are two that appear to be undisputable:

- Turbulence in the core is anisotropic.
- There is no inertial range, in which transfers of kinetic energy between scales occurs without loss.

Anisotropy of fluid motions is due to the combined action of the Coriolis and Lorentz forces. Together, they act to inhibit fluid motions perpendicular to the plane defined by $\mathbf{\Omega}$ and $\mathbf{B}$, resulting in so-called pancake-shaped flow structures, elongated in that plane. Since the Lorentz force is dominant, it is an important factor in the transfer of kinetic energy between spatial scales of motion; since that force is dissipative, the transfer involves loss of kinetic energy (Moffatt, 1967, 1978; Davidson, 2000; see Section 6.4). While core turbulence has some similarities to geostrophic flows (which are strongly affected by the conservative Coriolis force; e.g., see Charney, 1971; Rhines, 1979; Pedlosky, 1987; Cambon *et al.*, 1997; Cambon, 2001), the absence of an inertial range in core turbulence makes the two types of flows quite distinct.

6.1.5 Dynamic Regions in the Outer Core

The entire outer core is close to a well-mixed state, with composition and specific entropy nearly constant, independent of position (these variables change slowly with time). The small, spatially dependent deviations from this state are dynamically important and are the main focus of studies of core structure and dynamics, including the geodynamo problem. As a result of these deviations, there may exist as many as three dynamic regions in the outer core: a plume region at the bottom, a 'well-mixed' region in the middle, and a stable region at the top, as illustrated in **Figure 1**.

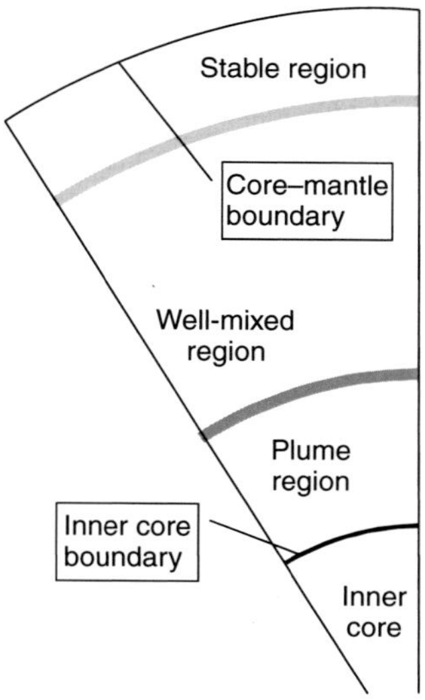

Figure 1 A cut-away view of the core, showing a plume region at the bottom of the outer core, a well-mixed region in the interior, and a stable region at the top. The radial extent of these regions is uncertain. The stable region may be well mixed by penetrative convective motions.

6.1.5.1 The plume region

The buoyancy released at the ICB due to solidification of the inner core creates very vigorous convection in the outer core (see [26] and discussion following). High-Rayleigh-number convective motions that are driven by a source of buoyancy at a boundary invariably consist of narrow rapid flows away from the boundary and slow, broad flows toward that boundary. The narrow, rapid motions upward from the ICB are likely to be small-scale plumes (i.e., elongated and flattened pancake-shaped structures). It follows that, near the bottom of the outer core, the area fraction, f, occupied by upwelling fluid is likely to be quite small; see **Figure 2**.

The plume region is characterized by the number, n, of plumes touching a given spherical surface located a distance r from Earth's center and the fraction, f, of area on that surface which they occupy. Both f and n are likely to be functions of r. The evolution of f and n with height above the ICB depends on the tendencies of plumes to entrain material, to merge, or to break up. Constraints on these variations are quantified in Section 6.5.2.

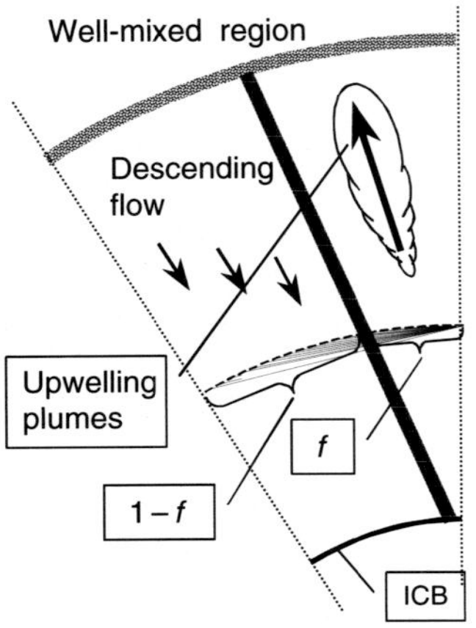

Figure 2 A cartoon of the plume region at the bottom of the outer core, showing upwelling plumes and descending flow occupying area fractions on spherical surfaces of magnitude *f* and 1 – *f*, respectively, separated by the dark hatched line. The magnitude of *f* increases with radial distance from the center of the Earth and the top of the plume region is characterized by *f* being about 1/2. The upwelling and descending parts are shown separately for clarity; in reality, these are intermingled.

Close to the ICB, f is almost certainly very small (reflecting the general property of vigorous convection described previously). Where $f \ll 1$, it is reasonable to model plumes in isolation or to consider only two-plume interactions. The structure and scaling of the plume region are quantified in Sections 6.4 and 6.5.

6.1.5.2 The well-mixed interior region

It is likely that in the bulk of the outer core convective motions are turbulent, with buoyant and dense plumes of fluid ascending and descending, all the while interacting vigorously. The motions are strongly affected by Coriolis and Lorentz forces and, as in the plume region, very likely are pancake-shaped (Braginsky and Meytlis, 1990); see Section 6.4. Nearly all parametrizations of core turbulence are focused on this region. Current approaches to parametrization are summarized in Section 6.7.

6.1.5.3 A stable region beneath the CMB?

It is possible that fluxes of heat or composition tend to make the region at the top of the outer core dynamically stable. If the flux of heat to the mantle is less than that conducted down the adiabat, the excess heat will tend to produce a stabilizing thermal gradient near the top of the outer core. Similarly an upward flux of light material, either due to pressure diffusion or the upwelling of compositionally buoyant plumes, may tend to create a stabilizing compositional gradient near the top of the outer core (Braginsky, 1999).

Even if the fluxes of heat and composition tend to produce stabilizing gradients at the top of the outer core, penetrative convective motions originating in the convectively unstable interior of the outer core may maintain a well-mixed state all the way to the CMB. Further, it is likely that the strength of the heat transferred to the mantle varies with location on the CMB, so that the stable region is confined to 'patches' at the top of the outer core where the heat flow from core to mantle is relatively small.

If a stable layer does exist, it is incapable of significant dynamo action, which requires vigorous radial motion. Consequently it has received relatively little attention and will not be considered further in this chapter.

6.2 Governing Equations

It is likely that convective motions in the outer core are driven by density differences of both thermal and compositional origin. As will be seen below, these density differences are very small perturbations on a reference state that is hydrostatic, adiabatic (having uniform specific entropy), and well mixed (having uniform composition).

Following Braginsky and Roberts (1995), thermodynamic variables in the outer core will be expressed as the sum of a reference-state portion, denoted by subscript 'a', plus a small convective portion, denoted by a subscript 'c'. (The velocity, having no reference-state portion, is written without subscript.) The myriad of variables introduced in this chapter are summarized in **Table 1**.

6.2.1 Reference State

The reference state satisfies the equations of state $\rho_a C_p \nabla T_a = \alpha T_a \nabla p_a$ and $K_S \nabla \rho_a = \rho_a \nabla p_a$ plus the hydrostatic and well-mixed equations $\nabla p_a = -\rho_a \mathbf{g}_a$ and $\nabla c_a = \mathbf{0}$, where T is temperature, p pressure, ρ density, α coefficient of thermal expansion, C_p specific heat at constant pressure, K_S adiabatic incompressibility, and c mass fraction of light constituent (called the composition in the following). Note that the Gruneisen parameter, $\gamma = \alpha K_S / \rho_a C_p$, is dimensionless and of unit order in the outer core.

Table 1 Notation and definition of symbols

Symbol	Name	Dimensions and magnitude
a, b, c	Components of the dimensionless perturbation magnetic field (Section 6.4 only); see [32]	
A_p	Dimensionless horizontal area of a plume; see [45]	
$\mathbf{B}$	Magnetic field vector	$\mathrm{Wb\,m^{-2}}$
$\mathbf{B}_0$	Large-scale magnetic field vector	$\mathrm{Wb\,m^{-2}}$
B_0	Typical magnitude of the magnetic field	$\approx 3\times10^{-3}\,\mathrm{Wb\,m^{-2}}$
c	Mass fraction of light constituent	
$\dot{c}_a$	Rate of increase of light material in the outer core; see [11] and [77]	$\sim 9\times10^{-20}\,\mathrm{s^{-1}}$
C	Co-density	
$\widehat{C}$	Scaled co-density; see [34]	
$\dot{C}$	Time rate of change of co-density; see [72]	$\sim 5\times10^{-20}\,\mathrm{s^{-1}}$
C_0	Typical magnitude of co-density	See [60]
C_p	Specific heat at constant pressure	$\mathrm{J\,kg^{-1}\,K^{-1}}$
C_{pe}	Effective specific heat of outer core	$\approx 1700\,\mathrm{J\,kg^{-1}\,K^{-1}}$
$\bar{C}_p$	Mass averaged specific heat	$\approx 800\,\mathrm{J\,kg^{-1}\,K^{-1}}$
C_{gr}	Effective specific heat for gravitational energy; see [85]	$\approx 300\,\mathrm{J\,kg^{-1}\,K^{-1}}$
C_{lh}	Effective specific heat for latent heat; see [86]	$\approx 600\,\mathrm{J\,kg^{-1}\,K^{-1}}$
D	Material diffusivity	$\approx 7\times10^{-9}\,\mathrm{m^2\,s^{-1}}$
f	Fraction of area occupied by plumes	See [59]
F_S	Divergence of flux of specific entropy	$\mathrm{W\,K^{-1}}$
F_C	Divergence of flux of composition	$\sim 3\times10^{-21}\,\mathrm{s^{-1}}$
$\mathbf{F}$	Body force	N
$\mathbf{F}_\nu$	Body force due to viscosity	N
$\mathbf{F}_{\nu m}$	Body force due to molecular viscosity	N
$\mathbf{g}_a$	Gravity vector in the reference state	$\mathrm{m\,s^{-2}}$
$\bar{g}$	Mass-average magnitude of gravity in the outer core	$=8.56\,\mathrm{m\,s^{-2}}$
g	Acceleration of gravity	$\mathrm{m\,s^{-2}}$
G	Smoothing operator; see [76]	
h	Depth or distance	m
$\boldsymbol{J}$	Buoyancy flux	$\mathrm{m^4\,s^{-3}}$
$\boldsymbol{J}_{icb}$	Buoyancy flux at the base of the outer core	$\sim \{15 \rightarrow 125\}\,\mathrm{m^4\,s^{-3}}$
$\boldsymbol{J}_c$	Buoyancy flux due to composition	$\mathrm{m^4\,s^{-3}}$
k	Thermal conductivity	$\approx 28\,\mathrm{W\,m^{-1}\,K^{-1}}$
K_S	Adiabatic incompressibility	$\mathrm{N\,m^{-2}}$
L	Typical length scale; also see [30]	See [62]
L_H	Latent heat of fusion	$\mathrm{J\,kg^{-1}}$
m	Mass	kg
M	Mass within radius r	kg
M_c	Mass of core	$=1.94\times10^{24}\,\mathrm{kg}$
M_{ic}	Mass of inner core	$=9.84\times10^{22}\,\mathrm{kg}$
M_{oc}	Mass of outer core	$=1.84\times10^{24}\,\mathrm{kg}$
$\dot{M}$	Convective mass flux	$\mathrm{kg\,s^{-1}}$
$\dot{M}_{ic}$	Rate of growth of inner-core mass	$\sim 3.5\times10^{6}\,\mathrm{kg\,s^{-1}}$
n	Number of plumes on a spherical surface	See [58]
n_B	Number of plumes threading a line of force	
N	Brunt Väisälä frequency	$\mathrm{s^{-1}}$
p_a	Pressure in the reference state	$\mathrm{N\,m^{-2}}$
P	Scaled dynamic pressure; see [33]	W
P_c	Power due to compositional convection	W
q	Heat source per unit mass	$\mathrm{W\,kg^{-1}}$
$q*$	Sum of internal heat sources; see [14]	$\mathrm{W\,kg^{-1}}$
q_{od}	Specific rate of Ohmic decay	$\mathrm{W\,kg^{-1}}$
q_{rd}	Specific rate of radioactive heating	$\mathrm{W\,kg^{-1}}$
Q	Heat flux from core	W
Q_{cd}	Heat flux down the adiabat	W

(Continued)

Table 1 (Continued)

Symbol	Name	Dimensions and magnitude
Q_{ic}	Heat from inner core	W
Q_{lh}	Heat released by solidification	W
Q_{oh}	Ohmic heating	W
Q_{rd}	Heat released by radioactive decay	W
Q_{sc}	Heat flux due to secular cooling	W
$\mathbf{r}$	Radial position	m
r	Radius	m
$\bar{r}$	Mass-averaged radius	2.76×10^6 m
s	Specific entropy	J kg^{-1} K^{-1}
$\dot{s}_a$	Time rate of change of specific entropy in the reference state; see [12]	W kg^{-1} K^{-1}
$\dot{s}_*$	Sources of specific entropy; see [13]	$\approx -3.7 \times 10^{-16}$ W kg^{-1} K^{-1}
t	Time	S
T	Temperature	K
$\bar{T}$	Mean temperature in outer core	$\approx$ 4350 K
$\mathbf{u}$	Fluid velocity	m s^{-1}
$\tilde{u}$	Smoothed velocity; see [76]	m s^{-1}
u, v, w	Plume velocity components; see [31]	m s^{-1}
U	Typical speed based on Coriolis force; see [18]	m s^{-1}
U_p	Typical speed of plumes; see [44]	See [61]
V_{ic}	Inner-core volume	$= 7.6 \times 10^{18}$ m^3
$\dot{V}_p$	Volume flux of plume material	$\sim 4 \times 10^9$ m^3 s^{-1}
$\dot{V}_{ic}$	Rate of growth of inner-core volume	m^3 s^{-1}
W	Speed of downwelling	m s^{-1}
x	Coordinate direction normal to the plane of **B** and $\mathbf{\Omega}$	m
x_{od}	Dimensionless Ohmic heating	Q_{od} 10^{-12} W^{-1}
x_{rd}	Dimensionless radioactive heating	Q_{rd} 10^{-12} W^{-1}
x_T	Dimensionless rate of cooling	$\approx [d\bar{T}/dt] 3 \times 10^{15}$ s K^{-1}
y	Third Cartesian coordinate	m
z	Coordinate aligned with $\mathbf{\Omega}$	m
α	Coefficient of thermal expansion	K^{-1}
α_s	Entropic expansion coefficient; $= \alpha T C_p$	$\approx 7 \times 10^{-5}$ K kg J^{-1}
β	Dimensionless distance along a field line; see [41]	
Γ	See [87]	
δ	Fractional jump in density due to change in composition across the inner-core boundary	$\approx$0.05
ΔC_c	Jump in composition at the ICB; see [78]	
η_m	Magnetic diffusivity	≈ 2 m^2 s^{-1}
θ	Colatitude	radian
κ	Thermal diffusivity; $= k/\rho C_p$	m^2 s^{-1}
$\bar{\kappa}$	Mass-averaged thermal diffusivity	$\approx 3 \times 10^{-6}$ m^2 s^{-1}
λ	Angle between **B** and $\mathbf{\Omega}$; see [32]	radian
μ	Chemical potential	J kg^{-1}
$\hat{\mu}$	Chemical potential gradient	$\approx 4.4 \times 10^7$ J kg^{-1}
μ_m	Magnetic permeability	$= 4\pi \times 10^{-7}$ Wb^2 s^2 m kg^{-1}
ξ, η, ζ	Plume coordinates; see [35]	
Π	Dynamic pressure	N m^{-2}
ϕ	Local longitude	radian
ρ_a	Density of the reference state	kg m^{-3}
$\bar{\rho}$	Mass-averaged density of the reference state	$= 1.09 \times 10^4$ kg m^{-3}
$\bar{\rho}_{ic}$	Mass-averaged density of inner core	kg m^{-3}
τ	dimensionless time; see [35]	
τ_{ic}	Age of the inner core	s
τ_{oc}	Mean turnover time of the outer core	s
ν	Kinematic viscosity	$\approx 5 \times 10^{-7}$ m^2 s^{-1}
$\bar{\psi}$	Mean gravitational potential of outer core	$\approx 10^7$ J kg^{-1}
$\mathbf{\Omega}$	Rotation vector	s^{-1}

(Continued)

Table 1 (Continued)

Symbol	Name	Dimensions and magnitude
Ω	Rotation rate	$=7.29\times10^{-5}\,\mathrm{s}^{-1}$
ICB	Inner-core boundary	
CMB	Core–mantle boundary	
Subscript 'a'	Adiabatic reference state	
Subscript 'c'	Convective perturbation	
E	Ekman number	[21]
R_e	Reynolds number	[3]
Λ	Lorentz number	[19]
R_o	Rossby number	[20]
R_m	Magnetic Reynolds number	[22]
R_a	Rayleigh number	[26]
S_{ch}	Schmidt number	[28]
P_r	Prandtl number	[29]

Magnitudes preceded by = are reasonably well known, and typically have two significant digits; magnitudes preceded by ≈ are accurate to about one significant digit (or a bit less); magnitudes preceded by ~ are rough approximations, with likely errors of 100%, and possibly more. A range of values is denoted by two numbers in curly brackets, separated by an arrow.
An additional dimensionless parameter is the Gruneisen parameter γ $(\alpha K_S/\rho_a C_P)$.

Much of our information of core structure and composition comes from teleseismic measurements of the speed of sound, $\sqrt{K_S/\rho_a}$.

6.2.2 Convective Equations

The equations governing convective motions in the outer core are (*see* Chapter 5)

$$\partial\mathbf{u}/\partial t + (\mathbf{u}\cdot\nabla)\mathbf{u} + 2\boldsymbol{\Omega}\times\mathbf{u} = -\nabla\Pi + C\mathbf{g}_a + (\mathbf{B}\cdot\nabla)\mathbf{B}/\rho_a\mu_m + \mathbf{F}_v \quad [6]$$

$$\nabla\cdot(\rho_a\mathbf{u}) = 0 \quad [7]$$

$$\partial\mathbf{B}/\partial t + (\mathbf{u}\cdot\nabla)\mathbf{B} = (\mathbf{B}\cdot\nabla)\mathbf{u} - \nabla\times(\eta_m\nabla\times\mathbf{B}) \quad [8]$$

$$\nabla\cdot\mathbf{B} = 0 \quad [9]$$

$$C = -\alpha_s s_c - c_c \quad [10]$$

$$\partial c_c/\partial t + (\mathbf{u}\cdot\nabla)c_c = -F_c - \dot{c}_a \quad [11]$$

and

$$\partial s_c/\partial t + (\mathbf{u}\cdot\nabla)s_c = -F_s - \dot{s}_a + \dot{s}_* \quad [12]$$

where $\mathbf{u}$ is the fluid velocity, $\boldsymbol{\Omega}$ rotation vector, Π dynamic pressure, $\mathbf{g}_a$ reference-state gravity, C co-density, $\mathbf{B}$ magnetic field vector, $\mathbf{F}_v$ viscous force, μ_m magnetic permeability, η_m magnetic diffusivity, $\alpha_s = (\alpha T/C_p)_a$ entropic expansion coefficient, F_c divergence of flux of composition, $\dot{c}_a = \mathrm{d}c_a/\mathrm{d}t$ temporal rate of increase of light material in the outer core (due to the growth of the inner core), s specific entropy (entropy per unit mass), F_s divergence of flux of specific entropy, $\dot{s}_a = \mathrm{d}s_a/\mathrm{d}t$ temporal rate of increase of specific entropy in the reference state, and

$$\dot{s}_* = q_*/T_a \quad [13]$$

represents the specific-entropy source due to heat sources (per unit mass) arising from radioactive decay, q_{rd}, Ohmic dissipation, q_{od}, and conduction of heat down the adiabat:

$$q_* = q_{rd} + q_{od} + (k/\rho_a)\nabla^2 T_a \quad [14]$$

with k being (molecular) thermal conductivity. Note that C is defined to be negative in buoyant plumes and that conduction of heat is a sink rather than a source of heat.

According to Roberts and Glatzmaier (2000), $\dot{c}_a \approx 9\times10^{-20}\,\mathrm{s}^{-1}$ and $\dot{s}_a \approx -3.7\times10^{-16}\,\mathrm{W\ kg^{-1}\ K^{-1}}$. It is estimated in Appendix 1 that $\dot{s}_* = (2.6x_{rd} + 2.6x_{oh} - 7.5)\times10^{-16}\,\mathrm{W\ kg^{-1}\ K^{-1}}$ where the rates of total heating ($Q = \int q\mathrm{d}m$) in the core due to radioactivity and Ohmic decay have been quantified by $x_{rd}\,10^{12}$ W and $x_{oh}\,10^{12}$ W, respectively.

By using the variables Π and C, several complications (involving the perturbations of density by pressure and gravity by density) have been sidestepped, although not ignored. Two advantages of using specific entropy rather than temperature in the density equation of state [10], are that the reference state has uniform specific entropy and that α_s is more nearly constant across the outer core than is the thermal expansion coefficient, α $[(\alpha_s)_{icb}/(\alpha_s)_{cmb} = 0.78$ while $\alpha_{icb}/\alpha_{cmb} = 0.57]$. Using data from Stacey and Davis (2004), $\alpha_s \approx \{8.3, 6.5\} \times 10^{-5}\,\mathrm{K\,kg\,J^{-1}}$ at the {top, bottom} of the outer core.

In the case that only molecular processes act, $\mathbf{F}_v$ is given by [2],

$$F_c = -D[\nabla^2 c_c - \nabla\cdot(\rho_a \delta \mathbf{g}_a/\hat{\mu})] \qquad [15]$$

and

$$F_s = -k(\nabla^2 T_c)/\rho_a T_a \qquad [16]$$

where D is material diffusivity, $\delta = \partial\mu/\partial p$, $\hat{\mu} = \partial\mu/\partial c$, and μ is the chemical potential difference between the two constituents (with core material being modeled as a binary alloy of metal and nonmetal; see Loper and Roberts, 1981).

For molten iron alloys near the melting point, $D \approx 7 \times 10^{-9}\,\mathrm{m^2\,s^{-1}}$ (Chalmers, 1964; Poirier, 1988), while Loper and Roberts (1981) estimated that a plausible range for $\rho_a\delta$ is between 1 and 2.5 and that $\hat{\mu} \approx 4.4 \times 10^7\,\mathrm{J\,kg^{-1}}$. The magnitude of $\nabla^2 c_c$ is difficult to estimate, but it is likely not to exceed that of the second term in the square bracket on the right-hand side of [15]; that second term is dominated by the divergence of gravity. Altogether, $F_c \sim 3 \times 10^{-21}\,\mathrm{s^{-1}}$; this is negligibly small compared with $\dot{c}_a$, which has been estimated following [14]. The value of thermal conductivity in the outer core is somewhat uncertain. Stacey and Anderson (2001) estimated k to vary from $46\,\mathrm{W\,m^{-1}\,K^{-1}}$ at the CMB to $63\,\mathrm{W\,m^{-1}K^{-1}}$ just above the ICB. However, Stacey (personal communication, 2005) now prefers a value $k = 28\,\mathrm{W\,m^{-1}\,K^{-1}}$, with little variation with depth. The magnitude of $\nabla^2 T_c$ is also difficult to estimate, but for any reasonable estimates, $F_s \ll \dot{s}_a$.

To a good approximation α_s may be treated as a constant and F_s and F_c neglected. Now eqns [11] and [12] may be combined to yield an evolution equation for the co-density:

$$\partial C/\partial t + (\mathbf{u}\cdot\nabla)C = \alpha_s[\dot{s}_a - \dot{s}_*] + \dot{c}_a \qquad [17]$$

Taking $\alpha_s \approx 7 \times 10^{-5}\,\mathrm{K\,kg\,J^{-1}}$ and using previous estimates, the forcing term on the right-hand side of [17] is roughly $[12 - 1.8\,(x_{rd} + x_{oh})] \times 10^{-20}\,\mathrm{s^{-1}}$, where x_{rd} and x_{oh} were introduced in the discussion following [14].

The unparametrized form of the governing equations contain four molecular diffusivities, for momentum (ν), material (D), heat ($\kappa = k/\rho C_p$), and magnetic flux (η_m). While ν, D, and κ are small and thus appear to require parametrization, η_m is sufficiently large that parametrization is neither needed nor desirable.

6.3 Parameters and Scaling

In this section, the dimensionless parameters associated with the governing equations are presented and discussed, beginning in Section 6.3.1 with those arising in the momentum and magnetic-diffusion equations and concluding in Section 6.3.2 with those arising in the equations governing variation of composition and entropy.

6.3.1 Momentum and Magnetic Diffusion Equations

As noted in Section 6.1.4, the force balances in [6] are far different from those in the traditional form of the momentum equation, [1]. Whereas the dominant balance in [1] is between pressure and inertia with the viscous force typically being small, the dominant balance in [6] appears to involve Coriolis, Lorentz, pressure, and buoyancy forces, with inertia and viscosity being small, although possibly not negligible. By custom for rapidly rotating systems, the Coriolis force is used as the reference force, against which other forces are measured. Most of these comparisons are quantified by dimensionless parameters, considered below. The exceptions are the pressure and buoyancy forces, which, when compared with the Coriolis force, yield scales for the dynamic pressure and velocity. The pressure scale is not of interest, while the characteristic velocity, U, is normally given by

$$U = C_0\bar{g}/2\Omega \qquad [18]$$

where $\Omega = \|\mathbf{\Omega}\| = 7.29 \times 10^{-5}\,\mathrm{s^{-1}}$, $\bar{g} = 8.56\,\mathrm{m\,s^{-2}}$ is the mass-averaged magnitude of gravity in the outer core, and C_0 is a characteristic amplitude of C. (Here and below, an overbar denotes an average (of a variable but well-determined quantity) within the outer core, while a subscript '0' denotes a typical or characteristic value (of a quantity of somewhat uncertain magnitude). Also, an equal sign implies accuracy to roughly two significant digits, $\approx$ implies one significant digit (or a bit less), and $\sim$ implies likely errors of 100% or possibly more.) The velocity deduced from secular variation of the geomagnetic field (e.g., see figure 4.1 of Merrill *et al.*, 1996), $0.38°\,\mathrm{yr^{-1}} \approx 7 \times 10^{-4}\,\mathrm{m\,s^{-1}}$, is produced by this scaling with $C_0 \approx 1.2 \times 10^{-8}$. It is shown in Section 6.4 that the rise speed of small-scale plumes can be greater than indicated by this velocity scale.

The Lorentz, inertial, and (molecular) viscous forces are quantified by the Elsasser, Rossby, and Ekman numbers, respectively:

$$\Lambda \equiv B_0^2/\eta_m \mu_m \bar{\rho} 2\Omega \qquad [19]$$

$$R_o \equiv U/2\Omega L \qquad [20]$$

and

$$E \equiv \nu/2\Omega L^2 \qquad [21]$$

where B_0 is a typical magnitude of the magnetic field in the outer core and $\bar{\rho}$ is the mean density. In addition, an important measure of dynamo action is the magnetic Reynolds number:

$$R_m \equiv UL/\eta_m \qquad [22]$$

which quantifies the relative importance of induction and Ohmic dissipation. In the core $\eta_m \approx 2\ m^2\ s^{-1}$; this parameter is invariably treated as a constant.

For large-scale motions in the core, E is very small and Λ is roughly of unit order; $\Lambda = 1$ if $B_0 = \sqrt{\eta_m \mu_m \bar{\rho} 2\Omega} \approx 2 \times 10^{-3} \text{Wb m}^{-2}$. One physical interpretation of the Elsasser number is that it is the ratio of the Joule damping time, $\eta_m \mu_m \bar{\rho}/B_0^2$, to the period of rotation of Earth, $2\pi\Omega$. A unit-order Elsasser number implies a large Lorentz force, capable of damping flow structures (such as free vortices) on timescales of the order of 1 day. This illustrates one difficulty in maintaining the dynamo; if the magnetic field is to be maintained efficiently, it must be configured so that this damping is not strong. It is likely that the Elsasser number in the core is large; see Section 6.5.3.

While the magnitudes of dimensionless numbers are illustrative of the relevant force balances, they do not tell the entire story. In addition to magnitude, forces have three other relevant attributes: conservative/dissipative/source, linear/nonlinear, and degrees of freedom. Conservative forces, such as inertia, Coriolis, and pressure, redistribute kinetic energy but do not degrade it to heat. Dissipative forces, such as Lorentz (coupled with finite electrical resistivity) and viscous, act to dissipate kinetic energy. The buoyancy force acts as a source of kinetic energy, counterbalancing dissipation. A force is nonlinear if it involves a dependent variable more than once (e.g., inertia) or if it involves a dependent variable that is governed by a nonlinear equation (Lorentz, buoyancy). Degrees of freedom refers to the structure of a force vector. The forces that have three degrees of freedom and are fully three-dimensional (3-D) are inertia and (molecular) viscous. The Coriolis and Lorentz forces involve cross products which constrain them to have two degrees of freedom. Buoyancy acts in a fixed direction and so has one degree of freedom. Pressure, being the gradient of a scalar, also has only one degree of freedom. These properties are summarized in **Table 2**. Note that the viscous term is the only dissipative force having three degrees of freedom.

The limitation in the degrees of freedom of the Coriolis force results in the well-known Taylor–Proudman Theorem:

$$(\mathbf{\Omega}\cdot\nabla)\mathbf{u} = 0 \qquad [23]$$

This constraint is valid only at dominant order and only if the Coriolis force is much larger than the Lorentz force (i.e., if $\Lambda \ll 1$). By the same token, the limitation in the degrees of freedom of the Lorentz force (in the case that $\Lambda \gg 1$) impresses a similar structure on the velocity, but now with invariance in the direction of $\mathbf{B}$ rather than $\mathbf{\Omega}$. These two constraints, taken together, imply that fluid motions in Earth's outer core are pancake shaped (Braginsky and Meytlis, 1990; see Section 6.4).

As seen in **Table 2**, the momentum equation contains three nonlinear terms, representing the Lorentz, inertial, and buoyancy forces. The buoyancy force is nonlinear by virtue of the advection term in [17]; this is a primary source of nonlinearity in the mathematical problem governing convective motions in the outer core. The noninertial force is proportional to the Rossby number, defined by [20], and is small for flows which are sufficiently rapid or of sufficiently large scale that $R_o \ll 1$. It is shown in Section 6.4 that

Table 2 Summary of forces in outer-core dynamics

Force	Dominant or secondary	Conservative, dissipative, or source	Linear or nonlinear	Degrees of freedom
Coriolis	Dominant	Conservative	Linear	2
Lorentz	Dominant	Dissipative	Nonlinear	2
Pressure	Dominant	Conservative	Linear	1
Buoyancy	Dominant	Source	Nonlinear	1
Inertia	Secondary	Conservative	Nonlinear	3
Viscous	Secondary	Dissipative	Linear	3

the inertial force is likely to be important in the dynamics of buoyant plumes and that the Lorentz force and accompanying magnetic diffusion equation may be linearized when considering buoyant plumes having small spatial and/or velocity scales such that $R_m \ll \Lambda$. In this limit, at dominant order in powers of R_m, equations [6] and [8] may be expressed as

$$\partial\mathbf{u}/\partial t + (\mathbf{u}\cdot\nabla)\mathbf{u} + 2\boldsymbol{\Omega}\times\mathbf{u} = -\nabla\Pi + C\mathbf{g}_a + (\mathbf{B}_0\cdot\nabla)\mathbf{B}/\rho_a\mu_m + \mathbf{F}_v \quad [24]$$

and, with constant magnetic diffusivity,

$$(\mathbf{B}_0\cdot\nabla)\mathbf{u} = -\eta_m\nabla^2\mathbf{B} \quad [25]$$

where $\mathbf{B}_0$ is the large-scale magnetic field (to be treated as a constant). Buoyancy is a source of kinetic energy, through convective instability, only if the density distribution is favorable (e.g., heavy fluid above light). The tendency for, and strength of, convective instability is quantified by the Rayleigh number which, in the present case, is of the form

$$R_a = C_0\bar{g}h^3/\nu\bar{\kappa} \quad [26]$$

where $\bar{\kappa}$ is the appropriate (thermal or compositional) mass-averaged molecular diffusivity and h is a vertical distance. Instability occurs when denser fluid underlies lighter and R_a exceeds a critical value of roughly 1000. Typically R_a is very large in the outer core. For example, with $\bar{\kappa} \equiv k/\bar{\rho}\bar{C}_p$ and using $\bar{\rho} = 1.09\times10^4$ kg m^{-3} and $\bar{C}_p \approx 800$ J kg^{-1}K^{-1} (Stacey and Davis, 2004) plus $k = 28$ W m^{-1} K^{-1} (F. Stacey, personal communication, 2005), $\bar{\kappa} \approx 3\times10^{-6}$ m^2 s^{-1}. If the outer core ($h \approx 2.2\times10^6$ m) has a heavy-over-light co-density contrast of, say, $C_0 \sim 10^{-9}$ with $\nu \approx 5\times10^{-7}$ m^2 s^{-1}, $\bar{g} \approx 8.56$ m s^{-2}, then $R_a \sim 6\times10^{22}$, indicating extreme convective instability. With compositional diffusivity, the Rayleigh number is even larger. On the other hand, if the outer core had a light-over-heavy co-density contrast, the Brunt Väisälä frequency would be

$$N = \sqrt{\bar{g}C_0/h} \quad [27]$$

With previous estimates, $N \approx 6\times10^{-8}$ s$^{-1} \approx 2$ yr^{-1}. (Note that this estimate is for illustration only; it is not suggested here that the bulk of the outer core is stratified.)

6.3.2 Composition and Specific Entropy

Equations [11] and [12] with the diffusive forms of the flux divergences [15] and [16], when nondimensionalized, yield two dimensionless parameters quantifying the relative strengths of advection and diffusion: the Schmidt and Peclet numbers are

$$S_{ch} = UL/D \quad [28]$$

and

$$P_r = UL/\bar{\kappa} \quad [29]$$

6.4 Scaling and Structure of Plumes

A dominant factor in structure of plumes in the outer core is the constraint imposed by rotation, quantified by [23] in the limit that buoyancy and Lorentz forces are negligibly small. If $\Lambda \geq O(1)$, this constraint is modified by the action of the Lorentz force in such a way that convective motions are facilitated. For example, the critical Rayleigh number for thermal convection in a rapidly rotating fluid decreases as the effect of the magnetic field increases (Chandrasekhar, 1961).

It is generally believed that the dynamo is in the strong-field regime (Roberts *et al.*, 2003; *see* Chapter 8), characterized by $\Lambda \gg 1$ in the bulk of the core. This suggests that, by adopting a suitable size and shape, the rise speed of plumes can exceed the magnitude predicted by [18]. On the other hand, the rise speed is limited by the condition that the Rossby number not exceed unity; only flows having a duration less than 1 day can have a large Rossby number. That is, the maximum characteristic rise speed is $2\Omega L$, where L is the thickness (i.e., smallest linear dimension) of the plume. In order that the characteristic speed of form [18] (with the mean value of gravity replaced by the local value) not exceed this maximum, the characteristic lengthscale must satisfy $L \geq C_0 g/4\Omega^2$. Detailed scaling of plume flow (presented below) reveals that L exceeds this lower bound in proportion to the Elsasser number; that is,

$$L = \frac{C_0 g\Lambda}{4\Omega^2} \quad [30]$$

Adopting this scaling, the velocity, magnetic field, and buoyancy may be expressed as (H. Shimizu, private communication, 2006)

$$\mathbf{u} = U[u\hat{\mathbf{x}} + \Lambda v\hat{\mathbf{y}} + \Lambda w\hat{\mathbf{z}}] \quad [31]$$

and

$$\mathbf{B} = B_0[\sin(\lambda)\hat{\mathbf{y}} + \cos(\lambda)\hat{\mathbf{z}}] + B_0 R_m\left[\frac{1}{\Lambda}a\hat{\mathbf{x}} + b\hat{\mathbf{y}} + c\hat{\mathbf{z}}\right], \quad [32]$$

$$\Pi = \frac{C_0^2 g^2}{4\Omega^2}\Lambda^2 P \quad [33]$$

$$C = C_0\widehat{C} \quad [34]$$

where x is a coordinate direction normal to the plane of **B** and **Ω**, z is aligned with the rotation axis (see **Figure 3**), C_0 is the magnitude of the buoyancy, and U is given by [18] with the mean magnitude of gravity replaced by the local value. The coordinates and time may be nondimensionalized as

$$\{x, y, z, t\} = \left\{L\xi, L\Lambda\eta, L\Lambda\zeta, \frac{1}{2\Omega}\Lambda\tau\right\} \quad [35]$$

It may be seen from [32] that the perturbation magnetic field is small provided that $R_\mathrm{m} \ll \Lambda$. Assuming this condition to be satisfied, the governing equations are, at dominant order in powers of Λ^{-1},

$$-v + \frac{\partial P}{\partial \xi} = 0 \quad [36]$$

$$\frac{Dv}{D\tau} + u + \frac{\partial P}{\partial \eta} + \sin(\theta)\sin(\phi)\widehat{C} = \sin(\lambda)\frac{\partial b}{\partial \beta} \quad [37]$$

$$\frac{Dw}{D\tau} + \frac{\partial P}{\partial \zeta} + \cos(\theta)\widehat{C} = \sin(\lambda)\frac{\partial c}{\partial \beta}, \quad [38]$$

$$\frac{\partial u}{\partial \xi} + \frac{\partial v}{\partial \eta} + \frac{\partial w}{\partial \zeta} = 0 \quad [39]$$

and

$$-\sin(\lambda)\frac{\partial}{\partial \beta}\{u, v, w\} = \frac{\partial^2}{\partial \xi^2}\{a, b, c\} \quad [40]$$

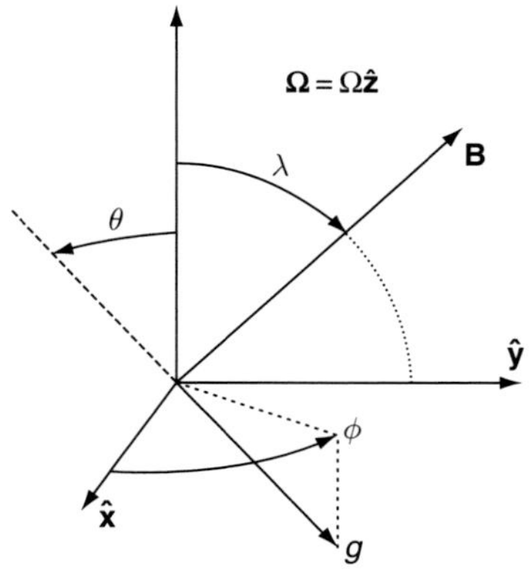

Figure 3 A depiction of the relative orientation of rotation, magnetic field and gravity vectors, and associated angles.

where θ is colatitude, ϕ is local longitude, and

$$\frac{\partial}{\partial \beta} = \frac{\partial}{\partial \eta} + \frac{\cos(\lambda)}{\sin(\lambda)}\frac{\partial}{\partial \zeta} \quad [41]$$

is a derivative in the direction of the large-scale magnetic field.

Equations [36]–[40], together with a suitably nondimensionalized version of [17], constitute a set of eight equations for eight unknowns: a, b, c, u, v, w, P, and $\widehat{C}$. The forcing function for this set of homogeneous equations is in the initial or boundary condition for $\widehat{C}$. Energy is provided to the flow by the buoyancy terms $\widehat{C}$ in [37] and [38] and is dissipated by the Lorentz terms in the same equations.

Although the aspect ratio of plumes governed by this set of equations is the same as that proposed by Braginsky and Meytlis (1990), there are some differences between the two approaches. Whereas Braginsky and Meytlis conceived of transient plumes arising from an unstable background state and dissipating before reaching a fully nonlinear state, the above equations govern possibly stable plumes surrounded by core fluid that is otherwise convectively stable.

In the absence of any instability, steady source of buoyancy at the ICB would produce a plume of indefinite length. However, it is likely that any such plume is unstable (e.g., see Eltayeb and Loper, 1997; Classen *et al.*, 1999; Eltayeb *et al.*, 2005), leading to the production of a sequence of plumes governed by the equations presented above. It is not known at the present time whether such plumes are stable structures or are prone to further instability, but the scaling presented in the following section suggests that they are unstable.

6.5 Dynamics of the Plume Region

This section considers the dynamics of an ensemble of plumes within the plume region near the base of the outer core. This region is characterized by narrow, relatively isolated (having area fraction $f \ll 1$) regions of rapid upwelling (plumes) embedded in a broad region of downward flow. Due to the secular evolution of composition and entropy in the outer core, the descending fluid is stably stratified. Thus the structure of the plume region is similar to that encountered in the box-filling mode (Turner, 1975). The magnitudes and fluxes within the buoyant plumes are quantified in Section 6.5.1 and the dynamic stratification of the downwelling regions is quantified in Section 6.5.2.

6.5.1 Plume Flux

Dynamics of plumes may be cast in terms of the buoyancy flux, $\mathcal{J}$:

$$\mathcal{J} = C_0 g \dot{V}_p \qquad [42]$$

where total upward volume flux of plume material is

$$\dot{V}_p \sim 4\pi r^2 U_p f = A_p U_p n \qquad [43]$$

In this expression, n is the number of plumes which occupy an area fraction f of a spherical surface a distance r from Earth's center,

$$U_p = C_0 g \Lambda / 2\Omega \qquad [44]$$

and A_p is the horizontal area of a single plume:

$$A_p \sim \frac{L^2 \Lambda}{\cos(\theta)} \qquad [45]$$

Note that $U_p = \Lambda U$, with U given by [18] with $\bar{g}$ replaced by g, and $U_p = 2\Omega L$ with L given by [30].

If a steady plume were to rise through fluid that is neutrally buoyant, its buoyancy flux would be constant, independent of height, in spite of entrainment of ambient fluid.

Equations [42]–[45], together with [30], may be combined to yield expressions for C_0, U_p, and L in terms of $\mathcal{J}$ and other factors. The most uncertain of the other factors are f and n. Since these are related to $\mathcal{J}$ by

$$n\Lambda^2 \mathcal{J} = 128\pi^2 r^4 \Omega^3 \overline{\cos(\theta)} f^2 \qquad [46]$$

where an overbar denotes an average, the expressions for C_0, U_p, and L may take a number of forms. For example, elimination of n yields

$$C_0 \sim \sqrt{\frac{1}{2\Lambda f}} \left[\frac{1}{rg} \sqrt{\frac{\Omega \mathcal{J}}{\pi}} \right] \sqrt{\frac{1}{2\Lambda f}} C_* \qquad [47]$$

$$U_p \sim \sqrt{\frac{\Lambda}{2f}} \left[\frac{1}{2r} \sqrt{\frac{\mathcal{J}}{\pi \Omega}} \right] \sqrt{\frac{\Lambda}{2f}} U_* \qquad [48]$$

and

$$L \sim \sqrt{\frac{\Lambda}{2f}} \left[\frac{1}{4\Omega r} \sqrt{\frac{\mathcal{J}}{\pi \Omega}} \right] \sqrt{\frac{\Lambda}{2f}} L_* \qquad [49]$$

Alternately, elimination of f (setting $\overline{\cos(\theta)} = 1/2$) yields

$$C_0 \sim \frac{2\Omega}{g\Lambda} \left[\frac{\Omega \mathcal{J}}{n} \right]^{1/4} \qquad [50]$$

$$U_p \sim \left[\frac{\Omega \mathcal{J}}{n} \right]^{1/4} \qquad [51]$$

and

$$L \sim \frac{1}{2\Omega} \left[\frac{\Omega \mathcal{J}}{n} \right]^{1/4} \qquad [52]$$

Although the factors f and n are poorly known and quite variable, they do have firm limits: $f \leq 1$ and $n \geq 1$. A slightly more stringent limit applies to f, on the assumption that in the well-mixed region upwelling and downwelling plumes occupy similar areas, $f \leq 1/2$. These imply the following limits:

$$\frac{16\pi^2 r^4 \Omega^3}{\mathcal{J}\Lambda^2} \geq n \geq 1 \qquad [53]$$

$$\frac{1}{2} \geq f \geq \frac{\Lambda}{16\pi r^2 \Omega} \sqrt{\frac{\mathcal{J}}{\Omega}} \qquad [54]$$

$$\frac{1}{rg} \sqrt{\frac{\Omega \mathcal{J}}{\pi \Lambda}} \leq C_0 \leq \frac{2\Omega^{5/4} \mathcal{J}^{1/4}}{g\Lambda} \qquad [55]$$

$$\frac{1}{2r} \sqrt{\frac{\mathcal{J}\Lambda}{\pi \Omega}} \leq U_p \leq (\Omega \mathcal{J})^{1/4} \qquad [56]$$

and

$$\frac{1}{4\Omega r} \sqrt{\frac{\mathcal{J}\Lambda}{\pi \Omega}} \leq L \leq \frac{\mathcal{J}^{1/4}}{2\Omega^{3/4}} \qquad [57]$$

These scaling expressions are local (not involving radial derivatives) and thus are valid for plumes rising through surroundings that are not neutrally buoyant. However, $\mathcal{J}$ is not independent of radius (i.e., distance from the center of Earth) in that case.

In estimates [53]–[57], the left-hand limits apply to the well-mixed region (with $f = 1/2$), with right-hand limits possibly approached in the plume region, where $f \ll 1$. A plausible range of values for the buoyancy flux at the base of the outer core is $\mathcal{J}_{icb} \sim 15 \rightarrow 125\ \mathrm{m^4\,s^{-3}}$ (see [94] and discussion following). For purposes of illustration, consider $\mathcal{J} \approx 50\ \mathrm{m^4\,s^{-3}}$. An estimate for Λ comes from dynamo modeling; a plausible range of the typical magnetic field obtained by numerical simulations (Glatzmaier, personal communication, 2006) is between 0.01 and 0.05 $\mathrm{Wb\,m^{-2}}$, giving a plausible range $25 < \Lambda < 626$. For purpose of illustration let $B_0 = 0.03\ \mathrm{Wb\,m^{-2}}$, which yields $\Lambda = 225$. Near the base of the outer core, $r = 1.3 \times 10^6$ m and $g = 4.5\ \mathrm{m\,s^{-2}}$, while in the interior, the mean values are appropriate: $\bar{r} = 2.76 \times 10^6$ m and $\bar{g} = 8.56\ \mathrm{m\,s^{-2}}$. With these numerical values, [53]–[57] become

$$7 \times 10^7 \geq n \geq 1 \qquad [58]$$

$$0.5 \geq f \geq 3 \times 10^{-5} \quad [59]$$

$$9.6 \times 10^{-11} \leq C_0 \leq 3.5 \times 10^{-8} \quad [60]$$

$$1.3 \times 10^{-3}\,\mathrm{m\,s^{-1}} \leq U_p \leq 0.25\,\mathrm{m\,s^{-1}} \quad [61]$$

and

$$8.7\,\mathrm{m} \leq L \leq 1700\,\mathrm{m} \quad [62]$$

It is likely that $\mathcal{J}$ becomes small near the top of the outer core, in which case the above numerical estimates need to be modified appropriately. Further, the plume scaling does not apply within a stable region near the top of the outer core (if such a region exists). In particular, note that the velocity limits in [61] cannot be directly compared with the velocity estimated from secular variation because the value of $\mathcal{J}$ employed above is not representative of conditions near the top of the outer core.

Within the plume region near the base of the outer core, the fact that number of plumes is proportional to the square of f suggests two things. First, within the plume region, plumes are few in number, large, and vigorous. Second, as the plumes rise, they become more numerous, smaller, and weaker. This implies that the plumes are unstable, breaking up as they rise.

6.5.2 Relative Magnitudes of Forces and Fluxes

With these scalings, the relative magnitude of the viscous force is quantified by the Ekman number, written as

$$E = \frac{16\pi r^2 \Omega^2 \nu}{\mathcal{J}} f \sim 0.02f \quad [63]$$

and the magnitude of the perturbation magnetic field produced by an individual plume, relative to the applied field, is

$$\frac{R_m}{\Lambda} = \frac{\mathcal{J}}{16\pi r^2 \Omega^2 \eta_m f} \sim \frac{1.2 \times 10^{-5}}{f} \quad [64]$$

The numerical estimates are suitable for the middle of the outer core. The appearance of the factor Λ in the denominator on the left-hand side of [64] is due to the fact that for plumes the relative magnitude of advection relative to diffusion is given by

$$\frac{(\mathbf{u}\cdot\nabla)}{\nabla^2} \sim \frac{2\Omega L^2}{\Lambda} \sim \frac{2.4 \times 10^{-5}}{f} \quad [65]$$

It follows that the Schmidt and Prandtl numbers for plume flow are

$$S_{ch} = \frac{\mathcal{J}}{16\pi r^2 \Omega^2 D f} \sim \frac{3400}{f} \quad [66]$$

and

$$P_r = \frac{\mathcal{J}}{16\pi r^2 \Omega^2 \bar{\kappa} f} \sim \frac{8}{f} \quad [67]$$

The numerical estimates in [63], [64], [66], and [67] employ the molecular values of viscosity and magnetic diffusivity; see the table of notation. It is clear that molecular viscosity, thermal diffusion, and material diffusion are small for all possible values of f, while magnetic induction in a single plume may become appreciable if $f \leq 1.2 \times 10^{-5}$. (Recall that the equations presented in Section 6.4 are valid only if magnetic induction is negligible.)

6.5.3 Plume Dynamo Action

It is not known whether individual small-scale buoyant plumes contribute positively to dynamo action. Assuming that they do, the combined and cumulative effect of an ensemble of small-scale buoyant plumes on the dynamo process is unknown, but it has the potential to be important. In fact, it was found in Section 6.5.2 that the effective magnetic Reynolds number of individual plumes may be of unit order provided $f \leq 1.2 \times 10^{-5}$. If the dynamo action of individual plumes were additive, the total effect would be quantified by $n_B R_m/\Lambda$, where n_B is the number of plumes threaded by a given line of force. Given that the lateral extent of individual plumes is of order L and assuming that they occupy a fraction f of a given line of force having length h, it follows that $n_B = fh/L$. Using this, [49] and [64] give

$$n_B \frac{R_m}{\Lambda} = \left[\frac{h}{4\pi\Omega\eta_m}\right]\sqrt{\frac{2\pi f\mathcal{J}}{\Omega\Lambda}} \quad [68]$$

With previous numerical estimates, this equation predicts the potential for significant dynamo action if $h\sqrt{f} > 8 \times 10^5$ m. Again it should be emphasized that these quantitative estimates, and the assumption of additive dynamo effect, are very uncertain. However, since $f \leq 1$ by definition, it appears that plumes can provide significant dynamo action only if the extent of a given line of force is larger than the depth of the core. This can be the case; in the strong-field limit, the field is dominated by a strong toroidal field which is predominantly aligned in the zonal direction.

6.5.4 Stratification in Downwelling Regions

Due to the secular evolution of the outer core, successive layers of fluid descending toward the ICB are progressively more buoyant, making the core in such regions stably stratified. The structure of plumes near the bottom of the outer core is likely to be affected by this stratification, which is quantified in this subsection.

Local stratification is typically quantified by the Brunt–Väisälä frequency of the form

$$N \simeq \sqrt{g \mid \mathrm{d}C/\mathrm{d}r \mid} \qquad [69]$$

Near the base of the outer core, $g \approx g_{\mathrm{icb}}$, while the vertical gradient of co-density, using [10], is

$$\frac{\mathrm{d}C}{\mathrm{d}r} = -\alpha_s \frac{\mathrm{d}s_c}{\mathrm{d}r} - \frac{\mathrm{d}c_c}{\mathrm{d}r} \qquad [70]$$

(The derivative of α_s is assumed small.) At the ICB, $\alpha_s \approx 6.5 \times 10^{-5}$ K kg J^{-1}.

Assuming that the descending motion is locally steady and vertical and writing $\mathbf{u} = -W\hat{\mathbf{r}}$, where $\hat{\mathbf{r}}$ is a unit vector pointing upward and using [11] (with $F_c = 0$) and [12] (with $F_s = 0$), [70] becomes

$$\mathrm{d}C/\mathrm{d}r = \dot{C}/W \qquad [71]$$

where

$$\dot{C} \equiv \alpha_s \dot{s}_* - \alpha_s \dot{s}_a - \dot{c}_a = [1.7x_{\mathrm{rd}} + 1.7x_{\mathrm{oh}} - 11.5] \times 10^{-20} \mathrm{s}^{-1} \qquad [72]$$

and the numerical estimates following [14] have been used. It is unlikely that the rate of heating due to radioactivity or Ohmic dissipation is sufficiently large to make this gradient positive. Assuming, for example, that $x_{\mathrm{rd}} + x_{\mathrm{oh}} \approx 3$ (i.e., the rate of internal heating due to radioactive decay and Ohmic heating is 3×10^{12} W), $\dot{C} \approx 5 \times 10^{-20} \mathrm{s}^{-1}$. By conservation of volume, in the limit that $f \ll 1$, the speed of downwelling is given by

$$W = 2\Omega L f \sim \sqrt{2f\Lambda}\left[\frac{1}{4r}\sqrt{\frac{\mathcal{J}}{\pi\Omega}}\right] \sim \sqrt{2f\Lambda}\, 4 \times 10^{-5} \mathrm{m\,s}^{-1} \qquad [73]$$

where [48] has been used. Using $\Lambda = 225$, $W \sim \sqrt{2f}$ 6×10^{-4}m s^{-1} . Combining [71] and [73] gives

$$\frac{\mathrm{d}C}{\mathrm{d}r} \sim \frac{1}{\sqrt{2f\Lambda}}\left[4r\dot{C}\sqrt{\frac{\pi\Omega}{\mathcal{J}}}\right] \qquad [74]$$

and, recalling that $\mathrm{d}C/\mathrm{d}r$ and $\dot{C}$ are negative, the Brunt–Väisälä frequency due to compositional stratification is

$$N = \sqrt{g\left|\frac{\mathrm{d}C}{\mathrm{d}r}\right|} \sim \sqrt{4rg|\dot{C}|\sqrt{\frac{\pi\Omega}{\mathcal{J}2f\Lambda}}} \sim \frac{10^{-7}}{(f\Lambda)^{1/4}}\,\mathrm{s}^{-1} \qquad [75]$$

This translates to an oscillation period of $(f\,\Lambda)^{1/4}$ 0.32 years or, if $\Lambda = 225$, $f^{1/4}$ 0.25 years. Further, if f were to achieve the smallest value quoted in the paragraph following [54], the period of oscillation would be about 5 days. While the magnitudes of these estimates are somewhat uncertain, they demonstrate that a significant degree of stratification is theoretically possible, and that strength of stratification increases as f decreases; that is, the stratification of descending material becomes progressively stronger as it approaches the ICB.

6.6 Cascades and Transfers of Energy in Core Turbulence

Perhaps the greatest difference between classic turbulence and turbulence in Earth's core is in the cascade of energy. In classic turbulence, kinetic energy is fed in at large scales and it cascades through the so-called inertial range to smaller scales, until viscous dissipation transforms it to heat. In contrast, in the core, the cascade process involves the Lorentz force which is dissipative at all scales. That is, there is nothing like an inertial range in which energy is transferred without loss. In particular, Siso-Nadal and Davidson (2004) have shown that the structure of a vortex in a homogeneous rotating hydromagnetic fluid evolves rapidly (on a scale of days) to a shape that minimizes the effects of the Coriolis and Lorentz forces. It is very likely that flows driven by buoyancy will have this same general characteristic of minimizing these forces. A likely shape is one flattened in the plane of $\mathbf{\Omega}$ and $\mathbf{B}$; this is the pancake shape identified by Braginsky and Meytlis (1990; see Section 6.4).

The existence of the inertial range also relies on an input of kinetic energy at the largest scales of motion, often by mechanical means, such as forced convection. If instead the energy of fluid motion is supplied by buoyancy forces (as is the case for most geophysical flows), the situation may be quite different. In order to generate large-scale structures, there must be a significant and reasonably efficient cascade of energy up the spatial scale. This is known to occur in the atmosphere and oceans due to the constraint imposed by the Coriolis force, coupled with the extreme aspect ratio of the fluid bodies (having depth much smaller than lateral extent; e.g., see Starr, 1968). It is not clear whether a similar process will act in the outer core, which is subject to the action of the Lorentz force and which is a deep fluid.

Given that the dynamo has a limited energy supply (*see* Chapter 2), it is apparent that the dynamo must avoid flows and structures that dissipate energy strongly on small scales. One way to accomplish this is to have the energy fed in at the largest dynamically possible scale, with virtually no cascade of energy to smaller scales. Another is to have the small-scale flow structures configured in a way to minimize Ohmic dissipation. At this point, our knowledge of the dynamics of small-scale flow structures is inadequate to determine which, if either, alternative is preferred.

6.7 Approaches to Parametrization of Turbulence

Numerical simulations of convection and dynamo action applicable to Earth's core are forced to operate in regions of parameter space far removed from core conditions. There are two reasons for this: limitations of speed and storage in current computers and the need for numerical stability. A good review of these limitations is given by Glatzmaier (2002). The first of these limitations will be with us for a very long time, while the second can be remedied by the advent of better numerical schemes.

Given the wide disparity in scales between the observed magnetic field ($\sim 10^6$ m) and dissipation of kinetic energy by molecular viscosity (<1 m), it is clear that some sort of parametrization of the smaller scales is necessary in numerical simulations of core dynamics and the dynamo process. Less clear are the processes to be parametrized and the proper form of the parametrizations. For example, scalar dissipative parametrizations are poor approximations to nonisotropic and nondiffusive processes.

6.7.1 The Need for Parametrization

The full set of governing equations consist of four evolution equations plus a diagnostic equation to be solved for two vectors (velocity and magnetic field) and three scalars (pressure, composition, and entropy). The molecular form of these equations contain four diffusivities: η_m, ν, κ, and D. However, the relatively large value of magnetic diffusivity permits the Lorentz force and the equation for the magnetic field to be linearized for small-scale flows, effectively obviating the need for parametrization. That leaves fluxes of momentum, material, and heat to be parametrized.

Consider first the momentum equation, [24], and the need to parametrize viscous momentum flux. The magnitude of molecular viscosity is very small, so that parametrization appears inevitable. However, it first must be determined whether parametrization is necessary. With current numerical models, parametrization of the viscous force is necessary to achieve numerical stability. However, there are numerical schemes on the horizon (such as WENO; see Jiang and Wu, 1999) that are numerically convergent with very small viscosity. Once such schemes are adopted, this requirement will disappear.

The question to be addressed now is whether, and under what circumstances, the viscous force is important and must be parametrized in numerical simulations. Viscosity is well known to be important in thin layers, such as the Ekman and Stewartson layers. Current numerical simulations explicitly resolve the Ekman layer, and this imposes a significant constraint on the choice of grid spacing and magnitude of viscosity. This limitation could be removed by the use of Ekman–Hartmann compatibility conditions; see Loper (1970). Further, when hydromagnetic effects are present, they dominate the Stewartson-layer structure, making the viscous force irrelevant; see Hollerbach (1996). In sum, there appears to be no need to parametrize small-scale viscous structures such as boundary layers.

What is rather less well known is the role of the viscous force in determining large-scale structure in flows driven by buoyancy forces, as is the case in Earth's core. Shimizu and Loper (1997) showed that, no matter how small the molecular viscosity, the viscous force is important in determining the scale of rotating hydromagnetic flow structures driven by buoyancy. This importance may be understood by realizing that both the Coriolis and Lorentz forces are anisotropic and provide no constraint on the shearing structure of flows that lie in the plane of $\mathbf{\Omega}$ and $\mathbf{B}$; the viscous force must provide that constraint. The extent of the flow structures which involve viscosity is quite large, and may not be realized in the outer core, which is of finite size.

The parametrized form of turbulent diffusion of entropy and material should mirror the structure of the relevant fluid motions. That is, the effective diffusivities should be anisotropic, having a structure such as that proposed by Braginsky and Meytlis (1990), as described in the following subsection.

6.7.2 Diffusive Parametrizations

The simplest form of parametrization is that dictated by molecular processes, using greatly enhanced coefficients, often called eddy diffusivities. "All models of the geodynamo crudely approximate… turbulent transport as a simple, isotropic, homogeneous, diffusive process modeled after molecular diffusion but with an enhanced 'turbulent diffusivity'." (Glatzmaier, 2002). This approach has been surprisingly successful, producing earth-like dynamos in numerical simulations. However, these simulated dynamos require a large amount of power – more than the Earth is likely to be able to supply. The diffusive coefficients used in current 3-D numerical simulations are sufficiently large that the resolved scales of motion are stable and the flow fields are in fact laminar. Glatzmaier (2005) has noted that turbulent motions are likely to be significantly different than laminar, so it is important to reduce the diffusivities and run 3-D models in the turbulent regime.

A variant on the eddy diffusivity approach is to use hyperdiffusivities, in which the magnitude of the diffusion coefficient increases as the lengthscale decreases. This has the advantage of reducing the amount of dissipation at larger scales, while providing numerical stability. The utility of this approach has been summarized by Glatzmaier (2002).

The influence of Coriolis and Lorentz forces is nonisotropic and a better parametrization will correspondingly entail nonisotropic (tensorial) eddy diffusivities. A heursitic model has been proposed by Braginsky and Meytlis (1990; summarized in appendix C of Braginsky and Roberts, 1995). In this model, the scalar viscosity is replaced by a tensor with principal axes locally aligned with $\mathbf{\Omega}$ and $\mathbf{B}$ such that the in-plane viscosity is of the same magnitude as magnetic diffusion (i.e., on the order of $1\ \mathrm{m^2\,s^{-1}}$), while that normal to the plane is smaller by a factor 10^{-3}. The parametrizations of diffusion of heat and material are the same as for viscosity.

6.7.3 Alternative Parametrizations

A fundamental shortcoming of all diffusive parametrizations is their inability to simulating nondiffusive processes that alter the scale and structure of flow and field. As a result, diffusive parametrizations tend to overpredict dissipation of magnetic and kinetic energy and misrepresent the dynamical structures. Several approaches have been suggested to circumvent this limitation, including the similarity and alpha models. Common features of these models are a smoothing operation and a closure assumption.

The similarity model is based on a premise and an assumption. The premise is that the dominant subgrid scales are those just below the resolution of the large-eddy simulation (LES) at hand. The assumption is that the structure of largest unresolved scales is similar to that of the smallest resolved scales. The development of the similar model entails the application of a series of smoothing operations of the form

$$\tilde{\mathbf{u}}(\mathbf{r}, t) = \int G(\mathbf{r}-\mathbf{r}')\mathbf{u}(\mathbf{r}', t)\mathrm{d}\mathbf{r}' \qquad [76]$$

where the smoothing operator, G, is typically Gaussian in form. Closure is achieved by assuming that the fluctuations to be parametrized are similar in form to those of the smallest resolved scales. Initial implementation of this method appeared promising; it clearly outperforms eddy-diffusive models (Buffett, 2003). However, recent results indicate that this method is of limited utility; its ability to accurately predict the structure of the subgrid processes seems to decrease fairly rapidly with scale separation.

The alpha model employs smoothing of small-scale fluctuations on a Lagrangian trajectory, with the result expressed in the usual Eulerian framework. The smoothing process can be cast in the form of [76]. The parameter alpha, which gives the method its name, is a lengthscale separating large-scale, active motions from small-scale passive motions. The formulation is closed, for example, by using Taylor's hypothesis of 'frozen-in' turbulence. For details, see Holm (2002). Like the similarity model, the alpha model is much in its infancy and it has yet to be demonstrated that it is capable of resolving small-scale structures accurately.

6.8 Unresolved Issues and Future Directions

The goal of numerical simulations of core dynamics and the geodynamo is to operate in parameter regimes where the results are relevant to Earth's core. Initial results with crude parametrizations, involving large eddy diffusivities, have been very promising, but these results are for parameters far from 'Earth like'. In order to make progress, a combination of advances needs to be made, including the use of larger and faster machines, the use of better algorithms (such as WENO, that do not need dissipation for numerical stability), and the use of better parametrizations of small-scale structures near boundaries (such as the Ekman or Ekman–Hartmann layer) and in the interior of the core (associated with turbulence and energy cascades).

Bigger and faster machines will be developed at a fairly predictable pace, but given the disparity between currently resolved scales and the smallest dynamic scales in the core, there is little hope of solving the geodynamo completely by direct numerical simulation (i.e., by brute force). Major advances can be made fairly quickly and economically by adopting new numerical schemes that do not need large dissipations for numerical stability, and parametrizations that obviate the need to resolve viscous boundary layers. However, the problem of parametrizing the unresolved scales will remain for some time to come.

Regarding turbulence and energy cascades, relatively little is known of 'mesoscale' dynamics, on scales having small magnetic Reynolds, Ekman, Peclet, and Schmidt numbers. Key questions that remain to be resolved include:

- On what scale is energy fed into convective motions by buoyancy forces?
- Is there a significant cascade of mechanical energy to larger spatial scales?
- With a linearized momentum equation, how does the energy cascade occur?
- On what scales is mechanical energy degraded to heat?
- Are there configurations of flow and field for which energy losses during cascade are small?
- Does the core adopt these loss-minimization structures?
- What is the energy requirement of the geodynamo?
- Is viscosity important in determining core flow structures or energy dissipation?

Once answers to these questions have been obtained, then dynamically consistent parametrizations may be devised and implemented.

Appendix 1: Core Evolution and Buoyancy Flux

This appendix contains essential background material for the main body of the chapter quantifying the evolution of the composition and entropy in the outer core. First the secular evolution of composition and specific entropy in the outer core are quantified. Sources of specific entropy are quantified next. Finally, the sources are combined to quantify the buoyancy flux.

Evolution of Composition

It follows from conservation of the heavy and light constituents of the core that the rate of evolution of the composition of the outer core is given by

$$\dot{c}_{\mathrm{a}} = \delta \dot{M}_{\mathrm{ic}} / M_{\mathrm{oc}} \qquad [77]$$

where δ is the fractional jump in density due to change of composition across the ICB, M_{oc} is the mass of the outer core, and $\dot{M}_{\mathrm{ic}} = \mathrm{d}M_{\mathrm{ic}}/\mathrm{d}t$ is the rate of growth of the solid inner core. A reasonable estimate based on eigenfrequencies of normal modes of Earth oscillation, is $\delta \approx 0.05$ (Masters and Gubbins, 2003). The magnitude of $\dot{M}_{\mathrm{ic}}$ is uncertain and somewhat controversial, but there is little doubt that the inner core is growing, and that $\dot{c}_{\mathrm{a}} > 0$. The estimate $\dot{c}_{\mathrm{a}} \approx 9 \times 10^{-20}\ \mathrm{s}^{-1}$ by Roberts and Glatzmaier (2000) corresponds to a growth rate $\dot{M}_{\mathrm{ic}} = 3.5 \times 10^{6}\ \mathrm{kg\,s}^{-1}$; if the inner core grew at this constant rate, its age would be 0.9 Ga.

The secular change of the reference-state composition acts as a volumetric sink of buoyant material within the outer core in eqn [11] governing the convective perturbation, c_{c}. A counterbalancing source is provided at the ICB by the solidification process. A fluid parcel gains light material as it passes close to the ICB (or as it penetrates the uppermost layers of the mushy inner core), with the change given by

$$(\Delta c_{\mathrm{c}})_{\mathrm{icb}} = \delta \dot{M}_{\mathrm{ic}} / (\dot{M})_{\mathrm{icb}} = \dot{c}_{\mathrm{a}} \tau_{\mathrm{oc}} \qquad [78]$$

where $\dot{M}$ is the convective mass flux (being a function of radius), $(\dot{M})_{\mathrm{icb}}$ is that flux evaluated at the ICB, and $\tau_{\mathrm{oc}} = M_{\mathrm{oc}} / (\dot{M})_{\mathrm{icb}}$ is the mean turnover time of the outer core.

Written in terms of the buoyancy flux, [78] is

$$(\mathcal{J}_{\mathrm{c}})_{\mathrm{icb}} = \delta [g/\rho]_{\mathrm{icb}} \dot{M}_{\mathrm{ic}} \approx \delta g_{\mathrm{icb}} \dot{V}_{\mathrm{ic}} \qquad [79]$$

where $\mathcal{J}_{\mathrm{c}} = c_{\mathrm{c}} g \dot{V}$ is the buoyancy flux due to composition, $\dot{V}_{\mathrm{ic}} = \dot{M}_{\mathrm{ic}} / \bar{\rho}_{\mathrm{ic}}$ is the rate of growth of the volume of the inner core, and $\bar{\rho}_{\mathrm{ic}} (\approx \rho_{\mathrm{icb}})$ is the mean density of the inner core. The rate of growth of the inner core is somewhat uncertain; one way to parametrize it is to write $\dot{V}_{ic} = V_{ic}/\tau_{ic}$, where V_{ic} ($=7.6 \times 10^{18}\,m^3$) is the current volume of the inner core and τ_{ic} would be the age of the inner core if its volume had increased at a constant rate. Estimates of τ_{ic} vary from about 0.5 to 4.0 Ga ($\tau_{ic} \approx 1.6 \rightarrow 13 \times 10^{16}$ s), giving $(\mathcal{J}_c)_{icb} \approx 13 \rightarrow 105\,m^4\,s^{-3}$. This is a relatively modest buoyancy flux; black smokers and other sources of buoyancy in the ocean typically have far greater magnitudes.

Assuming that there are no significant sources or sinks of material at the CMB, the variation of this buoyancy flux with radius in the outer core is given by

$$\mathcal{J}_c = (\mathcal{J}_c)_{icb}[M_c - M]/M_{oc} \qquad [80]$$

where $M(r)$ is the mass within a sphere of radius r and M_c ($\sim 1.94 \times 10^{24}$/kg) is the mass of the core. The buoyancy flux is related to the power, P_c, released by the rearrangement of matter in the outer core:

$$P_c = \int_{r_{icb}}^{r_{cmb}} \mathcal{J}_c \rho\, dr \qquad [81]$$

Using data in Stacey and Davis (2004), plus the estimated range of τ_{ic} given above, $P_c \approx 0.24 \rightarrow 1.9 \times 10^{12}$ W. This power is sufficient to drive the dynamo in the strong-field regime (Roberts *et al.* (2003) and *see* Chapter 2), even in the absence of a contribution from the specific entropy. However, if most of this energy is dissipated by small-scale motions, the amount available to sustain the large-scale field will be correspondingly smaller.

Evolution of Specific Entropy

The specific entropy of the outer core decreases with time due to the progressive cooling (Gubbins *et al.*, 2003):

$$\dot{s}_a \approx -Q_{sc}/\bar{T}M_c \qquad [82]$$

where

$$\bar{T} = \frac{1}{M}\int_{core} T\rho\, dV \qquad [83]$$

is the mean temperature of the core ($\bar{T} \approx 4350$ K) and Q_{sc} is the heat flux from core to mantle resulting from secular cooling of the core. The other source of heat from radioactive decay is, in effect, of external origin and so does not contribute to the secular entropy change. The contribution to [82] due to the rearrangement of material associated with the growth of the inner core is negligibly small.

In addition to the sensible cooling of the core, secular cooling includes the latent heat and gravitational energy released by the growth of the inner core; altogether

$$Q_{sc} = C_{pe}M_c[-d\bar{T}/dt] \qquad [84]$$

where $C_{pe} = \bar{C}_p + C_{gr} + C_{lh}$ is the effective specific heat of the core, $\bar{C}_p$ ($\approx 800\,J\,kg^{-1}\,K^{-1}$) is the mean (mass averaged) specific heat, and the effective specific heats for gravitational energy and latent heat are given by

$$C_{gr} = \left[\frac{M_{ic}}{\Gamma M_{oc}}\right]\frac{[\Delta\rho]\bar{\psi}}{\rho\bar{T}} \approx 300\frac{J}{kg\ K} \qquad [85]$$

and

$$C_{lh} = \left[\frac{M_{ic}}{\Gamma M_c}\right]\frac{L}{\bar{T}} \approx 600\frac{J}{kg\ K} \qquad [86]$$

where M_{ic} ($\approx 9.84 \times 10^{22}$ kg) and M_{oc} ($\approx 1.84 \times 10^{24}$ kg) are the masses of the inner and outer cores, $M_c = M_{oc} + M_{ic}$, $\bar{\psi}$ ($\approx 10^7\,J\,kg^{-1}$) is the mean gravitational potential of the outer core, and

$$\Gamma = \left[\frac{\alpha\rho_a g_a r}{3\rho_a C_p + 6K_S\alpha^2 T}\right]_{icb} \qquad [87]$$

is a measure of the rate of advance (increase of radius with time) of the intersection of the adiabat with the melting curve as the core cools; K_S is the adiabatic incompressibility. With previous estimates, $\Gamma \approx 0.02$.

Note that with these estimates, a relatively modest rate of cooling of, say, $1.14 \times 10^{-15}\,K\,s^{-1}$ ($=36\,K\,Ga^{-1}$) releases heat at a rate sufficient to equal that conducted down the adiabat at the CMB (3.6×10^{12} W with $k = 28\,W\,m^{-1}\,K^{-1}$). Note that $C_{pe}M_c \approx 10^{11}$ W Ga K^{-1}, so that a change of cooling rate of 10 K Ga^{-1} corresponds to a change of heat flux of 10^{12} W. For purposes of illustration, suppose that 5×10^{12} W are released by cooling of the core, corresponding to a cooling rate of 50 K Ga^{-1}. With a convective turnover time of 3700 years, the core cools by about $\Delta T = 2 \times 10^{-4}$ K each convective cycle.

The cooling rate of the core is somewhat uncertain. This may be quantified by writing $d\bar{T}/dt = -x_T\ 10\,K\,Ga^{-1} \approx -x_T\ 3 \times 10^{-16}\,K\,s^{-1}$, so that a value $x_T = 1$ corresponds roughly to 10^{12} W of core cooling. Altogether, $\dot{s}_a \approx -1.2x_T \times 10^{-16}\,W\,kg^{-1}\,K^{-1}$. The value ($\dot{s}_a \approx -3.7 \times 10^{-16}\,W\,kg^{-1}\,K^{-1}$) estimated by Roberts and Glatzmaier (2000) is achieved if the rate of heat loss from the core is 3.4×10^{12} W.

Specific-Entropy Sources

The variation of specific entropy within the outer core is more complicated than that of composition, for two reasons. First, [12] contains source terms other than that due to secular evolution of the reference state, and second, the specific entropy is changed due to exchanges of heat at the CMB as well as at the ICB. In this subappendix, first the three volumetric entropy sources that contribute to $\dot{s}_*$ (see [13]) are estimated close to the ICB, then the boundary sources at the ICB and CMB are estimated.

Volumetric sources

It is reasonable to assume that Ohmic heating is uniformly distributed (by mass) in the entire core so that

$$q_{\mathrm{oh}} = Q_{\mathrm{oh}}/M_{\mathrm{c}} \qquad [88]$$

It is likely, due to segregation induced by the solidification process, that radioactive heating is concentrated in the outer core, so that

$$q_{\mathrm{rd}} = Q_{\mathrm{rd}}/M_{\mathrm{oc}} \qquad [89]$$

There is substantial uncertainty in each of the three volumetric sources. It is uncertain whether there is a significant amount of radioactive heating in the core. The efficiency of the dynamo mechanism and the associated amount of Ohmic heating is not well constrained. While the radial structure of the adiabat is well constrained, the value of thermal conductivity is controversial. The uncertainties in heat sources will be accommodated by writing $Q_{\mathrm{rd}} = x_{\mathrm{rd}}\ 10^{12}$ W and $Q_{\mathrm{oh}} = x_{\mathrm{oh}}\,10^{12}$ W. The value of thermal conductivity will be taken as $k = 28\ \mathrm{W\,m^{-1}\,K^{-1}}$ (F. Stacey, personal communication, 2005), with $\nabla^2 T_{\mathrm{a}} \approx -7\times10^{-10}\ \mathrm{K\,m^{-2}}$. Altogether, near the ICB, $\{q_{\mathrm{rd}},\ q_{\mathrm{oh}},\ k[\nabla^2 T_{\mathrm{a}}]/\rho_{\mathrm{a}}\} = \{5.4x_{\mathrm{rd}},\ 5.2x_{\mathrm{oh}},\ -15\}\times 10^{-13}\mathrm{W\,kg^{-1}}$ and with $T_{\mathrm{icb}} \approx 5000$ K $\dot{s}_* = [2.7x_{\mathrm{rd}} + 2.6x_{\mathrm{oh}} - 7.5]\times10^{-16}\ \mathrm{W\,kg^{-1}\,K^{-1}}$.

Boundary sources

At the ICB, latent heat and inner-core sources increase s_{c}, while at the CMB, heat loss to the mantle (over and above that carried by conduction) decreases s_{c}. These changes are quantified by

$$(\Delta s_{\mathrm{c}})_{\mathrm{icb}} = [Q_{\mathrm{lh}} + Q_{\mathrm{ic}}]/(MT_{\mathrm{a}})_{\mathrm{icb}} \qquad [90]$$

and

$$(\Delta s_{\mathrm{c}})_{\mathrm{cmb}} = [Q_{\mathrm{cd}} - Q]/(\dot{M}T_{\mathrm{a}})_{\mathrm{cmb}} \qquad [91]$$

where

$$Q_{\mathrm{lh}} = L_{\mathrm{H}}\dot{M}_{\mathrm{ic}} \qquad [92]$$

is the rate of latent heat release due to solidification, Q_{cd} is the rate of transfer of heat by conduction down the adiabat, Q is the total rate of heat transfer from core to mantle, and L_{H} is the latent heat of fusion. The rate, Q_{ic}, of heat transfer from the inner core is due to those fractions of core secular cooling, Q_{sc}, and Ohmic dissipation, Q_{oh}, that occur within the inner core (with radioactive heating in the inner core being assumed small). The former is proportional to the potential temperature, giving

$$Q_{\mathrm{ic}} = \frac{M_{\mathrm{ic}}}{M_{\mathrm{c}}}\left[\frac{\bar{T}_{\mathrm{ic}}}{\bar{T}_{\mathrm{c}}} Q_{\mathrm{sc}} + Q_{\mathrm{oh}}\right] = 0.057 Q_{\mathrm{sc}} + 0.05 Q_{\mathrm{oh}} \qquad [93]$$

In a rough calculation, the effect of Q_{ic} may be ignored.

Buoyancy Flux

The changes of both composition and specific entropy at the ICB make the rising fluid buoyant and drive vigorous convection near the base of the outer core. Without loss of generality, the co-density of the descending fluid at the base of the outer core may be set equal to zero. The co-density of the fluid rising from the ICB is related to the buoyancy flux, $\mathcal{J}$, at the base of the outer core by

$$\mathcal{J}_{\mathrm{icb}} = \left[Cg\dot{V}\right]_{\mathrm{icb}} = -\left[L_H\left(\alpha/C_p\right)_{\mathrm{icb}} + \delta\right] g_{\mathrm{icb}}\dot{V}_{\mathrm{ic}} \qquad [94]$$

Using data from Gubbins *et al.* (2003) and Stacey and Davis (2004), $L_{\mathrm{H}}\ (\alpha/C_p)_{\mathrm{icb}} \approx 0.01$. Recalling that $\delta \approx 0.05$, it appears that change of composition is several times more important that latent heat in creating buoyancy at the ICB. Using previous estimates of τ_{ic} and other parameters, $\mathcal{J}_{\mathrm{icb}} \approx 15 \rightarrow 125\ \mathrm{m^4\,s^{-3}}$.

Barring significant mass exchanges at the CMB, the buoyancy at the top of the mantle is dominated by the change of specific entropy. If $Q > Q_{\mathrm{cd}}$, this change drives convection near the top of the outer core.

References

Braginsky SI (1999) Dynamics of the stably stratified ocean at the top of the core. *Physics of the Earth and Planetary Interiors* 111: 21–34.

Braginsky SI and Meytlis VP (1990) Local turbulence in the Earth's core. *Geophysical and Astrophysical Fluid Dynamics* 755: 71–87.

Braginsky SI and Roberts PH (1995) Equations governing convection in Earth's core and the geodynamo. *Geophysical and Astrophysical Fluid Dynamics* 79: 1–97.

Buffett BA (2003) A comparison of subgrid-scale models for large-eddy simulations of convection in the Earth's core. *Geophysical Journal International* 153: 753–765.

Cambon C (2001) Turbulence and vortex structures in rotating and stratified flows. *European Journal of Mechanics B* 20: 489–510.

Cambon C, Mansour NN, and Godeferd FS (1997) Energy transfer in rotation turbulence. *Journal of Fluid Mechanics* 337: 303–332.

Chalmers B (1964) *Principles of Solidification*. New York: Wiley.

Chandrasekhar S (1961) *Hydrodynamic and Hydromagnetic Stability*. Oxford: Clarendon Press.

Charney JG (1971) Geostrophic turbulence. *Journal of Atmospheric Science* 28: 1087–1095.

Classen S, Hempel M, and Christensen U (1999) Blob instability in rotating compositional convection. *Geophysical Research Letters* 26: 135–138.

Davidson PA (2000) *An Introduction to Magnetohydrodynamics*, 422pp. Cambridge: Cambridge University Press.

Eltayeb IA, Hamza EA, Jervase JA, Krishnan E, and Loper DE (2005) Compositional convection in the presence of a magnetic field. Part II. Cartesian plume. *Proceedings of the Royal Society of London* 461: 2605–2633.

Eltayeb IA and Loper DE (1997) On the stability of vertically oriented double diffusive interfaces. Part 3. Cylindrical interface. *Journal of Fluid Mechanics* 353: 45–66.

Glatzmaier GA (2002) Geodynamo simulations – How realistic are they? *Annual Review of the Earth and Planetary Science* 30: 237–257.

Glatzmaier GA (2005) Planetary and stellar dynamos: Challenges for next generation models. In: Soward AM, Jones CA, Hughes DW, and Weiss NO (eds.) *Fluid Dynamics and Dynamos in Astrophysics and Geophysics*, pp. 331–357. Boca Raton: CRC Press.

Gubbins D, Alfe D, Masters G, Price GD, and Gillan MJ (2003) Can the Earth's dynamo run on heat alone? *Geophysical Journal International* 155: 609–622.

Hollerbach R (1996) Magnetohydrodynamic shear layers in a rapidly rotating plane layer. *Geophysical and Astrophysical Fluid Dynamics* 82: 237–253.

Holm DD (2002) Lagrangian averages, averaged Lagrangians, and the mean effects of fluctuations in fluid dynamics. *Chaos* 12: 518–530.

Jiang G-S and Wu CC (1999) A high-order WENO finite difference scheme for ideal magneto-hydrodynamics. *Journal of Computational Physics* 150: 561–594.

Loper DE (1970) General solution for the linearized Ekman-Hartmann layer on a spherical boundary. *Physics of Fluids* 13: 2995–2998.

Loper DE and Roberts PH (1981) A study of conditions at the inner core boundary of the earth. *Physics of the Earth and Planetary Interiors* 24: 302–307.

Masters G and Gubbins D (2003) On the resolution of density within the Earth. *Physics of the Earth and Planetary Interiors* 140: 159–167.

Merrill RT, McElhinney MW, and McFadden PL (1996) *The Magnetic Field of the Earth: Paleomagnetism, the Core and the Deep Mantle*, 531pp. San Diego: Academic Press.

Moffatt HK (1967) On the suppression of turbulence by a uniform magnetic field. *Journal of Fluid Mechanics* 28: 571–592.

Moffatt HK (1978) *Magnetic Field Generation in Electrically Conducting Fluids*, 343pp. Cambridge: Cambridge University Press.

Pedlosky J (1987) *Geophysical Fluid Dynamics* 2nd edn., 710pp. New York: Springer.

Poirier JP (1988) Transport properties of liquid metals and viscosity of the Earth's core. *Geophysical Journal* 92: 99–105.

Rhines PB (1979) Geostrophic turbulence. *Annual Review of Fluid Mechanics* 11: 401–441.

Roberts PH and Glatzmaier GA (2000) Geodynamo theory and simulations. *Review of Modern Physics* 72: 1081–1123.

Roberts PH, Jones CA, and Calderwood AR (2003) Energy fluxes and Ohmic dissipation in the earth's core. In: Jones C, Soward A, and Zhang K (eds.) *Earth's Core and Lower Mantle*. London: Taylor and Francis.

Shimizu H and Loper DE (1997) Time and length scales of buoyancy-driven flow structures in a rotating hydromagnetic fluid. *Physics of the Earth and Planetary Interiors* 104: 307–329.

Siso-Nadal F and Davidson PA (2004) Anisotropic evolution of small isolated vortices within the core of the Earth. *Physics of Fluids* 16: 1242–1254.

Stacey FD and Anderson OL (2001) Electrical and thermal conductivities of Fe-Ni-Si alloy under core conditions. *Physics of the Earth and Planetary Interiors* 124: 153–162.

Stacey FD and Davis PM (2004) High pressure equations of state with applications to the lower mantle and core. *Physics of the Earth and Planetary Interiors* 142: 137–184.

Starr VP (1968) *Physics of Negative Viscosity Phenomena*, 255pp. New York: McGraw-Hill.

Turner JS (1975) *Buoyancy Effects in Fluids*. Cambridge: Cambridge University Press.

7 Rotational Dynamics of the Core

A. Tilgner, University of Göttingen, Göttingen, Germany

7.1	**Introduction**	208
7.1.1	Motivation	208
7.1.2	Equations of Motion	209
7.1.3	Inertial Oscillations in Infinitely Extended Fluids	211
7.1.4	Ekman Layers	213
7.2	**Inertial Oscillations in Spherical Shells**	215
7.2.1	The Mathematical Problem	216
7.2.2	Ray Geometry	217
7.2.3	Numerical Simulations	218
7.3	**Precession**	219
7.3.1	Solutions of the Inviscid Equation of Motion	220
7.3.2	Viscous Effects	221
7.3.3	Experiments	223
7.3.4	Numerical Solutions	224
7.3.4.1	Laminar flows	225
7.3.4.2	Instability	228
7.4	**Tides**	234
7.5	**Interaction with Buoyancy and Magnetic Fields**	236
7.6	**Summary and Outlook**	239
References		241

Nomenclature

a	long semi-axis of a spheroid (m)
c	short semi-axis of a spheroid (m)
d	distance between inner and outer boundary of a spherical shell (m)
e_i, e_o	ellipticities of inner and outer boundaries of a spheroidal shell
$e = 1 - c/a$	ellipticity
E	Ekman number
E_a	kinetic energy contained in $\mathbf{u}_a$
$\mathbf{k}$	wave vector
m	azimuthal wave number
r, θ, φ	spherical polar coordinates, with θ the colatitude
r_i, r_o	radii of inner and outer boundaries of a spherical shell
$\mathbf{u}$	velocity
$\mathbf{u}_a$	velocity field antisymmetric with respect to reflection at the origin
$\mathbf{u}_s$	velocity field symmetric with respect to reflection at the origin
$x, y, z,$	Cartesian coordinate system with z along rotation axis
$\hat{\mathbf{z}}$	unit vector along z, $\boldsymbol{\omega}_D = \omega_D \hat{\mathbf{z}}$
α	angle between rotation and precession axes
ν	kinematic viscosity, $(m^{-2}\,s^{-1})$
ω	angular frequency of inertial modes
ω_D	diurnal rotation rate ($\approx 2\pi/24\,h$), (s^{-1})
ω_F	rotation vector of fluid in a precessing container
$\boldsymbol{\Omega}$	precession vector
Ω	Ω_P/ω_D
Ω_P	rate of precession (s^{-1})

7.1 Introduction

7.1.1 Motivation

If the mantle was a perfectly rigid shell rotating at constant angular velocity, the core would rotate with the same angular velocity about the same axis because of viscous entrainment. This chapter deals with perturbations of the simple rigid rotation of the core. The most important perturbations are due to variations of the rotation of the mantle and deformation of the core–mantle boundary (CMB).

There are at least two areas of geophysics for which the rotational dynamics of the core are relevant. First, motions in the core act back on the mantle and affect its rotation (*see* Chapter 12). Investigation of Earth's rotation as a whole obviously requires study of the rotational dynamics of the core in particular. Second, the core motions also generate the magnetic field of the Earth. Flow induced by perturbations of the rotation of the core may contribute to the geodynamo or distort a magnetic field generated by convectively driven flow.

Convection, both thermal and compositional, is a plausible driving mechanism for core motion and ultimately the geodynamo. The structure of buoyancy-driven flows is well documented and it is in particular established that these flows can produce a dynamo effect. But the energy budget is tight and it could well be that buoyancy is insufficient to power the geodynamo (*see* Chapter 2). At any rate, precession of the rotation axis of the Earth and tidal deformation of the CMB are two effects of which we know that they exist, whereas we do not know with certainty whether the core is convecting, which makes it indispensable to study the response of the rotating core to precession and tides. These two mechanisms share many features, so that precession will be treated in detail, and the ideas developed in this context will then be transposed to the effects of tidal deformation. Both mechanisms have connections with celestial mechanics. A full treatment would require that we also take into account, for example, variations of the lunar orbit which result from dissipation in the core. But so far, only simpler models have been studied in which the motion or shape of the CMB is prescribed and the response of the core is inquired.

Figure 1 summarizes the problem under study. The mantle rotates about the geographic axis with the rate of diurnal rotation ω_D. The axis of (retrograde) precession is perpendicular to the ecliptic and forms an angle of 23.5° with the geographic axis. The shape of the CMB is assumed spherical or spheroidal (with the geographic axis as axis of symmetry). **Table 1** collects some relevant numbers.

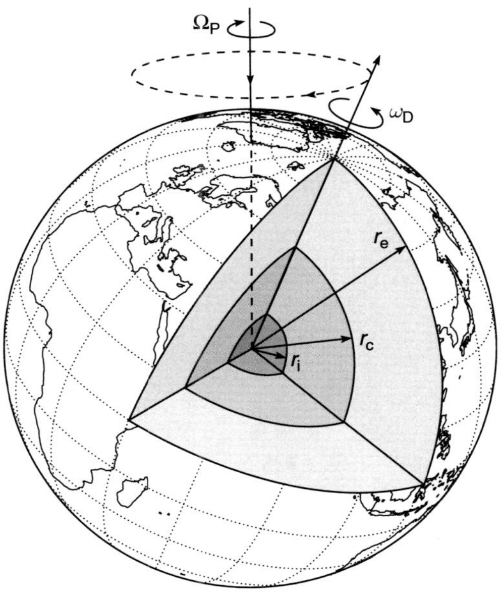

Figure 1 Interior structure of the Earth. The radii of the inner core, r_i, of the CMB, r_c, and of the Earth's surface, r_e, are indicated together with the rates of diurnal rotation ω_D and precession Ω_P.

Other processes have been suggested as possible perturbation of the core's rotation such as earthquakes (Aldridge, 1975). These are certainly less potent than precession and tides. Even the nutations can be neglected compared with precession (Bullard, 1949). There is also the possibility that the inner core and the

Table 1 Some properties of the Earth

r_i	1220 km
r_c	3480 km
r_e	6370 km
e_{IC}	1/414
e_{CMB}	1/393
Ω_P/ω_D	-1.07×10^{-7}
α	23.5°
ν	$10^{-6}\,m^2\,s^{-1}$
σ	$5 \times 10^5\,S\,m^{-1}$

The average radii of the inner core, r_i, the CMB, r_c, and the entire Earth, r_e, together with the ellipticities of the inner core and the CMB, e_{IC} and e_{CMB}, which are taken from Matthews *et al.* (1991b). The ellipticities are defined with the minor and major axes *c* and *a* as $e = 1 - (c/a)$. The Earth precesses in retrograde direction about the normal to the ecliptic with a period $2\pi/\Omega_P$ of 25 700 yr, the diurnal rotation period $2\pi/\omega_D$ equals 1 day. α is the angle between the rotation axis and the normal to the ecliptic. Estimates for the viscosity ν vary widely; see Poirier (1988). The conductivity σ is from Knittle and Jeanloz (1986).

mantle do not rotate about the same axis or at the same angular velocity which would force a complicated flow in the outer core. There is no undisputed seismological evidence for a differential rotation between the inner core and the mantle (Laske and Masters, 1999; Souriau and Poupinet, 2000) and gravitational torque acts to resist it (Buffett, 1996). Even if a differential rotation exists, it can only happen as a consequence of motions in the liquid core, so that differential rotation cannot be viewed as a driving mechanism in its own right, and it will not be covered in this chapter. A good starting point for the literature on the fluid dynamics of differential rotation is Hollerbach (1998).

As for any other fluid flow, it will be useful to distinguish laminar response from unstable and chaotic flows in which large-scale motion disintegrates into smaller eddies. It is at present not known which flow regime the Earth's core is in. A laminar flow draws little energy from the motion of the mantle, whereas unstable flow can dissipate orders of magnitude more power. Only if the flows driven by precession or tides are unstable can the rotational dynamics of the core make a significant contribution to the dynamo process.

Most of this chapter considers the core to be neutrally buoyant. This allows us to look at the peculiarities of rotational effects in isolation and to separate them from buoyancy effects. However, barring an extraordinary coincidence, the core is either stably stratified or convecting. It is important to check which properties of the rotational dynamics may be modified by buoyancy. This will be done towards the end in Section 7.5.

The topic covered in this chapter suffers from a paucity of directly useful observational data. In fact, **Table 1** contains virtually all relevant numbers, that is, the geometry of the core, the astronomical data on the precession of the Earth's axis, and the presumed viscosity and electrical conductivity of the core material. Because no direct observation of the core flow is possible, we have to try to deduce that flow from a combination of theory, numerical simulations, and laboratory experiments. Possible connections with observations to be exploited in the future are compiled in the concluding section.

The introductory section presents some general properties of rotating fluids which will be used later in the chapter. The second section is devoted to inertial modes. These are oscillatory motions in rotating bodies of fluid. Precession excites directly one particular of these modes (the spin-over mode, also called the tilt-over mode) and may excite many more, for instance, through instabilities or interaction with topography. In fact, any perturbation of rotating fluid which is small enough to be described by linearized equations of motion excites superpositions of inertial modes.

The third section is the largest in this chapter and collects our knowledge about precession-driven flow stemming from theory, experiment, and numerical simulation. The closely connected problem of tidally driven flows is treated in Section 7.4. The interaction with buoyancy and magnetic fields, as well as the dynamo effect, are discussed in Section 7.5 before conclusions and an outlook is given in the Section 7.6.

7.1.2 Equations of Motion

Since we will restrict ourselves to incompressible fluid, the Navier–Stokes equation is the appropriate starting point. This section shows how to use this equation if we want to allow for a variable rotation rate of the mantle.

Two different frames of reference are useful in formulating the equations of motion. The first one is attached to the boundaries and will be called the mantle frame, the second one is chosen such that the precession and rotation axes are stationary and will be called the precession frame; see **Figure 2** for definitions. A similar choice has to be made for the tidal problem, where one can opt either for the mantle frame or for a frame in which the tidal bulge is stationary.

We start by deriving the Navier–Stokes equation for an incompressible fluid in an arbitrary frame of reference rotating with the instantaneous angular frequency $\boldsymbol{\omega}_{\text{ref}}$ relative to inertial space and ask which acceleration is experienced by a fluid particle at position $\mathbf{r}$. The most algorithmic way of obtaining the relation between the acceleration measured in the rotating and an inertial frame consists in writing the Euler matrix describing the transformation between the two frames and taking its time derivatives. A more intuitive and shorter derivation uses the relation between time derivatives of vectors in the two frames (Goldstein, 1980) $(\partial/\partial t)_{\text{inert}} = (\partial/\partial t)_{\text{rot}} + \boldsymbol{\omega}_{\text{ref}}\times$. The subscripts specify whether the time dependences are viewed from the inertial or the rotating system. It follows that

$$\left(\frac{\partial}{\partial t}\mathbf{r}\right)_{\text{inert}} = \left(\frac{\partial}{\partial t}\mathbf{r}\right)_{\text{rot}} + \boldsymbol{\omega}_{\text{ref}} \times \mathbf{r}$$

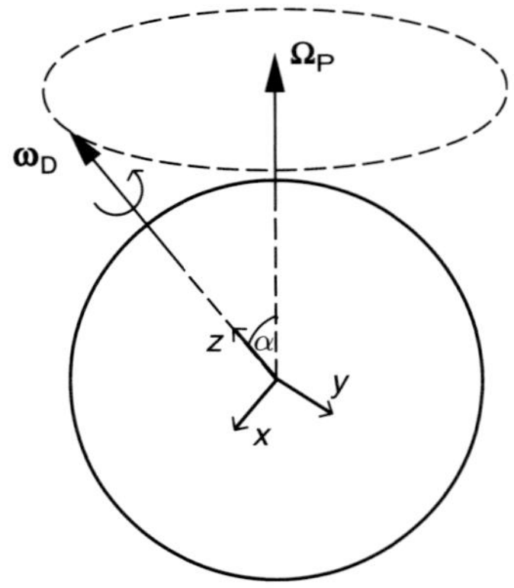

Figure 2 The z-axis is chosen to coincide with the axis of diurnal rotation ω_D, that is, the axis connecting geographic North and South Poles. This axis precesses around the precession axis Ω_P. The mantle frame is attached to the CMB. In this frame of reference, the boundaries are at rest but the precession axis is revolving around the z-axis. The precession frame is chosen such that both ω_D and Ω_P are independent of time, but the boundaries move with velocity $\omega_D \times \mathbf{r}$, where $\mathbf{r}$ is the position vector of a point on the boundary measured from the origin. Viewed from an inertial system, the tripod x, y, z of the precession frame is precessing around Ω_P such that Ω_P always remains in the x, z-plane.

and

$$\left(\frac{\partial^2}{\partial t^2}\mathbf{r}\right)_{\text{inert}} = \left(\frac{\partial^2}{\partial t^2}\mathbf{r}\right)_{\text{rot}} + 2\boldsymbol{\omega}_{\text{ref}} \times \left(\frac{\partial}{\partial t}\mathbf{r}\right)_{\text{rot}} + \boldsymbol{\omega}_{\text{ref}} \times (\boldsymbol{\omega}_{\text{ref}} \times \mathbf{r}) + \left(\frac{\partial}{\partial t}\boldsymbol{\omega}_{\text{ref}}\right)_{\text{inert}} \times \mathbf{r}$$

The second and third terms on the right hand side correspond to the Coriolis and centrifugal forces familiar from uniformly rotating systems, the fourth term is an additional force appearing in non-uniformly rotating frames generally referred to as the Poincaré force. The acceleration in the inertial frame can be deduced from the Navier–Stokes equation for an incompressible fluid with velocity $\mathbf{u}_{\text{inert}} = \left(\frac{\partial}{\partial t}\mathbf{r}\right)_{\text{inert}} = \mathbf{u}_{\text{rot}} + \boldsymbol{\omega}_{\text{ref}} \times \mathbf{r}$:

$$\left(\frac{\partial^2}{\partial t^2}\mathbf{r}\right)_{\text{inert}} = -\frac{1}{\rho}\nabla P + \nu\nabla^2(\mathbf{u}_{\text{rot}} + \boldsymbol{\omega}_{\text{ref}} \times \mathbf{r}) + \mathbf{f}_{\text{inert}} = -\frac{1}{\rho}\nabla P + \nu\nabla^2\mathbf{u}_{\text{rot}} + \mathbf{f}_{\text{inert}}$$

$$\nabla \cdot \mathbf{u}_{\text{inert}} = \nabla \cdot \mathbf{u}_{\text{rot}} = 0$$

ν denotes the viscosity of the fluid, ρ its density, and P the pressure. $\mathbf{f}_{\text{inert}}$ is a volume force (e.g., the gravitational attraction exerted by the Moon and the Sun) measured in the inertial frame. The above relations are now assembled to yield the Navier–Stokes equation in the mantle frame. The mantle rotates with a diurnal rate $\boldsymbol{\omega}_D$ whose precession is described by the precession vector $\boldsymbol{\Omega}_P$ such that $\boldsymbol{\omega}_{\text{ref}} = \boldsymbol{\omega}_D + \boldsymbol{\Omega}_P$ and $\left(\frac{\partial}{\partial t}\boldsymbol{\omega}_{\text{ref}}\right)_{\text{inert}} = \boldsymbol{\Omega}_P \times \boldsymbol{\omega}_D$. Dropping subscripts, the equation of motion becomes

$$\frac{\partial}{\partial t}\mathbf{u} + (\mathbf{u}\cdot\nabla)\mathbf{u} + 2(\boldsymbol{\omega}_D + \boldsymbol{\Omega}_P) \times \mathbf{u} = \nu\nabla^2\mathbf{u} - \nabla p - (\boldsymbol{\Omega}_P \times \boldsymbol{\omega}_D) \times \mathbf{r} + \mathbf{f} \quad \text{(mantle frame)} \qquad [1]$$

with a reduced pressure p which contains the centrifugal term and the factor $1/\rho$. The no-slip boundary condition at the outer boundary is $\mathbf{u} = 0$.

In the precession frame on the other hand, $\boldsymbol{\Omega}_P$ and $\boldsymbol{\omega}_D$ are stationary such that $\boldsymbol{\omega}_{\text{ref}} = \boldsymbol{\Omega}_P$ and the Poincaré force disappears, so that the equation of motion reads

$$\frac{\partial}{\partial t}\mathbf{u} + (\mathbf{u}\cdot\nabla)\mathbf{u} + 2\boldsymbol{\Omega}_P \times \mathbf{u} = \nu\nabla^2\mathbf{u} - \nabla p + \mathbf{f} \quad \text{(precession frame)} \qquad [2]$$

subject to the condition $\mathbf{u} = \boldsymbol{\omega}_D \times \mathbf{r}$ at the outer boundary for no-slip boundaries. Analytical work is generally easier in this second frame, but numerical simulations are easier in the first frame because $\mathbf{u}$ is small in this frame. The nonlinear term is then less important and the Courant-Friedrichs-Levy (CFL) condition due to advection is less severe, allowing a larger time step.

The most important point about [1] and [2] is that if $\mathbf{f}$ represents gravitational attraction, $\mathbf{f}$ derives from a potential and can thus be absorbed in the pressure gradient. Gravitational forces have no influence on the incompressible flow in the cavity of a body whose motion is prescribed. If we consider as a given that the Earth executes precessional motion, we do not need to bother about gravitational forces and torques exerted on the core. Equations [1] and [2] with $\mathbf{f} = 0$ describe the flow in a container precessing on a laboratory bench. Direct application to the Earth's core requires several assumptions, chiefly that density variations of the fluid can be ignored (which excludes buoyancy forces) and that all material properties remain constant throughout the liquid core. Magnetic fields will be considered in Section 7.5.

It is convenient to rescale all variables so that the equation becomes nondimensional. In the following sections and chapters, time is measured in units of $1/\omega_D$ and the length scale d is chosen either as the radius of the outer boundary or as the gap size between the inner core and the outer boundary if the inner core is taken into account. In these units, the Navier–Stokes equation becomes

$$\frac{\partial}{\partial t}\mathbf{u} + (\mathbf{u}\cdot\nabla)\mathbf{u} + 2(\hat{\mathbf{z}} + \boldsymbol{\Omega}) \times \mathbf{u} = E\nabla^2\mathbf{u} - \nabla p - (\boldsymbol{\Omega} \times \hat{\mathbf{z}}) \times \mathbf{r} \quad \text{(mantle frame)} \qquad [3]$$

where hats denote unit vectors and $\Omega = \Omega_P/\omega_D$. In order to avoid a proliferation of symbols, velocity, length, and time are denoted by the same letters in [1] as their nondimensional counterparts in [3]. The Ekman number E is given by $E = \nu(\omega_D d^2)^{-1}$ and measures in order of magnitude the ratio of viscous to Coriolis forces. Another nondimensional number of interest is the Rossby number $V(\omega_D d)^{-1}$, where V is a typical velocity, which estimates the ratio of the nonlinear to the Coriolis term. If this number is small, as is the case in many applications, the nonlinear term may be omitted from [3]. We will use this approximation, for example, for the study of inertial modes in Sections 7.1.3 and 7.2 and for the study of boundary layers in Section 7.1.4. In the precession frame (in which the Rossby number is never small) the nondimensional equation of motion is

$$\frac{\partial}{\partial t}\mathbf{u} + (\mathbf{u}\cdot\nabla)\mathbf{u} + 2\boldsymbol{\Omega} \times \mathbf{u} = E\nabla^2\mathbf{u} - \nabla p \quad \text{(precession frame)} \qquad [4]$$

7.1.3 Inertial Oscillations in Infinitely Extended Fluids

A rotating fluid supports oscillatory motion, so called inertial waves, inertial oscillations, or inertial modes. As for all types of waves, boundaries select eigenmodes and resonances for a finite fluid volume. Inertial modes of spherical shells are the topic of Section 7.2. It is easiest to explore the distinguishing features of inertial waves in an infinite domain, which we will do in this section.

Consider an unbounded ideal fluid rotating about the z-axis disturbed by motions small enough so that nonlinearities can be neglected in the co-rotating frame. The nondimensional equation in this frame of reference reads (see [4])

$$\frac{\partial}{\partial t}\mathbf{u} + 2\hat{\mathbf{z}} \times \mathbf{u} = -\nabla p, \quad \nabla\cdot\mathbf{u} = 0 \qquad [5]$$

We eliminate the pressure by taking the curl of [5]:

$$\frac{\partial}{\partial t}\nabla \times \mathbf{u} - 2\frac{\partial}{\partial z}\mathbf{u} = 0 \qquad [6]$$

Taking again the curl of [6] one finds with [5]

$$\frac{\partial}{\partial t}\nabla^2\mathbf{u} + 2\frac{\partial}{\partial z}\nabla \times \mathbf{u} = 0 \qquad [7]$$

$\nabla \times \mathbf{u}$ can now be eliminated by cross-differentiation between [6] and [7]:

$$\frac{\partial^2}{\partial t^2}\nabla^2\mathbf{u} + 4\frac{\partial^2}{\partial z^2}\mathbf{u} = 0 \qquad [8]$$

The last equation admits solutions in the form of plane waves $\mathbf{u} \propto e^{i(\mathbf{k}\cdot\mathbf{r}-\omega t)}$ provided that

$$\omega = \pm 2\hat{\mathbf{k}}\cdot\hat{\mathbf{z}} \qquad [9]$$

This dispersion relation shows that $-2 \le \omega \le 2$ and that ω depends on the direction of propagation of the wave, but not on its wavelength. The phase velocity $\mathbf{c}_p$ is given by

$$\mathbf{c}_p = \pm 2\frac{\hat{\mathbf{k}}\cdot\hat{\mathbf{z}}}{|\mathbf{k}|}\hat{\mathbf{k}} \qquad [10]$$

and the group velocity $\mathbf{c}_g$ is

$$\mathbf{c}_g = \pm 2\frac{\hat{\mathbf{k}} \times (\hat{\mathbf{z}} \times \hat{\mathbf{k}})}{|\mathbf{k}|} \qquad [11]$$

It follows that phase and group velocities are orthogonal; see **Figure 3**. Their orientations are not

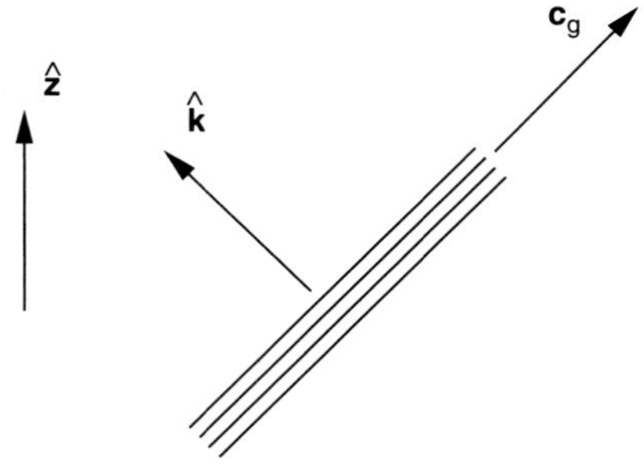

Figure 3 A wave packet indicated by a few wave crests is created by the superposition of plane waves with wave vectors parallel to $\hat{\mathbf{k}}$ and angular frequency $\omega = 2\hat{\mathbf{k}}\cdot\hat{\mathbf{z}}$. The envelope of the packet therefore extends to infinity perpendicular to $\hat{\mathbf{k}}$ but is localized in the direction of $\hat{\mathbf{k}}$. Individual wave crests propagate along $\hat{\mathbf{k}}$, but energy is transported in the direction of the group velocity $\mathbf{c}_g$ perpendicular to $\hat{\mathbf{k}}$, so that the envelope of the wave packet does not change in time. Reprinted from Tilgner A (2000) Oscillatory shear layers in source driven flows in an unbounded rotating fluid. *Physics of Fluids* 12: 1101–1111.

independent because $\mathbf{c}_p \cdot \hat{\mathbf{z}} = \pm 2|\hat{\mathbf{k}} \cdot \hat{\mathbf{z}}|^2/|\mathbf{k}|$ and $\mathbf{c}_g \cdot \hat{\mathbf{z}} = \pm 2|\hat{\mathbf{k}} \times \hat{\mathbf{z}}|^2/|\mathbf{k}|$, so that $\mathrm{sgn}(\mathbf{c}_p \cdot \hat{\mathbf{z}}) = \mathrm{sgn}(\mathbf{c}_g \cdot \hat{\mathbf{z}})$. The peculiar dispersion relation [9] opens the possibility to superpose waves at a particular frequency ω such that they form a localized packet around a surface with normal $\hat{\mathbf{n}}$ satisfying $\hat{\mathbf{n}} \cdot \hat{\mathbf{z}} = \pm\omega/2$. Since the group velocity is perpendicular to $\hat{\mathbf{n}}$, the wave motion will stay on that surface. The wave packet will not broaden as time goes on either, because the velocity field of the packet at any time t is given by

$$\int f(k)\mathrm{e}^{\mathrm{i}(\hat{\mathbf{n}}k \cdot \mathbf{r} - \omega t)}\mathrm{d}k = \mathrm{e}^{-\mathrm{i}\omega t} \int f(k)\mathrm{e}^{\mathrm{i}\hat{\mathbf{n}}k \cdot \mathbf{r}}\mathrm{d}k$$

where $f(k)$ represents the amplitude of the plane wave with wave-vector $k\hat{\mathbf{n}}$. Identical spatial distributions are therefore periodically recovered. In an infinite fluid, the wave motion can be localized in an arbitrarily thin region around the surface. Combinations of waves can be found which superpose to periodically form jets, shear layers, or a discontinuity in any higher derivative of the velocity field. These singularities can only occur in a periodic motion on surfaces with the appropriate inclination with respect to the rotation axis.

When inertial waves are reflected off a solid wall, their oscillation frequency stays constant, so that the angle between the wave-vector $\mathbf{k}$ and the rotation axis $\hat{\mathbf{z}}$ is conserved (unlike the familiar situation with light or acoustic waves where the angle between the wave vector and the normal to the reflecting surface is conserved (Phillips, 1963; Greenspan, 1968)). The same reflection law must also hold for internal shear layers since these can be represented as a superposition of waves. Some consequences of this unusual reflection law will be explored in Section 7.2.

A recurrent theme of the following sections is the appearance of inclined internal shear zones in rotating fluids. The words shear zone are meant to encompass simple shear layers as well as jets, for example. Because of their omnipresence, it is worthwhile to dwell more on the reasons for their existence. It is illuminating to look at the canonical shear layer across which pressure and normal velocity are continuous, but the tangential velocity is not. This model is the standard starting point for an elementary investigation of the Kelvin–Helmholtz instability in nonrotating fluids (Landau and Lifshitz, 1987) in which the surface of discontinuity can have an arbitrary orientation in space. In a rotating system on the contrary, the orientation of such a surface of discontinuity is not arbitrary because the Coriolis forces on both sides of the discontinuity are different and need to be balanced. Choose Cartesian coordinates x', y', z' such that $\hat{\mathbf{z}}'$ is normal to the surface of discontinuity and forms the angle ϑ with the rotation axis given by $\hat{\mathbf{z}} = \cos\vartheta\hat{\mathbf{z}}' + \sin\vartheta\hat{\mathbf{x}}'$. Denote by $[\cdot]$ the jump experienced by the quantity in square brackets when the plane spanned by the x' and y' axes is crossed. From $[p] = 0$ follows $[\partial p/\partial x'] = [\partial p/\partial y'] = 0$. Together with $[u_{z'}] = 0$, the x' and y' components of the equation of motion [5] become

$$\frac{\partial}{\partial t}[u_{x'}] = 2\cos\vartheta[u_{y'}]$$

$$\frac{\partial}{\partial t}[u_{y'}] = -2\cos\vartheta[u_{x'}]$$

If velocities vary in time like $\mathrm{e}^{\mathrm{i}\omega t}$, this system of equations admits nontrivial solutions only for

$$\omega = \pm 2\cos\vartheta = \pm 2\hat{\mathbf{z}} \cdot \hat{\mathbf{z}}' \qquad [12]$$

which is identical with the condition derived above if we think of the shear layer as a superposition of inertial waves with wavevectors parallel to $\hat{\mathbf{z}}'$.

There is a third, mathematical approach. It formalizes the preceding approaches and is only touched upon here because it depends on a whole body of theory on partial differential equations (*see* chapters 5 and 6 in vol. 2 of Courant and Hilbert (1968)). Within this general theory, the surfaces on which discontinuities may occur are called characteristic surfaces or characteristics for short. Equation [5] is classified as a hyperbolic equation for $|\omega| < 2$ which implies that inertial waves and internal layers only exist for $|\omega| < 2$. Characteristic surfaces degenerate to lines parallel to the z-axis for $\omega = 0$. The angle between the characteristic surfaces and the rotation axis increases with ω until these surfaces are perpendicular to the z-axis for $|\omega| = 2$. Characteristic surfaces disappear for $|\omega| > 2$ because the equation of motion is then of the so-called elliptic type, which means that flow with $|\omega| > 2$ do not allow for wave propagation and instead resemble potential flow in nonrotating systems. The distinguishing features of rotating flows only appear in the frequency range $|\omega| < 2$. For instance, a disk oscillating in a rotating tank generates inertial waves if the oscillation period of the disk is at least half the rotation period of the tank. If the disk oscillates faster, a flow is set up around the disk which is related to the potential flow around an oscillating disk in a nonrotating fluid (Reynolds, 1962a, 1962b).

Axisymmetric internal shear layers form conical surfaces around the rotation axis. In the limit $\omega = 0$, the cones become cylinders and occur at the rim of Taylor columns. In the case $\omega = 0$, [6] yields the stronger condition $\partial \mathbf{u}/\partial z = 0$ known as the Taylor–Proudman theorem: steady motion in the limit of small Ekman and Rossby numbers is independent of the z-coordinate. Steady flow (e.g., around obstacles) acquires columnar structure composed of Taylor columns aligned with the z-axis. Cylindrical, time-independent shear layers have been visualized experimentally numerous times. A few pictures exist also of oscillatory shear layers and their reflections from solid walls (Greenspan, 1968; Oser, 1958; Beardsley, 1970; McEwan 1970).

7.1.4 Ekman Layers

This section looks at motions for which the nonlinear term is still negligible in the mantle frame but we reinstate the viscous term. Starting from a solution of the inviscid equation of motion, the viscous corrections are to a first approximation confined to thin boundary layers which accommodate the no-slip boundary conditions. These so-called Ekman layers are of a well-known type omnipresent in rotating fluids.

Boundary layers of course also appear in non-rotating flows and involve a balance between viscous stresses, pressure gradients, and advection. In a rotating system, the Coriolis force enters into consideration and we may deduce from dimensional analysis that it introduces an intrinsic length scale, namely $(\nu/\omega_D)^{1/2} = E^{1/2}d$ with the definitions used in [3], which in most practical circumstances is small compared with the characteristic size of the fluid volume, that is, $E^{1/2} \ll 1$ ($E \approx 10^{-15}$ in the Earth's core). This length scale determines the thickness of the boundary layers at small Rossby numbers, that is, when advection is negligible. A peculiarity of Ekman layers is that they usually cause a flow inside the bulk directed perpendicular to the boundaries. This phenomenon is known as Ekman suction or Ekman pump. In time-dependent flows, the boundary layer approximation from which these conclusions are derived is not valid over the entire surface but breaks down at critical latitudes, as we shall see below. The radial component of velocity is much larger at critical latitudes than elsewhere on the boundary. The steep velocity gradients around critical latitudes extend in the form of internal layers or jets across the entire flow. Since we meet these layers so often throughout this chapter, it is worthwhile going through the mathematics showing that strong flows exist around critical latitudes. The section culminates in **Figure 5** which shows the singularity at critical latitudes in one special case.

An elementary treatment of Ekman layers in stationary flows can be found in many textbooks on fluid mechanics (Acheson, 1990; Kundu and Cohen, 2002). But we are mainly concerned with time-dependent flows in this chapter. A general treatment of Ekman layers in this case is still possible but rather abstract (Greenspan, 1968). The classical procedure is a formal asymptotic expansion in powers of $E^{1/2}$. It starts with a solution of the inviscid equation and adds to this solution a correction which is significant only near the boundaries. The boundary flow is connected to the bulk flow by matched asymptotics. The boundary layer then requires a correction to the bulk flow which in turn modifies the boundary layer at a higher order in E, etc. More and more refined solutions are obtained iteratively. To simplify matters, we deduce here the dominant contribution to the boundary layer from a more intuitive approach. We will treat the particular example of the boundary layers of the so-called spin-over mode which is central to precession-driven flow and which we will also meet in Section 7.2.

Assume that the CMB and the inner core are spherical. The inviscid equation of motion then allows the liquid core to spin about an arbitrary axis at an arbitrary angular velocity because there is no coupling between the core and the boundaries. For small but finite viscosity, the interior motion remains close to a rigid rotation. We regard the interior motion as given. Boundary layers form, whose structures we are about to compute. For finite viscosity, the inner rotation needs to be maintained by some forcing. We will later see that precession is able to do so.

In the mantle frame, introduce a local coordinate system θ', φ', z' attached to the inner-core boundary at colatitude θ with $\hat{\mathbf{z}}'$ pointing into the fluid (**Figure 4**). For definiteness the equations are derived for the inner core, analogous equations after changes of signs hold at the outer boundary. The local coordinate system is actually redundant for spherical boundaries but helps generalization to other boundary shapes. The rotation of the liquid core is given by $\boldsymbol{\omega} \times \mathbf{r}$. We will specify $\boldsymbol{\omega}$ by $\boldsymbol{\omega} = \omega(\sin t\,\hat{\mathbf{x}} + \cos t\,\hat{\mathbf{y}})$. Note that $\boldsymbol{\omega}$ is stationary in an inertial frame of reference. This solution obviously violates the no-slip boundary conditions.

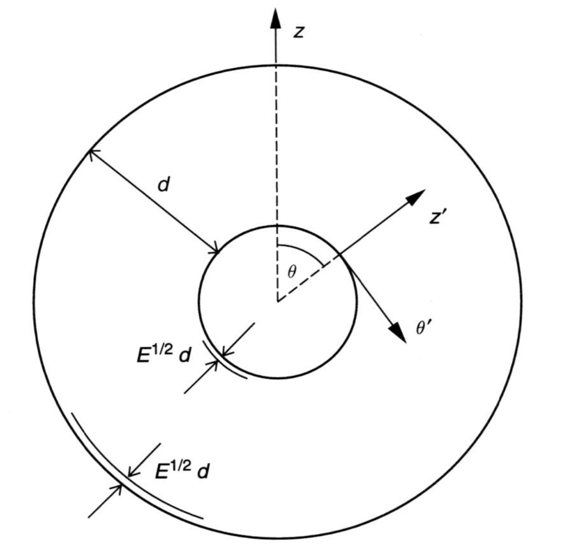

Figure 4 Definition of a local coordinate system. $\hat{\mathbf{z}}'$ is normal to the surface and is directed radially for a spherical core. Boundary layers of thickness $E^{1/2}d$ form, where $d = r_o - r_i$ is the gap size of the shell.

Let us restrict ourselves to situations in which the Rossby number is small so that the advection term can be dropped from the momentum equation [3]. We now look for a correction $\mathbf{v}$ such that $\mathbf{u} = \boldsymbol{\omega} \times \mathbf{r} + \mathbf{v}$ satisfies the no-slip boundary conditions, such that $\mathbf{v}$ has sizeable magnitude only close to the boundaries, and such that $\mathbf{v}$ is itself a solution of the linearized equation of motion

$$\frac{\partial}{\partial t}\mathbf{v} + 2\hat{\mathbf{z}} \times \mathbf{v} = E\nabla^2\mathbf{v} - \nabla\tilde{p} \qquad [13]$$

where $\nabla\tilde{p}$ is a pressure gradient which will also be confined to the boundary layer. We anticipate from the above dimensional arguments that the thickness of the boundary layer is $\mathcal{O}(E^{1/2})$. In that layer, the variation of velocity with z' is much more rapid than with θ' or φ'. The diffusion term $E\nabla^2\mathbf{v}$ then simplifies and becomes $E\partial^2\mathbf{v}/\partial z'^2$.

The condition $\nabla \cdot \mathbf{v} = 0$ implies (see [20]) that $\partial v_{z'}/\partial z'$ (which for a boundary layer of thickness $E^{1/2}$ is in order of magnitude $E^{-1/2}v_{z'}$) is equal to derivatives of $v_{\theta'}$ and $v_{\varphi'}$ with respect to the horizontal coordinates (which are in order of magnitude $v_{\theta'}$ or $v_{\varphi'}$) so that $v_{z'} = \mathcal{O}(E^{1/2}v_{\theta'}, E^{1/2}v_{\varphi'})$ and $v_{z'}$ is negligible compared with $v_{\theta'}$ and $v_{\varphi'}$ in the Coriolis term which becomes $2\boldsymbol{\Omega} \times \mathbf{v} \approx 2\cos\theta\, v_{\varphi'}\hat{\boldsymbol{\theta}}' + 2\cos\theta\, v_{\theta'}\hat{\boldsymbol{\varphi}}' - 2\sin\theta\, v_{\varphi'}\hat{\mathbf{z}}'$. Finally, it is useful to introduce the variable $\zeta = E^{-1/2}z'$ so that the balance between the viscous and Coriolis terms in the boundary layers becomes apparent. We now find from [13] using $E\partial^2/\partial z^2 = \partial^2/\partial\zeta^2$

$$\begin{aligned}
\frac{\partial}{\partial t}v_{\theta'} - 2\cos\theta v_{\varphi'} &= \frac{\partial^2}{\partial\zeta^2}v_{\theta'} + \frac{1}{r_i}\frac{\partial}{\partial\theta'}\tilde{p}\\
\frac{\partial}{\partial t}v_{\varphi'} + 2\cos\theta v_{\theta'} &= \frac{\partial^2}{\partial\zeta^2}v_{\varphi'} + \frac{1}{r_i\sin\theta}\frac{\partial}{\partial\varphi'}\tilde{p}\\
\frac{\partial}{\partial t}v_{z'} - 2\sin\theta v_{\varphi'} &= \frac{\partial^2}{\partial\zeta^2}v_{z'} + E^{-1/2}\frac{\partial}{\partial\zeta}\tilde{p}
\end{aligned} \qquad [14]$$

together with $\nabla \cdot \mathbf{v} = 0$. The boundary-layer flow is driven by the boundary conditions that $\mathbf{v} = -\boldsymbol{\omega} \times \mathbf{r}$ for $\zeta = 0$ (so that the total velocity is zero on the boundary) and $\mathbf{v} = 0$ for $\zeta \to \infty$ (in order to recover the inviscid solution at the edge of the boundary layer). This implies $\tilde{p} =$ const. for $\zeta \to \infty$ so that we may choose $\tilde{p} = 0$ for $\zeta \to \infty$, and the z'-component of [14] shows that $\tilde{p} = \mathcal{O}(E^{1/2})$. The pressure can therefore be neglected in the θ' and φ' components of the momentum equation and we arrive at

$$\frac{\partial}{\partial t}\begin{pmatrix} v_{\theta'} \\ v_{\varphi'} \end{pmatrix} + 2\cos\theta\begin{pmatrix} -v_{\varphi'} \\ v_{\theta'} \end{pmatrix} = \frac{\partial^2}{\partial\zeta^2}\begin{pmatrix} v_{\theta'} \\ v_{\varphi'} \end{pmatrix} \qquad [15]$$

More explicitly, the boundary condition at $\zeta = 0$ is

$$\begin{aligned}
v_{\theta'}(\zeta = 0) &= -\frac{1}{2}r_i\,\omega(\mathrm{e}^{\mathrm{i}t}\mathrm{e}^{\mathrm{i}\varphi} + \text{c.c.})\\
v_{\varphi'}(\zeta = 0) &= -\frac{1}{2}r_i\,\omega\cos\theta(\mathrm{i}\mathrm{e}^{\mathrm{i}t}\mathrm{e}^{\mathrm{i}\varphi} + \text{c.c.})
\end{aligned} \qquad [16]$$

where c.c. stands for complex conjugate. Equation [15] is solved by the ansatz $v_{\theta',\varphi'} = V_{\theta',\varphi'}\mathrm{e}^{\mathrm{i}t} + \text{c.c.}$ and $v_\pm = V_{\theta'} \pm \mathrm{i}V_{\varphi'}$ leading to

$$\mathrm{i}v_\pm \pm \mathrm{i}2\cos\theta v_\pm = \frac{\partial^2}{\partial\zeta^2}v_\pm \qquad [17]$$

These equations admit solutions of the form $v_\pm = v_{\pm 0}\mathrm{e}^{-\alpha_\pm\zeta}$ with

$$\alpha_\pm = \sqrt{\mathrm{i}(1 \pm 2\cos\theta)} \qquad [18]$$

where the square root is defined such that $\mathrm{Re}\{\alpha_\pm\} > 0$ and the boundary condition at infinity is fulfilled automatically. The boundary conditions at $\zeta = 0$ determine the final solution:

$$\begin{aligned}
v_{\theta'} &= -\frac{1}{2}r_i[(1-\cos\theta)\mathrm{Re}\{\mathrm{e}^{-\alpha_+\zeta}\mathrm{e}^{\mathrm{i}(\varphi+t)}\} + (1+\cos\theta)\\
&\quad \times \mathrm{Re}\{\mathrm{e}^{-\alpha_-\zeta}\mathrm{e}^{\mathrm{i}(\varphi+t)}\}]\\
v_{\varphi'} &= -\frac{1}{2}r_i[(1-\cos\theta)\mathrm{Im}\{\mathrm{e}^{-\alpha_+\zeta}\mathrm{e}^{\mathrm{i}(\varphi+t)}\} - (1+\cos\theta)\\
&\quad \times \mathrm{Im}\{\mathrm{e}^{-\alpha_-\zeta}\mathrm{e}^{\mathrm{i}(\varphi+t)}\}]
\end{aligned} \qquad [19]$$

$v_{z'}$ is determined by the requirement that $\nabla \cdot \mathbf{v} = 0$. Since $r_i \gg E^{1/2}d$, the radial coordinate does not vary much across the boundary layer and the representation of $\nabla \cdot \mathbf{v}$ in spherical coordinates simplifies to yield

$$\frac{\partial}{\partial z'} v_{z'} = -\frac{1}{r_i \sin\theta}\left(\frac{\partial}{\partial\theta}\sin\theta\, v_{\theta'} + \frac{\partial}{\partial\varphi} v_{\varphi'}\right) \qquad [20]$$

This equation together with the boundary condition $v_{z'}(z'=0)=0$ fully specifies $v_{z'}$ which will in general be different from zero at the edge of the boundary layer at $\zeta \to \infty$. The nonzero $v_{z'}$ constitutes precisely the Ekman suction and is given by the integral of the right-hand side of [20]:

$$\begin{aligned} E^{-1/2} v_{z'}(\zeta \to \infty) = {} & \frac{1-\cos 2\theta}{2\sin\theta}\mathrm{Re}\left\{\left(\frac{1}{\alpha_+}-\frac{1}{\alpha_-}\right)\mathrm{e}^{\mathrm{i}(\varphi+t)}\right\} \\ & + \sin\theta\frac{1-\cos\theta}{2}\mathrm{Re}\left\{\frac{\mathrm{i}}{\alpha_+^3}\mathrm{e}^{\mathrm{i}(\varphi+t)}\right\} \\ & - \sin\theta\frac{1+\cos\theta}{2}\mathrm{Re}\left\{\frac{\mathrm{i}}{\alpha_-^3}\mathrm{e}^{\mathrm{i}(\varphi+t)}\right\} \end{aligned} \qquad [21]$$

There is a singularity for $1 \pm 2\cos\theta = 0$ where either α_+ or α_- vanishes: $v_{z'}$ diverges at the critical latitudes of $\pm 30°$ ($\theta = 60°$ or $120°$). $E^{-1/2} v_r$, which is equal to $E^{-1/2} v_{z'}$ in spherical geometry is plotted in **Figure 5** against θ in the planes $\varphi + t = 0$ and $\varphi + t = \pi/2$. The first plane contains $\boldsymbol{\omega}$, whereas the second one is perpendicular to $\boldsymbol{\omega}$.

The singularity is of course not physical. The real boundary layer near the critical latitudes varies rapidly in both the z' and the θ' directions and is therefore much more difficult to treat. A complete

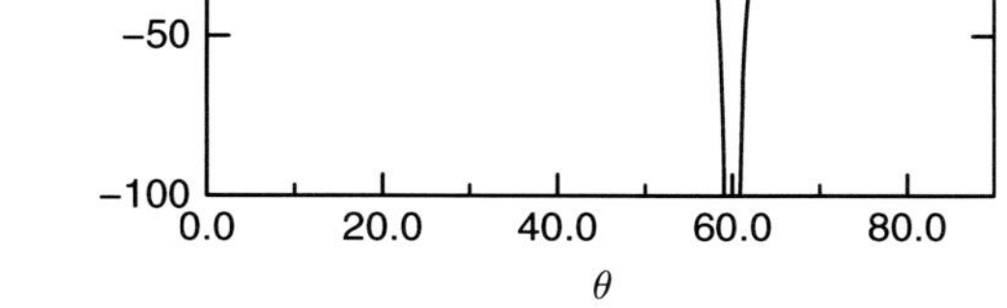

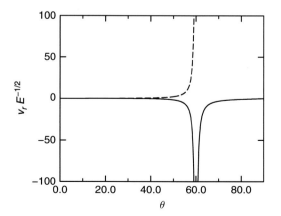

Figure 5 $v_r E^{-1/2}$ at the edge of the inner boundary layer of a spherical shell in the planes $\varphi + t = \pi/2$ (continuous) and $\varphi + t = 0$ (dashed) as a function of θ in degrees. Both curves are identical for $\theta > 60°$. At the outer boundary, v_r has the opposite sign.

theory is still lacking, but Roberts and Stewartson (1963) have shown that one obtains a self-consistent scaling for a boundary layer of thickness $\mathcal{O}(E^{2/5})$ with a lateral extent of $\mathcal{O}(E^{1/5})$ in the θ' direction. The singularity then disappears, but $v_{z'}$ near critical latitudes is still much larger than the $\mathcal{O}(E^{1/2})$ Ekman suction found at a distance from the critical latitudes.

The form of v_r in **Figure 5** suggests that a salient feature of the interior correction will be jets (in the plane $\varphi + t = \pi/2$) or shear layers (in the plane $\varphi + t = 0$) emanating from critical latitudes. It turns out that these shear zones traverse the entire fluid volume. v_r is time dependent in the rotating frame of reference and the shear layer will form an angle with the rotation axis as given by [12]. These shear layers will appear both in experiments and numerical simulations later in this chapter.

The calculation in this section was made for the case of differential rotation, but similar boundary layer effects must occur for any other perturbation of a rotating fluid. Boundary layers of the Ekman type rely on a balance between the Coriolis and viscous forces. Ekman layers can be substituted with the type of shear layers which can also exist in the bulk of a rotating fluid at places where the characteristic surfaces are tangential to the boundaries. In the case studied in this section, the inviscid solution depends on time as $\mathrm{e}^{\mathrm{i}t}$ in the mantle frame. Characteristic surfaces then form an angle of 30° with the rotation axis ($\theta = 60°$ in [12]) and are tangent to spherical boundaries at latitudes of $\pm 30°$, precisely where the Ekman boundary layers break down. The same break down occurs for any other time-dependent flow at critical latitudes determined by the frequency of the flow.

7.2 Inertial Oscillations in Spherical Shells

Section 7.1.3 has shown that plane waves can propagate in a rotating unbounded fluid. In a finite volume, inertial oscillations still occur, though not in the form of plane waves. An investigation of inertial modes in general is of interest because these oscillations are a useful starting point for the stability analysis of precession or tidally driven flows (Kerswell, 1993), and in some cases of thermal convection (Zhang, 1993, 1995). It has also been claimed that such modes have been detected with superconducting gravimeters following strong earthquakes (Melchior and Ducarme, 1986; Aldridge and Lumb, 1975), but see Zürn *et al.* (1987) for a critical comment.

More generally, we want to be able to handle arbitrary small perturbations of the rotating core. If the perturbation is small, we can use linearized equations of motion and proceed as for any linear oscillator: the response of the linear system is a superposition of eigenmodes of the system. If we know the eigenmodes, we can easily compute the effect of arbitrary forcings. This is the main motivation for studying inertial modes.

7.2.1 The Mathematical Problem

Consider a uniformly rotating body of fluid. If this motion is subject to small perturbations, the nonlinear advection term can be dropped from the Navier–Stokes equation in the rotating frame which becomes in nondimensional form

$$\frac{\partial}{\partial t}\mathbf{u} + 2\hat{\mathbf{z}} \times \mathbf{u} = -\nabla p + E\nabla^2\mathbf{u}, \quad \nabla \cdot \mathbf{u} = 0 \quad [22]$$

together with the boundary condition that $\mathbf{u} = 0$ on the boundaries. Searching for eigenmodes, we look for solutions in the form $\mathbf{u}(\mathbf{r}, t) = e^{i\omega t}\mathbf{v}(\mathbf{r})$ which leads to

$$i\omega\mathbf{v} + 2\hat{\mathbf{z}} \times \mathbf{v} = -\nabla\Phi + E\nabla^2\mathbf{v}, \quad \nabla \cdot \mathbf{v} = 0 \quad [23]$$

with $\mathbf{v} = 0$ on the boundaries and $p(\mathbf{r}, t) = e^{i\omega t}\Phi(\mathbf{r})$. Equation [23] constitutes a mathematically well posed eigenvalue problem. The eigenvalues ω have an imaginary part because of viscous damping.

We are mostly interested in very small Ekman numbers so that we are tempted to put $E = 0$ in eqn [23]. The order of the PDE then changes and we have to relax the boundary conditions. We can only impose that the velocity component normal to the boundaries vanishes in an ideal fluid.

Following general fluid dynamic experience we could naively expect that the solutions for E strictly zero and E different from zero but small, are very similar except for thin boundary layers next to solid boundaries. It turns out that in a rotating fluid, ignoring the viscous term leads to great difficulties. In mathematical terms, the problem for $E = 0$, even with the relaxed boundary conditions, is ill posed and in general does not have smooth solutions. The origin of the mathematical problem lies in an overspecification of the boundary conditions. Smooth solutions can only exist for particular shapes of the boundary. For a general treatment of hyperbolic partial differential equations and their admissible boundary conditions, see in particular Courant and Hilbert (1968).

A full sphere (as opposed to a spherical shell) is an example of a geometry which leads to smooth eigenfunctions. Simple methods for evaluating the eigenvalues and eigenmodes are presented in Greenspan (1968). More recently, even explicit formulae have been found for this geometry (Zhang *et al.*, 2004). The complete spectrum of eigenfrequencies is known for a full sphere but not for a spherical shell. Viscous effects can be included to lowest order by computing a boundary layer structure as in Section 7.1.4 together with its effects on the interior flow (see Greenspan, 1968). It is not even necessary to include details of the boundary layer structure at critical latitudes. The area around critical latitudes in which the boundary layer differs from a standard Ekman layer, and the in- or outflow from the boundary layer, scale with E in such a way that their contribution to the total dissipation vanishes for $E \to 0$. Formulae obtained with these approximations are in excellent agreement with experiments (Aldridge and Toomre, 1969) and simulations (Hollerbach and Kerswell, 1995).

The situation is entirely different for a spherical shell. The only smooth inviscid eigenmodes known for this geometry are purely toroidal, that is, they have zero radial velocity so that they are identical with modes of a full sphere. A most important mode of this class is the spin-over mode or tilt-over mode. Imagine a rotating spherical container full of fluid rotating about the same axis as the container. The spin-over mode is excited by impulsively changing the axis of rotation of the container. This operation may be viewed as an elementary step of precession. An ideal fluid is completely unaffected by the motion of the spherical boundaries, so that the fluid (viewed from an inertial frame of reference) continues rotating about its initial axis after the axis of the container has been tipped over. Viewed from a frame of reference attached to the boundaries, however, the axis of rotation of the fluid rotates about the axis of the container with a period equal to the rotation period of the container. With the choice of units leading to [23] this period is 2π. The fluid motion viewed from the mantle frame can be written as being proportional to $\omega_a(\sin t\hat{\mathbf{x}} + \cos t\hat{\mathbf{y}}) \times \mathbf{r} + \omega_z \cdot \hat{\mathbf{z}} \times \mathbf{r}$ with arbitrary ω_a and ω_z. The last term is time independent and an obvious solution of the equation of motion. The bracket can be transformed to $(\sin t\hat{\mathbf{x}} + \cos t\hat{\mathbf{y}}) = e^{i(\varphi + t)}(r/2) \cdot (\hat{\boldsymbol{\theta}} + i\cos\theta\,\hat{\boldsymbol{\varphi}}) + \text{c.c.} = \mathbf{v}_{\text{so}} + \text{c.c.}$ where c.c.

denotes complex conjugate. $\mathbf{v}_{so}$ is the velocity field of the spin-over mode. The spin-over mode is an inertial oscillation with angular frequency equal to 1. Indeed, it is straightforward to verify that

$$\frac{\partial}{\partial t}\mathbf{v}_{so} + 2\hat{\mathbf{z}} \times \mathbf{v}_{so} = \nabla(e^{i(\varphi+t)}\frac{i}{2}r^2 \sin\theta\cos\theta)$$

so that $\mathbf{v}_{so}$ is a solution of [23] for the eigenfrequency $\omega = 1$.

In the following two sections, we will see how the difficulties of the inviscid problem transpire into the solutions of the linearized Navier–Stokes equation.

7.2.2 Ray Geometry

We have seen in Section 7.1.3 that shear layers can exist in rotating fluids. We expect to find some of those in geometries where no smooth eigenfunctions exist. If we allow for no-slip boundary conditions, we more precisely expect such layers to emanate from critical latitudes. What happens at locations where such a layer meets another part of the boundary?

We have seen in Section 7.1.3 that inertial waves of wave-vector $\mathbf{k}$ have the angular frequency $\omega = \pm 2|\mathbf{k} \cdot \hat{\mathbf{z}}|/|\mathbf{k}|$, independent of $|\mathbf{k}|$. The group velocity is perpendicular to the phase velocity and directed along the characteristics of [22] for $E = 0$. The characteristics form the angle ϑ_r with the z-axis such that $\tan\vartheta_r = \pm(4/\omega^2 - 1)^{-1/2}$. Waves excited at the angular frequency ω can therefore superpose to form rays inclined at angle ϑ_r and every shear layer may be represented as a superposition of inertial waves. These waves reflect at boundaries, so that rays or shear layers must do the same. The reflection is very peculiar. Because ω fixes the inclination with respect to the rotation axis, and because an oscillation frequency is not modified by a reflection, rays are reflected such that the angle enclosed with the rotation axis remains unchanged.

What happens after multiple reflections? Closed cycles exist at some frequencies in spherical shells. The most important example for us is the case $\omega = 1$ which is relevant for the spin-over mode. **Figure 6** shows for a shell with $\eta = r_i/r_o = 0.35$ three independent closed cycles. They all start from critical latitudes. Two rays can emanate from the inner core and one from the CMB. All three lead to closed cycles.

Figure 7 shows numerical simulations of the spin-over mode in a spherical shell with a very small inner

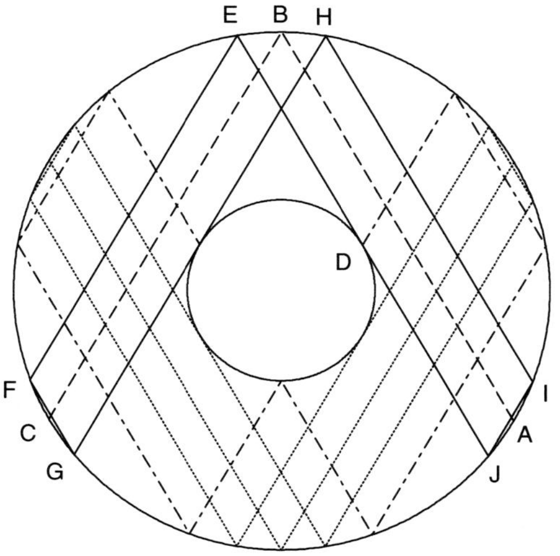

Figure 6 Ray patterns in a spherical shell with $r_i/r_o = 0.35$ for $\omega = 1$. The rotation axis is vertical in this and the following pictures showing meridional cross-sections. All rays shown start from critical latitude. The solid line (DEFGHIJ) starts tangentially from the inner core, and the dashed line (ABC) starts from the outer boundary. The dotted line results from the previous two after reflection about the equatorial plane. The dot dashed line runs along the second characteristic going through the northern critical latitude on the inner core. Its symmetric counterpart is not included in the figure.

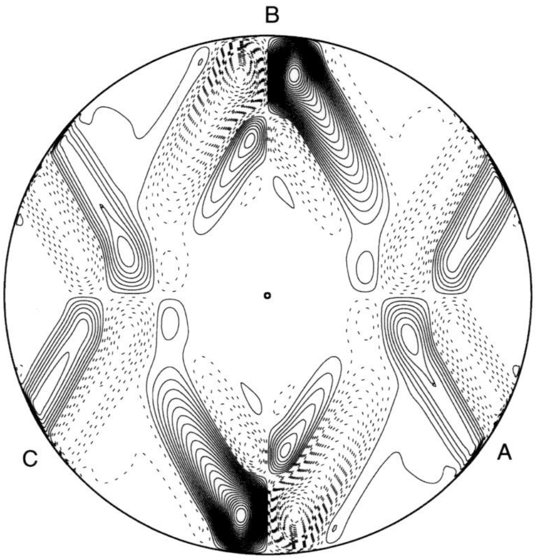

Figure 7 Radial velocity of the spin-over mode for $E = 10^{-6}$ in the meridional plane $\varphi + t = 0$ in the notation of Section 7.1.4. In this and all following contour plots, continuous and dashed contours lines indicate positive and negative values of the plotted quantity, respectively. The letters label the same points as in **Figure 6**.

core ($r_i/r_o = 0.01$). The closed cycle connected with the CMB at critical latitude is very prominent. Additional features appear in a shell of Earth like geometry ($r_i/r_o = 0.35$, see **Figure 8**). The ray tangential to the inner core gives rise to a pattern which is clearly recognizable in both **Figure 6** and the simulations. On the other hand, the other ray which could come from the critical latitude of the inner core does not show in the numerical results. It is not known at present why one is excited and the other is not. Another view at the layers caused by the inner core is offered by **Figure 21**.

More interestingly, the unusual reflection law opens the possibility for rays to converge after multiple reflection to caustics, limit cycles or attractors, examples of which are shown in **Figure 9**. This process is demonstrated in **Figure 10**. A ray has been started in poleward direction from the inner sphere at northern critical latitude and reflected 1500 times on the boundaries of a spherical shell with $r_i/r_o = 0.35$. For the next 500 reflections, a dot is placed in **Figure 10** at the colatitude θ at which a reflection off the outer boundary occurs. This procedure is repeated for different ω with a step size $\Delta\omega = 0.002$. For some ω the ray path covers more or less the entire shell, whereas for other ω, the ray reaches a periodic orbit.

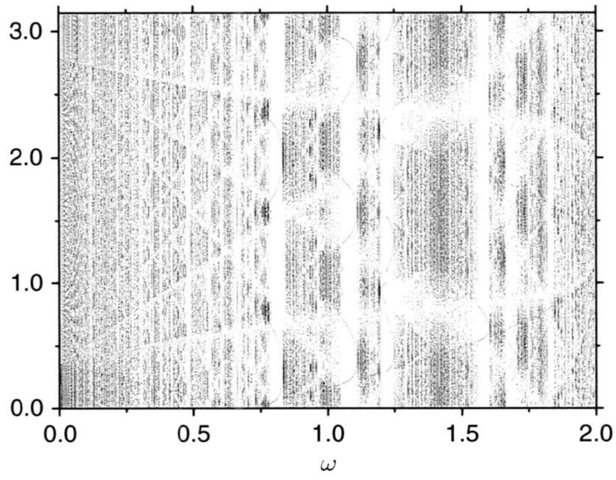

Figure 10 Ray reflections on the outer sphere at colatitude θ as a function of ω.

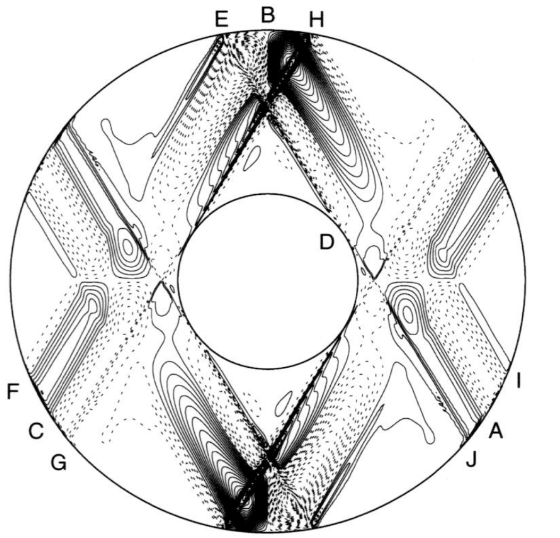

Figure 8 Radial velocity of the spin-over mode in a spherical shell for the same Ekman number and in the same meridional plane as in **Figure 7**. The letters label the same points as in **Figure 6**.

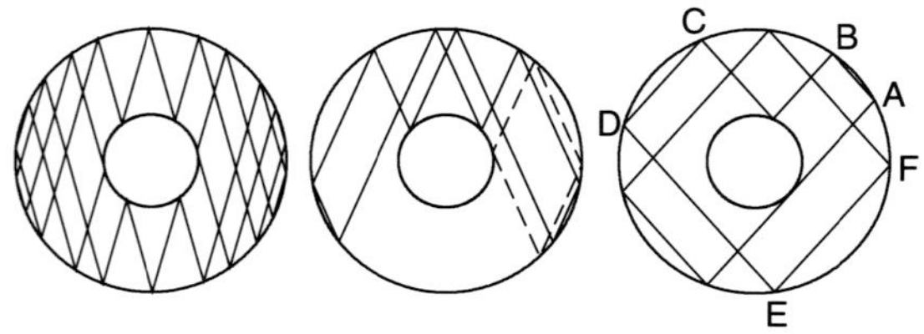

Figure 9 Ray attractors in a shell with $r_i/r_o = 0.35$ for $\omega = 0.532$ (left), $\omega = 0.81$ (center, the dashed line indicates a second attractor), and $\omega = 1.322$ (right). Adapted from Tilgner A (1999a) Driven inertial oscillations in spherical shells. *Physical Review E* 59: 1789–1794.

Note that the shear layer pattern of the spin-over mode in **Figure 6** is a closed circuit but not an attractor. Rays starting at nearby critical latitudes are not focused after multiple reflection onto this closed path.

Similar considerations apply to gravity waves in stably stratified fluids. They have the same dispersion relation as inertial waves and can thus also be found on attractors (Maas and Lam, 1995; Maas *et al.*, 1997).

7.2.3 Numerical Simulations

Apart from simulations of the spin-over mode, there are also numerical solutions for axisymmetric modes which are not toroidal. These modes offer a vast terrain for exploration in as far as properties of internal shear layers are concerned. These simulations either directly solve an eigenvalue problem or reproduce the excitation mechanism used in the experiments by Aldridge and Toomre (1969): the rotation rate of a spherical container is modulated sinusoidally in time. The Ekman pumps at the boundary then act at the frequency of the modulation and excite eigenmodes at that frequency. In mathematical terms, consider a spherical shell of gap d filled with fluid of viscosity ν which rotates about the z-axis with angular velocity $\Omega_0 + \Omega_1 \cos \tilde{\omega} t$. Using for units of time and length $1/\Omega_0$ and d, respectively, one obtains in the frame rotating at the rate Ω_0 about the z-axis eqn [22] as the nondimensional equation of motion. The Ekman number is now defined by $E = \nu/\Omega_0 d^2$. The no-slip boundary conditions require for the fluid velocity $\mathbf{u}$ that $\mathbf{u} = (\Omega_1/\Omega_0) \cdot \cos(\omega t)\hat{\mathbf{z}} \times \mathbf{r}$ at the inner and outer boundaries of radii r_i and r_o with $r_o - r_i = 1$ and $\omega = \tilde{\omega}/\Omega_0 \cdot \Omega_1/\Omega_0$ can be arbitrarily set equal to 1. Only axisymmetric flow is excited by the driving mechanism considered here so that temporal and spatial dependences are separated by the ansatz $\mathbf{u} = \mathrm{Re}\{\mathbf{v}(r, \theta)\mathrm{e}^{\mathrm{i}\omega t}\}$, where Re denotes the real part.

Figure 11 presents a study of the Ekman number dependence at $\omega = 1.32$. The internal layers narrow with decreasing E but retain a finite width so that perfect agreement with the geometric ray construction is not expected. Indeed, the simple pattern of internal layers at $\omega = 1.32$ (**Figure 11**) corresponds well to the attractor at $\omega = 1.322$ (**Figure 9**) but not to the one at exactly $\omega = 1.32$.

Many open questions remain concerning the behavior in the limit $E \to 0$. Ongoing research attempts to establish a relation between the attractors of inviscid fluids to the spectrum of eigenfrequencies of flows with small but finite Ekman numbers (Rieutord *et al.*, 2001). Another task for future research is to determine the stability properties of the shear layers.

Finally, note another important point about the role of critical latitudes. So far, we considered them to be special because the Ekman layers break down at these locations. This is the only mechanism available to generate an internal shear layer in the spin-over mode of a full sphere. But it turns out that for general modes, or even for the spin-over mode at the inner core of a spherical shell, internal shear layers preferentially form at critical latitudes even in inviscid fluids. Purely geometric reasons, rather than viscous effects, already single out the critical latitudes. This point has been studied in detail by Stewartson and Rickard (1969) and Rieutord *et al.* (2001) for discontinuities tangential to the inner sphere in inviscid inertial modes of a spherical shell.

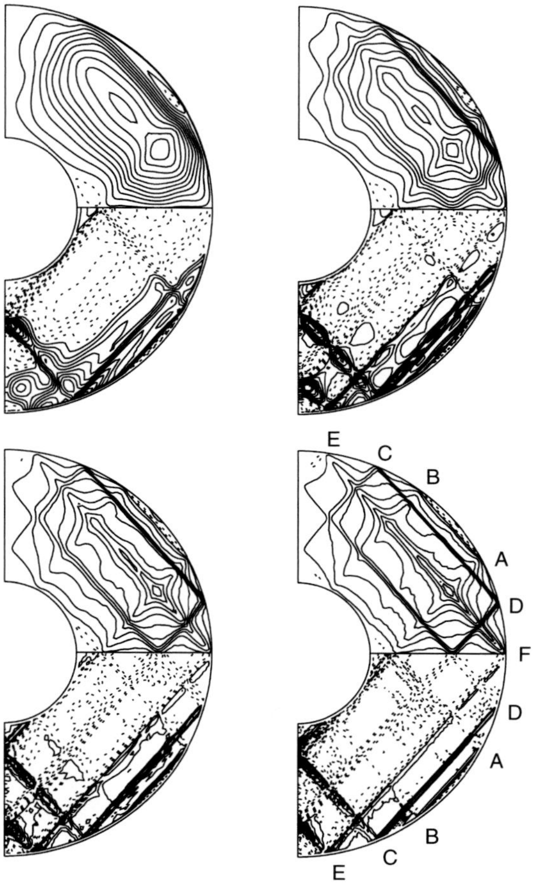

Figure 11 Flow patterns in a shell with $r_i/r_o = 0.35$ driven at $\omega = 1.32$ for Ekman numbers 10^{-5} (top left), 10^{-6} (top right), 10^{-7} (bottom left), and 10^{-8} (bottom right). The upper half of each panel shows meridional streamlines, the lower half azimuthal velocity u_φ. The Ekman layers have been removed from the plots of u_φ for clarity. Streamlines are shown at a time at which the instantaneous poloidal energy is maximum, u_φ is shown a quarter cycle of the driving force earlier. The rotation axis is vertical. A few points are labeled as in **Figure 9**. The remaining points of the attractor nearly coincide with the labeled ones. Adapted from Tilgner A (1999a) Driven inertial oscillations in spherical shells. *Physical Review E* 59: 1789–1794.

7.3 Precession

In this section, we consider the precessional motion of the mantle as given and ask for the response of the liquid core. In situations where this matters, we will assume that the CMB and the inner core rotate at identical angular velocities about identical axes. The relevant equations of motion are [3] and [4].

7.3.1 Solutions of the Inviscid Equation of Motion

Without precession, the fluid settles to a motion of uniform rotation in unison with the rotation of the container. Such a flow has constant vorticity. When the container starts to precess, the liquid tends to maintain its initial motion due to inertia. Viscous and pressure torques exerted by the boundaries on the fluid act to align the vorticity of the flow with the rotation axis of the mantle. As the mantle's axis is continuously moving in inertial space due to precession, this alignment is never quite reached. Let us assume that the flow maintains a spatially uniform but time-dependent vorticity throughout its evolution. Uniform vorticity flows are indeed solutions of the inviscid equation of motion and are commonly called Poincaré flows. These solutions are derived using the Lagrangian formalism in Poincaré (1910) (parts of which are translated into English in the final pages of the book by Lamb (1932)). A more pedestrian approach is used here (Sloudsky, 1895; Zharkov *et al.*, 1996).

Consider flows which depend linearly on the Cartesian coordinates *x*, *y*, *z*. To make such a flow fit into an ellipsoid of revolution determined by $(x/a)^2 + (y/a)^2 + (z/c)^2 = 1$, stretch the coordinates to transform the ellipsoid into a sphere, assume the flow is a solid body rotation in that sphere, and transform the coordinates back. In symbols, the stretched velocity vector $(u_x/a,\ u_y/a,\ u_z/c)$ is given by the vector product of any vector $(\tilde{\omega}_x,\ \tilde{\omega}_y,\ \tilde{\omega}_z)$ with $(x/a,\ y/a,\ z/c)$:

$$\mathbf{u} = \left(\tilde{\omega}_y z\frac{a}{c} - \tilde{\omega}_z y,\ \tilde{\omega}_z x - \tilde{\omega}_x z\frac{a}{c},\ \tilde{\omega}_x y\frac{c}{a} - \tilde{\omega}_y x\frac{c}{a}\right) \qquad [24]$$

It is easy to verify that **u** is solenoidal and does not penetrate the CMB (or an inner core if it has the same ellipticity as the CMB), but it satisfies neither no-slip nor stress-free boundary conditions and can therefore only be used as a solution of the inviscid Euler equation. These flows have spatially constant vorticity and the rotation of the fluid $\boldsymbol{\omega}_\mathrm{F}$ is given by

$$\boldsymbol{\omega}_\mathrm{F} = \frac{1}{2}\nabla \times \mathbf{u} = \frac{1}{2}\left(\left(\frac{a}{c} + \frac{c}{a}\right)\tilde{\omega}_x,\ \left(\frac{a}{c} + \frac{c}{a}\right)\tilde{\omega}_y,\ 2\tilde{\omega}_z\right) \qquad [25]$$

Since the ellipsoid is assumed symmetric about the *z*-axis we are free to choose a coordinate system such that $\Omega_y = 0$. The curl of the nonlinear term in the Navier–Stokes equation becomes

$$\begin{aligned}\nabla \times [(\nabla \times \mathbf{u}) \times \mathbf{u}] &= -[(\nabla \times \mathbf{u})\cdot\nabla]\mathbf{u} \\ &= -\left(\tilde{\omega}_z\tilde{\omega}_y\left(\frac{a}{c} - \frac{c}{a}\right),\ \tilde{\omega}_z\tilde{\omega}_x\left(\frac{c}{a} - \frac{a}{c}\right),\ 0\right)\end{aligned} \qquad [26]$$

and the Coriolis term can also be expressed in terms of $\tilde{\omega}_x$, $\tilde{\omega}_y$, and $\tilde{\omega}_z$. Inserting these expressions into [4] with *E* set to zero and eliminating the $\tilde{\omega}_x$, $\tilde{\omega}_y$, and $\tilde{\omega}_z$ in favor of $\boldsymbol{\omega}_\mathrm{F}$ yields the time evolution equation for $\boldsymbol{\omega}_\mathrm{F}$:

$$\begin{aligned}&\frac{\partial}{\partial t}\omega_{\mathrm{F}x} - \frac{a^2 - c^2}{a^2 + c^2}\omega_{\mathrm{F}z}\omega_{\mathrm{F}y} - \frac{2a^2}{a^2 + c^2}\omega_{\mathrm{F}y}\Omega_z = 0 \\ &\frac{\partial}{\partial t}\omega_{\mathrm{F}y} + \frac{a^2 - c^2}{a^2 + c^2}\omega_{\mathrm{F}z}\omega_{\mathrm{F}x} - \omega_{\mathrm{F}z}\Omega_x + \frac{2a^2}{a^2 + c^2}\omega_{\mathrm{F}x}\Omega_z = 0 \\ &\frac{\partial}{\partial t}\omega_{\mathrm{F}z} + \frac{2c^2}{a^2 + c^2}\omega_{\mathrm{F}y}\Omega_x = 0\end{aligned} \qquad [27]$$

Assume that precession is started at $t = 0$ from a state of uniform rotation. At early times $\omega_{\mathrm{F}z} \approx 1$ and therefore $\omega_{\mathrm{F}x}, \omega_{\mathrm{F}y} = \mathcal{O}(\Omega)$, so that the *z*-component of [27] yields more precisely $\omega_{\mathrm{F}z} = 1 - \mathcal{O}(\Omega^2)$.

Let us now consider on the contrary the steady case reached for $t \rightarrow \infty$. The *z*-component requires for $\frac{\partial}{\partial t}\boldsymbol{\omega}_\mathrm{F} = 0$ that $\omega_{\mathrm{F}y} = 0$ ($\Omega_x = 0$ is a trivial case of no precession at all). This means that $\hat{\mathbf{z}}$, $\boldsymbol{\Omega}$, and $\boldsymbol{\omega}_F$ all lie in the same plane. The *x*-component of [27] is thus automatically fulfilled and the *y*-component can be rewritten as

$$\omega_{\mathrm{F}x}\left(\frac{a^2 - c^2}{a^2 + c^2}\omega_{\mathrm{F}z} + \frac{2a^2}{a^2 + c^2}\Omega_z\right) = \omega_{\mathrm{F}z}\Omega_x \qquad [28]$$

For any $\omega_{\mathrm{F}z}$ different from $\Omega_z 2a^2/(c^2 - a^2)$, an $\omega_{\mathrm{F}x}$ can be found such that the resulting flow satisfies the Euler equation, nonlinear term included. Viscous boundary layers are necessary in order to select a solution. It is however plausible from physical intuition (and it is confirmed below by solutions of the full Navier–Stokes equation) that $\omega_{\mathrm{F}z}$ stays on the order $1 - \mathcal{O}(\Omega^2)$ during the evolution from the initial to the final steady state because that state departs little from the initial motion undisturbed by precession. As an alternative argument, it is plausible that $\omega_{\mathrm{F}z}$ is maximum for $\Omega = 0$ because both prograde and retrograde precession tilt the rotation axis away from its initial position. This again requires $\omega_{\mathrm{F}z} = 1 - \mathcal{O}(\Omega^2)$. With that assumption, $\omega_{\mathrm{F}x}$ can be computed to first order in Ω by setting $\omega_{\mathrm{F}z} = 1$ in [28].

Equations [27] and [28] can be expressed in a coordinate-independent form sometimes perceived as more elegant (Malkus, 1971):

$$\boldsymbol{\omega}_\mathrm{F} = (\boldsymbol{\omega}_\mathrm{F}\hat{\mathbf{z}})\hat{\mathbf{z}} - \frac{(\boldsymbol{\omega}_\mathrm{F}\hat{\mathbf{z}})(2 + \eta)}{(\boldsymbol{\omega}_\mathrm{F}\hat{\mathbf{z}})\eta + 2(1 + \eta)(\boldsymbol{\Omega}\hat{\mathbf{z}})}(\boldsymbol{\Omega} \times \hat{\mathbf{z}}) \times \hat{\mathbf{z}} \qquad [29]$$

with $\eta = a^2/c^2 - 1$. The full solution [24] is the solid body rotation plus a potential flow which is necessary to accommodate the boundary condition:

$$\mathbf{u} = \boldsymbol{\omega}_\mathrm{F} \times \mathbf{r} + \nabla\phi \qquad [30]$$

$$\phi = \frac{-(\boldsymbol{\omega}_\mathrm{F}\hat{\mathbf{z}})\eta}{(\boldsymbol{\omega}_\mathrm{F}\hat{\mathbf{z}})\eta + 2(1+\eta)(\boldsymbol{\Omega}\hat{\mathbf{z}})}(\boldsymbol{\Omega} \times \hat{\mathbf{z}}) \cdot \mathbf{r}(\hat{\mathbf{z}} \cdot \mathbf{r}) \qquad [31]$$

Figure 12 summarizes the main geometrical properties of the Poincaré solution: streamlines lie on ellipsoidal surfaces defined by $(x/a)^2 + (y/a)^2 + (z/c)^2 = \text{const.}$ In addition, streamlines are confined to planes perpendicular to $(c\tilde{\omega}_x, c\tilde{\omega}_y, a\tilde{\omega}_z)$, which is different from $\boldsymbol{\omega}_\mathrm{F}$. Streamlines are therefore ellipses with identical ellipticities lying in these planes. The normal to these planes is also different from the line running through the centers of the ellipses.

Application to Earth's parameters ($\Omega_z = \cos 23.5°/(26\,000 \cdot 365)$, $\Omega_x = \sin 23.5°/(26\,000 \cdot 365)$, $c/a = (399/400)$) of [28] with $\omega_{\mathrm{F}z}$ set to 1 leads to $\omega_{\mathrm{F}x} = 1.7 \times 10^{-5}$. The angle between the rotation axes of the mantle and the core is therefore 1.7×10^{-5} rad according to [28]. This simple differential rotation leads to rather complicated particle paths of core fluid relative to the mantle (Vanyo, 2004).

7.3.2 Viscous Effects

We have seen in the previous section that the inviscid equation of motion does not uniquely determine a solution. If viscosity is taken into account, one needs to find solutions which satisfy the no-slip boundary conditions. The first analytical attempts to include viscous effects represented the full solution as a Poincaré solution modified near the boundaries by a viscous boundary layer (Stewartson and Roberts, 1963; Roberts and Stewartson, 1965). In a linear theory which assumes zero Rossby number and small Ekman number, a particular orientation of the Poincaré flow is selected. It also becomes apparent within this approach that the boundary layers at critical latitudes differ from the viscous layers found elsewhere on the boundary, as was first noticed in Bondi and Lyttleton (1953).

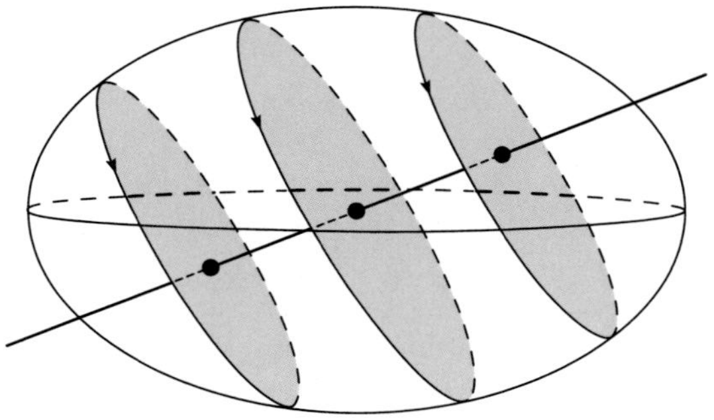

Figure 12 Sketch of the Poincaré flow. Three streamlines are shown together with the normal to the planes to which streamlines are confined.

Busse (1968) extended the previous theory by including nonlinear effects and determined the flow from an expansion in Ekman and Rossby numbers. The nonlinear effects introduce two novelties: first, the critical latitudes at which the Ekman layers break down now appear at angles of 60° with respect to the rotation axis of the fluid, not the axis of the container. Second, modifications of the Poincaré solution in the interior of the fluid in the form of a differential rotation appear. This correction is a second-order effect in the sense that its amplitude is proportional to the square of the precession rate. Crucially, however, the correction contains a singularity in the limit of zero Ekman number: in a spherical container of radius 1, it diverges at a distance cos 30° from the rotation axis of the fluid. One therefore does not recover the Poincaré solution from the full Navier–Stokes equation for an Ekman number tending to zero. This effect originates in a nonlinear interaction within the boundary layer itself (Greenspan, 1969).

The divergence found in the theory has its counterpart in the real system in the form of a cylindrical shear layer coaxial with the rotation axis of the fluid. Experiments have revealed a shear layer at the location predicted by theory and also additional weaker layers (see Section 7.3.3).

A very simple model introduced by Vanyo and Likins (1972) illustrates how viscous forces select a particular solution. Assume for simplicity that the boundaries are spherical with radii r_i and r_o for the inner core and the outer surface, respectively, and that the core and the mantle execute the same rotational and precessional movement. The Poincaré flow is then a solid body rotation $\boldsymbol{\omega}_\mathrm{F} \times \mathbf{r}$ with an arbitrary $\boldsymbol{\omega}_\mathrm{F}$. In the precession frame, [4] needs to be solved subject to the boundary conditions $\mathbf{u} = \hat{\mathbf{z}} \times \mathbf{r}$ at $r = r_\mathrm{i}, r_\mathrm{o}$. The simple idea used in Vanyo and Likins (1972) consists in subsuming the viscous effects in a frictional force per unit area proportional to $(\hat{\mathbf{z}} - \boldsymbol{\omega}_\mathrm{F}) \times \mathbf{r}$ acting at the boundaries. In dimensional units, the prefactor is given by $\rho\nu/h$, where h stands for the thickness of the boundary layers which will have to be chosen empirically. The phenomenological viscous force is simply proportional to the differential rotation between the bulk fluid and the boundaries. There is no symmetry argument which requires the viscous torque to be parallel to $(\hat{\mathbf{z}} - \boldsymbol{\omega}_F)$ since $\boldsymbol{\Omega}$ introduces another independent direction. The approximation is nonetheless attractive if we view the boundary layers as a lubricant between the interior fluid and the boundaries. Operating with $\int dV\ \mathbf{r} \times$ on [4] yields within the framework of this phenomenology

$$\frac{d}{dt}\boldsymbol{\omega}_F = \boldsymbol{\omega}_F \times \boldsymbol{\Omega} + \gamma(\hat{\mathbf{z}} - \boldsymbol{\omega}_F), \quad \gamma = 5\frac{\nu}{\omega_D\, dh}\frac{r_o^4 + r_i^4}{r_o^5 - r_i^5} \qquad [32]$$

In the stationary state, $\boldsymbol{\omega}_F$ is given by

$$\boldsymbol{\omega}_F = \frac{\gamma^2}{\gamma^2 + \Omega^2}(\hat{\boldsymbol{\Omega}} \times \hat{\mathbf{z}}) \times \hat{\boldsymbol{\Omega}} - \frac{\gamma\Omega}{\gamma^2 + \Omega^2}\hat{\boldsymbol{\Omega}} \times \hat{\mathbf{z}} + \cos\alpha\,\hat{\boldsymbol{\Omega}} \qquad [33]$$

Independently of the choice of h one finds for the stationary state $\boldsymbol{\omega}_F\hat{\mathbf{z}} = \omega_F^2$ and $\boldsymbol{\omega}_F \cdot (\Omega\hat{\boldsymbol{\Omega}} \times \hat{\mathbf{z}}) < 0$. The latter inequality shows that viewed from an inertial frame, the axis of the fluid lags behind the axis of the shell in the precessional motion. At infinitesimal Ω, $\boldsymbol{\omega}_F$ is orthogonal to $\boldsymbol{\Omega}$ and $\hat{\mathbf{z}}$. As $|\Omega|$ is increased, $\boldsymbol{\omega}_F$ gradually aligns with $\hat{\boldsymbol{\Omega}}$.

In order to get rid of any adjustable parameters like h, one needs to actually compute the structure of the boundary layers. Busse (1968) determines $\boldsymbol{\omega}_F$ for a spheroid with semi-axes a and c. This calculation is fairly involved and only the result is quoted here:

$$\frac{\boldsymbol{\omega}_F}{\omega_F^2} = \hat{\mathbf{z}} + \frac{A\hat{\mathbf{z}} \times (\boldsymbol{\Omega} \times \hat{\mathbf{z}}) + B(\hat{\mathbf{z}} \times \boldsymbol{\Omega})}{A^2 + B^2} \qquad [34]$$

with $A = 0.259(E/\omega_F)^{1/2}/a + (1 - c/a)\omega_F^2 + \boldsymbol{\Omega}\hat{\mathbf{z}}$ and $B = 2.62(E\cdot\omega_F)^{1/2}/a$, $E = \nu(\omega_D d^2)^{-1}$. The z-component of [34] reproduces the equation $\boldsymbol{\omega}_F\hat{\mathbf{z}} = \omega_F^2$ already obtained above. Equation [34], strictly speaking, obtains for a full spheroid with semi-axes a and c without an inner core. However, eqn [32] shows that the torque exerted by the inner core on the fluid in a spherical shell with $r_i/r_o = 0.35$ is only 2% of the total frictional torque, so that [34] can be reasonably applied to the Earth's core. Equation [34] for $E < 10^{-7}$ reproduces the angle of 1.7×10^{-5} already found at the end of Section 7.3.1. According to [34] the locus of $\boldsymbol{\omega}_F$ approximately sweeps half of a circular cone as Ω is varied; see **Figure 13**.

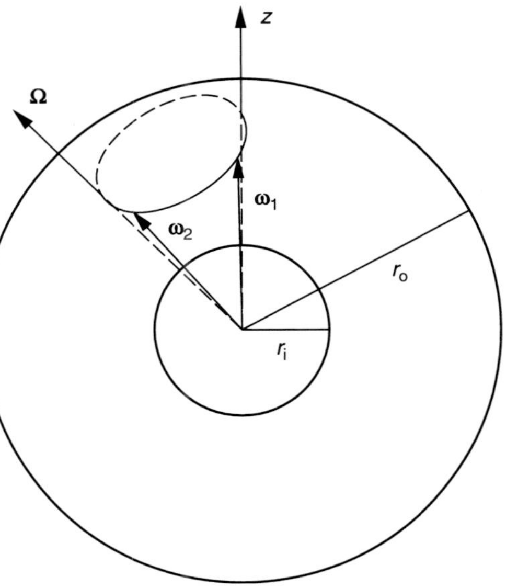

Figure 13 A spherical shell in prograde precession. For infinitesimal precession rates Ω, the rotation of the interior fluid $\boldsymbol{\omega}_1$ is nearly aligned with the z-axis. Viewed from the mantle frame, the fluid rotates with the angular velocity $\boldsymbol{\omega}_1 - \hat{\mathbf{z}}$ which is perpendicular to the plane containing $\hat{\mathbf{z}}$ and $\boldsymbol{\Omega}$. $\boldsymbol{\omega}_2$ shows a possible rotation vector of the fluid for large Ω. As Ω is increased, the rotation vector varies along a curve as shown and aligns with $\boldsymbol{\Omega}$ in the limit of large precession rates. For retrograde precession, the rotation vector of the fluid varies along the dashed curve from $\hat{\mathbf{z}}$ to $\boldsymbol{\Omega}$.

Noir *et al.* (2003) arrive at [34] by considering the balance of torques acting on the fluid. They obtain the frictional torque from the viscous decay rate of the spin-over mode computed in Greenspan (1968) which allows them to circumvent the most tedious part of the algebra presented in Busse (1968).

Viscous effects ultimately determine how much energy is dissipated by precession. This energy must at least be on the order of 10^{11} *W* if precession drives the geodynamo. The dissipated energy must be drawn from the rotational motion of the Moon and the Earth. Néron de Surgy and Laskar (1997) have computed the long-term evolution of the Earth including a frictional term of the same form as in [32]. It turns out that a precessional dissipation of nearly 10^{12} *W* is still compatible with geophysical constraints on the rotation rate.

Precessional forcing certainly has changed during Earth's history. The obliquity was not constant, though there is controversy about how much it varied (Williams, 1993; Néron de Surgy and Laskar, 1997). The more dramatic variation in precessional forcing comes from the change in distance between the Earth and the Moon. Precessional forcing was of course stronger when the Moon was closer to the Earth.

7.3.3 Experiments

Laboratory experiments have historically been our most important tool to gain insight into precession-driven flow. Even with todays computational capabilities, experiments are still superior in several aspects, especially concerning the accessible range of Ekman numbers.

Until recently, experimental diagnosis of precession-driven flow was restricted to torque measurements and optical visualization. **Figure 14** shows an experimental apparatus which sets an ellipsoidal container into precessional motion. It consists of basically two superposed turntables. Co-rotating cameras allow visual observation of dye tracers or flakes suspended in the fluid (normally water). The torque exerted by the motors in order to maintain the precessional motion of the container gives a direct measure of the energy dissipated in the fluid.

Figure 15 summarizes visual observations. At low precession rates, nested cylindrical shear layers appear which are approximately coaxial with the rotation axis of the fluid. Suspended flakes preferentially align with the shear in these layers but are randomly oriented in regions of low shear. The orientation of the particles is revealed by how they scatter incident light: viewed perpendicular to a plane of illumination, zones of high shear appear bright. When the precession rate is increased, these shear layers seem to develop wave-like instabilities as indicated in **Figure 15** and the flow ultimately becomes turbulent at high-enough precession rates. The visualization method of course does not allow us to tell whether the shear layers cause the instability or whether their deformation is simply an indicator of an instability of different origin. A sudden increase in energy dissipation, and hence in the torque applied by the motors, accompanies the onset of turbulence (Malkus, 1968). A hysteresis loop is observed in the torque when the precession rate is lowered from a value above the onset of turbulence in an ellipsoid of ellipticity 1/10, but none appears in an ellipsoid of ellipticity 1/400 (Vanyo, 1991). The onset of the

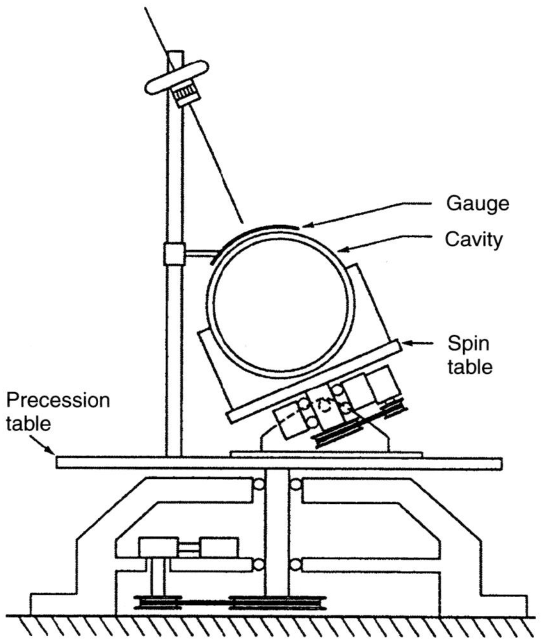

Figure 14 Experimental setup. A transparent ellipsoidal cavity rotates and precesses. A camera pointed along the rotation axis and mounted on the precession table takes pictures of the flow through a transparent gauge. Adapted from Vanyo J (1991) A geodynamo powered by luni-solar precession. *Geophysical and Astrophysical Fluid Dynamics* 59: 209–234.

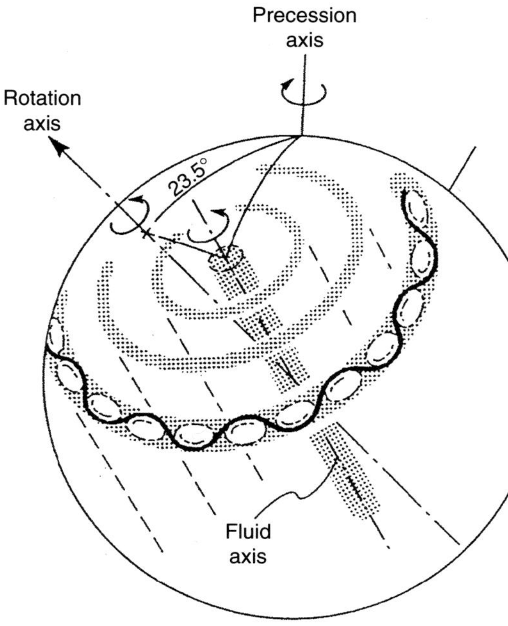

Figure 15 Sketch of visual observations. Adapted from Vanyo J, Lods D, and Wilde P (1994) Precessing mantle, liquid core and solid core interactions. In: *The 4th SEDI Symposium, Abstract Book*, pp. 62–64. Whistler, BC: SEDI.

wave-like instability occurred in one experiment (Malkus, 1968) at $E = 3.6 \times 10^{-6}$ and $\Omega_P/\omega_D = 1/60$ in a container of ellipticity 1/25. There is no systematic data available on how this onset depends on the control parameters.

More detail of the structure of the axisymmetric shear layers is given in **Figure 16** where the temporal evolution of an initially straight dye line in the equatorial plane is shown. The bulk of the fluid moves in retrograde direction and the strongest deformation of the dye streak occurs at a radius of about 0.866 (the radius of the boundary being 1). A cylinder at this radius connects the northern to the southern critical latitude ($\cos 30° = 0.866$).

Vanyo *et al.* (1995) contains numerous photographs of dye-tracer visualizations. It transpires that vortices reminiscent of a Kelvin–Helmholtz instability appear on the shear layers. There is also an axial velocity component inside these vortices, so that the flow in the vortices is helical. The gross structure of the helical velocity is not unlike that of buoyancy-driven flow, which adds to the suspicion that precession-driven flow might act as a dynamo.

The orientation of the velocity in the most prominent shear layer, and the location of that layer, in Malkus' experiment is compatible with the nonlinear theory of Busse. It is not clear at present where the other shear layers come from.

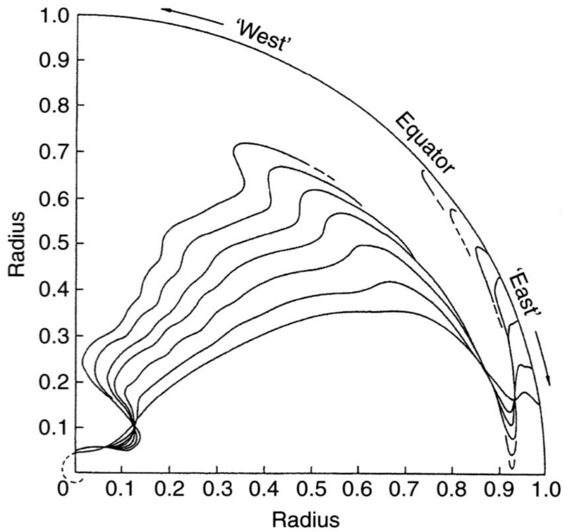

Figure 16 Deformation of a dye streak in the equatorial plane at successive moments in time. The view is from the South Pole; west and east are equivalent to retrograde and prograde, respectively. Reprinted from Malkus V (1968) Precession of the Earth as the cause of geomagnetism. *Science* 160: 259–264.

It is also surprising at first that there is no indication in visualizations of the inclined time-dependent shear layers which have been the topic of Section 7.2 (in particular **Figures 6** and **7**). It is likely that the response time of the suspended flakes is too long to show these layers. Indeed, every flake particle traverses during one rotation period a layer of one sign and the corresponding layer of opposite sign (because the spin-over mode has an azimuthal wave number of 1) so that averaged over one rotation period, there is no reorientation of the particle. Acoustic Doppler anemometry on the other hand confirmed the existence of oblique shear layers. Noir *et al.* (2001) compared the experimental velocity profiles with numerical simulations and found agreement for both the shape of the profile as well as the dependence of the velocities on the control parameters. They concluded that in the Earth's core, the velocity in these layers originating from the CMB is approximately $6 \times 10^{-6}\,\mathrm{m\,s^{-1}}$.

A most basic characterization of precession-driven flow is the orientation of the fluid axis. The first systematic data have been collected by Vanyo and Likins (1971) who compared their results with eqn [33]. A problem with [33] is that it predicts behavior which is symmetric between prograde and retrograde precession, which is not supported by observation. Pais and LeMouël (2001) re-examined Vanyo's data by fitting them to the theory of Busse, that is [34], and claimed good agreement. Another study by Noir *et al.* (2003) showed that discrepancies between experiment and theory remain which are larger than the experimental errors. Noir *et al.* (2003) started experimental determinations of the fluid axis of their own using three different methods: tracer particles, pressure measurements, and acoustic Doppler anemometry. Equation [34] is not expected to be universally applicable because it is derived using series developments which are valid in the limit of small Ekman numbers, small precession rates, and small ellipticities. Accordingly, Noir *et al.* (2003) observed deviations at some parameters but found general agreement, even at parameters for which it could not be expected. In particular, there is no reason to doubt that [34] describes very well the behavior of the Earth's core. Numerical simulations in the next section confirm this conclusion.

7.3.4 Numerical Solutions

Experimental progress primarily hinges on the development of new measurement techniques. Anemometry has never been an easy task, but it becomes especially cumbersome in a rotating or even precessing flow. Numerical simulations are attractive in the present context because all quantities of interest can easily be extracted from the data. On the other hand, we cannot simulate the lowest Ekman numbers achievable in experiments.

The most general geometry simulated to date is that of a spheroidal shell in which the minor axes of the inner and outer boundaries coincide with the common rotation axis of the boundaries. Most of the calculations presented here are for equal ellipticities of the inner and outer boundaries so that we state the mathematical problem for this case.

Consider incompressible fluid of kinematic viscosity ν in an ellipsoidal shell rotating with angular frequency ω_D about the z-axis. The shell furthermore executes precessional motion characterized by the precession vector $\Omega_p \hat{\mathbf{\Omega}}_p$ (hats denote unit vectors). The boundaries of the shell are given by

$$\frac{x^2}{a^2} + \frac{y^2}{a^2} + \frac{z^2}{c^2} = 1 \qquad [35]$$

$$\frac{x^2}{(\eta a)^2} + \frac{y^2}{(\eta a)^2} + \frac{z^2}{(\eta c)^2} = 1 \qquad [36]$$

$\eta < 1$ and both boundaries have the same ellipticity $e = 1 - c/a$. Units of length and time are chosen as $(1-\eta)a$ and $1/\omega_D$, respectively. The equation of motion for the velocity $\mathbf{u}(\mathbf{r}, t)$ reads in a frame of reference attached to the shell:

$$\frac{\partial}{\partial t}\nabla \times \mathbf{u} + \nabla \times \{(2(\hat{\mathbf{z}} + \mathbf{\Omega}) + \nabla \times \mathbf{u}) \times \mathbf{u}\} = E\nabla^2\nabla \times \mathbf{u} + 2\hat{\mathbf{z}} \times \mathbf{\Omega} \qquad [37]$$

$$\nabla \cdot \mathbf{u} = 0 \qquad [38]$$

The Ekman number E is defined by $E = \nu(\omega_D(1-\eta)^2 a^2)^{-1}$ and $\mathbf{\Omega} = \Omega_p/\omega_D \, \hat{\mathbf{\Omega}}_p$. The precession axis $\hat{\mathbf{\Omega}}$ forms the angle $\alpha (0 < \alpha < \pi/2)$ with the z-axis and is time dependent in the chosen system of reference:

$$\hat{\mathbf{\Omega}} = \sin\alpha \cos t\, \hat{\mathbf{x}} - \sin\alpha \sin t\, \hat{\mathbf{y}} + \cos\alpha\, \hat{\mathbf{z}} \qquad [39]$$

The boundary conditions require that $\mathbf{u} = 0$ at $r = r_i, r_o$. For spherical boundaries, $c = a$, $e = 0$, and $\eta = r_i/r_o$, where r_i and r_o denote the radii of the inner and outer boundary, respectively.

7.3.4.1 Laminar flows

The rotation axis of the fluid, $\boldsymbol{\omega}_F$, is predicted by [34] in the limit of small precession rates, Ekman numbers, and ellipticities. If we want to compare the prediction to numerical simulations, we first need a precise definition of the fluid axis. In the theory, the flow outside the boundary layers is assumed to have constant vorticity. In a sphere, such a flow corresponds to a solid body rotation. We thus seek to extract from the simulated flow a component which most closely resembles a solid body rotation. This turns out to be quite straightforward because of the numerical method that is usually employed (Tilgner, 1999d). Indeed, it is convenient to numerically solve the Navier–Stokes equation in spherical geometry after decomposing the velocity field $\mathbf{u}$ into poloidal and toroidal scalars, ϕ and ψ, by

$$\mathbf{u} = \nabla \times \nabla \times (\phi \hat{\mathbf{r}}) + \nabla \times (\psi \hat{\mathbf{r}}) \qquad [40]$$

and to decompose each scalar into spherical harmonics:

$$\phi = r \sum_{l=1}^{\infty} \sum_{m=-l}^{l} \phi_l^m(r,t) P_l^m(\cos\theta) e^{im\varphi}$$
$$\psi = r^2 \sum_{l=1}^{\infty} \sum_{m=-l}^{l} \psi_l^m(r,t) P_l^m(\cos\theta) e^{im\varphi} \qquad [41]$$

A solid body rotation of the fluid at radius r is entirely determined by the coefficients $\psi_1^1(r)$ and $\psi_1^0(r)$. The rotation of the fluid at radius r in the mantle frame is given by

$$\boldsymbol{\omega}(r) = -2\mathrm{Re}\{\psi_1^1(r)\}\hat{\mathbf{x}} + 2\mathrm{Im}\{\psi_1^1(r)\}\hat{\mathbf{y}} + \psi_1^0(r)\hat{\mathbf{z}} \qquad [42]$$

Re{} and Im{} denote the real and imaginary parts of the quantity in curly brackets. An average rotation is then defined by $< \boldsymbol{\omega} > = \frac{1}{V}\int \boldsymbol{\omega}(r) dV$, where V is the volume of the shell.

We are now in a position to compare numerical and analytical results. $\boldsymbol{\omega}_F$ determined from [34] with $a = c = r_o$ is given in the precession frame so that we need to compare it with $< \boldsymbol{\omega} > + \hat{\mathbf{z}}$. **Figure 17** shows a polar diagram in which the directions of $< \boldsymbol{\omega} > + \hat{\mathbf{z}}$ and $\boldsymbol{\omega}_F$ are given. This plot is a two-dimensional (2-D) projection of **Figure 13**. The locus of the directions visited by $< \boldsymbol{\omega} > + \hat{\mathbf{z}}$ when varying Ω depends on α but not on E according to [34]. This is well verified by the numerical results. Predicted and computed directions differ by angles comparable with the variations of direction of $\boldsymbol{\omega}(r)$ with r and are therefore on the order of the deviations within which a fluid rotation vector can be determined from the numerical data.

Figure 18 compares the kinetic energy of the flow in the mantle frame with the predicted value $1/2 \int (\boldsymbol{\omega}_F - \hat{\mathbf{z}})^2 dV$. The agreement is fair and the fractional error would appear even smaller in the precession frame because most of the rotational energy is subtracted out in transforming to the mantle frame, where deviations from solid body

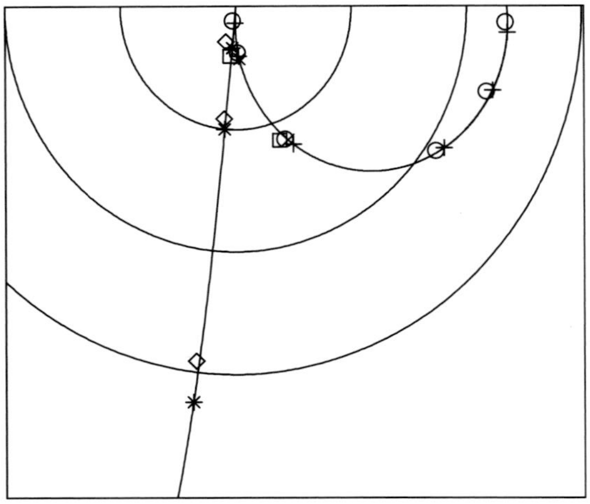

Figure 17 Polar plot showing the orientation of the average rotation of the fluid in a reference frame in which the axes of precession and rotation of the shell are stationary. The thin circles are located at 10°, 20°, and 30° from the North Pole. The thick lines show the locus of the directions of the rotation axis of the fluid predicted by [34] for $\alpha = 90°$ (left line) and $\alpha = 23.5°$ (right line), the latter ending at the top right of the figure at the position of the precession axis. Retrograde corresponds to clockwise in this figure. The symbols indicate $E = 10^{-4}$, $\alpha = 23.5°$ (circles); $E = 10^{-5}$, $\alpha = 23.5°$ (squares); and $E = 10^{-4}$, $\alpha = 90°$ (diamonds). The values of Ω can be deduced from **Figure 18** where more data for the same runs are shown; the points close to the pole are for small $|\Omega|$. Near to every data point obtained from direct simulation is another symbol showing the direction calculated with [34] for the same parameter set: crosses belong to circles, x to squares, and stars to diamonds. Reprinted from Tilgner A (1999b) Magnetohydrodynamic flow in precessing spherical shells. *Journal of Fluid Mechanics* 379: 303–318.

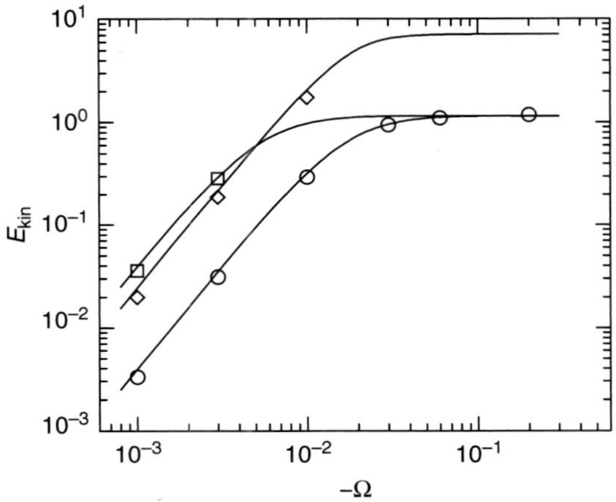

Figure 18 Kinetic energy E_{kin} of the flow in the mantle frame as a function of the (retrograde) precession rate Ω. The symbols indicate $E = 10^{-4}$, $\alpha = 23.5°$ (circles); $E = 10^{-5}$, $\alpha = 23.5°$ (squares); and $E = 10^{-4}$, $\alpha = 90°$ (diamonds). The solid lines show the kinetic energies deduced from [34]. Adapted from Tilgner A (1999b) Magnetohydrodynamic flow in precessing spherical shells. *Journal of Fluid Mechanics* 379: 303–318.

rotation are emphasized. The agreement also improves with decreasing E. More comparisons are in Tilgner and Busse (2001) for spherical shells, and Lorenzani and Tilgner (2001) for spheroidal shells.

Experiments have revealed oblique conical and cylindrical shear layers. The inclined shear layers are those contained in the spin-over mode and already seen in **Figures 7** and **8**. One of the cylindrical shear layers is predicted by the theory of Busse (1968). These cylinders are coaxial with the rotation axis of the fluid. In order to distill them from the numerical data, we have to plot the zonal component of the velocity axisymmetric about $\boldsymbol{\omega}_F$. This is done in **Figure 19**. The shear layers in the zonal velocity well known from experiments clearly appear in the plots. The exact location of the shear layers depends on both the boundary geometry and the Ekman number. The most precise comparison is possible in **Figure 20** in which the profiles of the zonal velocity in the equatorial plane perpendicular to ω_F are plotted as a function of radius. The strongest shear layer connects the critical latitudes, and the positions of the strongest prograde and retrograde jets correspond exactly to those given in figure 3 of Malkus (1968). The position of the smaller extrema depends on E and a perfect agreement with Malkus figure cannot be expected for this part of the profile. As can be deduced from **Figure 20**, the prograde jet becomes stronger with decreasing E. According to Busse (1968), a singularity should develop in the limit $E \to 0$. The Ekman number dependence of the maximum zonal velocity does not follow any simple law valid for the entire interval $10^{-4} < E < 10^{-6}$ but is compatible with a scaling in $E^{-3/10}$ for $10^{-5} < E < 10^{-6}$. Using the scaling in $E^{-3/10}$ down to the Ekman number of the Earth, one finds a velocity of 3×10^{-5} m s^{-1} for the prograde jet inside the core (Noir *et al.*, 2001).

Figure 20 contains additional structure in the zonal wind which compares favorably with Malkus' experiment. While we do not understand the origin of these other shear layers, the simulations show that they are not experimental artefacts and that their explanation is hidden in the equation of motion [37].

The simulations, as well as the experiments, had no significant inner core. The effect of an Earth like inner core has been investigated by Lorenzani and Tilgner (2001). The main cylindrical shear layer is too distant from the inner core to be noticeably

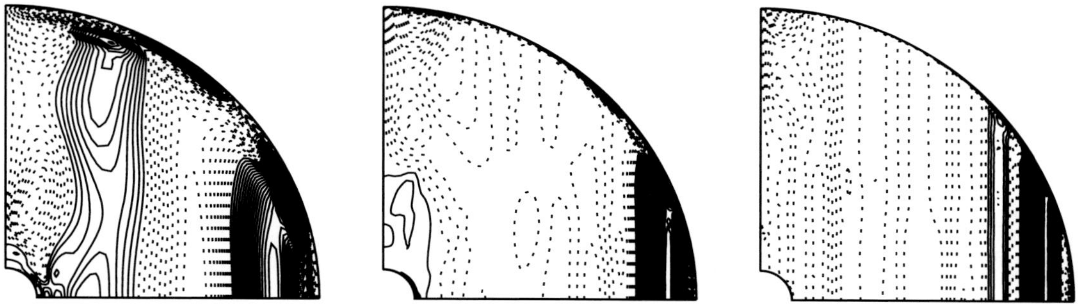

Figure 19 Zonal flow component $u_{\varphi'}$ (where φ' denotes the azimuthal angle with respect to the $\boldsymbol{\omega}_F$-axis) axisymmetric about $\boldsymbol{\omega}_F$, after subtraction of the average rotation of the fluid. $\boldsymbol{\omega}_F$ is pointing upwards in all panels. $\eta = 0.1$, $e = 0.04$, $\Omega = -10^{-5}$, and $\alpha = 30°$ in all cases. From left to right, the Ekman number is 10^{-4}, 10^{-5}, and 10^{-6}. Adapted from Lorenzani S and Tilgner A (2001) Fluid instabilities in precessing spheroidal cavities. *Journal of Fluid Mechanics* 447: 111–128.

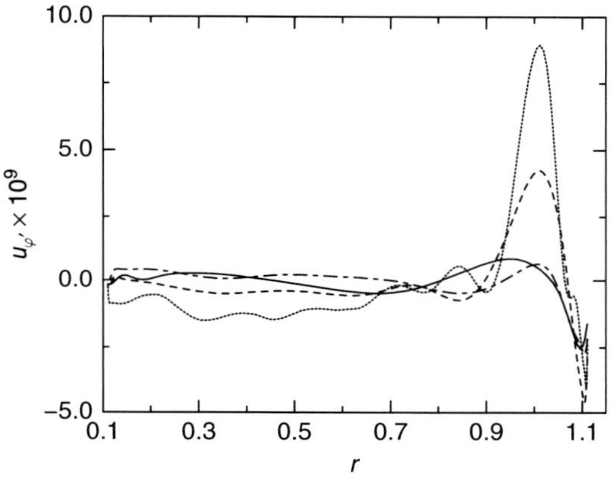

Figure 20 $u_{\varphi'}$ from **Figure 19** in the equatorial plane as a function of r for $\Omega = -10^{-5}$, $\alpha = 30°$ and $\eta = 0.1$. The line styles indicate $e = 0.04$, $E = 10^{-4}$ (solid); $e = 0.04$, $E = 10^{-5}$ (dashed); $e = 0.04$, $E = 10^{-6}$ (dotted); and $e = 0$, $E = 10^{-5}$ (dot dashed). For this last case, $u_{\varphi'}$ has been divided by 100 in order to make the curve fit into the figure. Adapted from Lorenzani S and Tilgner A (2001) Fluid instabilities in precessing spheroidal cavities. *Journal of Fluid Mechanics* 447: 111–128.

modified by it, but a new cylindrical shear layer emerges near the critical latitudes of the inner core.

In as far as the inclined shear layers are concerned, we have reliable theory about the scaling of the velocity with Ekman number and precession rate for the layer connected with the CMB. This scaling yields a velocity of $6 \times 10^{-6}\,\mathrm{m\,s^{-1}}$ when applied to the Earth's core (Noir *et al.*, 2001).

We are less advanced for the layers associated with the inner core and have to rely on numerics for that case. Tilgner (1999c) reports simulations with free-slip boundary conditions at the CMB and no-slip conditions at the inner core. This isolates the layer connected to the inner core, which is shown

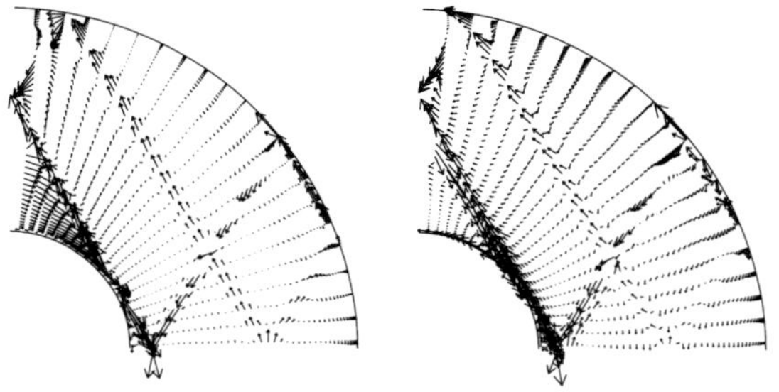

Figure 21 Velocity in meridional planes for $e_i = e_o = 1/400$, $E = 3 \times 10^{-6}$ and stress-free outer and no-slip inner boundaries. $\hat{\mathbf{z}}$ is pointing upwards, the axis pointing to the right is $\hat{\boldsymbol{\Omega}}$ in the left panel and $\hat{\mathbf{z}} \times \hat{\boldsymbol{\Omega}}$ in the right panel. Velocities at equal arrow length are larger in the right panel by a factor 1.24. The mean solid body rotation has been subtracted and the Ekman layer at the inner core has been removed from the plot. Adapted from Tilgner A (1999c) Non-axisymmetric shear layers in precessing fluid ellipsoidal shells. *Geophysical Journal International* 136: 629–636.

in **Figure 21**. The intensity of these shear zones decreases with decreasing E, but the extrapolation to $E = 10^{-15}$ is problematic. The numerical results do not behave according to any simple power law. The Ekman number of the simulation is presumably not small enough yet for the asymptotic regime. One is thus reduced to give an upper bound for the core flow: the radial velocity in the main shear zone will certainly be less than $6.9 \times 10^{-5}\,\mathrm{m\,s^{-1}}$.

The Poincaré solution in a shell not only violates no-slip boundary conditions, it even violates the no-penetration boundary condition if inner and outer boundaries have different ellipticities. Based on the experience of Section 7.2, which has shown that internal layers typically appear even in inviscid solutions in all but the simplest geometries, we may expect that a difference in ellipticities also brings internal layers into existence.

Simulations of the linearized equations of motion confirm this expectation. This time, the velocities on the characteristic surfaces tangent to the inner core are increasing with decreasing E because one obtains true discontinuities in the inviscid flow which are smeared out at finite Ekman number. Simulations yield directly a lower bound for the corresponding layer in the Earth's core of 3.4×10^{-7} m s^{-1}. We have no theory for the Ekman number dependence. The best power law fit to the numerical data suggests a scaling as $E^{-0.17}$. Assuming this is correct, one obtains velocities of 1.2×10^{-5} m s^{-1} for the Earth's core (Tilgner, 1999c). All the different mechanisms therefore seem to generate velocities in the shear zones in the range 10^{-7}–10^{-5} m s^{-1}, which is well below flow velocities deduced from secular variation.

The mechanism based on different ellipticities of the inner core and the CMB is a rather general effect of topography. In fact, the Poincaré flow over topography at the CMB of a height of some 10 km may force larger velocities than those found here, but this possibility remains to be investigated.

7.3.4.2 *Instability*

Instabilities generally break symmetries of the underlying flow. The laminar precession-driven flow is symmetric with respect to the origin in the sense that $\mathbf{u}(\mathbf{r}) = -\mathbf{u}(-\mathbf{r})$. It is therefore useful to separate the full velocity field into symmetric and antisymmetric components, such that $\mathbf{u} = \mathbf{u}_a + \mathbf{u}_s$, with $\mathbf{u}_s = (\mathbf{u}(\mathbf{r}) - \mathbf{u}(-\mathbf{r}))/2$ and $\mathbf{u}_a = (\mathbf{u}(\mathbf{r}) + \mathbf{u}(-\mathbf{r}))/2$. Adding and subtracting the equations for $\frac{\partial}{\partial t}\nabla \times \mathbf{u}(\mathbf{r})$ and $\frac{\partial}{\partial t}(\nabla \times \mathbf{u})(-\mathbf{r})$, and noting the fact that the curl of a vector is a pseudovector so that $\nabla \times \mathbf{u}_a = \frac{1}{2}[(\nabla \times \mathbf{u})(\mathbf{r}) - (\nabla \times \mathbf{u})(-\mathbf{r})]$ and $\nabla \times \mathbf{u}_s = \frac{1}{2}[(\nabla \times \mathbf{u})(\mathbf{r}) + (\nabla \times \mathbf{u})(-\mathbf{r})]$ one obtains

$$\frac{\partial}{\partial t}\nabla \times \mathbf{u}_s + \nabla \times [(\nabla \times \mathbf{u}_s) \times \mathbf{u}_s + (\nabla \times \mathbf{u}_a) \times \mathbf{u}_a] + 2\nabla \times ((\hat{\mathbf{z}} + \boldsymbol{\Omega}) \times \mathbf{u}_s) = E\nabla^2\nabla \times \mathbf{u}_s - 2\boldsymbol{\Omega} \times \hat{\mathbf{z}} \qquad [43]$$

$$\frac{\partial}{\partial t}\nabla \times \mathbf{u}_a + \nabla \times [(\nabla \times \mathbf{u}_s) \times \mathbf{u}_a + (\nabla \times \mathbf{u}_a) \times \mathbf{u}_s] + 2\nabla \times ((\hat{\mathbf{z}} + \boldsymbol{\Omega}) \times \mathbf{u}_a) = E\nabla^2\nabla \times \mathbf{u}_a \qquad [44]$$

together with $\nabla \cdot \mathbf{u}_a = \nabla \cdot \mathbf{u}_s = 0$ and $\mathbf{u}_a = \mathbf{u}_s = 0$ on the boundaries. Only $\mathbf{u}_s$ is forced by precession and $\mathbf{u}_a = 0$ is always a possible solution. An instability of that solution is necessary in order to obtain $\mathbf{u}_a \neq 0$. The energy E_a contained in the velocity field $\mathbf{u}_a$ is therefore a convenient indicator for the onset of instability. In order to localize instabilities spatially, it is helpful to also introduce the density $\epsilon_a(r)$ of the energy contained in $\mathbf{u}_a$ at radius r:

$$\epsilon_a(r) = \frac{1}{4\pi}\int_0^{\pi} d\theta \sin\theta \int_0^{2\pi} d\varphi \frac{1}{2}\mathbf{u}_a^2 \qquad [45]$$

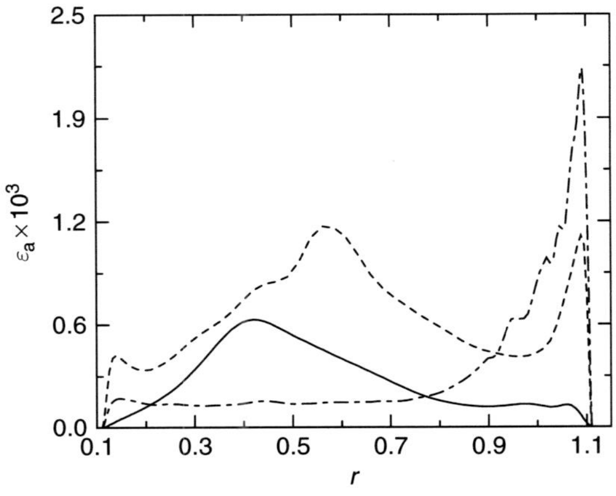

Figure 22 The energy density ϵ_a of the antisymmetric components (eqn [45]) as a function of r for $E = 5 \times 10^{-4}$, $e = 0.1$, $\alpha = 80°$, $\Omega = -0.08$ (solid); $E = 10^{-4}$, $e = 0.04$, $\alpha = 30°$, $\Omega = -0.05$ (dashed); and $E = 10^{-4}$, $e = 0.1$, $\alpha = 80°$, $\Omega = -0.08$ (dot dashed). The values for the last curve have been divided by 10. $\eta = 0.1$ for all cases. Adapted from Lorenzani S and Tilgner A (2001) Fluid instabilities in precessing spheroidal cavities. *Journal of Fluid Mechanics* 447: 111–128.

Three different situations are represented in **Figure 22**. Either only the bulk has become unstable, or only the boundary, or both. One instability does not seem to affect the other.

7.3.4.2.(i) *Boundary layer instability*

Let us first consider the instability localized in the boundary layer. **Figure 23** gives an impression of the unstable boundary layer flow. The comparatively small lateral length scales of the motions excited by the boundary layer instability become difficult to resolve at more extreme parameters. The development of this instability and ensuing numerical instabilities turn out to be the most serious obstacle on the way to high precession rates at low E. The instabilities of time-independent Ekman layers have already been studied in detail, both numerically and experimentally (Tatro and Mollo-Christensen, 1967; Lilly, 1966). In these studies, the Reynolds and Rossby numbers of the boundary layer emerged as the

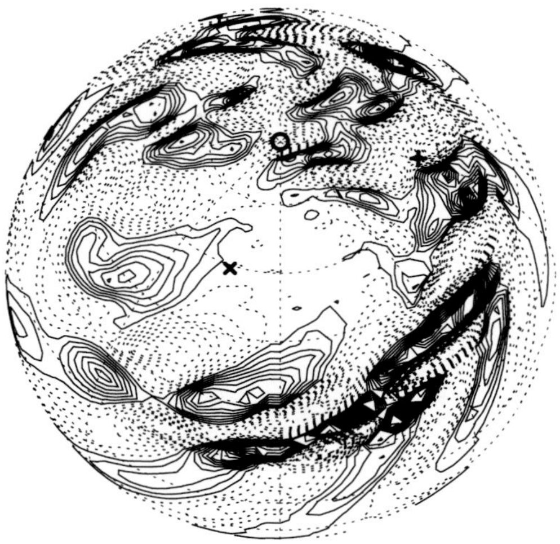

Figure 23 u_{ar} on a surface separated by 0.015 from the outer boundary for $e=0.04$, $\eta=0.1$, $E=10^{-4}$, and $\Omega=-0.03$. The North Pole is marked by a circle, the fluid axis by an x, and the precession axis by a +. Reprinted from Lorenzani S and Tilgner A (2001) Fluid instabilities in precessing spheroidal cavities. *Journal of Fluid Mechanics* 447: 111–128.

relevant parameter separating stable from unstable flows. In the present simulation, let us use as the boundary layer thickness the distance from the boundary at which the absolute value of the radial velocity averaged over spheroidal surfaces reaches a maximum. The layer thickness is approximately $1.4 \times E^{1/2}$ in all cases. Based on this thickness and the maximum tangential velocity, the Reynolds numbers of the boundary layer have been computed. The Rossby numbers of the simulated boundary layers lie in between 0.1 and 0.5 and have been estimated as $v_b[2(1+\Omega\cos\alpha)]^{-1}$ where v_b is the maximum tangential velocity at the edge of the boundary layer and $1+\Omega\cos\alpha$ is the total rotation rate about the z-axis. The critical Reynolds number lies somewhere in between 50 and 100, which falls into the range quoted by Tatro and Mollo-Christensen (1967) for time-independent Ekman layers.

Even though the Reynolds number of the boundary layer surely exceeded its critical value in some precession experiments, the Ekman layer instability has never been noted in experiments which indicates that it stays localized in thin boundary layers also at the parameters typical of experiments. The boundary layers are so thin that an instability within those layers is not easily detectable experimentally.

In the Earth's core, the boundary layers are likely to be unstable. The angle between the Earth's axis of figure and the rotation axis of the core is 1.7×10^{-5} which leads to velocities at the CMB of up to $4.3\,\mathrm{mm\,s^{-1}}$. This is one order of magnitude larger than the velocities deduced from the secular variation of the magnetic field. The precessional velocity field varies with the period of 1 day and the mantle of course screens variations of the magnetic field on that timescale. Nonetheless, these large velocities have an influence on the boundary layer dynamics. Based on a critical Reynolds number for the boundary layer of 100 and using $1.4 \times E^{1/2}$ for the layer thickness, the Earth's Ekman layer must be unstable if the Ekman number of the Earth's core is less than 3.5×10^{-14}, which is the case according to current estimates.

7.3.4.2.(ii) Instabilities of the bulk flow Let us now turn to instabilities of the bulk flow. It is numerically challenging to demonstrate these instabilities. If the boundary layer becomes unstable first, one has to resolve the small-scale structures which then appear. This considerably slows down the computation. It is therefore best to find sets of parameters where the boundary layers stay stable. The first examples which have been reported were for comparatively large Ekman numbers, which then also implies a large Poincaré force to drive the flow vigorously enough. A strong driving is achieved by choosing a large precession rate and a large angle between rotation and precession axes. A large precession rate by itself is not useful because a plateau is reached in the kinetic energy of the flow with increasing $|\Omega|$ (see **Figure 18**) and at yet larger $|\Omega|$, other inertial modes may become excited (Manasseh, 1992). An angle of about 60° between rotation and precession axes must usually be chosen for computations.

As an example, **Figure 24** shows E_a as a function of E at a fixed $|\Omega|$. E_a equals zero for the basic flow, $E_a \neq 0$ indicates the onset of a new flow.

The simulations of bulk instabilities in precessional flow (Tilgner and Busse, 2001; Lorenzani and Tilgner, 2001, 2003) lead to global instability which invades the entire fluid volume. It is impossible to connect the origin of instability by mere visual inspection to some particular feature of the underlying laminar flow. We therefore need a more incisive instrument than visualization.

It will prove very useful to spectrally decompose the flow field into components of different azimuthal wave numbers m'. The prime denotes a coordinate

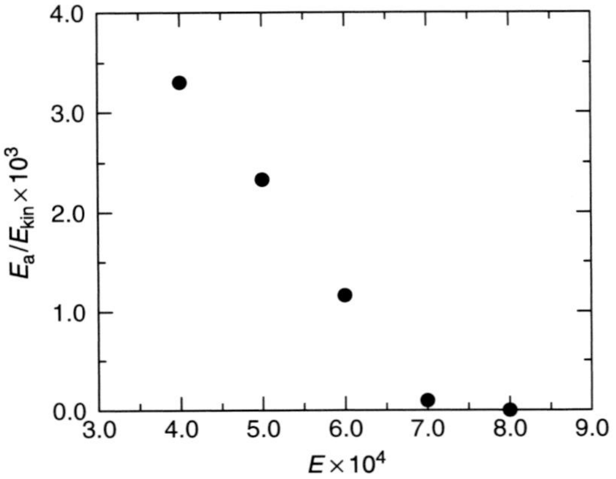

Figure 24 The ratio of the energy contained in the antisymmetric components and the total energy, E_a/E_{kin}, as a function of the Ekman number E for $\Omega = -0.2$ and $\alpha = 60°$ in a spherical shell with $r_i/r_o = 0.01$.

system in which the azimuth φ' is measured with respect to the axis of the fluid, not the axis of the container. In order to see why this makes sense, consider the equation governing the linear stability of an ideal fluid driven by precession in the mantle frame:

$$\frac{\partial}{\partial t}\nabla \times \mathbf{u}' + 2\nabla \times (\hat{\mathbf{z}} \times \mathbf{u}') + \nabla \times [(\nabla \times \mathbf{u}') \times \mathbf{u}_0] + \nabla \times [(\nabla \times \mathbf{u}_0) \times \mathbf{u}'] = 0 \qquad [46]$$

The total velocity **u** is the sum of the velocity perturbation $\mathbf{u}'$ and the basic flow field $\mathbf{u}_0$. If the third and fourth terms are small compared with the Cariolis term, they can be treated as a perturbation, the solution of the unperturbed problem being the inertial eigenmodes. Let us assume that some particular feature of the basic flow triggers the instability and that this feature is characterized by a single wave number m_0'. Cylindrical shear layers for example are axisymmetric with respect to ω_F and thus have $m_0' = 0$. In this case, the coefficients in [46] are independent of φ'. In addition, [46] is linear in $\mathbf{u}'$ so that $\mathbf{u}'$ must have a dependence on φ' in $e^{im_1'\varphi'}$. Axisymmetric shear layers excite unstable flow characterized by a single wave number m_1'.

However, the spin-over mode is a flow with $m_0' = 1$. It cannot excite an instability with a single wave number. In the lowest order of a perturbation calculation (outlined in Section 7.5), it must excite a pair of eigenmodes of the unperturbed problem. One speaks of a triad resonance when this happens because three entities compose the total flow: the laminar flow $\mathbf{u}_0$, which is itself an inertial mode if we identify $\mathbf{u}_0$ with the spin-over mode, and the two modes excited by the instability. Let these modes have wave numbers m_a' and m_b' and eigenfrequencies ω_a and ω_b. These modes have time and azimuthal dependences given by $e^{i(m_a'\varphi' - \omega_a t)}$ and $e^{i(m_b'\varphi' - \omega_b t)}$. They are coupled by the perturbation only if $|m_a' - m_b'| = 1$. The instability thus consists of at least one pair of inertial modes with wave numbers differing by one. Otherwise, the perturbation couplings, that is, the third and fourth terms in [46], would introduce new wave numbers. [46] can impossibly be satisfied if $\mathbf{u}'$ consists of a single mode and if $\mathbf{u}_0$ is characterized by an m_0' different from zero. For arbitrary m_0', the condition $|m_a' - m_b'| = m_0'$ has to be met. A look at a spectrum as a function of m' thus directly gives a hint as to what has caused an instability.

Figure 25 shows an example. The velocity fields corresponding to unstable states have been transformed to the primed coordinate system in which the z'-axis points along $\boldsymbol{\omega}_F$. The field has then been decomposed into spherical harmonics in this new system and the energy contained in the different spectral components has been plotted against the azimuthal wave number m'. In the example shown in **Figure 25**, a dominant contribution to $\mathbf{u}_a$ occurs at $m' = 7$ and 8, whereas $\mathbf{u}_s$ differs little from its shape in stable solutions. It is expected that there is a nonzero contribution from all wave numbers because the spectrum in **Figure 25** is for a saturated state and

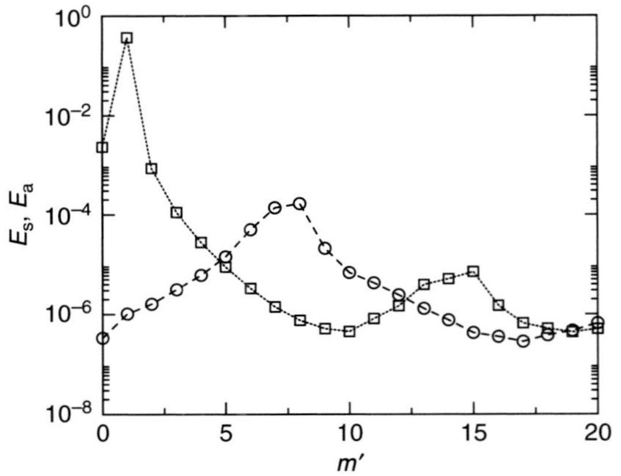

Figure 25 Energy contained in the modes with wave number m' as a function of m' for $\eta = 0.1$, $e = 0.04$, $\alpha = 30°$, $\Omega = -0.03$, and $E = 10^{-4}$. Antisymmetric (circles) and symmetric (squares) contributions are shown separately. Adapted from Lorenzani S and Tilgner A (2001) Fluid instabilities in precessing spheroidal cavities. *Journal of Fluid Mechanics* 447: 111–128.

not for an exponentially growing unstable motion. In addition, the prediction that only two wave numbers are populated is only true if the terms involving $\mathbf{u}_0$ are infinitesimally small. Nonlinear interactions generate a small peak in the spectrum of $\mathbf{u}_s$ at twice the wave number at which the spectrum of $\mathbf{u}_a$ peaks.

The structure of the instability for the set of parameters used in **Figure 25** is clarified in **Figures 26** and **27** and summarized in **Figure 28**. **Figures 26** and **27** are snapshots. As time goes on, the $m' = 7$ and $m' = 8$ patterns rotate independently of each other about the z'-axis. **Figure 26** shows $u_{az'}$ and u_{ar} in the plane perpendicular to ω_F. $m' = 7$ appears in u_{ar} whereas $m' = 8$ dominates $u_{az'}$. The two sets of rolls overlap but are centered at different radii. **Figure 27** shows cylindrical cuts at distances from the fluid axis corresponding to these two radii. For $m' = 7$ one finds columnar vortices symmetric about the equatorial plane of the primed coordinate system, whereas the $m' = 8$ vortices are antisymmetric about this plane.

Figure 28 reproduces these and additional observations in a sketch. Two sets of columnar vortices centered at different distances from the fluid axis exist, with wave numbers differing by one. The entire pattern is antisymmetric with respect to reflection at the origin. Individual rolls in the wave with odd wave number therefore have equal vorticity in the Northern and Southern Hemispheres. Rolls belonging to the wave with even wave number on the contrary have opposite vorticities in both hemispheres. The designations North and South refer of course to the primed coordinate system. The axial and azimuthal components in the outer roll pattern are of comparable magnitude, whereas $u_{az'}$ is small in the component with odd m'. $u_{az'}$ reaches its extremal values in between the outer vortices. The same sketch is valid for other parameters, except that the values of m' change.

The axisymmetric shear layers which are so conspicuous in experiments have $m'_0 = 0$. They cannot be the origin of the instability in **Figure 25**. The inviscid spin-over mode also has $m'_0 = 0$ in a sphere, but its viscous counterpart (with boundary layers, internal shear layers, etc.) has $m'_0 = 1$ and triggers the instability in **Figure 25**. Two additional features are important in ellipsoidal containers. The first is the ellipticity of the streamlines of the Poincaré solutions sketched in **Figure 12**. If one considers the unperturbed problem to be the eigenvalue problem in a sphere, and if one restricts oneself to small ellipticities of the container so that the elliptic deformation of the streamlines can be treated perturbatively, one will find instabilities involving two modes with wave numbers differing by 2. The second source of instability is introduced by the fact that the elliptical streamlines of the Poincaré flow lie in parallel planes and that there is shear between these planes because the line joining the centers of the ellipses is not perpendicular to the plane of the streamlines. This shear couples modes with wave numbers differing by 1. These last two instabilities occur even if viscosity is strictly zero. For this reason, they are classified as inertial instabilities and are more accessible to analytical treatment. Stability

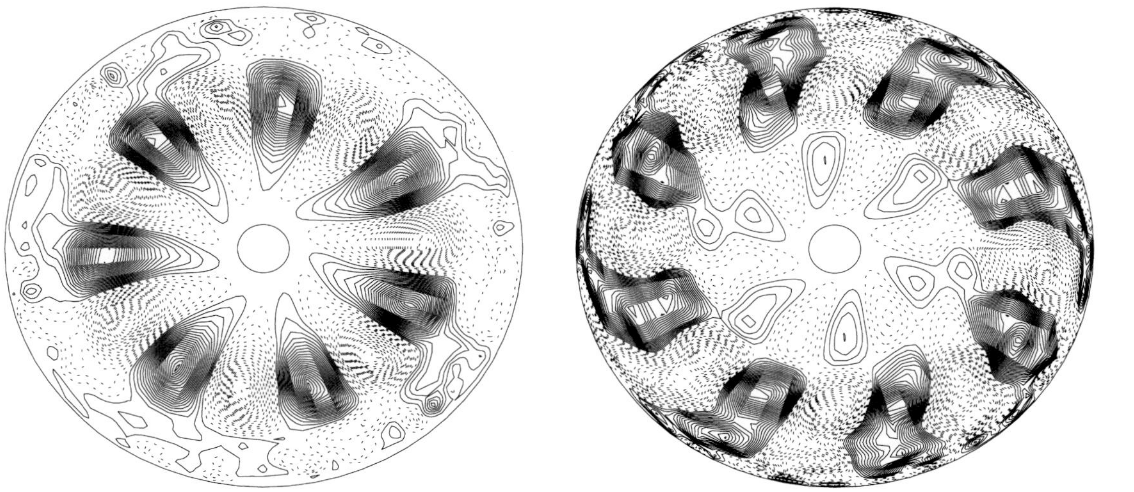

Figure 26 u_{ar} (left) and $u_{az'}$ (right) in the plane perpendicular to ω_F for the same parameters as in **Figure 25**. Adapted from Lorenzani S and Tilgner A (2001) Fluid instabilities in precessing spheroidal cavities. *Journal of Fluid Mechanics* 447: 111–128.

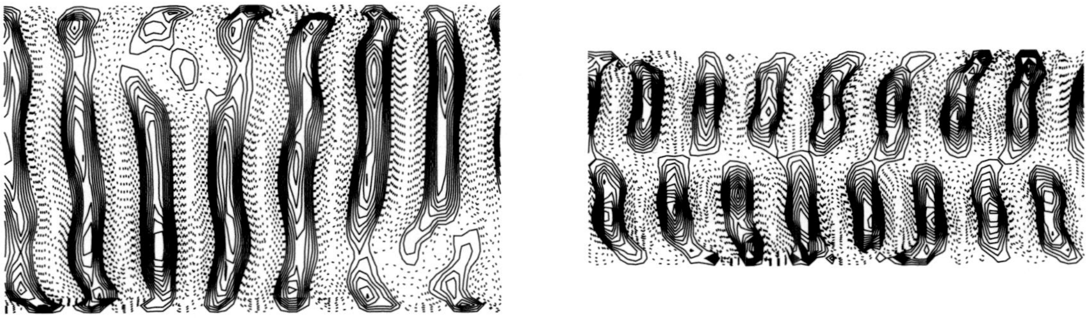

Figure 27 u_{ar} on cylindrical surfaces at distances 0.638 (left) and 0.869 (right) from the ω_F-axis for the same case as in **Figures 25** and **26.** φ' runs from $-\pi$ to π in going from left to right, and θ' is given on the ordinate. Adapted from Lorenzani S and Tilgner A (2001) Fluid instabilities in precessing spheroidal cavities. *Journal of Fluid Mechanics* 447: 111–128.

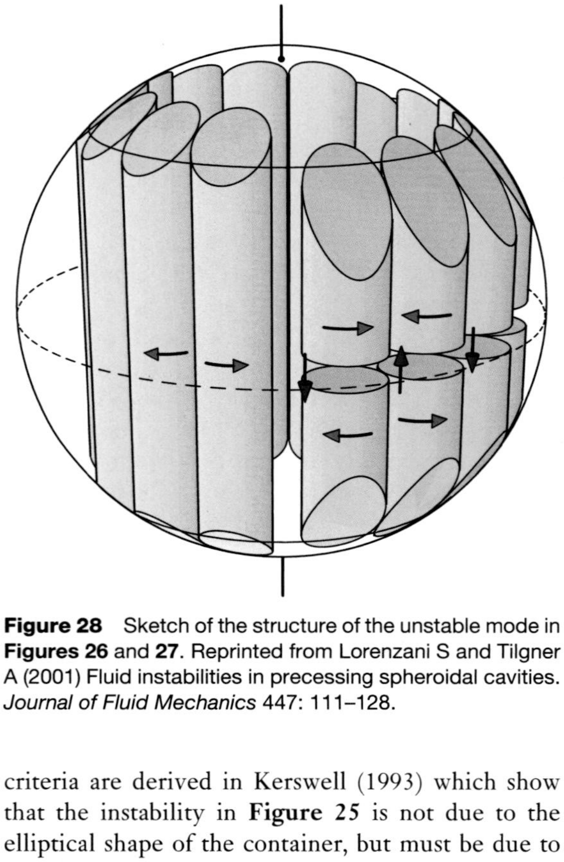

Figure 28 Sketch of the structure of the unstable mode in **Figures 26** and **27.** Reprinted from Lorenzani S and Tilgner A (2001) Fluid instabilities in precessing spheroidal cavities. *Journal of Fluid Mechanics* 447: 111–128.

criteria are derived in Kerswell (1993) which show that the instability in **Figure 25** is not due to the elliptical shape of the container, but must be due to viscous corrections to the spin-over mode.

7.3.4.2.(iii) Saturation and turbulence In order to isolate the two inertial instabilities just mentioned, one must get rid of the viscous boundary layers. This has been done by Lorenzani and Tilgner (2003) by applying free-slip boundary conditions. In this case a new phenomenon occurs which is commonly known as resonant collapse: the growing instability does not reach a saturated state but starts to oscillate instead (**Figure 29**). During collapse, a laminar large-scale inertial mode suddenly decays into small-scale turbulence. The small scales draw energy from the large scales which they dissipate. Once enough energy has been dissipated, the flow becomes laminar and the same instability as before grows once more only to decay into turbulence again. This cycle repeats indefinitely. During the growth phases, the two excited modes have the same qualitative features as those summarized in **Figure 28**. Beyond a certain amplitude of these modes, collapse can occur and E_a returns to small values. **Figure 29** shows an example for parameters far enough beyond the onset of instability so that the variations of the energy are not periodic any more. **Figure 30** demonstrates the appearance and disappearance of small-scale structures in the course of the oscillations.

Laboratory experiments were mostly done with full ellipsoids, but for geophysical applications, the effect of an inner core is of interest. In a simulation with the same parameters as above except for an inner core with $\eta = 0.35$, we expect the inner core to cause little change to the resonance conditions because the inertial modes observed without inner core reach their largest amplitude outside the region now occupied by the inner core. And indeed, the numerical simulation reveals an initial instability with the same pair as before ($m' = 3$ and 4) and the final stage is dominated by the pair $m' = 1$ and 3. However, no collapses occur in this particular case, but small scales remain permanently excited in a statistically stationary state at the end of the run (**Figure 31**).

Lorenzani and Tilgner (2003) have also simulated as closely as possible the experiments of Malkus (1968). It appears from these simulations that the

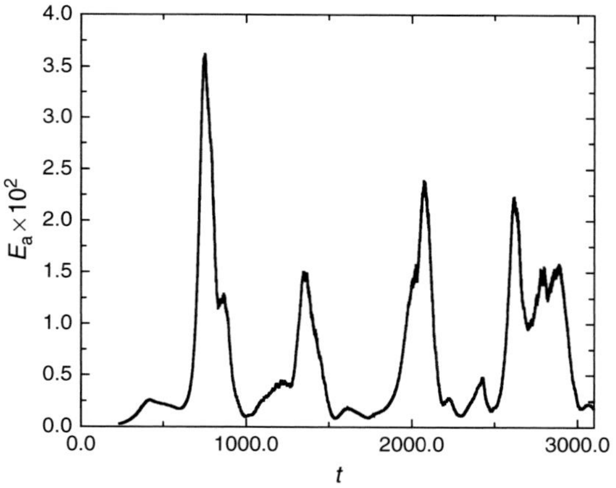

Figure 29 Time evolution of the energy E_a contained in the velocity components antisymmetric with respect to reflection at the origin for $e = 0.06$, $\alpha = 30°$, $\Omega = -0.14$, and $E = 5 \times 10^{-5}$. Adapted from Lorenzani S and Tilgner A (2003) Inertial instabilities of fluid flow in precessing spheroidal shells. *Journal of Fluid Mechanics* 492: 363–379.

vigorous motions observed in the experiments by Malkus (1968) are likely to be inertial rather than viscous instabilities. This does not exclude the possibility of a viscous instability appearing first at low precession rates. In one case, Malkus has observed a spectacular transition from a chaotic to a fully turbulent flow accompanied by a hysteresis effect (see Section 7.3.3). Simulations have shown that the sudden increase in turbulence fluctuations and the hysteresis have nothing to do with the onset of a new instability but are rather a manifestation of a reorientation of the vorticity of the underlying Poincaré flow. Turbulence appears at a precession rate at which a reorientation of the fluid axis occurs, too. Quite surprisingly, eqn [34] may have three coexisting solutions, two of which are stable (see Noir *et al.*, 2003). The presence of several solution branches readily explains the hysteresis effect. When the orientation of the basic flow changes abruptly, the strain in the basic flow changes equally abruptly and causes the transition from a more to a less stable flow. We will not consider these effects in any detail here

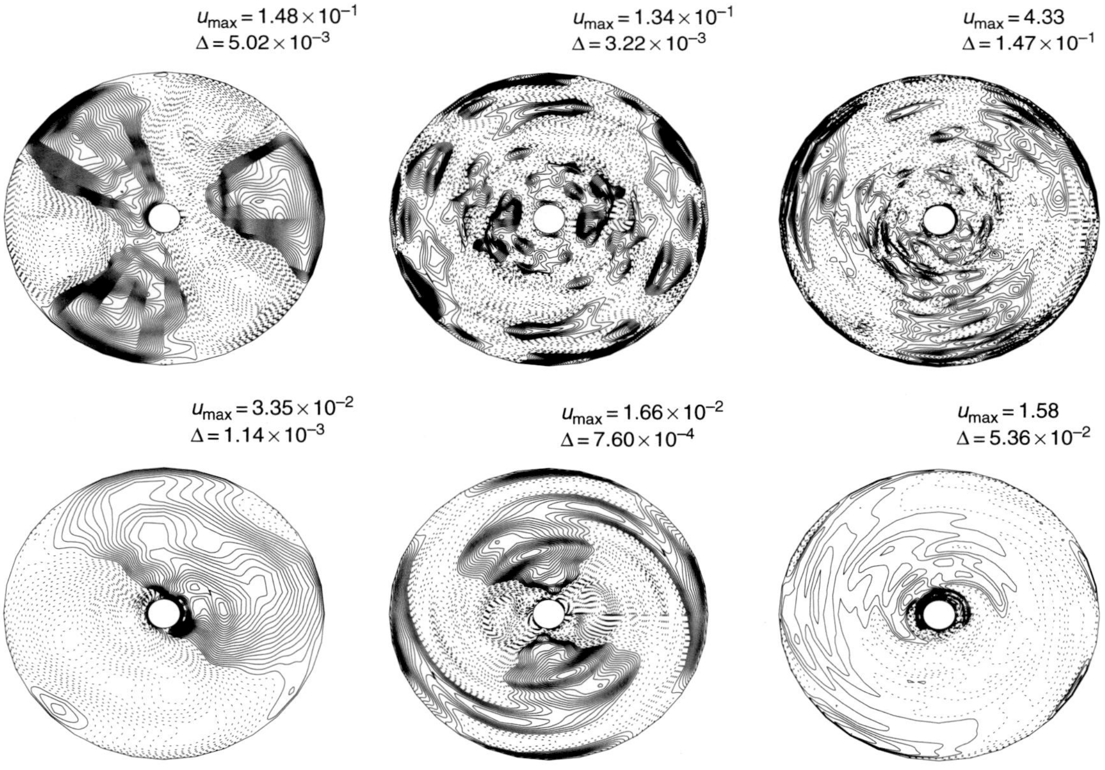

Figure 30 Pictures of the flow at the times of the first maximum (top) and the first minimum (bottom) in **Figure 29**. The different panels show u_{ar} (left), $u_{az'}$ (middle) and $(\nabla \times \mathbf{u}_a)_{z'}$ (right). Adapted from Lorenzani S and Tilgner A (2003) Inertial instabilities of fluid flow in precessing spheroidal shells. *Journal of Fluid Mechanics* 492: 363–379.

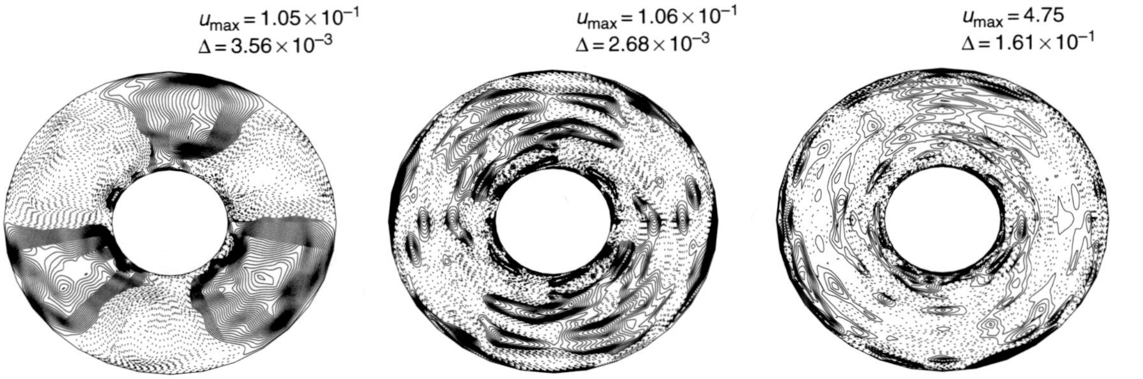

Figure 31 u_{ar} (left), $u_{az'}$ (middle), $(\nabla \times \mathbf{u}_a)_{z'}$ (right) for $e = 0.06$, $\alpha = 30°$, $\Omega = -0.14$, $E = 5 \times 10^{-5}$, and $\eta = 0.35$. Adapted from Lorenzani S and Tilgner A (2003) Inertial instabilities of fluid flow in precessing spheroidal shells. *Journal of Fluid Mechanics* 492: 363–379.

because a catastrophic reorientation as observed in Malkus' experiment is not relevant to the Earth. For $E = 10^{-15}$, $\alpha = 23.5°$, and $e = 1/400$, a similar transition occurs only at $\Omega \approx -10^{-3}$ whereas the actual precession rate of the Earth is $\Omega \approx -10^{-7}$.

7.3.4.2.(iv) Extrapolation to the Earth's core Two of the instabilities mentioned in this section can be extrapolated to the Earth. First, we expect the Ekman layers in the core to be unstable as mentioned above. Then there are the inertial instabilities. The theory of Kerswell (1993) has been verified by simulations and can be applied to the core. Unfortunately, according to that theory, the state of precession-driven flow in the Earth's core is uncertain because commonly accepted values of the viscosity of the core put the flow close to its stability limit. Fluid viscosity is one of the least well-constrained material properties of the core. Let us assume that the core is unstable. The core could then undergo resonant collapses. When collapse occurred in numerical simulations, the growth rate after the collapse was within a factor of 2–3 from the growth rate the same modes have during a linear growth phase starting from a Poincaré solution as described by perturbation theory. Using the upper bound for the growth rate of an inertial instability given by Kerswell (1993) applied to Earth's numbers gives a growth rate of 20 000 yr^{-1}. Viscosity acts to slow this growth. If collapses play a role in the Earth's core, they could manifest themselves in variations of the magnetic field with a time constant of 20 000 years or longer. There are some indications from paleomagnetic observations for variations which one naively expects in the presence of collapses (Aldridge and Baker, 2003).

On the other hand, we are not able, at present, to make reasonable guesses concerning the fate of the instabilities due to the axisymmetric shear layers or viscous corrections to the spin-over mode. In all the simulations made so far, the $m' = 1$ deviations from a flow with uniform vorticity in the basic state outweigh the $m' = 0$ deviations. However, as the Ekman number is decreased, the viscous corrections contributing to the $m' = 1$ deviations diminish, whereas the axisymmetric shear layer connecting the critical latitudes becomes more and more singular. An instability of that shear layer is thus plausible at low E. One can determine the typical velocity difference in the $m' = 1$ component from visualizations and construct a Rossby number from it. It turns out that this Rossby number increases with decreasing Ekman number, so that the mechanism related to the $m' = 1$ components should remain effective even at the lower Ekman numbers at which the axisymmetric shear layer possibly becomes unstable, too. Since we do not know what the stability criterion of these layers is, we are not yet able to conclude about the stability of the core flow of the Earth.

7.4 Tides

In addition to precession, tidal deformation of the CMB disturbs the rotation of the core flow. Comparatively little space of this chapter is devoted to tidally excited flows despite their potential geophysical significance. This is due to the fact that many ideas on precession-driven flows can be transposed with little modification to tidal flows so that there is not very much new material to present in this section.

We will consider a simple model in which the unperturbed CMB is spherical. Tidal forces deform this boundary into an ellipsoid with the long axis pointing toward the Moon. In this section, all examples will relate to the geometry in which the tidal elongation is perpendicular to the rotation axis, which corresponds to an equatorial tide. Precession is neglected so that tidal effects can be studied in isolation. The orientation of the axis of rotation of the core is not modified by tides. But the streamlines of the core flow, which are circular in the unperturbed core viewed from inertial space, become elliptical due to tidal deformation. Elliptical streamlines also appeared in precession-driven flow (see **Figure 12**) and gave rise to instability. The stability considerations will therefore be very similar for tidal and precessional flows.

The deformation of the boundary in an equatorial tide is characterized by an azimuthal wave number $m=2$. Suppose the tidal companion is fixed in inertial space. The flow then varies in the mantle frame with a frequency twice the rotation frequency, that is, $\omega=2$ in our dimensionless units. $\omega<2$ if the tidal companion is in prograde rotation.

Let us first look at internal shear layers which may modify the inviscid picture of **Figure 12**. For $\omega=2$, the group velocity of inertial waves is perpendicular to the rotation axis and critical latitudes are exactly at the poles. We therefore do not expect a pattern of multiple reflected internal layers as in the precessional case. But nonlinear interactions may start an axisymmetric shear layer at critical latitudes, which in the tidal case degenerates to a line joining the poles. Suess (1971) built an experiment to confirm the existence of such a shear zone. In this experiment, a flexible water-filled sphere was set into rotation. The sphere was deformed into an ellipsoid by a ring surrounding the equator of the sphere. The ring was fixed in the laboratory frame. The experiment confirmed the existence of a shear zone connecting the poles and revealed a second, albeit much weaker, axisymmetric shear layer at half the radius of the sphere. The origin of this second layer is unknown.

We proceed further in the analogy with precession-driven flow and investigate the stability of the flow, ignoring the shear zones just discussed. For small ellipticities we can again apply the perturbation expansion of Section 7.3.4.2 (see also Section 7.5). This analysis tells us that tidal distortion couples in triad resonances two inertial modes whose azimuthal wave numbers and frequencies differ by 2. There is also a more intuitive explanation for the instability of tidal flow. Imagine looking at the flow along the rotation axis from inertial space. One then sees elliptical streamlines. Let us represent the elliptical streamlines in a plane $z=$ const. with major axes along x and y by the stream function $\Psi=(1+\delta)x^2+(1-\delta)y^2$, where δ measures the deformation from the undeformed stream function x^2+y^2. The corresponding velocity field, $\nabla\times(\Psi\hat{\mathbf{z}})$, fits into any container whose boundaries in every plane $z=$ const. are ellipses (such as ellipsoids or ellipsoidal cylinders) with a ratio of long and short semi-axes of $[(1+\delta)/(1-\delta)]^{1/2}$. Ψ may be rewritten in the more suggestive form

$$\Psi=(x^2+y^2)+\delta(x^2-y^2) \qquad [47]$$

The first bracket is the stream function in absence of tidal deformation. The second term, proportional to δ, describes hyperbolic streamlines with axes oriented at 45° of the x and y axes.

Perturbing vorticity is amplified along the stretching direction of the hyperbolic streamlines. The effect of the stretching is most easily recognized in a deformed cylindrical vortex (instead of a vortex encased in an ellipsoid) infinitely extended along the z-axis with the stream function given by [47]. It is a simple exercise to verify that this vortex is unstable with respect to a perturbation of the form $(-\hat{\mathbf{x}}+\hat{\mathbf{y}})\times\mathbf{r}$ which grows at rate $\delta/2$. Indeed, the equation governing the stability of the basic flow $\mathbf{u}_0=\nabla\times(\Psi\hat{\mathbf{z}})$, namely [46] in an inertial frame of reference,

$$\frac{\partial}{\partial t}\nabla\times\mathbf{u}'+\nabla\times[(\nabla\times\mathbf{u}')\times\mathbf{u}_0] + \nabla\times[(\nabla\times\mathbf{u}_0)\times\mathbf{u}']=0 \qquad [48]$$

is solved by the perturbation $\mathbf{u}'=(-\hat{\mathbf{x}}+\hat{\mathbf{y}})\times\mathbf{r}e^{t\cdot\delta/2}$. This is a solid body rotation stationary in inertial space. In other words, this is a flow of spatially constant vorticity.

In ellipsoids, the simplest unstable mode is also a constant vorticity flow. A potential flow added to the solid body rotation accommodates the boundary conditions. The growth rate of the constant vorticity flow in the ellipsoid depends on the relative sizes of the axes of the ellipsoid. When the vorticity is constant in inertial space, its axis is rotating in the mantle frame with $\omega=1$. This unstable mode can be represented in the mantle frame as a superposition of the spin-over mode, which varies as $e^{-i(\varphi - t)}$, and its complex conjugate, which varies in $e^{i(\varphi - t)}$. These two modes fulfill the selection rules on wave number and frequency for a triad resonance stated in the next section. Other pairs of modes satisfying the selection rules can be excited as well, but their vorticity is directed parallel to the $(-\hat{\mathbf{x}} + \hat{\mathbf{y}})$ direction only in the time average. Computations of stability limits and many more fluid dynamic aspects of elliptical instabilities are reviewed in detail in Kerswell (2002).

Several experiments have investigated the instability of elliptically deformed rotating flows, either in cylindrical or in spheroidal geometry (Malkus, 1989; Gledzer and Ponomarev, 1992; Aldridge *et al.*, 1997; Seyed-Mahmoud *et al.*, 2004; Eloy *et al.*, 2000; Lacaze *et al.*, 2004, 2005). These experiments use one of two methods to realize the required deformation. Most use deformable containers in the same fashion as Suess (1971) (see **Figure 32**). Gledzer and Ponomarev (1992) used an interesting alternative. They spun up fluid in an ellipsoid or a cylinder with ellipsoidal cross-section and abruptly stopped the rotation of the container. The fluid continues to rotate due to its inertia and forms a vortex with elliptical stream lines. This vortex is prone to elliptical instability, an example of which is shown in **Figure 33**.

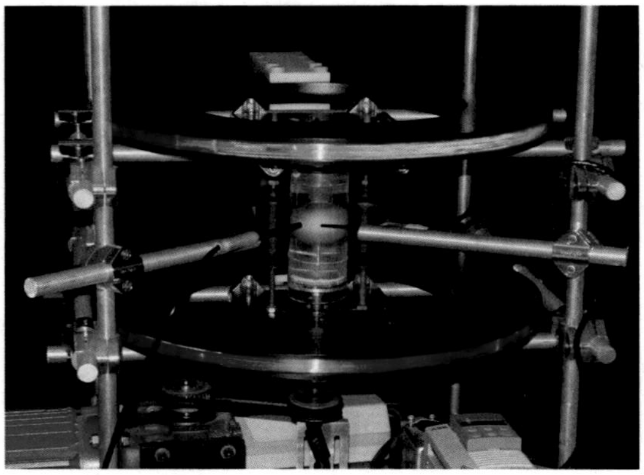

Figure 32 Experimental apparatus for the study of tidal instabilities in use at the 'Institut de Recherche sur les Phénomènes hors Equilibre' in Marseille, France. The fluid under study is in the container of opaque and ellipsoidal appearance in the center of the picture, which is mounted on a rotating shaft, and compressed and deformed by two gray cylindrical rollers with their axis parallel to the rotation axis. The rollers are stationary in the laboratory frame. A pair of Helmholtz coils allows the study of induction effects if Gallium is used as fluid. Courtesy of L Lacaze and P Le Gal.

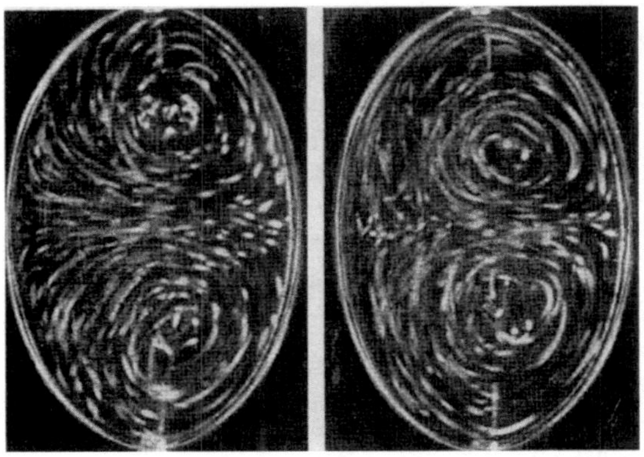

Figure 33 Elliptical instability in an ellipsoid. Streamlines are made visible with suspended particles. A flow with two eddies is excited by the instability. The eddies are time dependent as evidenced by the two panels which show the flow a short time interval apart. Reprinted from Gledzer E and Ponomarev V (1992) Instability of bounded flows with elliptical streamlines. *Journal of Fluid Mechanics* 240: 1–30.

In all cases, excellent agreement with theory has been obtained. After instability, the system settles in a state of one of three possible types, again in analogy with precession. Either a stationary state is reached, or the flow ends in a turbulent and statistically stationary state, or the flow undergoes resonant collapses. There is no understanding or systematic data on which of these three possibilities is realized under which circumstances.

The good agreement between theory and experiment concerning the stability criterion of tidal flow gives us confidence that we can apply this theory to the Earth. From the rotation rate of the Earth and the tidal deformation of the CMB, one computes a growth-rate for instabilities in inviscid fluid of approximately 1/7000 yr) (Aldridge *et al.*, 1997). This growth rate is lowered to $10^{-6}\,\mathrm{yr}^{-1}$ by viscous damping if the Ekman number of the core is 0.3×10^{-15}. If magnetic dissipation is also taken into account, it is unclear whether the Earth's core is unstable to tidal deformation (Kerswell, 1994), depending mostly on what the precise Ekman number of the core is.

7.5 Interaction with Buoyancy and Magnetic Fields

Up to now we have only considered neutrally buoyant fluid. This has greatly simplified the analysis and has allowed us some direct comparison with experiments. Eventually we will have to incorporate thermal and magnetic effects for the geophysical application. Relatively little work has been done in this direction until today so that we are left with many uncertainties. This section reviews the open questions and a few plausible but unproven answers.

As far as thermal effects are concerned, we have to distinguish between the assumptions that the core is convecting, or that it is stably stratified. Let us consider the latter possibility first. Stable stratification generally suppresses fluid motion, at least along the density gradient. But it also allows internal waves, which combine with pure inertial waves to form gravitoinertial waves. These waves have properties similar to inertial waves. In particular, discontinuities form on characteristic surfaces, which however are not conical any more (Dintrans *et al.*, 1999). The inviscid spin-over mode has no radial component and is thus unaffected by thermal stratification in a sphere. Laminar precession-driven flow cannot differ in any significant way in stable and neutral fluids, so that the instability mechanism through triad resonances can also operate in a stably stratified fluid by coupling gravitoinertial waves. Radial stratification is no impediment to instability (Kerswell, 1994), but in order to reach equal radial velocity, much more energy must be pumped into the flow compared with the neutral case because of the potential energy stored in the deformation of the background stratification.

If on the other hand the core is convecting, the time-averaged temperature distribution equals the adiabatic profile, so that the fluid is on average neutrally buoyant, but perturbations like precession now act on a medium which is not at rest. This could modify the precession-driven flow through the nonlinear term in the Navier–Stokes equation. The Rossby number (already introduced in Section 7.1.2) estimates in order of magnitude the ratio of the nonlinear and Coriolis terms and is $U/(\omega_D L)$ for a motion of typical velocity U and size L in the core which rotates at the rate ω_D. This number is about 10^{-5} for the whole core, using for U velocities deduced from secular magnetic variation, and is thus very small. In a first step, let us completely neglect the nonlinear term. Within this approximation, flows driven by different agents superpose linearly. Large-scale structures of inertial modes are therefore unchanged by convection. In particular, the orientation of the fluid axis is unaffected by convection. Dropping the nonlinear term from the Navier–Stokes equation eliminates both the axisymmetric shear layers and the instabilities. We therefore reinstate that term and treat it as a small perturbation. Within this framework, the nonlinear interaction of the (non-axisymmetric) viscous spin-over mode with itself yields an axisymmetric driving term (a Reynolds stress in the language of fluid mechanics) which drives the axisymmetric shear layers. The presence of convective motion adds new contributions to the total axisymmetric driving force but does not remove the one set up by precession. The cylindrical shear layers typical of precession-driven flow thus cannot disappear because of a background of convective motion. At the small scales of the shear zones, L in the expression for the Rossby number becomes small so that the Rossby number can be larger than 1. Significant interaction between convection and precession thus may occur at small scales, in which for example convective motion acts like an eddy diffusivity which broadens shear zones.

We will next show that inertial instabilities of Poincaré flow also survive in the presence of convection. To see this, it is helpful to have a more formal look at the perturbation calculation mentioned in Section 7.3.4.2. Let us start from [46], the linearized inviscid equation of motion valid in the mantle frame for a velocity perturbation $\mathbf{u}'$ on top of a basic flow field $\mathbf{u}_0$, so that the total velocity is $\mathbf{u}_0 + \mathbf{u}'$. We consider the third and fourth terms to be perturbations compared with the Coriolis term and denote them formally as $\epsilon\mathcal{V}\mathbf{u}'$, where $\mathcal{V}$ is a linear operator acting on $\mathbf{u}'$ and ϵ is a book-keeping parameter which allows us to keep track of the different orders of a perturbation calculation. The uncurled eqn [46] becomes

$$\frac{\partial}{\partial t}\mathbf{u}' + 2\hat{\mathbf{z}} \times \mathbf{u}' + \nabla p' + \epsilon\mathcal{V}\mathbf{u}' = 0 \qquad [49]$$

in which $\nabla p'$ is a pressure gradient. Inertial modes with frequency ω_j, which we compactly write as $e^{i\omega_j t}\boldsymbol{\Phi}_j(\mathbf{r})$, and their associated pressure p_j, are solutions of the unperturbed ($\epsilon = 0$) problem:

$$i\omega_j\boldsymbol{\Phi}_j(\mathbf{r}) + 2\hat{\mathbf{z}} \times \boldsymbol{\Phi}_j(\mathbf{r}) + \nabla p_j(\mathbf{r}) = 0 \qquad [50]$$

The $\boldsymbol{\Phi}_j$ are mutually orthogonal (Greenspan, 1968). With the appropriate normalization of the $\boldsymbol{\Phi}_j$, the integral over the fluid volume $\int \boldsymbol{\Phi}_k^*\boldsymbol{\Phi}_j\,dV = \delta_{j,k}$. We write the solution of [49] as a linear combination of eigenmodes

$$\mathbf{u}' = \sum_j c_j(t)e^{i\omega_j t}\boldsymbol{\Phi}_j(\mathbf{r}), \quad p' = \sum_j c_j(t)e^{i\omega_j t}p_j(\mathbf{r}) \qquad [51]$$

and pose an expansion of the coefficients c_j in ϵ as

$$c_j = c_j^{(0)} + \epsilon c_j^{(1)} + \epsilon^2 c_j^{(2)} + \cdots \qquad [52]$$

Order ϵ^0 yields $\frac{\mathrm{d}}{\mathrm{d}t}c_j^{(0)} = 0$. We obtain at order ϵ^1 by projecting [49] on mode $\mathbf{\Phi}_k$ and multiplying by $\mathrm{e}^{-\mathrm{i}\omega_k t}$:

$$\frac{\mathrm{d}}{\mathrm{d}t}c_k^{(1)} = -\sum_j c_j^{(0)}\mathrm{e}^{\mathrm{i}(\omega_j-\omega_k)t}\int \mathbf{\Phi}_k^* \mathcal{V}\mathbf{\Phi}_j \mathrm{d}V \qquad [53]$$

If we remember that $\mathbf{\Phi}_j$ depends on azimuth as $\mathrm{e}^{-\mathrm{i}m_j\varphi}$ we can derive the selection rule on azimuthal wave number for a triad resonance used in Section 7.3.4.2 from the requirement that the integral in [53] must be different from zero. Suppose that $\mathbf{u}_0$, and hence $\mathcal{V}$, depends on azimuth as cos $m_0\varphi$. The integral on the right-hand side of [53] contains the term $\mathrm{e}^{\mathrm{i}m_k\varphi}\cos(m_0\varphi)\mathrm{e}^{-\mathrm{i}m_j\varphi}$ whose integral over φ is different from zero only if $|m_k - m_j| = m_0$. A similar reasoning provides us with a condition on the eigenfrequencies of the two coupled inertial modes. Suppose that $\mathbf{u}_0$, and hence $\mathcal{V}$, depends on time as cos $\omega_0 t$. The right-hand side of [53] then contains two terms, with time dependences $\mathrm{e}^{\mathrm{i}(\omega_j-\omega_k+\omega_0)t}$ and $\mathrm{e}^{\mathrm{i}(\omega_j-\omega_k-\omega_0)t}$. There is resonance, that is, $c_k^{(1)}$ grows to infinity, if $|\omega_j - \omega_k| = \omega_0$. Equation [53] is linear in $\mathcal{V}$ or $\mathbf{u}_0$. If $\mathbf{u}_0$ contains more than one frequency, only one of them can couple any two inertial modes. It is for this reason that we do not expect convective motion to modify the onset of instabilities found in Section 7.3.4.2: in the mantle frame, the precession-driven basic flow has $\omega_0 = 1$, that is it varies on a diurnal timescale. Convection on the contrary is believed to vary on much longer timescales of decades or longer. For instance, the turnover time is approximately 1 ky for an eddy spanning the gap between the inner core and the CMB with a flow velocity of 0.1 mm s^{-1}. An instability involving a triad resonance between precession-driven flow and two inertial modes will thus be unaffected by convection simply because convection is far off that resonance. The same holds true for tidally driven motion. If the tidal companion is fixed in space, the tidal disturbance is characterized by $\omega_0 = 2$ in the above equations, which is again faster than convective timescales.

As the amplitudes of the unstable modes grow according to [53], the coupling term in [49] will eventually become large and higher orders of perturbation come into play. The unstable modes end in a saturated state or in resonant collapse. At this stage, noteworthy interaction between buoyancy and precession or tides is possible and will have to be investigated in the future.

The influence of magnetic fields has been much less studied. These studies involve some arbitrariness since we do not know the magnetic field inside the core. Oblique shear layers also exist in the magnetohydrodynamic case but they acquire a more oscillatory structure than before. The inertial waves forming the internal shear layers couple to the magnetic field and send off magnetic waves which occupy a much larger fraction of the fluid volume than the shear layers. At strong-enough magnetic fields, these layers also change orientation. This has been observed in numerical simulations in spherical shells (Tilgner, 1999b) and more accurately in infinitely extended fluids (Tilgner, 2000). Kerswell (1994) investigated the damping effect of magnetic fields on instability of tidal flow and concluded that, for Earth's parameters, the magnetic damping is of comparable magnitude as the viscous damping.

Finally, we may wonder whether precession-driven flow itself is capable of generating a magnetic field. The inviscid constant vorticity flow has streamlines confined to parallel planes. Efficient dynamo action requires additional flow components which can be provided by Ekman pumps or instabilities. In particular, the instability sketched in **Figure 28** looks promising because it is not unlike the helical columnar structure familiar from rotating convection. Simulations by Tilgner (2005) have shown that both mechanisms indeed modify the constant vorticity flow such that it can sustain magnetic fields at magnetic Reynolds numbers on the order of the magnetic Reynolds number of the Earth's core. Ekman pumps by themselves are too weak for field generation in the Earth. It has already been shown by Loper (1975) and Rochester *et al.* (1975) that energy considerations exclude a geodynamo if the core flow is laminar. But unstable precessional flow remains a plausible alternative to convection-driven flow to explain the geodynamo.

However, precessional flow varies on a diurnal timescale in the mantle frame so that one could expect the magnetic field to do the same, in which case it would be mostly screened because of the finite conductivity of the mantle. Surprisingly, magnetic field variation in one simulation available so far occurs mostly on a timescale slower than diurnal. In addition, the dipole moment, which lies near the equatorial plane in that simulation, exhibits random sudden reorientations reminiscent of reversals (see **Figure 34**). Future research will have to tell whether this behavior is typical, whether it persists in the presence of convectively driven motion (in which

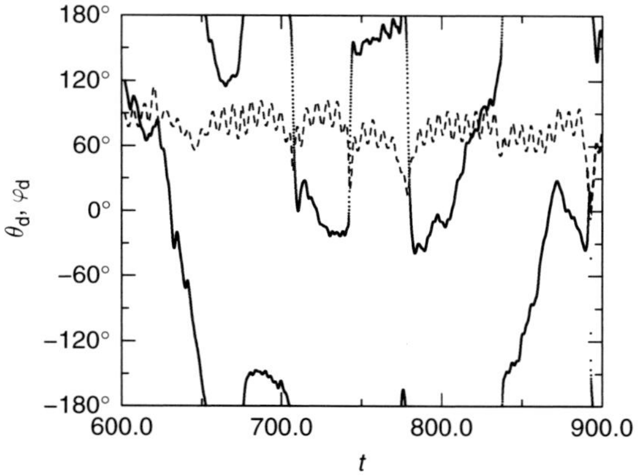

Figure 34 Time series of the polar angles θ_d, the angle between the dipole moment and the rotation axis (dashed line), and φ_d, the angle between the dipole moment and an arbitrary meridian (densely dotted line, appearing continuous for most of the time), for a simulation of a precession-driven dynamo. Reprinted from Tilgner A (2005) Precession driven dynamos. *Physics of Fluids* 17: 034104.

case precession might act as a trigger for reversals in dynamos driven by buoyancy), and whether precession-driven dynamos can also produce axial dipoles.

7.6 Summary and Outlook

This chapter has dealt with perturbations of uniform rotation of the liquid core. Because of a lack of direct observations, we have to use a combination of theory, numerical simulations, and laboratory experiments to guide our intuition about core flow. Section 7.2 allowed for arbitrary perturbations as long as the response of the core can be described by linearized equations of motion, that is the excited amplitude must be small. The following two sections were concerned with perturbations of arbitrary amplitude but of specific type: precession of the axis of rotation of the Earth and tidal deformation of the CMB. The primary response of the core to precession is a rotation of the core about an axis which forms an angle of 1.7×10^{-5} with the axis of rotation of the mantle. This leads to velocities of the core relative to the mantle of up to 4.3 mm s^{-1}, which is much larger than velocities inferred from magnetic secular variation, and large enough to cause instability of the boundary layer at the CMB. A summary of numerical values which appeared in Sections 7.3 and 7.4 is given in **Table 2**.

A uniformly rotating core is an assumption routinely used in the interpretation of geodetic observations (Matthews *et al.*, 1991a, 1991b). In this context, it is enough to get the core's angular momentum right. Simulations presented in this chapter show that while even the simplest analytical models accurately predict the core's rotation vector, the detailed structure of the actual flow looks quite different from a flow with constant vorticity because the constant vorticity flow does not satisfy all boundary conditions and is prone to instabilities in an ellipsoidal container.

All the flows found in this chapter are marked by internal shear zones. There are two origins for internal shear layers. One is the breakdown of Ekman layers. Boundary layers in rotating fluids are generally characterized by a balance between Coriolis and viscous forces. Exceptionally, a different equilibrium is possible when the Coriolis force is balanced by inertia. The scalings of the boundary layer thickness and velocities are then different from the usual Ekman layer. This occurs where shear zones tangential to the boundaries exist. The local accident in the boundary layers starts internal shear layers. These layers become weaker for decreasing Ekman number at equal bulk rotational energy of the fluid. The situation is different for those shear layers which are due to the geometry of the boundaries (e.g., inner and outer boundaries of different ellipticities). These layers become more pronounced with decreasing Ekman number and must turn into singularities in the limit of zero Ekman number.

Precession and tides provide non-axisymmetric forcing which excites directly a non-axisymmetric flow. Through nonlinear interaction, the flow acquires an axisymmetric component which typically contains cylindrically shear layers. These layers are very prominent in experimental visualizations.

The velocities in these shear zones are, under all reasonable assumptions, small compared with flow velocities deduced from magnetic secular variation. It is therefore possible that these layers are of little geophysical relevance, unless they are hydrodynamically unstable and excite stronger flow through instabilities. Early experiments nourished the intuition that axisymmetric shear layers are at the origin of instabilities in precession-driven flow. This impression may simply have arisen because of the visualization method employed in the experiments. The condition for stability of these layers is presently unknown and we cannot conclude about their stability in the core.

Table 2 A summary of best estimates obtained in Sections 7.3 and 7.4 for different characteristics of precessing and tidal core flows

Angle between fluid axis and geographic axis	1.7×10^{-5}
Maximal velocity of core fluid relative to mantle	$4.3\,\mathrm{mm\,s^{-1}}$
Velocity in inclined shear layers, connected to the CMB at critical latitudes	$6 \times 10^{-6}\,\mathrm{m\,s^{-1}}$
Velocity in inclined shear layers, connected to the inner core at critical latitudes	$1.2 \times 10^{-5}\,\mathrm{m\,s^{-1}}$
Axisymmetric prograde jet on the cylindrical surface joining the critical latitudes on the CMB	$3 \times 10^{-5}\,\mathrm{m\,s^{-1}}$
Growth rate of inertial instabilities excited by precession	$<20\,\mathrm{ky^{-1}}$
Growth rate of inertial instabilities excited by tides	$<7\,\mathrm{ky^{-1}}$

Note that we do not have any estimate for the axisymmetric shear layer attached to the inner core excited by precession.

Several instability mechanisms have been identified in numerical simulations. First, there is the boundary layer instability alluded to above. This instability is confined to the boundary layers and does not modify the flow globally. The bulk of the flow is affected by the second mechanism, which is the triad resonance in which the basic flow excites a pair of inertial modes. Several variants of this mechanism have been found, depending on what component of the flow triggers the instability. It can be either internal flow caused by viscous interactions at the boundaries, or it can be distortions of the solid body rotation of the core introduced by the shape of the boundaries. Our understanding of the latter type of excitation is good enough so that we can extrapolate numerical and analytical results to Earth's parameters. It turns out that both precessional and tidal excitation are close to the stability limit. Neither mechanism is clearly above or below the threshold for instability. Any prediction on the stability limit requires knowledge of the viscosity of the core. However, viscosity is one of the least constrained properties of the core.

A phenomenon commonly observed in experiments is the resonant collapse, in which cyclically unstable modes grow and suddenly decay because they draw energy from the basic flow which then becomes too weak to sustain the unstable modes. There is no theory yet telling us under which conditions these collapses occur instead of the flow simply reaching a stationary state. We therefore do not know whether to expect such collapses in the core.

The hydrodynamic stability of the core flow is crucial in connection with the geodynamo. A simple rotation cannot generate a magnetic field. The rotation must be perturbed in a nontrivial way. Convection flows notoriously are dynamos, but unstable precession-driven flows too are.

Most models reviewed in this chapter have neglected buoyancy. Viewed from a frame of reference attached to the mantle, convective flows occur on a much slower timescale than precessional or tidal flow. For this reason, the stability properties of precessional or tidal flows are basically unaltered by coexisting convection.

It can be hoped that simulations coupled with observations of the Earth will allow us to decide which of the above mechanisms is relevant for the core. For instance, measurements of tidal torques put upper bounds on the energy dissipated in the core due to tidal forcing. Current estimates yield approximately 100 GW for the dissipation in the solid Earth (Ray *et al.*, 1996, 2001), an unknown fraction of which occurs in the mantle. If the full 100 GW were available in the core, it could be enough to drive the geodynamo. At any rate, the dissipation is large enough so that we cannot exclude that tidally driven flow contributes significantly to the core dynamics.

Similarly, dissipation of precession-driven flow must show in changes of orbital parameters of the Earth and the Moon. Changes of the rotation rate of the Earth do not provide us with stringent constraints on precessional dissipation (see Section 7.3.2), and there is no published work on the effect this dissipation has on the Moon, which leaves room for future investigation.

Apart from geodetic and astronomical observations, the magnetic field may give interesting indications. The internal shear zones introduced by precession are too weak to be of direct relevance for the geomagnetic field (Busse, 1968; Pais and LeMouël, 2001). But if precession or tides are important, variations of the magnetic field and orbital parameters correlate. It has been claimed that correlations indeed exist between orbital parameters and paleomagnetic intensity and inclination (Channell *et al.*, 1998; Yamazaki and Hirokuni, 2002). However, these claims have been disputed (Valet,

2003). A more robust piece of data is the sequence of reversals. This sequence also seems to reveal a preferred interval between reversals of 100 ky (Consolini and DeMichelis, 2003). The period of 100 ky equals a period at which the eccentricity of the orbit of the Earth changes so that a small modulation of the precession rate with the same period is expected. However, it is far from clear that the amplitude of this modulation is large enough to have any direct effect on the geodynamo. Finally, the magnetic field may carry a signature of resonant collapses, such as slow increases in field intensity followed by sudden decreases. An analysis of the paleomagnetic record by Aldridge and Baker (2003) suggests that events of this type indeed occur.

The sheer magnitude of the flow velocity of the core relative to the mantle induced by precession shows that precession cannot be ignored in a description of core dynamics. Many open questions remain concerning the consequences precession or tides may have on the stability of the bulk of core flow and the geodynamo. Future progress in theory, simulation, and observations will have to tell us how rotational dynamics of the core interact with the geodynamo.

References

Acheson D (1990) *Elementary Fluid Dynamics*. Oxford, UK: Clarendon Press.

Aldridge K (1975) Inertial waves and the earths outer core. *Geophysical Journal of the Royal Astronomical Society* 42: 337–345.

Aldridge K and Baker R (2003) Paleomagnetic intensity data: A window on the dynamics of earths fluid core? *Physics of the Earth and Planetary Interiors* 140: 91–100.

Aldridge K and Lumb L (1975) Inertial waves identified in the Earths fluid outer core. *Nature* 325: 421–423.

Aldridge K, Seyed-Mahmoud B, Henderson G, and van Wijngaarden W (1997) Elliptical instability of the Earths fluid core. *Physics of the Earth and Planetary Interiors* 103: 365–374.

Aldridge K and Toomre A (1969) Axisymmetric inertial oscillations of a fluid in a rotating spherical container. *Journal of Fluid Mechanics* 37: 307–323.

Beardsley R (1970) An experimental study of inertial waves in a closed cone. *Studies in Applied Mathematics* 49: 187–196.

Bondi H and Lyttleton R (1953) On the dynamical theory of the rotation of the Earth: The effect of precession on the motion of the liquid core. *Proceedings of the Cambridge Philosophical Society* 49: 498–515.

Buffett B (1996) Inner core rotation: A test at the worldwide scale. *Geophysical Research Letters* 23: 3803–3806.

Bullard E (1949) The magnetic field within the Earth. *Proceedings of the Royal Society of London A* 197: 433–453.

Busse F (1968) Steady fluid flow in a precessing spheroidal shell. *Journal of Fluid Mechanics* 33: 739–751.

Channell J, Hodell D, McNamus J, and Lehman B (1998) Orbital modulation of the Earths magnetic field intensity. *Nature* 394: 464–468.

Consolini G and DeMichelis P (2003) Stochastic resonance in geomagnetic polarity reversals. *Physical Review Letters* 90: 058501.

Courant R and Hilbert D (1968) *Methoden der Mathematischen Physik*. Berlin, Germany: Springer.

Dintrans B, Rieutord M, and Valdettaro L (1999) Gravito-inertial waves in a rotating stratified sphere or spherical shell. *Journal of Fluid Mechanics* 398: 271–297.

Eloy C, Le Gal P, and Le Dizès S (2000) Experimental study of the multipolar vortex instability. *Physical Review Letters* 85: 3400–3403.

Gledzer E and Ponomarev V (1992) Instability of bounded flows with elliptical streamlines. *Journal of Fluid Mechanics* 240: 1–30.

Goldstein H (1980) *Classical Mechanics*. Reading, MA: Addison-Wesley.

Greenspan H (1968) *The Theory of Rotating Fluids*. Cambridge, UK: Cambridge University Press.

Greenspan H (1969) On the non-linear interaction of inertial modes. *Journal of Fluid Mechanics* 36: 257–264.

Hollerbach R (1998) What can the observed rotation of the Earths inner core reveal about the state of the outer core? *Geophysical Journal International* 135: 564–572.

Hollerbach R and Kerswell R (1995) Oscillatory internal shear layers in rotating and precessing flows. *Journal of Fluid Mechanics* 298: 327–339.

Kerswell R (1993) The instability of precessing flow. *Geophyscial and Astrophysical Fluid Dynamcis* 72: 107–144.

Kerswell R (1994) Tidal excitation of hydromagnetic waves and their damping in the Earth. *Journal of Fluid Mechanics* 274: 219–241.

Kerswell R (2002) Elliptical instability. *Annual Review of Fluid Mechanics* 34: 83–113.

Knittle E and Jeanloz R (1986) High-pressure metallization of FeO and implications for the Earths core. *Geophysical Research Letters* 13: 1541–1544.

Kundu P and Cohen I (2002) *Fluid Mechanics*. San Diego, CA: Academic Press.

Lacaze L, Le Gal P, and Le Dizès S (2004) Elliptical instability in a rotating spheroid. *Journal of Fluid Mechanics* 505: 1–22.

Lacaze L, Le Gal P, and Le Dizès S (2005) Elliptical instability of the flow in a rotating shell. *Physics of the Earth and Planetary Interiors* 151: 194–205.

Lamb H (1932) *Hydrodynamics*. Cambridge, UK: Cambridge University Press.

Landau L and Lifshitz E (1987) *Fluid Mechanics*. Oxford, UK: Pergamon Press.

Laske G and Masters G (1999) Limits on differential rotation of the inner core from an analysis of the Earths free oscillations. *Nature* 402: 66–69.

Lilly D (1966) On the instability of Ekman boundary flow. *Journal of the Atmospheric Sciences* 23: 481–494.

Loper D (1975) Torque balance and energy budget for the precessionally driven dynamo. *Physics of the Earth and Planetary Interiors* 11: 43–60.

Lorenzani S and Tilgner A (2001) Fluid instabilities in precessing spheroidal cavities. *Journal of Fluid Mechanics* 447: 111–128.

Lorenzani S and Tilgner A (2003) Inertial instabilities of fluid flow in precessing spheroidal shells. *Journal of Fluid Mechanics* 492: 363–379.

Maas L, Benielli D, Sommeria J, and Lam F-P (1997) Observation of an internal wave attractor in a confined, stably stratified fluid. *Nature* 388: 557–561.

Maas L and Lam F-P (1995) Geometric focusing of internal waves. *Journal of Fluid Mechanics* 300: 1–41.

Malkus V (1968) Precession of the Earth as the cause of geomagnetism. *Science* 160: 259–264.

Malkus V (1971) Do precessional torques cause geomagnetism? In: Roberts PH and Stewartson K (eds.) *Mathematical Problems in the Geophysical Sciences, Lectures in Applied Mathematics*, vol. 14, pp. 207–228. Providence, RI: American Mathematical Society.

Malkus V (1989) An experimental study of global instabilities due to the tidal (elliptical) distortion of a rotating elastic cylinder. *Geophysical and Astrophysical Fluid Dynamics* 48: 123–134.

Manasseh R (1992) Breakdown regimes of inertia waves in a precessing cylinder. *Journal of Fluid Mechanics* 243: 261–296.

Matthews P, Buffet B, Herring T, and Shapiro I (1991a) Forced nutations of the Earth: Influence of inner core dynamics. 1: Theory. *Journal of Geophysical Research* 96: 8219–8242.

Matthews P, Buffet B, Herring T, and Shapiro I (1991b) Forced nutations of the Earth: Influence of inner core dynamics. 2: Numerical results and comparisons. *Journal of Geophysical Research* 96: 8243–8257.

McEwan A (1970) Inertial oscillations in a rotating fluid cylinder. *Journal of Fluid Mechanics* 40: 603–640.

Melchior P and Ducarme B (1986) Detection of inertial gravity oscillations in the Earths core with a superconducting gravimeter at Brussels. *Physics of the Earth and Planetary Interiors* 42: 129–134.

Néron de Surgy O and Laskar J (1997) On the long term evolution of the spin of the Earth. *Astronomy and Astrophysics* 318: 975–989.

Noir J, Brito D, Aldridge K, and Cardin P (2001) Experimental evidence of inertial waves in a precessing spheroidal cavity. *Geophysical Research Letters* 28: 3785–3788.

Noir J, Cardin P, Jault D, and Masson J-P (2003) Experimental evidence of non-linear resonance effects between retrograde precession and the tilt-over mode within a spheroid. *Geophysical Journal International* 154: 407–416.

Noir J, Jault D, and Cardin P (2001) Numerical study of the motions within a slowly precessing sphere at low Ekman number. *Journal of Fluid Mechanics* 437: 283–299.

Oser H (1958) Experimentelle Untersuchung über harmonische Schwingungen in rotierenden Flüssigkeiten. *Zeitschrift Fur Angewandte Mathematik Und Mechanik* 38: 386–391.

Pais M and LeMouël J (2001) Precession-induced flows in liquid-filled containers and in the Earths core. *Geophysical Journal International* 144: 539–554.

Phillips O (1963) Energy transfer in rotating fluids by reflection of inertial waves. *Physics of Fluids* 6: 513–520.

Poincaré H (1910) Sur la précession des corps déformables. *Bulletin Astronomique* 27: 321–356.

Poirier J (1988) Transport properties of liquid metals and viscosity of the Earths core. *Geophysical Journal* 92: 99–105.

Ray R, Eanes R, and Chao B (1996) Detection of tidal dissipation in the solid Earth by satellite tracking and altimetry. *Nature* 381: 595–597.

Ray R, Eanes R, and Lemoine F (2001) Constraints on energy dissipation in the Earths body tide from satellite tracking and altimetry. *Geophysical Journal International* 144: 471–480.

Reynolds A (1962a) Forced oscillations in a rotating liquid (I). *Zeitschrift Fur Angewandte Mathematik Und Physik* 13: 460–468.

Reynolds A (1962b) Forced oscillations in a rotating liquid (II). *Zeitschrift Fur Angewandte Mathematik Und Physik* 13: 561–572.

Rieutord M, Georgeot B, and Valdettaro L (2001) Inertial waves in a rotating spherical shell: Attractors and asymptotic spectrum. *Journal of Fluid Mechanics* 435: 103–144.

Roberts P and Stewartson K (1963) On the stability of a MacLaurin spheroid of small viscosity. *Astrophysical Journal* 137: 777–790.

Roberts P and Stewartson K (1965) On the motion of a liquid in a spheroidal cavity of a precessing rigid body. II. *Proceedings of the Cambridge Philosophical Society* 61: 279–288.

Rochester M, Jacobs J, Smylie D, and Chong K (1975) Can precession power the geomagnetic dynamo? *Geophysical Journal of the Royal Astronomical Society* 43: 661–678.

Seyed-Mahmoud B, Aldridge K, and Henderson G (2004) Elliptical instability in rotating spherical fluid shells: Application to earths fluid core. *Physics of the Earth and Planetary Interiors* 142: 257–282.

Sloudsky T (1895) De la rotation de la terre supposée fluide á son intérieur. *Bulletin de la Société Impériale des Naturalistes* 9: 285–318.

Souriau A and Poupinet G (2000) Inner core rotation: A test at the worldwide scale. *Physics of the Earth and Planetary Interiors* 118: 13–27.

Stewartson K and Rickard J (1969) Pathological oscillations of a rotating fluid. *Journal of Fluid Mechanics* 35: 759–773.

Stewartson K and Roberts P (1963) On the motion of a liquid in a spheroidal cavity of a precessing rigid body. *Journal of Fluid Mechanics* 17: 1–20.

Suess S (1971) Viscous flow in a deformable rotating container. *Journal of Fluid Mechanics* 45: 189–201.

Tatro P and Mollo-Christensen E (1967) Experiments on Ekman layer instability. *Journal of Fluid Mechanics* 28: 531–543.

Tilgner A (1999a) Driven inertial oscillations in spherical shells. *Physical Review E* 59: 1789–1794.

Tilgner A (1999b) Magnetohydrodynamic flow in precessing spherical shells. *Journal of Fluid Mechanics* 379: 303–318.

Tilgner A (1999c) Non-axisymmetric shear layers in precessing fluid ellipsoidal shells. *Geophysical Journal International* 136: 629–636.

Tilgner A (1999d) Spectral methods for the simulation of incompressible flows in spherical shells. *International Journal for Numerical Methods in Fluids* 30: 713–724.

Tilgner A (2000) Oscillatory shear layers in source driven flows in an unbounded rotating fluid. *Physics of Fluids* 12: 1101–1111.

Tilgner A (2005) Precession driven dynamos. *Physics of Fluids* 17: 034104.

Tilgner A and Busse F (2001) Fluid flows in precessing spherical shells. *Journal of Fluid Mechanics* 426: 387–396.

Valet J-P (2003) Time variations in geomagnetic intensity. *Reviews of Geophysics* 41: 4.

Vanyo J (1991) A geodynamo powered by luni-solar precession. *Geophysical and Astrophysical Fluid Dynamics* 59: 209–234.

Vanyo J (2004) Core–mantle relative motion and coupling. *Geophysical Journal International* 158: 470–478.

Vanyo J and Likins P (1971) Measurement of energy dissipation in a liquid-filled, precessing, spherical cavity. *Transactions of the ASME Journal of Applied Mechanics* 38: 674–682.

Vanyo J and Likins P (1972) Rigid-body approximations to turbulent motion in a liquid-filled, precessing, spherical cavity. *Transactions of the ASME Journal of Applied Mechanics* 39: 18–24.

Vanyo J, Lods D, and Wilde P (1994) Precessing mantle, liquid core and solid core interactions. In: *The 4th SEDI Symposium, Abstract Book*, pp. 62–64. Whistler, BC: SEDI.

Vanyo J, Wilde P, Cardin P, and Olson P (1995) Experiments on precessing flows in the earths liquid core. *Geophysical Journal International* 121: 136–142.

Williams G (1993) History of the Earths obliquity. *Earth-Science Reviews* 34: 1–45.

Yamazaki T and Hirokuni O (2002) Orbital influence on earths magnetic field: 100,000 year periodicity in inclination. *Science* 295: 2435–2438.

Zhang K (1993) On coupling between the Poincaré equation and the heat equation. *Journal of Fluid Mechanics* 268: 211–229.
Zhang K (1995) On coupling between the Poincaré equation and the heat equation: Non-slip boundary condition. *Journal of Fluid Mechanics* 284: 239–256.
Zhang K, Liao X, and Earnshaw P (2004) On inertial waves and oscillations in a rapidly rotating spheroid. *Journal of Fluid Mechanics* 504: 1–40.
Zharkov V, Molodensky S, Brzezinsky A, Groten E, and Varga P (1996) *The Earth and Its Rotation*. Heidelberg, Germany: Wichmann.
Zürn W, Richter B, Rydelek P, and Neuberg J (1987) Comment: Detection of inertial gravity oscillations in the Earths core with a superconducting gravimeter at Brussels. *Physics of the Earth and Planetary Interiors* 49: 176–178.

8 Numerical Dynamo Simulations

U. R. Christensen and J. Wicht, Max-Planck-Institut für Sonnensystemforschung, Katlenburg-Lindau, Germany

8.1	**Introduction**	245
8.2	**Basic Formulation of the MHD Dynamo Problem**	247
8.2.1	Fundamental Ingredients	247
8.2.2	Basic Equations and Nondimensional Parameters	248
8.2.3	Boundary Conditions and Treatment of Inner Core	249
8.2.3.1	Mechanical conditions	249
8.2.3.2	Magnetic boundary conditions and inner-core conductivity	250
8.2.3.3	Thermal boundary conditions and distribution of buoyancy sources	250
8.2.4	Miscellaneous Modifications	251
8.2.4.1	Restricted spherical geometry	251
8.2.4.2	Modified inertial terms	251
8.2.4.3	Compressible models	251
8.2.4.4	Dynamos in Cartesian geometry	252
8.3	**Numerical Approaches**	252
8.3.1	Spectral Methods	252
8.3.1.1	Poloidal/toroidal decomposition	252
8.3.1.2	Spherical harmonic representation	253
8.3.1.3	Radial representation	253
8.3.1.4	Spectral equations	254
8.3.1.5	Time integration	255
8.3.1.6	Evaluation of nonlinear terms	256
8.3.1.7	Boundary conditions and inner core	256
8.3.1.8	Numerical truncation, hyperdiffusion, and subgrid scale models	258
8.3.2	Local Methods	259
8.4	**Model Results**	261
8.4.1	Fundamental Aspects	261
8.4.1.1	General properties of standard models: Weakly versus strongly driven dynamos	261
8.4.1.2	Variations and non-dipolar dynamos	264
8.4.1.3	Mechanism of magnetic field generation	266
8.4.1.4	Taylor state and torsional oscillations	268
8.4.1.5	Scaling laws for dynamos	270
8.4.2	Comparison with the Geomagnetic Field	272
8.4.2.1	CMB field morphology	272
8.4.2.2	Magnetic power spectra	274
8.4.2.3	Secular variation	275
8.4.2.4	Comparison with the paleomagnetic field	276
8.4.2.5	Mantle influence on the dynamo	277
8.5	**Perspectives**	278
References		279

8.1 Introduction

It is believed that the Earth's magnetic field is generated by a self-sustained homogeneous dynamo operating in the liquid part of the Earth's iron core. Thermal and compositional buoyancy forces drive a convective circulation of the electrically conducting fluid. Assuming that a magnetic field already exists, electrical currents are induced in the moving fluid. When the magnetic field associated with these currents has the strength and geometry suitable for the induction process so that there is no further need for a field of external origin, we speak of a self-sustained dynamo. Technical dynamos in cars or power plants are self-sustained, but rely for their operation on the guiding of the electrical currents by the complex arrangements of wires. In contrast, the geodynamo operates in a sphere whose conductivity is approximately uniform. In the decades from 1950 to 1990, dynamo theory has shown that homogeneous dynamo action is possible in principle and has elucidated the basic requirements (*see* Chapter 3). However, models that reproduce the detailed properties of the geomagnetic field had to wait until the numerical simulation of magnetohydrodynamic (MHD) flow has become mature enough.

While previous models had shown that three-dimensional self-consistent geodynamo simulations are possible in principle, the year 1995 brought a final breakthrough. Glatzmaier and Roberts (1995a, 1995b) and Kageyama and Sato (1995) showed that such models can indeed explain several properties of the geomagnetic field, including reversals. In the following decade, a large number of dynamo models have been published. In many of them, the magnetic field properties closely match those of the geomagnetic field in terms of spatial spectra and magnetic field morphology, secular variation, and sometimes the characteristics of dipole reversals. Scientists progressively use dynamo models as a tool to explain specific properties of the geomagnetic field even though some conditions in the models differ strongly from those in the Earth's core. All models assume a far too large viscosity in order to suppress small-scale turbulence which cannot be resolved with the available computational means. On the other hand, the magnetic Reynolds number, Rm, which describes the ratio of advection of magnetic field to magnetic diffusion, is only of order 10^3 in the geodynamo. This value is moderate in comparison to that for dynamos in other astrophysical systems and is accessible to direct numerical simulation. The ability to run numerical models at the correct value of the magnetic Reynolds number is probably the key for the success of geodynamo modeling. Whether the suppression of small flow scales seriously corrupts the essential physics of the dynamo process remains an open question.

In this chapter we review the progress in geodynamo modeling since 1995. Earlier reviews that (also) addressed numerical dynamo simulations have been given by Fearn (1998), Dormy *et al.* (2000), Jones (2000), Busse (2000), Zhang and Schubert (2000), Kono and Roberts (2002), Glatzmaier (2002) and Rüdiger and Hollerbach (2004, chapter 2). They usually concentrated on specific aspects of dynamo modeling, such as the comparison with geomagnetic field properties, the fundamental physical processes operating in dynamos, or the approximations and limitations of present models and the prospects to overcome them in the future. Here we try to give a fairly complete account on the results of dynamo modeling with regard to all these issues. We will assume that the reader is familiar with the basic concepts of MHDs, rotating fluids, and dynamo theory, which are discussed in earlier chapters (*see* Chapters 3 and 5) and in some of the above-mentioned review papers.

We restrict the discussion to direct numerical simulations of the full MHD dynamo problem. In this case, the coupled equations of magnetic induction and of flow of a conducting fluid influenced by the electromagnetic Lorentz force are solved. Aside from the MHD dynamo models, there are several simpler approaches to the dynamo problem that have been extensively studied in the past, but which received less attention for geodynamo applications after reliable solutions to the full dynamo problem have become available. We briefly mention the concepts of kinematic dynamo models, mean-field theory, and magnetoconvection without discussing the results in detail. In the kinematic dynamo problem, the fluid flow is analytically prescribed and unaffected by the magnetic field. Kinematic dynamos (*see* Chapter 3, section 3) have played an important role for establishing that spherical homogeneous dynamos are possible. More recently, they are only occasionally studied in the context of the geodynamo (e.g., Gubbins and Sarson, 1994; Willis and Gubbins, 2004). The so-called mean-field dynamo models (*see* Chapter 3, section 5) are based on a separation of scales, both for the flow **u** and the magnetic field **B**. The interaction of small unresolved scales of **u** and **B** plays an important role for generating a large-scale magnetic field. Since the details of the small-scale turbulence are poorly known, simple parametrizations are employed to describe this induction process. Today, the main application of mean-field dynamo modeling is in astrophysics, where the magnetic Reynolds number is very large, and we will address it only in so far as it provides a context for understanding magnetic field generation mechanisms in MHD models. The term magnetoconvection (*see* Chapter 5, section 5) describes a system that is governed by the full set of MHD equations for a convection-driven flow. However, the magnetic field is not self-sustained but is imposed by boundary conditions, although it can be modified by the flow. Magnetoconvection modeling is employed to understand how electromagnetic forces affect the flow in cases where the magnetic field cannot be expected to be self-sustained or where a specific configuration of the magnetic field is desired.

In Section 8.2 we present the fundamental MHD equations in a simple but probably adequate form for modeling the geodynamo. We proceed to describe the various modifications of the basic equations and their boundary conditions that have been employed by different authors. In Section 8.3 we discuss numerical schemes for solving the MHD equations in a sphere. Readers less interested in the technical aspects of dynamo modeling can skip this section. Most modelers have used some variant of a spectral transform method, in which the variables are expanded in spherical harmonic functions in the angular variables. Nonlinear terms in the equations are calculated on grid points, which requires repeated transformations between the spectral and the grid representation. Other numerical methods based on a purely local description are slowly emerging. In the main part of this chapter, we try to give a rather complete review of the modeling results obtained since 1995. Section 8.4 consists of two main parts. First we discuss the more fundamental questions addressed in geodynamo simulations, such as the mechanism of magnetic field generation or the influence of the various nondimensional model parameters. In the second part we discuss models geared to explain different properties of the geomagnetic field. In Section 8.5 we end the discussion by trying to give a perspective of how dynamo modeling may develop in the foreseeable future.

8.2 Basic Formulation of the MHD Dynamo Problem

8.2.1 Fundamental Ingredients

There is a general consensus that a successful self-consistent model of the geodynamo requires several basic ingredients.

1. A plausible mechanism for driving the flow of an electrically conducting fluid must be part of the model. In almost all published models, this is assumed to be thermal and/or compositional convection (but see Tilgner (2005) and (*see* Chapter 7, section 5) for a dynamo model driven by precession).
2. Cowling's theorem (*see* Chapter 3, section 3.5) prohibits simplified two-dimensional solutions to the dynamo problem, hence the fluid flow and the magnetic field must be modeled in three dimensions. Spherical geometry is essential to reproduce the global, dipole-dominated magnetic field of the Earth.
3. The model system must be rotating because Coriolis forces are essential for structuring the flow in a way that is conducive for dynamo action.

Other properties of the Earth's core are usually assumed not to play a primary role. In most models, the effects of compressibility on the density and the radial thermal structure of the core are ignored. Instead, the Boussinesq approximation is applied, in which density changes due to temperature or compositional variations enter only through a buoyancy term in the Navier–Stokes equation (*see* Chapter 5, section 2.5). In this respect, models for the geodynamo differ strongly from models of stellar dynamos. All material properties are taken to be constant. Most models consider a rotating fluid shell bounded at the inner radius r_i and outer radius r_o (see **Figure 1**). The region outside the shell (the mantle) is usually assumed to be insulating while the solid inner core is modeled as an electrical conductor with the same conductivity as the outer core. However, in many simple models, the inner core is treated as an insulator.

One of the simplest dynamo models that can be constructed along these lines has been used to define a benchmark for numerical codes (Christensen *et al.*, 2001). In the following section, we use the setup and

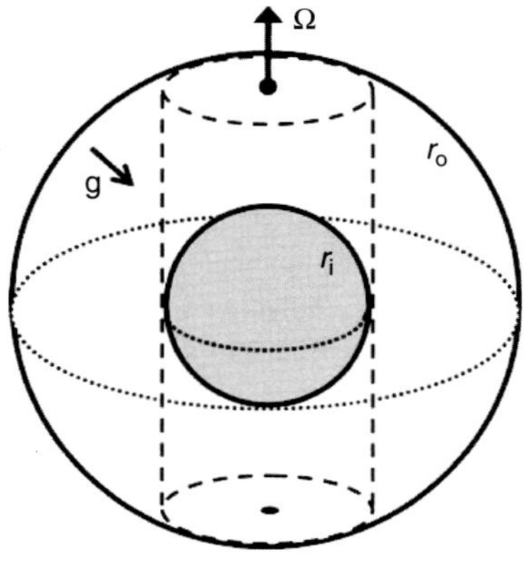

Figure 1 Basic geometry for a dynamo operating in a fluid shell rotating with angular frequency Ω. Broken lines show the inner-core tangent cylinder.

the definitions of the benchmark dynamo as starting point. We then proceed to discuss various modifications used in other dynamo models.

8.2.2 Basic Equations and Nondimensional Parameters

Following the considerations from above, the standard set of equations consists of the Navier–Stokes equation, including the Coriolis and Lorentz force terms, the condition that the velocity field $\mathbf{u}$ is divergence free, a transport equation for temperature T or composition, and the induction equation for the magnetic field $\mathbf{B}$. The latter results from Maxwell's equations, ignoring displacement currents, and Ohm's law. The physical background of these equations is discussed for example in Gubbins and Roberts (1987), Braginsky and Roberts (1995) or (*see* Chapter 5, section 2). The equations must be complemented by appropriate boundary conditions. A simple choice is that $\mathbf{u}$ vanishes on the boundaries (impenetrable no-slip boundaries), T is fixed on the outer and inner boundary to T_o and $T_o + \Delta T$, respectively, and $\mathbf{B}$ matches appropriate fields for $r < r_i$ and $r > r_o$. These external fields must have a form that excludes sources outside the dynamo region (*see* Chapter 3, section 3.1).

Most modelers treat the equations in nondimensional form, with some notable exceptions where physical variables are used (Glatzmaier and Roberts, 1995a, 1995b, 1997). However, there is no unique or generally accepted way to scale the equations. As a consequence, different sets of nondimensional control parameters are employed. For convection-driven dynamos, there are basically four independent control parameters. In the definition of the dynamo benchmark study, the fundamental length scale is the thickness $D = r_o - r_i$ of the spherical shell, time t is scaled by the viscous diffusion time D^2/ν (with ν the kinematic viscosity), temperature is scaled by ΔT, and magnetic induction by $(\rho\mu\lambda\Omega)^{1/2}$ (with ρ the density, μ the magnetic permeability, λ the magnetic diffusivity, and Ω the rotation rate). This leads to the following set of nondimensional equations:

$$E\left(\frac{\partial \mathbf{u}}{\partial t} + \mathbf{u}.\nabla\mathbf{u}\right) + 2\hat{\mathbf{z}} \times \mathbf{u} + \nabla\Pi = E\nabla^2\mathbf{u} + \mathrm{Ra}\frac{\mathbf{r}}{r_o}T + \frac{1}{\mathrm{Pm}}(\nabla \times \mathbf{B}) \times \mathbf{B} \qquad [1]$$

$$\frac{\partial \mathbf{B}}{\partial t} - \nabla \times (\mathbf{u} \times \mathbf{B}) = \frac{1}{\mathrm{Pm}}\nabla^2\mathbf{B} \qquad [2]$$

$$\frac{\partial T}{\partial t} + \mathbf{u}\cdot\nabla T = \frac{1}{\mathrm{Pr}}\nabla^2 T + \epsilon \qquad [3]$$

$$\nabla\cdot\mathbf{u} = 0,\ \nabla\cdot\mathbf{B} = 0 \qquad [4]$$

The unit vector $\hat{\mathbf{z}}$ indicates the direction of the rotation axis. We assume that gravity is a linear function of radius, $g(r) = g_o\, r/r_o$ where g_o is the value on the outer boundary. Π is the nonhydrostatic pressure and ϵ the volumetric heating rate. The four nondimensional control parameters are the Ekman number,

$$E = \frac{\nu}{\Omega D^2} \qquad [5]$$

a modified Rayleigh number,

$$\mathrm{Ra} = \frac{\alpha g_o \Delta T D}{\nu\Omega} \qquad [6]$$

the Prandtl number,

$$\mathrm{Pr} = \frac{\nu}{\kappa} \qquad [7]$$

and the magnetic Prandtl number,

$$\mathrm{Pm} = \frac{\nu}{\lambda} \qquad [8]$$

Here, κ is thermal diffusivity and α is thermal expansivity.

Variants of the nondimensional equations and control parameters result from different choices for the fundamental scales. For the length scale, often r_o is chosen instead of D. Other natural scales for time are the magnetic or the thermal diffusion time, or the rotation period. For scaling magnetic field strength, a choice of different options is available. The prefactor of 2, which is retained in the Coriolis term in eqn [1], is often incorporated into the definition of the Ekman number. When the thermal boundary conditions do not involve a fixed temperature contrast, temperature must be scaled in some other way, which leads to different definitions of the Rayleigh number. The conventional Rayleigh number for a fixed temperature contrast,

$$R = \frac{\alpha g_o \Delta T D^3}{\kappa\nu} \qquad [9]$$

is related to the modified Rayleigh number used here by $\mathrm{Ra} = RE/\mathrm{Pr}$. Although the control parameters and the place where they appear in the nondimensional equations differ for the various ways of scaling, they can usually be written as simple combinations of the four parameters that have been introduced above. However, in particular, the conversion between different forms of the Rayleigh number may involve complicated numerical factors. Kono and Roberts (2001, 2002) discuss in some detail different possible definitions of the key control parameters and the transformations between them.

Characteristic properties of the solution are often expressed in terms of nondimensional numbers. In the context of the geodynamo, the two most important ones are the magnetic Reynolds number Rm and the Elsasser number Λ. Usually the rms values of the velocity u_{rms} and of the magnetic field B_{rms} inside the spherical shell are taken as characteristic values. The magnetic Reynolds number,

$$\mathrm{Rm} = \frac{u_{\mathrm{rms}} D}{\lambda} \quad [10]$$

can be considered as a measure for the flow velocity and describes the ratio of advection of the magnetic field to magnetic diffusion. Other characteristic nondimensional numbers related to the flow velocity are the (hydrodynamic) Reynolds number,

$$\mathrm{Re} = \frac{u_{\mathrm{rms}} D}{\nu} \quad [11]$$

which measures the ratio of inertial forces to viscous forces, and the Rossby number,

$$Ro = \frac{u_{\mathrm{rms}}}{\Omega D} \quad [12]$$

a measure for the ratio of inertial to Coriolis forces. The Elsasser number,

$$\Lambda = \frac{B_{\mathrm{rms}}^2}{\mu_o \lambda \rho \Omega} \quad [13]$$

measures the ratio of Lorentz to Coriolis forces and is equivalent to the square of the nondimensional magnetic field strength in the scaling chosen here.

Typical values for the control parameters and the nondimensional numbers characterizing the flow and magnetic field in the Earth's core and in dynamo simulations are listed in **Table 1**. Here the Rayleigh number is normalized by its critical value $\mathrm{Ra_c}$ for the onset of convection without a magnetic field.

8.2.3 Boundary Conditions and Treatment of Inner Core

8.2.3.1 Mechanical conditions

In the simplest model, the fluid shell is treated as a container with rigid, impenetrable, and co-rotating walls. This implies that within the rotating frame of reference all velocity components vanish at r_o and r_i. Kuang and Bloxham (1997) argued that, because the Ekman number has a far larger value in the model than it has in the core, the no-slip boundary leads to excessively large Ekman-layer effects, and they suggest that it is preferable to eliminate the viscous Ekman layer entirely. Hence they replace the condition of zero horizontal velocity by one of vanishing viscous shear stresses (free-slip condition). The influence of mechanical boundary condition is discussed in Section 8.4.1.2.

There is no *a priori* reason why the inner core should co-rotate with the mantle, and some models allow for differential rotation of the inner core and mantle with respect to the reference frame (e.g. Glatzmaier and Roberts, 1995a, 1996b; Kuang and Bloxham 1997). The change of rotation rate is determined from the net torque, to which viscous forces, electromagnetic forces, and gravitational forces between density heterogeneities in the mantle and the inner core contribute (*see* Chapter 10, section 7). Neglecting the gravitational torque, Glatzmaier and Roberts (1995b, 1996b) found that the inner core rotates a few degrees per year by coupling to a mean prograde (eastward) flow near the inner core boundary. Following this modeling result, seismic evidence for differential rotation of similar order has been claimed (Song and Richards, 1996; Su *et al.*, 1996), but later work showed that differential rotation amounts to only a fraction of a degree per year (e.g., Laske and Masters, 1999; Zhang *et al.*, 2005). Subsequent dynamo simulations also showed much smaller values of inner-core rotation, in particular when gravitational coupling of the mantle and inner core was taken into account (Buffett and Glatzmaier, 2000; Wicht, 2002).

Table 1 Order of magnitude of parameters in the core and in dynamo models

	Control parameters				Characteristic numbers			
	Ra/Ra$_c$	**E**	**Pm**	**Pr**	**Rm**	**Re**	**Ro**	**Λ**
Core	$\gg 1$	10^{-15}–10^{-14}	10^{-6}–10^{-5}	0.1–1	10^2–10^3	10^8–10^9	$\approx 10^{-7}$	0.1–10
Weakly driven models	1–10	$>10^{-4}$	>1	1	40–100	<30	10^{-2}–10^{-1}	0.3–10
Strongly driven models	10–50	10^{-6}–10^{-4}	10^{-1}–10^3	0.025–10^3	10^2–10^3	<2000	3×10^{-4}–10^{-2}	0.1–100

8.2.3.2 Magnetic boundary conditions and inner-core conductivity

In the simplest models, the magnetic field inside the fluid shell matches continuously to a potential field in both the exterior and the interior regions. When a spectral numerical method based on spherical harmonic functions is adopted, these conditions are easy to satisfy (Section 8.3.1.7). This is not the case when local numerical methods are used (Section 8.3.2). In several of these models a condition of vanishing horizontal magnetic field components has been used (Kageyama and Sato, 1995, 1997a; Li *et al.*, 2002; Harder and Hansen, 2005). Allowing only for a finite radial field component at the boundary (sometimes called the 'quasi-vacuum condition') has no physical foundation. However, the solution obtained with this condition seems qualitatively similar to that obtained with the appropriate condition for a truly insulating exterior region (Harder and Hansen, 2005). The 'quasi-vacuum condition' may therefore be acceptable as a numerical convenience in studies where details of the external field structure are of no interest.

Treating the inner core as an insulator is obviously not realistic either, and in a number of models it has the same electrical conductivity as the fluid shell. In this case, an equation equivalent to eqn [2] must be solved for the inner core, where the velocity field simply describes the solid-body rotation of the inner core with respect to the reference frame. At the inner core boundary, a continuity condition for the magnetic field and the horizontal component of the electrical field apply. Results from a simplified mean-field dynamo model (Hollerbach and Jones, 1993) suggested that the finite conductivity of the inner core plays an essential role for stabilizing the dipole polarity of the dynamo field, which would reverse far more frequently in the absence of a conducting inner core. However, in a suite of MHD models including some in which the dipole field reverses, Wicht (2002) found only minor differences between cases with insulating and with conducting inner core when everything else was equal. Calculations by Stanley and Bloxham (2006) suggest that the importance inner-core conductivity for strengthening the axial dipole component increases when the relative size of the inner core is larger than in the present Earth.

In some models (Glatzmaier and Roberts, 1995a; Kuang and Bloxham, 1999), the mantle is not treated as an insulator, but a thin layer of moderate conductivity is assumed at its bottom. The main purpose is to allow for mechanical coupling of the mantle and the fluid core by electromagnetic torques. The effects of the conducting layer on the dynamo process itself has not been studied systematically, but is presumably small.

8.2.3.3 Thermal boundary conditions and distribution of buoyancy sources

In many geodynamo models, convection is driven by an imposed fixed temperature contrast between the inner and outer boundaries, setting $\epsilon = 0$ in eqn [3]. This condition is used for simplicity and has no physical basis. In the present Earth, convection is thought to be driven by a combination of thermal and compositional buoyancy, the latter arising from the release of the light alloying element(s) upon solidification of the inner core (Loper, 1978; Lister and Buffett, 1995). The heat loss from the core is controlled by the convecting mantle, which effectively imposes a condition of fixed heat flux at the core–mantle boundary (CMB) on the dynamo. The heat flux is spatially and temporally variable. The timescale is that of mantle convection, that is, several tens to a hundred million years, hence much longer than relevant timescales for the dynamo process. Geodynamo models that employ a heat flux condition on the outer boundary (e.g., Glatzmaier and Roberts, 1996a; Kuang and Bloxham, 1999) therefore ignore the temporal variation, but in some models the spatial variability is accounted for (see Section 8.4.2.5). Sources of heat are the release of latent heat of inner-core solidification, possible radioactive elements in the core (Gessmann and Wood, 2002; Rama Murthy *et al.*, 2003), and the secular cooling of the outer and inner core, which can effectively be treated like a heat source (*see* Chapter 2).

The rate of inner-core growth and the associated compositional flux from the inner core boundary depend directly on the heat loss to the mantle (Lister and Buffett, 1995). The only model which explicitly included both compositional and thermal driving of convection is that by Glatzmaier and Roberts (1996a) (used also in subsequent publications by these authors). The distinction between compositional and thermal buoyancy becomes dynamically important when the respective diffusivities differ. While the molecular diffusivities for heat and concentration in the Earth's core differ strongly, usually the effective (turbulent) values are assumed to be similar. For equal values of the diffusivities, the light element concentration and the temperature, weighted with their contributions to the buoyancy force, can be combined into a single variable which is sometimes termed 'co-density'. It satisfies an equation that is formally equivalent to eqn [3] (Braginsky and Roberts, 1995; Sarson *et al.*, 1997; Kutzner and Christensen, 2004). Hence, there is no fundamental distinction between compositional and thermal convection. The governing equations are usually written in terms of temperature, but with the understanding that compositional convection is also implied. The main distinction lies in the distribution of sources and sinks of the buoyant agent. For pure thermal convection powered by secular cooling and radiogenic heat, the sources are volumetrically distributed, and the CMB represents the sink. In the case of pure compositional convection, the inner core boundary is the source, and the sink is distributed throughout the outer core (negative ϵ in eqn [3], representing the homogeneous mixing of the excess concentration of light element into the fluid body). The influence of the mode of driving convection on the magnetic field is discussed in Section 8.4.1.2.

8.2.4 Miscellaneous Modifications

8.2.4.1 Restricted spherical geometry

Some models do not treat the dynamo problem for the full spherical shell, but assume some longitudinal symmetry, for example, twofold or fourfold symmetry, in order to reduce the computational requirements. In terms of a spherical harmonic expansion, this means that only functions of even order m, or only those with $m=0, 4, 8, \ldots$, respectively, are retained. A comparison of dynamos with and without longitudinal symmetry suggests that global properties of the dynamo, such as the mean magnetic energy density, are little affected by the symmetry assumption (Christensen *et al.*, 1999). Such models are less suitable to study, for example, magnetic reversals, because they do not allow for an equatorial dipole component.

More severe restrictions are used in the so-called 2½-dimensional models, which have been used as a numerically cheap alternative to fully three-dimensional dynamo models by some authors (Jones *et al.*, 1995). These models are fully resolved in the radial and latitudinal direction, but in azimuthal direction they take only the axisymmetric components and a single non-axisymmetric harmonic mode into account. The mutual interaction of axisymmetric and non-axisymmetric terms is accounted for, but in order to reduce the three-dimensional problem to a quasi two-dimensional one the contribution of certain nonlinear interaction terms to the non-axisymmetric parts must be neglected. Some aspects of the solutions agree qualitatively to those found in three-dimensional models at similar parameter values (Sarson *et al.*, 1998; Sarson and Jones, 1999).

8.2.4.2 Modified inertial terms

In some models, the inertial term in eqn [1] has been neglected (Glatzmaier and Roberts, 1995a, 1995b) or has been treated in a simplified way by retaining only its axisymmetric component (Glatzmaier and Roberts, 1996a; Kuang and Bloxham, 1997, 1999). The reason for ignoring inertial forces is, aside from the simplification, that they are considered to be much smaller in the Earth's core than the dominant Coriolis and Lorentz forces. The rationale for retaining axisymmetric inertial terms is that for differential motion of cylinders coaxial to the rotation axis, inertia is believed to be the only way to balance in the Earth's core residual electromagnetic torques acting on these cylinders, which gives rise to torsional oscillations (see Section 8.4.1.4). However, while inertial forces may perhaps play a small role under conditions of the Earth's core, it is not clear that this is also true for the nominal parameter values of the dynamo models. Unfortunately, no direct comparison of the same model with and without (full) inertial terms is available. An open question is if current dynamo models containing the (full) inertia terms give too much influence to inertia compared with the geodynamo.

8.2.4.3 Compressible models

While most geodynamo models assume the flow to be incompressible and employ the Boussinesq approximation, those by Glatzmaier and Roberts (1996a) (and subsequent models of these authors) take the density stratification and other non-Boussinesq effects, such as the dissipative contributions to the heat equation, into account. In their model, the anelastic approximation is used, which differs from the fully compressible formulation by suppressing sound waves (*see* Chapter 5, section 2.1). Because these anelastic dynamo models differ also in other conditions from published Boussinesq models, it is not clear if fluid compressibility has a significant influence on the dynamo solutions. In stars, the expansion/contraction of a rising/sinking parcel of fluid is, through the action of the Coriolis force, the principal source of helicity, which plays an essential role for the dynamo mechanism. It is not clear whether this effect can play a significant role in the Earth's fluid core, where the density changes by only 23%, but this question deserves further attention.

8.2.4.4 Dynamos in Cartesian geometry

Spherical geometry is essential for reproducing the appropriate topology of the magnetic field and has a strong influence on the shape of the flow. However, for plane layer geometry, more efficient numerical methods are available, which allow simulations at more realistic parameter values, for example, lower values of the Ekman number. Although Cartesian models cannot represent direct models of the geodynamo, they allow to address fundamental questions concerning the force balance and the dynamical regime. Plane-layer dynamos have been studied by St. Pierre (1993), Rotvig and Jones (2002), and Stellmach and Hansen (2004) at Ekman numbers (defined by eqn [5]) as low as 10^{-6}. Stellmach and Hansen (2004) find in their dynamos that at an Ekman number of order 10^{-4} the presence of a magnetic field has only a weak influence on the scale of the flow. At $E = 10^{-6}$, it promotes significantly larger flow scales than occur for nonmagnetic convection, as expected for the so-called strong-field dynamo regime (*see* Chapter 3, section 6.3) and (*see* Chapter 4, section 7). Furthermore, they find a robust example for a subcritical dynamo, where the self-sustained magnetic field is necessary for convection to occur at all (see also St. Pierre, 1993). In spherical shell dynamos, these effects have not been demonstrated so far.

8.3 Numerical Approaches

We will focus on outlining the principles of spectral methods that have been employed for most of the dynamo simulations to date. The quest to compute at more realistic parameter values makes massive parallel computing inevitable. Several researchers have therefore developed codes based on local methods that promise a better performance on multiprocessor systems. Section 8.3.2 offers a brief overview of these new developments.

8.3.1 Spectral Methods

Spectral approaches are well established for solving differential equations and were first applied to the dynamo problem by Bullard and Gellman (1954). The originally employed Galerkin approaches (Bullard and Gellman, 1954; Zhang and Busse, 1989, 1990) have been superseded by the more efficient pseudospectral methods. For geodynamo simulations, these have been introduced by Glatzmaier and Roberts (1995a) (based on earlier developments for modeling stellar dynamos (Glatzmaier, 1984)) and have been adopted subsequently by many other dynamo modelers.

In this approach, the unknowns are expanded into complete sets of functions in radial and angular directions. Chebyshev polynomials and spherical harmonic functions are the common choice. This allows to express all partial derivatives analytically. Employing orthogonality relations of spherical harmonic functions and using collocation or finite differencing methods in radius then lead to algebraic equations that are integrated in time with a mixed implicit/explicit scheme. The nonlinear terms and the Coriolis force are evaluated in grid space rather than in spectral space. Although this approach requires costly numerically transformations between the two representations, the resulting decoupling of all spherical harmonic modes leads to a net gain in computational speed. Before explaining these methods in more detail, we introduce the poloidal/toroidal decomposition.

8.3.1.1 Poloidal/toroidal decomposition

Representing the magnetic field in spherical coordinates (r, θ, ϕ) by a poloidal and a toroidal part,

$$\mathbf{B}(r, \theta, \phi) = \nabla \times \nabla \times [\hat{\mathbf{r}}\, g(r, \theta, \phi)] + \nabla \times [\hat{\mathbf{r}}\, h(r, \theta, \phi)] \quad [14]$$

guarantees that its divergence vanishes (eqn [4]). Three unknown field components are replaced by two scalar fields, the poloidal potential g and the toroidal potential h. This decomposition is unique, aside from an arbitrary radial function $f(r)$ that can be added to g or h without affecting $\mathbf{B}$. An analogous decomposition represents the flow field $\mathbf{u}$ by poloidal and toroidal potentials v and w in the Boussinesq approximation assumed here.

The two scalar potentials of a divergence-free vector field can be extracted from the radial component and the radial component of its curl:

$$\hat{\mathbf{r}} \cdot \mathbf{B} = -\Delta_H g \quad [15]$$

$$\hat{\mathbf{r}} \cdot (\nabla \times \mathbf{B}) = -\Delta_H h \quad [16]$$

The operator Δ_H denotes the horizontal part of the Laplacian:

$$\Delta_H = \frac{1}{r^2 \sin\theta} \frac{\partial}{\partial \theta} \sin\theta \frac{\partial}{\partial \theta} + \frac{1}{r^2 \sin\theta} \frac{\partial^2}{\partial^2 \phi} \quad [17]$$

8.3.1.2 Spherical harmonic representation

Spherical harmonic functions Y_{lm} are a natural choice for the horizontal expansion in colatitude θ and longitude ϕ:

$$Y_{lm}(\theta, \phi) = P_{lm}(\cos\theta)\mathrm{e}^{\mathrm{i}m\phi} \quad [18]$$

l and m denote degree and order, respectively; P_{lm} is an associated Legendre function. Different normalizations are in use. Here we adopt a complete normalization, so that the orthogonality relation reads

$$\int_0^{2\pi} \mathrm{d}\phi \int_0^{\pi} \sin\theta\, \mathrm{d}\theta\, Y_{lm}(\theta, \phi) Y_{l'm'}(\theta, \phi) = \delta_{ll'}\delta_{mm'} \quad [19]$$

As an example, we give the spherical harmonic representation of the magnetic poloidal potential $g(r, \theta, \phi)$ truncated at degree and order L,

$$g(r, \theta, \phi) = \sum_{l=0}^{L} \sum_{m=-l}^{l} g_{lm}(r)\, Y_{lm}(\theta, \phi) \quad [20]$$

with

$$g_{lm}(r) = \frac{1}{\pi} \int_0^{\pi} \mathrm{d}\theta \sin\theta\, g_m(r, \theta)\, P_{lm}(\cos\theta) \quad [21]$$

$$g_m(r, \theta) = \frac{1}{2\pi} \int_0^{2\pi} \mathrm{d}\phi\, g(r, \theta, \phi)\, \mathrm{e}^{-\mathrm{i}m\phi} \quad [22]$$

The potential $g(r, \theta, \phi)$ is a real function, so that $g^*_{lm}(r) = g_{l,-m}(r)$. The asterisk denotes the complex conjugate. Thus, only coefficients with $m \geq 0$ have to be considered. The same kind of expansion is made for the toroidal magnetic potential, the velocity potentials, pressure, and temperature.

Equations [21] and [22] define a two-step transform from the longitude/latitude representation to spherical harmonic representation $(r, \theta, \phi) \rightarrow (r, l, m)$. Equation [20] formulates the inverse procedure $(r, l, m) \rightarrow (r, \theta, \phi)$. Fast Fourier transforms (FFTs) can be employed in longitudinal direction, requiring (at least) $N_\phi = 2L + 1$ evenly spaced grid points ϕ_i. Several authors have developed fast Legendre transforms that could be employed in latitudinal direction (Lesur and Gubbins, 1999; Spotz and Swarztrauber, 2001; Healy, *et al.*, 2004). While not as efficient at FFTs, they could nevertheless help to further speed up dynamo calculations, a potential that still needs to be exploited. All spectral codes to date rely on Gauss–Legendre quadrature (Abramowitz and Stegun, 1984) for evaluating integral [21]:

$$g_{lm}(r) = \frac{1}{N_\theta} \sum_{j=1}^{N_\theta} w_j g_m(r, \theta_j)\, P_{lm}(\cos\theta_j) \quad [23]$$

where θ_j are the N_θ Gaussian quadrature points defining the latitude grid, and w_j are the respective weights (Abramowitz and Stegun, 1984). Prestored values of the associated Legendre functions at grid points θ_j provide the inverse transform [20] (Spotz and Swarztrauber, 2001). Generally, $N_\phi = 2N_\theta$ is taken, which provides isotropic resolution in the equatorial region (**Figure 2**). Choosing $L = [\min(2N_\theta, N_\phi) - 1]/3$ prevents aliasing errors.

8.3.1.3 Radial representation

Different approaches have been employed for representing the radial variation of the unknowns. In some codes, a spectral representation is used only for the

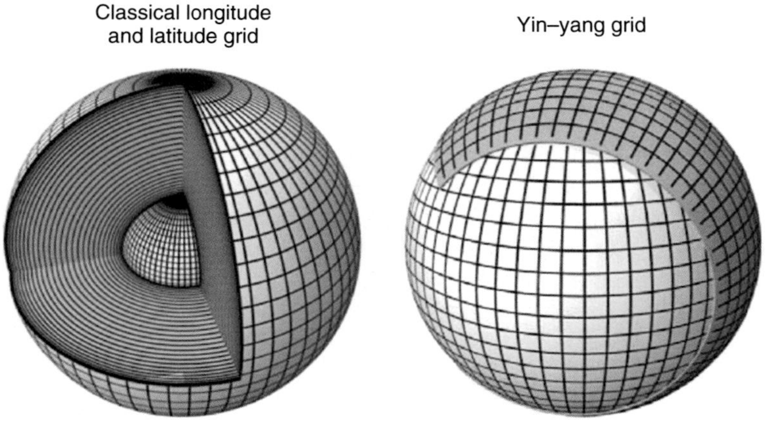

Figure 2 Comparison of the classical longitude/latitude grid and the yin–yang grid that has been used in conjunction with finite difference methods by Kageyama and Yoshida (2005). The latter avoids the highly inhomogeneous coverage in the polar region by reusing the low-latitude band to cover the whole spherical surface.

dependence on the angular variables and a finite difference scheme is employed in the radial direction (e.g., Kuang and Bloxham, 1999). Here we concentrate on schemes that also expand radial dependencies into complete sets of functions. Series of trigonometric functions (Zhang and Busse, 1988; Wicht and Busse, 1997) allow to automatically fulfill the boundary condition, but Chebyshev polynomials $C(x)$ are a more common choice. The polynomial of degree n is defined by

$$C_n(x) = \cos(n \arccos(x)), \quad -1 \leq x \leq 1 \qquad [24]$$

Truncating the radial expansion of the poloidal magnetic potential at degree N reads

$$g_{lm}(r) = \sum_{n=0}^{N} g_{lmn}\, C_n(r) \qquad [25]$$

with

$$g_{lmn} = \frac{2-\delta_{n0}}{\pi} \int_{-1}^{1} \frac{\mathrm{d}x\, g_{lm}(r(x))\, C_n(x)}{\sqrt{1-x^2}} \qquad [26]$$

The Chebyshev definition space ($-1 \leq x \leq 1$) is usually mapped linearly onto radius ($r_\mathrm{i} \leq r \leq r_\mathrm{o}$) by

$$x(r) = 2\frac{r-r_\mathrm{i}}{r_\mathrm{o}-r_\mathrm{i}} - 1 \qquad [27]$$

Nonlinear mapping can be employed to modify the radial dependence of the grid-point density (Tilgner, 1999).

When choosing the N_r extrema of $C_{N_r}-1$ as radial grid points,

$$x_k = \cos\left(\pi \frac{(k-1)}{N_r-1}\right), \; k = 1, 2, \ldots, N_r \qquad [28]$$

the values of the Chebyshev polynomials at these points are simply given by cosine functions:

$$C_{nk} = C_n(x_k) = \cos\left(\pi \frac{n(k-1)}{N_r-1}\right) \qquad [29]$$

This particular choice of grid points has two advantages. For one, the grid points become denser toward the inner and outer radius and better resolve boundary layers. In addition, FFTs can be employed to switch between grid representation [26] and Chebychev representations [25], rendering this procedure a fast Chebyshev transform. Choosing $N_r > N$ provides radial dealiasing. **Figure 2** shows the distribution of grid points, which becomes denser at the shell boundaries and toward the rotation axis. While the former is a desired effect, the latter may lead to numerical problems (see Section 8.3.2).

8.3.1.4 Spectral equations

We have now introduced the necessary tools for deriving the spectral equations. Taking the radial components of the Navier–Stokes equation [1] and the induction equation [2] provides evolution equations for the poloidal potentials $v(r, \theta, \phi)$ and $g(r, \theta, \phi)$. The radial component of the curl of these equations provides evolution equations for the toroidal counterparts $w(r, \theta, \phi)$ and $h(r, \theta, \phi)$. Expanding all potentials in spherical harmonics and Chebyshev polynomials, multiplying with Y^*_{lm}, and integrating over spherical surfaces (while making use of the orthogonality relation [19]) results in equations for the expansion coefficients v_{lmn}, g_{lmn}, w_{lmn}, h_{lmn}, p_{lmn} and T_{lmn}, respectively.

The two evolution equations for the poloidal and toroidal flow potentials are

$$\begin{aligned} &E\frac{l(l+1)}{r^2}\left[\left(\frac{\partial}{\partial t} + \frac{l(l+1)}{r^2}\right)C_n - C''_n\right] v_{lmn} \\ &+ C'_n\, p_{lmn} - Ra\frac{r}{r_o}\, C_n T_{lmn} \\ &= \int \mathrm{d}\Omega\, Y^*_{lm}\, \hat{\mathbf{r}}\cdot\mathbf{F} \end{aligned} \qquad [30]$$

and

$$\begin{aligned} &E\frac{l(l+1)}{r^2}\left[\left(\frac{\partial}{\partial t} + \frac{l(l+1)}{r^2}\right)C_n - C''_n\right] w_{lmn} \\ &= \int \mathrm{d}\Omega\, Y^*_{lm}\, \hat{\mathbf{r}}\cdot\nabla\times\mathbf{F} \end{aligned} \qquad [31]$$

Here, $\mathrm{d}\Omega$ is the spherical surface element. We use the summation convention for the Chebyshev index n. Radial derivatives of Chebyshev polynomials are denoted by primes. The action of the horizontal Laplacian [17] on spherical harmonics has been expressed analytically by

$$\Delta_\mathrm{H} Y_{lm} = -\frac{l(l+1)}{r^2} Y_{lm} \qquad [32]$$

We note that the terms on the left-hand side of eqns [30] and [31], resulting from the viscous term, the buoyancy term and the explicit time derivative, completely decouple in spherical harmonic degree and order. Terms that do not decouple, namely Coriolis force, Lorentz force, and advection of momentum, are collected on the right-hand side of eqns [30] and [31] into the forcing term **F**:

$$\mathbf{F} = -2\hat{\mathbf{z}}\times\mathbf{u} - E\,\mathbf{u}\cdot\nabla\mathbf{u} + \frac{1}{\mathrm{Pm}}(\nabla\times\mathbf{B})\times\mathbf{B} \qquad [33]$$

Resolving **F** into potential functions is not required. Its numerical evaluation is discussed in Section 8.3.1.6.

The pressure remains an additional unknown in the equation for the poloidal potential [30]. Hence, one more equation involving v_{lmn} and p_{lmn} is required. It is obtained by taking the horizontal divergence of the Navier–Stokes equation [1], providing

$$E\frac{l(l+1)}{r^2}\left[\left(\frac{\partial}{\partial t}+\frac{l(l+1)}{r^2}\right)C'_n - C't''_n - \frac{2l(l+1)}{r^3}C_n\right]v_{lmn} + \frac{l(l+1)}{r^2}C_n p_{l\,mn} = -\int d\Omega\, Y^*_{lm}\nabla_H\cdot\mathbf{F} \quad [34]$$

As an alternative, eqn [30] can be replaced by an equation obtained by applying the operator ($\hat{\mathbf{r}}\cdot\nabla\times\nabla\times$) to the Navier–Stokes equation. This procedure provides an equation for the poloidal flow potential that eliminates the pressure. However, it also increases the radial order of the differential equation, which may lead to additional numerical complications (Hollerbach, 2000; Tilgner, 1999).

The equations for the poloidal and toroidal magnetic field coefficients read

$$\frac{l(l+1)}{r^2}\left[\left(\frac{\partial}{\partial t}+\frac{1}{\mathrm{Pm}}\frac{l(l+1)}{r^2}\right)C_n - \frac{1}{\mathrm{Pm}}C''_n\right]g_{lmn} = \int d\Omega\, Y^*_{lm}\hat{\mathbf{r}}\cdot\mathbf{D} \quad [35]$$

and

$$\frac{l(l+1)}{r^2}\left[\left(\frac{\partial}{\partial t}+\frac{1}{\mathrm{Pm}}\frac{l(l+1)}{r^2}\right)C_n - \frac{1}{\mathrm{Pm}}C''_n\right]h_{lmn} = \int d\Omega\, Y^*_{lm}\hat{\mathbf{r}}\cdot\nabla\times\mathbf{D} \quad [36]$$

with the dynamo term

$$\mathbf{D} = \nabla\times(\mathbf{u}\times\mathbf{B}) \quad [37]$$

The last spectral equation concerns the temperature field and is given by

$$\frac{1}{\mathrm{Pr}}\left[\left(\mathrm{Pr}\frac{\partial}{\partial t}+\frac{l(l+1)}{r^2}\right)C_n - \frac{2}{r}C'_n - C''_n\right]T_{lmn} = -\int d\Omega\, Y^*_{lm}\mathbf{u}\cdot\nabla T \quad [38]$$

We have now derived a full set of equations [30], [31], [35], [36], [34], and [38], each describing the evolution of a single spherical harmonic mode of the six unknown fields (assuming that the terms on the right-hand side are given). Each equation couples $N+1$ Chebyshev coefficients for a given spherical harmonic mode (l, m). Typically, a collocation method is employed to solve for the Chebyshev coefficients. This means that the equations are required to be exactly satisfied at $N-1$ radial grid points defined by eqns [27] and [28]. Excluded are the points $r=r_i$ and $r=r_o$, where the boundary conditions provide additional constraints on the set of Chebyshev coefficients (see Section 8.3.1.7).

8.3.1.5 Time integration

The strong nonlinearities in the equations generally demand either higher-order time-stepping methods (Kageyama and Yoshida, 2005), multistep algorithms (Glatzmaier, 1984; Clune *et al.*, 1999; Hollerbach, 2000), and/or iterative schemes (Harder and Hansen, 2005; Hejda and Reshetnyak, 2003). Implicit time-stepping schemes offer increased stability and allow for larger time steps (Zhang and Busse, 1988; Wicht and Busse, 1997; Hejda and Reshetnyak, 2003). However, fully implicit approaches have the disadvantage that the nonlinear terms couple all spherical harmonic modes. The potential gain in computational speed is therefore lost at higher resolution, where one very large matrix has to be dealt with rather than a set of much smaller ones. Similar considerations hold for the Coriolis force, one of the dominating forces in the system and therefore a prime candidate for implicit treatment. However, the Coriolis term couples modes (l, m, n) with ($l+1$, m, n) and ($l-1$, m, n) and also couples poloidal and toroidal flow potentials. An implicit treatment of the Coriolis term therefore also results in a much larger (albeit sparse) inversion matrix. Most authors (Glatzmaier, 1984; Clune *et al.*, 1999; Matsui and Okuda, 2004) consequently prefer a mixed implict/explicit algorithm. Here, we describe the one suggested by Glatzmaier (1984).

Nonlinear and Coriolis terms, collected on the right-hand side of eqns [30], [31], [35], [36], [34], and [38], are treated explicitly with a second-order Adams–Bashforth scheme. Terms collected on the left-hand side are time-stepped with an implicit modified Crank–Nicolson algorithm. While the equations are coupled radially, they decouple for all spherical harmonic modes. But note that the poloidal flow potential eqn [31] and the pressure eqn [34] are coupled for a given spherical harmonic mode.

As an example, we derive the time-stepping equation for the poloidal magnetic potential of degree l and order m, denoting the explicit nonlinear term at radial grid point r_k with

$$D_{klm}(t) = \int d\Omega\, Y^*_{lm}\hat{\mathbf{r}}\cdot\mathbf{D}(t, r_k, \theta, \phi) \quad [39]$$

After discretization of the partial time derivative, $\partial g_{lmn}/\partial t = (g_{lmn}(t+\delta t) - g_{lmn}(t))/\delta t$, where δt, is the time step, we can formulate the left-hand side of eqn [35] as matrix multiplication. The matrices **A** and **G** are defined by

$$A_{kn} = \frac{l(l+1)}{r_k^2}\frac{1}{\delta t}C_{nk} \quad [40]$$

$$G_{kn} = \frac{l(l+1)}{r_k^2}\frac{1}{\mathrm{Pm}}\left(\frac{l(l+1)}{r_k^2}C_{nk} - C''_{nk}\right) \quad [41]$$

where $C_{nk} = C_n(r_k)$. The matrices depend on l but not on m. Advancing time from t to $t+\delta t$ is then a matter of solving

$$(A_{kn} + \alpha G_{kn})\, g_{lmn}(t+\delta t) = (A_{kn} - (1-\alpha)G_{kn})\, g_{lmn}(t) + \frac{3}{2}D_{klm}(t) - \frac{1}{2}D_{klm}(t-\delta t) \quad [42]$$

The classical Crank–Nicholson scheme is obtained for $\alpha = 0.5$, but Glatzmaier (1984) reports that a slightly larger weight of $\alpha = 0.6$ helps to stabilize the time integration. Since the stability requirements limiting δt will usually change during a computational run, the time step should be adjusted accordingly. The matrix **G** remains unchanged, but **A** has to be updated whenever δt is changed. This, in turn, requires a new triangulation of matrix $A_{kn} + \alpha G_{kn}$, which is then stored for subsequent time steps until the next adjustment of δt is in order. Courant's condition (Press *et al.*, 2003) offers a guideline concerning the size of δt demanding that δt should be smaller than the advection time between two grid points. Strong Lorentz forces require an additional stability criterion that is obtained by replacing the flow speed by Alfvén's velocity in a modified Courant criterion (Christensen *et al.*, 1999). The explicit treatment of the Coriolis force requires that the time step is limited to a fraction of the rotation period, which may be the relevant criterion at low Ekman number when flow and magnetic field remain weak. Nonhomogeneous grids and other numerical effects generally require an additional safety factor in the choice of δt.

8.3.1.6 Evaluation of nonlinear terms

Detailing the evaluation of the Coriolis force term and the nonlinear terms is beyond the scope of the chapter, but we outline a few key issues. The nonlinear terms can in principle be calculated in spherical harmonic space. Several early dynamo simulations followed this approach and a few more recent ones still do (Bullard and Gellman, 1954; Zhang and Busse, 1988, 1989, 1990; Wicht and Busse, 1997; Phillips and Ivers, 2003), but the respective representations are cumbersome to formulate and numerically very expensive to evaluate when the number of involved modes becomes large. The vast majority of recent codes therefore relies on pseudospectral (or spectral transform) methods, where the nonlinear terms are evaluated in grid space rather than in spectral space. However, the spatial derivatives required in the nonlinear terms are best calculated in spectral space. This is done by multiplication with im for the derivative with respect to ϕ, by precomputing radial derivatives of the Chebyshev polynomials at grid points, by using the expression [32] for the action of the horizontal Laplacian, and by employing recurrence relations (Abramowitz and Stegun, 1984) for the latitudinal derivatives of the Legendre polynomials. The pseudospectral method requires back-and-forth transformation between the spherical harmonic and the grid representation at each time step. In particular, the relatively slow Legendre transforms make this the most time-consuming part of pseudospectral codes.

8.3.1.7 Boundary conditions and inner core

Since the system of equations is formulated on a radial grid, boundary conditions can simply be satisfied by replacing the collocation equation at grid points r_i and r_o with appropriate expressions. The condition of zero radial flow on the boundaries implies

$$C_n(r)v_{lmn} = 0 \text{ at } r = r_\mathrm{i}, r_\mathrm{o} \quad [43]$$

Note that the summation convection with respect to radial modes n is used again. The no-slip condition requires that the horizontal flow components also have to vanish, provided the two boundaries are at rest. This condition is fulfilled when

$$C_n'(r)v_{lmn} = 0 \text{ at } r = r_\mathrm{i}, r_\mathrm{o} \quad [44]$$

and

$$C_n(r)w_{lmn} = 0 \text{ at } r = r_\mathrm{i}, r_\mathrm{o} \quad [45]$$

for all harmonic modes (l, m). The two conditions [43] and [44] replace the poloidal flow potential eqn [30] and the pressure eqn [34], respectively, at the collocation points r_i and r_o.

If the inner core and/or mantle are allowed to react to torques, a condition based on conservation of angular momentum replaces condition [45] for mode $(l=1, m=0)$ (Hollerbach, 2000):

$$\mathbf{I}\frac{\partial}{\partial t}w = \Gamma \quad [46]$$

The tensor **I** denotes the moment of inertia of inner core or mantle, ω is the mantle or inner-core rotation rate relative to that of the reference frame, and Γ is the respective torque. Viscous, magnetic, and gravitational torques have been considered (Xu and Szeto, 1994; Buffett, 1997; Buffett and Glatzmaier, 2000; Aurnou *et al.*, 1996; Wicht, 2002). Generally, it is assumed that the gravitational interaction between the oblateness of the Earth's mantle and the polar flattening of the inner-core causes very large restoring forces that minimize any deviations between the mantle and inner core rotation axes (Xu and Szeto, 1998; Buffett, 1996). All authors therefore consider only inner-core rotation about the polar axis.

Free-slip boundary conditions require that the viscous stress vanishes, which in turn implies that the nondiagonal components $e_{r\phi}$ and $e_{r\theta}$ of the strain-rate tensor vanish. Translated to the spectral representation, this requires

$$\left[C_n''(r)-2\frac{1}{r}C_n'(r)\right]v_{lmn}=\left[C_n'(r)-2\frac{1}{r}C_n(r)\right]w_{lmn}=0 \quad [47]$$

Magnetic boundary conditions at the interface with an insulating mantle or insulating inner core are similarly implemented. The toroidal magnetic field cannot enter any insulator and therefore has to vanish at the boundary

$$C_n(r)h_{lmn} = 0 \text{ at } r = r_\text{i} \text{ and/or } r = r_\text{o} \quad [48]$$

Matching conditions for the poloidal magnetic field with a source-free external potential field require that the following equations are satisfied at the boundary grid points:

$$C_n'(r)g_{lmn}-C_n(r)\frac{l+1}{r}g_{lmn} = 0 \text{ at } r = r_\text{i} \quad [49]$$

$$C_n'(r)g_{lmn}-C_n(r)\frac{l}{r}g_{lmn} = 0 \text{ at } r = r_\text{o} \quad [50]$$

If the inner core is modeled as an electrical conductor, a simplified dynamo equation has to be solved in which the fluid flow is replaced by the solid-body rotation of the inner core. The latter is described by a single toroidal flow mode ($l=1$, $m=0$). The resulting nonlinear terms can be expressed by a simple spherical harmonic expansion, where the superscript I denotes values in the inner core and ω_I its differential rotation rate:

$$\int \mathrm{d}\Omega\, Y_{lm}^{*}\, \hat{\mathbf{r}}\cdot\nabla\times\left(\mathbf{u}^\text{I}\times\mathbf{B}^\text{I}\right) = -\mathrm{i}\omega_\text{I} m\frac{l(l+1)}{r^2}g_{lm}^\text{I}(r) \quad [51]$$

$$\int \mathrm{d}\Omega\, Y_{lm}^{*}\, \hat{\mathbf{r}}\cdot\nabla\times\nabla\times\left(\mathbf{u}^\text{I}\times\mathbf{B}^\text{I}\right) = -\mathrm{i}\omega_\text{I} m\frac{l(l+1)}{r^2}h_{lm}^\text{I}(r) \quad [52]$$

The expensive back-and-forth transformations between spherical harmonic and grid representation are therefore not required for advancing the inner core magnetic field in time.

In the inner core, the magnetic potentials are again conveniently expanded into Chebyshev polynomials. The Chebyshev variable x spans the whole diameter of the inner core, so that grid points are dense near the inner-core boundary but sparse in the center. The mapping is given by

$$x(r) = \frac{r}{r_\text{i}}, \quad -r_\text{i} \le r \le r_\text{i} \quad [53]$$

Each point in the inner core is thus represented twice, by grid points (r, θ, ϕ) and $(-r, \pi-\theta, \phi+\pi)$. Since both representations must be identical, this imposes a symmetry constraint that can be fulfilled when the radial expansion comprises only polynomials of even order (Hollerbach, 2000):

$$g_{lm}^\text{I}(r) = \left(\frac{r}{r_\text{i}}\right)^{l+1}\sum_{i=0}^{M-1} g_{lm\,2i}^\text{I}\,C_{2i}(r) \quad [54]$$

An equivalent expression holds for the toroidal potential in the inner core. FFTs can again by employed efficiently for the radial transformations, using the M extrema of $C_{2M-1}(r)$ with $x > 0$ as grid points.

The sets of spectral magnetic field equations for the inner and the outer core are coupled via continuity equations for the magnetic field and the horizontal electric field. Continuity of the magnetic field is assured by (1) continuity of the toroidal potential, (2) continuity of the poloidal potential, and (3) continuity of the radial derivative of the latter. Continuity of the horizontal electric field demands (4) that the radial derivative of the toroidal potential is continuous, provided that the horizontal flow and the electrical conductivity are continuous at the interface. These four conditions replace the spectral equations ([35] and [36]) on the outer-core side and equations ([51] and [52]) on the inner-core side. Employing free-slip conditions or allowing for electrical conductivity differences between inner and outer core leads to more complicated and even nonlinear matching conditions (Schubert and Zhang, 2001).

8.3.1.8 Numerical truncation, hyperdiffusion, and subgrid scale models

It is numerically unfeasible to resolve the whole spectrum of different length scales and timescales present in the geodynamo. For example, turbulent eddies in the core can be expected to have a size down to 10 m, whereas the grid spacing in the currently best resolved models is several tens of kilometers. Dynamo models are therefore forced to run with unrealistically high viscous and thermal diffusivities. The magnetic diffusivity is the only exception, where realistic values can be used. The increased diffusivities damp smaller scales that cannot be represented numerically. The important question is whether the effects we strive to model are still correctly represented in the resolved scales.

In a spectral code, the resolution is determined by the truncation level (L, N) of the expansion. We can give the truncation a physical interpretation by assuming that the diffusion of the unresolved scales is practically infinite. Spectra of magnetic and kinetic energy are often taken to judge if the resolution is sufficient. The energy should clearly decay from the dominant mode toward the tail of the spectrum. A decay over several orders a magnitude is regarded desirable, but tests with different levels of truncation have indicated that the general properties of the solution do not change significantly when the drop in the energy spectra exceeds a factor of 50 (Christensen *et al.*, 1999; Kutzner and Christensen, 2002).

Several dynamo models have invoked hyperdiffusivities to ensure numerical stability at values of the Ekman number and the Rayleigh number that would normally require much higher levels of truncation than could be afforded (Glatzmaier and Roberts, 1995a, 1995b, 1996a; Kuang and Bloxham, 1997, 1999; Sakuraba and Kono, 1999; Stanley and Bloxham, 2004; and some other papers of these authors). This means that the Laplacians in some or all of the diffusive terms in eqns [1]–[3] are replaced by an operator which more strongly damps small spatial scales in the flow, magnetic field, or temperature field, respectively. Hyperdiffusion is generally implemented in the spherical harmonic representation by amplifying the diffusive term by a factor that depends on the spherical harmonic degree l; for example, in case of the equations for the magnetic potentials ([35] and [36]), the prefactor Pm^{-1} to the diffusive terms is replaced for $l > l_0$ by

$$\left(1 + a\left(l - l_0\right)^b\right)\mathrm{Pm}^{-1} \qquad [55]$$

a and b are positive constants, and l_0 is a threshold value below which the regular diffusivity applies.

In some implementations, the increase of diffusivity with harmonic degree is quite severe. For example, Glatzmaier and Roberts (1995a, 1995b) employed $b=3$, $a=0.075$, and $l_0=0$, which means that the diffusivity exceeds the regular value by a factor of 100 at $l=11$. Kuang and Bloxham (1997, 1999) use a milder form of hyperdiffusion with $b=2$, $a=0.05$, and $l_0=5$, for which the diffusivity has grown 10-fold at $l=20$. The effect of the hyperdiffusivity is to damp smaller scales more gradually compared to the simple truncation of the spectrum (e.g., Glatzmaier, 2002). The argument for using scale-dependent hyperdiffusivities is that the smaller resolved scales in the solution are expected to interact more strongly with the unresolved scales than the larger components. It is assumed that the main effect of this interaction is to transfer energy to the unresolved scales, which is parametrized by enhancing the diffusivities for the smaller resolved scales.

However, the use of hyperdiffusivity has been criticized on various grounds. First, there is no guideline how to choose the values (a, b, l_0) so that they properly represent the effects which they are supposed to parametrize. Hence the use of hyperdiffusivity adds complexity and an element of arbitrariness to the model. Second, in its usual form, hyperdiffusivity is anisotropic in the sense that the value of diffusivity is sensitive to the horizontal scale but is independent of the scale in the radial direction. While this effect is partly desired to allow for the formation of thin viscous boundary layers (Ekman layers) at the inner and outer boundaries, there is no physical reason for this anisotropy. While the use of hyperviscosity allows, for example, simulations at lower values of the Ekman number than would be possible with the use of regular viscosity, one must keep in mind that the nominal value of the Ekman number refers to the viscosity experienced by the largest scale in the flow (sometimes this value has been termed the 'headline' Ekman number). Intermediate scales in the flow may actually be more relevant for the dynamo. The effective Ekman number at these scales is larger than the nominal value and is in fact of an order where simulations are possible with regular viscosity.

Zhang and Jones (1997) found for convection in the presence of an imposed magnetic field in the linear (marginal stability) regime strong differences between cases with regular diffusivities and with hyperdiffusivities. Grote *et al.* (2000a) compared self-sustained dynamo models at moderately low Ekman number with and without hyperdiffusion and observed in some of their models large differences in the overall character of the generated magnetic field, also at large scales. They found that the toroidal magnetic field is generated by the α-effect in their dynamos with regular diffusivities, but by an ω-effect in the corresponding hyperdiffusive dynamos (see Section 8.4.1.3 for these effects). Roberts and Glatzmaier (2000) have run their original hyperdiffusive model (Glatzmaier and Roberts, 1996a) with regular diffusivities, using much higher spatial resolution. As expected, the power in the magnetic energy spectrum drops much more gently with increasing harmonic degree without hyperdiffusivity, but the spectrum is also changed at the low end. With regular diffusivity the dipole component is less dominant compared to other low-order multipole components (see also Kono and Roberts, 2002). However, because the model with regular diffusivity could only be run for a small fraction of a magnetic diffusion time and was branched off from the case with hyperdiffusion shortly after a reversal occurred, it remains unclear if the relative weakness of the dipole is a persistent difference between the models. In a systematic parameter study of dynamos with regular diffusivities, Christensen *et al.* (1999) found that some aspects of the solution, for example, the relative strength of the toroidal magnetic field inside and outside the inner-core tangent cylinder (see also Section 8.4.1.1), become similar to those observed in the hyperdiffusive dynamos by Glatzmaier and Roberts when the Ekman number and Rayleigh number approach the nominal values of the latter dynamos. It remains an open question to what extent hyperdiffusivities affect not only the small-scale components of the solution but also change the overall character of the dynamo. Most modelers that use hyperdiffusivities consider them as a 'necessary evil' (Glatzmaier, 2002), which should be abandoned as more powerful computational resources allow for high spatial resolution, or which should be replaced by a more sophisticated parametrization of subgrid-scale processes that has a better basis.

Nonlinear interactions in the dynamo are responsible for the transfer of energy between different length scales. Turbulence theory is mostly concerned with a cascade that transfers the energy from the large scale of the energy injection to the small scale where energy is ultimately dissipated away. But the cascade can also work in the opposite direction, particularly when strong rotational constraints play a role, transferring energy into scales that are larger than the scales where energy is fed into the system. Reynolds stresses (nonlinear momentum advection associated with small flow scales) that drive strong zonal flows, such as those observed at the surface of the gas planets, are a prominent example for this inverse cascade (Christensen, 2001; Heimpel *et al.*, 2005). Subgrid methods try to parametrize the energy transfer between numerically resolved scales and scales smaller than the chosen grid (see Matsui and Buffett (2005) and references therein and Chapter 6, section 7.2). The same general considerations apply to the action of nonlinear terms in the dynamo equation [2] and the temperature equation [3]. Information from the nonlinear interactions amongst the resolved scales is extrapolated to estimate the energy exchange between resolved and unresolved scales. Matsui and Buffett (2005) apply and compare several such methods of different complexity to a convection-driven dynamo in a rotating plane layer (see also Matsushima, 2006). The results are promising, but their applicability to more realistic dynamos in spherical shells needs to be proven.

8.3.2 Local Methods

In recent years, several authors have invested in developing new dynamo codes based on methods that we summarize under the term 'local' here: finite differences, finite elements, finite volume or control volume, and spectral element methods. There are several reasons for local approaches.

1. They permit a more flexible decomposition of the spherical shell allowing for local grid refinements that are adapted to a specific problem.
2. Some local methods are simpler to implement and all are more straightforward to parallelize than spectral schemes.
3. Local methods avoid the need for performing Legendre transforms at every time step, which are computationally expensive when the number of modes becomes large.

Spherical coordinates and longitude/latitude grids are the obvious choice for solving problems in spherical domains. However, the grid is highly anisotropic at higher latitudes where cell sizes decrease rapidly toward the poles. This anisotropy can cause numerical instabilities, and painfully small time steps may be required in some schemes to fulfill Courant's criterion near the rotation axis (Hejda and Reshetnyak, 2003). This is less of a problem when convection close to the rotation axis stays weak but can become an issue at higher Rayleigh numbers.

Local methods, on the other hand, allow for very flexible domain decompositions, which is the very reason for their success in complex industrial applications. However, complicated decompositions, nonorthogonal subdomains, and locally varying coordinate systems are penalized with higher code complexity and additional numerical costs (Harder and Hansen, 2005). Choosing a decomposition is a question of finding the best compromise for the specific problem and the specific geometry.

Finite element methods decompose the spherical shell into a number of subvolumes or cells (Matsui and Okuda, 2004; Chan *et al.*, 2001). Galerkin methods are applied to each individual cell, expanding the unknowns in linear (Matsui and Okuda, 2004), quadratic (Chan *et al.*, 2001), or higher-order functions. Hybrid solutions, using Fourier expansions in longitude and a spectral element decompositions in the meridional plane (Fournier *et al.*, 2005), have been implemented for thermal convection in a rotating sphere, and can be extended to the dynamo problem. Using higher-order spectral representation within the meridional cells defines the so-called spectral element schemes (Fournier *et al.*, 2004, 2005). All finite element applications circumvent possible instabilities in the polar region by providing a nearly even coverage of the spherical surface.

Finite difference methods approximate derivatives by differences of functional values on neighboring grid points. Dynamo models by Kuang and Bloxham (1999) and Dormy *et al.* (1998) apply finite differences in radial direction only. Earlier implementations of the finite difference method in all three dimensions (Kageyama *et al.* 1993; Kageyama and Sato, 1995, 1997a) retained the latitude/longitude grid. FFT-based low pass filters were employed to circumvent numerical instabilities in the polar region due to the grid anisotropy. A newer and more elegant approach manages to combine advantages of the classical decomposition with the flexibility of finite difference applications in a so-called yin–yang grid shown in **Figure 2**. The yin subgrid uses a longitude/latitude decomposition at low to intermediate latitudes where the coverage is nearly isotropic but omits a strip in longitude. The same grid geometry is used again in the yang counterpart to cover higher latitudes as well as the remaining longitudes. Two simple rotations about the z-axis and an equatorial axis convert yin to yang. Both sections overlap, implying some computational overhead of about 6% at high resolution.

Another way to circumvent the pole-instability problem are finite-volume schemes that are employed in conjunction with finite element differencing (Hejda and Reshetnyak, 2003; Harder and Hansen, 2005). These methods combine a decomposition into volume cells with staggered grids, and rather than solving the differential equations directly, they are expressed in conservative form by integration over each volume cell. Gauss' theorem is then employed to convert the integrals into fluxes over the cell boundaries. Temperature and pressure are evaluated in cell centers while normal flow and magnetic field components are computed on cell surfaces. The volume of each cell controls its relative importance in the solution, thus rendering cells close to the pole almost negligible in a longitude/latitude grid.

Spectral methods offer two significant advantages: derivatives are easily calculated with high accuracy, and, most notably in the context of dynamo simulations, the formulation of magnetic boundary conditions at the interface with an external insulator (eqns [48]–[50]) is straightforward. A larger number of grid points (meaning a larger number of unknowns) is required in local methods to gain accuracies comparable with pseudospectral schemes (Christensen *et al.*, 2001). As far as the magnetic boundary condition is concerned, local methods either apply artificial boundary conditions (see Section 8.2.3.2), or explicitly solve for the field in the outer region. Since the insulating region is purely diffusive, significantly less computational effort is necessary than for solving the dynamo equation in the shell. The additional numerical costs seem acceptable (Matsui and Okuda, 2004; Kageyama and Yoshida, 2005).

Most local codes do not employ the poloidal–toroidal decomposition [14] of flow and magnetic field. When primitive variables, that is, velocity and magnetic field components, are used, continuity and $\nabla \cdot \mathbf{B} = 0$ have to be enforced explicitly. The three components of the Navier–Stokes equation and the continuity equation form a system of four equations to determine four unknowns, three flow components and the pressure. The latter can be 'corrected' to enforce $\nabla \cdot \mathbf{u} = 0$ in Boussinesq codes or continuity of mass flux in compressible schemes (Harder and Hansen, 2005). Along this line, an auxiliary pseudo-pressure gradient ∇q can be incorporated into the dynamo equation. Solving $\nabla^2 q = 0$ at each time step in conjunction with a starting condition $\nabla \cdot \mathbf{B} = 0$ guarantees that $\mathbf{B}$ remains divergence free at all times. Note that this numerical trick does not affect the curl of the dynamo equation. It is thus compliant with Maxwell's equations (Chan *et al.*, 2001; Harder and Hansen, 2005), and simply takes advantage of the inherent gauge freedom.

The expected parallelization efficiency is one of the main incentives for using local rather than spectral methods. Parallelization of a spectral code comes at the cost of a significant communication overhead since the whole solution has to be broadcasted twice at every time step, each processor communicating with all other processors (Clune *et al.*, 1999). An efficient parallelization of spectral codes is more demanding than for local schemes, but is nevertheless possible, at least on parallel computers of medium size and high communication bandwidth. Reported parallelization efficiencies are 84% on 64 processors of an IBM p575 shared-memory system, where memory access rather than interprocessor communication is the limiting factor, about 80% on 128 processors of a distributed memory T3E with a highly optimized code (Clune *et al.*, 1999), and 66% on 512 processors of the Earth Simulator, which has a mixed shared/distributed architecture (Takahashi *et al.*, 2005).

For codes based on local methods, parallelization is a simple matter of assigning a separate subdomain of the sphere to each processor. Differential operators require information from neighboring cells only, which amounts to much less overall communication than for spectral methods. A peak performance of 46% has been reported on 4096 processors for the yin–yang finite element code running on the Earth Simulator (Kageyama and Yoshida, 2005). Could such a performance be reached with a pseudospectral code? Comparing performances of different codes on different architectures is a daunting task, in particular when absolute runtime or computer availability are also considered. It seems fair to say that, while local methods have the potential of running more efficiently than spectral methods on very large parallel systems with thousands of processors, spectral methods can still perform well on systems more typically available to most researchers, with tens to a few hundreds of processors.

8.4 Model Results

Numerical dynamo models have been studied for different purposes. In some cases, the main aim was to understand the physical principles that shape the fluid flow and lead to the generation of a magnetic field. Other models aimed at reproducing some properties of the geomagnetic field and elucidating the specific conditions for their occurrence. Often these two aims have been combined. In this section, we first treat the more fundamental aspects and then proceed to a detailed comparison of model results with the geomagnetic field.

8.4.1 Fundamental Aspects

8.4.1.1 General properties of standard models: Weakly versus strongly driven dynamos

We will first consider the 'standard' model setup and choice of boundary conditions adopted by the majority of modelers. It comprises no-slip boundaries, lack of internal heating, and regular (scale-independent) diffusivities. These dynamos often generate a magnetic field dominated by the axial dipole. By 'weakly driven dynamos' we mean primarily that the Rayleigh number does not far exceed the critical value for the onset of (nonmagnetic) convection, typically by a factor of less than 10. 'Strongly driven dynamos' are more than 10 times supercritical. The difference between these two groups goes beyond that in Rayleigh number. Depending on the convective vigor, there are restrictions on the possible choices of other control parameters (see **Table 1**) when one aims at obtaining a dynamo with a dipolar field. This point is discussed below and in Section 8.4.1.2. Of course, a gradual transition exists between weakly and strongly driven dynamos and there is no strict dichotomy.

A sufficiently large value of the magnetic Reynolds number $\mathrm{Rm} \approx 50$ is needed for self-sustained dynamo action. In weakly driven models, this can only be achieved when the magnetic Prandtl number is significantly larger than 1. In many weakly driven models, a fairly high value of the Ekman number ($E > 10^{-4}$) has been assumed. The flow pattern and the magnetic field are relatively large-scaled in such cases and high spatial resolution is not required.

One of the simplest models had been selected for a benchmark comparison study, and its properties have been quantitatively verified by a number of independent numerical codes (Christensen *et al.*, 2001). It is one of the few known dynamo solutions that is quasi-stationary in the sense that the only time dependence consists of a steady shift of the pattern in longitude. The magnetic Reynolds number is only 39, one of the lowest values for which a self-sustained dynamo has been found. **Figure 3** shows various aspects of the solution. It exhibits a fourfold symmetry in longitude and mirror symmetry with respect to the equator. Even a slight change in the control parameters breaks the symmetry of the solution and leads to a chaotic time variation. However, aside from the larger complexity and presence of smaller spatial scales, the general pattern of the flow and the magnetic field are broadly similar to those of the benchmark solution in many models with weak driving for a rather broad range of control parameters (e.g., Kageyama and Sato, 1997a; Christensen *et al.*, 1999).

When characterizing the flow, the region inside and outside the tangent cylinder must be distinguished. The tangent cylinder is an imaginary cylinder parallel to the rotation axis that touches the inner core at the equator (**Figure 1**). In the weakly driven models, the flow is organized in nearly geostrophic convection columns outside the tangent cylinder (**Figure 4**). The term 'geostrophic' refers to a flow that is controlled by the balance of Coriolis and pressure gradient force, and whose velocity field does not ideally change in the direction of the rotation vector. The particle motion in these columns is helical: superimposed on the vortex motion around the column there is a (non-geostrophic) flow along the column axis, which converges toward the equatorial plane in cyclonic vortices and diverges away from the equatorial plane in anticyclonic vortices. In the case of weakly driven convection, there is little motion inside the tangent cylinder.

The magnetic field is strongly dominated by the axial dipole in most models with weak driving. On the outer boundary it is concentrated into flux bundles centered near 60° latitude, which coincide with downwelling flow below the boundary (**Figures 3(a)** and **3(b)**). Weaker inverse flux spots, whose polarity is opposite to the dominant polarity in their hemisphere, occur in pairs north and south of the equator. In the interior of the shell, the axisymmetric part of the poloidal magnetic field has a typical bull-eye pattern and the axisymmetric toroidal field consists of two flux bundles of opposite polarity outside the inner core tangent cylinder near the equatorial plane (**Figure 3(c)**). The axisymmetric flow (**Figure 3(d)**) is weak.

In strongly driven models, the control parameters are somewhat closer to Earth values, although still different by orders of magnitude (see **Table 1**). Christensen *et al.* (1999) and Christensen and Aubert (2006) found that at sufficiently high Rayleigh number standard dynamos no longer generate a dipole-dominated magnetic field. The value of Ra/Ra_c at which dipolar dynamos break down rises with the inverse Ekman number. Hence strongly driven dynamos require a low Ekman number, typically less than 10^{-4}, when the emphasis is on models with a dipolar field that can be compared to the geodynamo. The magnetic Prandtl number can be around 1 or smaller in strongly driven models. The magnetic Reynolds number, which in the weakly

(a) (b) (c) (d)

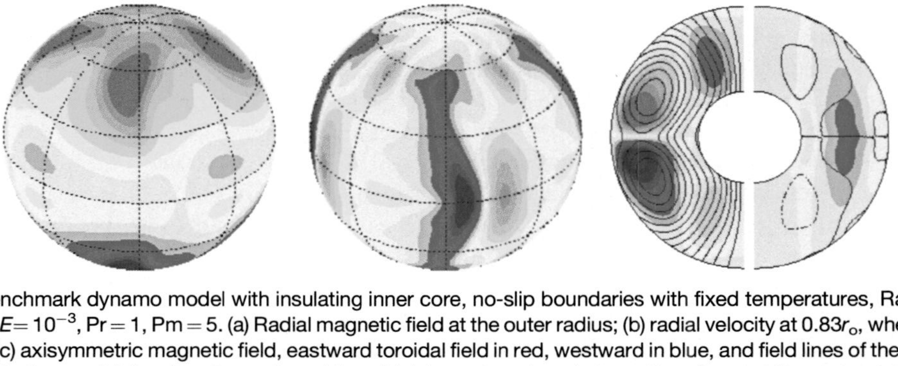

Figure 3 Benchmark dynamo model with insulating inner core, no-slip boundaries with fixed temperatures, Ra = 100 (1.8 × critical), $E = 10^{-3}$, Pr = 1, Pm = 5. (a) Radial magnetic field at the outer radius; (b) radial velocity at $0.83r_o$, where red stands for upwelling; (c) axisymmetric magnetic field, eastward toroidal field in red, westward in blue, and field lines of the poloidal field; (d) axisymmetric flow, reddish colors for eastward flow, bluish westward, and streamlines for meridional circulation.

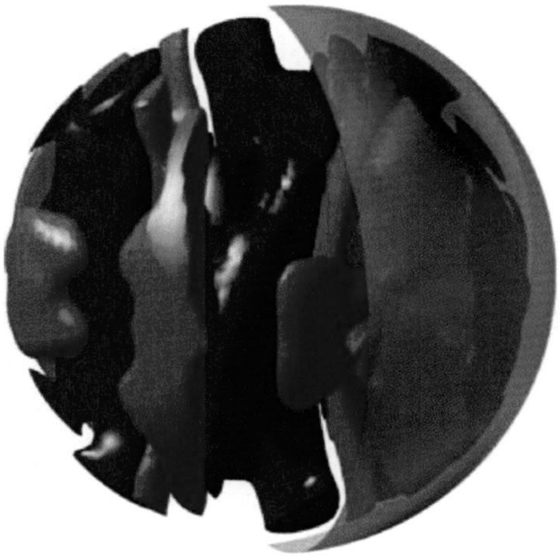

Figure 4 Isosurfaces of positive (red) and negative (blue) vorticity $\omega_z = \hat{\mathbf{z}} \cdot \nabla \times \mathbf{u}$ for the benchmark dynamo.

driven models is typically of order 100 or less, reaches values up to 1000 and therefore matches the likely Earth value (Christensen and Tilgner, 2004). Such models require larger computational resources and fewer of these cases have been published (e.g., Christensen and Tilgner, 2004; Takahashi *et al.*, 2005; Christensen and Aubert, 2006). A special case is models employing hyperdiffusivities (see Section 8.3.1.8), which use strong convective driving and a low value of the Ekman number, but suppress small spatial scales.

With regular diffusivites the strongly driven solutions contain much more small-scale structure than the weakly driven models (**Figure 5**). This is also reflected in the magnetic energy spectrum (**Figure 6**), which in the strongly driven case has a flat distribution from degree 4 to approximately degree 20 before it starts to fall off. In the benchmark case, the spectral power drops rapidly with harmonic degree beyond $l = 8$. Nonetheless, the dipole contribution to the magnetic energy stands out in both cases. Also, some morphological aspects of the flow and magnetic field pattern are not dissimilar in the strongly driven models from those in cases with weak driving. This comprises the organization of the flow outside the tangent cylinder in nearly geostrophic columns, the concentration of magnetic flux on the outer boundary in high-latitude patches, and the occurrence of inverse flux spots at low latitude. An important difference is that at higher Rayleigh number vigorous flow occurs also inside the tangent cylinder. Upwelling flow is often found at or nearby the polar axis (Glatzmaier and Roberts, 1995b; Olson *et al.*, 1999; Sreenivasan and Jones, 2005, 2006a). Due to a thermal wind mechanism (which means a flow governed by a balance of buoyancy and Coriolis forces), the warm rising plume is associated with a westward vortex flow near the outer shell boundary and an eastward vortex near the inner core boundary. **Figure 5(d)** shows this general pattern but also that some complexity exists in the details. The magnetic flux escaping from the fluid shell inside the tangent cylinder region is relatively weak. In the strongly driven cases, the axisymmetric toroidal magnetic field is more pronounced inside than outside the tangent cylinder. Outside the tangent cylinder, it is dominated by a pair of azimuthal flux bundles near the equatorial plane, as it is in the weakly driven models.

In both weakly and strongly driven models, the Elsasser number Λ, describing the strength of the magnetic field, is found to be roughly of order 1. The strength of Earth's field at the CMB corresponds to $\Lambda \approx 0.1$, and reasonable assumptions lead to the conclusion that the mean field strength inside the core is an order of magnitude larger. Hence, on an

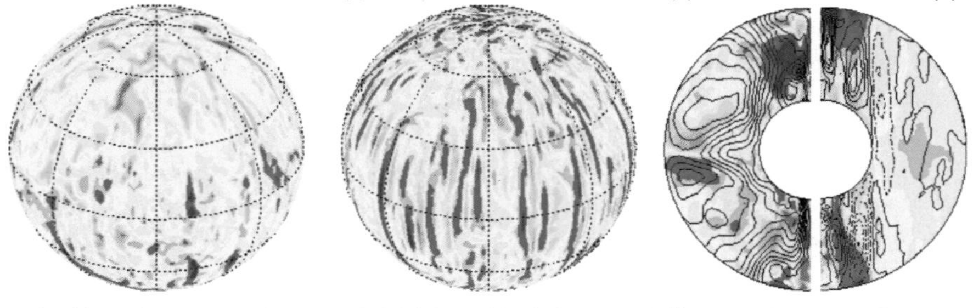

Figure 5 Strongly driven dynamo model, Ra = 3600 (42 × critical), $E = 3 \times 10^{-5}$, Pr = 1, Pm = 2.5. Snapshot of (a) radial magnetic field at the outer radius, (b) radial velocity at $0.93\,r_o$, (c) axisymmetric magnetic field, (d) axisymmetric flow. Details as in **Figure 3**.

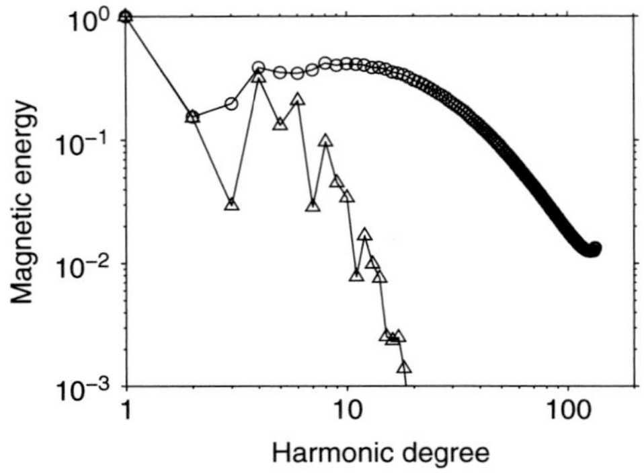

Figure 6 Spectrum of magnetic energy inside the conducting shell as function of harmonic degree for the benchmark model (triangles) and the strongly driven case shown in **Figure 5** (circles). In the latter case, a time-averaged spectrum is shown. The spectra have been normalized by the $l = 1$ component.

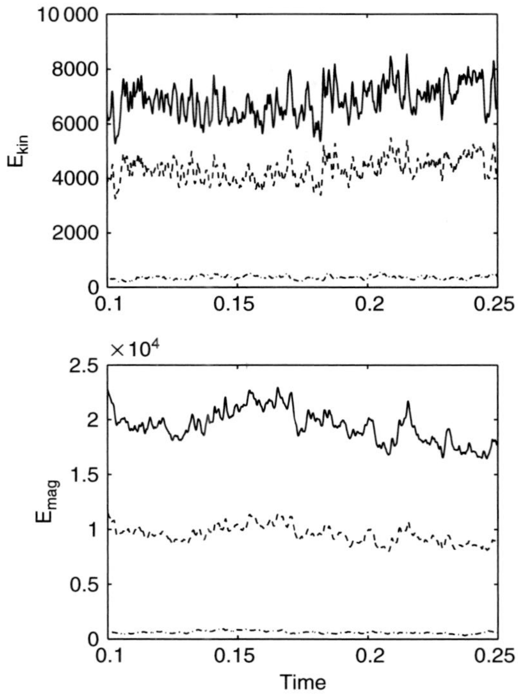

Figure 7 Time series of the kinetic energy density E_{kin} and magnetic energy density E_{mag} for the case shown on **Figure 5**. Full lines indicate total energies, broken lines the toroidal part, and dash-dotted lines the axisymmetric toroidal contribution.

order of magnitude scale, the models reproduce the geomagnetic field strength. However, the actual values of Λ in different models, particularly the strongly driven ones, cover a fairly large range from $\Lambda = 0.1$ to $\Lambda = 100$ (Christensen and Aubert, 2006).

In strongly driven dynamos, the flow and the magnetic field fluctuate chaotically with time (**Figure** 7). The partitioning of kinetic or magnetic energy into the toroidal and poloidal and the axisymmetric and non-axisymmetric components is of interest with a view on the field generation mechanism. When the standard spectral method is employed, it is easily calculated and has been reported by several authors (e.g., Grote *et al.*, 1999; Christensen *et al.*, 1999; Simitev and Busse, 2005). In the types of dynamos discussed above, the magnetic energy is rather evenly divided between the toroidal and poloidal field. The axisymmetric contribution is large in the weakly driven dynamos; in the benchmark case, it contributes 88% to the poloidal field and 55% to the toroidal field. These figures drop with increasing magnetic Reynolds number and are 16% (poloidal) and 7% (toroidal) for the case shown in **Figures 5** and 7. The axisymmetric toroidal magnetic field is usually weaker than the axisymmetric poloidal field. Most of the kinetic energy is contributed by the non-axisymmetric toroidal (~60%) and non-axisymmetric poloidal (~30%) parts, which reflects the dominance of columnar convection. The axisymmetric toroidal flow (differential rotation) contributes typically of the order 10% or less (but see next section). The energy in the axisymmetric poloidal flow (meridional circulation) is usually miniscule.

8.4.1.2 Variations and non-dipolar dynamos

Some published dynamos differ in their magnetic field pattern or flow structure from the cases described above. According to the symmetry properties of the magnetic field with respect to the equatorial plane, a dipolar and a quadrupolar family of dynamos are distinguished (Zhang and Gubbins, 1993). In the former case, the B_r and the B_ϕ components are antisymmetric about the equator, and the B_θ component is symmetric. These symmetries are reversed for the quadrupolar family. Although the field of an axial dipole and of an axial quadrupole show the respective symmetries, the classification does not necessarily imply the dominance of one or the other harmonic contribution. When we speak of a dipolar dynamo (in contrast to a dynamo belonging to the dipole family), we mean that it has a dominant axial dipole component. Weakly driven dynamos models often obey one or the other symmetry. The benchmark dynamo, for example, belongs to the dipolar symmetry family. Strongly driven models are not perfectly symmetric.

Ishihara and Kida (2000) report dynamo solutions dominated by the equatorial dipole, which have been studied more extensively by Aubert and Wicht (2004). They are found at Rayleigh numbers just beyond the onset of dynamo action and are members of the quadrupolar dynamo family. When the Rayleigh number is increased to slightly more supercritical values the equatorial solution becomes unstable. The axial dipole solution representing the complementary symmetry family takes over. The increase in meridional circulation associated with the larger Rayleigh number seems to favor the axial dipole configuration (Aubert and Wicht, 2004; Tilgner, 2004).

At a rather large Ekman number of 10^{-2}, Wicht and Olson (2004) found dynamos with a strong meridional circulation and lack of geostrophic columns, which exhibit quasi-periodic dipole reversals. But the more standard type with columnar flow seems to exist also at such large values of the Ekman number (Katayama *et al.*, 1999).

Grote *et al.* (2000b) and Simitev and Busse (2005) performed an extensive survey of parameter space for dynamos with stress-free boundaries, varying in particular the Prandtl number and the magnetic Prandtl number in a wide range (see also Grote and Busse, 2001; Busse and Simitev, 2005a, 2005b). The flow in these models is driven partly by internal heat sources and has usually a columnar structure. While some of their dynamos, particularly those at high Pr and high Pm, have magnetic fields dominated by a stable axial dipole component, they also find dynamos with a different magnetic field geometry. At moderate to low values of the magnetic Prandtl number, the dynamos have predominantly quadrupolar magnetic field symmetry or are of a hemispherical type, where the magnetic field generation process operates only in the Northern or Southern Hemisphere. The field structure in these dynamos frequently oscillates in a nearly periodic manner. This is also the case for some of their dipolar dynamos, which may show periodic reversals (Busse and Simitev, 2005b). Busse and Simitev (2006) explain the periods found in their oscillating models by a simple mean-field dynamo wave theory.

In models of nonmagnetic rotating convection with stress-free boundaries, a strong zonal flow is excited, which carries most of the kinetic energy. Grote and Busse (2001) describe so-called relaxation oscillation, in which convective flow is virtually absent most of the time. Nearly periodic bursts of convection occur when the zonal flow is suffiently diminished by viscous friction. The energy of the zonal flow is replenished by the associated inertial effects (Reynolds stresses), which in turn suppresses the convection (see also Christensen, 2002). However, in the presence of a self-sustained magnetic field, the zonal flow is inhibited and usually no longer dominates over other flow components. Relaxation oscillations are suppressed and convection becomes more efficient in terms of the net heat transport (Grote and Busse, 2001). The dynamo model of Kuang and Bloxham (1997, 1999), employing stress-free boundaries, is exceptional in the sense that a very strong zonal flow, which carries 80% of the kinetic energy, coexists with a strong magnetic field whose poloidal part is dominated by the axial dipole component.

In models with no-slip boundaries and magnetic Prandtl number of order 1, Kutzner and Christensen (2000, 2002) find that driving convection by a significant contribution of internal heating favors dynamos creating a quadrupolar or more complex poloidal magnetic field dominated by smaller scales. However, dipole-dominated solutions for internally heated dynamos can also be found, in particular at high values of the magnetic Prandtl number (Sakuraba and Kono, 1999; Grote *et al.*, 2000b). For otherwise similar parameter values, dipolar dynamos are preferred when the source of buoyancy is concentrated at the inner core boundary, as in the case of compositional convection. However, even in the latter case, the dipolar dynamos give way to those with a small-scaled field when convection is driven very strongly at a high supercritical Rayleigh number (Kutzner and Christensen, 2002). **Figure 8** illustrates the magnetic field structure on the outer boundary of such a dynamo.

Sreenivasan and Jones (2006b) study the influence of inertial forces by co-varying in a set of dynamo models the parameters in such a way that the relative weight of the inertial term in the governing equations changes, whereas the weight of all others forces remains unaffected. Strengthening the inertial forces leads to weaker and less dipolar magnetic fields. Christensen and Aubert (2006) demonstrate that the role of inertial forces is rather universal in controlling the type of dynamo solution. Balancing the inertial term against the Coriolis term in the Navier–Stokes equation, they define as characteristic parameter measuring their ratio a local (scale-dependent) Rossby number $Ro_l = u_{rms}/(\Omega l)$, where l is the

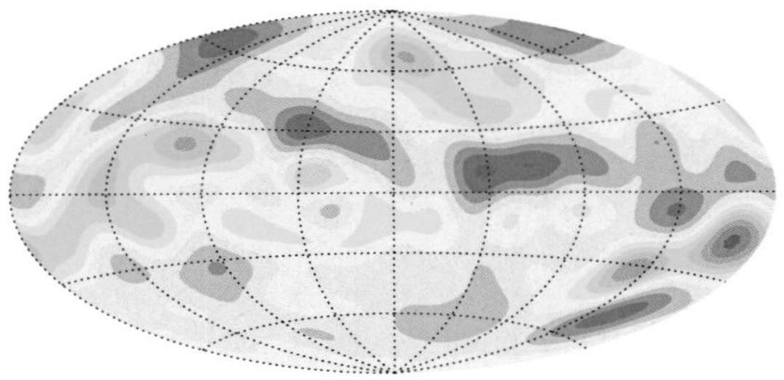

Figure 8 Snapshot of the magnetic field at the outer boundary of a dynamo model with Ra = 3600, $E = 3 \times 10^{-5}$, Pm = 0.5, and Pr = 1. Only the large-scale part up to harmonic degree and order 12 is shown. The dipole contribution is weak. When either Pm is increased to 1 or Ra decreased to 2400, the axial dipole contribution to the field becomes dominant.

characteristic length scale of the flow. They show that, independent of the values of the various control parameters, dipolar dynamos are obtained when Ro_l is below a critical value of about 0.12, and non-dipolar ones above this value (for cases with no-slip boundaries and no internal heat sources). Olson and Christensen (2006) elaborate this point further for a larger class of dynamos and propose a relation linking Ro_l to the fundamental control parameters. Based on this scaling law, they suggested that Ro_l in the geodynamo is close to the transition point between the dipolar and the non-dipolar regime, and that this may be the reason why the Earth shows stochastic reversals.

Inertial forces become larger when convection is driven more strongly. Their adverse influence on dipole field generation can be balanced by enhancing the rotational force. This explains why strongly driven dipolar dynamos need a low value of the Ekman number. It also explains why dipolar dynamos at low magnetic Prandtl number, which obviously need strong driving in order to reach a supercritical value of the magnetic Reynolds number, are only found when the Ekman number is also low. Christensen *et al.* (1999) and Christensen and Aubert (2006) determined that the minimum value of the magnetic Prandtl number at which dipolar dynamos exist varies with the Ekman number as

$$\mathrm{Pm}_{\min} \simeq 450\, E^{3/4} \qquad [56]$$

8.4.1.3 Mechanism of magnetic field generation

Several authors have analyzed numerical dynamo solutions in order to clarify the fundamental mechanisms by which the magnetic field is maintained. The term 'mechanism' can be understood in different ways. Often it refers to the process that generates axisymmetric poloidal field from axisymmetric toroidal field and vice versa, with the mediation of non-axisymmetric field components. Conceptually, this analysis follows closely the mean-field dynamo theory (*see* Chapter 3, section 5), taking the axisymmetric components as the mean magnetic field and the non-axisymmetric parts as the fluctuating or small-scale part.

The mechanism is more easily elucidated in weakly driven dynamos, where the flow outside the tangent cylinder is organized in well-defined geostrophic columns with negative helicity in the Northern Hemisphere and positive helicity in the Southern Hemisphere. In columnar convection, the helicity (defined as $H = \mathbf{u} \cdot \nabla \times \mathbf{u}$) is coherent within each hemisphere, meaning a systematic correlation of local velocity and local vorticity. Kageyama and Sato (1997b) describe the mechanism of how the columnar convection cells convert an axisymmetric toroidal field, which is eastward in the Northern Hemisphere and westward in the Southern Hemisphere, into the poloidal field of an axial dipole. The process is illustrated in **Figure 9(a)**, showing schematically three steps in the evolution of an initially purely azimuthal field, indicated by two field lines. The field lines are advected by the flow, represented by two geostrophic columns and the superimposed ageostrophic circulation (indicated by half-circular arrows). They are distorted and twisted into a poloidal field geometry following the scheme of the classical Parker loop (*see* Chapter 3, section 2.6). An identical field geometry is generated in other pairs of convection columns, and overall the field has a significant mean meridional component with dipole geometry. In the terminology of mean-field electrodynamics, the axisymmetric poloidal field is generated by an α-effect, although the helical eddies in the flow are rather large-scaled in contrast to the usual assumption of scale separation in mean-field theory. This mechanism for poloidal field generation seems to operate in all published numerical dynamo models with columnar convection.

The axisymmetric toroidal field can, in principle, be produced by an α-effect acting on the dipole field, or through shearing it by differential rotation in the mean zonal flow (ω-effect). Depending on which mechanism dominates for the creation of the toroidal mean field, dynamos are classified as α^2-dynamos or $\alpha\omega$-dynamos (*see* Chapter 3, section

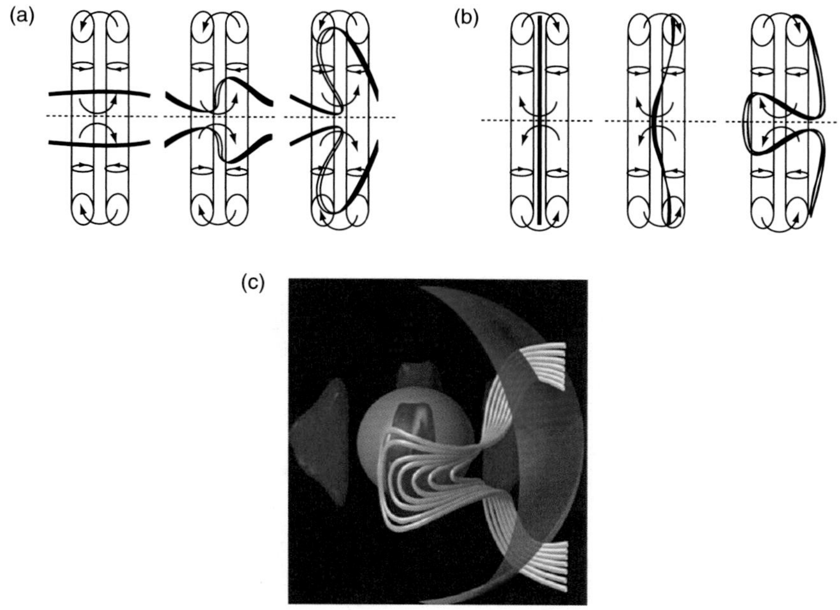

Figure 9 (a) Cartoon showing in a time sequence the conversion of azimuthal field into poloidal field by columnar convection. The broken line indicates the equatorial plane. Superimposed on the columnar circulation is an ageostrophic flow along the column axis and between the columns; this combination implies negative helicity in the Northern Hemisphere and positive helicity in the Southern Hemisphere. (b) Illustration of the generation of axisymmetric toroidal field from poloidal field by columnar convection. (c) Field line bundle in the benchmark dynamo model entering/emerging through one of the high-latitude flux concentrations on the outer boundary (compare **Figure 3**). The gray sphere in the center is the inner core and isosurfaces of anticyclonic vorticity $\omega_z < 0$ are shown in blue. Parts (a) and (b) are taken from Olson P, Christensen U, and Glatzmaier G A (1999) Numerical modeling of the geodynamo: Mechanisms of field generation and equilibration. *Journal of geophysical Research* 104, 10383–10404 and reproduced with permission of AGU.

5.2). The strong axisymmetric toroidal field inside the tangent cylinder that is found in many strongly driven dynamo models (e.g., Glatzmaier and Roberts, 1995a) is clearly due to shearing of poloidal field by the thermal wind circulation (compare **Figure 5**), i.e., an ω-effect. The origin of the toroidal field outside the tangent cylinder is more controversial and may indeed differ between dynamo models. Olson *et al.* (1999) suggested that it is created by the columnar flow through a mechanism similar to the one that creates the poloidal field. This is illustrated in **Figure 9(b)**, showing the evolution of an initially straight poloidal field line that is advected in the flow. The two nearly azimuthal segments of the field line in the final configuration, which run in opposite direction north and south of the equator, combine with the identical field configuration in other pairs of convection columns to form the global axisymmetric toroidal field. Note that the final field configuration in **Figures 9(a)** and **9(b)** is similar when it is properly phase-shifted and represents conceptually the actual field-line geometry found in the benchmark dynamo model (**Figure 9(c)**).

Schrinner *et al.* (2005) investigated the induction process in these simple dynamos in more detail in a mean-field context. They confirm that basically an α^2-mechanism operates, but they also show that simple mean-field models based on an isotropic α-effect are not adequate. Olson *et al.* (1999) studied the role of the ω-effect in their dynamo models by isolating the contribution to the generation of the mean axisymmetric toroidal field by shearing of the poloidal field in gradients of the azimuthal circulation. Because outside the tangent cylinder this contribution was found to be largely anti-correlated with the toroidal magnetic field, the ω-effect was ruled out as its main source.

However, for many other dynamo models, there is evidence that the ω-effect plays an essential role. Kageyama and Sato (1997a) found that their dynamo did not operate when the zonal flow was artificially suppressed, even though its contribution to the total kinetic energy is rather small. Simitev and Busse

(2005) monitored in their dynamo models several 'interaction integrals' which quantify the source for the various magnetic field components in terms of the interaction of different components of the velocity field and magnetic field. They find that the axisymmetric toroidal magnetic field is mainly sustained by the interaction of the axisymmetric toroidal flow with the axisymmetric poloidal field, that is, an ω-effect.

Ishihara and Kida (2000) studied the field generation mechanism from a different angle and identified regions where stretching of magnetic field lines converts kinetic energy into magnetic energy. They find that field line stretching is closely associated with the columnar helical flow illustrated in **Figure 9** and that the intense magnetic flux bundles generated in this way contribute to the axisymmetric poloidal and the axisymmetric toroidal field, which supports an α^2-mechanism. Buffett and Bloxham (2002) addressed the conversion of energy in the model of Kuang and Bloxham (1997, 1999) with stress-free boundaries, in which the axisymmetric toroidal flow component is dominant. They demonstrated that the source term for the axisymmetric toroidal magnetic energy closely matches a particular sink term for the axisymmetric toroidal kinetic energy, which represents the work done against Lorentz forces. This shows clearly that the toroidal magnetic field is generated by the ω-effect in this dynamo. The source term for the axial dipole part of the poloidal field, resulting from interactions of smaller-scaled velocity components with the magnetic field, was found to have its peak at flow scales for which the local magnetic Reynolds number is of order 1. This highlights that magnetic diffusion plays an essential role for the α-effect to convert magnetic field between the toroidal and poloidal components.

In summary, an α-effect associated with the helical flow in convection columns generates the axial dipole field in all models that do have a strong dipole. The role of differential rotation for the generation of the axisymmetric toroidal magnetic field seems to differ in various dynamo models. Somewhat surprisingly, this does not necessarily imply strong differences in the structure of the poloidal magnetic field that is observable outside the dynamo region.

8.4.1.4 Taylor state and torsional oscillations

The answer to the question if the force balance in dynamo models is essentially the same as in the geodynamo is crucial for judging how realistic the models are. The so-called Taylor state is a special consequence of the magnetostrophic regime, in which Coriolis, Lorentz, and pressure forces balance to first order. The geodynamo is commonly assumed to be in such a state, and the demonstration of proximity to a Taylor state is sometimes considered as the essential indicator for the correct force balance in a dynamo model.

Coriolis force and pressure gradient dominate the geostrophic dynamics of low-viscosity flows in fast rotating systems. The smallness of viscous forces and inertia in the Earth's core is expressed by the fact that Ekman and Rossby numbers are much smaller than 1 (**Table 1**). However, the contribution of both forces in the Navier–Stokes equation [1] is scale dependent (in contrast to the Coriolis force), so that these terms may be more important than the scale-independent parameters suggest (Christensen and Aubert, 2006; Olson and Christensen, 2006).

The Lorentz force enters the force balance as a first-order contribution when the Elsasser number is of order 1 like in the Earth's core. Cylinders that are concentric to the rotation axis play a particular role in the magnetostrophic regime, since neither pressure gradient nor Coriolis force contribute to the integrated azimuthal force on these particular surfaces (*see* Chapter 5, section 5.5). Thus, the integrated azimuthal Lorentz force on the so-called geostrophic cylinders can only be balanced by the small viscous and inertial effects. The respective Lorentz force integral has been named Taylor integral τ, in recognition of the fact that Taylor (1963) was the first to recognize the special significance of this force balance:

$$\tau = \frac{1}{\mathrm{Pm}} \int_{C(s)} \mathrm{d}S\, \hat{\phi} \cdot [(\nabla \times \mathbf{B}) \times \mathbf{B}] \qquad [57]$$

$C(s)$ denotes the geostrophic cylinders that depend on cylindrical radius s, and $\mathrm{d}S$ is a differential surface element.

While Taylor (1963) considered the case $\tau = 0$ in accordance with a magnetostrophic case where inertia and viscous forces are neglected all together, Braginsky (1970) reinstated boundary layer friction into the force balance. Assuming a linear velocity gradient over the Ekman layer that matches the geostrophic flow $u_\phi(s)$ with the condition $\mathbf{u} = 0$ on the boundary, the modified balance reads (Hollerbach, 1996)

$$\tau = E^{1/2} \frac{4\pi s}{(1 - s^2)^{1/4}} u_\phi(s) \qquad [58]$$

When one assumes that the azimuthal Lorentz force does not cancel significantly over $C(s)$, this equation constrains the nondimensional magnetic field strength $B = \Lambda^{1/2}$ to be

$$B = \mathcal{O}\left(\mathrm{Pm}^{1/2}E^{1/4}\right) \qquad [59]$$

This can be reconciled with the assumption that $B \approx 1$ only for large enough values of Pm and Ekman numbers $\mathrm{E} \geq \mathcal{O}(10^{-4})$. While many dynamo simulations are working in such a parameter regime, the geodynamo is obviously not.

Two further possibilities to nearly satisfy eqn [58] have been discussed in the literature. Model-Z dynamos, introduced by Braginsky (1975) (see also Braginsky and Roberts, 1987, 1994), minimize the Taylor integral with a configuration where the poloidal field lines are mostly aligned with the z-axis. They bend only close to the outer boundary in order to adjust to the outside potential field.

The option that has been examined more widely is the Taylor state, where Lorentz forces can be strong locally, but cancel over the geostrophic cylinder. A measure for the Taylorization, that is, for how well the Lorentz forces cancel, is

$$\mathrm{Tay} = \frac{\int_{C(s)} \mathrm{d}S\,\hat{\phi}\cdot[(\nabla\times\mathbf{B})\times\mathbf{B}]}{\left\{\int_{C(s)} \mathrm{d}S\left(\hat{\phi}\cdot[(\nabla\times\mathbf{B})\times\mathbf{B}]\right)^2\right\}^{1/2}} \qquad [60]$$

Taylor already realized that a perfect cancellation $\mathrm{Tay} = 0$ seems very unlikely for the time dependent flows expected in the Earth's core. Deviations from Taylor state would accelerate the cylinder, thereby producing azimuthal shear $\partial u_\phi/\partial s$. The shear creates azimuthal toroidal magnetic field by acting on poloidal field B_s. The Lorentz force associated with this newly created toroidal field opposes the acceleration according to Lenz' law, thereby trying to reinstate Taylor's condition [58]. The result are 'torsional oscillations' of the geostrophic cylinders around the Taylor state that are supposed to be the source of the fastest magnetic field changes in the Earth's core with a timescale around 50 years. They may also be the cause for geomagnetic jerks (Bloxham *et al.*, 2002) (*see* Chapter 4, section 5). Torsional oscillations travel as a wave inward and outward away from the cylindrical radius where they are excited and are ultimately damped by diffusive effects. Zatman and Bloxham (1997) suggest that torsional oscillations in the Earth's core can be identified in secular variation data. They proceeded to estimate the integrated field component B_s on geostrophic cylinders by inverting the force balance (Zatman and Bloxham, 1999), providing valuable information on the magnetic field inside the Earth's core, which is very hard to access otherwise (*see* Chapter 4, section 5.1).

Requirements for a numerical dynamo to work close to the Taylor regime are (1) negligible viscosity, preferably $E \ll 10^{-4}$ to rule out the viscously limited regime, and (2) negligible inertia except contributions of the form $\partial u_\phi(s)/\partial t$ that balance residual Lorentz torques on the cylinder. The ultimate test for the presence of a Taylor state is $|\mathrm{Tay}| \ll 1$.

The dynamo model of Kuang and Bloxham (1999) has been tailored with a magnetostrophic force balance and torsional oscillations in mind. Stress-free boundary conditions are assumed to minimize viscous damping. In addition, inertial terms are neglected in the Navier–Stokes equation [1] with the exception of axisymmetric contributions needed to describe torsional oscillations. While Kuang and Bloxham (1999) demonstrate that stress-free conditions indeed seem to favor a more 'magnetostrophic-like' force balance on cylinders, their dynamo models are not in the Taylor state. A close examination of their model at $Ro = 10^{-2}$ and $E = 10^{-4}$ (parameters rescaled to the definitions used here) has revealed that Reynolds stresses and viscous torques remain large and effectively balance the Lorentz torque on geostrophic cylinders (Dumberry and Bloxham, 2003). Busse and Simitev (2005a) present a model at $E = 2\times10^{-5}$ and low Prandtl number in which the acceleration of geostrophic cylinders is predominantly balanced by the Lorentz torque, indicative of torsional oscillations. Takahashi *et al.* (2005) claim that their simulations at $E = 2\times10^{-6}$ is in a quasi-Taylor state, but unfortunately stop at showing that the viscous torque on cylinders is much smaller than the Lorentz torque.

Dynamo simulations in a Cartesian box allow to go to small Ekman number at affordable numerical costs. Therefore, some authors have chosen this setup to explore fundamental questions like Taylorization despite the non-geophysical geometry. Rotvig and Jones (2002) report that their Cartesian simulations approach Taylor states at $E = 10^{-5}$, the (modified) Taylorization parameter going down to $|\mathrm{Tay}| = 10^{-2}$. Stellmach and Hansen (2004) report similar results for a magnetoconvection simulation at the same Ekman number, but find only incomplete Taylorization in dynamo models with Ekman numbers as low as 10^{-6}.

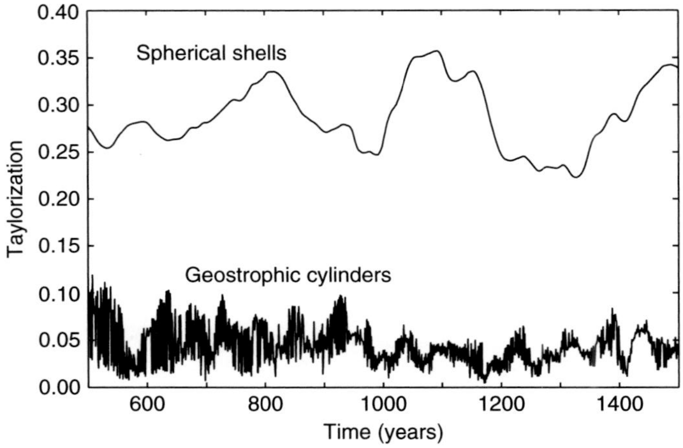

Figure 10 Taylorization parameter vs time for a dynamo model with Ra = 3000, $E = 3 \times 10^{-6}$, Pm = 0.5, Pr = 1. The rms value of Tay taken over all geostrophic cylinders or spherical shells is shown.

Wicht and Christensen (unpublished results) perform dynamo simulations at Ekman numbers down to $E = 3 \times 10^{-6}$ with a focus on potential torsional oscillations and Taylor states. They find that Taylorization with values as low as $|\mathrm{Tay}| = 10^{-2}$ is reached at their lowest Ekman number. However, while magnetic effects dominate the short-timescale force balance on geostrophic cylinders, other effects remain important, for example, Reynolds stress and time variability of B_s. In low Ekman number simulations, the magnetic field generally shows a lot of small-scale structure. In order to exclude a simple statistical fluctuation with alternating sign of Lorentz force on cyclindrical surface as the cause for the cancellation of Lorentz torques, Wicht and Christensen also calculated normalized torque integrals on (non-geostrophic) spherical surfaces $S(r)$, which replace $C(s)$ in eqn [60]. The same cancellation would be expected on both types of surfaces if it were a purely statistical effect. On spherical surfaces, the residual torques are an order of magnitude larger than on geostrophic cylinders, confirming that the dynamo works close to a Taylor state (**Figure 10**).

8.4.1.5 Scaling laws for dynamos

A fundamental question is how the characteristic properties of convection-driven dynamos in a rotating spherical shell (such as the flow velocity or the magnetic field strength) depend on the control parameters. Some heuristic scaling laws have been presented (e.g., Stevenson, 1979; Starchenko and Jones, 2002), but a well-founded and generally accepted theory is missing. Christensen and Tilgner (2004), Aubert (2005), and Christensen and Aubert (2006) took up the task to derive scaling laws, in a partly empirical way, from numerical solutions. Because the dynamo properties depend on four control parameters, this is a substantial enterprise which requires the calculation of a large number of results covering a sufficient region of the parameter space.

The essential step for obtaining simple scaling laws is the introduction of parameters that are independent of any of the three diffusivities that enter into the dynamo problem. As fundamental control parameter, Aubert (2005) and Christensen and Aubert (2006) (see also Christensen, 2002) defined a modified Rayleigh number that depends on the advected heat flux Q_{adv}.

$$\mathrm{Ra}_Q^* = \frac{1}{4\pi r_o r_i} \frac{\alpha g_o Q_{\mathrm{adv}}}{\rho c \Omega^3 D^2} \qquad [61]$$

The relation to the conventional Rayleigh R number (eqn [9]) and the modified Rayleigh number Ra (eqn [6]) is

$$\mathrm{Ra}_Q^* = \mathrm{Ra}(\mathrm{Nu}-1)\frac{E^2}{\mathrm{Pr}} = R(\mathrm{Nu}-1)\frac{E^3}{\mathrm{Pr}^2} \qquad [62]$$

where $\mathrm{Nu} = QD/(4\pi r_o r_i \rho c k \Delta T)$ is the Nusselt number, that is, the ratio of heat flow Q to the heat flow in a conductive state. Ra_Q^* is proportional to the (nondimensional) power generated by buoyancy forces. The characteristic properties of the solution also need to be scaled in a way that does not involve diffusivities. For example, the mean flow velocity is expressed by the Rossby number.

Aubert (2005) studied how the velocity of the mean zonal flow (the axisymmetric component of u_ϕ)

depends on Ra_Q^*. In the case of nonmagnetic rotating convection, he finds that the zonal flow is mainly driven by inertial effects (Reynolds stresses) and the corresponding Rossby number $\mathrm{Ro}_{\mathrm{zonal}}$ depends on both Ra_Q^* and the Ekman number. In contrast, in his dynamo cases, the zonal flow arises from a thermal wind effect and its amplitude is found to be a simple function of Ra_Q^*, independent of the parameters E, Pr, and Pm, that describe the influence of diffusivities:

$$\mathrm{Ro}_{\mathrm{zonal}} \simeq \mathrm{Ra}_Q^{*1/2} \qquad [63]$$

In a more extensive study, Christensen and Aubert (2006) derived scaling laws for the global velocity, described by the Rossby number Ro, the efficiency of heat transport, described by a modified Nusselt number $\mathrm{Nu}^* = (\mathrm{Nu} - 1)E\ \mathrm{Pr}^{-1}$, and the magnetic field strength, expressed by the Lorentz number $\mathrm{Lo} = B_{\mathrm{rms}}/[(\mu\rho)^{1/2}\Omega D]$. They found the latter parameter more useful for describing the magnetic field strength than the conventional Elsasser number [13]. Christensen and Aubert (2006) restricted their analysis to dynamos with a dipole-dominated magnetic field and found the following best-fitting scaling laws:

$$\mathrm{Nu}^* = 0.074\mathrm{Ra}_Q^{*0.53} \qquad [64]$$

$$\mathrm{Ro} = 0.85\mathrm{Ra}_Q^{*0.41} \qquad [65]$$

$$\mathrm{Lo}/f_{\mathrm{Ohm}}^{1/2} = 0.91\mathrm{Ra}_Q^{*0.34} \qquad [66]$$

Based on the idea that the magnetic field strength is determined by the power available to balance Ohmic dissipation, a correction factor $f_{\mathrm{Ohm}}^{1/2}$ has been introduced in eqn [66], where f_{Ohm} is the relative contribution of Ohmic dissipation to the total dissipation. f_{Ohm} is thought to be close to 1 in the core. The numerical results fit the scaling law for the modified Nusselt number very well. The fit for Rossby and Lorentz number is decent, the latter being shown in **Figure 11**.

Further discussion on these scaling laws can be found in Chapter 5, section 5.7. Olson and Christensen (2006) examined a broader set of models, also including non-dipolar dynamos, to find a scaling relation for the dipole moment as the main observable for magnetic fields of planets other than Earth. As long as the magnetic field is dipole dominated, they find that a Lorentz number that characterizes the mean dipole strength at the outer boundary of the dynamo region scales like $\mathrm{Lo}_{\mathrm{dip}} = \gamma\mathrm{Ra}_Q^{*1/3}$, with

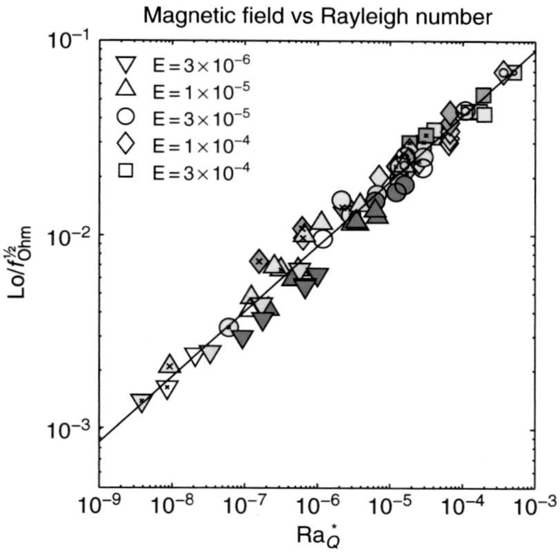

Figure 11 Lorentz number (nondimensional magnetic field strength) corrected for the fraction of Ohmic dissipation vs modified Rayleigh number for 72 dynamo models that generate a dipole-dominated magnetic field. The Ekman number is indicated by the shape of the symbol, the magnetic Prandtl number is color-coded ($\mathrm{Pm} > 3$, dark red; $1 < \mathrm{Pm} \leq 3$, light red; $\mathrm{Pm} = 1$, white; $0.3 \leq \mathrm{Pm} < 1$, light blue; $\mathrm{Pm} < 0.3$, dark blue). Prandtl numbers other than 1 are indicated by a large cross ($\mathrm{Pr} = 10$), small cross ($\mathrm{Pr} = 3$), small circle ($\mathrm{Pr} = 0.3$) or large circle ($\mathrm{Pr} = 0.1$) inside the main symbol. Adapted from Christensen UR, and Aubert J (2006) Scaling properties of convection-driven dynamos in rotation spherical shells and application to planetary magnetic fields. *Geophysical Journal International* 166: 97–114 with permission by Blackwell.

$\gamma = 0.1$–0.2. In the non-dipolar regime, the dipole field is at least 10 times weaker, with an abrupt transition in bottom-heated cases and a more gradual one for internally heated dynamos.

The fit of the numerical results for the Rossby number and the Lorentz number can be improved by assuming a weak additional influence of the magnetic Prandtl number (Christensen and Aubert, 2006). The Ekman number and the hydrodynamic Prandtl number do not seem to have a significant effect. The exponents for the dependence on the magnetic Prandtl number are small, 0.11 in case of the best fit to the Lorentz number. But the rather large extrapolation from model values of $\mathrm{Pm} \sim \mathcal{O}(1)$ to core values $\sim\mathcal{O}(10^{-6})$ implies that a Prandtl number dependence would nevertheless have a significant impact.

Using a somewhat more limited set of model results, Christensen and Tilgner (2004) derived a scaling law for the Ohmic dissipation time τ_{Ohm}, defined as magnetic energy divided by the rate of

Ohmic dissipation. They find it to depend on the magnetic Reynolds number according to

$$\tau_{\mathrm{Ohm}} = 1.74\,\tau_{\mathrm{dipole}}\,\mathrm{Rm}^{-1} \qquad [67]$$

where $\tau_{\mathrm{dipole}} = r_0^2(\pi^2\lambda)$ is the dipole decay time. Again, an additional dependence on the magnetic Prandtl number could somewhat improve the fit to the model data. Christensen and Tilgner (2004) also determined the Ohmic dissipation time for the Karlsruhe laboratory dynamo experiment, in which the magnetic Prandtl number has the small value of a liquid metal, and found it to agree much better with the scaling law eqn [67] than with a more complex one involving Pm. Therefore, they suggested that a weak dependence of the dynamo properties on the magnetic Prandtl number at values $\mathrm{Pm} \sim 1$ may disappear at $\mathrm{Pm} \ll 1$, so that the simpler Pm-independent scaling would represent conditions in the Earth's core reasonably well. Christensen and Tilgner (2004) also found that the characteristic timescale for the magnetic secular variation in numerical dynamo models is inversely proportional to the magnetic Reynolds number. They used this dependence to estimate a core value of $\mathrm{Rm} \approx 1000$ from the observed geomagnetic secular variation, which is equivalent to a flow velocity of $0.5\ \mathrm{mm\,s^{-1}}$.

Christensen and Aubert (2006) applied eqn [65] to estimate from the characteristic core flow velocity a modified Rayleigh number Ra_Q^* of order 10^{-13}. For this value eqn [66] predicts a plausible magnetic field strength in the core of order 1 mT. Using Jupiter's observed excess heat flow to obtain the Rayleigh number, the magnetic field strength in Jupiter's dynamo region is calculated to be of order 10 mT, which is also a plausible value. This agreement lends support to the applicability of the simple scaling laws to the geodynamo. The scaling laws that involve a dependence on the magnetic Prandtl number severely underpredict the field strength in the Earth's core. Clarifying the role of the magnetic Prandtl number and providing a better theoretical foundation for the scaling laws remains a task for the future.

8.4.2 Comparison with the Geomagnetic Field

Many dynamo simulations reproduce the first-order properties of the geomagnetic field: dipole dominance and the correct order of magnitude of global field strength. A more detailed comparison is thus necessary to assess how realistically the different simulations model the geodynamo process. We have to content ourself with comparing the field above the core surface, since there is little or no information about the geomagnetic field inside the core. This task is difficult for dynamo modelers and geomagnetists alike, since the Earth's magnetic field varies on a large range of timescales, reaching from reversal-rate variations over several tens of million years to geomagnetic jerks representing annual changes. In addition, details of the spatial structure of the field are known to some degree for the recent field, but are increasingly less resolved when going back in time. Since Chapter 9 is devoted to reversals and excursion, we concentrate on timescales from decadal to millennium variations.

In order to compare dimensionless simulations to geomagnetic properties, we have fixed the magnetic diffusion time D^2/λ to 122 kyr by assuming an electrical conductivity of $\sigma = 6 \times 10^5\,\mathrm{Sm}^{-1}$. This amounts to a dipole decay time of 29 kyr. The field strength is calculated by adopting Earth's rotation rate and a core density of $\rho = 1.1 \times 10^4\ \mathrm{kg\,m^{-3}}$.

8.4.2.1 CMB field morphology

How well do we know the geomagnetic field at the top of the dynamo region? Even today, with several satellites measuring the magnetic field with previously unknown precision, there are significant inherent limitations. While the Gauss decomposition (Merill *et al.*, 1998, chapter 2.2) separates magnetic fields of external and internal origin, it does not allow to distinguish between Earth's crustal field and the field produced by the dynamo process. Some separation of the two is possible nevertheless, based on two properties of the geomagnetic field. While the dynamo field varies on geologically short timescales, the crustal magnetization can be regarded stationary in comparison. Furthermore, the crustal field is of smaller scale, leaving the core field at scales below degree 14 basically uncontaminated (Maus *et al.*, 2006). Additional precaution is needed with downward continuing field measurements to the CMB, making it necessary to damp contributions beyond roughly $l = 12$ (Bloxham and Gubbins, 1985). Within these limitations, modern field models (Olsen, 2002; Maus *et al.*, 2006) provide a global representation of the present geomagnetic field at the CMB up to degree and order 14.

Figure 12 compares the radial magnetic field for the year 1990 in the GUFM model (Jackson *et al.*, 2000) with two dynamo simulations. While we have selected snapshots that closely match the

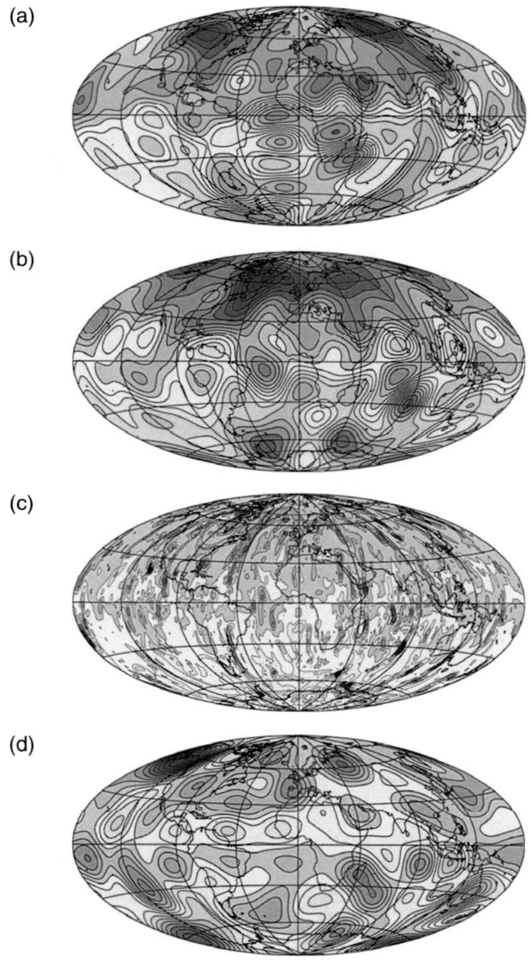

Figure 12 Comparison of the radial magnetic field at the CMB for the 1990 GUFM model (Jackson *et al.*, 2000) and two different dynamo simulations. (b) and (c) present a model at $E = 3 \times 10^{-5}$, $Ra = 42 \times Ra_c$, $Pm = 1$, and $Pr = 1$. Spectral filtering similar to the one employed in GUFM has been used in (b); (c) shows B_r at full numerical resolution up to spherical harmonic degree and order 133. The dipole is very stable in this model. (d) presents the filtered magnetic field of a reversing dynamo at $E = 3 \times 10^{-4}$, $Ra = 26 \times Ra_c$, $Pm = 3$, and $Pr = 1$.

geomagnetic field, most features are typical for dynamo models in the respective regimes. The longitude is arbitrary in the simulations. Continental contours have been included as a measure for relative size and orientation of magnetic structures. **Figures 12(b)** and **12(c)** depict a model at a relatively low Ekman number of $E = 3 \times 10^{-5}$ **Figure 12(c)** shows the full numerical resolution, while **Figure 12(b)** shows the same solution subject to a simulated 'crustal' filter that damps contributions beyond degree 12 and excludes contributions beyond degree 14. **Figure 12(d)** displays the filtered field for a reversing dynamo at $E = 3 \times 10^{-4}$ (Kutzner and Christensen, 2002; Wicht, 2002). We have selected a snapshot during a stable polarity interval.

There are many similarities between the geomagnetic field and the simulations apart from the obvious dipole dominance and the approximate agreement in the field strength: weaker field inside the tangent cylinder, strong normal polarity flux lobes close to the tangent cylinder, and pairwise inverse field patches around the equator. 'Normal' and 'inverse' refer to radial magnetic field directions that are inline or opposite to the dominant dipole direction. The reversing dynamo also possesses inverse field patches inside the tangent cylinder and at mid-latitudes in the Southern Hemisphere, very similar to today's field configuration. While these similarities are truly remarkable, the comparison also reveals discrepancies typical for many dynamo simulations. For example, in the reversing dynamo model, the dipole is significantly less pronounced than in the geomagnetic field (see also **Figure 13**). Another deviation concerns the field in the equatorial region. North–south aligned pairs of inverse magnetic field patches dominate the equatorial region in the simulations, in particular in the reversing numerical model, but we find mainly normal polarity patches in this region in today's and probably also the historic magnetic field (Finlay, 2005). Note that the simulation in **Figure 12(c)** shows many small-scale field structures described by spherical harmonic degrees beyond 14. These structures become even smaller for lower Ekman and larger Rayleigh numbers. Field patches that nicely resemble geomagnetic features when subjected to the crustal filter turn out to be blurred images of much narrower and significantly more complex small scale features.

In the dynamo simulations, flux maxima and minima have been associated with downwellings and upwellings in the convective flow (Christensen *et al.*, 1998; Olson *et al.*, 1999; Kutzner and Christensen, 2004). The convergent flow associated with downwellings collects magnetic field lines and thereby enhances the already present background field. The divergent flow connected to upwellings has the opposite effect. Intense flux lobes close to the tangent cylinder that are created by downwellings inside cyclonic convection columns and the low field strength around the pole caused by polar upwellings are prominent examples for these effects. In addition, upwellings create 'dipolar' pairs, consisting

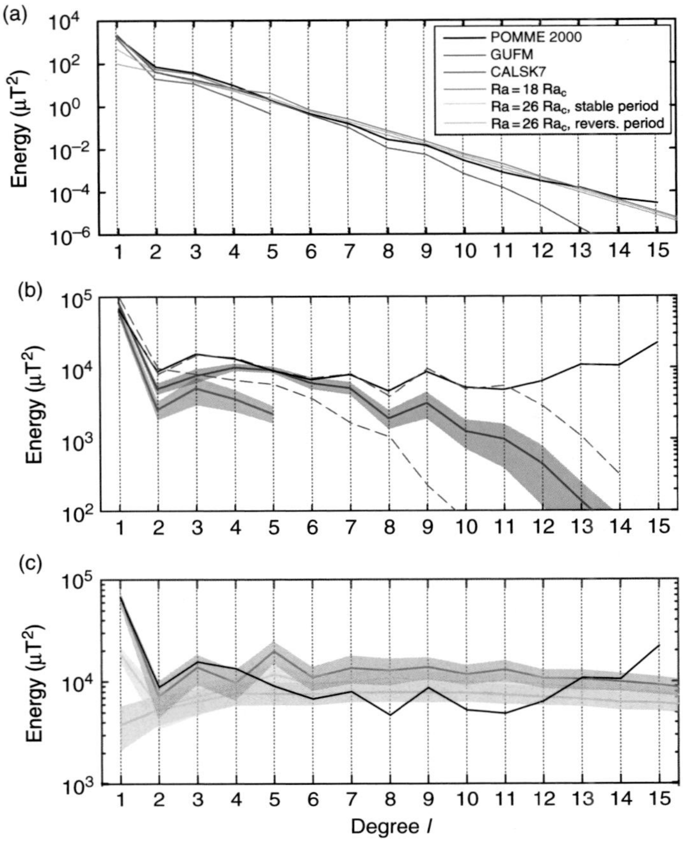

Figure 13 Comparison of time averaged geomagnetic power spectra for different geomagnetic field models and dynamo simulations. (a) Spectra at Earth surface. (b) Geomagnetic models at CMB. Colored bands in the width of the standard deviation indicate time variability. (c) Comparison of the geomagnetic POMME spectrum (black line) with spectra from dynamo simulations at the CMB. See text for more explanation.

of an inverse and a normal polarity patch, by pushing out horizontal field lines toward the exterior. Diffusion takes over close to the outer boundary to finally bring the pair to the CMB (Bloxham, 1986; Christensen *et al.*, 1998). The latter effect could be responsible for the inverse patches identified in the recent geomagnetic field.

8.4.2.2 Magnetic power spectra

The spatial resolution of geomagnetic field models decreases when we go back in time. Historic field models mainly rely on data from geomagnetic observatories and on measurements taken by mariners. They cover up to four centuries (Bloxham and Jackson, 1992; Jackson *et al.*, 2000) and provide Gauss coefficients up to degree 14 in 2.5-year intervals. However, the spatial resolution lies below degree 14 in the earlier epochs of these models due to decreasing data coverage, limited data quality, and the lack of intensity measurements prior to the 1830s. Archeomagnetic and paleomagnetic data allow to construct field models spanning several thousand years (Hongre *et al.*, 1998; Korte and Constable, 2003). The latest models cover the past 7000 years with a spatial resolution equivalent to degree $l = 5$ and and temporal resolution of 100 years (Korte *et al.*, 2005; Korte and Constable, 2005, 2006).

The time dependence of the resolution makes it difficult to decide which of today's smaller magnetic field features are typical and should therefore be replicated by a realistic dynamo model. Rather then comparing more snapshots of geomagnetic and simulated models for various epochs we proceed with analyzing magnetic power spectra W_l that describe the magnetic energy density carried by each spherical harmonic degree l ($W_l = (l+1)\sum_{m=0}^{l}(g_{lm}^2 + h_{lm}^2)$ where g_{lm}, h_{lm} are the Gauss coefficients). **Figure 13** compares magnetic power spectra of the POMME model for the year 2000 (Maus *et al.*, 2006) (black), time-averaged spectra of the historic GUFM model (Jackson *et al.*, 2000) (blue), the archeomagnetic CALSK7 model (Korte and Constable, 2005) (green), and geodynamo simulations. **Figure 13(a)** shows the spectra at a radial level equivalent to Earth's surface. Here, differences are hard to discern since all spectra decay similarly with the exception of the time-averaged GUFM model where the higher harmonics are obvious subject to stronger damping. **Figure 13(b)** visualizes the difference of the three geomagnetic field models at the CMB. Colored bands show the width of the respective standard deviations and illustrate time variability. We have added GUFM spectra for the two bounding years of this model, 1590 (lower dashed blue line) and 1990 (upper dashed blue line). The significantly stronger decay of the 1590 model reflects the inferior resolution in earlier years. Differences between the higher harmonic contributions in the 1990 GUFM and the 2000 POMME model indicate the stronger damping of the former. The degree 15 POMME model may be overly optimistic concerning the reliability of magnetic field coefficients beyond degree 12. The lower energy of higher moments in the CALSK7 model is possibly biased by low resolution and strong damping. The larger time variations, on the other hand, is a likely consequence of the longer averaging interval, 7500 years, instead of 400 years for GUFM (Korte and Constable, 2006).

Figure 13(c) compares time-averaged spectra from dynamo simulations with the POMME spectrum at the CMB. We have again selected models at parameters $E=3\times10^{-4}$, $\mathrm{Pr}=1$, and $\mathrm{Pm}=3$ for two different Rayleigh numbers: $\mathrm{Ra}=18\times\mathrm{Ra_c}$ (red, averaged over 64 ky), representing the stable dipole regime, and $\mathrm{Ra}=26\times\mathrm{Ra_c}$, representing the reversing case shown in **Figure 12**. Two spectra for the latter cover a stable polarity interval (yellow, averaged over 43 ky) and a period during which the dipole is weak and reverses several times (orange, averaged over 46 ky). The comparison demonstrates that dynamo simulations in the dipole-dominated regime (red and yellow) are well capable of reproducing the dipole moment and the power spectrum of the present geomagnetic field.

A specific feature reproduced by the dynamo simulations is the relatively low geomagnetic quadrupole contribution which lies clearly below the octupole contribution. Maxima at $l=3$ and $l=5$ in the simulation reflect the latitudinal differentiation into the regions inside and outside the tangent cylinder described in Section 8.4.1.1 and into the band of inverse field around the equator (**Figure 12**). Olson and Christensen (2002) report that the time-averaged octupole is significantly more pronounced in their numerical simulations than in paleomagnetic models (Section 8.4.2.4). However, the differentiation becomes weaker when the Rayleigh number is increased. The observable part of the model spectrum is essentially white during unstable periods (orange curve in **Figure 13**) with the exception of particularly low dipole and quadrupole contributions. The field seems to lose its large-scale coherence.

8.4.2.3 Secular variation

A simple global measure for the typical timescales associated with the field contribution of spherical harmonic degree l can be derived from the magnetic power spectrum W_l and the power spectrum of magnetic field variations $\dot{W}_l$:

$$\tau_l = \left(W_l/\dot{W}_l\right)^{1/2} \qquad [68]$$

Hongre *et al.* (1998) report that τ_l is typically of the order of a few hundred years in archeomagnetic data and decreasing with increasing degree. Olsen *et al.* (2006) find that the timescale varies like $\tau_l \propto l^{-1.35}$ in the present geomagnetic field. A similar behavior is found in dynamo simulations. Christensen and Olson (2003) find that τ_l roughly scales like l^{-1} for the non-dipole components. Examining a suit of dynamo models, Christensen and Tilgner (2004) derive a scaling, which relates τ_l to the inverse magnetic Reynolds number and is compatible with the values derived by Hongre *et al.* (1998):

$$\tau_l = 21.7\,\tau_{\mathrm{dipole}}\,l^{-1}\mathrm{Rm}^{-1} \qquad [69]$$

Equation [69] agrees with the idea that the magnetic field is mainly changed by advection.

Korte and Constable (2006) perform a spectral analysis in time for their CALSK7.2 model and find that the variation of all spherical harmonic contributions also contains timescales significantly longer than the advection time. Analyzing the dipole amplitude variation in a reversing dynamo model, Wicht (2002) also finds that the spectrum is dominated by periods much longer than the advection time.

Two more specific aspects of secular variation are the often-cited westward drift and possible eastward rotating vortices inside the tangent cylinder. The westward drift is a long-standing feature in the historic magnetic field and describes the motion of magnetic structures along the equator in the Atlantic hemisphere (see, e.g., Finlay and Jackson, 2003). Since this feature is so simple and persistent, its amplitude of about 0.2° per year is sometimes considered as the typical convective flow speed in the core. The faster eastward zonal flow inside the north polar cap of the tangent cylinder and its somewhat weaker southern counterpart have only been identified clearly in recent years (Olson and Aurnou, 1999; Hulot *et al.*, 2002; Amit and Olson, 2004). These vortices were first described by Glatzmaier and Roberts (1995b) and can be found in many dynamo simulations (e.g., Aubert, 2005; Sreenivasan and Jones, 2006a). Their origin is the thermal wind associated with hot plumes rising from the inner core boundary (see section 8.4.1.1).

While the fast eastward movement of magnetic features inside the tangent cylinder thus seems to be associated to a fluid flow, dynamo simulations suggest that this is less clear for the westward drift at lower latitudes as well as for some other secular variation features (Finlay, 2005). Westward as well as eastward drifts at low latitudes are present in several dynamo simulations (Kuang and Bloxham, 1997; Glatzmaier and Roberts, 1996a) and advection as well as wave propagation can be the causes (Rau *et al.*, 2000; Kono *et al.*, 2000; Christensen and Olson, 2003). Both mechanisms may contribute, one or the other being dominant at different times. Aubert (2005) analyzed time-averaged zonal flows for a suit of dynamo simulations (see also Section 8.4.1.5) and found it generally very similar to the flow inferred from present-day secular variation (Hulot *et al.*, 2002).

The secular variation in high Rayleigh number simulations (without strong hyperdiffusivity) offers a very mixed and complicated picture (Kono *et al.*, 2000; Rau *et al.*, 2000; Finlay, 2005). A recent analysis of a numerical dynamo simulation at $E = 3 \times 10^{-4}$ by Finlay (2005) suggests that inverse patches are formed by downwellings around the equator, ride with these downwellings on MAC waves (balance of magnetic, Archimedean, and Coriolis forces in the Navier–Stokes equation), can decouple from the wave, may be advected by westward or eastward jets, and finally dissolve again.

8.4.2.4 Comparison with the paleomagnetic field

Scarcity and uncertain dating of paleomagnetic data permit only statistical field models spanning the past 5 My (Constable and Parker, 1988; Kono and Tanaka, 1995; Johnson and Constable, 1995, 1997; Kelly and Gubbins, 1997; Constable and Johnson, 1999; Hatakeyama and Kono, 2002). Constable and Parker (1988) suggested that each Gauss coefficient may be treated as an independent normally distributed random variable, assuming that the dynamo works as a so-called gaint Gaussian process. The statistical paleomodel is then defined by mean and standard deviation of the Gauss coefficients, the latter serving as a measure for the amplitude of secular variation. This approach allows to disregard dating uncertainties of paleomagnetic data. Most statistical paleosecular models agree on that all mean coefficients vanish except for the dominating axial dipole contribution and a much weaker axial quadrupole that amounts to a few percent of the dipole strength. Constable and Parker (1988) proposed a particularly simple statistical secular variation model where the standard deviation depends on degree l only. This model was extended later to account for the observed higher dipole variability and the latitudinal dependence of directional data, which can be accommodated by a larger standard deviation of the degree $l = 2$ order $m = 1$ components (Kono and Tanaka, 1995; Johnson and Constable, 1995, 1997; Kelly and Gubbins, 1997; Hatakeyama and Kono, 2002).

Paleomagnetists commonly express directional data obtained at a specific site by the position (longitude, latitude) that a virtual geomagnetic North Pole (VGP) would assume if the field were purely dipolar. The appropriately calculated angular standard deviation (ASD) around the mean VGP position, also called VGP dispersion, is a measure for the amplitude of paleosecular variation at a given site. The ASD shows a very distinct latitude dependence (**Figure 14**), rising from about 12° at the equator to about 20° at higher latitudes (Kono *et al.*, 2000). This dependence could not be reproduced by the models of Glatzmaier *et al.* (1999), where no significant latitudinal variation is found. The models of Sakuraba and Kono (1999), Wicht (2005), and Christensen and Olson (2003) shows the correct variation but not the correct amplitude, the former two varying too strongly and the latter one too weakly.

Figure 14 displays ASD values for the two numerical dynamo models whose spectra are shown in **Figure 13**. We have averaged over time spans of 120 ky and 500 ky, respectively, for sites on a one-by-one degree grid. Paleomagnetic ASD values are based on data sets spanning 5 My with a much scarcer data coverage in time and space (Kono *et al.*, 2000).

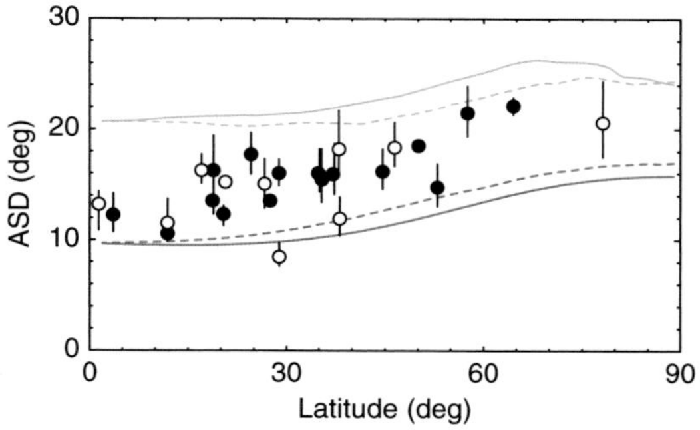

Figure 14 Azimullay-averaged angular standard deviation of the VGP position (ASD) after Kono and Roberts (2002) (symbols with error bars) and for the stable (red) and reversing (yellow) dynamos shown in **Figure 13**. Closed symbols and solid lines (open symbols, dashed lines) denote Northern Hemisphere (Southern Hemisphere) results. In the reversing simulation, transitional VGPs further away than 45° from either pole have been disregarded in the analysis.

The latitude dependence is nicely reproduced and the amplitudes embrace the paleomagnetic values, being too low for $\mathrm{Ra} = 18 \times \mathrm{Ra_c}$ and too high for $\mathrm{Ra} = 26 \times \mathrm{Ra_c}$.

Wicht (2005) speculates that the latitudinal ASD variation is a consequence of the increased convectional activity inside the tangent cylinder at higher Rayleigh numbers. The same effect also seems to be responsible for a larger number of excursions detected at higher latitude sites in his model. Such a behavior would be reflected in a pronounced fluctuation of mostly the axisymmetric magnetic field components. However, this is not compliant with the preferred paleomagnetic interpretation. Kono *et al.* (2000) perform a statistical analysis of the dynamo simulation by Sakuraba and Kono (1999) and find increased g_{21} and h_{21} standard deviations that describe the latitudinal ASD dependence in paleosecular variation models.

8.4.2.5 Mantle influence on the dynamo

Persistent non-axisymmetric features in the 5 My mean geomagnetic field are a matter of debate. Kelly and Gubbins (1997) identify flux concentrations under Canada and Siberia similar to those in the present field configuration. These structures can also be found in archeomagnetic models (Korte and Constable, 2006). Johnson and Constable (1997) argue that these features lie beyond the resolution of current paleomagnetic data compilations and suggest that the most convincing nonzonal structure is an anomalous radial field under the Pacific, where the magnetic equator bulges toward the north. These features can clearly be identified as an inclination anomaly, and this is supported by a large number of lava flow data from Hawaii. Probably related is the particularly low secular variation observed in the Pacific hemisphere in historic (Jackson *et al.*, 2000), archeomagnetic (Korte and Constable, 2006), and possibly also the paleomagnetic field. Because we expect that in the fluid core itself there is no way to break longitudinal symmetry, any persistent deviation from a purely latitudinal dependence in the long-term magnetic field and its time-dependent fluctuation should be due to an external influence on the dynamo. The most plausible one is that a heterogeneous lower mantle imprints an azimuthally varying boundary condition on the dynamo (*see* Chapter 12).

While the present and the historic magnetic spectrum show a particularly weak quadrupole contribution, this is not true for archeomagnetic and paleomagnetic time-averaged fields, where the axisymmetry quadrupole is perhaps the only significant non-dipole component. Axial dipole and quadrupole belong to the two different equatorial symmetry families (see Section 8.4.1.1). The two families are mixed by equatorially anti-symmetric flows. A particularly simple mechanism is the coupling of time-averaged axial dipole and quadrupole contributions by a time-averaged equatorially antisymmetric flow. However, this flow component should average to zero over time, unless the equatorial symmetry is broken in the dynamo model by some external influence. Inhomogeneous thermal outer-boundary conditions are one way to accomplish this (Zhang and Gubbins, 1992; Olson and Christensen, 2002; Bloxham, 2002; Christensen and Olson, 2003).

Olson and Christensen (2002) ran dynamo simulations with different imposed heterogeneous heat flux patterns, confirming that this is indeed a viable mechanism. The only one of their dynamo models that produced a significant time-averaged axial quadrupole used a boundary heat flux of degree $l = 1$ and order $m = 0$. They also performed simulations with a heat flow pattern inferred from seismic tomographic models of the lower mantle, translating higher seismic velocities into lower temperatures and hence higher-than-average heat flux. Such heat flux variations include a slight equatorial asymmetry but are generally dominated by the ($l = 2$, $m = 2$) pattern. The equatorial asymmetry was not strong enough to cause a significant axial quadrupole term in the time-average field. McMillan *et al.* (2001) arrive at the same conclusion based on similar simulations by Glatzmaier *et al.* (1999).

The strong azimuthal heat flux variation suggested by the lower-mantle seismic velocity anomalies may cause persistent non-axisymmetric magnetic field features. While McMillan *et al.* (2001) and Hulot and Bouligand (2005) find no statistically significant trace in the 'tomographic' simulation by Glatzmaier *et al.* (1999), Bloxham (2002) and Christensen and Olson (2003) identify non-axisymmetric features in their 'tomographic' dynamo models. In particular, they find reduced secular variation in an area of low heat flow underneath the Pacific that readily compares to the low secular variation in the historic, archeomagnetic, and paleomagnetic field. However, Christensen and Olson (2003) point out that the zonal variation in secular variation amplitude is quite weak in the models and significantly smaller than the latitudinal variation that features prominently in paleomagnetic directional ASD data.

We pointed out that downwelling in the center of cyclonic columns is responsible for longitudinal field variations in the form of flux concentration close to the tangent cylinder. This effect can be very time dependent at higher Rayleigh numbers, and not all cyclonic vortices intensify the field to the same degree. The number, intensity, and form of these flux patches change with time, sometimes resembling a configuration similar to the current or historic magnetic field, but the time-averaged field is axisymmetric as long as the CMB condition is homogeneous. Using a tomographic heat flux boundary condition helps to stabilize the field to a degree where the time-averaged field features two flux patches in the Northern Hemisphere at locations very similar to that inferred for the geomagnetic field (Olson and Christensen, 2002). However, these simulations also show an intense flux patch in the Southern Hemisphere that is not present in geomagnetic data.

8.5 Perspectives

Numerical geodynamo models have had a remarkable success in matching geomagnetic field properties, to an extent that was not dreamed of 15 years ago. They produce dipole-dominated magnetic fields of the right strength, show spatial power spectra and a field morphology at the CMB that resemble the observed ones and a reversal behavior similar to that deduced from paleomagnetism.

However, not all models perform equally well. Also, the values of fundamental control parameters cannot match Earth values, and their model values are usually chosen rather arbitrarily or according to numerical feasibility. The understanding of which parameter combinations lead to Earth-like dynamos and for what reasons has not kept pace with the surge in geodynamo modeling. First steps into this direction have been made by searching for scaling laws that relate characteristic dynamo properties to the control parameters. However, so far these laws are largely empirical, and their range of validity remains unclear. Simulations at more extreme parameter values, such as Ekman numbers $E < 10^{-6}$ and magnetic Prandtl numbers $\mathrm{Pm} \ll 1$, are needed to consolidate the scaling laws, requiring tremendous computational resources. New (local) methods may gain the advantage over standard spectral codes on massive parallel systems, but the potential of spectral methods seems not exhausted yet. We must develop a better theoretical understanding of the force balance in the dynamo models and compare it to the expected balance in the Earth's core. How important is viscous friction in the dynamo models and at which Ekman number will its influence become negligible? Do small unresolved scales act mainly as an energy sink, or do they play an important role for the induction process? Are there essential differences between dynamos operating at a magnetic Prandtl number of order 1 and those working at a very low value of Pm? To answer such questions and to put scaling laws on a firm theoretical basis remains a major task for the future.

In addition to better understand convection-driven dynamos in the most basic setup, the consequences of various 'real Earth' complications need to be explored. How does the interplay of compositional and thermal buoyancy (double-diffusive convection) affect the dynamo? Are Boussinesq models basically adequate, or do non-Boussinesq effects such as fluid compressibility or the redistribution of thermal energy by Ohmic heating have a large influence? What are the most appropriate boundary and coupling conditions at the CMB and the inner core boundary (e.g., is the flux of light element at the inner core boundary heterogeneous?), and how does this affect the dynamo? This list of open questions is certainly far from exhaustive.

Earth is not the only magnetic planet in the solar system. It would strengthen the confidence in the geodynamo models tremendously if it could be shown that the models can reproduce the field strength and geometry of other planets, after adaptation to the specific conditions in these objects. So far, a comparative dynamo theory covering the diversity of planetary magnetic fields is in its infancy.

References

Abramowitz M and Stegun I (1984) *Pocketbook of Mathematical Functions*. Thun, Switzerland: Harry Deutsch.

Amit H and Olson P (2004) Helical core flow from geomagnetic secular variation. *Physics of the Earth and Planetary Interiors* 147: 1–25.

Aubert J (2005) Steady zonal flows in spherical shell dynamos. *Journal of Fluid Mechanics* 542: 53–67.

Aubert J and Wicht J (2004) Axial versus equatorial dipolar dynamo models with implications for planetary magnetic fields. *Earth and Planetary Science Letters* 221: 409–419.

Aurnou JM, Brito D, and Olson PL (1996) Mechanics of inner core superrotation. *Geophysical Research Letters* 23: 3401–3404.

Bloxham J (1986) The expulsion of magnetic flux from the Earth's core. *Geophysical Journal of the Royal Astronomical Society* 87: 669–678.

Bloxham J (2002) Time-independent and time-dependent behaviour of high-latitude flux bundles at the core–mantle boundary. *Geophysical Research Letters* 29: 1854.

Bloxham J and Gubbins D (1985) The secular variation of Earth's magnetic field. *Nature* 317: 777–781.

Bloxham J and Jackson A (1992) Time-dependent mapping of the magnetic field at the core–mantle boundary. *Journal of Geophysical Research* 97: 19537–19563.

Bloxham J, Zatman S, and Dumberry M (2002) The origin of geomagnetic jerks. *Nature* 420: 65–68.

Braginsky SI (1970) Torsional magnetohydrodynamic vibrations in the Earth's core and variation in day length. *Geomagnetizm i Aeronomiya* 10: 1–8.

Braginsky SI (1975) Nearly axially symmetric model of the hydromagnetic dynamo of Earth. *Geomagnetizm i Aeronomiya* 15: 122–128.

Braginsky SI and Roberts PH (1987) A model-Z geodynamo. *Geophysical and Astrophysical Fluid Dynamics* 38: 327–349.

Braginsky SI and Roberts PH (1994) From Taylor-state to model-Z. *Geophysical and Astrophysical Fluid Dynamics* 77: 3–13.

Braginsky SI and Roberts PH (1995) Equations governing convection in Earth's core and the geodynamo. *Geophysical and Astrophysical Fluid Dynamics* 79: 1–97.

Buffett BA (1996) Gravitational oscillations in the length of day. *Geophysical Research Letters* 23: 2279–2282.

Buffett BA (1997) Geodynamic estimates of the viscosity of the Earth's inner core. *Nature* 388: 571–573.

Buffett BA and Bloxham J (2002) Energetics of numerical geodynamo models. *Geophysical Journal International* 149: 211–224.

Buffett BA and Glatzmaier GA (2000) Gravitational braking of inner-core rotation in geodynamo simulations. *Geophysical Research Letters* 27: 3125–3128.

Bullard EC and Gellman H (1954) Homogeneous dynamos and terrestrial magnetism. *Philosophical Transactions of the Royal Society of London A* 247: 213–278.

Busse FH (2000) Homogeneous dynamos in planetary cores and in the laboratory. *Annual Review of Fluid Mechanics* 32: 383–408.

Busse FH and Simitev R (2005a) Convection in rotating spherical fluid shells and its dynamo states. In: Soward AM, Jones CA, Hughes DW, and Weiss NO (eds.) *Fluid Dynamics and Dynamos in Astrophysics and Geophysics*, pp. 359–392. New York: Taylor and Francis.

Busse FH and Simitev R (2005b) Dynamos driven by convection in rotating spherical shells. *Astronomische Nachrichten* 326: 231–240.

Busse FH and Simitev R (2006) Parameter dependences of convection-driven dynamos in rotating spherical fluid shells. *Geophysical and Astrophysical Fluid Dynamics* 100: 341–361.

Chan KH, Zhang K-K, Zou J, and Schubert G (2001) A nonlinear, 3-D spherical α^2 dynamo using a finite element method. *Physics of the Earth and Planetary Interiors* 128: 35–50.

Christensen U and Olson P (2003) Secular variation in numerical geodynamo models with lateral variations of boundary heatflow. *Physics of the Earth and Planetary Interiors* 138: 39–54.

Christensen U, Olson P, and Glatzmaier GA (1998) A dynamo model interpretation of geomagnetic field structures. *Geophysical Research Letters* 25: 1565–1568.

Christensen U, Olson P, and Glatzmaier GA (1999) Numerical modelling of the geodynamo: A systematic parameter study. *Geophysical Journal International* 138: 393–409.

Christensen UR (2001) Zonal flow driven by deep convection in the major planets. *Geophysical Research Letters* 28: 2553–2556.

Christensen UR (2002) Zonal flow driven by strongly supercritical convection in rotating spherical shells. *Journal of Fluid Mechanics* 470: 115–133.

Christensen UR and Aubert J (2006) Scaling properties of convection-driven dynamos in rotating spherical shells and application to planetary magnetic fields. *Geophysical Journal International* 166: 97–114.

Christensen UR, Aubert J, Busse FH, *et al.* (2001) A numerical dynamo benchmark. *Physics of the Earth and Planetary Interiors* 128: 25–34.

Christensen UR and Tilgner A (2004) Power requirement of the geodynamo from Ohmic losses in numerical and laboratory dynamos. *Nature* 429: 169–171.

Clune TC, Eliott JR, Miesch MS, Toomre J, and Glatzmaier GA (1999) Computational aspects of a code to study rotating turbulent convection in spherical shells. *Parallel Computing* 25: 361–380.

Constable CG and Johnson CL (1999) Anisotropic paleosecular variation models: Implication for geomagnetic field observables. *Physics of the Earth and Planetary Interiors* 115: 35–51.

Constable CG and Parker RL (1988) Statistics of the geomagnetic secular variation for the past 5 m.y. *Journal of Geophysical Research* 93: 11569–11581.

Dormy E, Cardin P, and Jault D (1998) MHD flow in a slightly differentially rotating spherical shell, with conducting inner core, in a dipolar magnetic field. *Earth and Planetary Science Letters* 160: 15–30.

Dormy E, Valet J-P, and Courtillot V (2000) Numerical models of the geodynamo and observational constraint. *Geochemistry, Geophysics, and Geosystems* 1(10), doi:10.1029/2000GC0000062.

Dumberry M and Bloxham J (2003) Torque balance, Taylor's constraint and torsional oscillations in a numerical model of the geodynamo. *Physics of the Earth and Planetary Interiors* 140: 29–51.

Fearn DR (1998) Hydromagnetic flow in planetary cores. *Reports on Progress in Physics* 61: 175–235.

Finlay CC (2005) *Hydromagnetic Waves in Earth's Core and Their Influence on Geomagnetic Secular Variation*. PhD Thesis, School of Earth and Environment, University of Leeds.
Finlay CC and Jackson A (2003) Equatorially dominated magnetic field change at the surface of Earth's core. *Science* 300: 2084–2086.
Fournier A, Bunge H-P, Hollerbach R, and Vilotte J-P (2004) Application of the spectral-element method to the axisymmetric Navier–Stokes equation. *Geophysical Journal International* 156: 682–700.
Fournier A, Bunge H-P, Hollerbach R, and Vilotte J-P (2005) A Fourier-spectral element algorithm for thermal convection in rotating axisymmetric containers. *Journal of Computational Physics* 204: 462–489.
Gessmann CK and Wood BJ (2002) Potassium in the Earth's core?. *Earth and Planetary Science Letters* 200: 63–78.
Glatzmaier GA (1984) Numerical simulation of stellar convective dynamos. 1: The model and method. *Journal of Computational Physics* 55: 461–484.
Glatzmaier GA (2002) Geodynamo simulations – How realistic are they? *Annual Review of Earth and Planetary Sciences* 30: 237–257.
Glatzmaier GA, Coe RS, Hongre L, and Roberts PH (1999) The role of the Earth's mantle in controlling the frequency of geomagnetic reversals. *Nature* 401: 885–890.
Glatzmaier GA and Roberts PH (1995a) A three-dimensional convective dynamo solution with rotating and finitely conducting inner core and mantle. *Physics of the Earth and Planetary Interiors* 91: 63–75.
Glatzmaier GA and Roberts PH (1995b) A three-dimensional self-consistent computer simulation of a geomagnetic field reversal. *Nature* 377: 203–209.
Glatzmaier GA and Roberts PH (1996a) An anelastic evolutionary geodynamo simulation driven by compositional and thermal convection. *Physica D* 97: 81–94.
Glatzmaier GA and Roberts PH (1996b) Rotation and magnetism of Earth's inner core. *Science* 274: 1887–1891.
Glatzmaier GA and Roberts PH (1997) Simulating the geodynamo. *Contemporary Physics* 38: 269–288.
Grote E and Busse FH (2001) Dynamics of convection and dynamos in rotating spherical fluid shells. *Fluid Dynamics Research* 28: 349–368.
Grote E, Busse FH, and Tilgner A (1999) Convection-driven quadrupolar dynamos in rotating spherical shells. *Physical Review E* 60: 5025–5028.
Grote E, Busse FH, and Tilgner A (2000a) Effects of hyperdiffusivities on dynamo simulations. *Geophysical Research Letters* 27: 2001–2004.
Grote E, Busse FH, and Tilgner A (2000b) Regular and chaotic spherical dynamos. *Physics of the Earth and Planetary Interiors* 117: 259–272.
Gubbins D and Roberts PH (1987) Magnetohydrodynamics of the Earth's core. In: Jacobs JA (ed.) *Geomagnetism*, vol. 2, pp. 1–183. London: Academic Press.
Gubbins D and Sarson GR (1994) Geomagnetic field morphologies from a kinematic dynamo model. *Nature* 368: 51–55.
Harder H and Hansen U (2005) A finite-volume solution method for thermal convection and dynamo problems in spherical shells. *Geophysical Journal International* 161: 522–532.
Hatakeyama T and Kono M (2002) Geomagnetic field model for the last 5 My: Time-averaged field and secular variation. *Physics of the Earth and Planetary Interiors* 133: 181–215.
Healy DM, Jr., Kostelec PJ, and Rockmore D (2004) Towards safe and effective high-order Legendre transforms with applications to FFTs for the 2-sphere. *Advances in Computational Mathematics* 21: 59–105.
Heimpel M, Aurnou J, and Wicht J (2005) Simulation of equatorial and high-latitude jets on Jupiter in a deep convection model. *Nature* 438: 193–196.
Hejda P and Reshetnyak M (2003) Control volume method for the dynamo problem in the sphere with the free rotating inner core. *Studia Geophysica et Geodaetica* 47: 147–159.
Hollerbach R (1996) On the theory of the geodynamo. *Physics of the Earth and Planetary Interiors* 98: 163–185.
Hollerbach R (2000) A spectral solution of the magneto-convection equations in spherical geometry. *International Journal of Numerical Methods in Fluids* 32: 773–797.
Hollerbach R and Jones CA (1993) Influence of the Earth's inner core on geomagnetic fluctuations and reversals. *Nature* 365: 541–543.
Hongre L, Hulot G, and Khokholov A (1998) An analysis of the geomagnetic field over the past 2000 years. *Physics of the Earth and Planetary Interiors* 106: 311–335.
Hulot G and Bouligand C (2005) Statistical paleomagnetic field modelling and symmetry considerations. *Geophysical Journal International* 161: 591–602.
Hulot G, Eymin C, Langlais B, Mandea M, and Olsen N (2002) Small-scale structure of the geodynamo inferred from Ørsted and Magsat satellite data. *Nature* 416: 620–623.
Ishihara N and Kida S (2000) Axial and equatorial magnetic dipoles generated in a rotating spherical shell. *Journal of the Physical Society of Japan* 69: 1582–1585.
Jackson A, Jonders ART, and Walker MR (2000) Four centuries of geomagnetic secular variation from historical records. *Philosophical Transactions of the Royal Society of London A* 358: 957–990.
Johnson CL and Constable CG (1995) The time averaged geomagnetic field as recorded by lava flows over the past 5 million years. *Geophysical Journal International* 122: 489–519.
Johnson CL and Constable CG (1997) The time averaged geomagnetic field: Global and regional biases for 0–5 Ma. *Geophysical Journal International* 131: 643–666.
Jones CA (2000) Convection-driven geodynamo models. *Philosophical Transactions of the Royal Society of London A* 358: 873–897.
Jones CA, Longbottom AW, and Hollerbach R (1995) A self-consistent convection driven geodynamo model, using a mean field approximation. *Physics of the Earth and Planetary Interiors* 92: 119–141.
Kageyama A and Sato T (1995) Computer simulation of a magnetohydrodynamic dynamo. II. *Physics of Plasmas* 2: 1421–1431.
Kageyama A and Sato T (1997a) Dipole field generation by an MHD dynamo. *Plasma Physics and Controlled Fusion* 39: 83–91.
Kageyama A and Sato T (1997b) Generation mechanism of a dipole field by a magnetohydrodynamic dynamo. *Physical Review* E 55: 4617–4626.
Kageyama A, Watanabe K, and Sato T (1993) Simulation study of a magnetohydrodynamic dynamo: Convection in a rotating spherical shell. *Physics of Fluids B* 54: 2793–2805.
Kageyama A and Yoshida M (2005) Geodynamo and mantle convection simulations on the Earth Simulator using the Yin-Yang grid. *Journal of Physics Conference Series* 16: 325–338.
Katayama J, Matsushima M, and Honkura Y (1999) Some characteristics of magnetic field behavior in a model of MHD dynamo thermally driven in a rotating spherical shell. *Physics of the Earth and Planetary Interiors* 111: 141–159.
Kelly P and Gubbins D (1997) The geomagnetic field over the past 5 million years. *Geophysical Journal International* 128: 315–330.
Kono M and Roberts PH (2001) Definition of the Rayleigh number for geodynamo simulation. *Physics of the Earth and Planetary Interiors* 128: 13–24.

Kono M and Roberts PH (2002) Recent geodynamo simulations and observations of the geomagnetic field. *Reviews of Geophysics* 40: 1013.

Kono M, Sakuraba A, and Ishida M (2000) Dynamo simulation and palaeosecular variation models. *Philosophical Transactions of the Royal Society of London A* 358: 1123–1139.

Kono M and Tanaka H (1995) Mapping the Gauss coefficients to the pole and the models of paleosecular variation. *Journal of Geomagnetism and Geoelectricity* 47: 115–130.

Korte M and Constable CG (2003) Continuous global geomagnetic field models for the past 3000 years. *Physics of the Earth and Planetary Interiors* 140: 73–89.

Korte M and Constable CG (2005) Continuous geomagnetic field models for the past 7 millennia. 2: CALS7K. *Geochemistry, Geophysics, and Geosystems* 6: Q02H16.

Korte M and Constable CG (2006) Centennial to millennial geomagnetic secular variations. *Geophysical Journal International* 167: 43–52.

Korte M, Genevey A, Constable CG, Frank U, and Schnepp E (2005) Continuous geomagnetic field models for the past 7 millennia. 1: A new global data compilation. *Geochemistry, Geophysics, and Geosystems* 6: Q02H15.

Kuang W and Bloxham J (1997) An Earth-like numerical dynamo model. *Nature* 389: 371–374.

Kuang W and Bloxham J (1999) Numerical modeling of magnetohydrodynamic convection in a rapidly rotating spherical shell: Weak and strong field dynamo action. *Journal of Computational Physics* 153: 51–81.

Kutzner C and Christensen UR (2000) Effects of driving mechanisms in geodynamo models. *Geophysical Research Letters* 27: 29–32.

Kutzner C and Christensen UR (2002) From stable dipolar towards reversing numerical dynamos. *Physics of the Earth and Planetary Interiors* 131: 29–45.

Kutzner C and Christensen UR (2004) Simulated geomagnetic reversals and preferred virtual geomagnetic pole paths. *Geophysical Journal International* 157: 1105–1118.

Laske G and Masters G (1999) Limits on differential rotation of the inner core from an analysis of the Earth's free oscillations. *Nature* 402: 66–69.

Lesur V and Gubbins D (1999) Evaluation of fast spherical transforms for geophysical applications. *Geophysical Journal International* 139: 547–555.

Li J, Sato T, and Kageyama A (2002) Repeated and sudden reversals of the dipole field generated by a spherical dynamo action. *Science* 295: 1887–1890.

Lister JR and Buffett BA (1995) The strength and efficiency of thermal and compositional convection in the geodynamo. *Physics of the Earth and Planetary Interiors* 91: 17–30.

Loper DE (1978) Some thermal consequences of the gravitationally powered dynamo. *Journal of Geophysical Research* 83: 5961–5970.

Matsui H and Buffett BA (2005) Sub-grid scale model for convection-driven dynamos in a rotating plane layer. *Physics of the Earth and Planetary Interiors* 153: 108–123.

Matsui H and Okuda H (2004) Treatment of the magnetic field for geodynamo simulations using the finite element method. *Earth, Planets and Space* 56: 945–954.

Matsushima M (2006) Reexamination of a scale-similarity model for the subgrid-scale flux in the Earth's core. *Geophysical and Astrophysical Fluid Dynamics* 100: 363–377.

Maus S, Rother M, Stolle C, *et al.* (2006) Third generation of the Potsdam Magnetic Model of the Earth POMME. *Geochemistry, Geophysics, and Geosystems* 7: Q07008.

McMillan DG, Constable CG, Parker RL, and Glatzmaier GA (2001) A statistical analysis of magnetic fields from some geodynamo simulations. *Geochemistry, Geophysics, and Geosystems* 2, doi:10.1029 2000GC000130.

Merill RT, McElhinny MW, and McFadden PL (1998) *The Magnetic Field of the Earth*. Academic Press: San Diego, USA.

Olsen N (2002) A model of the geomagnetic field and its secular variation for epoch 2000 estimated from Ørsted data. *Geophysical Journal International* 149: 454–462.

Olsen N, Lühr H, Sabaka TJ, *et al.* (2006) CHAOS – A model of the earth's magnetic field derived from CHAMP, Ørsted and SAC-C magnetic satellite data. *Geophysical Journal International* 166: 67–75.

Olson P and Aurnou JM (1999) A polar vortex in the Earth's core. *Nature* 402: 170–173.

Olson P, Christensen U, and Glatzmaier GA (1999) Numerical modeling of the geodynamo: Mechanisms of field generation and equilibration. *Journal of Geophysical Research* 104: 10383–10404.

Olson P and Christensen UR (2002) The time-averaged magnetic field in numerical dynamos with non-uniform boundary heat flow. *Geophysical Journal International* 151: 809–823.

Olson P and Christensen UR (2006) Dipole moment scaling for convection-driven planetary dynamos. *Earth and Planetary Science Letters* 250: 561–571.

Phillips CG and Ivers DJ (2003) Strong field anisotropic diffusion models for the Earth's core. *Physics of the Earth and Planetary Interiors* 140: 13–28.

Press WH, Teukolsky SA, Vetterling WT, and Flannery BP (2003) *Numerical Recipes,* 2nd edn., vol. 1. Cambridge: Cambridge University Press.

Rama Murthy V, van Westrenen W, and Fei YW (2003) Experimental evidence that potassium is a substantial radioactive heat source in planetary cores. *Nature* 423: 163–165.

Rau S, Christensen U, Jackson A, and Wicht J (2000) Core flow inversions tested with numerical dynamo models. *Geophysical Journal International* 141: 485–497.

Roberts PH and Glatzmaier GA (2000) A test of the frozen-flux approximation using a new geodynamo model. *Philosophical Transactions of the Royal Society of London A* 358: 1109–1121.

Rotvig J and Jones CA (2002) Rotating convection-driven dynamos at low Ekman number. *Physical Review E* 66: 056308.

Rüdiger G and Hollerbach R (2004) *The Magnetic Universe*. Weinheim, Germany: Wiley-VCH.

Sakuraba A and Kono M (1999) Effect of the inner core on the numerical solution of the magnetohydrodynamic dynamo. *Physics of the Earth and Planetary Interiors* 111: 105–121.

Sarson GR and Jones CA (1999) A convection driven geodynamo reversal model. *Physics of the Earth and Planetary Ineriors* 111: 3–20.

Sarson GR, Jones CA, and Longbottom AW (1997) The influence of boundary region heterogeneities on the geodynamo. *Physics of the Earth and Planetary Ineriors* 101: 13–32.

Sarson GR, Jones CA, and Longbottom AW (1998) Convection driven geodynamo models of varying Ekman number. *Geophysical and Astrophysical Fluid Dynamics* 88: 225–259.

Schrinner M, Rädler KH, Schmitt D, Rheinhardt M, and Christensen U (2005) Mean-field view on rotating magnetoconvection and a geodynamo model. *Astronomische Nachrichten* 326: 245–249.

Schubert G and Zhang K (2001) Effects of an electrically conducting inner core on planetary and stellar dynamos. *Astrophysical Journal* 557: 930–942.

Simitev R and Busse FH (2005) Prandtl-number dependence of convection-driven dynamos in rotating spherical fluid shells. *Journal of Fluid Mechanics* 532: 365–388.

Song XD and Richards PG (1996) Seismological evidence for differential rotation of the earth' inner core. *Nature* 382: 221–224.

Spotz WF and Swarztrauber PN (2001) A performance comparison of associated Legendre projections. *Journal of Computational Physics* 168: 339–355.

Sreenivasan B and Jones CA (2005) Structure and dynamics of the polar vortex in the Earth's core. *Geophysical Research Letters* 32: L20301.

Sreenivasan B and Jones CA (2006a) Azimuthal winds, convection and dynamo action in the polar regions of planetary cores. *Geophysical and Astrophysical Fluid Dynamics* 100: 319–339.

Sreenivasan B and Jones CA (2006b) The role of inertia in the evolution of spherical dynamos. *Geophysical Journal International* 164: 467–476.

St. Pierre MG (1993) The strong-field branch of the Childress-Soward dynamo. In: Proctor MRE, Matthews PC, and Rucklidge AM (eds.) *Solar and Planetary Dynamos*, pp. 295–302. Cambridge: Cambridge University Press.

Stanley S and Bloxham J (2004) Convective-region geometry as the cause of Uranus' and Neptune's unusual magnetic fields. *Nature* 428: 151–153.

Stanley S and Bloxham J (2006) Numerical dynamo models of Uranus' and Neptune's magnetic fields. *Icarus* 184: 556–572.

Starchenko SV and Jones CA (2002) Typical velocities and magnetic field strengths in planetary interiors. *Icarus* 157: 426–435.

Stellmach S and Hansen U (2004) Cartesian convection-driven dynamos at low Ekman number. *Physical Review E* 70: 056312.

Stevenson DJ (1979) Turbulent thermal convection in the presence of rotation and a magnetic field: A heuristic theory. *Geophysical and Astrophysical Fluid Dynamics* 12: 139–169.

Su WJ, Dziewonski AM, and Jeanloz R (1996) Planet within a planet: Rotation of the inner core of the Earth. *Science* 274: 1883–1887.

Takahashi F, Matsushima M, and Honkura Y (2005) Simulations of a quasi-Taylor state geomagnetic field including polarity reversals on the Earth simulator. *Science* 309: 459–461.

Taylor JB (1963) The magneto-hydrodynamics of a rotating fluid and the Earth's dynamo problem. *Proceedings of the Royal Society of London A* 274: 274–283.

Tilgner A (1999) Spectral methods for the simulation of incompressible flows in spherical shells. *International Journal of Numerical Methods in Fluids* 30: 713–724.

Tilgner A (2004) Small scale kinematic dynamos: Beyond the α-effect. *Geophysical and Astrophysical Fluid Dynamics* 98: 225–234.

Tilgner A (2005) Precession driven dynamos. *Physics of Fluids* 15: 034104.

Wicht J (2002) Inner-core conductivity in numerical dynamo simulations. *Physics of the Earth and Planetary Ineriors* 132: 281–302.

Wicht J (2005) Palaeomagnetic interpretation of dynamo simulations. *Geophysical Journal International* 162: 371–380.

Wicht J and Busse FH (1997) Magnetohydrodynamic dynamos in rotating spherical shells. *Geophysical and Astrophysical Fluid Dynamics* 86: 103–109.

Wicht J and Olson P (2004) A detailed study of the polarity reversal mechanism in a numerical dynamo model. *Geochemistry, Geophysics, and Geosystems* 5: Q03H10.

Willis AP and Gubbins D (2004) Kinematic dynamo action in a sphere: Effects of periodic time-dependent flows on solutions with axial dipole symmetry. *Geophysical and Astrophysical Fluid Dynamics* 98: 537–554.

Xu S and Szeto AMK (1994) Gravitational coupling in the Earth's interior revisited. *Geophysical Journal International* 118: 94–100.

Xu S and Szeto AMK (1998) The coupled rotation of the inner core. *Geophysical Journal International* 133: 279–297.

Zatman S and Bloxham J (1997) Torsional oscillations and the magnetic field within the Earth's core. *Nature* 388: 760–763.

Zatman S and Bloxham J (1999) On the dynamical implications of models of B_s in the Earth's core. *Geophysical Journal International* 138: 679–686.

Zhang J, Song XD, Li YC, Richards PG, Sun XL, and Waldhauser F (2005) Inner core differential motion confirmed by earthquake waveform doublets. *Science* 309: 1357–1360.

Zhang K and Gubbins D (1992) On convection in the Earth's core driven by lateral temperature variation in the lower mantle. *Geophysical Journal International* 108: 247–255.

Zhang K and Gubbins D (1993) Convection in a rotating spherical fluid shell with an inhomogeneous temperature boundary condition at infinite Prandtl number. *Journal of Fluid Mechanics* 250: 209–232.

Zhang K and Jones CA (1997) The effect of hyperviscosity on geodynamo models. *Geophysical Research Letters* 24: 2869–2872.

Zhang K and Schubert G (2000) Magnetohydrodynamics in rapidly rotating spherical systems. *Annual Review of Fluid Mechanics* 32: 409–443.

Zhang KK and Busse FH (1988) Finite amplitude convection and magnetic field generation in a rotating spherical shell. *Geophysical and Astrophysical Fluid Dynamics* 44: 33–53.

Zhang KK and Busse FH (1989) Convection driven magnetohydrodynamic dynamos in rotating spherical shells. *Geophysical and Astrophysical Fluid Dynamics* 49: 97–116.

Zhang KK and Busse FH (1990) Generation of magnetic fields by convection in a rotating spherical fluid shell of infinite Prandtl number. *Physics of the Earth and Planetary Ineriors* 59: 208–222.

9 Magnetic Polarity Reversals in the Core

G. A. Glatzmaier and R. S. Coe, University of California, Santa Cruz, CA, USA

9.1	Observations	283
9.2	Models	288
9.3	Conclusions	295
References		295

9.1 Observations

The observational evidence for evaluating and improving geodynamo simulations comes from direct observations of the geomagnetic field over the past few centuries and from the paleomagnetism of magnetic materials, such as rocks, fired artifacts, adobe, and even pigments from paintings, over the rest of geologic time. Thus, for information about most geomagnetic behaviors we rely upon the results of paleomagnetic studies. These show us that the Earth's magnetic field has been dominantly dipolar in the past, as it is today. Moreover, to a reasonable approximation it averages to axial dipolar over a few tens of thousands of years (e.g., Merrill, 1996). This is demonstrated for the past 150 million years by the agreement between plate reconstructions based on marine magnetic anomalies and paleomagnetic apparent polar wander paths (Besse and Courtillot, 2002). The preponderance of paleomagnetic evidence suggests that dipolar dominance was typical earlier as well (McElhinny, 2004), perhaps even during much of the Pre-Cambrian, but Pre-Cambrian evidence is sparse and equivocal (Dunlop and Yu, 2004). To second order, however, a small but significant departure from dipolar of the time-averaged field has long been noted (Wilson, 1971). Averaged over the past 5 My, the departure can be represented as an axial quadrupole contribution that is 3–5% of, and the same sign as, the axial dipole (Johnson and Constable, 1997; McElhinny, 2004). Comparison of paleomagnetic and plate-tectonic continental reconstructions suggests that a 3% axial quadrupole component may have been typical over the past 200 My (Besse and Courtillot, 2002).

The other salient feature of the field is that it reverses polarity. Geologically speaking, reversals occur very quickly. Even for the most recent reversal (Matuyama–Brunhes), which occurred 0.78 Ma, the dating error of the most favorable basalt flows is comparable to the transition duration (Singer *et al.*, 2005). For this reason transition duration must be obtained from sedimentary records, using the stratigraphic thicknesses of the transition zones and estimates of the deposition rate. From 30 selected records of the last four reversals, Clement (2004) found that the average time for the field to change direction from one polarity to the other was 7000 years (**Figure 1**). These reversals, so defined, tended to occur more quickly at low latitudes than at mid-to-high latitudes, with individual durations ranging from 2000 to 12 000 years. This regionally varying behavior implies that the nondipole field played a large role.

In detail, reversal records display a wide variety of field behavior (Coe and Glen, 2004). Almost all exhibit large intensity drops during transition, but their directional behavior varies greatly. The most detailed, high-deposition-rate lava-flow records and sedimentary records show that at least some reversals are complex, with episodes of oscillatory and rapid field change (**Figures 2** and **3**). Often the field reverses briefly, but relapses to intermediate directions one or more times before finally attaining stable opposite polarity. This behavior complicates estimation of duration. For instance, if one includes an early unsuccessful swing to normal polarity in the most recent reversal transition, its duration is about 18 000 years (**Figure 4**), three times longer than if one considers that swing to be an unrelated precursor (Singer *et al.*, 2005). Inclusion of the precursory swing in direction seems reasonable, in light of the complex transition paths of **Figures 2** and **3**. Even excluding the precursor, high-resolution sediment records show that this transition is complex, with five or six large swings of the field direction (**Figure 5**) (Channell and Lehman, 1997). Some dynamo simulations have also produced comparably complex and long reversal transitions – for example, the second reversal during

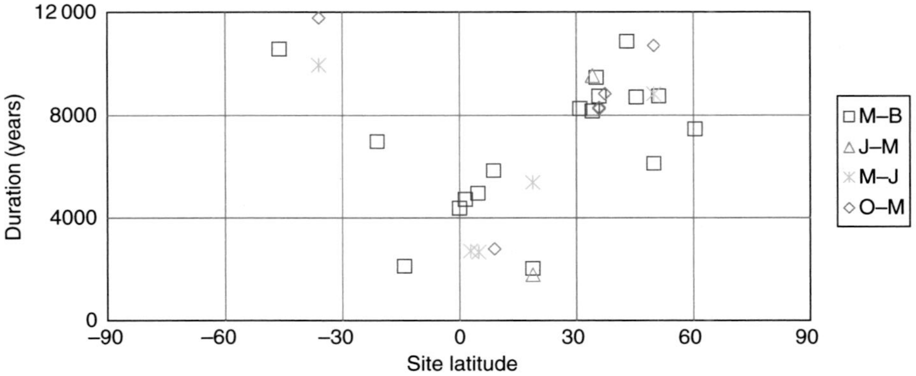

Figure 1 Duration of directional change from one stable polarity to the other during the last four reversals, as determined from transition-zone thickness and estimated deposition rate in sedimentary sections. Note that the duration so-defined increases with latitude. This method may underestimate the duration of complex reversals by designating early or late instability as unrelated pre- or post-transition excursions. MB, Matuyama–Brunhes; J–M, Jaramillo–Matuyama; M–J, Matuyama–Jaramillo; O–M, Olduvai–Matuyama. Adapted from Clement BM (2004) Dependence of the duration of geomagnetic polarity reversals on site latitude. *Nature* 428: 637–640.

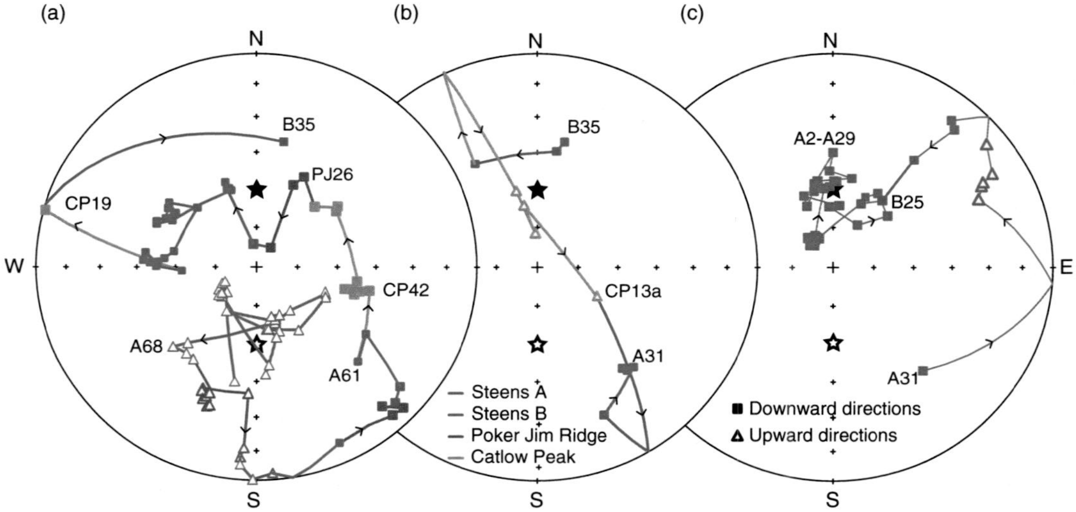

Figure 2 Composite directional record of the 16.6 Ma Steens Mountain reversal recorded in four sections of superposed lava flows in SE Oregon. Time progresses from left to right. Note the increased complexity of the transition path when new directions found at Steens B (Camps *et al.*, 1999), Poker Jim Ridge, and Catlow Peak (Jarboe *et al.*, 2005) are spliced into the original record of Mankinen *et al.* (1985) from Steens A and B sections. Open and closed stars are the reversed and normal axial dipole directions for that region.

the tomographic simulation of Glatzmaier *et al.* (1999) (see case 'h' of **Figure 14** and Coe and Glen (2004, figure 7 and plate 1)).

Nonetheless, various authors have discerned some statistical regularity in the directional behavior of reversals over the past 2–20 My: the tendency of transitional virtual geomagnetic poles (VGPs) to cluster in preferred longitudinal bands or patches on Australasia and the Americas and to avoid the Central Pacific Basin (Laj *et al.*, 1991; Clement, 1991; Hoffman, 1992; Hoffman and Singer, 2004; Love, 2000), though the statistical significance of this tendency has been disputed (Valet *et al.*, 1992; McFadden *et al.*, 1993; Prévot and Camps, 1993;

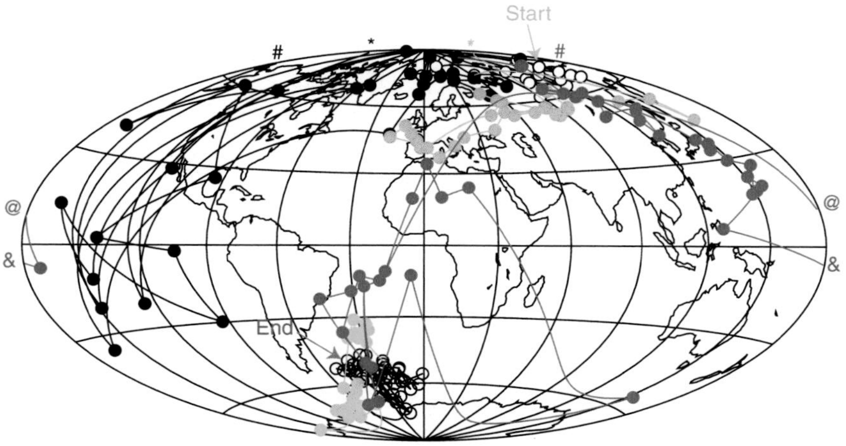

Figure 3 The Gauss–Matuyama (2.58 Ma) reversal record of VGPs recorded in sediments deposited in Searles Lake, California (Glen *et al.*, 1999b). Note the highly complex VGP path, with initial and final excursions in orange, multiple rapid oscillations in black, and main reversing phase including two large swings from high to equatorial latitudes in red.

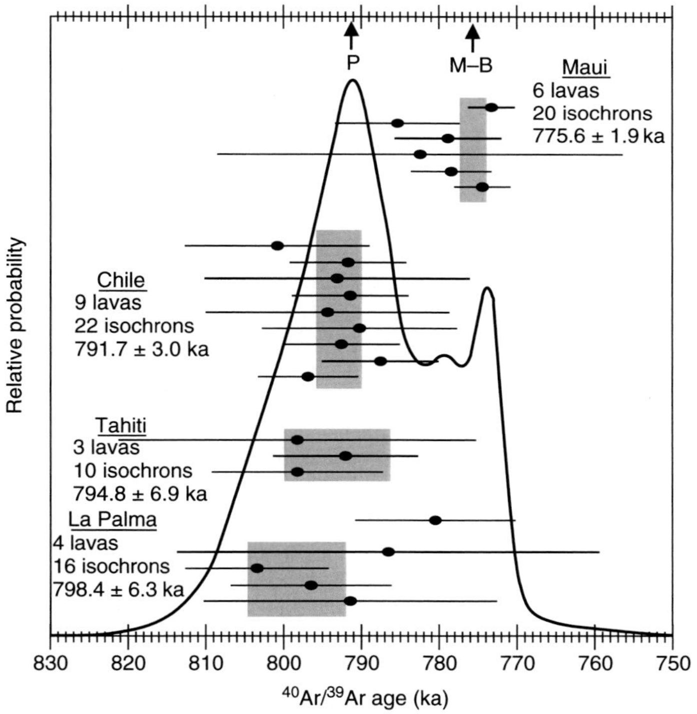

Figure 4 Ar/Ar ages of 23 transitionally magnetized lava flows from four widely spaced localities that record transitional field directions attributed to the most recent, Matuyama–Brunhes (M–B) reversal. Gray bands show the weighted mean age and 2σ uncertainty for each lava sequence. The ages of the lavas from Maui and the uppermost flow from La Palma correspond to the accepted age of the M–B reversal from sedimentary cores (Tauxe *et al.*, 1996), whereas the others correspond to what has been termed the M–B precursor (P) (Hartl and Tauxe, 1996). The probability density curve indicates that these lavas together span a minimum of 18 ky, a period about three times longer than the conventionally cited M–B duration. Adapted from Singer BS, Hoffman KA, Coe RS, *et al.* (2005) Structural and temporal requirements for geomagnetic field reversal deduced from lava flows. *Nature* 434: 633–636.

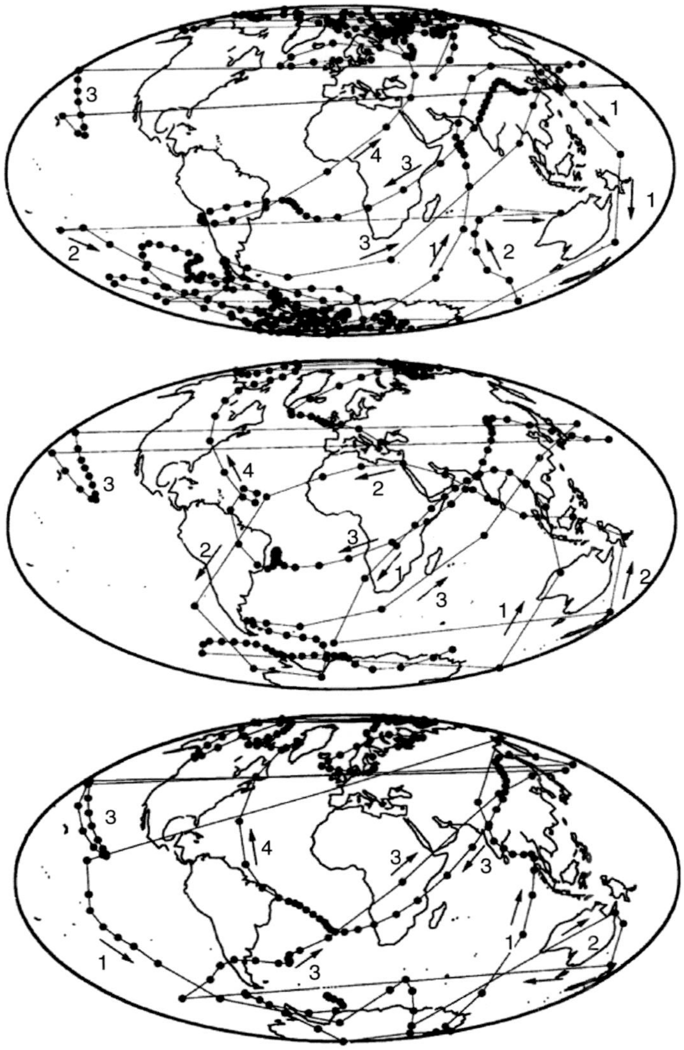

Figure 5 Three high-resolution records of VGPs spanning the most recent (Matuyama–Brunhes) reversal, from three separate cores of ODP site 984 drift sediments in the northern Atlantic. Note the complexity of the transition path, even though the precursor (cf. **Figure 4**) is not included. Adapted from Channell JET and Lehman B (1997) The last two geomagnetic polarity reversals recorded in high-deposition-rate sediment drifts. *Nature* 389: 712–715.

Merrill and McFadden, 1999). The long timescale suggests mantle influence, and the preferred areas do correlate with large-scale seismic tomography, overlying regions of higher than average P- and S-wave velocity (Laj *et al.*, 1991; Hoffman and Singer, 2004). Assuming that higher seismic velocity signifies regions with lower than average lower-mantle temperature, convective downwelling in the core could be localized there and concentrate poloidal flux lines. Transitional VGP preference would then be expected (Gubbins and Coe, 1993) and has in fact been produced by dynamo simulations employing appropriate heat-flux boundary conditions at the core–mantle boundary (CMB) (Coe *et al.*, 2000; Olson and Christensen, 2002; Kutzner and Christensen, 2004).

Earth's reversals occur aperiodically, in fact almost randomly, but the mean duration appears to change progressively over long time intervals (**Figure 6**). From 0 to 165 Ma the field reversed on average about 2 times per million years, but over 10 My intervals the average reversal rate varied from highs of at least 5 per million year to a low of 0. During the interval from 124 to 83 Ma, the

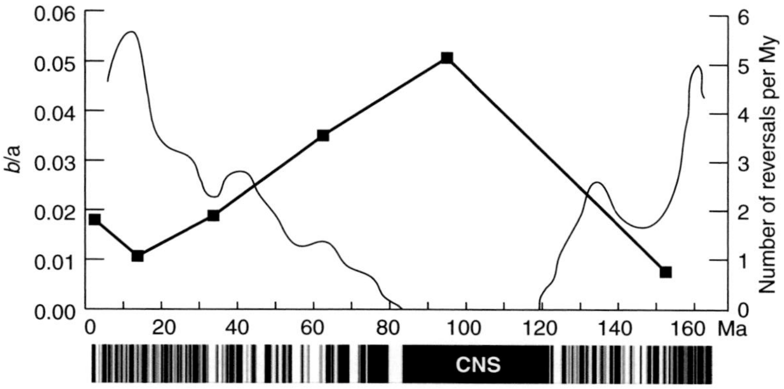

Figure 6 Smoothed reversal rate (thin wavy line) corresponding to normal and reversed chrons (black and white bars) of the geomagnetic polarity timescale. The rate goes to zero in the Cretaceous Normal Superchron, from 83 to 124 Ma. The squares denote *b/a*, a relative measure of the average ratio of the antisymmetric to symmetric parts of the geomagnetic field (excluding the axial dipole) as a function of time in the past, inferred from paleomagnetic secular variation using the results of McFadden *et al.* (1991). These demonstrate a strong inverse correlation between *b/a* and reversal rate. We note that recent studies (see Ogg (2004)) suggest even higher reversal rates from 155 to 165 Ma than shown here (~10 per My). Adapted from Coe RS and Glatzmaier GA (2006) Symmetry and stability of the geomagnetic field. *Geophysical Research Letters* 33: L21311 (doi:10.1029/2006GL027903).

so-called Cretaceous Normal Superchron (CNS), no true polarity reversals have been unequivocally demonstrated. From analysis of the latitudinal variation of paleomagnetic secular variation recorded by lava flows over the past 150 My, McFadden *et al.* (1991) proposed that average reversal rate is related to symmetry of the nonaxial dipole field: a higher degree of symmetry of the field about the equator, as expressed by its spherical harmonic gauss coefficients, correlates with more frequent reversals (**Figure 6**). For example, the equatorial dipolar and axial quadrupolar harmonics of the field are symmetric and the axial octupolar harmonic is antisymmetric (Merrill, 1996). Recent analysis of the Glatzmaier *et al.*'s (1999) dynamo simulations supports this idea: **Figure 7** shows that the most stable, nonreversing simulation (case 'e' of **Figure 14**) has far more energy associated with antisymmetric than with symmetric gauss coefficients (Coe and Glatzmaier, 2006). Furthermore, a

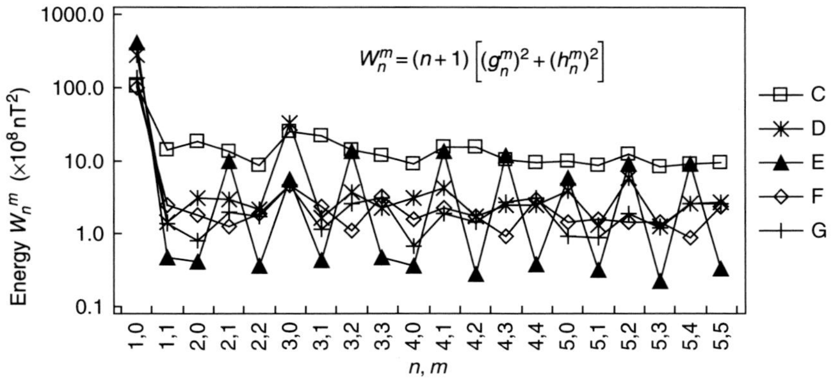

Figure 7 Time-averaged spatial energy density W_n^m at the CMB associated with each harmonic degree and order up to n, $m = 5,5$ for five of the simulations of Glatzmaier *et al.* (1999) that are shown in **Figure 14**. The terms associated with gauss coefficients g_n^m and h_n^m are antisymmetric about the equator when $(n+m)$ is odd and are symmetric about the equator when $(n+m)$ is even. Note that the most stable case, E, heavily favors antisymmetric energy terms (with $n+m$ odd), as does generally the second most stable case D. Adapted from Coe RS and Glatzmaier GA (2006) Symmetry and stability of the geomagnetic field. *Geophysical Research Letters* 33: L21311 (doi:10.1029/2006GL027903).

simulation with a solid inner core only one-quarter of its size today (Roberts and Glatzmaier, 2001) produced an even more antisymmetric nonaxial dipole, suggesting that reversals may have been much less common in the distant geologic past when Earth's inner core was smaller (*see* Chapter 2). Limited paleomagnetic evidence available from rocks older than 1 Ga appears to support this suggestion (Coe and Glatzmaier, 2006).

The CNS was clearly a time of exceptional field stability, though a few instances of short-lived reversed directions have been reported within it (Tarduno, 1990; Gilder *et al.*, 2003; Ogg *et al.*, 2004). These might represent aborted reversal attempts. Much more recently, during the past 0.78 My of today's normal polarity epoch when global simultaneity is easier to establish, there have been eight or more brief excursions of the field to low intensity and reversed or nearly reversed directions that have been detected around the globe (**Figure 8**) (Champion *et al.*, 1988; Guyodo and Valet, 1999; Lund *et al.*, 2006). They are generally distinguished from complete polarity reversals by their short duration, a few thousand to at most a few tens of thousands of years. In the paleomagnetic and seafloor magnetic anomaly records their classification is somewhat arbitrary (Acton *et al.*, 2006), but in dynamo simulations aborted reversals are those for which the field deep in the core maintains its original polarity throughout the time of intermediate and reversed directions at the surface of the Earth (**Figure 9**) (Glatzmaier *et al.*, 1999).

9.2 Models

Although the magnetic dipole reversal mechanism is still poorly understood, mathematically it is easy to see why there can be two oppositely directed magnetic basins of attraction. Given a solution to the set of equations that govern the thermodynamics and magnetohydrodynamics (MHD) for a convective dynamo (i.e., fluid flow, magnetic field, and thermodynamic variables), completely reversing the magnetic field everywhere would also satisfy the same set of equations; that is, reversing the sign of the magnetic field vector everywhere will not change the Lorentz force or Joule heating because the field is quadratic in these terms. It will also not affect the magnetic induction or conservation equations because the field, although linear in these two equations, appears in every term.

Reversals of the dipolar part of the magnetic field were originally studied using two-dimensional (2-D, axisymmetric) models of the mean magnetic fields. These models are kinematic, that is, the fluid flow is prescribed or parametrized instead of being part of the solution. Only the longitudinally averaged part of

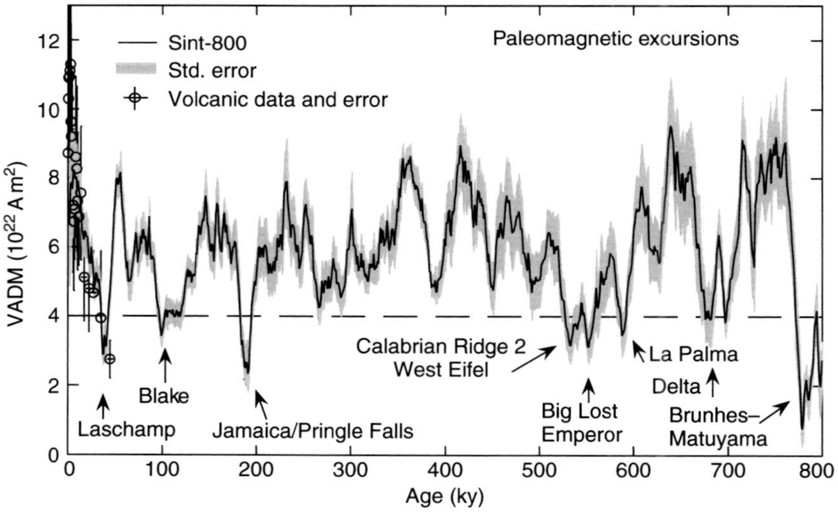

Figure 8 Excursions of the geomagnetic field since the last reversal (Brunhes–Matuyama), as indicated by deep minima in Earth's dipole moment (VADM). Results are from a global stack of records of relative paleointensity data derived from marine sedimentary cores, normalized to fit volcanic absolute paleointensity results. Adapted from Guyodo Y and Valet JP (1999) Global changes in intensity of the Earth's magnetic field during the past 800 kyr. *Nature* 399: 249–252.

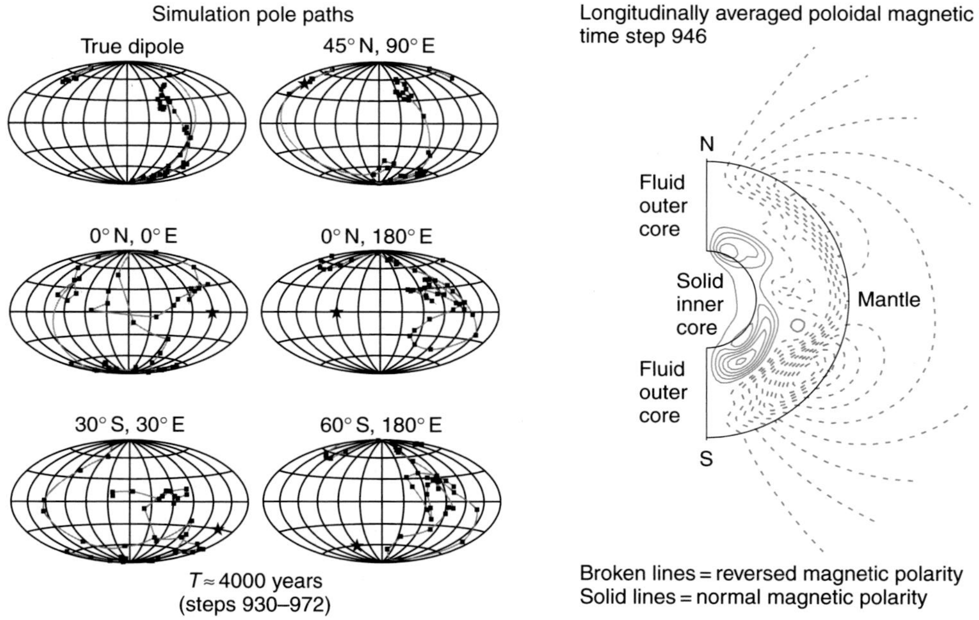

Figure 9 An aborted reversal during the case 'c' simulation of Glatzmaier *et al.* (1999) that occurs around 88 000 years in the record of **Figure 14(c)**. Left: True dipole path and VGP paths for five locations around the globe. Right: Longitudinally averaged poloidal flux in the outer core at the midpoint of the directional excursion shows that while the field reversed in most of the outer core and above, it retained the original normal polarity close to the inner core and within it. Note that each plotted Time Step represents about 100 years and 3500 numerical time steps.

the magnetic field is calculated (*see* Chapter 3). This method continues to be employed in the solar community to study the periodic reversals associated with the sunspot cycle (e.g., Bonanno *et al.*, 2006) and in the geophysics community to study the aperiodic paleomagnetic reversals (e.g., Giesecke *et al.*, 2005). Such models prescribe the two main ingredients for a self-sustaining dynamo: the 'alpha effect', which twists toroidal (longitudinally directed) magnetic field into poloidal (radially and latitudinally directed) field, and the 'omega effect', which shears poloidal field into toroidal field. The alpha effect can also act on poloidal field to generate (or destroy) toroidal field. The alpha prescription in these models parametrizes the effects of helical fluid flow and the omega prescription represents the effects of differential rotation (i.e., the variation of angular velocity with radius and latitude). Replacing these complicated, nonlinear, 3-D, time and spatially dependent processes with 2-D and usually time-independent functions greatly simplifies the problem. These studies formed a basis for understanding convective dynamos. The simplifications and assumptions built into these models greatly reduce the complexity and computational expense and therefore can provide good statistics because of the long times that can be simulated. Unfortunately with this approach one cannot reconstruct what the 3-D dynamics must be that maintains the prescribed alpha and omega effects. Discovering and understanding the details of the dynamo mechanisms requires a self-consistent solution of the full set of nonlinear 3-D equations that represent conservation of mass, momentum, energy, magnetic flux, and magnetic induction (*see* Chapters 5 and 8).

Early, self-consistent MHD models of the solar dynamo were developed by Gilman (1983) and Glatzmaier (1985). Instead of prescribing the alpha and omega effects parametrically, they solved a full set of MHD equations for the 3D time-dependent fluid flow, thermodynamics, and magnetic field in a rotating spherical shell that served as an analog of convection and magnetic field generation in the solar interior. The resulting solutions maintained a differential rotation at the surface, very similar to that observed on the Sun, and a magnetic field that continuously and periodically reversed as a 'dynamo wave' propagating in latitude, somewhat similar to the large-scale fields observed on the solar surface. However, the coarse spatial resolution, which was barely affordable at that time, forced the simulated fluid flows to be unrealistically laminar. Later it was found that the differential rotation below the surface predicted by these early models was not consistent with the profile inferred from helioseismolgy. Computers today provide much better spatial resolution, which allows the simulations to be at least weakly turbulent. This changes the style of the dynamics (from large convective cells to small convective plumes), the transport of angular momentum, and therefore the pattern of differential rotation, the generation of vorticity, and helicity, and ultimately the dynamo mechanism. However, the solar dynamo is still far from being well understood.

The original 3-D MHD geodynamo models were developed in the early 1990s. Surprisingly perhaps, the challenge then was not to produce reversals, but to stop them from occurring too frequently. These were strongly convective and rotationally dominant (high Rayleigh number, low Ekman number) dynamos that were quite unstable. In addition, the solid inner core (*see* Chapter 10) was originally approximated as an insulator (for numerical simplicity) because most people at that time felt the small inner core would have little effect on the dynamo in the outer fluid core. However, when Hollerbach and Jones (1993) showed that a finite-conducting inner core in their 2-D 'mean-field dynamo model' stabilizes their solution, Glatzmaier and Roberts (1995) made the inner core conducting in their 3-D MHD model and obtained a relatively stable, dipole-dominated dynamo (**Figure 10**) that, after a couple of magnetic diffusion times, produced an isolated dipole reversal (**Figure 11**). The large insulating inner core in their original (frequently reversing) model obstructed the flow; whereas their new conducting inner core provided an anchor for the field via magnetic torque, which resulted in a solution still time dependent but not continuously reversing. As mentioned above, tests with a much smaller conducting inner core (less flow obstruction) produced a stabler, much more antisymmetric magnetic field (Roberts and Glatzmaier, 2001; Coe and Glatzmaier, 2006), which suggests that the reversal frequency may have been very low in the distant past (*see* Chapter 2).

It has been more than a decade since that first spontaneous magnetic dipole reversal was found in a 3-D MHD self-consistent computer simulation of the geodynamo (Glatzmaier and Roberts, 1995). Since then several groups around the world have developed similar geodynamo models (*see* Chapter 8), some of which also produce magnetic reversals (Kida *et al.*, 1997; Kageyama *et al.*, 1999; Sarson and Jones, 1999; Kutzner and Christensen, 2002; Busse, 2002; Wicht, 2002; Takahashi *et al.*, 2005; Reshetnyak and Steffen, 2005). These simulations are less laminar than the early solar dynamo simulations. However, since no one yet can afford the computing resources to produce a geodynamo simulation close to the Earth's parameter regime, the simulations are still extremely crude approximations of the geodynamo. The various MHD geodynamo models employ different sets of parameters, boundary conditions, numerical methods, and spatial resolution (Glatzmaier, 2002); therefore, a detailed comparison is difficult. Yet many of the results have features quite similar to geomagnetic and paleomagnetic observations. Reviews have been written that qualitatively compare the results to the paleomagnetic reversal record (e.g., Dormy *et al.*, 2000; Kono and Roberts, 2002).

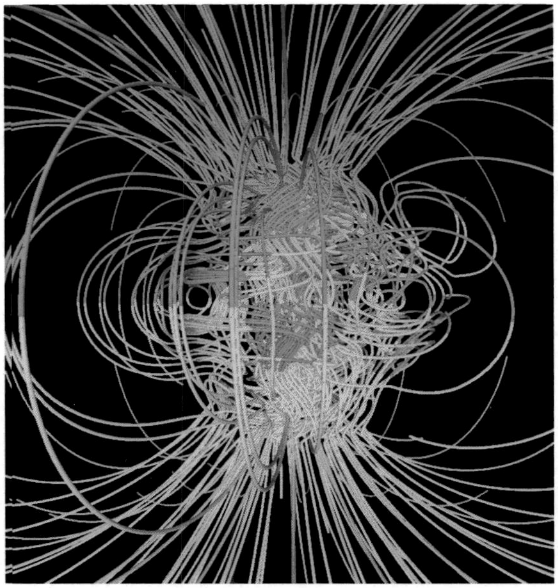

Figure 10 A snapshot of the simulated geomagnetic field produced by Glatzmaier and Roberts (1995). A set of magnetic lines of force illustrated the 3D structure of the field, which is intense and complicated inside the fluid core and smooth and dipole-dominated outside the core. The rotation axis of the model Earth is vertical in the illustration and yellow lines represent outward directed field and blue line represent inward directed field. The field is sheared around the 'tangent cylinder' to the inner-core equator.

Reversing MHD dynamo simulations fall loosely into three categories: (1) periodic dynamo waves, for which the field continuously reverses (Kida *et al.*, 1997; Busse, 2002), (2) reversals possibly triggered by large-scale plume events, fluctuations in the axisymmetric meridional circulation, or changes in kinetic and magnetic energies (Sarson and Jones,

Computer-simulated magnetic dipole reversal

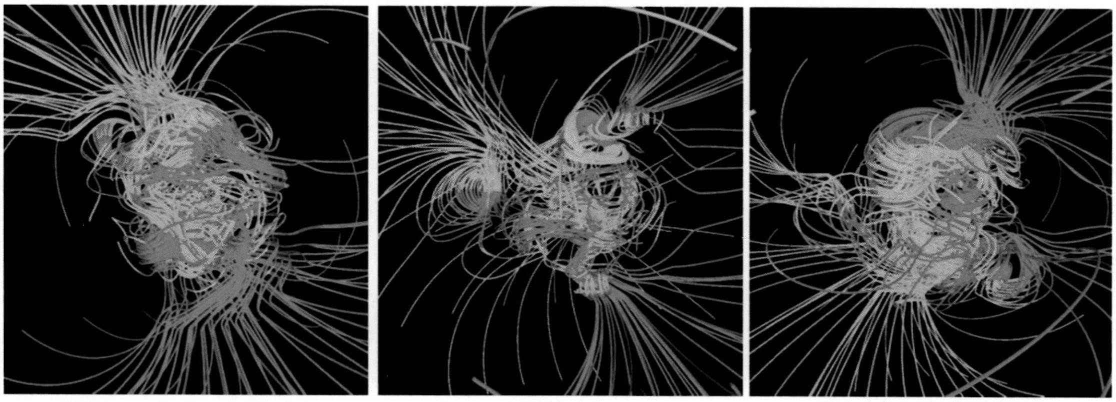

Figure 11 Three snapshots of a simulated magnetic field (as in **Figure 10**) at 500 years before the mid-point in the dipole reversal, at the mid-point and at 500 years after the mid-point.

1999; Kageyama *et al.*, 1999; Wicht and Olson, 2004; Reshetnyak and Steffen, 2005), and (3) those with relatively long epochs of varying durations between reversals of varying durations and with more frequent aborted reversals (Glatzmaier *et al.*, 1999; Kutzner and Christensen, 2002; Takahashi *et al.*, 2005). For the simulations of category (3) the reversals occur through the combined nonlinear action of many small-scale fluctuations.

The early solar dynamo models (Gilman, 1983; Glatzmaier, 1985) simulated the periodic and continuously occurring solar reversals as an alpha–omega dynamo wave. The more recent simulated reversals of Kida *et al.* (1997) and Busse (2002) also appear to be global-scale dynamo waves driven by laminar convection. However, as discussed above, paleomagnetic reversals do not continuously occur, that is, the epochs between reversals are long (~10 magnetic dipole diffusion times) compared with the typical duration of a reversal (~0.1–0.5 magnetic dipole diffusion time). (A geomagnetic dipole diffusion time is 20 000 years.) They are also aperiodic, that is, the lengths of these epochs are highly irregular (**Figure 6**). In addition, aborted reversals (or excursions) are much more frequent than full reversals (**Figure 8**). Single dipolar reversals among several aborted reversals have been simulated by Glatzmaier and Roberts (1995), Kageyama *et al.* (1999), and Sarson and Jones (1999).

It appears that the progression through the three categories of simulations is correlated with the vigor of the convection (i.e., the Rayleigh and Reynolds numbers) and the relative effect of rotation (i.e., the Ekman and Rossby numbers) (*see* Chapters 5, 3, 8, and 11). Assuming the rotation rate, electrical conductivity, heat flux, and dimensions of the Earth's core, the smaller the prescribed thermal and viscous diffusivities the more turbulent the flow and the closer these nondimensional numbers approach Earth-like values and the more Earth-like the reversals appear. However, the character and frequency of the reversals also likely depend on several other model specifications: the ratio of buoyancy and Coriolis forces, the pattern of the heat flux over the CMB, and the relative size of the solid inner core.

Wicht and Olson (2004), for example, studied a weak-field, slowly rotating, large-scale convective dynamo, which produced a series of quite regular (periodic) reversals with periods long relative to the duration of a reversal but short relative to a magnetic diffusion time (**Figure 12**). These results are more Earth-like than the continuously reversing dynamo-wave solutions because of their relatively long, stable epochs between the reversals; however, the process is still periodic, that is, the durations of the epochs do not vary randomly as the Earth's do (**Figure 6**). Reversed-polarity magnetic flux is generated when convective plumes twist the field; this reversed flux is then advected throughout the outer core by the dominant meridional circulation. However, Lorentz forces have very little effect on the flow in their simulations; that is, the solutions are nearly kinematic. The main advantage of this approach is the ease of analysis; that is, the reversal mechanism is large scale and the reversal frequency is high and constant. However, the reversal mechanism may not be very Earth-like.

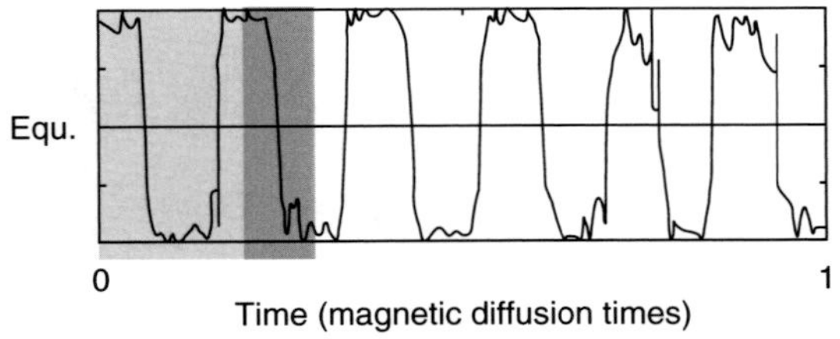

Figure 12 The magnetic pole latitude (at the surface) vs time (in magnetic diffusion times, here estimated as 80 000 years). Adapted from Wicht J and Olson P (2004) A detailed study of the polarity reversal mechanism in a numerical dynamo model. *Geochemistry, Geophysics, and Geosystems* 5, doi:10.1029/2003GC000602.

Kutzner and Christensen (2002) studied more strongly convective and more rotationally dominant cases. They found that convective dynamos with small Rayleigh numbers (i.e., small convective driving for a given rotation rate) tend to be stable and have a dominant axial dipole; large Rayleigh numbers produce frequently reversing dynamos and less dipolar fields. The cases that fall within a narrow region of parameter space between these two regimes have longer and more irregular times between reversals and therefore appear more Earth-like (**Figure 13**).

Glatzmaier *et al.* (1999) also chose strongly convective, rotationally dominant, cases to test the sensitivity of convective dynamos to the pattern of the heat-flux boundary condition on the CMB (**Figure 14**), presumably imposed by mantle convection. Their model

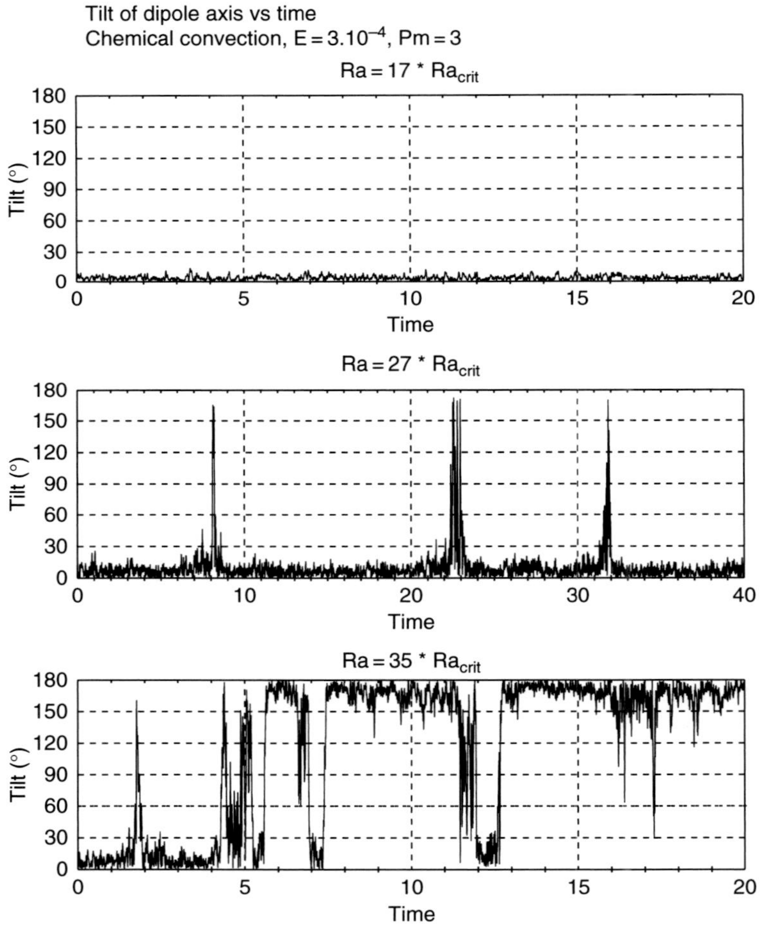

Figure 13 Three time series of a simulated field. The top row is for a low Rayleigh number, which is stable. The middle row is for a higher Rayleigh number, which has aborted reversals. The bottom row is for a still higher Rayleigh number, which shows dipole reversals and aborted reversals. Each row shows the dipole colatitude at the surface vs time (in magnetic diffusion times, here estimated as 80 000 years). Adapted from Kutzner C and Christensen U (2002) From stable dipolar to reversing numerical dynamos. *Physics of the Earth and Planetary Interiors* 121: 29–45.

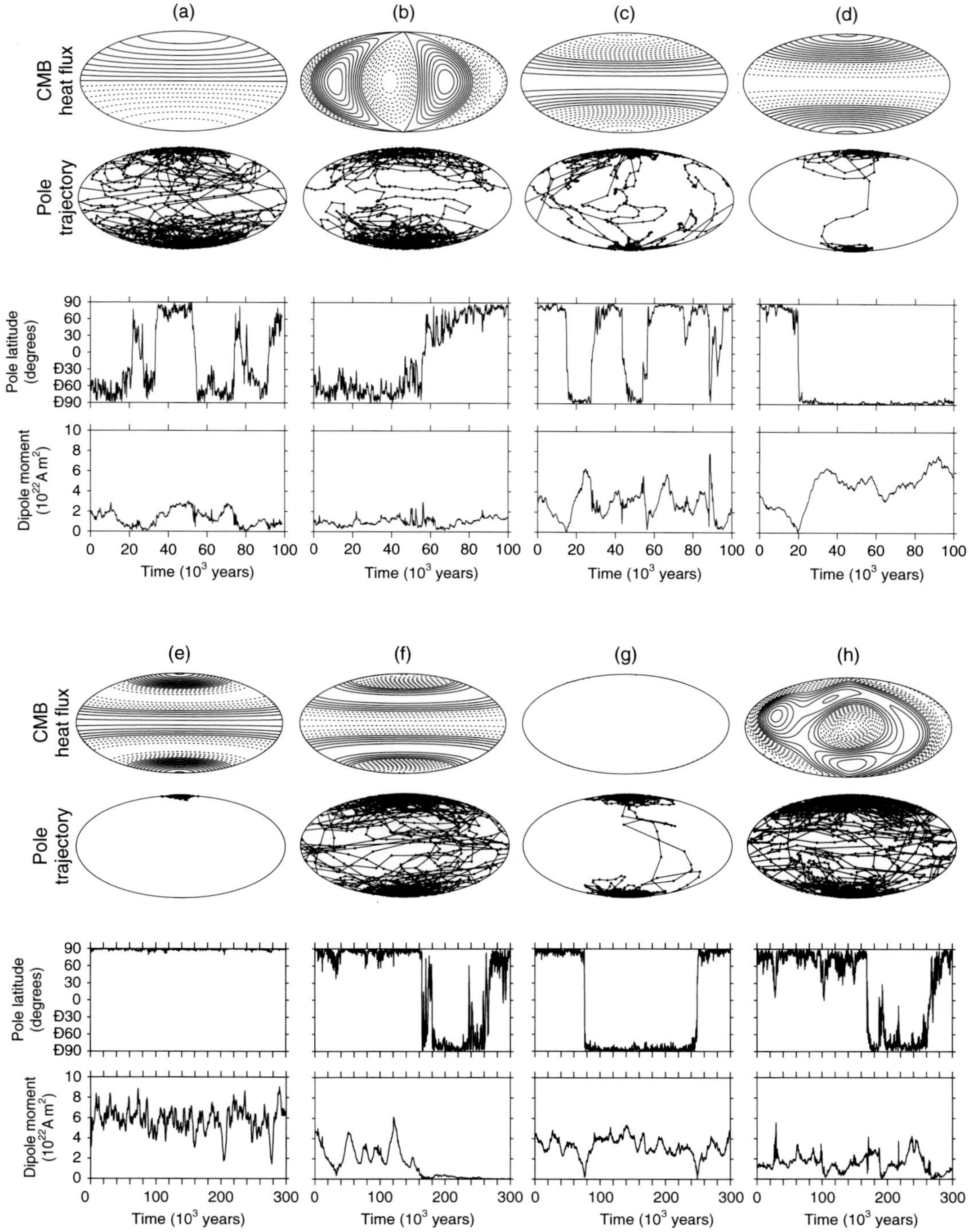

Figure 14 Eight dynamo simulations with different imposed patterns of radial heat flux at the CMB. The top row shows the patterns of CMB heat flux. Solid contours represent greater heat flux out of the core relative to the mean; broken contours represent less. Case 'g' has a uniform CMB heat flux and case 'h' has a pattern based on seismic tomography, assuming lower sound speed corresponds to warmer mantle and therefore smaller heat flux out of the core. The second row shows the trajectory of the south magnetic pole of the dipole part of the field outside the core, spanning the times indicated in the plots below; the marker dots are about 100 years apart. The plots in the third and fourth rows show the south magnetic pole latitude and the magnitude of the dipole moment (in units of 10^{22} A m^2) vs time (in units of 1000 years). Reproduced from Glatzmaier GA, Coe RS, Hongre L, and Roberts PH (1999) The role of the Earth's mantle in controlling the frequency of geomagnetic reversals. *Nature* 401: 885–890 with permission from Nature.

differs from all other geodynamo models by solving the equations of motion within the anelastic approximation instead of the Boussinesq approximation (*see* Chapters 5, 3, and 8). That is, in their model the variation of density with depth is taken into account and both compositional and thermal buoyancy are computed. In addition, more self-consistent thermal and compositional boundary conditions are applied at the inner-core boundary. They found that forcing greater heat flux through the CMB in the polar and equatorial regions, opposed to mid-latitude, produces a strong, stable, axial dipole, whereas the opposite CMB heat flux pattern produces frequent reversals (*see* Chapter 12). Of the eight cases they tested, their Earth-like tomographic heat flux case (h) and their homogeneous heat flux case (g) appear most Earth-like in terms of the long and irregular times between reversals and the relatively short and highly variable reversal durations. In addition, like the Earth, case 'h' has more frequent aborted reversals than successful reversals. The system seems to be continually trying to reverse via MHD instabilities but only after many attempts are the conditions favorable for the new polarity to continue to grow and the old polarity to be fully destroyed (**Figure 15**). This continuing process of trying to reverse occurs in these simulations at roughly the frequency at which aborted reversals occur in the Earth, which is similar to the frequency that regular full reversals occur in the Wicht and Olson (2004) simulations (**Figure 12**). However, unlike the reversals seen by Wicht and Olson (2004), many of these simulated reversals differ in terms of morphology and duration.

As computational resources continue to improve, simulations using much smaller (i.e., more realistic) diffusivities become possible. Takahashi *et al.* (2005) produced some of the most realistic geodynamo simulations to date by using much larger Rayleigh numbers and smaller Ekman numbers, that is, smaller, more realistic viscous, thermal, and magnetic diffusivities. Several dipole reversals occurred with a highly variable reversal frequency (**Figure 16**). They also show that the internal dynamics of the fluid core changes significantly when the viscous forces become much smaller than the Coriolis (rotational) and Lorentz (magnetic) forces, which underscores the importance of reaching the very low-viscosity (turbulent) regime if one wishes to simulate and understand the dynamo mechanism in the Earth's core.

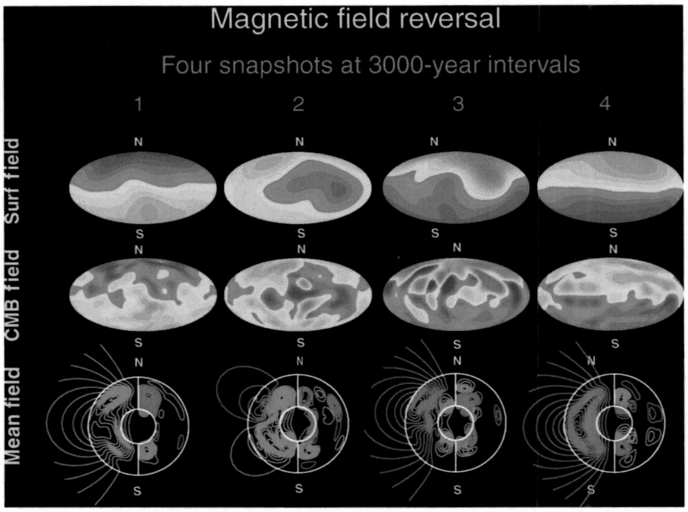

Figure 15 A sequence of snapshots of the longitudinally averaged magnetic field through the interior of the core and of the radial component of the field at the CMB and at what would be the surface of the Earth, displayed at roughly 3000-year intervals, spanning the first dipole reversal of case 'h' in **Figure 14**. In the plots of the average field, the small circle represents the inner-core boundary and the large circle is the CMB. The poloidal field is shown as magnetic field lines on the left-hand sides of these plots (blue is clockwise and red is counter-clockwise). The toroidal field direction and intensity are represented as contours (not magnetic field lines) on the right-hand sides (red is eastward and blue is westward). Hammer (equal area) projections of the entire CMB and surface are used to display the radial field (with the two different surfaces displayed as the same size). Reds represent outward-directed field and blues inward field. The surface field, which is typically an order of magnitude weaker, was multiplied by 10 to enhance the color contrast. Adapted from Glatzmaier GA, Coe RS, Hongre L, and Roberts PH (1999) The role of the Earth's mantle in controlling the frequency of geomagnetic reversals. *Nature* 401: 885–890.

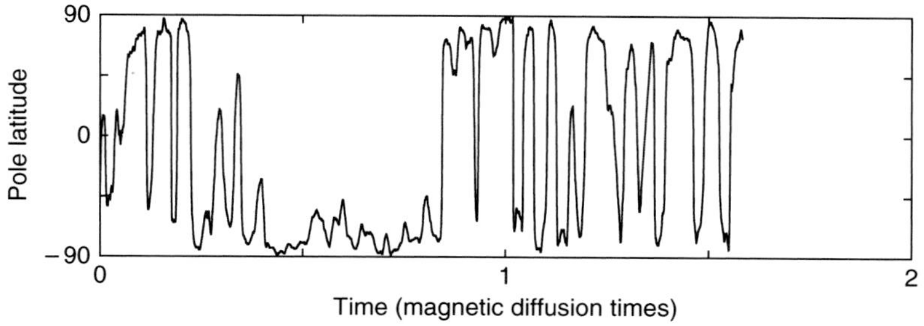

Figure 16 The magnetic pole latitude (at the surface) vs time (in magnetic diffusion times, here estimated as 200 000 years). Adapted from Takahashi F, Matsushima M, and Honkura Y (2005) Simulations of a Quasi–Taylor state geomagnetic field including polarity reversals on the Earth Simulator. *Science* 309: 459–461.

9.3 Conclusions

Our understanding of geomagnetic reversals has improved considerably over the years with paleomagnetic studies and geodynamo simulations. Paleomagnetic observations now provide considerable constraints on the time-averaged field and the character of reversals, some of which have been matched to first-order in some dynamo simulations. Nonetheless, more 'ground truthing' observations and more realistic simulations are needed to discover the details of the reversal mechanism in the Earth's core and what influences its range of variation.

In terms of observations, longer, more reliable, and more detailed records of paleomagnetic field behavior are needed, especially early in Earth's history. This of course will take considerable time and effort.

In terms of models, geodynamo simulations need to be more turbulent and rotationally dominant. Laminar flows are certainly easier to produce and analyze; but it is unlikely that the large-scale convection cells seen in such studies produce a reversal mechanism representative of that in the Earth's turbulent outer core. The same conclusion has been reached in the solar dynamo community. In addition, much longer simulations are needed to gather better statistics on the types and frequencies of reversals. Finally, the sensitivities to several model specifications, like density stratification, boundary conditions, compositional buoyancy, inner-core conductivity, and diffusion coefficients, need to be tested. These modeling goals all require significant amounts of computing resources and, as in the past, compromises will need to be made as we wait for computer hardware to improve. Support for this research was provided by NSF CSEDI program.

References

Acton G, Guyodo Y, and Brachfeld S (2006) The nature of a cryptochron from a paleomagnetic study of chron C4r.2r recorded in sediments off the Antarctic Peninsula. *Physics of the Earth and Planetary Interiors* 156: 213–222.

Besse J and Courtillot V (2002) Apparent and true polar wander and the geometry of the geomagnetic field over the last 200 Myr. *Journal of Geophysical Research* 107(6–1): 6–31.

Bonanno A, Elstner D, and Belvedere G (2006) Advection-dominated solar dynamo model with two-cell meridional flow a positive alpha-effect in the tachocline. *Astronomische Nachrichten* 327: 680–685.

Busse FH (2002) Convective flows in rapidly rotating spheres and their dynamo action. *Physics of Fluids* 14: 1301–1314.

Camps P, Coe RS, and Prevot M (1999) Transitional geomagnetic impulse hypothesis: Geomagnetic fact or rock-magnetic artifact? *Journal of Geophysical Research* 104: 17747–17758.

Champion DE, Lanphere MA, and Kuntz MA (1988) Evidence for a new geomagnetic reversal from lava flows in Idaho: Discussion of short polarity reversals in the Brunhes and late Matuyama polarity chrons. *Journal of Geophysical Research* 93: 11667–11680.

Channell JET and Lehman B (1997) The last two geomagnetic polarity reversals recorded in high-deposition-rate sediment drifts. *Nature* 389: 712–715.

Clement BM (1991) Geographical distribution of transitional VGPs – Evidence for non-zonal equatorial symmetry during the Matuyama–Brunhes geomagnetic reversal. *Earth and Planetary Science Letters* 104: 48–58.

Clement BM (2004) Dependence of the duration of geomagnetic polarity reversals on site latitude. *Nature* 428: 637–640.

Coe RS and Glatzmaier GA (2006) Symmetry and stability of the geomagnetic field. *Geophysical Research Letters* 33: L21311 (doi:10.1029/2006GL027903).

Coe RS and Glen JMG (2004) The complexity of reversals. In: Channell JET, Kent DV, Lowrie W, and Meert JG (eds.) *Geophysical Monograph Series, Vol. 145: Timescales of the Internal Geomagnetic Field*, pp. 221–232. Washington, DC: AGU.

Coe RS, Hongre L, and Glatzmaier GA (2000) An examination of simulated geomagnetic reversals from a palaeomagnetic perspective. *Philosophical Transactions of the Royal Society of London, Series A* 358: 1141–1170.

Dormy E, Valet J-P, and Courtillot V (2000) Numerical models of the geodynamo and observational constraints. *Geochemistry, Geophysical, and Geosystems* 1, doi:2000GC000062.

Dunlop DJ and Yu Y (2004) Intensity and polarity of the geomagnetic field during Precambrian time. In: Channell JET, Kent DV, Lowrie W, and Meert JG (eds.) *Geophysical Monograph Series, Vol. 145: Timescales of the Internal Geomagnetic Field*, pp. 85–100. Washington, DC: AGU.

Giesecke A, Rudiger G, and Elstner D (2005) Oscillating alpha-squared-dynamos and the reversal phenomenon of the global geodynamo. *Astronnomische Nachrichten* 326: 693–700.

Gilder S, Chen Y, Cogne JP, Tan XD, Courtillot V, Sun DJ, and Li YG (2003) Paleomagnetism of Upper Jurassic to Lower Cretaceous volcanic and sedimentary rocks from the western Tarim Basin and implications for inclination shallowing and absolute dating of the M-0 (ISEA?) chron. *Earth and Planetary Science Letters* 206: 587–600.

Gilman PA (1983) Dynamically consistent nonlinear dynamos driven by convection in a rotating spherical shell. II: Dynamos with cycle and strong feedbacks. *Astrophysical Journal Supplement* Series 53: 243–268.

Glatzmaier GA (1985) Numerical simulations of stellar convective dynamos. II: Field propagation in the convection zone. *Astrophysical Journal* 291: 300–307.

Glatzmaier GA (2002) Geodynamo simulations – How realistic are they? *Annual Review of Earth and Planetary Sciences* 30: 237–257.

Glatzmaier GA, Coe RS, Hongre L, and Roberts PH (1999) The role of the Earth's mantle in controlling the frequency of geomagnetic reversals. *Nature* 401: 885–890.

Glatzmaier GA and Roberts PH (1995) A three-dimensional self-consistent computer simulation of a geomagnetic field reversal. *Nature* 377: 203–209.

Glen JMG, Coe RS, and Liddicoat JC (1999) A detailed record of paleomagnetic field change from Searles Lake, California 2. The Gauss Matuyama polarity reversal. *Journal of Geophysical Research* 104: 12883–12894.

Gubbins D and Coe RS (1993) Longitudinally confined geomagnetic reversal paths from non-dipolar transition fields. *Nature* 362: 51–53.

Guyodo Y and Valet JP (1999) Global changes in intensity of the Earth's magnetic field during the past 800 kyr. *Nature* 399: 249–252.

Hartl P and Tauxe L (1996) A Precursor to the Matuyama/Brunhes transition-field instability as recorded in pelagic sediments. *Earth and Planetary Science Letters* 138: 121–135.

Hoffman KA (1992) Dipolar reversal states of the geomagnetic field and core mantle dynamics. *Nature* 359: 789–794.

Hoffman KA and Singer BS (2004) Regionally recurrent paleomagnetic transitional fields and mantle processes. In: Channell JET, Kent DV, Lowrie W, and Meert JG (eds.) *Geophysical Monograph Series, Vol. 145: Timescales of the Internal Geomagnetic Field*, pp. 233–243. Washington, DC: AGU.

Hollerbach R and Jones CA (1993) Influence of the Earth's core on geomagnetic fluctuations and reversals. *Nature* 365: 541–543.

Jarboe NA, Coe RS, Glen JM, and Renne PR (2005) Compilation of a composite geomagnetic polarity reversal path recorded in basalts erupted during initial Yellowstone hotspot volcanism. *EOS, Transaction of the American Geophysical Union, Fall Meeting Supplement* 85(52): GP21A–0016.

Johnson CL and Constable CG (1997) The time-averaged geomagnetic field: Global and regional biases for 0–5 Ma. *Geophysical Journal International* 131: 643–666.

Kageyama A, Ochi MM, and Sato T (1999) Flip–flop transitions of the magnetic intensity and polarity reversals in the magnetohydrodynamic dynamo. *Physical Review Letters* 82: 5409–5412.

Kida S, Araki K, and Kitauchi H (1997) Periodic reversals of magnetic field generated by thermal convection in a rotating spherical shell. *Journal of the Physical Society of Japan* 66: 2194–2201.

Kono M and Roberts PH (2002) Recent geodynamo simulations and observation of he geomagnetic field. *Reviews of Geophysics* 40: 1013 (doi:10.1029/2000RG000102).

Kutzner C and Christensen U (2002) From stable dipolar to reversing numerical dynamos. *Physics of the Earth and Planetary Interiors* 121: 29–45.

Kutzner C and Christensen UR (2004) Simulated geomagnetic reversals and preferred virtual geomagnetic pole paths. *Geophysical Journal International* 157: 1105–1118.

Laj C, Mazaud A, Weeks R, Fuller M, and Herrero-Bervera E (1991) Geomagnetic reversal paths. *Nature* 351: 447.

Love JJ (2000) Statistical assessment of preferred transitional VGP longitudes based on palaeomagnetic lava data. *Geophysical Journal International* 140: 211–221.

Lund S, Stoner JS, Channell JET, and Acton G (2006) A summary of Brunhes paleomagnetic field variability recorded in ocean drilling program cores. *Physics of the Earth and Planetary Interiors* 156: 194–204.

Mankinen EA, Prévot M, Grommé CS, and Coe RS (1985) The Steens Mountain (Oregon) geomagnetic polarity transition.1: Directional history, duration of episodes, and rock magnetism. *Journal of Geophysical Research* 90: 393–416.

McElhinny M (2004) Geocentric axial dipole hypothesis: A least squares perspective. In: Channell JET, Kent DV, Lowrie W, and Meert JG (eds.) *Geophysical Monograph Series, Vol. 145: Timescales of the Internal Geomagnetic Field*, pp. 1–12. Washington, DC: AGU.

McFadden PL, Barton CE, and Merrill RT (1993) Do virtual geomagnetic poles follow preferred paths during geomagnetic reversals? *Nature* 361: 342–344.

McFadden PL, Merrill RT, McElhinny MW, and LeeS (1991) Reversals of the Earth's magnetic field and temporal variations of the dynamo families. *Journal of Geophysical Research* 96: 3923–3933.

Merrill RT (1996) *The Magnetic Field of the Earth*, 531pp. San Diego, CA: Academic Press.

Merrill RT and McFadden PL (1999) Geomagnetic polarity transitions. *Reviews of Geophysics* 37: 201–226.

Ogg JG (2004) The Jurassic Period. In: Gradstein FM, Ogg JG, and Smith AG (eds.) *A Geologic Time Scale*, pp. 307–343. Cambridge: Cambridge University Press.

Ogg JG, Agterberg FP, and Gradstein FM (2004) The Cretaceous Period. In: Gradstein FM, Ogg JG, and Smith AG (eds.) *A Geologic Time Scale*, pp. 344–383. Cambridge: Cambridge University Press.

Olson P and Christensen UR (2002) The time-averaged magnetic field in numerical dynamos with nonuniform boundary heat flow. *Geophysical Journal International* 151: 809–823.

Prévot M and Camps P (1993) Absence of preferred longitude sectors for poles from volcanic records of geomagnetic reversals. *Nature* 366: 53–57.

Reshetnyak M and Steffen B (2005) A dynamo model in a spherical shell. *Numerical Methods and Programming* 9: 27–34(http://num-meth.srcc.msu.su).

Roberts PH and Glatzmaier GA (2001) The geodynamo, past, present and future. *Geophysical and Astrophysical Fluid dynamics* 94: 47–84.

Sarson GR and Jones CA (1999) A convection driven geodynamo reversal model. *Physics of the Earth and Planetary Interiors* 111: 3–20.

Singer BS, Hoffman KA, Coe RS, *et al.* (2005) Structural and temporal requirements for geomagnetic field reversal deduced from lava flows. *Nature* 434: 633–636.

Takahashi F, Matsushima M, and Honkura Y (2005) Simulations of a Quasi–Taylor state geomagnetic field including polarity reversals on the Earth Simulator. *Science* 309: 459–461.

Tarduno JA (1990) A brief reversed polarity interval during the cretaceous normal polarity superchron. *Geology* 18: 683–686.

Tauxe L, Herbert T, Shackleton NJ, and Kokk YS (1996) Astronomical calibration of the Matuyama–Brunhes boundary: Consequences for magnetic remanence acquisition in marine carbonates and the Asian loess sequences. *Earth and Planetary Science Letters* 140: 133–146.

Valet JP, Tucholka P, Courtillot V, and Meynadier L (1992) Palaeomagnetic constraints on the geometry of the geomagnetic field during reversals. *Nature* 356: 400–407.

Wicht J (2002) Inner-core conductivity in numerical dynamo simulations. *Physics of the Earth and Planetary Interiors* 132: 281–302.

Wicht J and Olson P (2004) A detailed study of the polarity reversal mechanism in a numerical dynamo model. *Geochemistry, Geophysics, and Geosystems* 5, doi:10.1029/2003GC000602.

Wilson RL (1971) Dipole offset: The time-averaged palaeomagnetic field over the past 25 million years. *Geophysical Journal Royal Astronomical Society* 22: 491–504.

10 Inner-Core Dynamics

I. Sumita, Kanazawa University, Kanazawa, Japan

M. I. Bergman, Simon's Rock College, Great Barrington, MA, USA

10.1	**Introduction**	299
10.2	**Core Composition, Phase Diagram, and Crystal Structure of Iron**	300
10.2.1	Composition of the Core	300
10.2.2	Phase Diagram of Iron Alloy	301
10.2.3	Crystal Structure of Iron	303
10.3	**Solidification of the Inner Core**	303
10.4	**Grain Size and Rheology in the Inner Core**	305
10.5	**Origin of the Inner-Core Elastic Anisotropy**	306
10.5.1	An Overview	306
10.5.2	Anisotropy of h.c.p. Iron	307
10.5.3	Dynamical Models	308
10.5.4	Solidification Texturing Models	309
10.6	**Origin of Other Inner-Core Seismic Structures**	310
10.6.1	Properties and Structure of the Inner-Core Boundary	310
10.6.2	Inner-Core Attenuation and Scattering	311
10.6.3	Hemispherical Variation of Seismic Velocity, Anisotropy, and Attenuation	312
10.6.4	The Deep Inner Core and the Inner-Core Transition Zone	312
10.7	**Inner-Core Rotation**	313
10.7.1	An Overview	313
10.7.2	Electromagnetic Coupling	313
10.7.3	Gravitational Coupling	313
10.7.4	Combined Coupling	314
10.8	**Summary**	314
References		315

10.1 Introduction

With a radius of 1200 km the Earth's inner core is only 30% smaller than its moon. Being the solid, central core of a planet, its dynamics are fascinating and unique. The inner-core boundary is a solid–liquid phase boundary, where the geotherm crosses the liquidus temperature of the iron alloy that comprises the Earth's core. The solidity of the inner core is a consequence of the melting temperature gradient being steeper than the geothermal gradient, as a result of the effects of pressure on the liquidus temperature. In smaller planets and satellites the entire core may be solid. As the Earth cools and solidification proceeds, the radius of the inner-core boundary increases with time. There is considerable uncertainty in the parameters required for thermal history calculations, but it is likely that the inner core is between 1 and 3 billion years old. Many of the dynamical processes that take place in the inner core are likely to be related to its growth, which in turn is strongly controlled by the pattern of heat transfer in the outer core, or in other words, the 'climate' of the outer core. Although the inner core, outer core, and mantle are all dynamically coupled, both thermally and mechanically, the effects of coupling are particularly important for the inner core because of its small volumetric size. This implies that the inner core can act as a recorder of the geodynamical processes occurring not only in the outer core, but also in the mantle.

The study of inner-core dynamics has been motivated by a wide range of observations and inferences, many coming from seismology. Geodynamics has attempted to explain the origin of these inner-core properties and attributes, such as elastic anisotropy and anomalously high attenuation. Another set of observations that has given us insight into the inner core comes from geomagnetism. For example, inner-core solidification has been proposed to be an important candidate for the energy source to drive the geodynamo, and it may affect long-term variations of the strength of the geomagnetic dipole. In addition, motion of the inner core is thought to affect geodetic observations of polar motion.

In this chapter a review of current understanding of the dynamical processes that operate in the inner core is presented. First, a review of current knowledge of the composition of the inner core and its phase relations is presented. These are important for estimating the compositional buoyancy that helps drive convection in the outer core (*see* Chapters 2 and 5), and for understanding the process through which the inner core is solidifying. Next, a discussion on the grain size and stress in the inner core, which are important in controlling the mechanism and timescale of deformation and hence the strain rate and viscosity, is presented. This is followed by a critical examination of models proposed for the origin of inner core elastic anisotropy, one of the most intriguing features of the inner core. These models require knowledge of the stable phase of iron under inner-core conditions and the single-crystal elasticity of the hexagonal close-packed phase of iron, which remain uncertain. Also reviewed here is the orgin of other properties inferred of the inner core, such as anomalously high attenuation and lateral heterogeneity. Finally, the mechanisms of inner-core superrotation, another consequence of the dynamical coupling between the inner core and the outer core, are discussed.

10.2 Core Composition, Phase Diagram, and Crystal Structure of Iron

This section examines the composition and phase diagram of the core, in particular, how they impact the solidification and deformation of the Earth's inner core. We also examine briefly the crystal structure of iron under inner-core conditions. See Jeanloz (1990) and Poirier (1994) for detailed reviews on the composition of the core, and Boehler (1996a, 1996b) for a review of the pressure–temperature phase diagram of iron (evidence for the existence of the various solid phases and the variation of the melting temperature with pressure).

10.2.1 Composition of the Core

The composition of the core remains "an uncertain mixture of all the elements" (Birch, 1952). The average density of the Earth, the solar abundance of the elements, and the presence of iron meteorites, thought to be remnants of planetoid cores, all point to the core containing elemental iron (Birch, 1952). Comparison with iron meteorites (Brown and Patterson, 1948; Buchwald, 1975) suggests that the core also contains about 8% nickel by mass. Because nickel has a density similar to that of iron (Stacey, 1992), its presence in the core is difficult to detect seismically, and furthermore, little theoretical or experimental work on the partitioning of nickel has been carried out.

More studies have been aimed at the composition of the less-dense alloying components in the core, and their partitioning upon solidification. Early static compression experiments extrapolated to core conditions (Bridgman, 1949; Birch, 1952) and shock compression experiments (McQueen and Marsh, 1966) showed that iron–nickel alloys under core pressures are about 6–10% more dense than the outer core (Bullen, 1949), implying that less-dense alloying components must be present. Recent calculations using first principles and molecular dynamics (Laio *et al.*, 2000) confirm that liquid iron at 330 GPa and 5400 K is about 6% more dense than the seismic Earth model PREM in the core (Dziewonski and Anderson, 1981), and calculations based on lattice potential theory lean toward a 10% deficit (Shanker *et al.*, 2004.

Iron meteorites contain FeS (troilite) inclusions, suggesting that FeS may alloy with iron in the core (Mason, 1966), and experiments show that sulfur has an affinity for liquid iron–nickel alloys, at least in the range 2–25 GPa and 2073–2623 K (Li and Agee, 2001). Iron–sulfur–silicon liquids have been found to be miscible above 15 GPa at 2343 K (Sanloup and Fei, 2004). Cosmochemical arguments have been made that the mantle is depleted in sulfur relative to its chondritic abundance by two orders to magnitude, and that the missing sulfur could reside in the core (Murthy and Hall, 1970). However, since sulphur is volatile, this line of argument has been questioned (Ringwood, 1977).

FeO is absent from iron meteorites because at atmospheric pressure it is soluble by less than 0.1% by mass in solid iron, and its solubility increases only slightly above the liquidus (Ringwood, 1977). However, its solubility increases with temperature and pressure as it becomes metallic, so that it could be present in the core (Ringwood and Hibberson, 1990). Other possible alloying components include iron silicate (Fe_xSi_{1-x}), Fe_3C, and FeH (MacDonald and Knopoff, 1958; Wood, 1993; Fukai, 1984; Okuchi, 1997; Lin *et al.*, 2002), all likely to be metallic under core pressures. Abundance and volatility during condensation from the nebula, the mode of core formation, and solubility in iron during core formation presumably determined the composition of less-dense elements in the core. See McDonough (2003) for a comprehensive discussion of light elements in the core.

The inner core is more dense than the outer core, by about 600 kg m^{-3} according to the seismic Earth model PREM (Dziewonski and Anderson, 1981), and 550 ± 50 kg m^{-3} according to a model by Masters and Shearer (1990). Part of this difference, perhaps 200 kg^{-3}, can be attributed to iron being more dense in the solid phase (Stacey 1992; Laio, *et al.*, 2000). The rest is attributed to the less-dense alloying components partitioning into the outer core during solidification of the core. This partitioning has important consequences for the thermodynamics (*see* Chapter 2), fluid dynamics (*see* Chapter 5), and metallurgy of the core (see Section 10.3).

The seismically inferred density increase at the inner–outer core boundary, above that due to the phase transition, can be explained if the core is an iron-rich alloy. However, unless the solidus slope is vertical in a phase diagram, some nonzero fraction of the less-dense alloying components will fractionate into the inner core as it solidifies. Jephcoat and Olson (1987) extrapolated the density of pure iron to inner-core conditions and found that it is too dense to explain the inner-core seismic data, suggesting the inner core also contains some light alloy, equivalent to 3–7% sulfur by mass. Similarly, Laio *et al.* (2000) has calculated that solid iron under the conditions at the inner–outer core boundary is 2–3% more dense than given by PREM (Dziewonski and Anderson, 1981), and Lin *et al.* (2005) confirmed this experimentally by studying the sound velocities of pure iron at high pressure and temperature. Stixrude *et al.* (1997) calculated that the mass fraction of the less-dense element depends on which less-dense element and compound is present (for instance, FeS vs FeS_2). However, all estimates indicate the density jump at the inner–outer core boundary exceeds the 200 kg m^{-3} expected for the phase transition alone, implying that light elements are segregating into the outer core.

Geochemists have examined the question of radioactive isotopes in the core, particularly U^{238}, Th^{232}, and K^{40}. Although the presence of radioactive heat had long been proposed as a source of energy for the geodynamo (Bullard, 1950; Gubbins *et al.*, 1979), the presence of heat-producing isotopes has long been questioned, as there is little experimental evidence for their solubility in liquid iron under core conditions (McDonough, 2003). Paleomagnetic evidence suggests that the Earth's magnetic field is at least 3 billion years old (McElhinny and Senanayake, 1980). Since the compositional buoyancy provided by inner-core solidification helps drive the dynamo (*see* Chapter 2), the inner core may be as old as the dynamo. However, thermal calculations (Labrosse *et al.*, 2001) suggest that such an old inner core is incompatible with the rate at which the inner core is currently growing. A resolution to this problem may be the presence of K^{40} in the core (Gessmann and Wood, 2002; Murthy *et al.*, 2003), which, with a half-life of 1.25 billion years, would have been an important heat source earlier in the Earth's history, allowing a slow rate of inner-core growth. Another possibility is that the power requirements of the dynamo are less than previously thought (Christensen and Tilgner, 2004), mitigating the need for an old inner core and hence radioactive heat sources. In light of all these uncertain possibilities the presence of radioactive elements in the core, and the age of the inner core, remain open questions.

10.2.2 Phase Diagram of Iron Alloy

Considerable work has been done in determining the alloy phase diagram of the core and the crystal structure of iron under inner-core pressure and temperature conditions. In a binary system with end members A and B, where each pure solid exists only in a single solid phase (crystal structure), and where A and B are miscible in the liquid phase, as is likely in the outer core (Alfe *et al.*, 1999; Helffrich and Kaneshima, 2004), three types of phase diagrams are possible. Which type of phase diagram exists for a given system is determined by the minimum in Gibbs free energy

$$G = H - TS \qquad [1]$$

where H is the enthalpy, T is the temperature, and S is the entropy. If the crystal structures are the same, and the atomic sizes are similar, then A and B may be completely miscible in the solid state as well as the

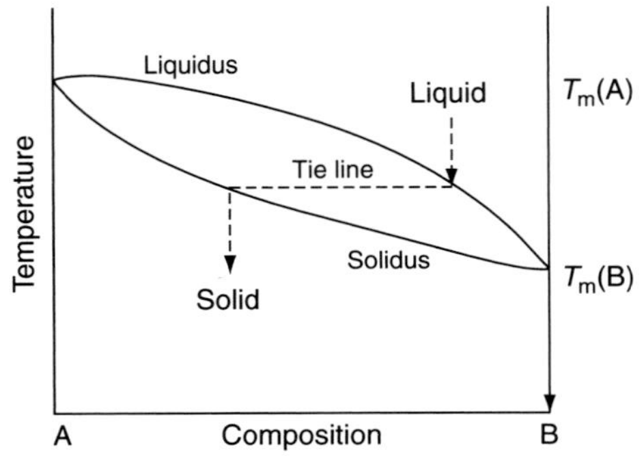

Figure 1 Phase diagram for a binary system (A,B) exhibiting complete solubility in both the liquid and solid phases (the enthalpy of mixing, ΔH, equals zero), so that both the liquid and solid phases are ideal solutions. The liquidus represents the temperature at which liquid of a given composition solidifies. A horizontal tie line at that temperature connects to the solidus, which gives the composition of the solid. The space between the curves represents the width of the phase loop, which is a measure of the compositional difference between the liquid and solid.

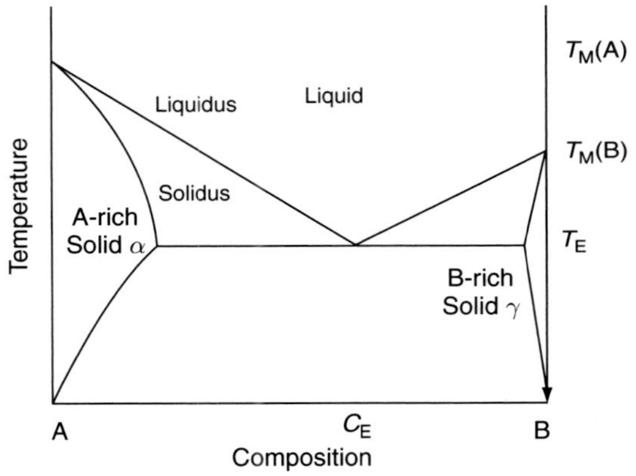

Figure 2 Phase diagram for a binary system exhibiting complete solubility in the liquid phase, but where $\Delta H > 0$ in the solids, so that the solids are real solutions. Because $\Delta H > 0$ in the solids, the system remains liquid to lower temperatures than the melting temperature of pure A or pure B (T_M(A) or T_M(B)). ΔH is sufficiently positive that the miscibility gap in the solid phase extends up to the liquid phase. Such a phase diagram represents a eutectic system. Two solid phases, A-rich α and B-rich γ, exist at all temperatures beneath the eutectic temperature T_E, the lowest temperature at which liquid can be in equilibrium with solid. At T_E, the composition of the liquid is C_E.

liquid state ($\Delta H = 0$ upon mixing). Liquid and solid are then both ideal solutions, and only two phases, liquid and solid, are possible in the temperature-composition field (**Figure 1**).

If the change in enthalpy upon mixing in the solid state is non-zero, then the solid is a real solution. If $\Delta H > 0$, atoms A and B are incommensurate, either because of a differing crystal structure, atomic size, or both. In this case it is energetically favorable for the mixture to remain a liquid to a temperature below that of the pure solid melting temperatures. Accompanying the melting point depression is a miscibility gap in the solid at low temperatures such that two solid phases co-exist. If ΔH is large enough, the miscibility gap can extend up to the liquid phase (Porter and Easterling, 1992). In this situation, the lower entropy associated with two separate solid phases, an A-rich α-phase and a B-rich β-phase (rather than randomly mixed atoms in a single phase), is made up for by the lower enthalpy that results from A and B existing primarily in separate phases.

Such a system exhibits a eutectic phase diagram (**Figure 2**), and there exists a eutectic composition that has a minimum melting temperature. Due to the entropic contribution, B will have nonzero solubility in the A-rich α-phase, and vice versa, but such a solid solution is not ideal. Examples of this type of phase behavior are ubiquitous, including model laboratory systems sodium chloride–water, ammonium chloride–water, lead–tin, and zinc–tin.

In a binary system with $\Delta H > 0$ it is also possible for stable solid phases other than those of pure A and pure B to exist. When such intermediate phases are present (as is common with iron alloys), a more complex phase diagram results. Other solidification reactions such as peritectic solidification can result. However, the essential solidification structure of the alloy is similar to that of a eutectic system (Porter and Easterling, 1992).

If $\Delta H < 0$, it is energetically favorable for a solid phase to form, so that a maximum in the melting temperature occurs for some intermediate composition. Such a solid is known as an ordered alloy because the atoms arrange themselves in a particular structure known as a superlattice. Ordered alloys often occur only within a certain compositional range. There is no experimental evidence that any of the possible core alloy components form an ordered alloy with iron under inner-core conditions.

At low pressures Fe–FeS forms a eutectic with a melting point depression of 600 K from that of iron. However, Boehler (1996a) has presented evidence showing that the melting point depression decreases with increasing pressure, based on experiments reaching pressures up to 62 GPa, still much lower than the 330 GPa at the inner–outer core boundary.

This suggests that Fe–FeS may form an ideal solid solution at inner-core pressures. This does not preclude an iron-enriched inner core, provided the liquidus and solidus are sufficiently separated in composition (the phase loop in **Figure 1** is wide), and the melting temperature of iron is higher than that of FeS, at core pressures. The latter may well be the case (Boehler, 1992; Anderson and Ahrens, 1996). However, Alfe *et al.* (2000) found from *ab initio* calculations that the concentration of sulfur in the solid state is very nearly that of the liquid state. Such a similarity in composition would seem to have difficulty explaining the excess density jump at the inner–outer core boundary. (The similarity indicates that the liquidus and solidus are close in composition. While this suggests an ideal solid solution, a eutectic system could still be a possibility if a significant fraction of the inner core has not yet cooled to the eutectic temperature (Fearn *et al.*, 1981).)

On the other hand, FeO alone may not be able to explain the presence of a less-dense component in the inner core. Sherman (1995), also using first-principles calculations, showed that the concentration of oxygen in the solid state is very low at inner-core pressures. However, Alfe *et al.* (1999) suggested that the concentration could be higher for other assumed crystal structures, and Stixrude *et al.* (1997) were able to explain the inner-core density with FeO present in solid solution with pure iron rather than as a mixture of separate phases.

Since the composition of the core remains uncertain, the phase diagram is also uncertain, but in spite of these uncertainties it is likely that a binary phase diagram captures the salient features of core solidification.

10.2.3 Crystal Structure of Iron

Understanding the solid phase(s) of pure and alloyed iron under inner-core conditions is central to interpreting the seismic anisotropy of the inner core (see Section 10.5). **Figure 3** summarizes the phase boundaries of iron. Under atmospheric pressure and room temperature, pure iron takes on a body-centered cubic (b.c.c.) crystal structure (the α-phase). At higher temperatures it undergoes a phase transformation to a face-centered cubic (f.c.c.) crystal structure (the γ-phase), and before melting it undergoes an entropy-driven phase transformation to another b.c.c. crystal structure (the δ-phase). At high pressure, pure iron transforms to a phase with a higher packing density, the hcp ε-phase. It has generally been thought that this is the stable phase of iron under inner-core conditions, though there have been some reports of a double h.c.p. phase (the β-phase; Saxena *et al.*, 1993). However, this experimental inference has been questioned (Vocadlo *et al.*, 1999; Kubo *et al.*, 2003). It has also been suggested that at the high temperatures of the inner-core, ε-iron could transform to a b.c.c. phase (Ross *et al.*, 1990; Matsui and Anderson, 1997; Belonoshko *et al.*, 2003). The role of impurities on the stable phases of iron under inner-core conditions is also under scrutiny (Vocadlo *et al.*, 2003).

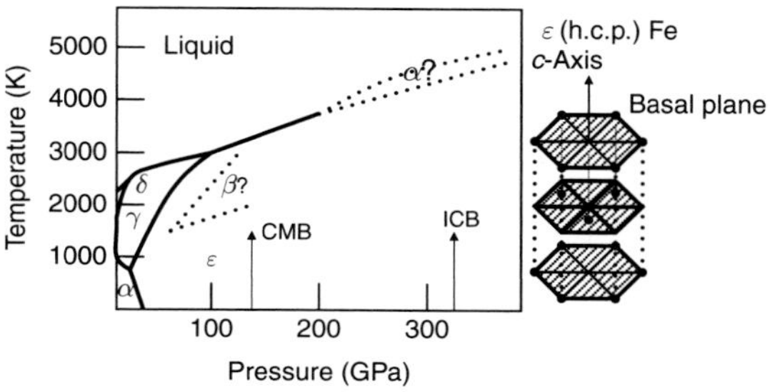

Figure 3 Summary of the phase boundaries of iron. The uncertainty of the boundaries increases with temperature and pressure, and the existence of the double h.c.p. β and b.c.c. α' phases is uncertain. To the right is the crystal structure of the h.c.p. ε phase.

10.3 Solidification of the Inner Core

Although hottest at the center, the inner core is solidifying directionally outwards due to the effect of pressure on the liquidus slope. A flat solid–fluid interface can become morphologically unstable (Mullins and Sekerka, 1963, 1964) due to constitutional supercooling (Rutter, 1958; Porter and Easterling, 1992). This occurs when rejection of solute during solidification of an alloy leads to a solute boundary layer in the fluid with a scale thickness D/V, where D is the solute diffusivity and V is the inner-core growth velocity. If the gradient of the local freezing temperature (the liquidus)

$$\mathrm{d}T_{\mathrm{L}}/\mathrm{d}z = \Delta T/(D/V) \qquad [2]$$

where ΔT is the temperature difference between the liquidus and solidus, exceeds the actual temperature gradient in the fluid, $\mathrm{d}T/\mathrm{d}z$, then freezing is predicted ahead of the flat solid–fluid interface (see **Figure 4**). This results in the growth of a solid perturbation into the fluid.

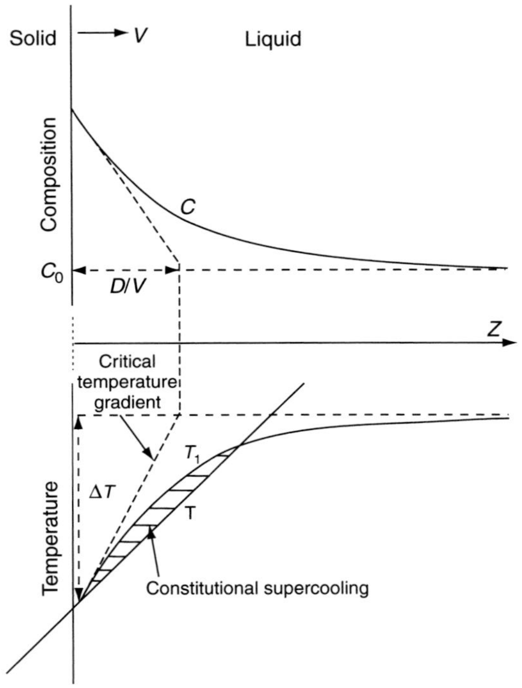

Figure 4 Dendritic growth in directionally solidifying alloys can occur when the liquid is constitutionally supercooled. Adjacent to the solid is a solute boundary layer of scale thickness D/V, where D is the solute diffusivity and V is the growth rate of the solid. The solute concentration C in the boundary layer is enriched relative to its value C_0 far from the solid. This enrichment depresses the equilibrium freezing temperature T_L by an amount ΔT from its value far from the solid. If the temperature T in the liquid is less than T_L, then solid is predicted ahead of the solid/liquid interface, a condition known as constitutional supercooling. The criterion for constitutional supercooling is that the temperature gradient in the liquid, $dT/dz < \Delta T/(D/V)$. Adapted from Porter DA and Easterling KE (1992) *Phase Transformations in Metals and Alloys*. London: Chapman and Hall.

Morphological instability due to constitutional supercooling typically results in dendritic growth of solute-poor solid, with interdendritic, solute-rich pockets. Dendrites oriented close to the direction of heat flow grow most rapidly (dendrites also grow in particular crystallographic directions, the key to solidification texturing). Directionally solidified alloys exhibit columnar crystals, elongated in the direction of dendritic growth (Porter and Easterling, 1992). The mushy zone is the mixed solid–fluid region during dendritic growth. Mushy zones are ubiquitous features in directionally solidifying metallic alloys, organic systems, and aqueous salt solutions on both sides of the eutectic, though they are less common in silicates because of the low entropy associated with faceted dendrites (Jackson, 1958; Miller and Chadwick, 1969).

Loper and Roberts (1981) extended the condition for morphological instability to include the effect of pressure variations on the liquidus, estimating that the inner-core growth rate is nearly 500 times supercritical. In essence, the argument for morphological instability of the inner-core boundary rests on the small temperature gradient dT/dz near the inner-core boundary, in spite of the inner-core growth velocity V also being very small. Fearn *et al.* (1981) suggested that the center of the Earth could exceed the eutectic temperature, so that the entire inner core would be in a mushy state.

Loper and Roberts (1978, 1980) also examined the thermodynamics of a slurry, which is a general mixed phase region. They considered an inner core formed by precipitation of heavy, iron-rich particles downwards. Shimizu *et al.* (2005) later considered the possibility that the inner core might in part grow by precipitation rather than dendritic growth, so that the region near the inner–outer core boundary might be a slurry layer rather than a mushy layer. They found, however, that morphological instability is likely for all realsitic values of the liquidus concentration slope.

Morse (1986, 2002) has questioned the prediction of a mushy inner core on other grounds – the assumption that heat and solute near the inner–outer core boundary are removed only by diffusion. Loper and Roberts (1981) ignored the effect of convection, arguing that it must become small near the boundary, but Morse (1986, 2002) suggests that convection at the base of the outer core reduces the solute buildup, making the inner-core growth rate nearly five orders of magnitude less than that needed for morphological instability.

Section 10.5.4 summarizes the seismic evidence for solidification structures in the inner core. There is also meteoritic evidence, although this evidence is ambiguous because of the relative lack of data and the large extrapolation from a meteorite to a planetary core. Iron meteorites such as those from the Cape York shower exhibit compositional gradients that are too large to result from the general fractionation of a planetoid core, but instead are more likely due to microsegregation between secondary and tertiary dendrite arms (Esbensen and Buchwald, 1982; Haack and Scott, 1992). Moreover, the FeS (troilite) nodules in meteorites are elongated and oriented, suggesting interdendritic pockets of melt during directional solidification (Esbensen and Buchwald, 1982).

10.4 Grain Size and Rheology in the Inner Core

The rheology of iron at core conditions governs the style and rate of deformation in the inner core. For timescales longer than the Maxwell relaxation time

$$\tau = \eta / G \quad [3]$$

the inner core can be considered to behave as a viscous fluid. Here η is the viscosity and G is the rigidity. Current estimates of inner-core viscosity span a large range, of the order 10^{13}–10^{21} Pa s, but even if the upper limit applies the inner core is less viscous than the mantle, because it is closer to its melting temperature. With a rigidity of the order of 10^{11} Pa (Dziewonski and Anderson, 1981), τ is between 10^2 and 10^{10}s, which is far shorter than mantle timescales. Constraints from rheology come from theoretical and empirical laws of mineral physics, and from elastic and anelastic properties of seismic waves at various frequencies. Here, we review the viscous properties followed by the elastic and anelastic properties.

Viscosity relates the stress σ to the strain rate $d\varepsilon/dt$. The general form of the strain rate in a solid can be expressed as

$$d\varepsilon/dt = A(DGb)/(kT)(b/d)^p(\sigma/G)^n \quad [4]$$

where D is the diffusion coefficient, b is the Burgers vector, k is Boltzmann's constant, T is temperature, d is the grain size, and p, n, and A are dimensionless constants (Van Orman, 2004).

Deformation occurs under imposed stress either by the diffusive movement of vacancies and atoms, or from slipping of lattice by dislocations (Poirier, 1985). Both of these mechanisms of deformation coexist, but the mechanism that gives the larger strain rate under a given stress becomes the predominant one. At high stress conditions, power law creep ($p = 0$, $n = 3$–5) gives the largest strain rate under a given stress. The nonlinear dependence of strain rate on stress results because the dislocation density increases with applied stress.

At low-stress conditions, both dislocation (Harper–Dorn creep) and diffusion creeps are possible. For large grain size, a Newtonian-type dislocation creep (Harper–Dorn creep, with $p = 0$, $n = 1$) dominates over diffusion creep. A Newtonian dependence is interpreted to arise from dislocation density that is independent of stress. One interpretation for the stress-independent dislocation creep was provided by Wang *et al.* (2002), who suggested that dislocation density results from the sum of two stresses: the applied stress and the Peierl's stress required to move the dislocations. For low-stress conditions Peierl's stress dominates, and the dislocation density becomes independent of applied stress. For small grain size, a Newtonian diffusion creep becomes faster. This is because the smaller grain size yields a larger spatial gradient of density of vacancies or atoms, which causes the deformation. There are two types of diffusion creeps: one where diffusion occurs within the lattice (Nabarro–Herring creep: $p = 2$, $n = 1$) and the other where diffusion occurs along the grain boundaries (Coble creep: $p = 3$, $n = 1$). The activation energy for grain boundary diffusion is smaller than that for lattice diffusion (Poirier, 1985), so for the high-temperature conditions of the inner core, Nabarro–Herring creep is more important.

These regimes of deformation are summarized in the form of deformation mechanism maps for various materials of interest (Frost and Ashby, 1982; also see **Table 1** for a general summary of the various high-temperature deformation mechanisms). Deformation mechanism maps are often constructed as a function of homologous temperature, T/T_m, (where T_m is the melting temperature) and normalized stress, σ/G. Compilation of data have shown that materials with the same crystal

Table 1 High temperature deformation mechanisms

	$p = 0$ (grain size independent)	$p = 2$	$p = 3$
$n = 1$(Newtonian)	Harper–Dorn creep (dislocation density independent of stress) Low stress Large grain size	Nabarro–Herring creep (diffusion within lattice) Low stress	Coble creep (diffusion along grain boundaries) Low stress Small grain size
$n = 3$–5	Power law creep (dislocation density increases with stress) High stress		

structure and similar bonding have similar deformation mechanism maps (Frost and Ashby, 1982), so h.c.p. metals can be used as an analog of h.c.p. iron to estimate the constants in eqn [4]. Homologous temperature of the inner core is high (>0.9), so we are left with estimating the magnitude of stress and grain size in order to estimate the viscosity of the inner core.

Several studies have sought to identify and quantify sources of stress in the inner core (see also Section 10.5). For instance, Jeanloz and Wenk (1988) considered a nonadiabatic large-scale degree-one flow in the inner core. They assumed a temperature variation of about 1 K with a length scale of 1000 km, and estimated that a buoyancy stress of about 10^4 Pa would arise. Together with the deformation mechanism of iron (Frost and Ashby, 1982), they estimated a viscosity of 10^{10}–10^{16} Pa s. This leads to a very large strain rate of the order of $10^{-9}\,s^{-1}$, or an equivalent flow velocity of about $10^{-3}\,m\,s^{-1}$, which is unrealistically fast.

Yoshida *et al.* (1996) considered an anisotropic inner-core growth and obtained a strain rate of the order of $10^{-18}\,s^{-1}$. Such a small strain rate is a result of a very slow growth rate of the inner core. They considered power law creep and Nabarro–Herring (diffusion) creep as possible candidates. From evaluating the stress needed to cause this strain rate, they showed that diffusion creep occurs for a grain size less than 6 m, and a power law creep occurs if the grain size is larger than 6 m, though there are uncertainties in the parameter values used. As an additional condition, they assumed that dynamic recrystallization governs the grain size, which yields a smaller grain size for larger stress. Using these constraints, they obtained a grain size of 5 m, which is marginally in the diffusion creep regime, and a viscosity of 3×10^{21} Pa s. In another model, they applied a Maxwell relaxation model to seismic attenuation data, and estimated the viscosity to be greater than 10^{16} Pa s.

Bergman (1998, 2003) examined the issue of grain size in the inner core. One constraint on grain size comes from assuming that the frequency-dependent inner core attenuation (see Section 10.5.4 of this chapter) arises from scattering (e.g., Cormier *et al.*, 1998; Cormier and Li, 2002). For this case, the grain size becomes large, of the order of a seismic body wavelength, that is, 100 m to 1 km (Bergman, 1998). It is also possible to estimate the grain size from extrapolation of dendrite size obtained from laboratory experiments (although by six orders of magnitude!). Due to the slow cooling rate in the inner core, the grain size estimate is of the order of a few hundred meters (Bergman, 2003).

Van Orman (2004) argued that Harper–Dorn creep is the dominant deformation mechanism in the inner core. This mechanism occurs at low stress and large grain size conditions, a regime not considered by Yoshida *et al.* (1996). For the low stress level in the inner core, Van Orman (2004) showed that power law creep is unlikely, and that Harper–Dorn creep would occur for grain sizes exceeding 10–30 μm. In this likely case, the estimated viscosity becomes 10^{10}–10^{12} Pa s, which is quite small. The deformation mechanism in the inner core thus remains unclear.

It is important to note that the above estimates are for solid h.c.p. iron. The presence of melt within the inner core would drastically reduce the viscosity, so that the above estimates should be taken as an upper limit. In addition, the presence of melt would promote diffusion of atoms along grain boundaries, and cause pressure solution-type creep (which is similar to Coble creep), thus resulting in a grain-size dependence.

10.5 Origin of the Inner-Core Elastic Anisotropy

10.5.1 An Overview

Inner-core elastic anisotropy, where P-waves travel about 3% faster in the polar direction than parallel to the equatorial plane, with a larger degree of anisotropy in the deeper inner core, is well established from seismic body waves and free oscillations. The most plausible interpretation, based on the analogy with mantle anisotropy, is that it results from the lattice-preferred orientation of the h.c.p. iron that comprise the inner core. Alternatively, it could be due to shape-preferred orientation of melt pockets (Singh *et al.*, 2000). Seismology has also shown that there is an attenuation anisotropy, where P-waves in the polar direction are more attenuated (see Section 10.5.4; Creager, 1992; Song and Helmberger, 1993; Souriau and Romanowicz, 1996, 1997; Yu and Wen, 2006).

A variety of models have been proposed to explain the origin of inner-core elastic anisotropy. We can classify them into two categories. The first category consists of dynamical models, which assume the crystals align as a result of plastic deformation within the inner core. The second set of models assume that the inner-core texture forms during solidification from the liquid outer core. **Table 2** lists the advantages and drawbacks of each of the models. In the next subsection we summarize the anisotropic properties of h.c.p. iron and its style of slip under dislocation creep, followed by a

Table 2 Models for inner-core anisotropy

Mechanism	Reference	Advantages	Drawbacks
Deformation models			
Inner-core thermal convection	Jeanloz and Wenk (1988)	Analogy with mantle	Inner core not likely to be thermally convecting
Preferential equatorial solidification due to outer core convection	Yoshida, *et al*. (1996)	Explains relation of fast axis with spin axis, preferential solidification observed in lab experiments	Low stress levels so that texture may take age of IC to develop
Radial component of Maxwell stress	Karato (1999)	Anisotropy reflects magnetic field, larger stress level	Stress likely to be balanced by other stresses
Azimuthal component of Maxwell stress	Buffett and Wenk (2001)	Can sustain flow	Requires *c*-axis to be slow, depth dependence?
Solidification models			
Paramagnetic susceptibility	Karato (1993)	Novel	Iron under inner-core conditions not likely to be paramagnetic
Anisotropic heat flow due to outer core convection	Bergman (1997)	Observed in lab, simple depth dependence, may also explain attenuation anisotropy	Requires *c*-axis to be fast, effects of deformation?

None of the models by themselves can easily explain latitudinal variations, or abrupt changes with depth.

review of the two categories of models for understanding inner-core elastic anisotropy.

10.5.2 Anisotropy of h.c.p. Iron

The anisotropy of h.c.p. iron at inner-core conditions was first estimated from analog h.c.p. metals. Wenk *et al.* (1988) considered Ti at room conditions to be a good analog for the inner core because it has a similar c/a ratio and ratio of linear compressibility as h.c.p. iron. They concluded that P-waves should be faster in the *c*-axis direction. Sayers (1989) analyzed several h.c.p. metals and arrived at the same conclusion. These estimates were based on room temperature values. Bergman (1998) argued, based on the temperature dependence of the elastic constants of h.c.p. metals, that the magnitude of anisotropy of h.c.p. iron would become even larger at core temperatures.

In recent years, first-principles calculations and high-pressure experiments have made direct determination of elastic constants possible (Steinle-Neumann *et al.*, 2003). Stixrude and Cohen (1995) calculated elastic constants of h.c.p. iron at core pressures (but 0 K), again finding the *c*-axis fast, by about 4% for compressional waves. These calculations were then extended to high temperatures by Steinle-Neumann *et al.* (2001), who found that the sense of P-wave anisotropy reversed from that at 0 K, so that the *a*-axis of a single crystal is 10–12% faster than the *c*-axis at high temperature. According to Steinle-Neumann *et al.* (2001), the seismic inference of a 3% fast P-wave velocity in the polar direction can thus be explained by 1/3 alignment of *a*-axes (or basal planes) in the polar direction. However, calculations by Gannarelli *et al.* (2003) did not find a reversal of the sense of anisotropy. Experimental determination of the elastic constants by Mao *et al.* (1998, 1999) inferred a bulk modulus and rigidity similar to previous values, but with the fast axis lying at an angle intermediate between the *c*- and *a*-axes.

If the preferred orientation of h.c.p. iron occurs by dislocation creep, a critical issue is whether the predominant slip plane is basal or prismatic. Poirier and Price (1999) applied the criterion derived by Legrand (1984) to determine whether basal or prismatic slip dominates in h.c.p. iron. To distinguish between the various slip regimes, Legrand (1984) defined a parameter R as

$$R = (c_{66} \cdot \gamma_B)/(c_{44} \cdot \gamma_P) \quad [5]$$

where c_{66} and c_{44} are elastic constants and γ_B and γ_P are the stacking fault energies in the basal and prism planes, respectively. Legrand (1984) showed that when $R < 1$ the primary slip system is basal and when $R > 1$ it is prismatic. Using *ab initio* calculations, Poirier and Price (1999) calculated the stacking fault energies for h.c.p. iron, and using the published elastic constants for 0 K h.c.p. iron, they found that R becomes 0.37–0.43, indicating that h.c.p. iron slips predominantly on the basal plane. This result is also obtained when the elastic constants of Steinle-Neumann *et al.* (2001) are used.

Wenk *et al.* (2000b) conducted high-pressure deformation experiments of h.c.p. iron in a diamond anvil, at pressures of up to 220 GPa. By imposing a uniaxial nonhydrostatic stress, they found that *c*-axes align parallel to the axis of the diamond cell. Comparing with the results from plasticity theory, they found that basal slip is dominant in determining the overall preferred orientation even if prismatic slip is favored over basal slip.

10.5.3 Dynamical Models

The key issues for a dynamical origin of the preferred orientation are the source and pattern of stress, and the mechanism of alignment. The first of the dynamical models was by Jeanloz and Wenk (1988), who proposed that preferred orientation of h.c.p. iron is caused by inner-core convection driven by internal heat sources. By evaluating the radiogenic heat generation in the inner core, they concluded that the Rayleigh number in the inner core is supercritical. They showed that if a spherical harmonic degree $l = 1$ convection exists in the inner core, then the resulting simple shear flow in the equatorial region yields a *c*-axis alignment of about 45° from the polar direction. Averaging such a crystal alignment, and using the elastic constants of Ti, Jeanloz and Wenk obtained a fast P-wave in the polar direction. There are, however, several issues with this model. First, the amount of radiogenic heat is highly uncertain so that thermal convection in the inner core may not occur. Thermal history calculations generally do not yield thermal convection in the inner core because the geothermal gradient in the inner core becomes smaller than the adiabat (due to the slow growth rate of the inner core), which allows sufficient time for the inner core to cool (Sumita *et al.*, 1995; Yoshida *et al.*, 1996; Yukutake, 1998). Also, this model does not explain why a degree 1 convection pattern with a symmetry axis corresponding to the rotation axis should be preferred. Later, numerical studies of thermal convection in the inner core were carried out by Weber and Machetel (1992), and by Wenk *et al.* (2000a). In the latter study, preferred orientation of h.c.p. iron under a polar downwelling and an equatorial upwelling was calculated, which resulted in a fast P-wave in the polar direction.

An explanation for an inner-core flow pattern with an axis coinciding with the rotation axis was given by Yoshida *et al.* (1996). They argued that the inner-core growth should be of zonal degree 2 because of the columnar convection in the outer core, which transports heat more efficiently from low latitudes of the inner core. This causes the inner core to grow faster near the equator (see **Figure 5**). Experiments conducted by

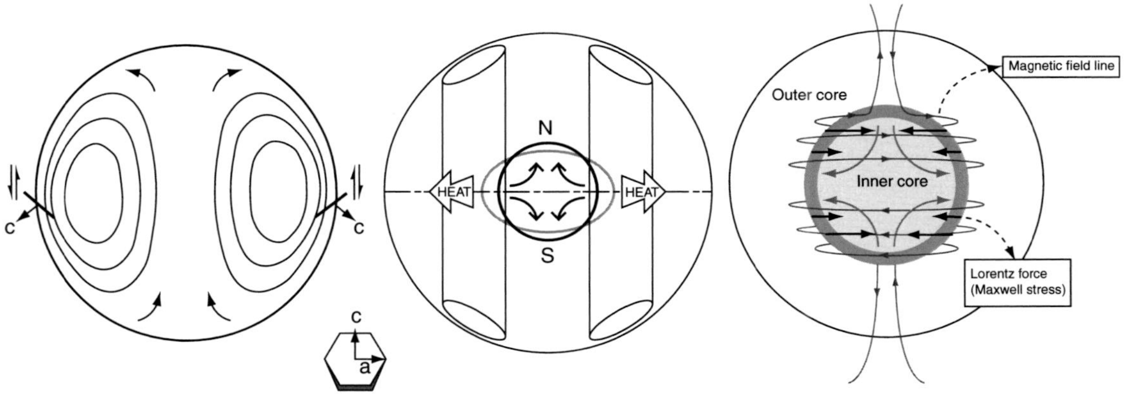

Figure 5 Some dynamical models exhibiting a radial flow in the inner core. In these models such flow is considered to be responsible for the preferred orientation of crystals. (a) An internally heated inner core can convect. A degree 1 mode is the first unstable mode. Adapted from Jeanloz R (1990) The nature of the Earth's core. *Annual Review of Earth and Planetary Sciences* 18: 357–386. (b) Convection in a rapidly rotating, spherical fluid shell such as the outer core is more efficient at transporting heat perpendicular to the rotation axis. This leads to an inner core that solidifies more oblately than the gravitational equipotential. A solid-state flow results, and the stress may lead to a recrystallization texture. Adapted from Yoshida S, Sumita I, and Kumazawa M (1996) Growth model of the inner core coupled with outer core dynamics and the resulting elastic anisotropy. *Journal of Geophysical Research* 101: 28085–28103. (c) Maxwell stresses resulting from the magnetic field in the outer core can squeeze the inner core, causing a flow. Here a toroidal magnetic field is shown to cause such flow. In this model, the flow pattern depends on the magnetic field pattern of the outer core. Adapted from Karato S (1999) Seismic anisotropy of the Earth's inner core resulting from flow induced by Maxwell stresses. *Nature* 402: 871–873.

Bergman *et al.* (2005) show this effect in a rotating hemisphere of salt water solidifying from the center. However, because of the density difference between the inner and outer cores, the inner core must deform isostatically to maintain its spherical shape, and an overall stress field with uniaxial tension in the polar direction results. Using the elastic constants of Stixrude and Cohen (1995) and Kamb's theory of preferred orientation by recrystallization under stress (Kamb, 1959), Yoshida *et al.* (1996) showed that this model can explain the observed inner-core anisotropy. Sumita and Yoshida (2003) used the same model with the elastic constants of Steinle-Neumann *et al.* (2001) and found that *a*-axes aligning in the polar direction can explain the sense of anisotropy. One problem with this model is that the stress resulting from anisotropic growth is quite small due to the slow growth rate of the inner core. As a result, preferred orientation would take a geologically long time (on the order of 10^9 years) to develop. The applicability of Kamb's theory under the presence of dislocations is also uncertain.

Karato (1999) argued that Maxwell stresses resulting from the geomagnetic field generated in the outer core could also induce flow in the inner core, by radially squeezing the inner core at the ICB. In this model, a toroidal magnetic field of $B = 10^{-2} - 10^{-1}$ T results in Maxwell stresses $\sigma = 10^2 - 10^4$ Pa. Since the induced flow pattern in the inner core is controlled by the magnetic field pattern in the outer core, this model predicts that the inner-core anisotropy reflects the magnetic field pattern. However, Buffett and Bloxham (2000) questioned whether the Maxwell stresses can actually drive a radial flow in the inner core, or whether they are simply balanced by other stresses. They considered the balance between magnetic, pressure and buoyancy stresses, and showed that these stresses would not equilibrate because of the incompatibility between thermodynamic and hydrostatic equilibrium conditions at the ICB. As a result, they predict that viscous flow would occur, but they also showed this flow to be weak and confined near the ICB. As a result, they concluded that the radial component of the Maxwell stress is unsuitable for causing preferred orientation.

Buffett and Wenk (2001) used the elastic constants by Steinle-Neumann *et al.* (2001), and considered the Lorentz force tangential to the inner core. They calculated the resulting anisotropy that forms under an inner-core shear flow. The resulting stress in this model becomes quite small, of the order of several Pa. Assuming both basal and prismatic slips to occur for h.c.p. iron, they showed that the *c*-axes become parallel to the equatorial plane, which is consistent with the seismic inferences, if the *c*-axis of h.c.p. iron under inner-core conditions is indeed the slow direction. Since the Maxwell stress decreases with the distance from the inner-core boundary, the alignment becomes weak with depth, so that an explanation is needed to produce a stronger anisotropy with depth. Another issue is that the stress field depends strongly on the morphology of the magnetic field, which is not known well.

To summarize, the condition of thermodynamic equilibrium at the ICB is important in evaluating the flow pattern in the inner core, as the analysis of Buffett and Bloxham (2000) showed. Since the flow pattern in the outer core, the solidification at the ICB, and the pattern and magnitude of Maxwell stresses are all coupled, a self-consistent model of inner-core anisotropy needs to properly incorporate all of these effects.

10.5.4 Solidification Texturing Models

The second category of models to explain the origin of inner-core anisotropy invoke solidification. Karato (1993) proposed that h.c.p. iron may have a paramagnetic susceptibility and would align during solidification under a magnetic field, hence resulting in a preferred orientation. However, this model predicts larger anisotropy near the ICB, which is inconsistent with the observations. Furthermore, theoretical studies show h.c.p. iron to be nonmagnetic at core pressures (Söderlind *et al.*, 1996; Steinle-Neumann *et al.*, 1999).

A solidification texturing during directional cooling was shown by Bergman (1997) to be a possible mechanism for understanding inner-core elastic anisotropy. The directional cooling occurs perpendicular to the rotation axis, as a result of the pattern of convection in the outer core, as suggested by Yoshida *et al.* (1996), later demonstrated in experiments by Bergman *et al.* (2005). Solidification experiments on tin-rich alloys exhibit dendrites that grow in a particular crystallographic direction controlled by the direction of heat flow, resulting in elastic anisotropy. This result was extended to h.c.p. zinc alloys by Bergman *et al.* (2000). Hence, the direction of heat flow becomes a preferred crystallographic direction (**Figure 6**). The geometry of the crystal growth results in an increase in anisotropy with depth. The effect of fluid flow on solidification texturing of h.c.p. zinc alloys was examined by Bergman *et al.* (2003). In analogy with the solidification of h.c.p. sea ice, the experiments showed that the fluid flow causes

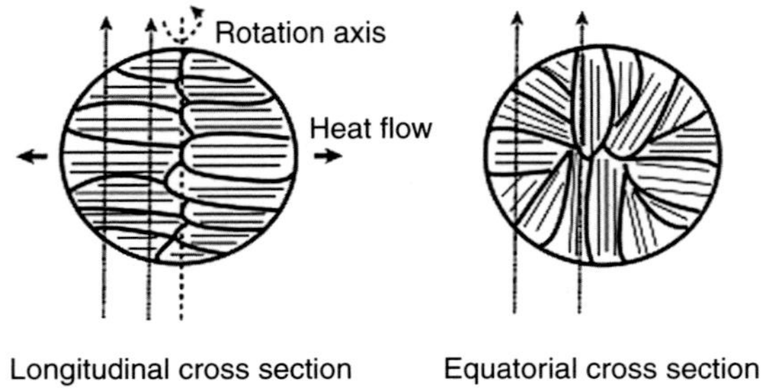

Figure 6 A solidification texturing model resulting in a preferred orientation of crystals. Heat flow perpendicular to the rotation axis leads to dendritic growth in the cylindrically radial direction. The heavier lines represent columnar crystals, the lighter lines primary dendrites. The left panel represents a longitudinal cross section, the right panel an equatorial cross section. North–south seismic rays (left panel), represented by the dotted arrows, are always perpendicular to the growth direction of dendrites. The component of rays perpendicular to the rotation axis (right panel) that is parallel to the growth direction of dendrites increases with turning depth in the inner core. This is the origin of the depth dependence associated with solidification texturing of the inner core. (Such geometric depth dependence becomes less strong for rays not turning on the equatorial plane.) Adapted from Bergman MI (1997) Measurements of elastic anisotropy due to solidification texturing and the implications for the Earth's inner core. *Nature* 389: 60–63.

transverse texturing, indicating that alignment is sensitive to the direction of fluid flow as well as that of heat flow.

Brito *et al.* (2002) conducted directional solidification experiments using liquid (pure) gallium, and measured the resulting anisotropy of the polycrystalline gallium. They found that for all cases crystals elongated parallel to the imposed thermal gradient that determined the direction of crystal growth, and that other factors such as turbulence and magnetic field have little influence. They also found that the preferred orientation of crystals is controlled primarily by the presence of seed grains, and that this strongly affects the amount of anisotropy. Seeding has been observed to have more of an effect on the texture of solidifying fresh water than salt water (Weeks and Wettlaufer, 1996). Unresolved issues with solidification texturing include whether conduction will dominate over convection as regards heat transport in the outer core, the role of post-solidification deformation in modifying the solidification texture, whether the observed depth dependence can be explained by the simple depth dependence predicted by geometry, and uncertainty as to the sense of fastest wave propagation in h.c.p. iron under inner-core conditions.

10.6 Origin of Other Inner-Core Seismic Structures

In this section we review other seismic structures of the inner core, and their geodynamical interpretations.

10.6.1 Properties and Structure of the Inner-Core Boundary

The inner-core boundary has been found to be seismically sharp, with a transition thickness of less than 5 km (Cummins and Johnson, 1988). Such a sharp inner-core boundary may be explained by a rapid increase in solid fraction of partial melt with depth, which might result from convection within the mushy layer (Loper, 1983) or from compaction (Sumita *et al.*, 1996). It has been proposed that the inner-core radius may depend on frequency, if the inner-core boundary is a diffuse boundary defined by the relaxation time equivalent to the period of a seismic wave (Anderson, 1983). However, this has not been substantiated, which indicates that the inner-core boundary is probably not a gradual transition from the outer core to the inner core.

We next consider structure near the inner-core boundary. At the high pressures of the Earth's core, the difference between the bulk modulus of the solid and liquid is small. As a result, the P-wave velocity jump at the inner-core boundary arises primarily from the finite rigidity of the inner core. The bottom most part of the outer core has been found to have a small to zero P-wave velocity gradient compared to PREM (Souriau and Poupinet, 1991; Song and Helmberger, 1992, 1995; Kaneshima *et al.*, 1994). One interpretation is that this is a layer with solid crystals that are not interconnected, such that the rigidity is zero. P-wave velocity in this layer is then given approximately by the Wood's formula (Mavko *et al.*, 1998)

$$V = (K_R/\rho)^{1/2} \qquad [6]$$

where K_R is the Reuss average of the bulk modulus of the solid–liquid composite, and ρ is the average density.

If we use the values at the top of the inner core for solid and bottom of the outer core for liquid, then we find that the average density increases more with solid fraction than does the average bulk modulus. As a consequence, P-wave decreases with solid fraction. For example, a 20% volumetric fraction of solid particles yields a velocity decrease of approximately

0.2%. Highly porous dendrites such as those observed in the solidification of ammonium chloride crystals, which does not allow shear waves to propagate, may be analogous to such a layer. Such a layer also results in attenuation. A frequency-dependent attenuation at the top of the outer core has been reported (e.g., Tanaka and Hamaguchi, 1993). A similar analysis for the bottom of the outer core may be used to constrain the solid fraction in this layer (Stevenson, 1983). An apparent absence of scattering in this layer, may also be used to constrain the length scale of dendrites.

10.6.2 Inner-Core Attenuation and Scattering

Knowledge of seismic attenuation in the inner core could give earth scientists considerable insight into the nature of the inner core, but at present the depth and frequency dependence of the seismic quality factor in compression Q_α or in shear Q_β remain uncertain (Masters and Shearer, 1990). At body-wave frequencies (1 Hz), Q_α in the upper inner core is quite low, about 200, increasing to 440–1000 deeper into the inner core (Cormier, 1981; Doombos, 1983; Souriau and Roudil, 1995). Using normal modes, Widmer *et al.* (1991) found Q_β to be even lower, about 120. Little depth resolution is available from normal-mode data.

Taken together these studies could indicate a frequency dependence of attenuation, or high attenuation in compression, but mineral physicists are not in agreement. Mao *et al.* (1998) found that the aggregate shear wave speed of h.c.p. iron extrapolated from 220 GPa and room temperature to inner-core conditions is about 15% greater than that in the inner core (Dziewonski and Anderson, 1981), suggesting near-melting softening. However, some studies (Jackson *et al.*, 2000; Laio *et al.*, 2000; Steinle-Neumann *et al.*, 2001) found a Poisson's ratio for h.c.p. iron under inner-core conditions comparable to that from PREM, suggesting that the high Poisson's ratio of the inner core is an intrinsic property of h.c.p. iron at high temperature.

Attenuation in the inner core can result from intrinsic relaxation and diffusion mechanisms, and from scattering. As seismic waves travel they cause an adiabatic pressure perturbation, which causes a temperature change and, potentially, freezing and melting. These cause thermal and compositional diffusion, and result in attenuation. Loper and Fearn (1983) derived an expression for frequency-dependent attenuation due to diffusion. Under this mechanism, attenuation increases with melt fraction, with peak attenuation at the timescale corresponding to thermal and compositional diffusion.

Singh *et al.* (2000) evaluated the attenuation for the case where liquid can flow in the interconnected space of a solid matrix, as a result of the pressure variations arising from the propagation of seismic waves. In the seismic frequency range of interest, their results showed an attenuation peak at a viscosity of around 250 Pa s. Likely values for the molecular viscosity of iron at core conditions (Section 10.4) are five orders of magnitude smaller than this, so that attenuation from this mechanism is negligibly small.

Another interesting finding is the coda waves following the inner-core-reflected wave (PKiKP) (e.g., Vidale and Earle, 2000). One interpretation is that these are caused by strong scatterers with a scale length of about 2 km in the outermost 300 km of the inner core (Vidale and Earle, 2000). They may alternatively be due to reverberation effects near the inner-core boundary (Poupinet and Kennett, 2004). Laboratory experiments indicate that there are at least three length scales associated with solidification that could be relevant to scattering: the grain size, the spacing between dendrites, and the spacing between the chimneys of the upwelling in the mushy layer (e.g., Tait *et al.*, 1992). Reverberation near the inner-core boundary might be caused by waves trapped in a low-Q mushy zone between the high-Q outer core and the deeper inner core.

Scattering can give rise to an apparent Q that is lower than the intrinsic material value. This has been explored by Cormier (1981) and Cormier and Li (2002) for 10–30 km wavelength body waves. Bergman (1998) and Cormier *et al.* (1998) also examined whether anisotropic scattering as a result of columnar crystal growth might also be responsible for an attenuation anisotropy (Creager, 1992; Souriau and Romanowicz, 1996, 1997; Oreshin and Vinnik, 2004; Yu and Wen, 2006), where waves traveling parallel to the fast rotation axis exhibit smaller amplitudes and more complex waveforms. Anisotropic scattering has been observed in directionally solidified h.c.p. zinc alloys (Bergman *et al.*, 2000). Like the elastic anisotropy, the attenuation anisotropy may also have a hemispherical variation. On the other hand, some studies have shown that there are regions without such attenuation anisotropy (Helffrich *et al.*, 2002).

10.6.3 Hemispherical Variation of Seismic Velocity, Anisotropy, and Attenuation

Next we turn to laterally heterogeneous structures in the inner core. Recent seismological studies have revealed that there exists a large-scale longitudinal variation in P-wave velocity and its anisotropy, which appears to be depth dependent. Using P-waves traveling in the east–west direction, it was found that the eastern hemisphere has a larger *V*p and also a smaller *Q* compared to the western hemisphere (Tanaka and Hamaguchi, 1997; Niu and Wen, 2001; Wen and Niu, 2002; Cao and Romanowicz, 2004), and using P-waves traveling in the polar direction, it was found that the Western Hemisphere has a larger anisotropy compared to the Eastern Hemisphere (e.g., Tanaka and Hamaguchi, 1997; Creager, 1999). Furthermore, Cao and Romanowicz (2004) found a hemispherical variation of the depth dependence of attenuation of P-waves. Their results indicate that *Q* in the Eastern Hemisphere increases with depth, whereas that in the Western Hemisphere decreases with depth. Since the outer core and deep inner core have a large *Q*, a minimum *Q* should exist somewhere near the ICB. They proposed that this minimum exists at a deeper depth in the Western Hemisphere compared to the Eastern Hemisphere.

Such hemispherical structure may be produced by lateral variation of the inner-core solidification rate, caused by the outer-core flow controlled by the thermally heterogeneous mantle (Sumita and Olson, 1999, 2002). More rapid solidification of the inner core in the Western Hemisphere would lead to a porous inner core because there is insufficient time to expel liquid by compaction, leading to a low *V*p there. Similarly, if the inner-core anisotropy results from preferential growth, the Western Hemisphere would have a larger anisotropy. On the other hand, long-term mantle control to explain hemispherical variations requires that the inner core be locked to the mantle (Buffett, 1996b), implying that seismologists are inferring an inner-core oscillation rather than a rotation. Future observations should resolve this issue. There may also be some positive feedback mechanism, for example, involving an anisotropic thermal conductivity, but this has not yet been studied in detail.

We also need to consider the origin of the hemispherical difference of the depth of minimum *Q* proposed by Cao and Romanowicz (2004). There are at least two candidates for the laterally variable minimum *Q* that arise from the variation of melt fraction. One results at a very low melt fraction at a temperature just above the eutectic. This is inferred from the acoustic measurements of a partially molten binary eutectic system (e.g., Spetzler and Anderson, 1968; Stocker and Gordon, 1975; Watanabe and Kurita, 1994). These experiments have shown that as the temperature is raised, *Q* abruptly decreases at the eutectic temperature, but recovers slightly at higher temperatures, before decreasing at yet higher temperatures. Another candidate for a minimum *Q* exists at a much higher melt fraction, since the liquid outer core has a high *Q*. However, these two candidates have caveats. If the cause of the minimum is a very low melt fraction, then there should be a stepwise increase of *Q* at a certain depth corresponding to the eutectic temperature. However, this has not yet been observed. On the other hand, if the minimum occurs at a higher melt fraction, then the melt fraction cannot be too high so as to inhibit shear wave propagation. It is uncertain whether such melt fraction can be realized within the inner core.

10.6.4 The Deep Inner Core and the Inner-Core Transition Zone

Finally, we consider the seismic structure of the deeper section of the inner core. Deeper in the inner core attenuation has been found to be smaller, which may be explained by a smaller liquid fraction or from a larger grain size. Seismic anisotropy is found to be larger in the deeper inner core (e.g., Creager, 2000), with a possible transition zone at about 200 km depth between the isotropic and anisotropic inner core (Song and Helmberger, 1998), and laterally variable. However, the sharpness of this transition seem to differ for different wavelengths and geographical locations (Leyton *et al.*, 2005).

Several possible explanations can be given for the cause of larger anisotropy with depth. A smaller anisotropy at shallow parts can arise from random alignment in the horizontal plane or from alignment such that the symmetry axis (*c*-axis for h.c.p.) is in the radial direction. The former can result from the slow kinetics of crystal alignment, the geometry of solidification texturing, or fluid flow in the outer core, and the latter from a principal stress axis in the radial direction at shallow depths. Such a stress field may arise from electromagnetic stresses or from compaction. Some observations even suggest that the deepest part of the inner core may be distinctly different from other parts of the inner core (Ishii and Dziewonski, 2002, 2003; Beghein and Trampert, 2003), perhaps suggesting a phase transition from h.c.p. to b.c.c. iron.

10.7 Inner-Core Rotation

10.7.1 An Overview

Inner-core rotation relative to the mantle was first proposed on theoretical grounds by Gubbins (1981), and then found to occur, on the order of degrees per year, in numerical dynamo models by Glatzmaier and Roberts (1995a, 1995b). Following the numerical simulations, Song and Richards (1996) inferred that the tilted symmetry axis of inner-core anisotropy is rotating eastward about the spin axis, at a rate of 1.1 deg yr^{-1}, and Su *et al.* (1996) found a rate as high as 3 deg yr^{-1}. However, using a lateral anisotropy gradient rather than the tilted symmetry axis, Creager (1997) limited the differential rotation to 0.2–0.3 deg yr^{-1}. In these studies there is a tradeoff between the rate of super-rotation and the level of anisotropy and symmetry axis tilt, or the anisotropy gradient. Vidale *et al.* (2000) examined the PKiKP coda from nuclear tests, which eliminates tradeoffs associated with event mislocations, to also infer a differential rotation rate of 0.2 deg yr^{-1}. However, using the splitting of core-sensitive normal modes, Laske and Masters (1999) preferred no rotation, though the data allows a maximum inner-core super-rotation of 0.2–0.3 degrees yr^{-1}. Numerical simulations with a better resolution and less hyper-diffusion also predict a modest prograde inner-core rotation of about 0.1 degrees yr^{-1}, which further decreases to 0.02 degree yr^{-1} when gravitational coupling is included (Buffett and Glatzmaier, 2000). Although there have been controversial seismological issues (Souriau and Poupinet, 2003), recent studies using earthquakes with similar waveforms (Zhang *et al.*, 2005) continue to support inner-core super-rotation. Here we review mechanisms that can cause as well as inhibit inner-core rotation.

10.7.2 Electromagnetic Coupling

Various torques can act on the inner core and these can make the inner core rotate. Electromagnetic torque was considered by Gubbins (1981) and was also the driving mechanism in the numerical models by Glatzmaier and Roberts (1995a, 1995b). Electromagnetic torque originates from the restoring force caused by the outer-core flow that stretches the magnetic field lines. It is expressed as (Rochester, 1962)

$$\Gamma = 1/\mu_0 \int (B_r B_\phi \, r \sin\theta \, dS) \qquad [7]$$

where μ_0 is the magnetic permeability, B_r and B_ϕ are the radial and azimuthal fields at the ICB. Quantitative estimates of inner-core rotation from electromagnetic torque were given by Gubbins (1981). He showed that when an electromagnetic torque of 10^{19} Nm is applied to the inner core, it would rotate and approach a steady angular velocity of about 0.13 deg yr^{-1}, and induce a toroidal magnetic field of about 80 gauss. He also showed that an oscillatory motion with a period of about 10 years is possible.

The mechanism of inner-core rotation in the numerical calculations by Glatzmaier and Roberts (1995a, 1995b) was analyzed in Glatzmaier and Roberts (1996). They showed that the electromagnetic coupling caused by the thermal wind within the inner-core tangent cylinder was the primary cause. Within the tangent cylinder, upwellings form above the polar region and downwellings near the equator. This meridional circulation causes eastward flow near the inner core and westward flow near the core–mantle boundary, by conservation of angular momentum. The eastward flow, coupled with the magnetic field, results in an electromagnetic coupling between the outer-core flow and the inner core, and spins up the inner core.

An analytical model of inner-core rotation driven by the thermal wind was also derived by Aurnou *et al.* (1996). Assuming that the temperature inside the tangential cylinder is higher than the temperature outside, they showed that the resulting eastward thermal wind, coupled with the magnetic field, can quantitatively explain the inner-core superrotation. This model was extended in Aurnou *et al.* (1998) by numerically calculating the resulting inner-core rotation from electromagnetic coupling for three different outer-core flow patterns. This work confirmed that the thermal wind can efficiently couple inner-core rotation to outer-core flow.

10.7.3 Gravitational Coupling

Mass anomalies in the mantle can deform the inner core. When the inner core rotates relative to the mantle, a misalignment of the inner core topography relative to the mantle mass anomaly causes a gravitational restoring force to the inner core.

This torque was shown to become quite large, of the order of 10^{21} N m by Buffett (1996b). Although there are uncertainties in the mass anomaly within the mantle, this estimate indicates that a gravitational torque can exceed the electromagnetic torque by orders of magnitude, and thus lock the inner core to the mantle to inhibit rotation. Such gravitational coupling between the mantle and the inner core can cause an exchange of angular momentum between them, and result in length-of-day variations (Buffett, 1996a, 1996b).

Gravitational torque can become small if the inner-core topography cannot become as high as that caused by the mass anomaly of the mantle. If this torque becomes smaller than the electromagnetic torque, the inner core may rotate. Buffett (1997) showed that it is possible to constrain the viscosity of the inner core if the inner core is rotating under the gravitational torque. One case is when the viscosity of the inner core is so small that the inner core can deform rapidly, and thus misalignment of the mantle mass anomaly and inner-core topography does not occur. The other case is when the viscosity of the inner core is so high that the inner core is slow to deform and thus the inner-core topography becomes small. For this case, however, because of high viscosity, once it becomes locked to the mantle, it would become unable to rotate again. From the above constraints, the inner-core viscosity was found to become less than 10^{16} Pa s or greater than 10^{20} Pa s.

10.7.4 Combined Coupling

Estimates of electromagnetic torque have large uncertainty because the strength of the toroidal field inside the core cannot be observed. Aurnou and Olson (2000) calculated the inner-core rotation under the combined effects of electromagnetic, gravitational, and viscous torques. Among these torques, the viscous torques was found to be very small. When the electromagnetic and gravitational torques are comparable, the inner-core rotation showed time-dependent features. A more general case of inner-core rotation, where the inner core was also allowed to tilt, was studied by Xu *et al.* (2000).

Incorporation of the combined effects of gravitational torque and viscous deformation of the inner core into a numerical geodynamo calculation was done by Buffett and Glatzmaier (2000), who demonstrated that the gravitational torque can significantly suppress inner-core rotation.

10.8 Summary

The inner core remains a difficult part of our planet to study. In every way it presents challenges: for mineral physicists, the extreme pressures and temperatures are difficult to achieve; for seismologists, the 5200 km of material above obfuscates inner-core signals; for geodynamicists, the wide variety and interaction of possible phenomena and the large uncertainty of physical parameters complicates interpretation. Nevertheless progress continues along multiple fronts: observational scientists will become yet better at extracting signals; experimentalists will continue to make technical improvements to achieve the extreme conditions of the inner core, and to design appropriate analog experiments; computational scientists will go where experimentalists cannot; and theoreticians will make progress in coupling data from all disciplines. Further progress towards understanding the inner core will require a multi- and interdisciplinary effort. We also hope and predict that geodynamicists will stimulate new observations, such as was the case with inner-core superrotation. Perhaps it should have been possible to have predicted inner-core elastic anisotropy prior to its discovery! Proposition of testable models by geodynamicists is one of the key directions to the study of the inner core.

It is always dangerous to make specific predictions about the directions a field of study will take, but we will nevertheless try. The composition of the light elements of the core is likely to remain uncertain–there are too many ways to put the puzzle together. This means that the phase diagram of the core will also remain uncertain. It does not mean, however, that there will be no progress toward understanding such important issues as the partitioning of elements between the inner and outer cores, and the light elements' effects on the melting temperature. On the other hand, it seems reasonable that we will make progress towards understanding the stable phases(s) of pure and alloyed iron under inner-core conditions. Likewise, we expect that the current uncertainty concerning the elastic constants of h.c.p. iron under inner-core conditions will get sorted out. Similarly, if there is another stable phase of iron in the inner core, we will determine its elastic properties.

Inner-core seismology is difficult, but hopefully not intractable. Over time we will come to some agreement concerning the elastic and anelastic properties of the inner core. Of particular interest will be the geographic distributions of the elastic anisotropy, the bulk attenuation, and the attenuation anisotropy. One thing seems certain: as we get a better picture of the inner core, it, like the rest of the earth above, is not simple and featureless. In some ways this makes it less interesting – there is no one simple physical mechanism that can explain all the data. Ultimately, though, it shows that the inner core too exhibits the rich array of phenomena that makes earth science challenging and interesting. We predict that geodynamicists will take up the challenge to cross disciplines, and come to understand the origin of inner core anisotropy, in all of its detail as well as other interesting seismic properties that we dig up. We also predict that the inner core may hold some of the keys to understanding long-standing problems in the earth sciences, such as magnetic field generation and reversal, and core thermodynamics.

References

Alfe D, Price GD, and Gillan MJ (1999) Oxygen in the Earth's core: A first-principles study. *Physics of the Earth and Planetary Interiors* 110: 191–210.

Alfe D, Gillan MJ, and Price GD (2000) Constraints on the composition of the Earth's core from *ab initio* calculations. *Nature* 405: 172–175.

Anderson DL (1983) A new look at the inner core of the Earth. *Nature* 302: 660.

Anderson WW and Ahrens T (1996) Shock temperatures and melting in iron sulfides at core pressures. *Journal of Geophysical Research* 101: 5627–5642.

Aurnou JM, Brito D, and Olson PL (1996) Mechanics of inner core super-rotation. *Geophysical Research Letters* 23: 3401–3404.

Aurnou JM, Brito D, and Olson PL (1998) Anomalous rotation of the inner core and the toroidal magnetic field. *Journal of Geophysical Research* 103: 9721–9738.

Aurnou J and Olson P (2000) Control of inner core rotation by electromagnetic, gravitational and mechanical torques. *Physics of the Earth and Planetary Interiors* 117: 111–121.

Beghein C and Trampert J (2003) Robust normal mode constraints on inner-core anisotropy from model space search. *Science* 299: 552–555.

Belonoshko AB, Ahuja R, and Johansson B (2003) Stability of the body-centered-cubic phase of iron in the Earth's inner core. *Nature* 424: 1032–1034.

Bergman MI (1997) Measurements of elastic anisotropy due to solidification texturing and the implications for the Earth's inner core. *Nature* 389: 60–63.

Bergman MI (1998) Estimates of the Earth's inner core grain size. *Geophysical Research Letters* 25: 1593–1596.

Bergman MI (2003) Solification of the Earth's core. In: Veronique D, Kenneth C, Stephan Z, and Shun-Ichiro K (eds.) *Geodynamics Series, 31: Earth's Core: Dynamics, Structure, Rotation*, pp. 105–127, (10.1029/31GD08), Washington, DC: American Geophysical Union.

Bergman MI, Agrawal S, Carter M, and Macleod-Silberstein M (2003) Transverse solidification textures in hexagonal close-packed alloys. *Journal of Crystal Growth* 255: 204–211.

Bergman MI, Giersch L, Hinczewski M, and Izzo V (2000) Elastic and attenuation anisotropy in directionally solidified (hcp) zinc and the seismic anisotropy in the Earth's inner core. *Physics of the Earth and Planetary Interiors* 117: 139–151.

Bergman MI, Macleod-Silberstein M, Haskel M, Chandler B, and Akpan N (2005) A laboratory model for solidification of Earth's core. *Physics of the Earth and Planetary Interiors* 153: 150–164.

Birch F (1952) Elasticity and constitution of the Earth's interior. *Journal of Geophysical Research* 57: 227–286.

Boehler R (1992) Melting of the Fe–FeO and Fe–FeS systems at high pressures: Constraints on core temperatures. *Earth and Planetary Science Letters* 111: 217–227.

Boehler R (1996a) Fe–FeS eutectic temperatures to 620 kbar. *Physics of the Earth and Planetary Interiors* 96: 181–186.

Boehler R (1996b) Melting temperature of the Earth's mantle and core: Earth's thermal structure. *Annual Reviews - Earth and Planatery Sciences* 24: 15–40.

Bridgman PW (1949) Linear compression to 30,000 kg/cm^2, including relatively incompressible substances. *Proceedings of the American Academy of Arts and Sciences* 77: 187–234.

Brito D, Elbert D, and Olson P (2002) Experimental crystallization of gallium: Ultrasonic measurements of elastic anisotropy and implications for the inner core. *Physics of the Earth and Planetary Interiors* 129: 325–346.

Brown H and Patterson C (1948) The composition of meteoritic matter. III-Phase equilibria, genetic relationships, and planet structures. *Journal of Geology* 56: 85–111.

Buchwald VF (1975) *Handbook of Iron Meteorites,* Vol. 1: Berkeley CA: University of California Press.

Buffett BA (1996a) Gravitational oscillations in the length of day. *Geophysical Research Letters* 23: 2279–2282.

Buffett BA (1996b) A mechanism for decade fluctuations in the length of day. *Geophysical Research Letters* 23: 3803–3806.

Buffett BA (1997) Geodynamic estimates of the viscosity of the Earth's inner core. *Nature* 388: 571–573.

Buffett BA and Bloxham J (2000) Deformation of Earth's inner core by electromagnetic forces. *Geophysical Research Letters* 27: 4001–4004.

Buffett BA and Glatzmaier GA (2000) Gravitational braking of inner-core rotation in geodynamo simulations. *Geophysical Research Letters* 27: 3125–3128.

Buffett BA and Wenk H-R (2001) Texturing of the Earth's inner core by Maxwell stresses. *Nature* 413: 60–63.

Bullard EC (1950) The transfer of heat from the core of the Earth. *Monthly Notices of the Royal Astronomical Society* 6: 36–41.

Bullen KE (1949) Compressibility–pressure hypothesis and the Earth's interior. *Monthly Notices of the Royal Astronomical Society. Geophysical Supplement* 5: 355–368.

Cao A and Romanowicz B (2004) Hemispherical transition of seismic attenuation at the top of the Earth's inner core. *Earth and Planetary Science Letters* 228: 243–253.

Christensen UR and Tilgner A (2004) Power requirement of the geodynamo from Ohmic losses in numerical and laboratory dynamos. *Nature* 429: 169–171.

Cormier VF (1981) Short-period PKP phase and the anelastic mechanism of the inner core. *Physics of the Earth and Planetary Interiors* 24: 291–301.

Cormier VF, Li X, and Choy GL (1998) Seismic attenuation of the inner core: Viscoelastic or stratigraphic? *Geophysical Research Letters* 25: 4019–4022.

Cormier VF and Li X (2002) Frequency-dependent seismic attenuation in the inner core 2. A scattering and fabric interpretation. *Journal of Geophysical Research* 107(B12): 2362 (doi:10.1029/2002JB001796).

Creager KC (1992) Anisotropy of the inner core from differential travel times of the phases PKP and PKIKP. *Nature* 356: 309–314.
Creager KC (1997) Inner core rotation rate from small-scale heterogeneity and time-varying travel times. *Science* 278: 1284–1288.
Creager KC (1999) Large-scale variations in inner-core anisotropy. *Journal of Geophysical Research* 104: 23127–23139.
Creager KC (2000) Inner core anisotropy and rotation. In: Karato S-I, Stixrude L, Lieberman R, *et al.* (eds.) *Mineral Physics and Seismic Tomography From the Atomic to Global Scale*, pp. 89–114. Washington, DC: American Geophysical Union.
Cummins P and Johnson LR (1988) Synthetic seismograms for an inner core transition of finite thickness. *Geophysical Journal* 94: 21–34.
Doombos DJ (1983) Observable effects of the seismic absorption band in the Earth. *Geophysical Journal of the Royal Astronomical Society* 75: 693–11.
Dziewonski AM and Anderson DL (1981) Preliminary reference Earth model. *Physics of the Earth and Planetary Interiors* 25: 297–356.
Esbensen KH and Buchwald VF (1982) Planet (oid) core crystallization and fractionation-evidence from the Agpalilik mass of the Cape York iron meteorite shower. *Physics of the Earth and Planetary Interiors* 29: 218–232.
Fearn DR, Loper DE, and Roberts PH (1981) Structure of the Earth's inner core. *Nature* 292: 232–233.
Frost HH and Ashby MF (1982) *Deformation Mechanism Maps*. 166 pp. Tarrytown, NY: Pergamon.
Fukai Y (1984) The iron–water reaction and the evolution of the Earth. *Nature* 308: 174–75.
Gannarelli CM, Alfe S, and Gillan MJ (2003) The particle-in-cell model for ab initio thermodynamics: Implications for the elastic anisotropy of the Earth's inner core. *Physics of the Earth and Planetary Interiors* 139: 243–253.
Gessmann CK and Wood BJ (2002) Potassium in the Earth's core? *Earth and Planetary Science Letters* 200: 63–78.
Glatzmaier GA and Roberts PH (1995a) A three-dimensional self-consistent computer simulation of a geomagnetic field reversal. *Nature* 377: 203.
Glatzmaier GA and Roberts PH (1995b) A three-dimensional convective dynamo solution with rotating and finitely conducting inner core and mantle. *Physics of the Earth and Planetary Interiors* 91: 63.
Glatzmaier GA and Roberts PH (1996) Rotation and magnetism of the Earth's inner core. *Science* 274: 1887–1891.
Gubbins D (1981) Rotation of the inner core. *Journal of Geophysical Research* 86: 11695–11699.
Gubbins D, Masters TG, and Jacobs JA (1979) Thermal evolution of the Earth's core. *Geophysical Journal of the Royal Astronomical Society* 59: 57–99.
Haack H and Scott ERD (1992) Asteroid core crystallization by inward dendritic growth. *Journal of Geophysical Research* 97: 14727–14734.
Helffrich G, Kaneshima S, and Kendall J-M (2002) A local, crossing-path study of attenuation and anisotropy of the inner core. *Geophysical Research Letters* 29(12): 1568 (doi: 10.1029/2001GL014059).
Helffrich G and Kaneshima S (2004) Seismological constraints on core composition from Fe-O-S liquid immiscibility. *Science* 306: 2239–2242.
Ishii M and Dziewonski AM (2002) The innermost inner core of the Earth: Evidence for a change in anisotropic behavior at the radius of about 300 km. *Proceedings of the National Academy of Sciences* 99(22): 14026–14030.
Ishii M and Dziewonski AM (2003) Distinct seismic anisotropy at the center of the Earth. *Physics of the Earths and Planetary Interiors* 140: 203–217.
Jackson I, Fitzgerald JD, and Kokkonen H (2000) High temperature viscoelastic relaxation in iron and its implication for the shear modulus and attenuation of the Earth's inner core. *Journal of Geophysical Research* 105: 23605–23634.
Jackson KA (1958) Mechanism of growth. In: Maddin R, *et al.* (ed.) *Liquid Metals and Solidification*, pp. 174–186. Cleveland: American Society for Metals.
Jeanloz R (1990) The nature of the Earth's core. *Annual Review of Earth and Planetary Sciences* 18: 357–386.
Jeanloz R and Wenk H-R (1988) Convection and anisotropy of the inner core. *Geophysical Research Letters* 15: 72–75.
Jephcoat A and Olson P (1987) Is the inner core of the Earth pure iron? *Nature* 325: 332–335.
Kamb WB (1959) Theory of preferred crystal orientation developed by crystallization under stress. *Journal of Geology* 67: 153–170.
Kaneshima S, Hirahara K, Ohtaki T, and Yoshida Y (1994) Seismic structure near the inner core-outer core boundary. *Geophysical Research Letters* 21(2): 157–160.
Karato S (1993) Inner core anisotropy due to the magnetic field induced preferred orientation of iron. *Science* 262: 1708–1711.
Karato S (1999) Seismic anisotropy of the Earth's inner core resulting from flow induced by Maxwell stresses. *Nature* 402: 871–873.
Kubo A, Ito E, and Katsura T, *et al.* (2003) *In situ* X-ray observation of iron using Kawai-type apparatus equipped with sintered diamond: Absence of β phase up to 44 GPa and 2100 K. *Geophysical Research Letters* 30: 1126 (doi: 10.1029/2002GL016394).
Labrosse S, Poirier J-P, and Le Mouel J-L (2001) The age of the inner core. *Earth and Planetary Science Letters* 190: 111–123.
Laio A, Bernard S, Chiarotti GL, Scandolo S, and Tosatti E (2000) Physics of iron at Earth's core conditions. *Science* 287: 1027–1030.
Laske G and Masters G (1999) Rotation of the inner core from a new analysis of free oscillations. *Nature* 402: 66–69.
Leyton F, Koper KD, Zhu L, and Dombrovskaya M (2005) On the lack of seismic discontinuities in the inner core. *Geophysical Journal International* 162: 779–786.
Legrand B (1984) Relations entre la structure el'ectronique et la facilite de glissement dans les metaux hexagonaux compacts. *Philosophical magazine* 49: 171–184.
Li J and Agee CB (2001) Element partitioning constraints on the light element composition of the Earth's core. *Geophysical Research Letters* 28: 81–84.
Lin J-F, Heinz DL, Campbell AJ, Devine JM, and Shen G (2002) Iron–silicon alloy in Earth's core? *Science* 295: 313–315.
Lin J-F, Sturhahn W, Zhao J, Shen G, Mao H-K, and Hemley RJ (2005) Sound velocities of hot dense iron: Brich's law revisited. *Science* 308: 1892–1894.
Loper DE (1983) Structure of the inner core boundary. *Geophysical and Astrophysical Fluid Dynamics* 22: 139–155.
Loper DE and Fearn DR (1983) A seismic model of a partially molten inner core. *Journal of Geophysical Research* 88: 1235–1242.
Loper DE and Roberts PH (1978) On the motion of an iron-alloy core containing a slurry I. General theory. *Geophysical and Astrophysical Fluid Dynamics* 9: 289–321.
Loper DE and Roberts PH (1980) On the motion of an iron-alloy core containing a slurry II. A simple model. *Geophysical and Astrophysical Fluid Dynamics* 16: 83–127.
Loper DE and Roberts PH (1981) A study of conditions at the inner core boundary of the Earth. *Physics of the Earth and Planetary Interiors* 24: 302–307.
MacDonald GJF and Knopoff L (1958) The chemical composition of the outer core. *Journal of Geophysics* 1: 1751–1756.

Mao HK, Shu J, Shen G, Hemley RJ, Li B, and Singh AK (1998) Elasticity and rheology of iron above 220 GPa and the nature of the Earth's inner core. *Nature* 396: 741–743.
Mao HK, Shu J, Shen G, Hemley RJ, Li B, and Singh AK (1999) correction Elasticity and rheology of iron above 220 GPa and the nature of the Earth's inner core. *Nature* 399: 280.
Mason B (1966) Composition of the Earth. *Nature* 211: 616–618.
Masters TG and Shearer PM (1990) Summary of seismological constraints on the structure of the Earth's core. *Journal of Geophysical Research* 95: 21691–21695.
Matsui M and Anderson OL (1997) The case for a body-centered cubic phase (α') for iron at inner core conditions. *Physics of the Earth and Planetary Interiors* 103: 55–62.
Mavko G, Mukerji T, and Dvorkin J (1998) *The Rock Physics Handbook*. Cambridge: Cambridge University Press.
McDonough WF (2003) Compositional model for the Earth's core. In: Holland HD and Turekian KK (eds.) *Treatise on Geochemistry, vol. 2: The Mantle and core*, pp. 547–568. Oxford: Pergamon.
McElhinny MW and Senanayake WE (1980) Paleomagnetic evidence for the existence of the geomagnetic field 3.5 Ga ago. *Journal of Geophysical Research* 85: 3523–3528.
McQueen RG and Marsh SP (1966) Shock-wave compression of iron–nickel alloys and the earth's core. *Journal of Geophysical Research* 71: 1751–1756.
Miller WA and Chadwick GA (1969) The equilibrium shapes of small liquid droplets in solid-liquid phase mixtures: Metallic h.c.p. and metalloid systems. *Proceedings of the Royal society of London Series A* 312: 257–276.
Morse SA (1986) Adcumulus growth of the inner core. *Geophysical Research Letters* 13: 1557–1560.
Morse SA (2002) No mushy zones in the Earth's core. *Geochimica et cosmochimica Acta* 66(12): 2155–2165.
Mullins WW and Sekerka RF (1963) Morphological stability of a particle growing by diffusion or heat flow. *Journal of Applied Physics* 34: 323–329.
Mullins WW and Sekerka RF (1964) Stability of a planar interface during solidification of a dilute binary alloy. *Journal of Applied Physics* 35: 444–451.
Murthy VR and Hall HT (1970) The chemical composition of the Earth's core: Possibility of sulfur in the core. *Physics of the Earth and Planetary Interiors* 2: 276–282.
Murthy VR, van Westrenen W, and Fei Y (2003) Experimental evidence that potassium is a substantial radioactive heat source in planetary cores. *Nature* 423: 163–165.
Niu F and Wen L (2001) Hemispherical variations in seismic velocity at the top of the Earth's inner core. *Nature* 410: 1081–1084.
Okuchi T (1997) Hydrogen partitioning into molten iron at high pressures: Implications for Earth's core. *Science* 278: 1781–1784.
Oreshin SI and Vinnik LP (2004) Heterogeneity and anisotropy of seismic attenuation in the inner core. *Geophysical Research Letters* 31: L02613 (doi:10.1029/2003GL018591).
Poirier J-P (1985) *Creep of Crystals*. New York: Cambridge University Press.
Poirier J-P (1994) Light elements in Earth's outer core; A critical review. *Physics of the Earth and Planetary Interiors* 85: 319–337.
Poirier J-P and Price GD (1999) Primary slip system of epsilon-iron and anisotropy of the Earth's inner core. *Physics of the Earth and Planetary Interiors* 110: 147–156.
Porter DA and Easterling KE (1992) *Phase Transformations in Metals and Alloys*. London: Chapman and Hall.
Poupinet G and Kennett BLN (2004) On the observation of high frequency PKiKP and its coda in Australia. *Physics of the Earth and Planetary Interiors* 146: 497–511.
Ringwood AE (1977) Composition of the core and implications for the origin of the Earth. *Geochemical Journal* 11: 111–135.
Ringwood AE and Hibberson W (1990) The system Fe–FeO revisited. *Physics and Chemistry Minerals* 17: 313–319.
Rochester MG (1962) Geomagnetic core–mantle coupling. *Journal of Geophysical Research* 67: 4833–4836.
Ross M, Young DA, and Grover R (1990) Theory of the iron phase diagram at Earth core conditions. *Journal of Geophysical Research* 95: 21713–21716.
Rutter JW (1958) Imperfections resulting from solidification. In: Maddin R, *et al.* (eds.) *Liquid Metals and Solidification*, pp. 243–262. Cleveland: American Society for Metals.
Sanloup C and Fei Y (2004) Closure of the Fe–S–Si liquid miscibility gap at high pressure. *Physics of the Earth and Planetary Interiors* 147: 57–65.
Sayers CM (1989) Seismic anisotropy of the inner core. *Geophysical Research Letters* 16: 270–276.
Saxena SK, Shen G, and Lazor P (1993) Experimental evidence for a new iron phase and implications for Earth's core. *Science* 260: 1312–1314.
Shanker J, Singh BP, and Srivastava SK (2004) Volume–temperature relationship for iron at 330 GPa and the Earth's core density deficit. *Physics of the Earth and Planetary Interiors* 147: 333–341.
Sherman DM (1995) Stability of possible Fe-FeS and Fe-FeO alloy phases at high pressure and the composition of the Earth's core. *Earth and Planetary Science Letters* 132: 87–98.
Shimizu H, Poirier J-P, and Le Mouel J-L (2005) On crystallization at the inner core boundary. *Physics of the Earth and Planetary Interiors* 151: 37–51.
Singh SC, Taylor MAJ, and Montagner J-P (2000) On the presence of liquid in Earth's inner core. *Science* 287: 2471–2474.
Söderlind P, Moriarty JA, and Wills JM (1996) First-principles theory of iron up to Earth-core pressures: Structural, vibrational and elastic properties. *Physical Review B-condensed Matter* 53: 14063–14072.
Song X and Helmberger DV (1992) Velocity structure near the inner core boundary from waveform modeling. *Journal of Geophysical Research* 97(B5): 6573–6586.
Song X and Helmberger DV (1993) Anisotropy of the Earth's inner core. *Geophysical Research Letters* 20: 285–288.
Song X and Helmberger DV (1995) A P wave velocity model of Earth's core. *Journal of Geophysical Research* 100(B7): 9817–9830.
Song XD and Helmberger DV (1998) Seismic evidence for an inner core transition zone. *Science* 282: 924–927.
Song XD and Richards PG (1996) Observational evidence for differential rotation of the Earth's inner core. *Nature* 382: 221–224.
Souriau A and Poupinet G (1991) The velocity profile at the base of the liquid core from PKP(BC+Cdiff) data: An argument in favour of radial inhomogeneity. *Geophysical Research Letters* 18(11): 2023–2026.
Souriau A and Poupinet G (2003) Inner core rotation: A critical appraisal. In: Dehant V, *et al.* (eds.) *Geodynamics Series, vol. 31: Earth's Core: Dynamics, Structure, Rotation*, pp. 65–82, (10.1029/31GD06), Washington, DC: AGU.
Souriau A and Roudil P (1995) Attenuation in the uppermost inner core from broad-band GEOSCOPE PKP data. *Geophysical Journal International* 123: 572–587.
Souriau A and Romanowicz B (1996) Anisotropy in inner core attenuation: A new type of data to constrain the nature of the solid core. *Geophysical Research Letters* 23(1): 1–4.
Souriau A and Romanowicz B (1997) Anisotropy in the inner core: Relation between P-velocity and attenuation. *Physics of the Earth and Planetary Interiors* 101: 33–47.
Spetzler H and Anderson DL (1968) The effect of temperature and partial melting on velocity and attenuation in a simple binary system. *Journal of Geophysical Research* 73: 6051–6060.

Stacey FD (1992) *Physics of the Earth,* 3rd edn. Brisbane: Brookfield Press.
Steinle-Neumann G, Stixrude L, and Cohen RE (1999) First-principles elastic constants for the hcp transition metals Fe, Co and Re at high pressure. *Physical Review B* 60: 791–799.
Steinle-Neumann G, Stixrude L, Cohen RE, and Gulseren O (2001) Elasticity of iron at the temperature of the Earth's inner core. *Nature* 413: 57–60.
Steinle-Neumann G, Stixrude L, and Cohen RE (2003) Physical properties of iron in the inner core. In: Dehant V, *et al.* (eds.) Geodynamics Series, 31: *Earth's Core: Dynamics, Structure, Rotation,* pp. 213–232, (10.1029/31GD10), Washington, DC: AUG.
Stevenson DJ (1983) Anomalous bulk viscosity of two-phase fluids and implications for planetary interiors. *Journal of Geophysical Research* 88(B3): 2445–2455.
Stixrude L and Cohen RE (1995) High-pressure elasticity of iron and anisotropy of the Earth's inner core. *Science* 267: 1972–1975.
Stixrude L, Wasserman E, and Cohen RE (1997) Composition and temperature of Earth's inner core. *Journal of Geophysical Research* 102: 24729–24739.
Stocker RL and Gordon RB (1975) Velocity and internal friction in partial melts. *Journal of Geophysical Research* 80: 4828–4836.
Su WJ, Dziewonski AM, and Jeanloz R (1996) Planet within a planet: Rotation of the inner core. *Science* 274: 1883–1887.
Sumita I and Olson P (1999) A laboratory model for convection in Earth's core driven by a thermally heterogeneous mantle. *Science* 286: 1547–1549.
Sumita I and Olson P (2002) Rotating thermal convection experiments in a hemispherical shell with heterogeneous boundary heat flux: Implications for the Earth's core. *Journal of Geophysical Research* 107: 10.1029/2001JB000548.
Sumita I and Yoshida S (2003) Thermal interactions between the mantle, outer and inner cores, and the resulting structural evolution of the core. In: Dehant V, Kenneth C, Shun-Ichiro K, *et al.* (eds.) *Geodynamics Series, 31: Earth's Core: Dynamics, Structure, Rotation* pp. 213–232 and (10.1029/31GD14), Washington: AGU.
Sumita I, Yoshida S, Hamano Y, and Kumazawa M (1995) A model for the structural evolution of the Earth's core and its relation to the observations. In: Yu kutake T (ed.) *The Earth's Central Part: Its Structure and Dynamics*, pp. 231–261. Toyko: Terra Scientific.
Sumita I, Yoshida S, Kumazawa M, and Hamano Y (1996) A model for sedimentary compaction of a viscous medium and its application to inner-core growth. *Geophysical Journal International* 124: 502–524.
Tait S, Jahrling K, and Jaupart C (1992) The planform of compositional convection and chimney formation in a mushy layer. *Nature* 359: 406–408.
Tanaka S and Hamaguchi H (1997) Degree one heterogeniety and hemispherical variation of anisotropy in the inner core from PKP(BC)–PKP(DF) times. *Journal of Geophysical Research* 102: 2925–2938.
Tanaka S and Hamaguchi H (1993) Degree one heterogeniety at the top of the Earth's core, revealed by SmKS Travel Times. In: Le Mouel JL, *et al.* (ed.) *Geophysical Monograph Series, 72: Dynamics of the Earth's Deep Interior and Earth Rotation*, pp. 127–134. Washington: IUGG and AUG.
Van Orman JA (2004) On the viscosity and creep mechanism of Earth's inner core. *Geophysical Research Letters* 31: L20606 (doi: 10.1029/2004GL021209).
Vidale JE and Earle PS (2000) Fine-scale heterogeneity in the Earth's inner core. *Nature* 404: 273–275.
Vidale JE, Dodge DA, and Earle PS (2000) Slow differential rotation of the Earth's inner core indicated by temporal changes in scattering. *Nature* 405: 445–448.
Vocadlo L, Brodholt J, Alfe D, Price GD, and Gillan MJ (1999) The structure of iron under the conditions of the Earth's inner core. *Geophysical Research Letters* 26: 1231–1234.
Vocadlo L, Alfe D, Gillan MJ, Wood IG, Brodholt JP, and Price GD (2003) Possible thermal and chemical stabilization of body-centered-cubic iron in the Earth's core. *Nature* 424: 536–539.
Wang JN, Wu JS, and Ding Y (2002) On the transitions among different creep regimes. *Materials Science and Engineering A* 334: 275–279.
Watanabe T and Kurita K (1994) Simultaneous measurements of the compressional-wave velocity and the electrical conductivity in a partially molten material. *Journal of Physics of the Earth* 42: 69–87.
Weber P and Machetel P (1992) Convection within the inner-core and thermal implications. *Geophysical Research Letters* 19: 2107–2110.
Weeks WF and Wettlaufer JS (1996) Crystal orientations in floating ice sheets. In: Arsenault RJ, Cole D, Gross T, *et al.* (eds.) *The Johannes Weertman Symposium*, pp. 337–350. Zimbabwe: The Minerals, Metals, & Materials Society.
Wen L and Niu F (2002) Seismic velocity and attenuation structures in the top of the Earth's inner core. *Journal of Geophysical Research* 107B11: 2273 (doi: 10.1029/2001JB000170).
Wenk H-R, Baumgardner JR, Lebenson RA, and Tome CN (2000a) A convection model to explain anisotropy of the inner core. *Journal of Geophysical Research* 105: 5663–5677.
Wenk H-R, Matthies S, Hemley RJ, Mao H-K, and Shu J (2000b) The plastic deformation of iron at pressures of the Earth's inner core. *Nature* 405: 1044–1047.
Wenk H-R, Takeshita T, Jeanloz R, and Johnson GC (1988) Development of texture and elastic anisotropy during deformation of hcp metals. *Geophysical Research Letters* 15: 76–79.
Widmer R, Masters G, and Gilbert F (1991) Spherically symmetric attenuation within the Earth from normal mode data. *Geophysical Journal International* 104: 541–553.
Wood BJ (1993) Carbon in the core. *Earth and Planetary Science Letters* 117: 593–607.
Xu S, Crossley D, and Szeto AMK (2000) Variations in length of day and inner core differential rotation from gravitational coupling. *Physics of the Earth and Planetary Interiors* 117: 95–110.
Yoshida S, Sumita I, and Kumazawa M (1996) Growth model of the inner core coupled with outer core dynamics and the resulting elastic anisotropy. *Journal of Geophysical Research* 101: 28085–28103.
Yu W and Wen L (2006) Inner core attenuation anisotropy. *Earth and Planetary Science Letters* 245: 581–594.
Yukutake T (1998) Implausibility of convection in the Earth's inner core. *Physics of the Earth and Planetary Interiors* 108: 1–13.
Zhang J, Song X-D, Li Y, Richards PG, Sun X, and Waldhauser F (2005) Inner core differential motion confirmed by earthquake waveform doublets. *Science* 309: 1357–1360.

11 Experiments on Core Dynamics

P. Cardin, Université Joseph-Fourier, Grenoble, France

P. Olson, Johns Hopkins University, Baltimore, MD, USA

11.1	**Introduction**	319
11.2	**Rotational Dynamics**	320
11.2.1	Formation of Taylor Columns	320
11.2.2	Viscous Layers	321
11.2.2.1	Ekman boundary layers	321
11.2.2.2	Stewartson layers	323
11.2.3	Nonlinear Effects	325
11.2.3.1	Nonlinear resonance of the Poincaré mode	325
11.2.3.2	Geostrophic zonal motions in precession experiments	325
11.2.3.3	Parametric instabilities	327
11.2.3.4	Quasi-geostrophic turbulence	327
11.3	**Thermal Convection in a Rapidly Rotating Spherical Shell**	328
11.3.1	Experimental Conditions	328
11.3.2	Onset of Convection	328
11.3.3	Developed Convective States	329
11.3.4	Liquid Metal Experiments	331
11.3.5	Zonal Flows	331
11.3.6	Heat Flux Measurements	332
11.3.7	Compositional Convection	332
11.4	**Magnetohydrodynamics**	332
11.4.1	Hartmann Layers	333
11.4.2	Internal Magnetic Layers	333
11.4.3	Magnetic Columns	335
11.4.4	Turbulence Under the Influence of a Magnetic Field	335
11.4.5	Magnetoconvection in Rotating Fluids	336
11.5	**Experimental Dynamos**	336
11.5.1	The Riga Dynamo	337
11.5.2	The Karlsruhe Dynamo	338
11.5.3	Turbulence and Dynamo Onset	339
11.5.4	Toward a Magnetostrophic Dynamo?	340
References		341

11.1 Introduction

Laboratory experiments on the dynamics of the Earth's core are usually designed to understand the basic fluid and solid mechanical processes that affect the geodynamo, the evolution of the core, and its interaction with the mantle. As a rule, their intent is not to reproduce the full complexity of the core, because laboratory fluid experiments, like numerical simulations, are subject to many limitations. First and foremost, the volume of the working fluids is limited. The linear dimensions in fluid experiments range from centimeters to a few meters at most, a full six orders of magnitude smaller than the linear dimensions of the core. Second, only a restricted range of physical properties of the working fluids is available. Water is generally the preferred fluid for nonmagnetic experiments, whereas gallium or sodium is used in situations where magnetic fields and electric currents are required. Both of these liquids have viscous, thermal, and magnetic diffusivities that are reasonably close to the core values. However, in spite of the overall match of their transport properties, there is still a mismatch between experiments using these fluids and the core.

Two important physical attributes that distinguish experiments on the core from experiments on the Earth's mantle, for example, are the background solid-body rotation and the magnetic field, which are critical in the core but are unimportant in the solid, poorly conducting silicate portions of the Earth. In laboratory experiments, the rotation and the magnetic field are generally imposed on the fluid and are limited to upper values around $\approx$1000 rpm and $\approx$1 T, respectively. These physical parameters, along with the diffusivities, enter into the definitions of two of the critical dimensionless numbers that characterize the dynamics of the core, the Ekman number E, and the Hartmann number Ha. Because of the dependence on length scale, rotation, and magnetic field strength, both of these numbers assume unrealistic values in experiments, even though they are sometimes closer to the core in experiments than in numerical simulations.

One major advantage that laboratory experiments offer is their ability to naturally include the nonlinear effects and the associated instabilities that accompany large-amplitude forcing in fluids. One major disadvantage is the difficulty of obtaining full global measurements of highly variable quantities. As a rule, experimental flows have to be understood using a finite number of locally based measurements. Under these circumstances it is often difficult to get a general understanding of the system without the help of theory or an accompanying numerical simulation. Lastly, it often is difficult to demonstrate asymptotic behavior using a single experimental apparatus, as it is generally impractical to vary the dimensionless parameters widely enough in a given apparatus. Because of this limitation, it is usually necessary to repeat even very successful experiments using different fluids and different setups, and then make careful intercomparisons of their results. The next section in this chapter will describe how the planetary rotation influences the flow in the core. The following section describes experiments on thermal convection in rapidly rotating spheres. The next sections treat the magnetic influence on the dynamics of electrically conducting fluids, including fluids with convection and rotation. As we shall see, too little is known from experiments of the dynamics in situations where all of these effects are present simultaneously. Finally, we review some of the recent results of experimental dynamos.

11.2 Rotational Dynamics

The Coriolis acceleration derived from the basic solid-body rotation of the Earth plays a major role in the dynamics of the liquid core. In this section, we examine several simple experiments that feature the Coriolis acceleration and give physical insight into how it influences fluid motion in the core, in situations where magnetic fields are not as important. Later we examine magnetic effects on fluid motion, and then we examine situations where the effects of rotation and magnetic fields are present simultaneously.

11.2.1 Formation of Taylor Columns

Consider the motion in a cylindrical vessel of water on a table rotating at angular velocity Ω (i.e., period of rotation T). The top of the water is a free surface and the bottom is a flat rigid boundary. Centered in the bottom boundary is a circular disk made of a porous material (cf. **Figure 1**). When solid-body rotation is established in the fluid, a uniform suction is suddenly started at the disk. As a result of the suction, water is removed from the tank, and the resulting inward, converging flow is deflected into a cyclonic vortex by the Coriolis effect. The growing vortex is seen in the pictures (**Figure 2**) by the azimuthal shear that mark its edges. This shear tends to align flakes suspended in the fluid in the azimuthal direction; these produce the black regions seen in the images. The progressive growth of the vortex along the direction of the axis of rotation gives birth to an elongated flow structure called a Taylor column (Taylor, 1923). In this case, the formation of the Taylor column lasts only a few periods of rotation. The mechanism for its growth is the propagation of inertial waves originating near the porous disk. The group velocity at which energy propagates by inertial waves is given by (*see* Chapter 7)

$$\mathbf{v}_g = \frac{2}{k^3}\mathbf{k} \times (\boldsymbol{\Omega} \times \mathbf{k})$$

Assuming the wave number $k \approx \pi/d$ and propagation perpendicular to the rotation axis, we find that the rise time is $Th/4d \approx 33\,\mathrm{s}$, in good agreement with the observations (**Figure 2**). We verified experimentally that the growth rate of the Taylor column is independent of both the height of the fluid and the diameter of the tank. However, we find that it is proportional to the diameter of the suction area. Although the formation of the column is the result

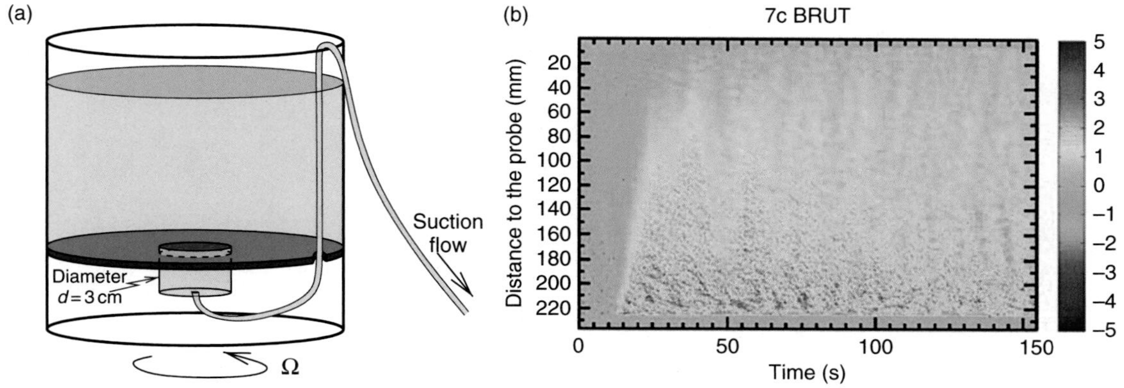

Figure 1 (a) Sketch of a cylindrical tank of water on a rotating platform. The diameter and the height of water are both $h = 20$ cm. The diameter of the porous disk at the base of the tank is $d = 3$ cm. The period of rotation of the turning table is $T = 9$ s. The water in the tank is allowed to spin up to a state of solid-body rotation and then a small amount of water is removed or injected through the porous disk. (b) Space-time evolution of the resulting axial component of the velocity, obtained using ultrasonic Doppler velocimetry from a probe located at the surface of the fluid and pointing towards the porous disk located at 220 mm. The color scale gives the axial velocity in mm s^{-1}.

of the propagation of inertial waves in the fluid, the final state of the flow is columnar and almost two-dimensional (2D) and is called 'geostrophic'.

The formation process can also be seen in the spatiotemporal diagram of the z-component of the velocity at the center of the Taylor column measured by ultrasonic Doppler velocimetry and shown in **Figure 1** (the velocimetry technique is described in Brito *et al.*, 2001). These measurements reveal many propagating structures which consist of superimposed inertial waves with different wave numbers. After the formation of the Taylor column, Doppler measurements show a linear dependence of the axial component and an invariance of the tangential component of the velocity with the height, as predicted for a geostrophic column. Applying these results to the outer core, the time needed to reach the final 2D columnar state is inversely proportional to the transverse length scale of the flow (i.e., the column diameter). Small-scale columns may therefore not be geostrophic. A typical column diameter that corresponds to this condition may be estimated by equating the turnover time of convective structure d/V to the time of formation of the column $Th/4d$. With $V = 10^{-3}$ m s^{-1}, $h = 10^6$ m, we find that scales lower than 5 km are not geostrophic in the Earth's core.

11.2.2 Viscous Layers

Following the formation of the Taylor column, the flow continues to evolve in time, although at a slower rate. The rate of evolution and the amplitude of the flow are no longer controlled by inertial waves, but instead become controlled by viscous effects concentrated in two types of shear layers: viscous boundary layers called Ekman layers and internal shear layers called Stewartson layers. Both layers may play an important role in the core.

11.2.2.1 Ekman boundary layers

Ekman layers are boundary layers in which there is a balance between the viscous force and the Coriolis acceleration. They are typically quite thin. For example, in the Taylor column experiment just described the thickness of a laminar Ekman layer is of order $\sqrt{\nu/\Omega}$, that is, $\approx$1 mm. Accordingly, it is generally very difficult to study them experimentally. In the Earth's core, the predicted thickness of the laminar Ekman layer is of the order of 1 m. Nevertheless, they have a finite range of stability, and in some situations they play an important role in the larger-scale dynamics.

Early experiments by Faller (1963) have identified two types of instabilities in laminar Ekman layers, which occur at local Reynolds numbers (defined in terms of the thickness of the Ekman layer and the velocity at the top of the layer) of 125 and 55 (called types II, respectively). Lingwood (1997) has explained these instabilities in term of convective instabilities, and the predictions of a linear stability theory based on this interpretation agree with the experimental results. According to these results, purely laminar Ekman layers in the core could be

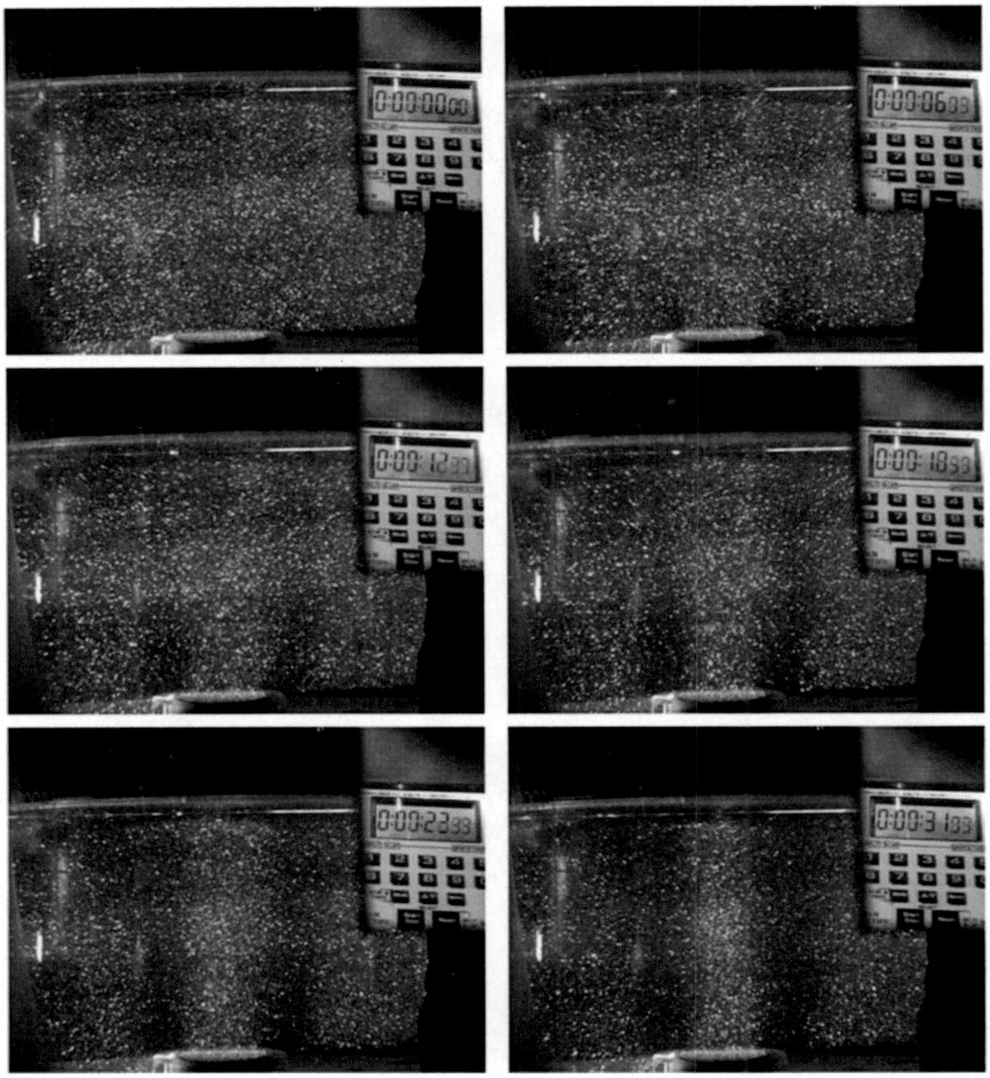

Figure 2 Visualization of the formation of a Taylor column in kalliroscope fluid. The injection rate is $7 \times 10^{-7}\,\mathrm{m^3\,s^{-1}}$ and the period of rotation of the platform is $T = 20\,\mathrm{s}$.

marginally unstable, because their local Reynolds numbers $(V/\sqrt{\nu\Omega})$ are around 100. Assuming a molecular viscosity of $\nu = 10^{-6}\,\mathrm{m^2\,s^{-1}}$ for the fluid in the outer core gives a typical velocity of $10^{-3}\,\mathrm{m\,s^{-1}}$ to trigger boundary-layer instabilities, slightly larger than velocities at the top of the core inferred from the geomagnetic field secular variation (see Eymin and Hulot, 2005).

Ekman boundary layers produce a pumping effect on the fluid outer core that acts perpendicular to the core–mantle and inner–outer core boundaries. Ekman pumping has its origin in the requirement for mass conservation in the Ekman boundary layer (Greenpan, 1968). Although the velocity induced by the Ekman pumping is quite small, a factor $E^{1/2}$ smaller than the free stream velocities parallel to the boundary, it nevertheless can have a large influence on geostrophic flows. Ekman pumping tends to suppress the jump of vorticity between the fluid and the boundary, for example. In rotating spin-up, where a fluid adjusts to changes in the rotation rate of its container, Ekman pumping is the mechanism that accelerates the core of the fluid to the new rotation rate. Using an ultrasonic Doppler velocimetry, we have measured azimuthal velocity in the bulk of a rotating sphere during a spin-up experiment and checked the linear theory of Greenspan (Greenspan, 1968) up to Rossby number ($\Delta\Omega/\Omega$) of 0.3 (Brito *et al.*, 2004). The Greenspan theory is valid for nonmagnetic planetary cores ($Ro \ll 1$) if one considers only geostrophic motions with longer periods than 1 day.

Ekman layers may include singularities, particularly on curved surfaces (Stewartson and Roberts, 1963). For stationary Ekman layers in spherical shells, the singularity occurs at the critical latitude of 0° in the equatorial region where the rotation vector is parallel to the layer. For oscillating flows with pulsations between 0 and 2Ω, where Ω is angular velocity of the rotating system, the critical latitude moves from the equator to the pole in both hemispheres, the amount depending on the pulsation frequency. Ekman layer singularities have been observed experimentally in precessing experiments (Noir *et al.*, 2001a) where they lead to the formation of internal shear layers in the fluid that are aligned along characteristic surfaces (Kerswell, 1995). Noir *et al.* (2001a) have measured these time-oscillating detached shear layers in a precessing experiment using Doppler velocimetry. They are generally invisible with flakes or dye because they oscillate at frequencies close to the container frequency (exactly at the container frequency in the Noir *et al.* (2001a) experiment). This internal shear layer consists of a main jet connected to the Ekman layer singularity and alternating jets parallel to the main one (Noir *et al.*, 2001b). Its spatiotemporal structure may be understood in terms of inertial waves propagating inside the fluid (*see* Chapter 7). Application to the liquid outer core, if we ignore the action of the magnetic field and other nonlinear effects, suggests that this mechanism will generate inertial waves of wavelength 20 km with typical velocity of $v = 6 \times 10^{-6}\,\mathrm{m\,s^{-1}}$.

11.2.2.2 Stewartson layers

Stewartson layers (Stewartson, 1957) are internal shear layers located along the edges of Taylor columns, as in **Figure 1**. Compared to Ekman boundary layers, Stewartson layers have a much larger thickness, proportional to $E^{1/4}L$, where $E = \nu/\Omega L^2$ is the Ekman number, ν the kinematic viscosity, Ω the rotation rate, and L a typical length scale. Their dynamics is complicated by the presence of a nested asymptotic sublayer of thickness $E^{1/3}L$. The classical experimental method for producing Stewartson shear layers is to apply a differential rotation to a disk in a cylinder of fluid in solid-body rotation. Hide and Titman (1967) studied the stability of these layers, and found that at a critical value of the Rossby number $\Delta\Omega/\Omega$ geostrophic vortices develop inside the Stewartson layer. They obtained very different results for positive versus negative Rossby numbers, a result that has been confirmed by new experiments (Früh and Read, 1999) but is still not well understood. Nevertheless, Früh and Read (1999) demonstrated that the onset of this instability is dictated by the local Reynolds number, which is a measure of the shear across the layer, and scales like $E^{1/4}L$ relative to the molecular dissipation ν/L^2. The critical value of this local Reynolds number is around 19.

We have studied the stability of Stewartson layers in a spherical Couette experiment (Schaeffer and Cardin, 2005) consisting of a rotating spherical shell with a differentially rotating inner core. **Figure 3** shows photographs of the Stewartson layer below and above the threshold of instability, respectively. Because of the spherical geometry, the unstable Stewartson layer generates Rossby waves. Generation of these Rossby waves (Hide, 1966) relies on the variation of the height of a column as the column moves radially away or toward the rotation axis. When a fluid column moves outward, its height

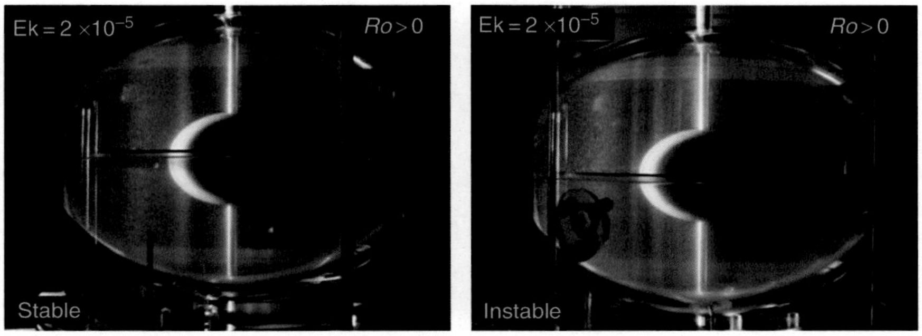

Figure 3 Instability of a Stewartson layer in a Couette shear flow between differentially rotating spheres (Schaeffer, 2004).

is reduced by the spherical outer boundary, and conservation of mass requires the cross-sectional area of the column to increase. By conservation of angular momentum, this causes a decrease in its vorticity, a consequence of the Kelvin Circulation Theorem. Symmetry considerations dictate that an inward motion generates a cyclonic vortex and outward motion generates an anticyclonic vortex. With a spherical outer boundary, these waves propagate in the prograde azimuthal direction. Columnar Rossby waves have been measured experimentally using an ultrasonic Doppler techniques. **Figure 4** shows spatio-temporal diagrams of the radial velocity in the equatorial plane. The Rossby wave fills the whole shell with spiral arms. It is propagating in time in the prograde direction. The onset of these waves is controlled by the critical Rossby number $(\Delta\Omega/\Omega)$, which in this geometry is of the order of $\beta E^{1/2}$ where $\beta = (1/h)(\mathrm{d}h/\mathrm{d}s)$, and where h is the height of the column and s the cylindrical radius. In terms of local Reynold number, this relationship can be written $\beta E^{-1/4}$ which explicitly contains a dependence on the global rotation rate, in contrast with the situation in the experiment by Früh and Read (1999) made in a flat geometry. In absence of a magnetic field, any variation of the period of the inner core larger than

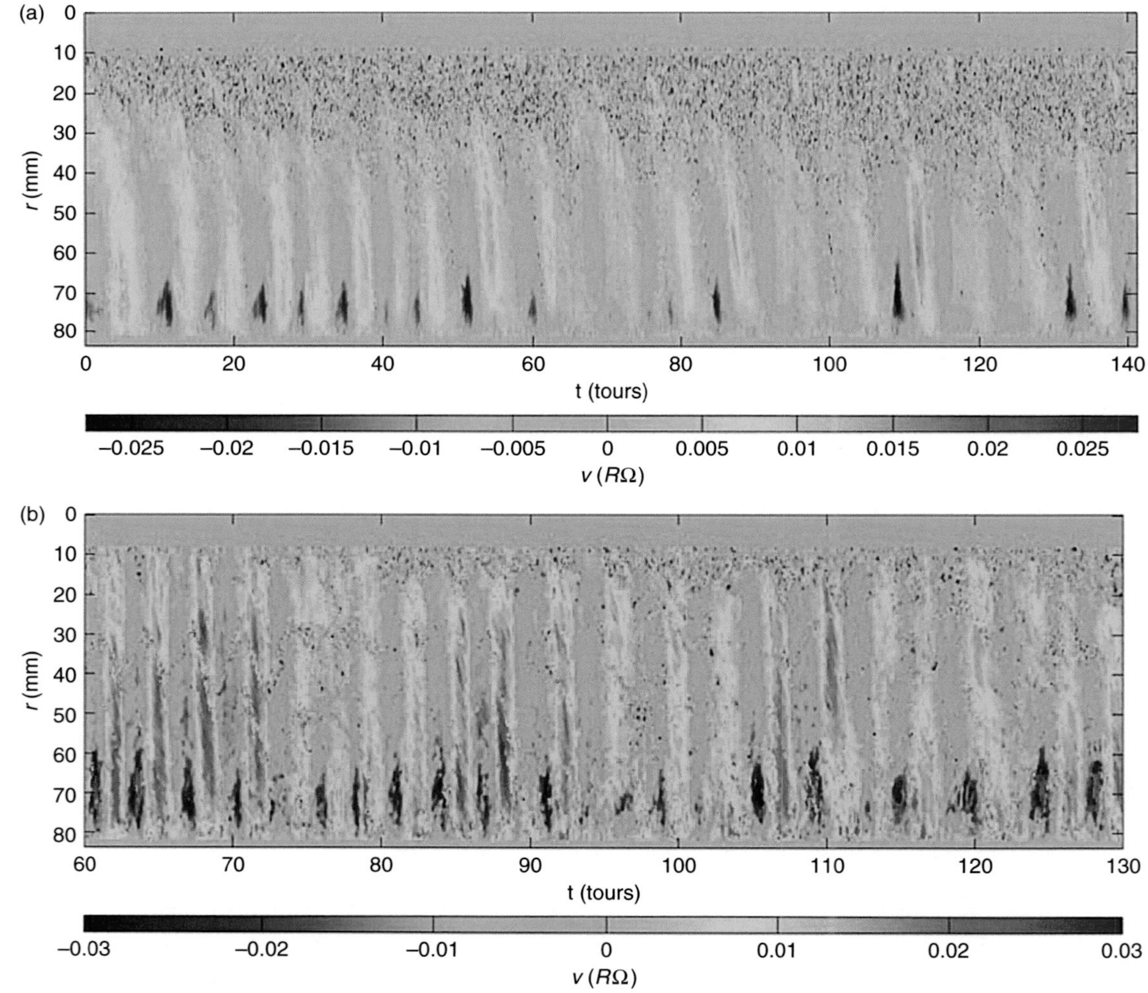

Figure 4 Propagation of a Rossby wave in a rotating spherical Couette flow. Time evolution of the radial velocity in the equatorial region (colors are indicative of the velocity amplitude). r is the distance from the rotation axis to the ultrasonic probe. The differentially rotating inner core is located at $r = 80$ mm. The velocity is plotted in nondimensional units of $R\Omega$. The Ekman number is $E = 2.1 \times 10^{-5}$. (a) $Ro = \Delta\Omega/\Omega = 0.17$ (just above critical). (b) $Ro = 0.68$.

10^{-2} s will generate Rossby waves at the tangent cylinder, but the presence of magnetic fields strongly affects the Stewartson layer at the tangent cylinder and also affects wave generation there (Dormy *et al.*, 1998).

11.2.3 Nonlinear Effects

Many nonlinear effects have been observed in rotating fluid experiments. One important source of the kinetic energy for nonlinear motions is precession. As described in detail in Chapter 7 the axis of rotation of the solid Earth precesses with a period of about 26 000 years, and the response of the fluid core to the precession consists of a solid-body rotation about an axis which lags the rotation of the solid by a small angle, the so-called Poincaré response, plus other, more complex flows.

11.2.3.1 Nonlinear resonance of the Poincaré mode

The pioneering work of Malkus (1968) was the first to verify experimentally that the linear response of the fluid to precession is component of solid-body rotation about an equatorial axis, as predicted by Poincaré (1910). The overall solid-body rotation of the fluid is slightly tilted away from the minor axis of the spheroid and the two axis form an angle θ which can be measured in laboratory (Vanyo *et al.*, 1995) and compared to the theoretical viscous correction to the Poincaré mode direction predicted by Busse (1968). In our experiments (Noir *et al.*, 2001b, 2003), we have observed jumps of the angle θ by varying continuously the retrograde precession forcing as shown in **Figure 5**. This effect can be understood as a nonlinear resonance between the gravest (lowest frequency) fundamental inertial mode in the spheroidal cavity and the precession.

In the course of its evolution, the physical parameters of the Earth's rotation have changed with time. Periodic resonances between the tilt-over mode and different nutations (annual, semiannual, etc.) should have produced large-amplitude excursions of the axis of rotation of the liquid core and affected global geodynamics (Greff-Lefftz and Legros, 1999). For these resonances, nonlinear theory reduces to the linear analysis used in Greff-Lefftz and Legros (1999), but for larger forcings (as, e.g., for the Bradley nutation) the nonlinear resonance condition applies.

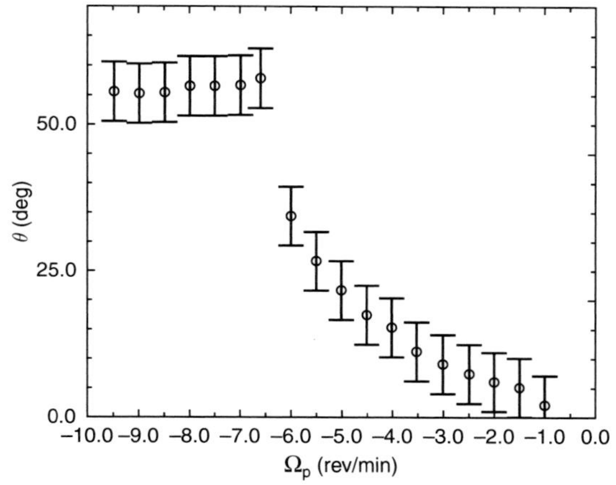

Figure 5 Rapid variations of the position of the axis of the Poincaré mode in a precessing experiment. The obliquity of the experimental spheroid is $\alpha = 20°$ and the rotation of the spheroid is 300 rpm. The angle between the minor axis of the spheroid and the Poincaré axis is plotted against the rotation rate Ω_p of the precession. Around the value $\Omega_p = -6.2$ rpm, a jump is observed. From Noir J, Jault D, Cardin P (2001a) Numerical study of the motions within a slowly precessing sphere at low Ekman number. *Journal of Fluid Mechanics* 437: 283–299.

11.2.3.2 Geostrophic zonal motions in precession experiments

Precession experiments exhibit very nicely the characteristic cylindrical flow structures of rotation-dominated fluids (**Figure 6**). The flake images in **Figure 7** show some of these structures. The shear zones in these images correspond to stationary geostrophic azimuthal motions superimposed on the Poincaré mode. These motions were measured quantitatively the first time by Malkus (1968) and his measurements have subsequently been confirmed. In our experiments, dye has been injected in the fluid and using a camera attached to spheroid, we followed the deformation of the dye patch. Typical photographs are shown in **Figure 8** and the results confirm Malkus' findings (**Figure 9**). Apart from small variations, the primary motion consists of a cylinder of fluid rotating faster than the Poincaré mode and located at a dimensionless radius of 0.9. Noir *et al.* (2001b) propose to explain these geostrophic motions as nonlinear effects in the Ekman layer singularity while Hollerbach and Kerswell (1995) proposed that this results from nonlinear interactions of inertial waves in the fluid. Spherical numerical simulation reproduces the cylindrical shear layers with the correct wavelength (see **Figure 9**) but the amplitude is incorrect by a factor 3. This

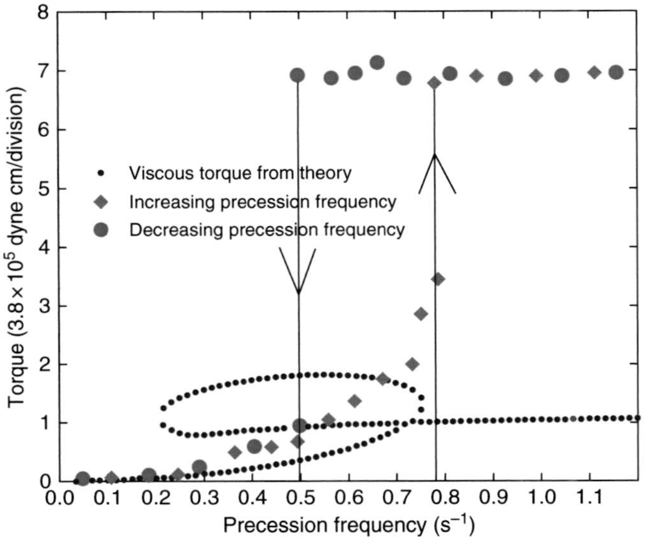

Figure 6 Torque variations along the rotating axis of the spheroid versus precession rate. Two jumps and an hysteresis are observed (Malkus, 1968). The viscous torque deduced from a torque balance is also plotted and may explain the occurrence of the jumps, although its amplitude is too small by a factor of 3.

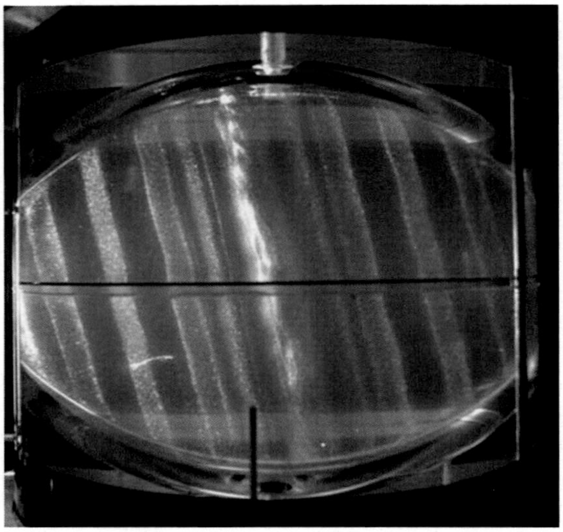

Figure 7 Side view of a meridional plane (illuminated by flakes reflection) of a precessing flow. The obliquity of the spheroid is $\alpha = 20°$, the rotation of spheroid is 300 rpm. The precessing table spins progradely with an angular velocity of 10 rpm. Cylindrical shear zones denote the presence of concentric geostrophic zonal flows (description of the experimental set-up in Noir *et al.* (2003)).

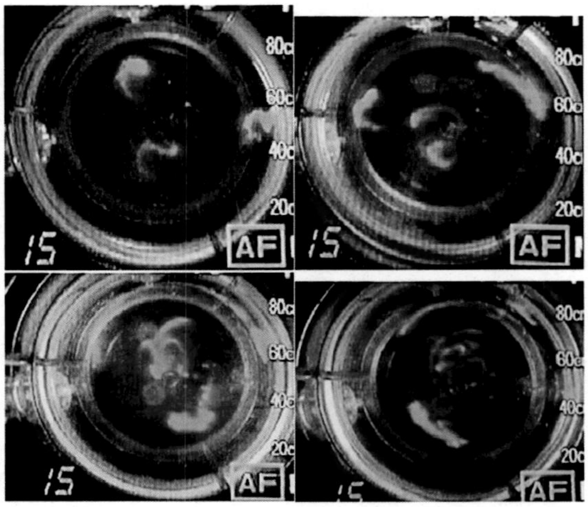

Figure 8 Top view of precessing spheroid with three patches of dye injected at latitudes 30°, 55°, and 70°. Four different snapshots are presented. The dye is transported by the geostrophic zonal motions and shows a double vortex in the center (description of the experimental set-up in Noir *et al.*, 2003).

discrepancy is still unsolved. Even with this discrepancy, these results allow us to estimate the amplitude ($\approx 10^{-5}$ ms^{-1}) and the size ($\approx$20 km) of the stationary geostrophic cylinders in the outer core, which are predicted to be located at 30° in absence of magnetic field (Noir *et al.*, 2001a). Geostrophic cylinders should also exist at the critical latitude of the inner core but they have not been described yet.

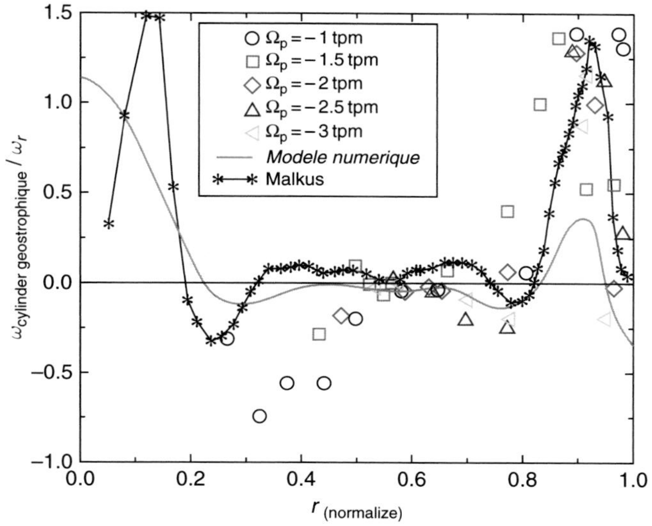

Figure 9 Amplitude of the geostrophic motions normalized by the angular velocity of the Poincaré mode versus the distance along the major axis of the spheroid for different forcings. The obliquity of the spheroid is $\alpha = 20°$ and its rotation is 212 rpm. The black solid line with stars shows the results of Malkus (1968) and the gray solid line shows the numerical results of Noir *et al.* (2001b).

11.2.3.3 Parametric instabilities

Elliptic and shear instabilities may be present in precession experiments, and also may be present in the core (Kerswell, 2002). These instabilities develop from a resonance between two inertial wave modes and a background shear flow. All these ingredients are present in tidally driven flows and precessing flows (*see* Chapter 7). As shown by Malkus (1989), in a cylindrical experiment the parameters can be tuned so as to get sudden, catastrophic events called elliptical instabilities. The elliptical instability first grows linearly (Kerswell, 2002) and then nonlinearly, destabilizing the background flow to small-scale turbulence, which grows until it reaches viscous saturation (Eloy *et al.*, 2000). Lacaze *et al.* (2004) have observed elliptical instabilities in spheroidal geometry. Their experimental findings agree with the theoretical prediction. But intermittent turbulent regimes similar to those observed in cylindrical symmetry have not yet been observed in precessing experiments. Peharps the two central vortices in the last photograph in **Figure 8** and the vertical oscillation of the axis of rotation of the fluid in **Figure 7** are the spherical equivalents of the structures observed in cylindrical geometry, but more experimental work is needed to confirm this. Without a better understanding of the turbulent breakdown of the spherical elliptic instabilities in precessing or tidally driven experiments, any application to planetary cores would be highly speculative (Aldridge and Baker, 2003). This subject deserves further investigation.

11.2.3.4 Quasi-geostrophic turbulence

As the forcing increases, rotating flows become turbulent, through instabilities originating in the fluid volume and by boundary-layer instabilities. Ultimately this turbulence becomes 3-D, but for moderate forcings, the turbulence remains nearly geostrophic. According to the Taylor–Proudman constraint, it is possible for columnar flow to become turbulent in planes perpendicular to the axis of rotation long before it becomes turbulent in the axial direction. Ekman pumping and topographic effects (or β-effect) both strongly influence this quasi-geostrophic turbulence, which differs from 2D turbulence as generally studied by physicists (Tabeling, 2002). For β-plane geometry flows, Rhines (1975) predicted a dominant turbulent scale $\sqrt{U/\beta}$ and scaling dependence for the spectra of kinetic energy. Zonal jets in giant planets are thought to be the result of the inverse cascade of kinetic energy from small scales to the so-called Rhines scale (k^{-5} law, where k is the wave number). Many experiments have tried to check Rhines' theory, using turbulence in β-plane flows. Here we present two of these. A group in

Texas led by H. Swinney (see for example Aubert *et al.*, 2002) studied the inverse cascade and the formation of zonal jets in a cylindrical β-tank with fluid injected at the bottom by small tubes. Using the large Coriolis Table at Grenoble, Read *et al.* (2004) studied the same process by injecting a spray of salty water onto the top of their 10 m rotating tank. Both experiments showed evidence of azimuthal axisymmetric flows, but neither was able to precisely measure the power-law variation of the spectra, which is often very difficult to retrieve experimentally. In a spherical shell, the β-effect varies with cylindrical radius and there is also a discontinuity at the tangent cylinder of the inner core. These variations cannot be neglected in deriving the expected Rhine's scale and may cause spherical shell turbulence to differ considerably from the character of β-plane turbulent flows. This is another topic that would benefit from additional experimental investigation.

11.3 Thermal Convection in a Rapidly Rotating Spherical Shell

11.3.1 Experimental Conditions

Apart from space experiments made in the microgravity environment of Earth's orbit (Hart *et al.*, 1986; Egbers *et al.*, 1999), all laboratory investigations of rotating spherical convection in planetary cores rely on the centrifugal force as a substitute for the radial gravity force. It is not difficult for centrifugal acceleration to dominate over laboratory gravity in a rotating apparatus. For a typical experimental size of 10 cm, a spin rate in excess of 100 rpm is required to fulfill this condition. It is perhaps surprising that the centrifugal acceleration can adequately model the spherical gravity of a planet for rotating fluid mechanics, but experiments have shown it can (Busse, 1970). The centrifugal acceleration varies linearly with the cylindrical radius, like the component of the central gravity perpendicular to the axis of rotation of a planet with a uniform density. The difference in direction (outward vs inward) between the two accelerations is taken into account by a reverse temperature gradient (cooler at the center) in the experiments, which leads to buoyancy (Archimedean) forces having their proper orientation in the experiment. The analogy is further strengthened because in a rotating planet the axial component of the central gravity is mainly balanced by pressure forces, a consequence of the Taylor–Proudman constraint. Thermal convection is produced experimentally in spherical shells by imposing thermal boundary conditions at the inner and outer boundary, the higher temperature on the outer boundary for the reasons given above (Carrigan and Busse, 1983). If the boundary temperatures are uniform, then the temperature varies on surfaces of constant centrifugal potential in the direction of the axis of rotation. This results in an ageostrophic buoyancy force perpendicular to the surface of the cylinder. A thermal wind (an ageostrophic azimuthal flow) is therefore produced to balance this buoyancy force in the experiment, an effect which is practically zero in the Earth's core. The amplitude of spurious flow is given by the thermal wind balance, which can be written as

$$\frac{\partial u_\phi}{\partial z} = \frac{1}{2}\alpha\Omega s \frac{\partial T}{\partial z}$$

where z is the coordinate along the axis of rotation, s the cylindrical radius, α the thermal expansion, and Ω the rotation rate of the sphere. The maximum velocity of this thermal wind is located close to the inner-core equator and is given by Carrigan and Busse (1983):

$$u_\phi = \frac{\alpha\Omega\,\Delta T}{2}$$

where ΔT is the imposed temperature difference between the inner and outer boundary. It represents a azimuthal flow of 2 mm s^{-1} for a rotation of 100 rpm with 1 K of temperature difference ($\alpha = 2 \times 10^{-4}$ K^{-1} for water).

11.3.2 Onset of Convection

Pioneering experiments by Busse and Carrigan (1976) have shown that the onset of thermal convection in a rapidly rotating sphere may be described by propagating thermal Rossby waves (Busse, 1970), which consist of columns of alternating positive and negative vorticity aligned with the axis of rotation and propagating in the prograde direction just outside the tangent cylinder of the inner core. Busse and his colleagues performed the first quantitative determination of the onset of convection in rotating annulus with a constant slope for bottom and top boundary (Busse and Carrigan, 1974) and a spherical shell, using water as the working fluid, and flake visualization (Carrigan and Busse, 1983). Their results agree relatively well with the theoretically predicted $E^{-1/3}$ asymptotic law for the critical Rayleigh number for convective onset at low Ekman number. Cordero and Busse (1992) determined the critical parameters for convective onset in a rotating hemispherical shell using temperature measurements. Laboratory gravity was used with the centrifugal acceleration in these experiments to generate paraboloidal surfaces of potential of gravity as close as possible to spherical surfaces in a 'Southern' Hemisphere geometry. They also took into account the thermal wind effect in determining the critical frequency of the thermal Rossby waves. In addition they observed the spiral structure of the convection columns and transitions to quasi-periodic states and chaos (Cordero, 1993). Following these initial studies, many discrepancies between theoretical, numerical, and experimental determinations of the onset of convection in rotating spherical shells were subsequently discovered. Using an approximate method to solve the linear stability problem, Jones *et al.* (2000) showed how to resolve some of the discrepancies between theoretical and numerical results. Dormy *et al.* (2004) completed the comparative study between numerical and analytical approaches, especially for the differential heating case. A close comparison with experimental data is often not easy but E. Dormy has computed the critical parameters for the Cordero and Busse (1992) experiment and found a critical frequency of ω_c/Ω of 2.2×10^{-3} instead of 2.6×10^{-3} found experimentally (E. Dormy, personal communication).

11.3.3 Developed Convective States

The convective flow remains columnar in a rotating sphere above the threshold for convective onset provided the amplitude of the buoyancy forces remains smaller than the Coriolis forces. Cardin and Olson (1994) have performed experiments up to 50 times the critical Rayleigh number in water. We observed that the critical Rossby waves disappear above the onset. Numerous columnar vortices fill the spherical shell with a quasi-geostrophic form of turbulent convection (**Figure 10**). In this regime, strong thermal plumes generate retrograde vortices at the inner-core boundary prolongated outward by a prograde spiraled ribbon-shaped street of retrograde vortices. The structures that characterize fully developed rotating convection have been clearly observed by Sumita and Olson (2000) in a hemispherical experiment with water. Just beyond the onset of convection, Rossby waves tend to fill the sphere, and the azimuthal number of convective features increases along a radius. Consequently, radial bifurcations or trees are observed in the field of flow (cf. **Figure 11**). This bifurcation is understood as an effect of the increase of the boundary slope (β-effect) with the cylindrical radius, which controls the azimuthal wave number of the convection. As the thermal forcing becomes stronger (more than 8 times critical), the Rossby waves are expelled toward the outer portion of the spherical shell, where the large slope strongly inhibits radial motion (**Figure 11**) as the innermost part of the convection

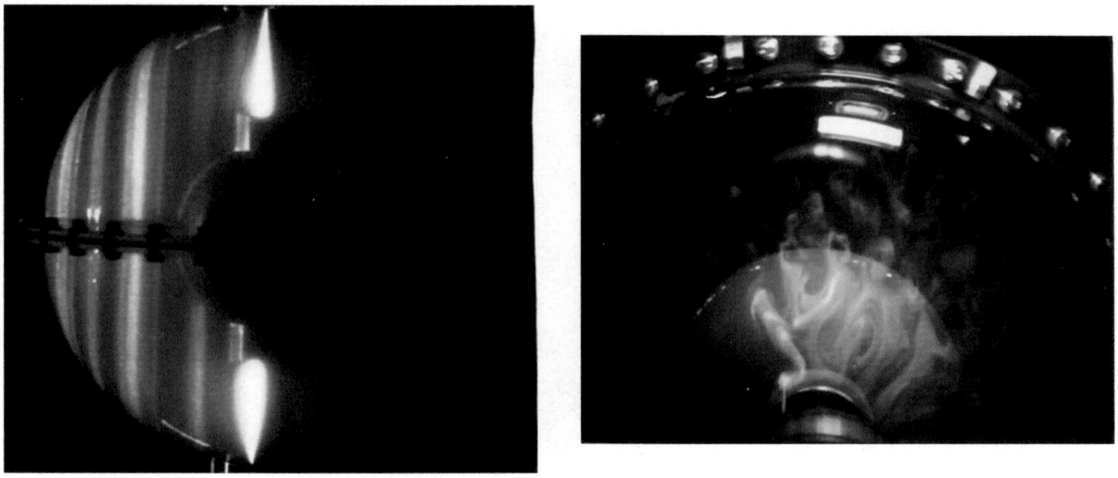

Figure 10 Experimental results on the structure of thermal convection at 50 times critical. *Left*: Side view of chaotic convection using a light beam and flakes inside the fluid. *Right*: Top view showing equatorial section of the columns and their organization using fluorescein dye released at the North Pole of the inner core. Rotation is anticlockwise. Reproduced from Cardin P and Olson P (1994) Chaotic thermal convection in a rapidly rotating spherical shell: Consequences for flow in the outer core. *Physics of the Earth and Planetary Interiors* 82: 235–259, with permission from Elsevier.

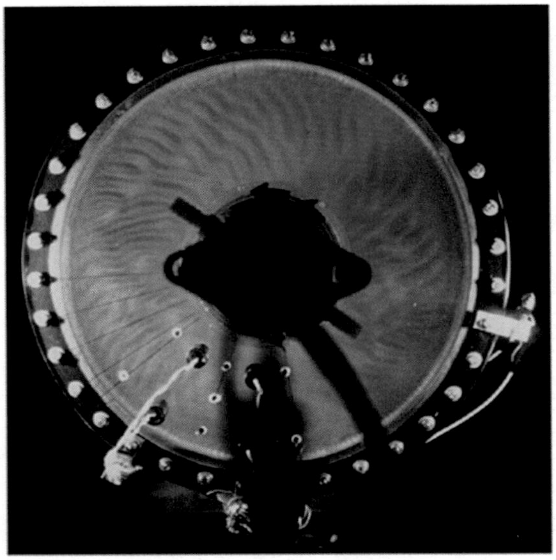

Figure 11 Experimental results on the structure of thermal convection at 6 times critical in a rapidly rotating hemispherical shell. Top view showing equatorial section of the columns seen by flakes reflection. Rotation is anticlockwise. Reproduced from Sumita I and Olson P (2000) Laboratory experiments on high Rayleigh number thermal convection in a rapidly rotating hemispherical shell. *Physics of the Earth and Planetary Interiors* 117: 153–170, with permission from Elsevier.

becomes more turbulent. This regime has been called 'dual convection'. Aubert *et al.* (2003) argued that this transition between inertial regimes and Rossby wave regimes is controlled by a local Reynolds number, the ratio of turnover time, and the Rossby wave period.

Quantitative measurements of the radial velocity in rapidly rotating convection have been made using the ultrasonic Doppler velocimetry (Brito *et al.*, 2001; Aubert *et al.*, 2001, 2003; Gillet *et al.*, 2007a). A typical space-time diagram is shown in **Figure 12**. Thermal convection tends to be stronger in the vicinity of the inner cylinder, for the reasons discussed above. The time average of the root mean square (rms) radial velocity shows a maximum close to the inner cylinder (**Figure 13**). Following Cardin and Olson (1994), Aubert *et al.* (2001) proposed that the maximum of measured radial velocity can be matched to a scaling law $(\alpha g Q / \rho C_p \Omega^3 D^4)^{2/5}$ where Q is the heat flux, α the thermal expansion, g the acceleration of gravity, ρ the density, C_p the heat capacity, Ω the rotation rate, and D the gap. This scaling results from a balance between Archimedean (buoyancy), inertial, and Coriolis forces, and some other assumptions. The quality and the range of the experimental data may allow for other scalings. The typical radial size of the columns in **Figure 12** is between 10 and 20 mm, not far from the critical size at the onset (15 mm) for these experimental parameters. It has proven to be difficult to measure any change in vortex size over the range of accessible experimental control parameters, although there is some expectation on theoretical grounds to observe larger vortices fed by the inverse cascade of energy in 2D turbulence as the Rayleigh number increases. Rhines (1975) also suggested that the inverse cascade stops at the scale $\propto \sqrt{U_{\text{conv}}/\beta}$ which corresponds to the balance between the Reynolds stresses and the vortex stretching in the vorticity equation. Higher-rotation-rate experiments are needed to study the evolution of the size of the vortices and other turbulent effects.

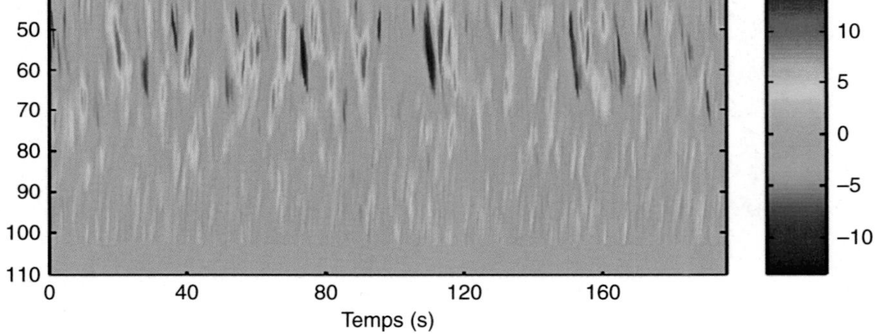

Figure 12 Experimental results on the structure of thermal convection for 40 times the critical Rayleigh number, with $E = 6.5 \times 10^{-6}$ and $P = 7$. Abcissa is time in s, ordinate is the depth in the spherical shell, outer sphere is at 110 mm, and the inner cylinder is located at 40 mm. Colors represent the radial velocity in mm s^{-1} (red is outward, blue inward). Strong anticyclones moving retrograde close the inner cylinder, appearing as a time oscillation (blue–red) (as in (110,50) region). For more details see Gillet *et al.* (2007a).

11.3.4 Liquid Metal Experiments

Using a liquid metal as the working fluid allows experimental study of thermal convection at low Prandtl number ($P = \nu/\kappa$, where ν is the kinematic viscosity and κ the thermal diffusivity) which is the regime of interest for core dynamics in terms of this parameter. Even if liquid metals (gallium and sodium) have a somewhat lower Prandtl number ($P \approx 0.025$) than expected for the liquid iron at core conditions ($P \approx 0.1$), they are a better model than water ($P \approx 7$), and they have the critical property that their thermal diffusivity is much larger than their viscosity. Liquid metal are not easy to use because they are opaque, and because of their very high thermal conductivity, high heat flux are needed to initiate the thermal convection. At Rayleigh numbers only 4 times critical, convective velocities are much larger than corresponding ones in water (Aubert *et al.*, 2001, 2003;Gillet *et al.*, 2007a), and Reynolds numbers as large as 2000 have been obtained (cf. **Figure 13**). Measured radial velocities seems to follow the scaling law described above for water (Aubert *et al.*, 2001), but the size of the vortices is larger in the gallium experiment and the inertial convection region (the vortices) occupy about two-thirds of the gap. The Rossby wave convection is pushed outward and may be responsible for large variations of the rms velocity in the profile (at large radius) shown in **Figure 13**.

11.3.5 Zonal Flows

Azimuthal mean flows are almost always observed in rotating spherical experiments, and they were detected in the early rotating annulus experiments in Busse and Hood (1982). Cardin and Olson (1994) and Sumita and Olson (2000) reported a slow retrograde azimuthal transport of dye injected around the inner core (cf. **Figure 10**) at Rayleigh numbers above a few times critical. Using ultrasonic measurements, Aubert *et al.* (2001) reported the first quantitative measurements of zonal flows for water and gallium. This study has been augmented by Gillet *et al.* (2007a) using more precise measurements (typical examples in **Figure 14**). A larger zonal flow is present in the gallium experiment compared to the water experiments. This difference is the result of much stronger convective velocities in 'low Prandtl number' liquid-metal experiment, which is combined with the nonlinear effect associated with the axisymmetric geostrophic flow. Just above the onset of convection we expect the zonal flow to increase quadratically with the convective velocity (as in the water experiment in Aubert *et al.* (2001). For larger forcings, Gillet *et al.* (2007a) proposed a 4/3-power-law scaling, based on the Rhines scales for the size of the zonal and non-axisymmetric flows. In case of gallium experiment, the presence of the large zonal flow strongly affects the organization of thermal convection, even just above the onset (Plaut and Busse, 2002).

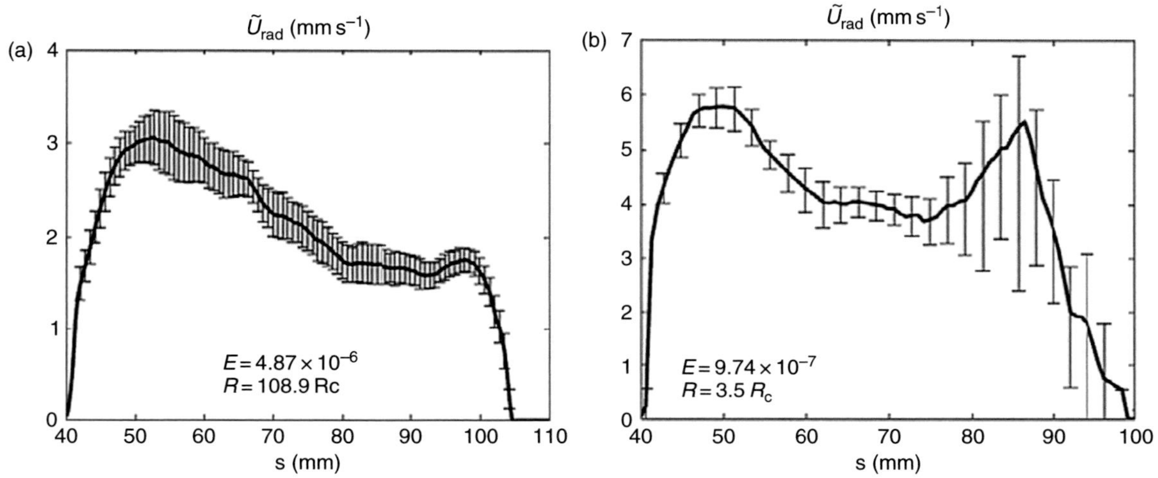

Figure 13 Experimental measurement of the radial convective velocity vs the radius. (a) In water for $Ra = 102\,Ra_c$, $E = 4.5 \times 10^{-6}$, $P = 7$. (b) In Gallium for $Ra = 3.5\,Ra_c$, $E = 7.9 \times 10^{-7}$, $P = 0.025$. For more details see Gillet *et al.* (2007a).

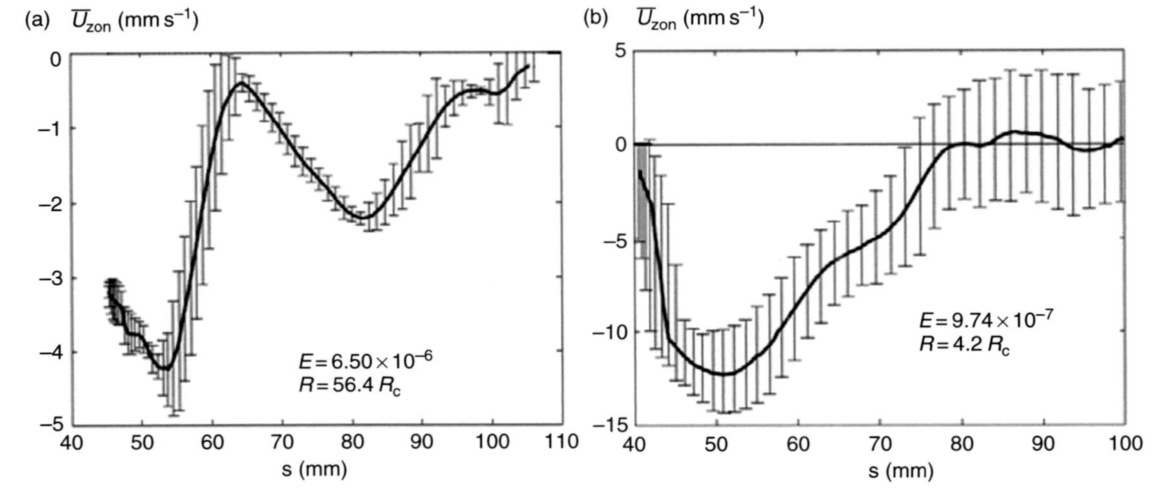

Figure 14 Experimental measurement of the zonal velocity vs the radius in rotating convection. (a) In water for $Ra = 102\,Ra_c$, $E = 4.5 \times 10^{-6}$, $P = 7$. (b) In Gallium for $Ra = 3.5\,Ra_c$, $E = 7.9 \times 10^{-7}$, $P = 0.025$. For more details see Gillet *et al.* (2007a).

Additional effects arise when the forcing is not uniform. Sumita and Olson (2002) studied the effect of a heterogeneous boundary heat flux on the dynamics. For some conditions, they observed the formation of large-scale spiraling front with a jet. The jet traverses the entire fluid shell, all the way from the outer to the inner boundary in some cases. The front is stationary with respect to the thermal boundary heterogeneity, and it separates two distinct types of large-scale flow which is superposed on a small-scale flow consisting of thermal convective vortices.

11.3.6 Heat Flux Measurements

Temperatures are generally easy to measure in experiments and can give much information on the dynamics (amplitude, fluctuations, transport, etc.). In this section we will focus on heat transfer measurements. Sumita and Olson (2000, 2003) performed a series of experiments with water and silicon oil to measure the ratio between the Nusselt number and the Rayleigh number in the hemispherical shell setup. The Nusselt number is measured by the difference of temperature of the cooling liquid at the entrance and exit of the inner sphere and the Rayleigh number is measured by the jump of temperature across the gap. J. Aurnou has compared these results with numerical results in **Figure 15**. He found a power law with a 0.4 exponent while numerical results fit a 0.55 power law. The dependence of the Prandtl number is not trivial to evaluate, and even if the numerical results seem to show little or no dependence on it, we know from many experiments that the dynamical regimes are very different in liquid metal versus water experiments, suggesting that Prandtl number differences may persist into the asymptotic regime.

11.3.7 Compositional Convection

Cardin and Olson (1992) modeled compositional convection in the outer core by the release of denser fluid (a mixture of water and sugar) at the inner spherical surface of a rotating spherical shell filled with pure water. They observed very thin, tenuous plumes, which moved radially from the inner boundary to the outer boundary. An analysis in terms of buoyancy fluxes (compositional vs thermal) led to the proposal that the dynamics are mainly dominated by thermal convective structures, in spite of the fact that the compositional buoyancy is larger.

11.4 Magnetohydrodynamics

We now consider the action of the Lorentz force on the dynamic of a nonrotating, electrically conducting fluid. The experiments of this type are substantially more difficult, as they entail the problems of handling liquid metals and the problems of producing and controlling strong magnetic fields.

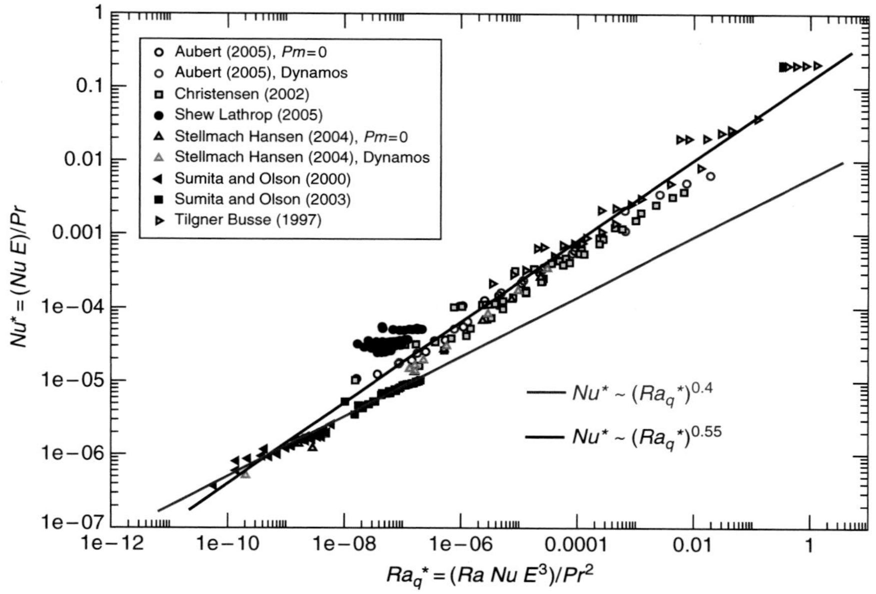

Figure 15 Modified Nusselt number vs modified Rayleigh number for experiments and numerical simulations. Different power laws have been found. J. Aurnou (personal communication).

The role of the magnetic forces in these experiments is somewhat analogous to the rotational constraint in the previous experiments, and the analogy extends to the formation of boundary layers and magnetic columns. Ultimately, however, we are interested in systems where both rotation and magnetic field are present.

11.4.1 Hartmann Layers

In presence of a magnetic field, diffusive magnetic layers called Hartmann layers form at the interfaces between materials with different electrical properties (Hartmann and Lazarus, 1937; Moreau, 1990). In a Hartmann layer of thickness $Ha^{-1}L = \sqrt{\mu\rho\eta\nu}/B$ the Lorentz forces are in balance with the viscous forces. Murgatroyd (1953) found that the friction factor in those Hartmann layers is a function of the ratio of the Reynolds number (*Re*) to the Hartmann number (*Ha*), which is equivalent to a local Reynolds number defined in terms of the thickness of the Hartmann layer. The stability of the viscomagnetic Hartmann layer has been studied experimentally by Moresco and Alboussière (2004) and the transition to turbulence (and vice versa, the transition to laminar flow) has been found to occur at a local Reynolds number of 380 ($\pm 10\%$) (see **Figure 16**). The experimentally determined critical value of the local Reynolds number is two orders of magnitude smaller than the one deduced from linear stability analysis, which implies that nonlinear effect plays an important role in the stability of Hartmann layers. It is also possible that the roughness of the boundary itself plays an important role to destabilize the Hartmann layers, which are very thin in many experimental situations (sometimes as thin as 3 μm).

The stability of the secondary type of Hartmann layer, the type that are oriented parallel to the magnetic field (with characteristic thickness $Ha^{-1/2}$ L), has been examined by Burr *et al.* (2000). Above the critical Reynolds number ($2000 < Re < 5000$), elongated vortices with their axes aligned with the imposed magnetic field appear above the boundary layer. In the absence of strong rotational effects, magnetic Hartmann boundary layers at the core–mantle boundary are expected to be stable, given the typical speeds of flow in the core inferred from geomagnetic secular variation.

11.4.2 Internal Magnetic Layers

Free-standing, detached diffusive magnetic layers may exist in the interior of an electrically conducting fluid. As with Stewartson shear layers, they occur in specific geometries such as a spherical shell, which makes them of interest for core dynamics. Dormy

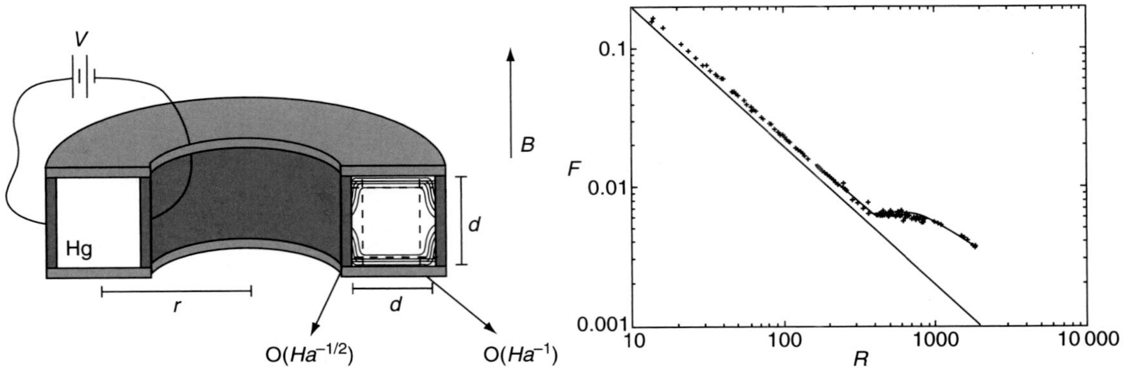

Figure 16 Experimental study of the stability of the Hartmann layers. A flow of liquid mercury ($d = 1$ cm) is produced by an imposed electric current in a imposed vertical magnetic field B (up to 13 T). The Hartmann boundary layer becomes instable at a local Reynolds number of 380. From Moresco P and Alboussière T (2004) Experimental study of the instability of the Hartmann layer. *Journal of Fluid Mechanics* 504: 167–181.

et al. (1998) found them numerically in a linear spherical Couette flow with an imposed dipolar magnetic field. As in Hartmann layers, the viscous and magnetic forces are in balance in the shear layer and this equilibrium generates a super-rotating toroidal jet embedded by the imposed magnetic field lines. The thickness of the jet is $Ha^{-1/2}L$ (Dormy *et al.*, 2002) and the amplitude of the super-rotation is limited. The electric boundary conditions can change the shape and the amplitude of this internal shear layers (Hollerbach, 2000). A super-rotating jet has been observed in the Deruiche Tourneur sodium (DTS) experiment in Grenoble (Nataf *et al.*, 2006). **Figure 17** shows Doppler measurements of the angular velocity in the spherical shell along an ultrasound ray crossing the outer sphere at (10° N latitude, 0° longitude) and (25° S latitude, 47° E longitude). The maximum in the measured angular velocity is larger than the imposed angular velocity of the inner core, which demonstrates the presence of super-rotation. A comparison with nonlinear numerical results (computed at lower forcing) confirms that the maximum of super-rotation is located close to the equator of the inner sphere in the experiments,

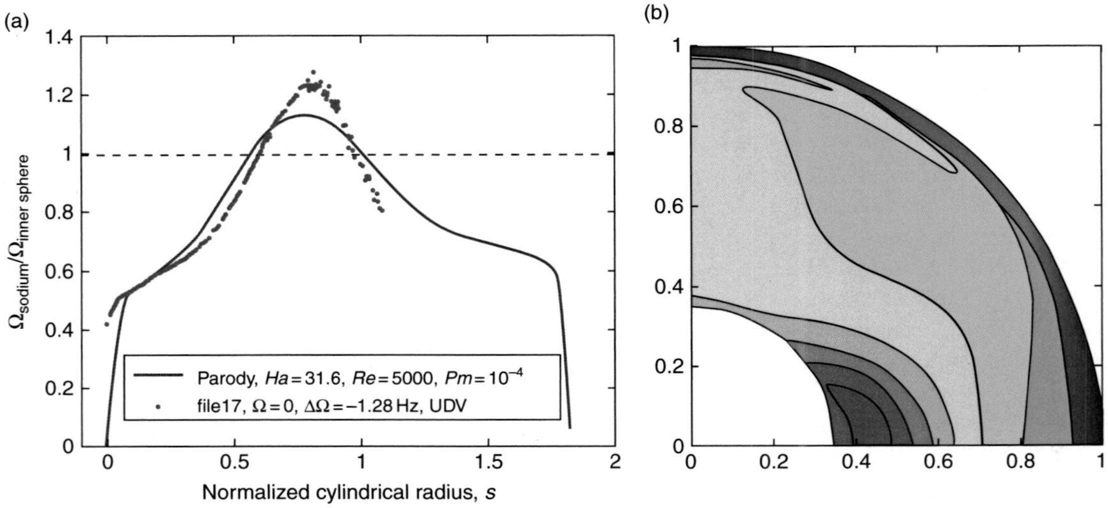

Figure 17 Experimental evidence of super-rotation in the DTS experiment. (a) Experimental and numerical profiles of angular velocity along a particular ray in the spherical shell. (b) Meridional map of angular velocity obtained by numerical means corresponding to numerical profile on the left. Contour is 0.1 the angular velocity of inner core, the maximum (red) is 1.18 the angular velocity of the inner sphere.

whereas it occurs close to magnetic field line attached to equator of the outer boundary in the calculations. Strong nonlinear effects are evidently responsible for shifting the super-rotation crescent inward where the magnetic field is the strongest.

11.4.3 Magnetic Columns

Through the Lorentz force, the presence of a magnetic field tends to make the flow field anisotropic, elongating the dynamic structures along the field lines. This elongation occurs by two different process – Joule dissipation and Alfvén wave propagation (Moreau, 1990; Loper and Moffatt, 1993; Shimizu and Loper, 1997). The effect of these two is to deform an initially isotropic perturbation (such as spherical blob) into a columnar structure, elongated parallel to the local magnetic field direction. The theory behind effect is fully described in Chapter 6. Any motion of a conducting fluid across the magnetic field lines induces eddy (Foucault) electric currents and also causes Joule dissipation, which tend to suppress such a transverse flow. Sreenivasan and Alboussière (2002) have performed experiments to study the formation of a magnetic column (a 2D vortex) from a 3D flow structure. They observed anisotropic decay of a vortex created by an impulse of electrical current. The vortex velocity is small enough to be in the regime where the induced magnetic field is negligible compared to the imposed magnetic field (i.e., the magnetic Reynolds number is less than unity). For strong magnetic fields (i.e., the interaction parameter N, measuring the ratio between Lorentz and nonlinear forces, is greater than one), they observed an exponential decay of the kinetic energy until the column reaches the other boundary. This is interpreted as the formation of a magnetic column by the pseudodiffusive process of Joule currents flowing in the direction of the magnetic field. At larger magnetic fields (large Lindquist number), we expect the formation of the columns to be driven by Alfvén waves; however, to our knowledge the experimental formation of magnetic columns by this particular mechanism has never been reported. In the core, elongated fluid columns can be formed by either of the two propagating wave processes we have described – fluid disturbances propagating as inertial waves along the direction of the rotation axis and by Alfvén waves propagating along the magnetic field lines. To compare the relative strength of these two processes in the core, we estimate the Lehnert number λ (Cardin *et al.*, 2002) which is the ratio of their respective group velocities:

$$\lambda = \frac{B}{\sqrt{\mu_0 \rho \Omega} L}$$

For small λ (as in the core), a rapid disturbance will generate quasi-geostrophic columns in a few days by inertial wave propagation. At longer periods of time, the motion will correspond to a balance between the Coriolis and the Lorentz forces (the magnetostrophic balance). It is possible that short-period events in the geomagnetic secular variation could be interpreted as motions of this type. Experiments are needed to confirm how this process might work in the core.

11.4.4 Turbulence Under the Influence of a Magnetic Field

Experiments on magnetohydrodynamic (MHD) turbulence have been made to study fundamental properties of the turbulence under the action of a strong magnetic field. Although there are many well-known applications of MHD turbulence in metallurgy and related subjects, the number of such experiments that have been applied to the core is relatively small. The statistical properties of MHD turbulence are often characterized by their magnetic and kinetic spectra. Kinetic energy spectra are deduced from time measurements of pressure while magnetic energy spectra are inferred from the variations of one component of the magnetic field. Often both measurements are made in the same location. Time spectra are sometimes converted into spatial spectra under the hypothesis of ergodicity (which is true only for homogeneous turbulence and could be wrong in the core), or the so-called Taylor assumption, which applies to turbulence transported in a mean flow. In experiments at small Rm and in the presence of a weak external magnetic field ($(\sigma B^2 R/\rho U) \ll 1$), the magnetic field can be considered as a passive vector in the flow, and magnetic energy spectra typically exhibit a $-11/3$ power exponent in the inertial regime in this situation (Bourgoin *et al.*, 2002), which is related to the classical Kolmogorov exponent ($-5/3$) of the kinetic energy spectra. For stronger magnetic fields (i.e., $(\sigma B^2 R/\rho U > 1)$) Alemany *et al.* (1979) found a -5 power exponent for the magnetic energy spectra and a -3 power exponent for the kinetic energy spectra in an experiment of free decay of grid-generated turbulence in mercury. The two distinct results at different strengths of the applied magnetic field were subsequently confirmed by Messadek and Moreau (2002).

11.4.5 Magnetoconvection in Rotating Fluids

When acting separately, both rotation and magnetic field tend to stabilize an electrically conducting fluid layer against thermal convection. However, in a classic paper Chandrasekar (1968) used marginal stability analysis to show that the stability to thermal convection can be reduced when both effects are present together. A local minimum in the critical Rayleigh number is possible when the Coriolis and the Lorentz forces are both large and are of comparable magnitude (i.e., when the ratio of these two forces, the Elsasser number, are of order 1). Nagakawa (1957) observed this local stability minimum in his study of an horizontal plane of liquid mercury heated from below and put in a strong vertical magnetic field. More recently, Aurnou and Olson (2001) have made a comprehensive set of thermal convective experiments in horizontal layer of liquid gallium subject to uniform rotation and a uniform vertical magnetic field. Aurnou and Olson (2001) focused on the relationship between Nusselt and Rayleigh numbers at various Ekman and Elsasser numbers. As expected, they found that the convection is inhibited by rotation and magnetic field acting separately, but they did not find regimes of enhanced convection under the dual action of rotation and magnetic field. This is possibly due to the limited parameter regime sampled in these experiments, in which the minimum instability was missing. Gillet *et al.* (2007b) performed magnetoconvection experiments in a rotating spherical shell with liquid gallium. Strong electrical currents on the axis of rotation were used to induce a toroidal magnetic field with a quasi-cylindrical geometry. Again, the magnetic field was evidently too weak in this experiment to observe the decrease of the critical Rayleigh number with the Elsasser number. The results of Gillet *et al.* (2007b) agree with the variations of the critical parameters deduced from a quasi-geostrophic numerical approach at low Elsasser number. The concept introduced by Chandrasekar (1968), that magnetic fields and rotation may simultaneously destabilize a fluid is potentially very important for the geodynamo and other planetary dynamos, as it opens the possibility of subcritical dynamo action in liquid cores. Experimental studies with larger magnetic fields would be valuable in this context, to investigate thermal convection at low Prandtl and magnetic Prandtl numbers in regimes where Coriolis and Lorentz forces are comparable and dominant in the system.

11.5 Experimental Dynamos

According to the magnetic induction equation (*see* Chapter 3), the magnetic Reynolds number ($\mathrm{Rm} = UL/\eta$, where U and L are typical velocity and length, and η the magnetic diffusivity) is the critical parameter to initiate dynamo action in an electrically conducting fluid. For velocity fields with suitable symmetry properties (i.e., fluid motions with helicity), kinematic dynamo studies have shown that critical magnetic Reynolds numbers between 10 and 100 are necessary to reach dynamo conditions (Dudley and James, 1989); *see* Chapter 3. Such magnetic Reynolds numbers are not easy to produce in a laboratory experiment. The best electrical conductor among the common fluids is molten sodium, with a magnetic diffusivity $\eta = 0.09\,\mathrm{m^2\,s^{-1}}$ at 120°C (see Nataf (2003) for physical properties). Even though its use requires special safety (and administrative!) procedures and also requires control at relatively high operating temperatures (the melting temperature of sodium is 98°C), it remains the only practical fluid to use in experimental dynamo modeling. To reach high magnetic Reynolds numbers in the relatively small volume of an experiment, very high fluid velocities are needed (more precisely, very high gradients of velocity are needed, since solid-body rotation does not count here). Numerically, velocities of order $10\,\mathrm{m\,s^{-1}}$ are needed even in relatively large-sized containers, 1 m or larger. As we have seen before, thermal convection velocities in laboratory experiment are too small for this, by two orders of magnitude or more, so a thermally convective dynamo experiment remains an impractical goal. Instead, the approach has been to achieve very high velocities in liquid sodium containers in which the fluid is driven mechanically by powerful motors. A large amount power (>100 kW) is needed to produce such high velocities, even allowing for the fact that the density of liquid sodium is relatively low. The main advantages that experiments offer for modeling dynamo action in planetary cores stem from their transport properties, which are closer to the Earth's core than most numerical dynamos. The magnetic Prandtl number of liquid sodium ($Pm = \nu/\eta \approx 10^{-5}$) is comparable to the magnetic Prandtl number of liquid iron. Compare these with present numerical dynamo models, which encounter difficulties with magnetic Prandtl numbers smaller than about 0.1 (*see* Chapter 8). It is easy to understand that small-scale feature and rapid time variations of the flow depend strongly on the values of the diffusion coefficients in the fluid. In addition, liquid metals tend to easily produce highly turbulent dynamos, since the Reynolds number is Pm^{-1} times greater than the magnetic Reynolds number. Presumably, this is one of the reasons why numerical modeling is limited to magnetic Prandtl numbers close to 1. At this stage, experimental dynamo modeling remains far removed from the conditions of the core, in terms of geometry, the source of motions, and the strength of planetary rotation. However, the today's dynamo experiments are just the beginning. The priority up until now has been to obtain self-generation; the next goal is to increase their geophysical realism.

11.5.1 The Riga Dynamo

In the year 2000, the Riga group headed by Prof. Agris Gailitis culminated a 30 year effort when they achieved self-excitation of a magnetic field in a liquid sodium experiment (Gailitis *et al.*, 2003). Their experimental apparatus was specially designed to produce the flows of the kinematic dynamo theory of Ponomarenko (1973) at relatively low critical magnetic Reynolds number (Gailitis *et al.*, 2003). As shown in **Figure 18**, the apparatus consists of three coaxial stainless steel cylinders, each 3 m long. Liquid sodium is accelerated downward by a helicoidal propeller located at the top of the inner cylinder and returns to the top by a vertical flow in the larger-diameter cylindrical shell. The liquid sodium in the outermost cylinder is at rest, and acts to increase the magnetic diffusion time of the system. At fast rotation rates of the propeller (>1900 rpm, corresponding to a magnetic Reynolds number of about 20 based on the radius of the inner cylinder), an oscillating self-sustained magnetic field was observed. Below and above the onset, the frequency of this magnetic field is very close to the kinematic dynamo model predictions (cf. **Figure 19**). The experimental growth rate below the onset of dynamo action (determined by the study of an imposed external magnetic field) is also very close to the appropriate prediction of 2D kinematic dynamo calculations (cf. **Figure 19**). Following a step increase in the velocity, the dynamo-induced magnetic field increases exponentially with time, with a growth rate in agreement with the numerical kinematic dynamo model predictions. The field grows until it reaches a saturation level of a few militeslas. Saturated states have been maintained for several minutes (compared to the magnetic diffusion time of order 1 s), leaving no doubt about the presence of a self-sustained magnetic field (Gailitis *et al.*, 2003).

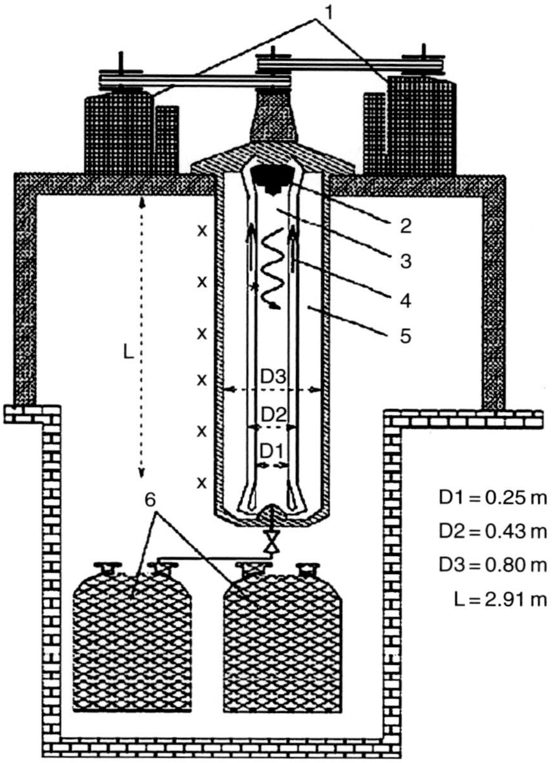

Figure 18 The Riga dynamo facility. Main parts of the apparatus include (1) two motors (55 kW each); (2) propeller; (3) helical flowregion; (4) back flow region; (5) sodium at rest; (6) sodium storagetanks; (∗) position of the flux-gate sensor; (x) positions of the six Hall sensors. Reprinted (figure) with permission from Gailitis A, Lielausis O, Dement'ev S *et al.* (2000) Detection of a flow induced magnetic field eigenmode in the Riga dynamo facility. *Physical Review Letters* 84: 4365–4368. Copyright (2000) by the American Physical Society.

The power dissipated in the dynamo regime is shown in **Figure 19**. There is a clear deviation from the Ω^3 law expected from simple hydraulic arguments and observed below the onset (Gailitis *et al.*, 2001). An increase of 10 kW is needed to sustain the dynamo and compensate the Joule dissipation of the electrical currents supporting the magnetic field. This level of power consumption matches the dissipation predicted for a magnetic field with the

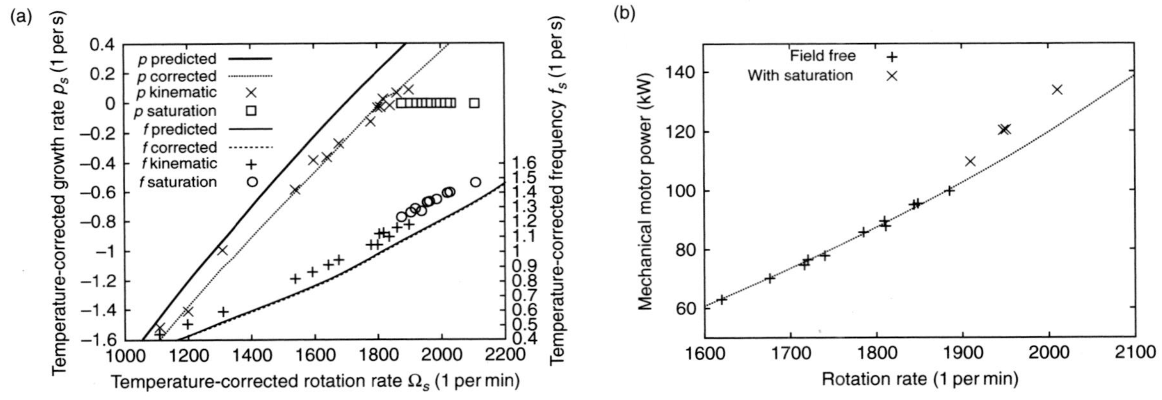

Figure 19 Results of the Riga dynamo experiment. (a) Measured growth rates *p* and frequencies *f* for different rotation rates in the kinematic and the saturation regime, compared with the numerical predictions. (b) Motor power below (+) and within (x) the dynamo regime. The motor power scales as the third power of the rotation rate (dashed line) of the propeller below the dynamo onset. (a) From Gailitis *et al.* (2003). Reprinted (figure) with permission from Gailitis A, Lielausis O, Platacis E, *et al.* (2001) Magnetic field saturation in the Riga dynamo experiment. *Physical Review Letters* 86: 3024–3027. Copyright (2001) by the American Physical Society.

geometry of the least-stable magnetic eigenmode and with the experimentally observed amplitude. In the saturated state, the azimuthal fluid flow is expected to be weaker away from the propeller, and this effect would dictate the level of saturation (Gailitis *et al.*, 2003). Velocity measurements are required to quantify the action of the Lorentz forces on the flow. Using ultrasound Doppler velocimetry, first tests have shown that the sodium that was expected to be at rest in outer cylinder is in fact in motion during dynamo action (F. Stefani, personal communication).

11.5.2 The Karlsruhe Dynamo

A second successful dynamo experiment was made in the year 2000, this time in Karlsruhe. Featuring a design inspired by the early kinematic dynamo models of Roberts (1972) and Busse (1992), the apparatus consists of an array of 52 stainless steel spin generators, as shown in **Figure 20**. Each spin generator consists of two parts: a central tube of 10 cm of diameter where the motion approximates a Poiseuille flow, and a surrounding outer cylinder where the liquid sodium follows a helical flow path constrained by a series of blades. Liquid sodium is circulated through these pipes by three remote MHD pumps with power 210 kW. This experiment has two different scales: the spin-generator scale and the modulus scale, which is 10 times larger. This scale separation makes it suitable for applications of dynamo mean-field theories (Rädler and Brandenburg, 2003). For

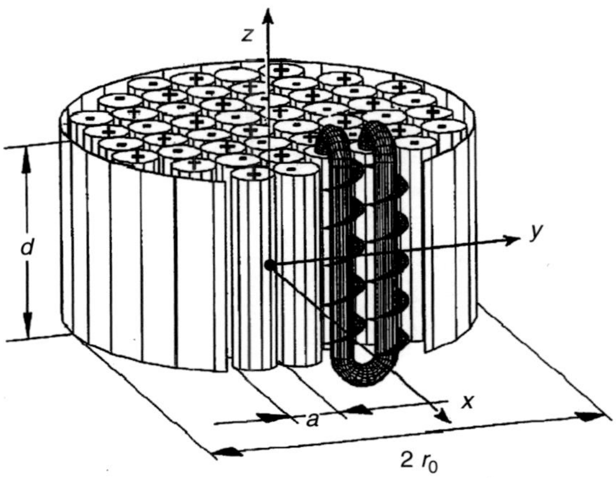

Figure 20 Sketch of the Karlsruhe dynamo apparatus. An array of 52 spin (helical flow) generators are enclosed inside a 1.7 m diameter stainless steel cylinder. Liquid sodium is pumped through the generator array. Dimensions: $a = 0.21$ m, $r_0 = 0.85$ m, $d = 0.703$ m. From Stieglitz R and Müller U (2001) Experimental demonstration of a homogeneous two-scale dynamo. *Physics of Fluids* 13: 561–564

volume rates larger than $\approx 100\ \mathrm{m^{-3}\,h^{-1}}$, Stieglitz and Müller (2001) observed a self-sustained stationary magnetic field of a few militeslas for long periods of time (more than 1 h) (Müller *et al.*, 2006). The experimental onset agrees quite well with kinematic dynamo results of Tilgner (1997) and with mean-field theory results of Rädler *et al.* (1998) as shown in **Figure 21** even if the experimental threshold is 10% lower than the predicted values.

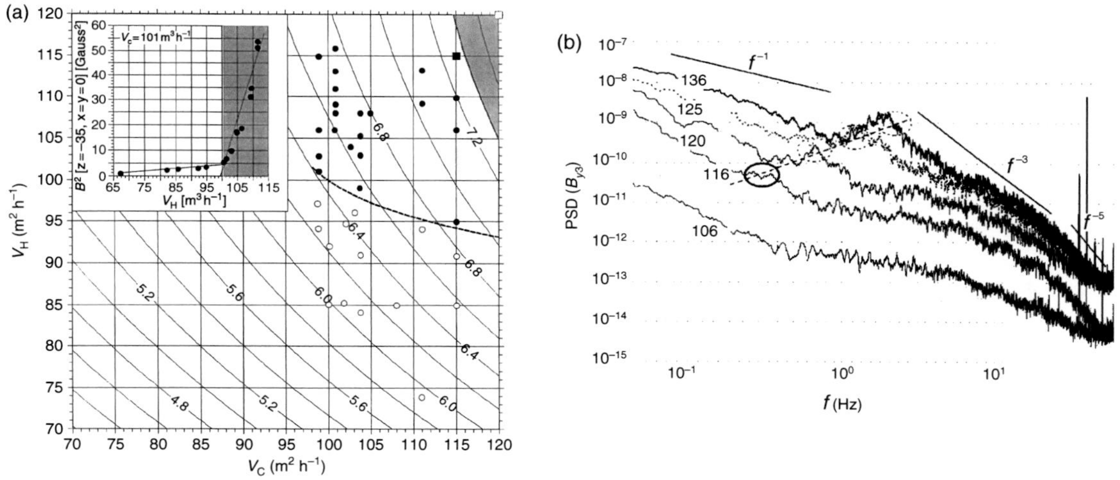

Figure 21 Results of the Karlsruhe dynamo experiment. (a) Stability diagram for the onset of dynamo action as a function of the central flow rate V_C and the helical flow rates V_H. The filled dots denote experimentally measured dynamo action, open dots denote nondynamo states. The dashed line indicates the experimentally obtained marginal stability curve, defined by a regression of the steep increase of the magnetic energy ($\approx B^2$) with V_H (see subgraph in the upper left). The gray-marked domain specifies the flow-rate domain for the existence of a dynamo as calculated by Radler. The isolines show the values of a modified magnetic Reynolds number according to the mean-field model of Radler. The two squared symbols indicate Tilgners calculations for onset of dynamo action. (b) Power spectral density (PSD) for the components B_y for five different volumetric flow rates of operation at the center of the modulus. $V_H = 100\,\mathrm{m}^{-3}\,h^{-1}$ and for five central volumetric flow rates $V_C = 106, 116, 120, 125, 136\,\mathrm{m}^{-3}\,\mathrm{h}^{-1}$. (a) From Stieglitz R and Müller U (2001) Experimental demonstration of a homogeneous two-scale dynamo. *Physics of Fluids* 13: 561–564. (b) From Müller U, Stieglitz R, Horanyi S (2004) A two-scale hydromagnetic dynamo experiment. *Journal of Fluid Mechanics* 498: 31–71.

Outside the modulus, the large-scale magnetic field is mainly a dipole with its axis perpendicular to the direction of the channels (Müller *et al.*, 2004), in agreement with numerical and theoretical models. As predicted, the magnetic field has a staircase structure in the direction of the channels, but two preferred directions of the equatorial dipole were found (instead of arbitrary many directions, as predicted theoretically). Müller *et al.* (2004) showed that these two preferred directions are controlled by an external magnetic field and permanent magnetic fields. The magnetic field along the central axis of the modulus has been measured. The geometry agrees quite well with the numerically predicted eigenmode, but some loss of symmetry has been observed (Müller *et al.*, 2006). In the center region, field intensities up to 40 mT have been measured, 10 times greater than the values of the magnetic field outside the device. Saturation of the magnetic field can be understood in this experiment as a back reaction of the Lorentz forces on the fluid flow. Müller *et al.* (2006) measured up to a 13% decrease on the amplitude of the velocity in the central pipe above the dynamo onset. More interestingly, they also reported an increase of the fluctuations of the velocity field around its mean (rms) value as the experiment reaches dynamo conditions. Generally in MHD the presence of the magnetic field reduces the turbulent fluctuations of the velocity field. The authors associate the fluctuations of velocity field with the fluctuations of the magnetic field, which become more and more energetic as the dynamo becomes more supercritical (Müller *et al.*, 2006). This effect can be seen in magnetic field spectra, which exhibit a maximum at moderate frequency (1 Hz). The frequency of this maximum shifts to higher values as the flow rate increases above the value for dynamo onset. These spectra show two trends around the peak, one with an f^{-1}-variation at lower frequency, another with an f^{-3}-variation at larger frequencies, and yet another f^{-5} dependency at very large frequencies.

11.5.3 Turbulence and Dynamo Onset

The magnetic Prandtl number is the ratio of the viscosity to the magnetic diffusivity of the fluid. The magnetic Prandtl number of liquid metals is usually very low, and its value in the outer core is not too different from liquid sodium, 5×10^{-6}. A magnetic Reynolds numbers of a few tens as required for dynamo action implies an ordinary Reynolds numbers larger than 10^7 in dynamo experiments, and also in the core. Consequently, dynamo experiments are necessarily turbulent, and for the same reason the geodynamo is also turbulent, although the turbulence in the outer core has special characteristics (*see* Chapter 6). The two successful dynamo experiments have overcome the turbulence problem by very precise control of the flow (constrained by guiding pipes, for example), reducing the effects of turbulence, which are usually undesirable. For instance, in the Riga dynamo, Gailitis *et al.* (2000) estimate the level of turbulence to be less than 10% and the corresponding increase of the critical magnetic Reynolds number only 1%. Similar levels of turbulence have been observed in Karlsruhe experiment (Müller *et al.*, 2006). On the other hand, dynamo experiments in Maryland (Peffley *et al.*, 2000; Sisan *et al.*, 2003), Cadarache (Bourgoin *et al.*, 2002) and Wisconsin (Spence *et al.*, 2006) are based on less constrained, more turbulent flows, and have not reached the critical value of the magnetic Reynolds number. The presence of turbulence in these less-constrained experiments has raised their effective critical magnetic Reynolds numbers above the predictions of numerical kinematic dynamos based on purely laminar flow (Marié *et al.*, 2003; Ravelet *et al.*, 2005; Nornberg *et al.*, 2006). All these experiments detected hydrodynamic turbulence using a weak magnetic field as a passive vector tracer of the flow. Nevertheless, their observation of an upward shift of the dynamo onset because of the turbulence needs to be understood better (Ponty *et al.*, 2005; Laval *et al.*, 2006). All of these experiments reveal how the flow induces magnetic fields from the interaction of the turbulence with the imposed magnetic field. For instance, the exponential decay rate of the magnetic field after shutdown provides information on the growth rate of the induced magnetic field (Peffley *et al.*, 2000) and gives some information on how far the experiment is from dynamo onset. Time-averaged large-scale flows also have been observed to induce magnetic fields in a different direction than the applied field, an example of the Ω (Lehnert, 1958; Brito *et al.*, 1995; Spence *et al.*, 2006) or Parker effect (Volk *et al.*, 2005). The induction effects due to small-scale turbulent fluctuations and their contribution to the mean induction, the so-called α-effect, has also studied experimentally. In PERM, Stepanov *et al.* (2006) investigated induction mechanisms in a screw-flow of gallium in a toroidal channel during a spin-down process. Their measurements yielded a low upper bound on the α-effect compared to the predictions from dimensional analysis. These low values were confirmed in the Gallium Van Karmán experiment in Lyon (R. Volk, personal communication). Still, the experimental result of the Wisconsin experiment seems to contradict these two other experiments, as Spence *et al.* (2006) advocate the presence of a turbulent electromotive force to generate an induced magnetic field parallel to the imposed dipolar field aligned with the axis of symmetry of their experiment. According to Cowling's theorem, this component cannot result from the interaction of the axisymmetric laminar part of the flow with the axisymmetric imposed magnetic field. In Oct 2006, as this chapter was being written, the Cadarache experiment has produced a self-sustained magnetic field. The Cadarache dynamo is made of a cylindrical vessel of 100 l of liquid sodium, with two counter-rotating propellers (a Van Karmán geometry). After many failed attempts like in the Maryland and Wisconsin experiments, a dynamo regime has been found for a particular set of conditions. The conditions include propellers made of ferromagnetic material (iron), a copper cylindrical inner boundary introduced to isolate the sodium in the outer part of the vessel, and an inner copper annulus placed between the two propellers in order to reduce large-scale turbulence. Presumably, the two US teams will be able to reproduce these results in their apparatuses.

11.5.4 Toward a Magnetostrophic Dynamo?

The first author's group in Grenoble has designed a liquid sodium experiment to study the properties of flow in the regime where both rotation and magnetic field play an important role in the dynamics (Cardin *et al.*, 2002). The experiment consists of a rotating spherical shell (with a core-like radius ratio of 0.35) with a differentially rotating inner core. The Couette flow in the spherical shell is subject to a strong dipolar magnetic field imposed by a permanent magnet located inside the inner core. We have found experimental evidence of super-rotation (Nataf *et al.*, 2006) that features an equatorial inner region rotating faster than the inner core. The induced magnetic field has been measured outside the spherical shell. Strong amplification of magnetic field has been observed when the inner-core rotation is opposite to the outer sphere. We have also detected propagating disturbances that travel along latitude lines in the sphere. A range of azimuthal wave numbers (from 2 to 6) and their corresponding propagation velocities have also been measured. More work is needed to determine if these disturbances correspond to the hydromagnetic waves described by Hide (1966), or if they have another source. The ultimate goal of this project is to reach the dynamo regime in a realistic geometry, by repeating these experiments in a sequence of larger spherical shells.

References

Aldridge K and Baker R (2003) Paleomagnetic intensity data: A window on the dynamics of Earth's fluid core? *Physics of the Earth and Planetary Interiors* 140: 91–100.

Alemany A, Moreau R, Sulem PL, and Frisch U (1979) Influence of an external magnetic field on homogeneous MHD turbulence. *Journal de Mécanique* 18: 277–313.

Aubert J, Brito D, Nataf H-C, Cardin P, and Masson J-P (2001) A systematic experimental study of rapidly rotating spherical convection in water and liquid gallium. *Physics of the Earth and Planetary Interiors* 128: 51–74.

Aubert J, Gillet N, and Cardin P (2003) Quasigeostrophic models of convection in rotating spherical shells. *Geochemistry, Geophysics, Geosystems* 4: 1052 (doi:10.1029/2002GC000456).

Aubert J, Jung S, and Swinney HL (2002) Observations of zonal flow created by potential vorticity mixing in a rotating fluid. *Geophysical Research Letters* 29: 1876–1879.

Aurnou JM and Olson PL (2001) Experiments on Rayleigh Bénard convection, magnetoconvection and rotating magnetoconvection in liquid gallium. *Journal of Fluid Mechanics* 430: 283–307.

Bourgoin M, Marié L, Pétrélis F, *et al.* (2002) Magnetohydrodynamics measurements in the von Kármán sodium experiment. *Physics of Fluids* 14: 3046.

Brito D, Aurnou J, and Cardin P (2004) Turbulent viscosity measurements relevant to planetary core–mantle dynamics. *Physics of the Earth and Planetary Interiors* 141: 3–8.

Brito D, Cardin P, Nataf H-C, and Marolleau G (1995) Experimental study of a geostrophic vortex of gallium in a transverse magnetic field. *Physics of the Earth and Planetary Interiors* 91: 77–98.

Brito D, Nataf H-C, Cardin P, Aubert J, and Masson J-P (2001) Ultrasonic Doppler velocimetry in liquid gallium. *Experiments in Fluids* 31: 653–663.

Burr U, Barleon L, Müller U, and Tsinober A (2000) Turbulent transport of momentum and heat in magnetohydrodynamic rectangular duct flow with strong sidewall jets. *Journal of Fluid Mechanics* 406: 247–279.

Busse F (1992) Dynamo theory of planetary magnetism and laboratory experiments. In: Friedrich A and Wunderlin A (eds.) Springer *Proceedings in physics 69; Evaluation of Dynamical Structures in Complex Systems*, pp. 197–207. Berlin: Springer.

Busse FH (1968) Steady fluid flow in a precessing spheroidal shell. *Journal of Fluid Mechanics* 33: 739–751.

Busse FH (1970) Thermal instabilities in rapidly rotating systems. *Journal of Fluid Mechanics* 44: 441–460.

Busse FH and Carrigan CR (1974) Convection induced by centrifugal buoyancy. *Journal of Fluid Mechanics* 62: 579–592.

Busse FH and Carrigan CR (1976) Laboratory simulation of thermal convection in rotating planets and stars. *Science* 191: 81–83.

Busse FH and Hood LL (1982) Differential rotation driven by convection in a rapidly rotating annulus. *Geophysical and Astrophysical Fluid Dynamics* 21: 59–74.

Cardin P, Brito D, Jault D, Nataf H-C, and Masson J-P (2002) Towards a rapidly rotating liquid sodium dynamo experiment. *Magnetohydrodynamics* 38: 177–189.

Cardin P and Olson P (1992) An experimental approach to thermochemical convection in the Earth's core. *Geophysical Research Letters* 19: 1995–1998.

Cardin P and Olson P (1994) Chaotic thermal convection in a rapidly rotating spherical shell: Consequences for flow in the outer core. *Physics of the Earth and Planetary Interiors* 82: 235–259.

Carrigan CR and Busse FH (1983) An experimental and theoretical investigation of the onset of convection in rotating spherical shells. *Journal of Fluid Mechanics* 126: 287–305.

Chandrasekar S (1968) *Hydrodynamic and Hydromagnetic Stability*. New York, USA: Clarendon Press.

Cordero S (1993) Experiments on convection in a rotating hemispherical shell: Transition to chaos. *Geophysics Research Letters* 20: 2587–2590.

Cordero S and Busse FH (1992) Experiments on convection in rotating hemispherical shells – Transition to a quasi-periodic state. *Geophysical Research Letters* 19: 733–736.

Dormy E, Cardin P, and Jault D (1998) MHD flow in a slightly differentially rotating spherical shell, with conducting inner core, in a dipolar magnetic field. *Earth and Planetary Science Letters* 160: 15–30.

Dormy E, Jault D, and Soward AM (2002) A super-rotating shear layer in magneto-hydrodynamic spherical Couette flow. *Journal of Fluid Mechanics* 452: 263–291.

Dormy E, Soward AM, Jones CA, Jault D, and Cardin P (2004) The onset of thermal convection in rotating spherical shells. *Journal of Fluid Mechanics* 501: 43–70.

Dudley ML and James RW (1989) Time-dependent kinematic dynamos with stationary flows. *Proceedings of the Royal Society of London A* 425: 407–429.

Egbers C, Brasch W, Sitte B, Immohr J, and Schmidt J-R (1999) Estimates on diagnostic methods for investigations of thermal convection between spherical shells in space. *Measurement Science and Technology* 10: 866–877.

Eloy C, Le Gal P, and Le dizès S (2000) Experimental study of the multipolar vortex instability. *Physical Review Letters* 85: 3400–3403.

Eymin C and Hulot G (2005) On core surface flows inferred from satellite magnetic data. *Physics of the Earth and Planetary Interiors* 152: 200–220.

Faller AJ (1963) An experimental study of the instability of the laminar Ekman boundary layer. *Journal of Fluid Mechanics* 15: 560–576.

Früh W-G and Read PL (1999) Experiments on a barotropic rotating shear layer. Part 1: Instability and steady vortices. *Journal of Fluid Mechanics* 383: 143–173.

Gailitis A, Lielausis O, Dement'ev S, *et al.* (2000) Detection of a flow induced magnetic field eigenmode in the Riga dynamo facility. *Physical Review Letters* 84: 4365–4368.

Gailitis A, Lielausis O, Platacis E, Gerbeth G, and Stefani F (2003) The Riga dynamo experiment. *Surveys in Geophysics* 24: 247–267.

Gailitis A, Lielausis O, Platacis E, *et al.* (2001) Magnetic field saturation in the Riga dynamo experiment. *Physical Review Letters* 86: 3024–3027.

Gillet N, Brito D, Jault D, and Nataf H-C (2007a) Experimental and numerical study of convection in a rapidly rotating spherical shell. *Journal of Fluid Mechanics* (in press).

Gillet N, Brito D, Jault D, and Nataf H-C, (2007b) Experimental and numerical study of magnetoconvection in a rapidly rotating spherical shell (in press).

Greenpan H (1968) *The Theory of Rotating Fluids*. Cambridge UK: Cambridge University Press.

Greff-Lefftz M and Legros H (1999) Core rotational dynamics and geological events. *Science* 286: 1707–1709.

Hart JE, Glatzmaier GA, and Toomre J (1986) Space-laboratory and numerical simulations of thermal convection in a rotating

hemispherical shell with radial gravity. *Journal of Fluid Mechanics* 173: 519–544.
Hartmann J and Lazarus F (1937) Experimental investigations on the flow of mercury in a homogeneous magnetic field. *Kongelige Danske Videnskabernes Selskab, Mathematisk-Fysiske Meddebelser* 15: 1–45.
Hide R (1966) Free hydromagnetic oscillations of the Earth's core and the theory of the geomagnetic secular variation. *Proceedings of the Royal Society of London A* 259: 615–650.
Hide R and Titman C (1967) Detached shear layers in a rotating fluid. *Journal of Fluid Mechanics* 29: 39–60.
Hollerbach R (2000) Magnetohydrodynamic flows in spherical shells. *LNP Vol. 549: Physics of Rotating Fluids* 549: 295.
Hollerbach R and Kerswell RR (1995) Oscillatory internal shear layers in rotating and precessing flows. *Journal of Fluid Mechanics* 298: 327–339.
Jones CA, Soward AM, and Mussa AI (2000) The onset of thermal convection in a rapidly rotating sphere. *Journal of Fluid Mechanics* 405: 157–179.
Kerswell RR (1995) On the internal shear layers spawned by the critical regions in oscillatory Ekman boundary layers. *Journal of Fluid Mechanics* 298: 311–325.
Kerswell RR (2002) Elliptical instabilities. *Annual Review of Fluid Mechanics* 34: 83–113.
Lacaze L, Le Gal P, and Le Dizès S (2004) Elliptical instability in a rotating spheroid. *Journal of Fluid Mechanics* 505: 1–22.
Laval J-P, Blaineau P, Leprovost N, Dubrulle B, and Daviaud F (2006) Influence of turbulence on the dynamo threshold. *Physical Review Letters* 96(20): 204503.
Lehnert B (1958) An experiment on axisymmetric flow of liquid sodium in a magnetic field. *Arkiv for Fysik* 13: 109–116.
Lingwood RJ (1997) Absolute instability of the Ekman layer and related rotating flows. *Journal of Fluid Mechanics* 331: 405–428.
Loper D and Moffatt K (1993) Small-scale hydromagnetic flow in the Earth's core: Rise of a vertical buoyant plume. *Geophysical and Astrophysical Fluid Dynamics* 68: 177–202.
Malkus WVR (1968) Precession of the Earth as the cause of geomagnetism. *Science* 160: 259–264.
Malkus WVR (1989) An experimental study of the global instabilities due to the tidal (elliptical) distortion of a rotating elastic cylinder. *Geophysical and Astrophysical Fluid Dynamics* 48: 123–134.
Marié L, Burguete J, Daviaud F, and Léorat J (2003) Numerical study of homogeneous dynamo based on experimental von Kármán type flows. *European Physical Journal B* 33: 469–485.
Messadek K and Moreau R (2002) An experimental investigation of MHD quasi-two-dimensional turbulent shear flows. *Journal of Fluid Mechanics* 456: 137–159.
Moreau R (1990) *Magnetohydrodynamics*. Dordrecht, NL: Kluwer Academic Publishers.
Moresco P and Alboussière T (2004) Experimental study of the instability of the Hartmann layer. *Journal of Fluid Mechanics* 504: 167–181.
Müller U, Stieglitz R, and Horanyi S (2004) A two-scale hydromagnetic dynamo experiment. *Journal of Fluid Mechanics* 498: 31–71.
Müller U, Stieglitz R, and Horanyi S (2006) Experiments at a two-scale dynamo test facility. *Journal of Fluid Mechanics* 552: 419–440.
Murgatroyd W (1953) Experiments on magneto-hydrodynamic channel flow. *Philosophical Magazine A* 44: 1348–1354.
Nagakawa Y (1957) Experiments on the instability of a layer of mercury heated from below and subject to the simultaneous action of a magnetic field and rotation. *Proceedings of the Royal Society of London A* 242: 81–88.
Nataf H-C (2003) Dynamo and convection experiments. In: Jones CA, Soward AM, and Zhang K (eds.) *Earth's Core and Lower Mantle*, pp. 153–179. London: Taylor and Francis.
Nataf H-C, Alboussière T, Brito D, *et al.* (2006) Experimental study of super-rotation in a magnetostrophic spherical Couette flow. *Geophysical and Astrophysical Fluid Dynamics* 100: 281–298 (doi:10.1080/03091920600718426).
Noir J, Brito D, Aldridge K, and Cardin P (2001b) Experimental evidence of inertial waves in a precessing spheroidal cavity. *Geophysical Research Letters* 28: 3785–3788.
Noir J, Cardin P, Jault D, and Masson J-P (2003) Experimental evidence of nonlinear resonance effects between retrograde precession and the tilt-over mode within a spheroid. *Geophysical Journal International* 154: 407–416.
Noir J, Jault D, and Cardin P (2001a) Numerical study of the motions within a slowly precessing sphere at low Ekman number. *Journal of Fluid Mechanics* 437: 283–299.
Nornberg MD, Spence EJ, Kendrick RD, Jacobson CM, and Forest CB (2006) Intermittent magnetic field excitation by a turbulent flow of liquid sodium. *Physical Review Letters* 97(4): 044503.
Peffley NL, Cawthorne AB, and Lathrop DP (2000) Toward a self-generating magnetic dynamo: The role of turbulence. *Physical Review E* 61: 5287–5294.
Plaut E and Busse F (2002) Low prandtl number convection in a rotating cylindrical annulus. *Journal of Fluid Mechanics* 464: 345–363.
Poincaré R (1910) Sur la précession des corps déformables. *Bulletin Astronomique* 27: 321–356.
Ponomarenko Y (1973) On the theory of hydromagnetic dynamos. *Zhurnal Prikladnoi Mekhaniki i Tekhnicheskoi Fiziki* 6: 47–51.
Ponty Y, Mininni PD, Montgomery DC, Pinton J-F, Politano H, and Pouquet A (2005) Numerical study of dynamo action at low magnetic Prandtl numbers. *Physical Review Letters* 94(16): 164502.
Rädler K-H, Apstein E, Reinhardt M, and Schüler M (1998) The Karlsruhe dynamo experiment, a mean field approach. *Studia Geophysica et Geodaetica* 42: 224–231.
Rädler K-H and Brandenburg A (2003) Contributions to the theory of a two-scale homogeneous dynamo experiment. *Physical Review E* 67(2): 026401.
Ravelet F, Chiffaudel A, Daviaud F, and Léorat J (2005) Toward an experimental von Kármán dynamo: Numerical studies for an optimized design. *Physics of Fluids* 17: 7104.
Read P, Yamazaki YH, Lewis SR, *et al.* (2004) Jupiters and Saturns convectively driven banded jets in the laboratory. *Geophysical Research Letters* 31: L22701 (doi:10.1029/2004GL02016).
Rhines P (1975) Waves and turbulence on a beta-plane. *Journal of Fluid Mechanics* 69: 417–443.
Roberts GO (1972) Dynamo action of fluid motions with two-dimensional periodicity. *Philosophical Transactions of the Royal Society of London* 271: 411–454.
Schaeffer N (2004) *Instabilités, turbulence et dynamo dans une couche de fluide cisaille en rotation rapide*. PhD Thesis, Université Joseph Fourier Grenoble.
Schaeffer N and Cardin P (2005) Quasigeostrophic model of the instabilities of the Stewartson layer in flat and depth-varying containers. *Physics of Fluids* 17: 4111.
Shimizu H and Loper DE (1997) Time and length scales of buoyancy-driven flow structures in a rotating hydromagnetic fluid. *Physics of the Earth and Planetary Interiors* 104: 307–329.
Sisan DR, Shew WL, and Lathrop DP (2003) Lorentz force effects in magnetoturbulence. *Physics of the Earth and Planetary Interiors* 135: 137–159.
Spence EJ, Nornberg MD, Jacobson CM, Kendrick RD, and Forest CB (2006) Observation of a turbulence-induced large scale magnetic field. *Physical Review Letters* 96(5): 055002.

Sreenivasan B and Alboussière T (2002) Experimental study of a vortex in a magnetic field. *Journal of Fluid Mechanics* 464: 287–309.
Stepanov R, Volk R, Denisov S, Frick P, Noskov V, and Pinton J-F (2006) Induction, helicity, and alpha effect in a toroidal screw flow of liquid gallium. *Physical Review E* 73(4): 046310.
Stewartson K (1957) On almost rigid rotation. *Journal of Fluid Mechanics* 3: 17–26.
Stewartson K and Roberts PH (1963) On the motion of a liquid in a spheroidal cavity of a precessing rigid body. *Journal of Fluid Mechanics* 17: 1–20.
Stieglitz R and Müller U (2001) Experimental demonstration of a homogeneous two-scale dynamo. *Physics of Fluids* 13: 561–564.
Sumita I and Olson P (2000) Laboratory experiments on high Rayleigh number thermal convection in a rapidly rotating hemispherical shell. *Physics of the Earth and Planetary Interiors* 117: 153–170.
Sumita I and Olson P (2002) Thermal convection experiments in a rotating hemispherical shell with heterogeneous boundary heat flux: Implications for the Earth's core. *Journal of Geophysical Research* 107: doi:10.1029/2001JB000548.
Sumita I and Olson P (2003) Experiments on highly supercritical thermal convection in a rapidly rotating hemispherical shell. *Journal of Fluid Mechanics* 492: 271–287.
Tabeling P (2002) Two-dimensional turbulence: A physicist approach. *Physics Reports* 362: 1–62.
Taylor G (1923) Experiments on the motion of solid bodies in rotating fluids. *Proceedings of the Royal Society of London A* 104: 213–218.
Tilgner A (1997) Predictions on the behaviour of the Karlsruhe dynamo. *Acta Astron mica et geophysica Univ Comenianae XIX* 19: 51–62.
Vanyo JP, Wilde P, Cardin P, and Olson P (1995) Experiments on precessing flows in the Earth's liquid core. *Geophysical Journal International* 121: 136–142.
Volk R, Odier P and Pinton J-F (2005) Fluctuation of magnetic induction in von kármán swirling flows. *Physics of Fluids* 18(8): 085105–085105-10.

12 Core–Mantle Interactions

B. A. Buffett, The University of Chicago, Chicago, IL, USA

12.1	Introduction	345
12.2	Thermal Interactions	345
12.3	Electromagnetic Interactions	349
12.4	Mechanical Interactions	352
12.5	Chemical Interactions	354
12.6	Conclusions	355
References		356

12.1 Introduction

Two giant heat engines operate inside the Earth. One powers plate tectonics and accounts for most of the geological phenomena we observe at the surface. The other operates in the core, where it continually sustains the Earth's magnetic field against persistent ohmic losses. These two heat engines are coupled, primarily through interactions at the boundary between the core and mantle. Transfer of heat, mass, momentum, and electric current across the boundary profoundly affects the dynamics and evolution of both regions on timescales ranging from days to hundreds of millions of years. On short timescales, we observe diurnal wobbles in the Earth's rotation which are strongly affected by relative motion between the core and the mantle (e.g., Mathews and Shapiro, 1992). This and other types of mechanical interaction influence the flow in the core (e.g., Moffatt and Dillon, 1976) and contribute to variations in the length of day on decadal timescales (*see* Chapter 4). On longer timescales, we expect variations in both the magnitude and spatial distribution of heat flow across the boundary. The resulting thermal interactions influence the vigor and pattern of convection in the core, and may alter the frequency of reversals (Glatzmaier *et al.*, 1999). Electromagnetic interactions are also possible, particularly if the lowermost mantle has a large electrical conductivity. Electric currents near the base of the mantle can generate both large- and small-scale magnetic fields, while the associated Lorentz force contributes to the mechanical interaction. More recent suggestions of chemical reactions between the core and mantle (Brandon and Walker, 2005) raise the intriguing but contentious suggestion that mass has been transferred between the two regions over geological time. In this chapter, we focus on the consequences of core–mantle interactions for processes in the core. We address a number of recent advances in our understanding of thermal, mechanical, electromagnetic, and chemical interactions, as they relate to the dynamics and evolution of the core. We also deal with the role of core–mantle interactions as a means of detecting deep-Earth processes at the surface.

12.2 Thermal Interactions

Heat flow across the core–mantle boundary (CMB) is a fundamental parameter for the evolution of the core. It controls the rate of cooling and solidification of the core, and determines the vigor of convection in the fluid outer core (*see* Chapter 2). Convection is driven by buoyancy from the boundaries of the outer core in response to cooling and inner-core growth. Chemical buoyancy arises through the exclusion of the light elements from the inner core (Braginsky, 1963), whereas thermal buoyancy is generated by latent heat release on solidification (Verhoogen, 1961) and by forming cold, dense fluid in the thermal boundary layer at the top of the core. Each of these buoyancy sources is paced by the CMB heat flow.

The magnitude of the CMB heat flow is dictated by the mantle. The large and relatively sluggish mantle imposes control over the loss of heat from the core. Estimates of the CMB heat flow are often obtained from inferences of the temperature jump across the thermal boundary layer on the mantle side of the interface (e.g., Buffett, 2002). Typical values of the heat flow are 6–12 TW, although our present state of knowledge is not sufficient to rule out higher or lower values. Such a broad range of values permits two distinct styles of convection in the core (see **Figure 1**). One style occurs when the CMB heat flow exceeds the conduction of heat along the adiabatic gradient at the top of the core (denoted by Q_{ad}). Convection transports the superadiabatic part of heat flow through the core, creating a thermal boundary layer on the core side of the boundary. This provides a source of cold, dense fluid that drives convection from the top down into the core. The other regime occurs when the CMB heat flow is less than Q_{ad}. In this case, the mantle is unable to remove the heat carried by conduction down the adiabat. Excess heat accumulates at the top of the core (Gubbins *et al.*, 1982) or is convectively mixed into the interior by chemical buoyancy (Loper, 1978). Either of these possibilities eliminates thermal buoyancy production at the top of the core, restricting the buoyancy production to the inner-core boundary region.

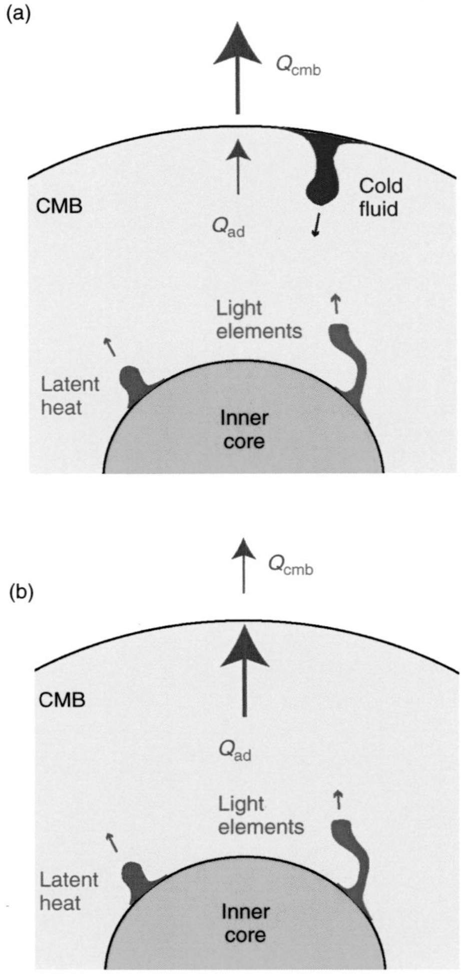

Figure 1 Two distinct styles of convection operate in the core, depending on the relative magnitude of the CMB heat flow, Q_{cmb}, and the heat flow conducted down the adiabat at the CMB, Q_{ad}. (a) A thermal boundary layer forms at the top of the core when $Q_{cmb} > Q_{ad}$. Cold and dense fluid at the CMB drives convection from the top. Latent heat and compositional buoyancy from the inner-core boundary drive convection from the bottom. (b) The thermal source of buoyancy at the top of the core disappears when $Q_{cmb} < Q_{ad}$. The excess heat that is carried to the CMB by conduction either accumulates at the top or is mixed back into the interior by compositional buoyancy.

Representative values for $Q_{ad} = 5–6$ TW lie within a plausible range of values for the CMB heat flow. This implies that either style of convection in the core is possible. It is also possible that the style of convection has changed over time. High CMB heat flow is likely in the early Earth, and it may be essential to power the dynamo prior to the formation of the inner core. A switch to bottom-driven convection becomes possible as the Earth cools and the inner core grows. We might even expect the style of convection to switch intermittently between the two regimes if the current heat flow is sufficiently close to Q_{ad}. Fluctuations in mantle convection can transiently shift the CMB heat flow above or below Q_{ad}, altering the convective regime.

Evidence for fluctuations in mantle convection are found in geological observations of changes in the rate of plate motions. Bunge *et al.* (2003) have recently incorporated geological estimates of plate motions into three-dimensional models of mantle convection in order to assess the consequences of changes in plate-spreading rates. Interpretation of observations from the last 120 million years suggests that spreading rates have generally decreased over this time, perhaps by more than 20% (Xu *et al.*, 2006; Lithgow-Bertelloni and Richards, 1998). While the magnitude of this is debated (Rowley, 2002), it is clear that changes in plate motion can cause thermal interactions with the core. **Figure 2** shows the predicted change in heat flow at the surface and CMB when plate motions are varied in the calculation by Bunge *et al.* (2003). (The plate motions over the past 120 My are cyclically repeated to extend the record back in time. The primary goal is to repeat the calculation with different initial conditions.) High heat flow at the surface coincides with times of rapid plate spreading. Changes at the CMB are felt about 50–60 My later, once the subducted material begins to arrive at the base of the mantle. The amplitude of the variation in CMB heat flow is 1 TW, or about 20% of the time-averaged value.

Changes in the CMB heat flow alter the supply of thermal and chemical buoyancy at the boundaries of

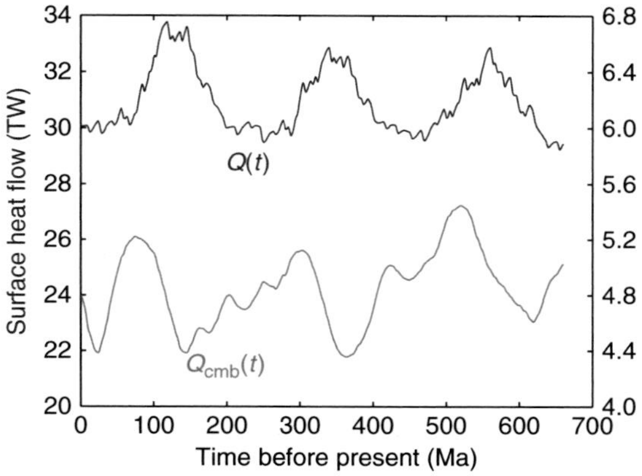

Figure 2 Heat flow at the surface, $Q(t)$, and the CMB, $Q_{\mathrm{cmb}}(t)$, from a numerical convection model with imposed plate motions (Bunge *et al.*, 2003). Plate motion histories over the past 120 My are cyclically repeated at earlier times to assess the response with different initial conditions. High heat flow at the surface coincides with times of faster than average plate spreading rates. Increases in heat flow at the CMB occur 50–60 My later.

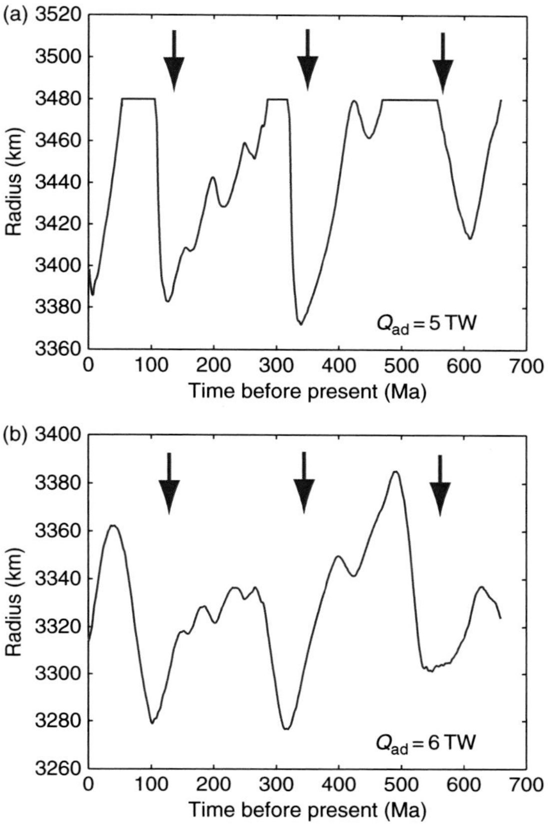

Figure 3 Changes in the radius of a stratified layer in the core in response to variations in CMB heat flow. (a) The stratified layer vanishes when the CMB heat flow exceeds the adiabatic heat flow Q_{ad}. Under these conditions, the radius of the stratified layer equals the radius of the core (e.g., 3480 km). The stratified layer reappears whenever the CMB heat flow drops below Q_{ad}. (b) A persistent stratified layer is present when $Q_{\mathrm{ad}} = 6\,\mathrm{TW}$, although the thickness of the layer varies with time. The arrows indicate the times when the heat flow at the surface is maximum.

the outer core. A stratified layer at the top of the core can develop when the heat flow drops below Q_{ad} because the mantle can no longer keep pace with the heat supplied by conduction along the adiabat. Excess heat accumulates at the top of the core unless convection in the core can entrain warm fluid back into the interior. In the absence of entrainment, we expect a warm, stratified layer to grow in response to changes in densities inside and below the stratified layer (Lister and Buffett, 1998). Heat accumulates in the stratified layer, causing the warm fluid to encroach down into the underlying convective region. However, light elements from the inner-core boundary accumulate in the convective region, so the stratified layer gradually becomes heavier in term of composition. This interplay between the thermal and chemical buoyancy ultimately limits the growth of the stratified layer.

Figure 3 shows two predictions for the thickness of the stratified layer using the model of Lister and Buffett (1998). The CMB heat flow $Q_{\mathrm{cmb}}(t)$ is taken from **Figure 2** and two representative values for Q_{ad} are considered. The value $Q_{\mathrm{ad}} = 5\,\mathrm{TW}$ causes an intermittent stratified layer to develop whenever $Q_{\mathrm{cmb}}(t)$ drops below Q_{ad}. A permanent stratified layer develops when $Q_{\mathrm{ad}} = 6\,\mathrm{TW}$, although the thickness of the layer fluctuates in time. In both cases, there is a lag between the time when Q_{cmb} is low and the time when the stratified layer reaches its maximum thickness. This time lag is set primarily by the thermal diffusion time for the stratified layer. We find a relatively thick stratified layer at 100–110 Ma, following a low CMB heat flow at 130 Ma. The low CMB heat flow (in this model) is a consequence of slow plate motions at the surface, 50–60 My earlier. The thickest layers are predicted to coincide with the time of maximum surface heat flow, although this result depends on the timescale for variations in the surface heat flow. The existence of a stratified layer in the core might be inferred from paleomagnetic observations of changes in the secular variation of the magnetic field, because a stable layer would filter magnetic fluctuations from the underlying convective region. Distinctive features in the field may also develop in the stratified layer through interactions with convective motion in the underlying region (Zhang and Schubert, 2000). More dramatic changes in the structure and dynamics of the core occur if the change in CMB heat flow is more substantial. For example, Nimmo and Stevenson (2000) suggest that a transition from plate tectonics to stagnant lid convection on Mars is sufficient to suppress convection throughout the Martian core, causing the termination of the magnetic field.

Lateral variations in heat flow at the CMB also affect the dynamics of the core. Cold slabs at the base of the mantle are expected to increase the local heat flow by increasing the local temperature gradient at the boundary. This effect seems unavoidable because the low viscosity of the core liquid maintains a nearly constant temperature over the boundary. Small adiabatic variations in temperature over kilometer-scale topography produce temperature variations of roughly 1 K, but these variations are small compared with temperature anomalies of several hundred Kelvins or more in the mantle. Consequently, variations in heat flow over the CMB can be inferred from the thermal structure on the mantle side of the boundary. Numerical simulations (Zhang and Gubbins, 1993, 1996; Sarson *et al.*, 1997) and experiments (Sumita and Olson, 1999, 2002) show that lateral variations in heat flow drive fluid motion in the core. In fact, several studies have sought to interpret core flows obtained from secular variation of the magnetic fields in terms of regional variations in heat flow (Bloxham and Gubbins, 1987; Kohler and Stevenson, 1990). Numerical simulations of convection with lateral variations in heat flow reveal a tendency to lock the pattern of convection to the pattern of heat flow at the boundary, although this typically occurs for a narrow range of parameter values. More commonly, the flow is highly time dependent. The source of the time dependence appears to be a consequence of switching the flow between the horizontal scale of the imposed boundary conditions and the natural scale of convection (Zhang and Gubbins, 1996; Sumita and Olson, 2002).

Similar conclusions are drawn from numerical geodynamo models (Sarson *et al.*, 1997; Olson and Christensen, 2002; Bloxham, 2002). Nonhomogeneous boundary conditions in geodynamo models yield persistent structure in the time-averaged flow (Olson and Christensen, 2002; Bloxham, 2002), although there can be substantial variation about the average (Christensen and Olson, 2003; Bloxham, 2002). The spatial pattern of the flow is expressed in the structure of the magnetic field. Models that use seismic heterogeneity at the base of the mantle to infer the local heat flow have been successful in producing a magnetic field with persistent nondipole structure (see **Figure 4**), similar to that deduced from paleomagnetic observations over the past few million years (Gubbins and Kelly, 1993; Johnson and Constable, 1997, 1998). However, the persistent nondipole structure is not prominent in individual snapshots of the solution; it emerges only after averaging the time-dependent part of the field. So far, it has not been possible to use thermal interactions to explain stationary features in the historical field over the past 300 years (Bloxham, 2002).

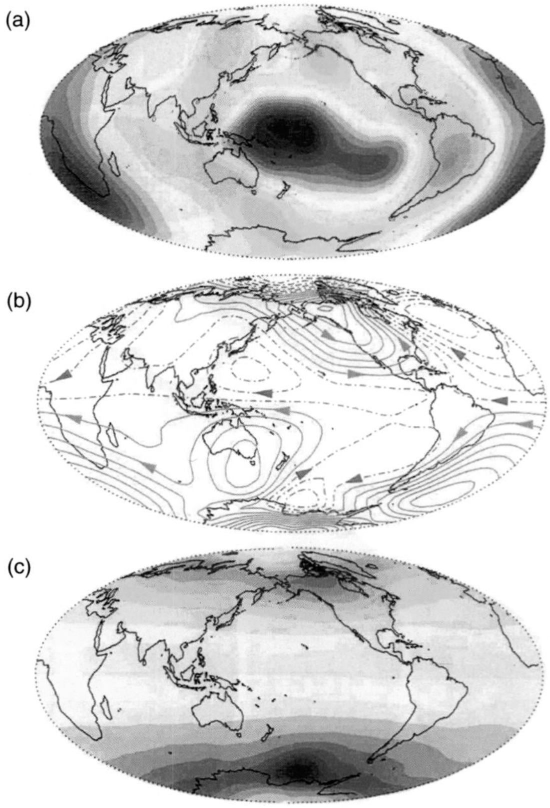

Figure 4 Lateral variations in Q_{cmb} induce flow near the top of the core and alter the time-averaged magnetic field from calculations of Olson and Christensen (2002). (a) Lateral variations in boundary heat flow are based on seismic models of velocity heterogeneity near the CMB. (b) Toroidal streamlines of steady flow below the CMB are due to imposed heat flow conditions. (c) Time-averaged radial magnetic field at CMB reveals nonzonal structure.

Spatial variations in heat flow may also have an important influence on magnetic reversals. The study of Glatzmaier *et al.* (1999) showed that different patterns of heat flow at the CMB dramatically alter the frequency of magnetic reversals. When the pattern of CMB heat flow is compatible with the natural pattern of convection in the outer core, there is a tendency to produce a stronger dipole field with less time dependence and fewer magnetic reversals. The opposite behavior is observed when the pattern of heat flow and the natural scale of convection are incompatible; in this case, much larger variations are observed in the field and reversals appear more frequently. Somewhat surprisingly, the most realistic behavior, in terms of time variations in the virtual geomagnetic pole (VGP), was obtained with uniform boundary conditions. Such a model cannot explain the persistent nondipole structure of the field, so it appears there are inconsistencies in our current understanding of thermal interactions. A related concern is the fact that dynamo models typically fail when the lateral variations in heat flow are comparable to the average (convective) heat flow (Olson and Christensen, 2002). Because a large part of the heat flow at the CMB is carried by conduction along the adiabat, the convective part of the average heat flow could be quite small (or even negative). Lateral variations in heat flow could easily exceed the average convective heat flow, which appears to doom the magnetic field in thermally driven dynamo models. It is not presently known if the dynamo models would still fail if convection was driven primarily by compositional buoyancy. Reconciling the behavior of the field with plausible thermal interactions remains an outstanding challenge.

Another consequence of thermal interactions involves the behavior of the field during a reversal. Compilations of VGPs from transition fields suggest that the VGPs are clustered into one of two preferred longitudes during a reversal (Laj *et al.*, 1991; Clement, 1991). In fact, it is possible for different sites to record different paths for the same reversal because the VGP location is based on the assumption of a dipole field. Non-dipole (multipole) components of the field also contribute to the direction of the field at a given site, shifting the VGP location from the position of the actual dipole axis. Different reversal paths emerge from different sites when the multipole components become prominent relative to the dipole. Kutzner and Christensen (2004) investigate the question of preferred reversal paths using a geodynamo model with lateral variations in CMB heat flow. Part of the time-averaged magnetic field in this study included an equatorial dipole, which defines a preferred orientation in longitude. VGP paths from a large number of sites on the surface sense this orientation, causing a preferred path during reversals. However, the scatter in VGP directions due to multipole components means that the preferred direction emerges only when a large number of sites are averaged. The question of whether preferred reversal paths can be inferred from a small number of sites at the surface remains a contentious issue, both for the models and the paleomagnetic observations. Additional questions arise because the models are still very far from Earth-like conditions. On the other hand, the signatures of thermal core–mantle interactions should be embedded in the flow and field at the top of the core. Identifying these signatures in both observations and models should be an important part of future progress.

12.3 Electromagnetic Interactions

Electric currents are induced in the lower part of the mantle as a consequence of dynamo action in the liquid core. Several types of interactions between the core and mantle are possible. One involves the force on current-carrying material, which can transfer momentum between the core and the mantle. This mechanism is commonly proposed to explain variations in the length of day over periods of several decades. A second type of interaction causes a distortion of the magnetic field as it diffuses through the mantle toward the surface. The importance of both effects depends on the electrical conductivity of the lower mantle. Laboratory-based estimates for mantle silicates and oxides, extrapolated to lower-mantle conditions, typically yield conductivities of $10\ \mathrm{S\ m^{-1}}$ or less (Xu *et al.*, 2000). Such low values are expected to yield relatively weak currents in the lower mantle and small electromagnetic core–mantle interactions. However, more substantial electromagnetic interactions are possible if the lowermost mantle is not composed entirely of silicates and oxides. Chemical reactions between the core and mantle have been proposed as a mechanism to incorporate iron alloys into the base of the mantle (Knittle and Jeanloz, 1989). The iron alloy may be a reaction product (Jeanloz, 1990) or a result of incorporating core material directly into the mantle (Poirier and LeMouel, 1992; Buffett *et al.*, 2000; Petford *et al.*, 2005; Kanda and Stevenson, 2006). More recent evidence of a new high-pressure phase of $MgSiO_3$ (Murakami *et al.*, 2004; Oganov and Ono, 2004) opens new possibilities for high conductivities in the lower mantle (Ono *et al.*, 2006). Regions of low conductivity may also arise from changes in the partition of iron between the dominant mineral components due to a transition in the spin state of iron in ferropericlase $(Mg_{0.83}, Fe_{0.17})O$ (Badro *et al.*, 2003).

One of the earliest suggestions of electromagnetic core–mantle interaction was motivated by the observation of small fluctuations in the length of day over periods of several decades (Bullard *et al.*, 1950; Rochester, 1962; Roden, 1963). Motion of the core relative to the mantle sweeps lines of magnetic field through the lower mantle, inducing a horizontal electric current. The resulting forces on the core and mantle act to oppose any relative motion. The axial component of the associated torque transfers angular momentum between the core and mantle. Stix and Roberts (1984) were the first to use detailed estimates of flow at the top of the core to determine the electromagnetic torque on the mantle. They predicted variations in the torque which were sufficient to explain the fluctuations in the length of day. However, these variations were superimposed on a steady torque, which had the undesirable effect of causing a steady change in the angular velocity of the core and the mantle. To avoid this effect, they proposed an additional (balancing) torque that arises when radial electric currents leak across the CMB. The existence of such a current is reasonable, although it cannot be constrained by surface observations. As a result, electromagnetic interactions provide a viable but unproven mechanism for explaining decadal variations in the length of day.

Holme (1998) revisited the question of electromagnetic core–mantle interactions by showing that the part of the core flow which contributes to the torque is not constrained by measurements of secular variation (also see Wicht and Jault, 1999). This means that the torque is a product of the assumptions used to resolve the nonuniqueness of the flow rather than by the data constraints. The strategy proposed by Holme (1998) was to determine the flow at the top of the core by fitting the observed secular variation subject to the condition that rearrangement of magnetic flux explains the length-of-day variations. (This calculation did not require any radial current across the CMB.) Plausible core flows were found to explain the variations in the length of day, provided the conductance of the lower mantle was roughly 10^8 S. The necessary conductance could be obtained with a 100 km layer at the base of the mantle and an average conductivity of $10^3\ S\ m^{-1}$, although other combinations of thickness and conductivity are possible.

A similar value for the conductance of the lower mantle was obtained from the study of the Earth's nutation (Buffett, 1992). Periodic variations in the direction of the Earth's rotation (nutations) are caused by the lunar and solar tides. The Earth's response to tidal torques includes a differential rotation of the core relative to the mantle, which alters the observed motion at the surface. Electromagnetic interactions alter this response by introducing a restoring force that opposes the differential rotation. The associated dissipation due to Ohmic loss is detected as a phase lag in the response relative to that predicted for an elastic Earth model (e.g., Mathews and Shapiro, 1992). Comparison of theoretical predictions (Wahr, 1981; Mathews *et al.*, 1991) and observations (Herring *et al.*, 1991) reveals a large discrepancy in the annual nutation, which is particularly sensitive to relative motion between the core and the mantle. This discrepancy could be eliminated by including electromagnetic interactions, as long as the conductance of the lower mantle is 10^8 S, comparable to the value required to explain length-of-day variations. However, there is one important distinction. Nutations involve nearly diurnal motion of the core relative to the mantle. (An annual nutation is defined by the period of the beat frequency between the nearly diurnal motion and exactly one cycle per day.) The associated skin depth for magnetic diffusion over diurnal periods limits electric currents to the immediate vicinity of the boundary, so the discrepancies in the nutations are most easily explained by high (10^5–$10^6\ S\ m^{-1}$) conductivities in a relatively thin (10^2–10^3) layer. A variety of mechanisms have been examined to explain (or dispute) such a conducting layer at the base of the mantle (Knittle and Jeanloz, 1991; Poirier and LeMouel, 1992, Buffett *et al.*, 2000, Petford *et al.*, 2005, Kanda and Stevenson, 2006). While improvements in nutation theory (Mathews *et al.*, 2002) and observations (Herring *et al.*, 2002) continue to support the existence of a conducting layer, other sources of dissipation have also been proposed. The most likely alternative is viscous dissipation in the liquid core (Buffett, 1992; Mathews and Guo, 2005; Deleplace and Cardin, 2006), although this explanation requires a viscosity which is roughly 4 orders of magnitude larger than recent estimates (de Wijs *et al.*, 1998; Zhang and Guo, 2000).

Another type of electromagnetic interaction arises when the electrical conductivity in the lower mantle varies laterally. Steady flow in the core sweeps the radial (poloidal) component of the magnetic field through the mantle, inducing local electric currents. In a uniform conductor, these currents produce a tangential (toroidal) component of the magnetic field. However, a heterogeneous distribution of conductivity channels current through the conducting regions. The added complexity in the current induces a poloidal field which can re-enforce the initial poloidal field. Busse and Wicht (1992) showed that self-sustaining dynamo action is possible in this case, even when the flow in the core is spatially uniform. In effect, the complexity of the conductivity structure replaces the need for complicated fluid motion in a uniform conductor. While the conditions required for self-sustaining dynamo action are unlikely to be realized in the Earth's core, it is possible to induce a small-scale radial field with an amplitude as large as 0.1 mT (Buffett, 1996c). This mechanism can also contribute to the large-scale field at the surface by distorting the unseen toroidal field to produce an observable poloidal component (Koyama *et al.*, 2001).

Electromagnetic interactions also accompany time variations in the amplitude of the magnetic field. The largest change in amplitude probably occurs during magnetic reversals. While the general structure of the field during a reversal is not well known, we do know that the axial part of the dipole field vanishes and reappears with the opposite polarity in as little as several thousand years (Clement, 2004). Rapid time variations in the axial dipole create large-scale electric fields in the lower mantle. When the distribution of electrical conductivity of the mantle is heterogeneous, the resulting currents generate a poloidal field (see **Figure 5**). The structure of the induced field depends on the spatial distribution of electrical conductivity. A large-scale pattern of electrical conductivity with a local conductance of 10^8 should produce an observable field during a magnetic reversal. Such a mechanism provides an attractive explanation for preferred reversal paths because the geographic location and structure of the induced field is fixed by the distribution of electrical conductivity in the lower mantle. Costin and Buffett (2004) explored this mechanism using a distribution of electrical conductivity that was based on estimates of topography on the CMB (dominantly a degree-2 pattern in spherical harmonics). The main part of the induced field had a nonzonal degree-3 pattern, which was superimposed on a decreasing axial dipole field to predict VGP paths. Preferred paths were caused by the nonzonal part of the field, although the actual location of the preferred paths depended on the location of the observing sites. Costin and Buffett (2004) used the location of sites in the sediment database of Clement (1991) to predict a clustering of reversal paths through the Americas and Asia (see **Figure 6**). This prediction was consistent with inferences drawn directly from the observations (Clement, 1991).

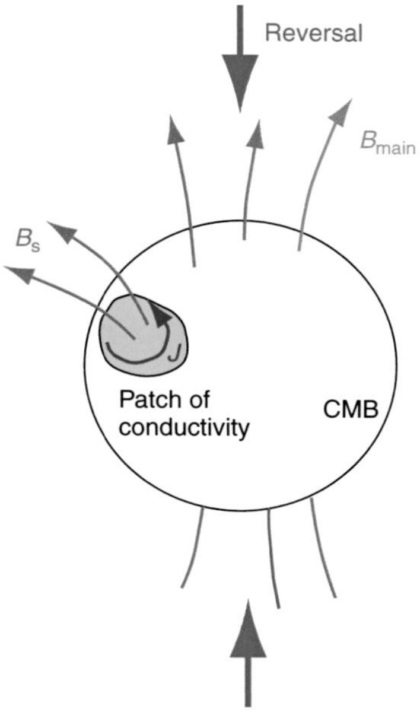

Figure 5 Time variations in the main magnetic field, B_{main}, during reversals induce electric currents J in the conductive regions at the CMB. The electric currents generate a secondary field, B_s, which is superimposed on B_{main}. The resulting distortion of B_{main} can give rise to preferred reversal paths because the location of the secondary field is fixed by the spatial distribution of electrical conductivity near the base of the mantle.

An interesting test of this prediction is afforded by the current rate of decrease in the dipole field, which is not too different from the average rate of decrease during a reversal (perhaps within a factor of 2 or 3). The induced field is both observable and consistent with historical observations of the field, although it is not currently possible to separate contributions from electromagnetic core–mantle interactions and the underlying variations that originate within the core. This ambiguity is a source of concern because the contribution of electromagnetic core–mantle interactions to the field at the surface can potentially complicate the way we interpret observations of secular variation in the magnetic field.

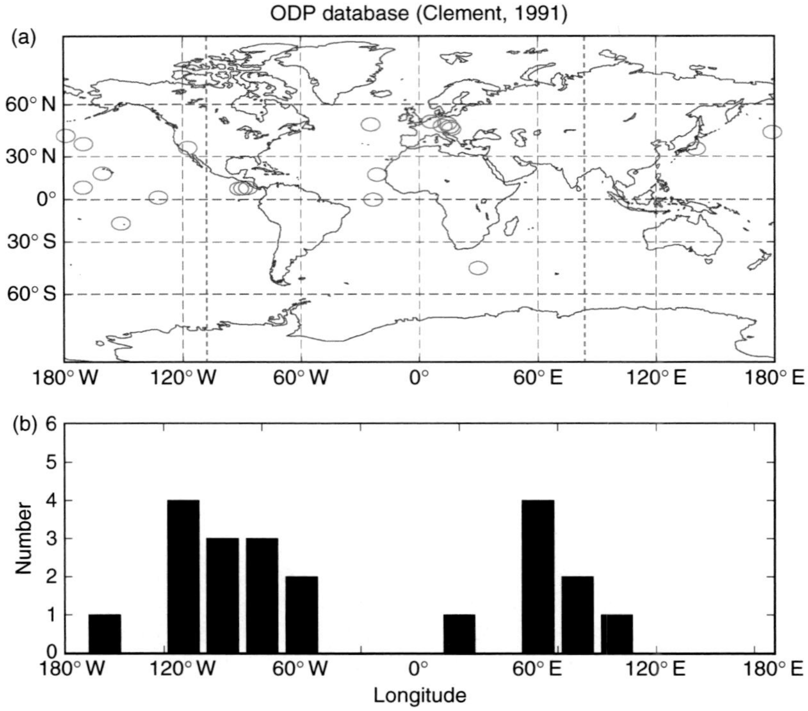

Figure 6 (a) Location of samples from the sediment database of Clement (1991) and (b) histogram of reversal path longitudes predicted by Costin and Buffett (2004). The spatial distribution of electrical conductivity is based on estimates of the large-scale topography on the CMB. The secondary field induced by this heterogeneous distribution of conductivity causes reversal paths to preferentially pass though the Americas and Asia.

12.4 Mechanical Interactions

Mechanical interactions are commonly invoked in the transfer of momentum between the core and the mantle. Most attention has focused on mechanisms that account for the observed variation in the length of day over timescales of several decades. Lorentz forces due to electromagnetic interactions offer one possible mechanism, although several other types of mechanical interactions have also been proposed. One such mechanism is due to flow of the core over topography on the CMB (Hide, 1969). Pressure differences on the leading and trailing sides of bumps on the boundary result in a torque that transfers angular momentum. When seismic estimates of the CMB topography first appeared in the literature (Morelli and Dziewonski, 1987; Forte and Peltier, 1991), it became possible to estimate the pressure (or topographic) torque using models of flow in the core (Hide, 1989; Jault and LeMouel, 1989). Most models of core flow are constrained by assuming a geostrophic force balance in the tangential direction (*see* Chapter 4). Under this assumption, the flow field recovered from observations of secular variation can be converted to a pressure field. When that pressure field was integrated over the boundary with known topography, the resulting torque was too large by several orders of magnitude to explain the observed variation in length of day (Jault and LeMouel, 1990). A small shift in the position of the topography could greatly reduce the torque, leading to speculations that the pattern of flow was locked by the topography to keep the torque small. Alternatively, it was possible to make small changes to the flow which eliminated the pressure torque entirely, but did not substantially alter the fit to the secular variation observations (Kuang and Bloxham, 1993). These results demonstrate the sensitivity of the calculation to small errors in either the topography or the flow models.

More problematic is the consistency of calculating a pressure torque using flow models that assume a geostrophic force balance at the top of the core. Geostrophic flow represents a balance between the Coriolis force and pressure gradient. Such a flow can have no time variation because the presence of accelerations implies a departure from geostrophic conditions. There can also be no net torque on the core (or the mantle) because a net torque also implies a change in momentum. Kuang and Bloxham (1997) used this result to show that the pressure torque due to a geostrophic flow must vanish. They did this by demonstrating that the torque associated with the Coriolis force is identically zero. The sum of the pressure and Coriolis torques must vanish if there is no change in momentum in a geostrophic flow, so it follows that the pressure torque is zero. Of course, it is possible that the flow at the top of the core is not geostrophic, so the pressure torque need not vanish. Jault and LeMouel (1999) argue that the geostrophic approximation would still provide a good approximation for the actual pressure field (and hence the pressure torque). However, it is doubtful that a reliable pressure field for mechanical coupling calculations can be recovered from the flow, given the sensitivity of the calculation to small errors.

Other estimates of the pressure torque have been obtained using models of idealized flow over boundary topography (e.g., Anufriev and Braginsky, 1975, 1977; Moffatt and Dillon, 1976). Pressure in the fluid is perturbed by the flow around the topography. The influence of this perturbation on the boundary topography produces the pressure torque. The amplitude of the pressure torque in these calculations varies as h^2, where h is the height of the topography. One factor of h arises because the perturbation depends on the height of the topography; the second factor of h arises because the integral over the surface for the pressure torque depends on the presence of topography. For boundary topography of a few kilometers at large spatial scales, the resulting pressure torque is probably too small to explain the observed variations in the length of day (e.g., Mound and Buffett, 2005). Similar conclusions have been obtained using numerical geodynamo models that include the influence of boundary topography (Kuang and Chao, 2001). An important aspect of the study by Kuang and Chao (2001) is that they avoid the use of idealized flow and make no assumptions about the structure of the magnetic field. The pressure at the boundary evolves in response to both the underlying convection and the presence of boundary topography. The fact that the pressure torque in these calculations is small would seem to resolve the question of pressure coupling.

Other consequences of fluid pressure on the CMB can be detected at the surface. For example, fluctuations in pressure produce observable changes in both gravity and surface topography (Fang *et al.*, 1996; Dumberry and Bloxham, 2004; Greff-Lefftz *et al.*, 2004). A typical pressure of 10^3 Pa changes on timescales of several decades as the flow evolves at the CMB. When the mantle responses elastically to the pressure change, the surface displacement can be a few millimeters and the change in the gravity field is within current limits of detection (Dumberry and Bloxham, 2004; Greff-Lefftz *et al.*, 2004). Recent interest in this process is motivated by observations of decadal changes in the elliptical part of the gravity field (Cox and Chao, 2002). While it appears that pressure changes in the core are too small to explain the observations of Cox and Chao (2002), current advances in space geodetic observations hold the promise of making the somewhat smaller core processes observable at the Earth's surface.

Gravitational interactions between the core and the mantle cause a different type of mechanical coupling. Convection in both regions creates and evolves density anomalies inside the Earth. Gravitational attraction of these density anomalies can result in a transfer of momentum (Jault and LeMouel, 1989; Rubincam, 2003). The distribution of density anomalies in the mantle can be inferred from seismic tomography models by assuming that variations in wave speed are due mainly to temperature (Dziewonski *et al.*, 1977; Hager *et al.*, 1985; Ricard and Vigny 1989). Unfortunately, a similar procedure is not feasible in the outer core, because the density anomalies are far too small to be detected seismically (Stevenson, 1987). An order of magnitude estimate for the density anomalies in the outer core suggests that gravitational interactions could be important (Jault and LeMouel, 1989). However, it is not presently possible to evaluate the torque without knowing the distribution of density anomalies in the outer core.

An alternative source of gravitational interaction occurs between density anomalies in the mantle and the inner core (Buffett, 1996a). The principle source of density heterogeneity in the inner core is caused by topography on the inner-core boundary, which can be estimated by assuming that the boundary tends to adjust over time toward an equipotential surface (*see* Chapter 10). (Density anomalies may also occur inside the inner core, but these are probably small because the inner core solidifies from the well-mixed outer core.) Density anomalies in the outer core are expected to be very small (e.g., $\delta\rho/\rho \approx 10^{-8}$), so the primary disturbance to equipotential surfaces in the core is probably caused by

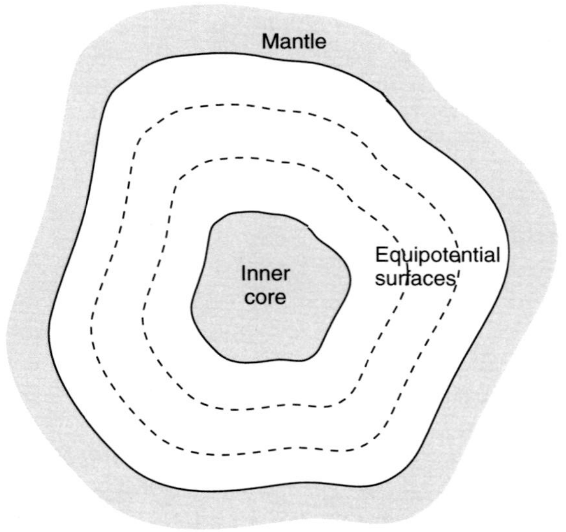

Figure 7 Schematic illustration of the heterogeneous Earth viewed on an equatorial cross section. Density anomalies in the mantle drive flow and perturb the CMB. Surfaces of constant potential inside the core are disturbed from axial symmetry by the combined effect of mantle density anomalies and boundary displacements. The inner core adjusts to a hydrostatic state by aligning the boundary with an equipotential surface. This state also minimizes the gravitational potential energy. Rotation of the inner core relative to its equilibrium position produces a mutual gravitational torque on the mantle and inner core.

density anomalies in the mantle. This means that the distribution of density anomalies in the mantle determines the shape of the inner core (see **Figure 7**). Estimates of the gravity perturbations from dynamic geoid studies (Forte *et al.*, 1994; Defraigne *et al.*, 1996) suggest that the peak-to-peak topography on the inner-core boundary is about 100 m or less. When the inner core rotates through the gravitational field of the mantle, a strong gravitational force acts to restore the inner core to its equilibrium position. An equal and opposite force on the mantle is about 2 orders of magnitude larger than the gravitational attraction due density anomalies in the outer core (Buffett, 1996a), primarily because the density anomalies associated with inner-core topography are much larger (e.g., $\delta\rho/\rho \approx 10^{-4}$). These large gravitational forces provide a plausible explanation for the decadal variations in length of day. The mechanism relies on a combination of electromagnetic and gravitational torques (Buffett, 1996b). Motions in the fluid outer core are tightly coupled to the inner core by electromagnetic forces on account of the high electrical conductivity on either side of the inner-core boundary. The inner core rotates in response to the strong electromagnetic torque, which transfers momentum to the mantle through the action of gravitational torques on the inner core. Tests of this mechanism in numerical geodynamo models yield torques on the mantle with a typical amplitude of 2×10^{18} N m (Buffett and Glatzmaier, 2000), which is sufficient to explain the variations in length of day. It is also possible to explain the length-of-day variations when the fluid motions in the outer core are restricted to the form of torsional oscillations (Mound and Buffett, 2005). These oscillations are of interest because the typical period is compatible with the timescale for variations in length of day (Braginsky, 1970, 1984). An independent test for gravitational interactions would involve the search for a free mode of oscillation due to the gravitational restoring force between the inner core and the mantle. Mound and Buffett (2006) present evidence for a 6 year oscillation in the length of day, which they attribute to gravitational interactions. Corroborating evidence could potentially be sought in time variations of the gravity field (Buffett and Glatzmaier, 2000).

12.5 Chemical Interactions

It is generally assumed that the core has been chemically isolated from the mantle since the time of core formation. The initial composition of the core is established by a series of processes that separate liquid iron into large enough volumes to inhibit chemical equilibration with the rest of the Earth (e.g., Stevenson, 1990; Rubie *et al.*, 2003). Subsequent transport of mass between liquid iron and silicates is assumed to be insignificant after the core forms. This conventional view is challenged by the studies of Knittle and Jeanloz (1989, 1991), which used diamond-anvil cell experiments to show that liquid iron reacts with silicates and oxides at high pressure and temperature. The reaction products are thought to include iron alloys, such as FeO and FeSi, and iron-depleted silicate minerals (Knittle and Jeanloz, 1989; Goarant *et al.*, 1992). Mantle minerals in direct contact with the core liquid should quickly establish chemical equilibrium, because reactions at high temperature are expected to be fast and convective mixing in the core is relatively efficient (the convective overturn time is probably of the order of 10^3 years). On the other hand, convection in the mantle exposes fresh surfaces of unreacted material more slowly. Jeanloz (1990) suggests that chemical reactions and transport of light elements into the core could have occurred over most of the Earth's history. Subduction of oxidized ocean crust may also drive reactions by continually altering chemical conditions at the base of the mantle (Walker, 2005). In either case, chemical reactions could alter the composition of the core by permitting a flux of mass between the mantle and the core.

Evidence for the core leaking back into the mantle is based on measurements of Os isotope ratios in some lavas associated with hot spots (Walker *et al.*, 1995; Brandon *et al.*, 1998). Enrichment in $^{186}Os/^{188}Os$ and $^{187}Os/^{186}Os$ relative to upper-mantle materials has been attributed to small additions of core material to the lower mantle. This core material is subsequently entrained in mantle plumes, where it contributes to lavas that are sampled at the surface (see Brandon *et al.* (2003) for a recent review). Only small amounts of core material are required to perturb the isotopic composition of the mantle because the concentration of Os in the core vastly exceeds that in the mantle, based on partitioning of Os between silicates and liquid metal. However, the observed enrichment in isotopic ratios require a source region with elevated ratios of Pt/Os and Re/Os in order to produce excess ^{186}Os and ^{187}Os through radioactive decay of ^{190}Pt and ^{187}Re. Given the long half-lives involved (e.g., 489 and 42 Ga), the elevated ratios of Pt/Os and Re/Os must be maintained for a very long time to produce the required isotopic enrichment.

Brandon *et al.* (1998) attribute the elevated ratios of Pt/Os and Re/Os in the liquid core to the growth of the inner core. The initial abundances of Pt, Re, and Os in the core should be close to the chondritic abundances because all of these elements are highly siderophile; most of the initial inventory enters the core with little fractionation. Solidification of the core removes more Os from the liquid core than either Re or Pt (Walker, 2000). This process increases Pt/Os and Re/Os ratios in the outer core above the chondritic ratios. Unfortunately, the current experimental estimates for the partitioning of Pt, Re, and Os between liquid and solid iron appear to require an old age for the inner core (Brandon *et al.*, 2003), which is incompatible with models for the thermal history of the core (Labrosse *et al.*, 2001; Buffett, 2002). A further complication has arisen with the recent measurement of tungsten isotopes in lavas that exhibit $^{186}Os/^{188}Os$ and $^{187}Os/^{188}Os$ enrichment (Schersten *et al.*, 2004). So far, no evidence for a core signature has been found in the tungsten isotopes, although the debate continues (Brandon and Walker, 2005).

Despite the controversy over isotopic signals from the core, it is difficult to escape the conclusion that some form of chemical interaction is inevitable. Differences in redox state imply that the core and the mantle cannot be in bulk equilibrium (Walker, 2005). Even a local equilibrium between the core and mantle minerals adjacent to the boundary is liable to be perturbed as the core cools. Changes in temperature alter the solubility of dissolved components, while growth of the inner core continually fractionates light elements into the outer core. Eventually, the liquid core may become supersaturated in light elements, causing an immisicible liquid phase to form (Ito *et al.*, 1995). Alternatively, the excess concentration of light elements may drive back reactions at the CMB, removing the excess light elements and precipitating mantle-like silicates (Buffett *et al.*, 2000). Walker (2005) has also discussed the possibility of electrochemical reactions at the CMB (Kavner and Walker, 2006) and the consequences of chemical shifts on the mantle side of the boundary. There is no shortage of mechanisms that can cause chemical interactions between the core and the mantle, and there are good reasons for suspecting that many of these operate within the Earth. The challenge lies in quantifying their importance for the evolution and dynamics of the planet.

12.6 Conclusions

Interactions between the core and the mantle take many forms. Thermal interaction involves changes in the rate and spatial distribution of heat flow across the CMB. The resulting changes in the vigor and pattern of convection in the core can dramatically affect the geodynamo and produce observable signals in the magnetic field at the Earth's surface. Electromagnetic interactions are the result of electric currents near the boundary. The consequences are entirely dependent on the electrical conductivity of the lower mantle. A conductance of 10^8 S for the lower mantle is sufficient to explain variations in the length of day and to contribute significantly to the induction of magnetic field inside the Earth. Mechanical interactions are most commonly attributed to the effects of fluid pressure acting on the CMB. While pressure torques are not thought to be the primary mechanism for transferring angular momentum between the core and the mantle, deformations due to fluid pressure on the CMB can cause observable changes in gravity and surface displacement. A more viable explanation for the variations in length of day at decadal periods involves gravitational interactions, primarily between the inner core and the mantle. Better observations of these gravitational interactions may provide new insights into the nonhydrostatic distribution of mass in the mantle. Chemical interactions of some form appear to be likely, although the details are presently unknown. Advances in high-pressure and high-temperature experiments may soon begin to fill in these details and provide unexpected surprises in the chemical evolution of the core.

References

Anufriev AP and Braginsky SI (1975) Influence of irregularities of the boundary of the Earth's core on the velocity of the liquid and on the magnetic field. *Geomagnetism and Aeronomy* (Engl. transl.) 15: 754–757.

Anufriev AP and Braginsky SI (1977) Influence of irregularites of the boundary of the Earth's core on fluid velocity and the magnetic field, II. *Geomagnetism and Aeronomy* 17: 78–82.

Badro J, Fiquet G, Guyot F, *et al.* (2003) Iron partitioning in Earth's mantle: Toward a deep lower mantle discontinuity. *Science* 300: 789–791.

Bloxham J and Gubbins D (1987) Thermal core–mantle interactions. *Nature* 325: 511–513.

Bloxham J (2002) Time-independent and time-dependent behaviour of high-latitude flux bundles at the core–mantle boundary. *Geophysical Research Letters* 29.

Bloxham J (2000) Sensitivity of the geomagnetic axial dipole to thermal core–mantle interactions. *Nature* 405: 63–65.

Braginsky SI (1963) Structure of F layer and reasons for convection in the Earth's core. *Doklady Akademii Nauk SSSR* 149: 1311–1317.

Braginsky SI (1970) Torsional magnetohydrodynamic vibrations in the Earth's core and variations in length of day. *Geomagnetism and Aeronomy* (Engl. transl.) 10: 1–8.

Braginsky SI (1984) Short-period geomagnetic secular variation. *Geophysical and Astrophysical Fluid Dynamics* 30: 1–78.

Brandon AD and Walker RJ (2005) The debate over core–mantle interaction. *Earth and Planetary Science Letters* 232: 211–225.

Brandon AD, Walker RJ, Morgan JW, Norman MD, and Prichard HM (1998) *Science* 280: 1570–1573.

Brandon AD, Walker RJ, Puchtel IS, Becker H, Humayun M, and Revillon S (2003) Os-186–Os-187 systematics of Gorgona Island komatiites: Implications for early growth of the inner core. *Earth and Planetary Science Letters* 206: 411–426.

Buffett BA (1992) Constraints on magnetic energy and mantle conductivity from the forced nutations of the Earth. *Journal of Geophysical Research* 97: 19581–19597.

Buffett BA (1996a) Gravitational oscillations in the length of day. *Geophysical Research Letters* 23: 2279–2282.

Buffett BA (1996b) A mechanism for decade fluctuations in the length of day. *Geophysical Research Letters* 23: 3803–3806.

Buffett BA (1996c) Effects of a heterogeneous mantle on the velocity and magnetic fields at the top of the core. *Geophysical Journal International* 125: 303–317.

Buffett BA and Glatzmaier GA (2000) Gravitational breaking of inner-core rotation in geodynamo simulations. *Geophysical Research Letters* 27: 3125–3128.

Buffett BA, Garnero EJ, and Jeanloz R (2000) Sediments at the top of the core. *Science* 290: 1338–1342.

Buffett BA (2002) Estimates of heat flow in the deep mantle based on the power requirements for the geodynamo. *Geophysical Research Letters* 29(12): 1566 (doi: 10.1029/2001GL014649).

Bullard EC, Freeman C, Gellman H, and Nixon J (1950) The westward drift of the Earth's magnetic field. *Philosophical Transactions of the Royal Society of London A* 132: 61–92.

Bunge H-P, Hagelberg CR, and Travis BJ (2003) Mantle circulation models with variational data assimilation: Inferring past mantle flow and structure from plate motion histories and seismic tomography. *Geophysical Journal International* 152: 280–301.

Busse FH and Wicht J (1992) A simple dynamo caused by conductivity variations. *Geophysical and Astrophysical Fluid Dynamics* 62: 135–144.

Christensen UR and Olson P (2003) Secular variation in numerical geodynamo models with lateral variations of boundary heat flow. *Physics of the Earth and Planetary Interiors* 138: 39–54.

Clement BM (1991) Geographical distribution of transitional VGPs: Evidence for non-zonal equatorial symmetry during the Matuyama–Bruhnes reveral. *Earth and Planetary Science Letters* 104: 48–58.

Clement BM (2004) Dependence of the duration of geomagnetic polarity reversals on site latitude. *Nature* 637–640.

Costin SO and Buffett BA (2004) Preferred reversal paths caused by a heterogeneous conducting layer at the base of the mantle. *Journal of Geophysical Research* 109: B06101 (doi:10.1029/2003JB002853).

Cox CM and Chao BF (2002) Detection of a large-scale mass redistribution in the terrestrial system since 1998. *Science* 297: 831–833.

Defraigne P, Dehant V, and Wahr JM (1996) Internal loading of an inhomogeneous compressible earth with phase boundaries. *Geophysical Journal International* 125: 173–192.

Deleplace B and Cardin P (2006) Viscomagnetic torque at the core mantle boundary. *Geophysical Journal International* 167: 557–566.

de Wijs GA, Kresse G, Vocadlo L, *et al.* (1998) The viscosity of liquid iron at the physical conditions of the Earth's core. *Nature* 392: 805–807.

Dubrovinsky L, *et al.* (2003) Iron–silica interaction at extreme conditions and the electrical conductivity of the lower mantle. *Nature* 422: 58–61.

Dumberry M and Bloxham J (2004) Variations in the Earth's gravity field caused by torsional oscillations in the core. *Geophysical Journal International* 159: 417–434.

Dziewonski AM, Hager BH, and O'Connell RJ (1977) Large-scale heterogeneities in the lower mantle. *Journal of Geophysical Research* 82: 239–255.

Fang M, Hager BH, and Herring TA (1996) Surface deformation caused by pressure changes in the fluid core. *Geophysical Research Letters* 23: 1493–1496.

Forte AM and Peltier WR (1991) Mantle convection and core–mantle boundary topography – explanations and implications. *Technophysics* 187: 91–116.

Forte AM, Woodward RL, and Dziewonski AM (1994) Joint inversion of seismic and geodynamic data for models of three-dimensional mantle heterogeneity. *Journal of Geophysical Research* 99: 21857–21877.

Gibbons SJ and Gubbins D (2000) Convection in the Earth's core driven by lateral variations in the core–mantle boundary heat flow. *Geophysical Journal International* 142: 631–642.

Glatzmaier GA, Coe RS, Hongre L, and Roberts PH (1999) The role of the Earth's mantle in controlling the frequency of geomagnetic reversals. *Nature* 401: 885–890.
Goarant F, Guyot F, Peronneau J, and Poirier JP (1992) High-pressure and high-temperature reactions between silicates and liquid iron alloys in the diamond cell studied by analytical electron microprobe. *Journal of Geophysical Research* 97: 4477–4487.
Greff-Lefftz M, Pais MA, and LeMouel JL (2004) Surface gravitational field and topography changes induced by the Earth's fluid core motions. *Journal of Geodesy* 78: 386–392.
Gubbins D, Thompson CJ, and Whaler KA (1982) Stable regions in the Earth's liquid core. *Geophysical Journal of the Royal Astronomical Society* 68: 241–251.
Gubbins D and Kelly P (1993) Persistent patterns in the geomagnetic field over the past 2.5 Myr. *Nature* 365: 829.
Hager BH, Clayton RW, Richards MA, Comer RP, and Dziewonski AM (1985) Lower mantle heterogeneity, dynamic topography and the geoid. *Nature* 313: 541–546.
Herring TA, Buffett BA, Mathews PM, and Shapiro II (1991) Forced nutations of the Earth: Influence of inner core dynamics, 3. Very long baseline interferometry data analysis. *Journal of Geophysical Research* 96: 8259–8273.
Herring TA, Mathews PM, and Buffett BA (2002) Modeling of nutation and precession: Very long baseline results. *Journal of Geophysical Research* 107: 2069.
Hide R (1969) Interaction between Earth's liquid core and mantle. *Nature* 222: 1055–1056.
Hide R and Malin SRC (1970) Novel correlations between the global features of the Earth's gravitational and magnetic fields. *Nature* 225: 605.
Hide R (1989) Fluctuations in the Earth's rotation and the topography of the core–mantle interface. *Philosophical Transactions of the Royal Society of London A* 328: 351–363.
Holme R (1998) Electromagnetic core–mantle coupling I. Explaining decadal changes in the length of day. *Geophysical Journal International* 132: 167–180.
Ito E, Morooka K, Ujike O, and Katsura T (1995) Reactions between molten iron and silicate melts at high pressure – Implications for the chemical evolution of the Earth's core. *Journal of Geophysical Research* 100: 5901–5910.
Jault D and LeMouel J-L (1989) The topographic torque associated with tangentially geostrophic motion at the core surface and inferences on the flow inside the core. *Geophysical and Astrophysical Fluid Dynamics* 48: 273–296.
Jault D and LeMouel J-L (1990) Core–mantle boundary shape – Constraints inferred from the pressure torque acting between the core and the mantle. *Geophysical Journal International* 101: 233–241.
Jault D and LeMouel J-L (1999) Comment on 'On the dynamics of topographical core–mantle coupling'. *Physics of the Earth and Planetary Interiors* 114: 211–215.
Jeanloz R (1990) The nature of the Earth's core. *Annual Review of Earth and Planetary Sciences* 18: 357–386.
Johnson CL and Constable CG (1997) The time-averaged geomagnetic field: Global and regional biases for 0–5 Ma. *Geophysical Journal International* 131: 643.
Johnson CL and Constable CG (1998) Persistently anomalous Pacific geomagnetic fields. *Geophysical Research Letters* 25: 1011–1014.
Kanda RVS and Stevenson DJ (2006) Suction mechanism for iron entrainment into the lower mantle. *Geophysical Research Letters* 33: L02310 (doi:10.1029/2005GL025009).
Kavner A and Walker D (2006) Core/mantle-like interactions in an electric field. *Earth and Planetary Science Letters* 248: 316–329.
Knittle E and Jeanloz R (1989) Simulating the core–mantle boundary – An experimental study of high-pressure reactions between silicates and liquid iron. *Geophysical Research Letters* 16: 609–612.
Knittle E and Jeanloz R (1991) The Earth's core–mantle boundary: Results of experiments at high pressure and temperature. *Science* 251: 1438–1443.
Kohler MD and Stevenson DJ (1990) Modeling core fluid motions and the drift of magnetic-field patterns at the CMB by using topography obtained by seismic inversion. *Geophysical Research Letters* 17: 1473–1476.
Koyama T, Shimizu H, and Utada H (2001) Possible effects of lateral heterogeneity in the D″ layer on electromagnetic variations of core origin. *Physics of the Earth and Planetary Interiors* 129: 99–116.
Kuang W and Bloxham J (1993) On the dynamics of topographical core–mantle coupling. *Geophysical and Astrophysical Fluid Dynamics* 72: 161–195.
Kuang W and Bloxham J (1997) On the dynamics of topographical core–mantle coupling. *Physics of the Earth and Planetary Interiors* 99: 289–294.
Kuang W and Chao BF (2001) Topographic core–mantle coupling in geodynamo modeling. *Geophysical Research Letters* 28: 1871–1874.
Kutzner C and Christensen UR (2004) Simulated geomagnetic reversals and preferred virtual geomagnetic pole paths. *Geophysical Journal International* 157: 1105–1118.
Labrosse S, Poirier JP, and LeMouel JL (2001) The age of the inner core. *Earth and Planetary Science Letters* 190: 111–123.
Laj C, Mazaud A, Weeks R, Fuller M, and Herrero-Bervera E (1991) Geomagnetic reversal paths. *Nature* 351: 447.
Lister JR and Buffett BA (1998) Stratification of the outer core at the core–mantle boundary. *Physics of the Earth and Planetary Interiors* 105: 5–19.
Lithgow-Bertelloni C and Richards MA (1998) The dynamics of Cenozoic and Mesozoic plate motions. *Reviews of Geophysics* 36: 27–78.
Loper DE (1978) Some consequences of a gravitationally powered dynamo. *Journal of Geophysical Research* 83: 5961–5970.
Mathews PM, Buffett BA, Herring TA, and Shapiro II (1991) Forced nutations of the Earth – Influence of inner core dynamics, 1. Theory. *Journal of Geophysical Research* 96: 8219–8242.
Mathews PM and Shapiro II (1992) Nutations of the Earth. *Annual Review of Earth and Planetary Sciences* 20: 469–500.
Mathews PM, Herring TA, and Buffett BA (2002) Modeling of nutation and precession: New nutation series for nonrigid Earth and insights into the Earth's interior. *Journal of Geophysical Research* 107 (doi:10.1029/2001JB000390).
Mathews PM and Guo JY (2005) Viscoelectromagnetic coupling in precession-nutation theory. *Journal of Geophysical Research* 110: B02402 (doi:10.1029/JB002915).
Moffatt HK and Dillon RF (1976) Correlation between gravitational and geomagnetic fields caused by interaction of core fluid motion with a bumpy core–mantle interface. *Physics of the Earth and Planetary Interiors* 13: 67–78.
Morelli A and Dziewonski AM (1987) Topography of the core–mantle boundary and lateral heterogeneity of the liquid core. *Nature* 325: 678–683.
Mound JE and Buffett BA (2005) Mechanisms of core–mantle angular momentum exchange and the observed spectral properties of torsional oscillations. 110: B08103 (doi:10.1029/2004JB003555).
Mound JE and Buffett BA (2006) Detection of a gravitational oscillation in length of day. *Earth and Planetary Sciences* 243: 383–389.
Murakami M, Hirose K, Kawamura K, Sata N, and Ohishi Y (2004) Post-perovskite phase transition in $MgSiO_3$. *Science* 304: 855–858.

Nimmo F and Stevenson DJ (2000) Influence of early plate tectonics on the thermal evolution and magnetic field of Mars. *Journal of Geophysical Research* 105(E5): 11969–11979.
Oganov AR and Ono S (2004) Theoretical and experimental evidence for a postperovskite phase of $MgSiO_3$ in Earth's D″ layer. *Nature* 430: 445–448.
Olson P and Christensen UR (2002) The time-averaged magnetic field in numerical dynamos with non-uniform boundary heat flow. *Geophysical Journal International* 151: 809–823.
Ono S, Oganov AR, Koyama T, and Shimizu H (2006) Stability and compressibility of the high-pressure phases of Al_2O_3 up to 200 GPa: Implications for the electrical conductivity of the base of the lower mantle. *Earth and Planetary Science Letters* 246: 326–335.
Petford N, Yuen D, Rushmer T, Brodholt J, and Stackhouse S (2005) Shear-induced material transfer across the core–mantle boundary aided by the post-pervoskite phase transition. *Earth, Planets and Space* 57: 459–464.
Poirier JP and LeMouel JL (1992) Does infiltration of core material into the lower mantle affect the observed geomagnetic field? *Physics of the Earth and Planetary Interiors* 99: 1–17; 73: 29–37.
Ricard Y and Vigny C (1989) Mantle dynamics with induced plate tectonics. *Journal of Geophysical Research* 94: 17543–17559.
Rochester MG (1962) Geomagnetic core–mantle coupling. *Journal of Geophysical Research* 67: 4853.
Roden RB (1963) Electromagnetic core–mantle coupling. *Geophysical Journal of the Royal Astronomical Society* 7: 361–374.
Rowley DB (2002) Rate of plate creation and destruction: 180 Ma to present. *GSA Bulletin* 114: 927–933.
Rubie DC, Melosh HJ, Reid JE, Liebske C, and Righter K (2003) Mechanisms of metal-silicate equilibration in the terrestrial magma ocean. *Earth and Planetary Science Letters* 205: 239–255.
Rubie DC, Gessman CK, and Frost DJ (2004) Partitioning of oxygen during core formation on the Earth and Mars. *Nature* 429: 58–61.
Rubincam DP (2003) Gravitational core–mantle coupling and the acceleration of the Earth. *Journal of Geophysical Research* 108(B7): 2338 (doi:10.1029/2002JB002132).
Sarson GR, Jones CA, and Longbottom AW (1997) The influence of boundary region heterogeneity on the geodynamo. *Physics of the Earth and Planetary Interiors* 101: 13–32.
Schersten A, Elliot T, Hawkesworth C, and Norman M (2004) Tungsten isotope evidence that mantle plumes contain no contribution from the Earth's core. *Nature* 427: 234–237.
Stevenson DJ (1987) Limits on lateral density and velocity variations in the Earth's outer core. *Geophysical Journal of the Royal Astronomical Society* 88: 311–319.
Stevenson DJ (1990) Fluid dynamics of core formation. In: Newson H and Jones JH (eds.) *The Origin of the Earth*, pp. 231–249. London: Oxford Press.
Stix M and Roberts PH (1984) Time-dependent electromagnetic core–mantle coupling. *Physics of the Earth and Planetary Interiors* 36: 49–60.
Sumita I and Olson P (1999) A laboratory model for convection in Earth's core driven by a thermally heterogeneous mantle. *Science* 286: 1547–1549.
Sumita I and Olson P (2002) Rotating thermal convection experiments in a hemispherical shell with heterogeneous boundary heat flux: Implications for the Earth's core. *Journal of Geophysical Research* 107.
Tapley BD, Bettadpur S, Ries JC, Thompson PF, and Watkins MM (2004) GRACE measurements of mass variability in the Earth system. *Science* 305: 503–505.
Verhoogen J (1961) Heat balance of the Earth's core. *Geophysical Journal of the Royal Astronomical Society* 4: 276–281.
Wahr JM (1981) The forced nutations of an elliptical, rotating, elastic and oceanless Earth. *Geophysical Journal of the Royal Astronomical Society* 64: 705–727.
Walker D (2000) Core participation in mantle geochemistry. *Geochimica et Cosmochimica* 64: 2897–2911.
Walker D (2005) Core–mantle chemical issues. *Canadian Mineralogist* 43: 1553–1564.
Walker RJ, Morgan JW, and Horan MF (1995) ^{187}Os enrichment in some plumes: Evidence for core–mantle interaction. *Science* 289: 819–822.
Wicht J and Jault D (1999) Constraining electromagnetic core–mantle coupling. *Physics of the Earth and Planetary Interiors* 111: 161–177.
Wicht J and Jault D (2002) Electromagnetic core–mantle coupling for laterally varying mantle conductivity. *Journal of Geophysical Research* 105: 23569–23579.
Xu X, Lithgow-Bertelloni C, and Conrad CP (2006) Global reconstructions of Cenozoic seafloor ages: Implications for bathymetry and sea level. *Earth and Planetary Science Letters* 243: 552–564.
Xu Y, Shankland TJ, and Poe BT (2000) Laboratory-based electrical conductivity in the Earth's mantle. *Journal of Geophysical Research* 105: 27865–27875.
Zhang K and Gubbins D (1993) Convection in a rotating spherical shell with an inhomogeneous temperature boundary condition: Condition at infinite Prandtl number. *Journal of Fluid Mechanics* 250: 209.
Zhang K and Gubbins D (1996) Convection in a rotating spherical fluid shell with an inhomogeneous temperature boundary condition at finite Prandtl number. *Physics of Fluids* 8: 1141–1148.
Zhang K and Schubert G (2000) Teleconvection: Remotely driven thermal convection in rotating stratified spherical layers. *Science* 290: 1944–1947.
Zhang YG and Guo GJ (2000) Molecular dynamics calculations of the bulk viscosity of liquid iron–nickel alloy and the mechanisms for the bulk attenuation of seismic waves in the Earth's outer core. *Physics of the Earth and Planetary Interiors* 122: 289–298.

Printed in the United States
By Bookmasters